T0188657

Foundations of Plasma Physics for Physicists and Mathematicians

Foundations of Plasma Physics for Physicists and Mathematicians

Geoffrey J. Pert
Department of Physics,
University of York, UK

This edition first published 2021

Registered Offices
John Wiley & Sons, Inc., 111 River Street, Hoboken, NJ 07030, USA
John Wiley & Sons Ltd, The Atrium, Southern Gate, Chichester, West Sussex, PO19 8SQ, UK

Editorial Office
The Atrium, Southern Gate, Chichester, West Sussex, PO19 8SQ, UK

For details of our global editorial offices, customer services, and more information about Wiley products visit us at www.wiley.com.

Wiley also publishes its books in a variety of electronic formats and by print-on-demand. Some content that appears in standard print versions of this book may not be available in other formats.

Library of Congress Cataloging-in-Publication Data

Names: Pert, (Geoffrey J.), author.
Title: Foundations of plasma physics for physicists and mathematicians / Geoffrey J. Pert.
Description: First edition. | Hoboken, NJ : Wiley, 2021. | Includes bibliographical references and index.
Identifiers: LCCN 2020045338 (print) | LCCN 2020045339 (ebook) | ISBN 9781119774259 (hardback) | ISBN 9781119774273 (adobe pdf) | ISBN 9781119774280 (epub)
Subjects: LCSH: Plasma (Ionized gases)
Classification: LCC QC718 .P375 2021 (print) | LCC QC718 (ebook) | DDC 530.4/4–dc23
LC record available at https://lccn.loc.gov/2020045338
LC ebook record available at https://lccn.loc.gov/2020045339

Cover Design: Wiley
Cover Images: Blue glowing plasma © sakkmesterke/Getty Images,
Laser-plasma interaction illustration - courtesy of Geoffrey J. Pert

Set in 9.5/12.5pt STIXTwoText by SPi Global, Chennai, India

C9781119774259_120321
Printed and bound by CPI Group (UK) Ltd, Croydon CR0 4YY

This book is dedicated with much appreciation to the mentors of my early career.
Peter Smy, Stuart Ramsden and Michael Woolfson
who all showed great faith in an unknown physicist
and whose guidance shaped my career as a professional scientist.
Also to
Malcolm Haines
who forgot more plasma physics than I ever knew.
Finally, but not least, to my longstanding friend and colleague
Greg Tallents
without his support, help, and encouragement this book
would have never seen the light of day.

Contents

Preface

Plasma often called 'the fourth state of matter' is the most abundant form of observable matter in the universe. Nonetheless plasma physics is a relatively new discipline, unknown before 1900. Ionisation studies originated in the experimental study of gas breakdown in strong electric fields by Paschen (1889). The inception of low temperature plasma physics may be considered to have been the result of the discovery by Thomson (1897) of the negatively charged electron in gas discharges in 1897, although plasma physics as we know it today did not evolve until the mid-1920s. Positively charged ions, essentially gas molecules which had lost an electron, were found soon after. In the early years of the twentieth century, gas discharges and arcs were empirically investigated. The related topics of breakdown of gases and ionisation and recombination dominated this field known as *ionised gases*. This activity becoming progressively more realistic during the early years of the twentieth century as the mobile role of molecules and particles became better understood; but still only in media of low ionisation, where the number of charged particles is small compared to that of the background gas. von Engel (1965) gives an interesting history of the study of ionised gases up to about to 1900. The early studies of breakdown, glow discharges and arcs are covered in detail by von Engel (1965), Cobine (1958), and Loeb (1955). By 1930 an essential understanding of the low temperature plasma had been developed exemplified by the two volume text (von Engel and Steenbeck 1932). At that time quantum physics was not well developed and essentially relatively simple classical models had to used. These yielded basic formal structures with functional relationships but with coefficients to be determined by experiment – an approach which has worked well and is still used today, exemplified by volumes such as Brown (1967). The field, now known as *low temperature plasma physics* has become increasingly developed (Raizer, 1997; Lieberman and Lichtenberg, 2005; Smirnov, 2015) as the earlier models have been progressively refined and improved. Although the basic theory has remained unchanged since the 1930s the detailed methodology has been refined and agreement between experiment and theory improved. In recent years a wide range of direct uses for both direct current and microwave discharges has been found such as gas laser pumping. Important commercial applications now include plasma processing, coating, etching, lighting, and lightning (Lieberman and Lichtenberg, 2005).

In the 1920s, the subject broadened after it became realised that the interior of stars for example must be at such high temperatures that they could contain only fully stripped bare atomic nuclei and electrons (Eddington, 1959). Astrophysical plasmas became an important subject area today embracing a much wider field than simply stellar interiors. Nowadays plasma physics embraces the whole gamut of astronomical bodies from aurora, the magnetosphere to the edge of black holes and led to many conceptual advances in subject (Alfvén, 1950). This area remains an active and important discipline in its own right. At this stage, the end of 1920s, we see the identification of the essential criteria defining plasma through two seminal works. Both involve the nature of the interactive force between charged particle in the plasma.

The first important conceptual step forward towards distinctive plasma physics was taken in 1923 by the publication of the paper by Debye and Hückel (1923) relating to the behaviour of electrolyte ions in solution. Considering the behaviour of positively and negatively charged particles with equal average charge density, Debye and Hückel argued that each ion of one charge is surrounded more closely by ions of the opposite charge thereby decreasing the field of the primary ion below that associated with Coulomb force law over distances in excess of a characteristic length, known as the Debye length (Section 1.3), which tends to zero as the particle temperature tends to zero.

The second key step was the identification of collective waves by Langmuir (1928) (and more fully Tonks and Langmuir 1929). It was proposed that cold plasma moved as a 'block', the individual particles oscillating in the field of their neighbours through the Coulomb force in a *plasma wave* Section 1.2. Again the medium is acting collectively through the Coulombic interparticle field, and *not* through the individual particle fields (see Section 6.2).

Realising that they were observing behaviour not described by conventional nomenclature Langmuir coined the name *plasma* in Langmuir (1928). More details on the origin of the name are given by Tonks (1967) and Mott-Smith (1971) who state that it arose from similarities with blood plasma as a transport fluid. In fact the word plasma stems from the Greek πλάσμα meaning a 'maleable substance' which is appropriate in view of the uncanny ability of the positive column of a glow discharge to fill the space available to it. The name was slow to become adopted but by the 1950s was widely used, particularly in connection with controlled fusion.

The following decade was characterised by the development of plasma kinetic theory.

Firstly Landau (1936) developed a form of Fokker–Planck equation Section 7.3 to account for the fact that most particle interactions are long range and therefore weak involving many particles simultaneously. In contrast to earlier work analogous to gases where the interaction was assumed short range and the Chapman–Enskog method, Section 4.11, used for fluids, had been incorrectly used to determine the hydrodynamic behaviour in plasma. This approach was followed by an improved more formal stochastic picture by Chandrasekhar (1943) (see Section 7.2) in an astrophysical context. The method was completed for application in plasma by Rosenbluth, MacDonald and Judd (1957), the final element being the introduction of the Coulomb logarithm by Cohen, Spitzer, and Routly (1950) following several early workers. The development of *magneto-hydrodynamics*, the plasma equivalent of fluid theory which treats ensembles of large numbers of particles averaged over velocity, was now complete.

The second development of kinetic theory was in the area of the microscopic distribution in which the particle motions are treated individually before a final averaging. Recognising the relative importance of collective behaviour over collisional, Vlasov (1938) introduced what is essentially a collisionless Boltzmann equation Section 6.2. This kinetic equation has proved very successful in dealing with the large number of plasma modes of oscillation. An important development of the theory due to Landau (1946) took account of the velocity resonance at the wave phase speed in warm plasma to develop damping Appendix 12.A. The approach has been widely used to identify micro-instability and damping.

These two approaches complete the underlying theoretical structure of plasma physics, individually appropriate for different problems. By about 1960 the formal structure of plasma physics was well established. Further development involved the development of theoretical pictures to investigate specific problems.

The successful achievement of man-made thermo-nuclear reactions in 1952 led to a marked increase in activity on the possibility of controlled reactions suitable for the generation of electrical power in the USA, UK, and USSR, which was mostly hidden behind security barriers. One of the first, and probably best known, was by John Lawson at Harwell, initially issued as a secret report (Lawson, 1955), but later a revised form in the open literature (Lawson, 1957), introducing the familiar Lawson criterion giving a necessary condition for useful thermo-nuclear power. His final conclusion (Lawson, 1955) remains as true today as it was 60 years ago.

> Even with the most optimistic possible assumptions, it is evident that the conditions for the operation of a useful thermonuclear reactor are very severe.

Nonetheless as we shall see in the course of this book, a major part of the road to controlled nuclear fusion has now been achieved.

Despite these problems, the possibility of such a device and its anticipated rewards led to surge in activity in plasma physics as it is was clear that the lowest temperatures require the working medium to be plasma. The general interest in the possibility of achieving fusion power using plasma as the medium has led to an enormous increase in the resources allocated to these problems and a wide variety of experimental device (Glasstone and Lovberg, 1960). As a consequence, the subject as a whole has moved rapidly forward with particular emphasis on conditions required for stable confined plasma, particularly toroidal geometries (tokamaks). Since 1970, laser generated plasma has offered an alternative route to fusion through inertial confinement.

It is a characteristic of any new field, that its development is likely to be reflected in the way plasma physics has developed over the years, namely in fits and starts. 1900–1940 the era of gas discharges, and collisional effects, 1940–1950 the start of what we might call the modern era where collective behaviour was being understood. 1950–1970 the age of major theoretical and experimental development when the basic understanding of waves and instabilities was developed. 1970–present the construction of large toroidal machines and detailed theory. plus laser-plasma interactions. This is reflected in the text books used. Several of the best date from 1960–1980, but have been updated. We have tried to reflect this historical growth in the structure of the book and included as many of the classic papers and books as possible. Consequently many of the papers cited may appear 'old-hat'.

The historical development is interesting as case study in itself as plasma the fourth state of matter was only recognised from about 1925 and its development as a discipline has been rapid. The period 1950–1970 was particularly important as due to security issues relating to fusion much of the work carried out in laboratories in the US, USSR, and UK was classified. A major event in 1956 was the lecture given by Kurchatov (1956) at Harwell revealing the progress made by Soviet workers and opening the security blanket.

Major declassification occurred as a result of the second UN conference on the Peaceful Uses of Atomic Energy at Geneva when fusion was included and lead to a sharing of results on Plasma Physics. Since then the subject aside from matters relating to H-bombs has lain outside the security blanket.

A key assessment by Lawson (1957) of the conditions necessary for controlled fusion to take place defined *ignition* as the state when more energy is released in the fusion reaction that is put into heating the fuel.[1]

---

1 Allowing for the conversion efficiency η from plasma energy to output, we equate the fusion energy release to the sum of initial energy of the particles $3kT$ and the bremsstrahlung heat loss $\alpha T^{1/2}\tau$ to the fusion energy release

$$\frac{\eta}{4(1-\eta)}\langle\sigma v\rangle\,\Delta E\tau = 3kT + \alpha T^{1/2}\tau \tag{1}$$

It turned out that on both sides of the 'iron curtain' the major experimental efforts had been on toroidal linear pinches, loosely based on the steady Bennett pinch (problem 7). Unfortunately, such plasmas are inherently unstable as we show in Section 16.2.2. As we shall see work in the USSR had gone farthest to mitigate these effects, but none were successful. None-the-less in the west and the USSR work continued to proceed relatively independently. In 1960, the US programme contained many different elements aside from pinches, which were conveniently surveyed in Glasstone and Lovberg (1960); the major effort being directed towards steady plasma in the *stellarator*. In the United Kingdom, the study of toroidal Z-pinches had been actively pursued since 1945 (Hendry and Lawson, 1993) ultimately leading to the construction of *ZETA* a large toroidal pinch.[2] USSR led the development of toroidal discharges initially proposed by Tamm (1990) and Shafranov (2001) which metamorphosed into *tokamaks*. By 1969, the Russians were demonstrating very promising results with their T3 Tokamak which had been distrusted in the United States on the basis of their unconventional diagnostics. Despite the cold war, a group from Culham, United Kingdom led by Derek Robinson and Nic Peacock were invited to Russia (Forrest, 2011) to independently check the ion temperatures. The Culham group were probably the world leaders on laser scattering at that time, a method recognised as being unequivocal. These experiments confirmed the Russian claims and a result tokamaks have become recognised as the most probable route to magnetically confined fusion with large machines being built in Europe (JET), the United states (PLT), and Japan (JT60). At present an international collaboration is constructing the ITER device to demonstrate break-even.[3]

Following de-classification in 1972, a second possible route to controlled thermo-nuclear fusion was opened with laser compression and *inertial confinement fusion* with the publication by Nuckolls, Wood, Thiessen and Zimmerman (1972), and shortly later by Clark, Fisher and Mason (1973). In this scheme, fusion takes place as a small pellet or shell of thermo-nuclear fuel is compressed by the ablation pressure and then blows apart due to its high temperature having being heated to 'burn' temperatures by a high power laser pulse. Since then the conditions necessary to achieve thermo-nuclear reactions have been progressively refined from these simple geometries as technical problems have arisen until today's designs involve complex targets and either direct heating by the laser or indirect by X-rays inside a small cavity (*hohlraum*) (Lindl, 1993). Large-scale laser

---

where n is the particle density, τ the containment time, and T the temperature; η is the efficiency parameter, the fusion reaction rate averaged over the particle velocity distribution, and α the bremsstrahlung emission coefficient. This occurred when the *Lawson criterion* is satisfied

$$n\,\tau > \frac{3\mathit{k}T}{\frac{\eta}{4(1-\eta)}\,\langle\sigma v\rangle\ \Delta E - \alpha T^{1/2}} \tag{2}$$

Taking representative values for η, α; and noting that the DT reaction requires the lowest temperature for ignition with minimum $\mathit{k}T \approx 20$ keV (Lawson 1957) identified the key criterion as $n\tau > 10^{20}$ m^{-3} s which with only the following slight modification has remained unchanged today (see Wesson, 2011). Later studies have indicated that triple product $n\ \tau\ T$ obtained under a constant pressure condition giving a minimum temperature for the DT reaction of 14 keV and triple product $nT\tau \approx 3 \times 10^{21}$ keV s m^{-3}.

2 Unfortunately ZETA was initially stigmatised by an unfortunate premature press release claiming the observation of controlled thermonuclear reactions, whereas in fact the measured neutrons were generated in the electric fields produced by instabilities. Despite this ZETA later carried out much novel and valuable work, including the serendipitous discovery of self-reversal in *reversed field pinches* (Bodin and Newton, 1980) and improved stability.

3 Author's note: I was fortunate to be attending the APS Plasma Division meeting when these results were announced and remember well the excitement caused. In retrospect looking back over 50 years it was possibly the most important result in that time, both scientifically and geopolitically, reconciling the Russian and American fusion programmes across cold war divisions. It also saw the death knell of the pathologically damaging anomalous Bohm type diffusion in fusion studies (see p. 35 in chapter 2).

facilities are required for successful implementation of this scheme and have therefore been constructed. The core of the *inertial confinement* scheme is essentially hydro-dynamic and is treated in some detail in our previous work (Pert, 2013, Ch.14). The interaction of the laser with plasma surface on the other hand has many features which lie within the ambit of this work and are treated here in Chapters 11 and 14.

One of the characteristics of plasma is the large number of degrees of freedom it possesses especially in the presence of a background magnetic field. As a result, there are a wide range of differing wave motions, both stable and growing, possible. Consequently, much of this discussion is concerned with waves of different types, both free and driven.

This text in two parts, reflecting this development of plasma physics. The first part, Chapters 1–8 guide the reader through the underlying theoretical structure of the subject. The remaining Chapters 9–14 deal with specific problems, principally associated with fusion

Chapter 1. Characteristic Collective Behaviour of Plasma

Chapter 2. Classical Behaviour for Plasma with Simple Collisions

Chapter 3. Single Particle Model – Motion of Charged Particles in Static Fields Neglecting Particle Interactions

Chapter 4. Fluid Theory and Hydrodynamic Equations Obtained by Integrating Basic Kinetic Theory Over the Particle Distribution with Only Short Range Collisional Interactions

Chapter 5. Review of the Properties of Waves in Dispersive and Anisotropic Media Including Geometrical Optics

Chapter 6. Development of Kinetic Effects When Collective Effects Dominate, Therefore Neglecting any Short Range Collisions – Vlasov Equation

Chapter 7. Plasma Theory and magnetohydrodynamic (MHD) Equations Obtained by Integrating Basic Kinetic Theory Over the Particle Distribution Contrasting the Difference Between Short Range Collisions (Gases) and Long Range Coulombic Interactions (Plasma)

Chapter 8. MHD Hydrodynamics for Plasma Where the Coulomb Force Is Strong Inducing Co-operative Behaviour Modifying Fluid Dynamics to Include the Role of the Magnetic Field on Charged Particles

Chapter 9. Simplification to MHD Flow Introduced by Infinite Conductivity – Absence of Collisional Effects

Chapter 10. Spectrum of Hydro-magnetic Acoustic Waves Found in Ideal MHD Plasma Including Shock Waves

Chapter 11. Spectrum of Waves in Cold (Zero Temperature) Plasma in Magnetic Fields

Chapter 12. Dielectric Properties of Waves in Warm Plasma Mainly in the Absence of a Magnetic Field

Chapter 13. Generalisation of the Classical Theory of Electro-magnetic Waves to Plasma and the Non-linear Interactions at High Beam Intensity

Chapter 14. Brief Introduction to Interaction Between a Laser Beam and Plasma Within the Classical Quasi-linear Regime, Where the Absorption/Thermal Conduction Are All Within the Constraints of 'Linear' Theory. Non-linear Regimes Needing Extensive Numerical Calculation Are Appropriate to Later Research Texts Such as Atzeni and ter Vehn (2004)

Chapter 15. MHD Theory of Stable Plasma Configurations, Particularly Toroidal Configurations

Chapter 16. MHD Theory of Instability in Plasma Configurations

Supplementary sections on the basic models of low temperature plasma, and the theory of complex variables and Laplace transforms are included to provide, if necessary, background material and revision.

The book was initially planned to be aimed at final year undergraduate and postgraduate students. However as work progressed, it became clear that the extended and complex nature of the subject introduced many new concepts and techniques from other areas which would be unfamiliar to the targeted students. The objective of the book therefore changed to provide the background material in both basic plasma physics and other related topics which would allow an easy access to complex research. The book is not intended to provide a review of the current state of plasma research, but to provide a ready source to the methodology and techniques used in that research, a sort of compendium in the basic physics underlying plasmas. Although the book is now aimed at postgraduate students, it remains very suitable for use as a basic final year undergraduate course in the methodology of plasma physics. It is the result of many years working in plasma physics research, especially laser-plasma interactions. Much of the material is based on different lecture courses given at the Universities of Alberta, Hull and York, over the years. The choice of topics included has ranged over different fusion orientated syllabi. In the interests of conciseness, much important (and interesting to the author) material in atomic and ionisation physics, astrophysics, magnetospheric physics, and low temperature industrially orientated plasma that would have been included in a full survey of the field has of necessity been omitted. However, the basic material contained in the book should have wide application.

July 2020

Geoff Pert FRS
University of York, York

1

Fundamental Plasma Parameters – Collective Behaviour

1.1 Introduction

Plasma is generally defined as *an electrically neutral, conducting gas*. The designation was introduced by Langmuir (1928) to describe the behaviour of the medium in an electrical discharge distant from the electrodes.[1] In this region, the gas contains a balanced number of positively and negatively electrically charged ions and electrons so that the total charge density is zero. Thus, if the ion density is n_i and charge is Ze, the electron density is $n_e = Z\, n_i$. Collisions are relatively infrequent, and the conductivity correspondingly high.

The behaviour of the plasma is determined by the motion of individual particles subject to electric and magnetic forces either externally applied or resulting from their charge. In fact, this is a complex problem as the particles themselves generate space charges and currents, and in consequence, electric and magnetic fields. Thus, the particles mutually interact through these self-generated fields.

This mutual interaction can take place in two ways due to the long-range nature of the fields. First, over short ranges, the particles act independently, and the interaction resembles the form of the conventional two particle collision found in gases. However, over longer distances, the particles behave collectively leading to group interactions such as a wave oscillations.

In all conventional plasmas, the ions and electrons can be treated as classical particles. The basic equations governing plasma are therefore Maxwell's equations for the fields combined with Newton's law of motion for the dynamics. Relativistic plasmas can be formed under exceptional conditions, but will not be treated here.

An immediate consequence of this mutual interaction is that a new form of longitudinal oscillation becomes possible. In contrast to sound waves in gases, where the forces are due to pressure and communicated by short range collisions, the force is due to the long-range nature of electrostatic forces. This oscillation introduces a characteristic frequency, the electron plasma frequency Π_e, at which the plasma can respond to temporal changes imposed upon it. The plasma cannot follow changes occurring over times $\sim \Pi_e^{-1}$. As a consequence, it may be anticipated that there is also a limiting distance over which spatial change can occur determined by the thermal velocity and the plasma frequency $\lambda_D \approx \overline{v}_e/\Pi_e$, where $\overline{v}_e = \sqrt{k T_e/m_e}$, m_e being the electron mass, k Boltzmann's constant, and T_e the electron temperature.

1 See Tonks (1967) for an account of the origin of the name.

Foundations of Plasma Physics for Physicists and Mathematicians, First Edition. Geoffrey J. Pert.

In this chapter and the next, we introduce the essential characteristics of dilute plasmas by considering some basic problems, which illustrate the collective behaviour of the constituent particles resulting from their electric charges. The electrostatic Coulomb force has a potential which varies as $1/r$, where r is the distance from the charge. The force is therefore effective at long ranges; as distances increase, the cumulative effect of the larger number particles balances the fall-off of the individual fields. This leads to a number of characteristic effects, of which we examine some of the simplest in this chapter: the space charge shielding of electric fields and longitudinal waves. The collision of particles by many charges simultaneously and their effects are deferred until the following Chapter 2.

A final important characteristic of plasma concerns the relative strength of the electrostatic interaction energy compared to thermal. If this ratio is small, the particles move freely and the correlation between particles is small; the plasma is *dilute*. The particle distribution is therefore approximately Maxwellian. In contrast if this ratio is large, particles are 'tied' to each other and the distribution highly correlated. The ultimate limit of this *strongly coupled plasma* behaviour can be seen in an ionic crystal lattice.

1.2 Cold Plasma Waves

Consider a cold (zero temperature) quasi-neutral plasma comprising ions of charge $Z\,e$ and electrons of charge $-e$ so that

$$Zn_{i0} = n_{e0}$$

The plasma is considered to be collision-free. Therefore, if the electrons are disturbed, the ions remaining stationary, a collective motion of the electrons is established by the internally generated fields. This contrast with a gas where collisions transmit the motion from element to element.

In one dimension, the motion of the particles takes place as planar slabs moving normal to their surface. Consider a block of electrons initially at x_0 of thickness δx_0 displaced at a distance X, the ions remaining stationary. As a result, there is an increase in the number of electrons at one end and a corresponding increase of an equal number of unbalanced positive charges at the other end of the displaced block of width X, which have been uncovered by the moving slab. Applying Gauss' theorem across the block, it is trivial to show that a uniform field

$$E_x = n_{e0} eX/\epsilon_0 \tag{1.1}$$

is generated on the plasma within the block. This field will accelerate the electrons back toward the ions

$$\frac{\mathrm{d}^2X}{\mathrm{d}t^2} = -\frac{eE_x}{m_e} = -\frac{n_{e0}\,e^2}{\epsilon_0\,m_e}X = -\Pi_e{}^2X \tag{1.2}$$

The electrons therefore oscillate about the ions with a frequency

$$\Pi_e = \sqrt{n_{e0}e^2/\epsilon_0\,m_e} \tag{1.3}$$

known as the *electron plasma frequency* or more commonly succinctly the *plasma frequency*.

Clearly, we may build up a more general bulk oscillation by summing over a series of slabs to allow for general forms of the initial displacement, i.e. as a wave with varying amplitude, provided the linear assumption is made that the ordering of the electrons remains unchanged. These *cold plasma waves* are non-propagating and are standing waves with zero group velocity.

The characteristic plasma frequency essentially represents the fastest speed at which electrons can respond. Therefore, we can propagate waves with frequency $\omega > \Pi_e$, but lower frequencies may be damped; in waves where $\omega < \Pi_e$, the electrons can move sufficiently rapidly to prevent the propagation of the wave.

1.2.1 Wave Breaking

Consider the slab whose ambient position is x_0 and is displaced by X and compare it with the neighbouring slab from $x_0 + \delta x_0$ which is displaced by $X + \delta X$. The spacing between the slabs

$$\begin{aligned}\delta x(t) &= [x(t) + \delta x(t)] - x(t) = [x_0 + \delta x_0 + X(t) + \delta X(t)] - [x_0 - X(t)] \\ &= \delta x_0 + \delta X(t)\end{aligned} \tag{1.4}$$

Clearly, the ordering of the slabs is maintained provided $\delta x(t) > 0$ if $X(t) > 0$ and *vice versa*, i.e.

$$\frac{\partial X(t)}{\partial x_0} > -1 \tag{1.5}$$

If this condition is not upheld, the electron sheets lose their integrity, the phases are mixed, and the wave damped. The wave energy is converted into disordered electron kinetic energy.

For a cold plasma wave differentiating Eq. (1.2) with respect to x_0, multiplying the result by $\partial X/\partial x_0$ and integrating yields

$$\Pi_e^{\,2}\left(\frac{\partial X}{\partial x_0}\right)^2 + \left(\frac{\partial \dot{X}}{\partial x_0}\right)^2 = \text{const} = W \tag{1.6}$$

so that if $W < \Pi_e^{\,2}$ at any time, the wave does not break.

The spatial position of the element x_0 is given by $x = X + x_0$, and the displacement $X(t)$ at time t by Eq. (1.2). Hence, the instantaneous electric field at point x at time t, $E(x, t)$ is easily obtained from Eq. (1.1). The behaviour of the wave at breaking is easily demonstrated by a simple example introduced by Dawson (1959). Consider a sinusoidal wave such that

$$X(0) = A\ \sin(k\, x_0) \quad \text{and} \quad \dot{X}(0) = 0$$

The electric field at time $t = 0$ is, therefore,

$$E(x, 0) = \frac{en_{e0}}{\epsilon_0} A \sin(k\, x_0) = E_{\text{max}}\ \sin(k\, x_0) \tag{1.7}$$

where $x = x_0 + A \sin(k\, x_0)$. It is easily established from Eq. (1.5) that this wave breaks if $A \geq 1/k$. The structure of the electric field as the amplitude of the wave varies is shown in Figure 1.1. It is clearly seen that wave breaking occurs if the amplitude of the wave $A \geq 1/k$ as predicted. Once the wave has broken the calculated field is multi-valued, which is, of course, unphysical. Since the maximum field intensity $E_{\text{max}} = en_{e0}A/\epsilon_0$, the amplitude of the critical field leading to wave breaking is

$$E_{\text{break}} = \frac{e\, n_{e0}}{\epsilon_0\, k} \tag{1.8}$$

which may be written in terms of the 'bounce frequency' $\omega_b = \sqrt{ekE_0/m}$ when E_0 is the intensity amplitude of the wave. At the wave breaking condition, $\Pi_e = \omega_b$.

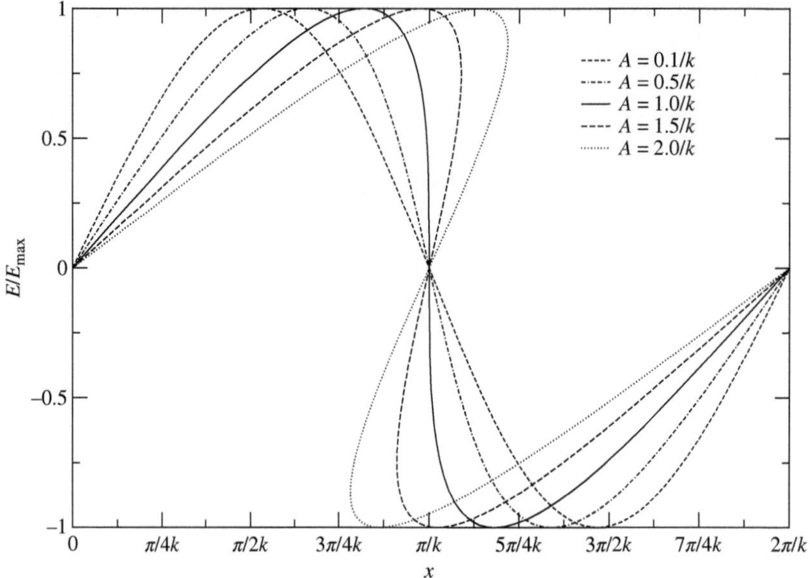

Figure 1.1 Plots of the electric field distribution as fractions of the peak field for the model plasma wave. For $A < 1/k$, it can be seen the field is well behaved, but for $A > 1/k$, it is multi-valued characteristic of breaking.

1.3 Debye Shielding

In a plasma, both ions and electrons are charged. As a consequence of the statistical distribution of the particles, there is a rapidly varying space charge field, the *micro-field*. The spatial and temporal fluctuations of this micro-field are determined by the random motion of the charged particles. In the presence of an applied electric field, the particles become polarised as in a dielectric, setting up a field which neutralises the applied one. The distance over which this space charge field is established is known as the *Debye length*. It is important to note that in contrast to a dielectric, the polarisation is dynamic, particles moving rapidly within the space charge, and the shielding is established as a statistical average over temporal fluctuations.

In particular, as a consequence of their negative charges, electrons tend to 'clump' around the positive charges of the ions. Provided the temperature is sufficiently high, the electrons are not bound to the ions but move freely, with slightly increased density in the neighbourhood of the ion. Provided the spatial range of the perturbation is extensive, a large number of particles are weakly slightly perturbed. Individual particles are averaged, and the space charge forms a continuum, established by the perturbed particle distribution.

Consider a test ion inserted into a plasma, which contains on average n_i^0 ions of charge Ze per unit volume and n_e^0 electrons with charge $-e$ per unit volume. The ions have an equilibrium thermal distribution with temperature T_i and the electrons T_e. Within the plasma, there are local variations of the average electric potential ϕ within which will be corresponding variations of the ion and electron densities. Since each particle is in thermal equilibrium with its own temperature, the local densities are given by the Boltzmann distribution

$$\begin{aligned} n_e &= n_e^0 \ \exp[e\,\phi/\kappa T_e] \\ n_i &= n_i^0 \ \exp[-Ze\,\phi/\kappa T_i] \end{aligned} \tag{1.9}$$

although overall charge neutrality holds for the average densities

$$Z\,n_i^0 = n_e^0 \tag{1.10}$$

Consider the field in the neighbourhood of the test ion with charge Ze at $r = 0$. Its positive charge will attract electrons and repel ions giving rise to a net charge imbalance locally. If this region of charge, separation is large compared to the inter-particle separation distance $n_i^{-1/3}$, the densities will show the net increase or decrease predicted by the Boltzmann equation (1.9) when time averaged. Over large distances and times, the inherent 'graininess' of the particles is averaged out to give a quasi-continuum distribution. The statistical distribution of the electrons will give a slowly varying potential, ϕ, described by Poisson's equation:

$$\nabla^2\phi = -\frac{1}{\epsilon_0}[Zn_ie - n_ee] \tag{1.11}$$

Near the ion, the field is spherically symmetric

$$\frac{1}{r^2}\frac{\mathrm{d}}{\mathrm{d}r}\left(r^2\frac{\mathrm{d}\phi}{\mathrm{d}r}\right) = -\frac{1}{\epsilon_0}n_e^0\left[\exp\left[\frac{-Ze\phi}{\mathcal{k}T_i}\right] - \exp\left[\frac{e\phi}{\mathcal{k}T_e}\right]\right] \tag{1.12}$$

and subject to the boundary conditions:

1. Near the ion, the field tends to that of the ion alone: $\phi = Ze/4\pi\epsilon_0 r$ as $r \to 0$.
2. Far from the ion, the field must vanish: $\phi \to 0$ as $r \to \infty$.

Since the field is small except close to the ion, we may expand the exponentials retaining only the leading terms

$$\frac{1}{r^2}\frac{\mathrm{d}}{\mathrm{d}r}\left(r^2\frac{\mathrm{d}\phi}{\mathrm{d}r}\right) = \frac{n_e^0e^2}{\epsilon_0}\left\{\frac{1}{\mathcal{k}T_e} + \frac{Z}{\mathcal{k}T_i}\right\}\phi = \frac{\phi}{\lambda_D^2} \tag{1.13}$$

where λ_D is the Debye length

$$\lambda_D^2 = \frac{\epsilon_0}{n_ee^2}\Big/\left\{\frac{1}{\mathcal{k}T_e} + \frac{Z}{\mathcal{k}T_i}\right\} \tag{1.14}$$

To solve this differential equation, substitute: $\phi = f/x$ and $x = r/\lambda_D$ to obtain

$$\frac{\mathrm{d}^2f}{\mathrm{d}x^2} = f \tag{1.15}$$

subject to $f(0) = 1$, and $f(x) \to 0$ as $x \to \infty$ whose solution is

$$f(x) \sim \exp(-x) \tag{1.16}$$

Hence, we see that the field at large distances from the ion is

$$\phi = \frac{Ze}{4\pi\epsilon_0 r}\exp\left(-\frac{r}{\lambda_D}\right) \tag{1.17}$$

and falls off more rapidly for distances $r \gtrsim \lambda_D$ than Coulombic due to additional electrons and depleted ions. The electrons therefore on a statistical basis screen the ion. At this statistical level, there is locally a higher-than average electron density reflecting weak electron correlation over distances $\sim \lambda_D$. Since the field falls off rapidly as $r \gtrsim \lambda_D$, the Coulomb field is screened, and its range reduced from ∞ to $\sim \lambda_D$.

It is often convenient to separate the Debye length into separate electron and ion components: λ_e and λ_i represent the electron and ion correlation lengths, respectively.

$$\left.\begin{aligned} \lambda_e^2 &= \frac{\epsilon_0 \kappa T_e}{n_e e^2} \\ \lambda_i^2 &= \frac{\epsilon_0 \kappa T_i}{Z^2 n_i e^2} \end{aligned}\right\} \quad \frac{1}{\lambda_D^2} = \frac{1}{\lambda_e^2} + \frac{1}{\lambda_i^2}$$

In many cases, screening is due to the electrons alone. The large ion mass ensures that they move much more slowly. As a result, the ions are nearly static and, as a consequence, tend to be more grainy in their averaged distribution; in particular, they cannot respond to rapidly oscillating fields. They generally play little role in screening. Consequently, the electron Debye length is generally used as a measure of the screening distance $\lambda_D \approx \lambda_e$. Note that the effect of a uniform background ion space charge is to neutralise some of the electron charge and therefore increase the Debye length.

The Debye length measures the penetration of electric fields into the plasma. However, as we have pointed out, the model leading to the Debye length is only valid if the number of particles in the Debye sphere ($n_i\ \lambda_D{}^3$) is large, or alternatively that the Debye length is much larger than the inter-particle separation. At distances $\gtrsim \lambda_D$, the plasma particles act collectively to smooth out electric field fluctuations, although as we shall show well organised coherent collective motions are possible. For distance less than λ_D, thermal fluctuations are insufficient to dominate space charge disturbances on that scale, and electric fields can be found.

The screening effect of the Debye shielding over lengths $> \lambda_D$ clearly marks a difference in the form of the interactions between particles. When the particle separation is small $d \gtrsim \lambda_D$, the particles clearly interact on a 'short range', i.e. on an individual basis as between gas molecules – *collisional interactions.* On the other hand, when the separation is large $d \gtrsim \lambda_D$, the particle no longer 'sees' individual scatterers but only a 'smoothed out' continuum of interacting elements – *coherent or collective interactions.*

Clearly, the plasma frequency is related to the Debye length as both represent terms over which fields may be damped by the plasma. In fact

$$\lambda_D\, \Pi_e = \sqrt{\frac{\kappa T_e}{m_e}} = \overline{v}_e \tag{1.18}$$

the electron thermal speed.

1.3.1 Weakly and Strongly Coupled Plasmas

The analysis (with its imperfection) which leads to Debye shielding is only valid if the number of electrons and ions in the Debye sphere is sufficiently large that the 'graininess' of the particle distribution is lost in the averaging. In this case, the plasma is weakly coupled, and the thermal energies are large compared to electrostatic energies. Such a plasma is called *dilute*, when

$$n\ \lambda_D{}^3 \gg 1 \tag{1.19}$$

If this condition is not obeyed, individual particles in the neighbourhood of an ion are strongly influenced by the field. In this case, the averaging we have used is no longer applicable, and individual particles are strongly influenced by their neighbours, i.e. strongly correlated and the plasma is dense: $\lambda_D \lesssim n_i^{-\frac{1}{3}}$. We can no longer express the distribution in terms of the one particle (Boltzmann) distribution function. Such plasmas are highly structured. For example an ionic crystal formed of positive and negative ions forms a well-structured lattice, provided the electrostatic forces of attraction are much larger than the thermal ones tending to destroy the structure. Similar effects are found in dusty plasma. Due to their large charge and (normally) low temperature, the ions are strongly correlated. The strength of ion correlation is defined by the *ion coupling coefficient*, which reflects the relative strength of the ion interaction energy to thermal energy:

$$\Gamma_i = \left\{\left(\frac{4\ \pi\ n_i^{-\frac{1}{3}}}{3}\right) \Big/ \sqrt{\left(\frac{\epsilon_0 kT_i}{Z^2 n_i e^2}\right)}\right\}^2 = \frac{Z^2\ e^2}{\epsilon_0\ R_0\ kT_i} \approx \frac{n_i^{-\frac{1}{3}}}{\lambda_D} \tag{1.20}$$

where $R_0 = (3/4\ \pi\ n_i)^{1/3}$ is the radius of the ion sphere. If $\Gamma_i > 1$, the ions in the plasma are strongly coupled. The term is the ratio of the electrostatic energy at the inter-particle separation distance to the thermal energy

$$\Gamma \approx \frac{\{Z^2 e^2/\epsilon_0 R_0\}}{\{kT_i\}} \rightarrow \frac{\text{Electrostatic potential energy between particles}}{\text{Thermal energy per particle}} \tag{1.21}$$

In contrast, electrons are rarely strongly coupled due to their higher temperature and smaller charge.

The plasma is strongly correlated when the electrostatic energy is greater than the thermal energy. Strongly coupled plasmas are unusual. It is difficult to make plasmas in the range $\Gamma \approx 1 \leftrightarrow 100$ for laboratory study. When strongly coupled, the overall particle distribution function can no longer be written in terms of a single particle distribution function as the higher-order terms are not small. Calculations of these systems are very difficult, and Monte-Carlo methods must often be used.

1.3.2 The Plasma Parameter

Closely related terms to the coupling coefficients are the *Debye number* – the number of electrons in the Debye sphere

$$N_D = \frac{4\ \pi}{3}\ n_e\ \lambda_D{}^3 \tag{1.22}$$

and the *plasma parameter* defined by the *Coulomb or Spitzer logarithm* (Eq. (2.8)) required for Rutherford scattering of electrons and ions (Section 2.1)

$$\Lambda = 4\pi n_e \lambda_D{}^3 \approx 4\ \pi\ \frac{(\epsilon_0\ kT_e)^{3/2}}{Z\ e^3\ n_e{}^{1/2}} \tag{1.23}$$

where Ze is the ion charge.

The plasma parameter has a further important scaling property. The momentum transfer cross section, derived from later Eq. (2.7), σ_d represents the cross section for momentum transfer from electrons to ions. Comparing the Debye length with the mean free path of electrons

$$n_i \sigma_d \lambda_D = \frac{Z\ e^3\ n_e{}^{1/2}}{(4\pi)^2\ \epsilon_0{}^{3/2}\ (kT_e)^{3/2} \ln \Lambda} = (4\ \pi\ \Lambda\ \ln \Lambda)^{-1} \tag{1.24}$$

Many plasmas are dilute, implying that $\Lambda \gg 1$ and that the mean free path is long compared to the Debye length. As a result, collisions often play only a small role in determining the dynamical behaviour of the particles. Interactions are due to long-range collective forces rather than short-range collisions as in a gas. In many cases, the plasma may be treated as collisionless, and the effect of collisions ignored.

1.4 Diffusion and Mobility

In low temperature, plasma ions and electrons drift through the background gas suffering many collisions. As a result, the movement of the particles is inhibited by collisions which destroy the directed motion only allowing particles to drift from collision to collision under the driving force. Such forces are typically either the partial pressure of the particle or an applied field, for example an electro-static field. In each case, however, the drag force due to collisions is the same.

1. In the first case, *diffusion*, the particle drift responds to the density gradient. Defining the particle flux Γ as the number of particles passing through unit area perpendicular to the drift per unit time, the diffusion flux is given by Fick's law $\mathbf{\Gamma}_D = -D\ \nabla n$, where D is the diffusion coefficient and n the particle density.
2. In second case, the presence of an electric field (or more generally any conservative field such as gravity) induces the particle drift at a rate $\mathbf{\Gamma}_E = \mu\ n\ \mathbf{E}$, where μ is the mobility.

The drift of ions through a background gas is a characteristic problem in gas discharge studies. Their behaviour is governed by induced charges between the ion and the gas molecule. A model of mobility under such forces was developed by Langevin in 1905, but lost for many years before being rediscovered after about 20 years (see McDaniel, 1964, p. 431).

1.4.1 Einstein–Smoluchowski Relation

As we have pointed out, the two processes are essentially similar, although the driving force is markedly different. As a result, we expect that the diffusion and mobility coefficients are related. This relationship was found independently by Einstein and Smoluchowski in 1905 in the course of their studies of Brownian motion. Consider a system in which the particles drift under the action of both diffusion and mobility in a conservative field with potential U, then the flux is the sum of that from both effects

$$\mathbf{\Gamma} = -D\ \nabla n - \mu\ n\ \nabla U \tag{1.25}$$

In equilibrium, the flux $\Gamma = 0$ everywhere, and therefore

$$D = -\mu \frac{n}{\mathrm{d}n/\mathrm{d}U} \tag{1.26}$$

But in thermal equilibrium, the particle density must satisfy Boltzmann's equation $n = A\ \exp(-U/\mathcal{k}T)$, where A is a constant related to the density profile. Clearly, $\mathrm{d}n/\mathrm{d}U = -n/\mathcal{k}T$ and as a result, we obtain the *Einstein–Smoluchowski relation*

$$D = \mu\ \mathcal{k}T \tag{1.27a}$$

When the conservative force is due to the particle charge q and the electric potential ϕ, the general potential $U = q\ \phi$ and the Einstein–Smoluchowski relation is modified to take account of the charge

$$D = \mu\ \mathcal{k}T/q \tag{1.27b}$$

Since the diffusion coefficient is positive, the mobility has the same sign as the charge.

1.4.2 Ambipolar Diffusion

We consider now the case when the density is such that Debye length is small compared to its scale length. Both electrons and ions with ion charge Z are drifting simultaneously, but the two particles of different polarity cannot separate, otherwise, substantial space charge fields are established; therefore, quasi-neutrality must be maintained $Z\, n_i \approx n_e$. Consequently, since the electron diffusion coefficient is much larger than the ion, an electric field is established between the two so that the mobility can reduce the electron flux and increase the ion to make the two equal

$$\Gamma_e = -D_e \frac{dn_e}{dx} - \mu_e\, n_e\, E \quad \text{and} \quad \Gamma_i = -D_i \frac{dn_i}{dx} + \mu_i\, n_i\, E \tag{1.28}$$

Since ambipolar diffusion is only important in cold plasma, where the ion charge $Z \approx 1$, we set $n_i = n_e$ and $\Gamma_i = \Gamma_e$, we obtain the joint *ambipolar diffusion* flux with coefficient D_a

$$\Gamma = -D_a \frac{dn}{dx} = -\frac{(D_i\, \mu_e + D_e\, \mu_i)}{\mu_e + \mu_i} \frac{dn}{dx} \tag{1.29}$$

Making use of the Einstein–Smoluchowski relation

$$D_a = D_i \frac{(1 + D_e/D_i\, \mu_i/\mu_e)}{(1 + \mu_i/\mu_e)} = D_i \frac{1 + T_e/T_i}{1 + \mu_i/\mu_e} \approx D_i\, (1 + T_e/T_i) \tag{1.30}$$

since in general $\mu_e \gg \mu_i$. It follows that the electron partial pressure is added to the ion driving the diffusion through the inter-particle electric field.

Ambipolar diffusion plays an important role in the behaviour of many low-temperature discharges such as glow discharges and the after-glow of decaying discharges. The nature and role of ambipolar diffusion was extensively reviewed by Phelps (1990).

1.5 Wall Sheath

At a planar wall at the edge of the plasma, an electric field is established which retards the electrons and accelerates the ions normal to the wall in the direction x. In the absence of the electric field, the net flux of electrons to the wall is given by the standard result from kinetic theory $\frac{1}{4}\, n_e\, \overline{v}_e$, where $\overline{v}_e = \sqrt{8\, k T_e/\pi\, m_e}$ is the mean thermal speed of the electrons.[2] In comparison, the flux of ions is $\frac{1}{4}\, n_i\, \overline{v}_i$. Since the ion mass is much larger than the electron, the ion thermal speed $\overline{v}_i \ll \overline{v}_e$ and the electron flux much larger than the ion. As a result, a negative surface charge builds on the wall which generates an electric field and reduces the electron flux to that of the ions.

Within the region of space charge, namely the sheath, the field is given by Poisson's equation

$$\nabla^2 V = -\frac{e}{\epsilon_0}(n_i - n_e) \tag{1.32}$$

2 Consider the flux of particles of speed v striking a small area δS at an angle θ to the normal. In time δt, the particles contained within a parallelepiped of volume $\delta S \times v\, \delta t\, \cos\theta$ will pass through the area. The probability of a particle having speed $v - (v + \delta v)$ is given by the distribution $\tilde{f}(v)\delta v$. The total number of particles of speed $(v, \delta v)$ moving at an angle between θ and $\theta + \delta\theta$ through δS is, therefore, $2\,\pi\, \sin\theta\, \delta S \times v\, \delta t\, \cos\theta\, \delta\theta \times n\, \tilde{f}(v)\, \delta v/4\,\pi$ since the velocities are uniformly distributed in angle. Noting that particles only strike the area if $v > 0$ and $0 < \theta < \pi/2$ and integrating over θ, the total number of particles striking unit area ($\delta S = 1$) per unit time ($\delta t = 1$), namely the flux

$$\text{Flux} = \frac{1}{2} \int_0^{\pi/2} \sin\theta\, \cos\theta\, d\theta \int_0^{\infty} n\, \tilde{f}(v)\, dv = \frac{1}{4} n\, \overline{v} \tag{1.31}$$

assuming the ions are singly charged. The electron density is given by Boltzmann's equation

$$n_e = n_0 \exp(e\,V/\mathcal{k}T_e) \tag{1.33}$$

since the electron drift velocity is always small compared to the thermal speed. However, this condition is not always satisfied for the ions, and the appropriate condition for their density must be identified.

1.5.1 Positively Biased Wall

If the wall is strongly biased positive so that the ions are retarded, Boltzmann's equation may be used for the ions also.

$$n_i = n_0 \exp(-e\,V/\mathcal{k}T_i) \tag{1.34}$$

Substituting in the sheath equation (1.32) leads directly to the planar form of Eq. (1.12) and hence the field

$$V = V_0 \exp(-x/\lambda_D) \tag{1.35}$$

where V_0 is the potential difference between the wall and the neutral plasma body at infinity. Clearly, the sheath width is equal to the Debye length.

1.5.2 Free Fall Sheath

If the absence of an applied field and when the mean free path is greater than the sheath width, the electron density is still given by Boltzmann's equation (1.33) due to the penetration into the field allowed by their velocity distribution. The ions in contrast free fall in the field with constant flux

$$n_i\,v_i = n_0\,v_0 \qquad \text{and} \qquad \frac{1}{2}\,m_i\,v_i^2 + e\,V = \frac{1}{2}\,m_i\,v_0^2 \tag{1.36}$$

where v_0 is the speed normal to the wall at which the ions enter the sheath. Hence, the sheath equation (1.32) becomes

$$\frac{d^2V}{dx^2} = \frac{n_0\,e}{\epsilon_0}\left\{\exp(e\,V/\mathcal{k}T_e) - 1/\sqrt{1 - 2\,e\,V/M_i\,v_0^2}\right\} \tag{1.37}$$

Substituting $\eta = -e\,V/\mathcal{k}T_e$ and $\xi = x/\lambda_D$, we obtain

$$\begin{aligned}\frac{d^2\eta}{d\xi^2} &= 1/\sqrt{1 + r\,\eta} - \exp(-\eta)\\ &\approx 1 - \frac{1}{2}\,r\,\eta - (1 - \eta) = \left(1 - \frac{r}{2}\right)\,\eta\end{aligned} \tag{1.38}$$

where $r = \mathcal{k}T_e/\frac{1}{2}m_i\,v_0^2$.

If $r > 2$, the solution oscillates, whereas if $r < 2$, one solution increases exponentially and cannot match the field into the bulk plasma. Hence, the sheath can only form if *Bohm's criterion* is obeyed

$$\frac{1}{2}m_i\,v_0^2 > \frac{1}{2}\mathcal{k}T_e \tag{1.39}$$

and the ions must have velocity greater than $\sqrt{\mathcal{k}T_e/m_i}$ on entering the sheath from the neutral plasma body.

Assuming the condition $r = 2$, we may calculate the potential drop to the wall since the ion and electron fluxes must be equal

$$n_0\ v_0 = n_0\ \sqrt{\kappa T_e/m_i} = \frac{1}{4}\ n_0\ \overline{v}_e\ \exp(-\eta_0) = \frac{1}{2}\ n_0\ \sqrt{\kappa T_e/2\ \pi\ m_e}\ \exp(-\eta_0)$$

Hence, substituting for v_0 from Bohm's criterion, we obtain

$$\eta_0 = -\frac{e\ V}{\kappa T_e} = \frac{1}{2}\ln\left(\frac{m_i}{2\ \pi\ m_e}\right) \approx 2.84 + \frac{1}{2}\ \ln\left(\frac{m_i}{m_p}\right) \tag{1.40}$$

where m_p is the proton mass. Clearly, the potential drop across the sheath is three times the electron temperature measured in eV.

The thickness of the sheath is obtained from

$$\frac{1}{2}\left(\frac{\mathrm{d}\eta}{\mathrm{d}\xi}\right)^2 = \frac{2}{r}\sqrt{1 + r\eta} + \exp(-\eta) + C$$

obtained by integrating Eq. (1.38). Integrating again

$$\begin{aligned} \xi_0 &= \int_0^{\eta_0} \left(1\Big/\frac{\mathrm{d}\eta}{\mathrm{d}\xi}\right)\,\mathrm{d}\eta \approx \int_0^{\eta_0} \frac{\mathrm{d}\eta}{\left[\frac{4}{r}\sqrt{(1 + r\eta) + 2\exp(-\eta) + 2C}\right]^{1/2}} \\ &\approx \frac{1}{2} r^{1/4} \int_0^{\eta_0} \frac{\mathrm{d}\eta}{\eta^{1/4}} \approx 2 \end{aligned} \tag{1.41}$$

The thickness of the sheath is, therefore, approximately $2\lambda_D$.

1.5.2.1 Pre-sheath

As we have seen above, the free fall sheath requires an ion flux into its edge for stability with speed $v_0 > \sqrt{\kappa T_e/m_i}$ which must be supplied from the quasi-neutral plasma bulk. The manner in which this flux is formed can be deduced as follows:

At the sheath edge all the ions are accelerated to the surface, where they recombine with electrons and are lost, i.e. there are no reflected particles. The ion distribution is therefore highly anisotropic; only that part with velocities directed from the plasma to the wall contributing, the backward half being missing. The ion density is therefore half that in the bulk.

The plasma in this pre-sheath region is quasi-neutral. Therefore, the electron density must also be half that in the bulk. Since the electron density is determined by the Boltzmann equation (1.33), this reduction must be accomplished by a potential drop at the sheath edge relative to the bulk

$$n_s/n_0 = \exp(e\ V_s/\kappa T_e) = 1/2 \quad \text{and} \quad e\ V_s/\kappa T_e = -\ln 2$$

This potential drop will provide an ion drift velocity $\sqrt{2\ e\ V_s/m_i} = \sqrt{2\ \ln 2\ \kappa T_e/m_i}$ which is sufficient to satisfy Bohm's criterion.

1.5.3 Mobility Limited Sheath

If the ionisation is low and the density high, then the mean free path is short compared to the sheath width. The motion of the ions and electrons is controlled by collisions, namely by diffusion and mobility. Thus, the ion flux, which is constant through the sheath

$$\Gamma_i = n_i v_i = -D_i \frac{\mathrm{d}n_i}{\mathrm{d}x} + n_i \mu_i E \tag{1.42}$$

where D_i is ion diffusion coefficient and μ_i the ion mobility – the ion drift speed per unit field. The electron density is again given by the Boltzmann term (1.33). The electric field E is given by the Maxwell equation

$$\frac{\mathrm{d}E}{\mathrm{d}x} = \frac{e}{\epsilon_0}(n_i - n_e) \tag{1.43}$$

Since the ion flux is constant, it must equal the drift from the plasma bulk to the wall, which can only be determined by a complete description of the overall plasma body. Thus, in this case, the sheath must be considered to be an integral part of the complete dynamic plasma system. We note the contrast with the free fall sheath.

2

Fundamental Plasma Parameters – Collisional Behaviour

As we noted in Chapter 1, the most characteristic plasma behaviour is associated with collective effects in which a large number of particles interact through the long-range Coulomb forces. In this chapter, we examine the behaviour of collisional effects, which are closely associated with the molecules of a classical gas, where the interactions are short-range collisions. In plasma, these also play an important role in damping wave motions and introducing entropy generation.

In plasma, the long-range behaviour of the Coulomb force again plays an important role. The effective range of the force is limited by the Debye length as over larger distances, collective effects dominate. In contrast to molecular forces in gases, these lengths are much larger than the inter-particle separation. Collisional interaction in plasma is therefore significantly different from gases in that instead of two particle interactions being the norm as with molecules, a charged particle interacts with many neighbouring particles simultaneously. Fortunately, the inverse square law of force associated with charges has some surprising but useful properties. As we shall see, the collisional behaviour of multi-particle collisions has the same theoretical form as that due to binary collisions. In addition, as the cross section for collision is identical for both classical and quantum collisions (apart from a logarithmic term), and consequently, collisional behaviour may be generally treated as classical (Landau and Lifshitz 1959).

2.1 Electron Scattering by Ions

2.1.1 Binary Collisions – Rutherford Cross Section

Consider the collision between a particle of charge q and mass m and a second of charge Q and mass M (Figure 2.1). When one particle is much heavier than the other $M \gg m$, the scatterer M may be taken as a stationary centre of force in which the lighter test particle is scattered by the Coulomb interaction between them. The force between the particles is along the line of separation, i.e. radially

$$\mathbf{F} = \frac{qQ}{4\pi\epsilon_0 r^2}\,\hat{\mathbf{r}} \tag{2.1}$$

of inverse square law form, $\hat{\mathbf{r}}$ being the unit vector in the direction of the inter-particle separation $\mathbf{r}$. The collision is elastic in that the speed after the collision is the same as that before v. The consequence of the collision is therefore a rotation of the velocity through the angle of scatter χ. The particle path is a hyperbola, symmetric about a line through the centre. The impact parameter b is defined as the distance between the projected incoming path and a line parallel to it through the centre.

Foundations of Plasma Physics for Physicists and Mathematicians, First Edition. Geoffrey J. Pert.

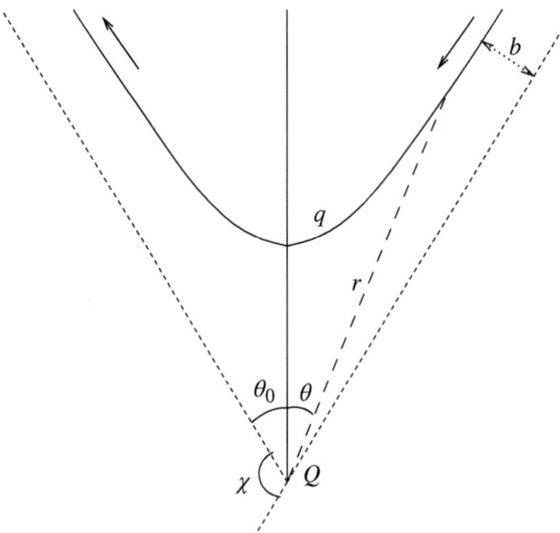

Figure 2.1 The Coulomb collision path of a light ion of charge q passing a heavy ion Q at impact parameter b giving a scattering angle χ.

By symmetry, the impulse communicated to the test particle m is along the line of symmetry and thus provides the change of momentum associated with scattering the particle through an angle χ

$$I = 2mv_0\cos\theta_0 = \int_{-\infty}^{\infty} F_z\,\mathrm{d}t = \frac{1}{4\pi\epsilon_0}\int_{-\infty}^{\infty}\frac{qQ}{r^2}\cdot\cos\theta\,\mathrm{d}t = \frac{1}{4\pi\epsilon_0}\int_{-\theta_0}^{\theta_0}\frac{qQ\cos\theta}{r^2\dot{\theta}}\,\mathrm{d}\theta \tag{2.2}$$

But since angular momentum is conserved, the force being central (i.e. radial), $mr^2\dot{\theta}$ is constant and

$$mr^2\dot{\theta} = mbv$$

Hence,

$$I = \frac{qQ}{4\pi\epsilon_0 bv}\int_{-\theta_0}^{\theta_0}\cos\theta\,\mathrm{d}\theta = \frac{qQ\sin\theta_0}{2\pi\epsilon_0 bv} \tag{2.3}$$

Since the angle of scatter $\chi = \pi - 2\theta_0$, we obtain the angle of scatter for an impact parameter b

$$\cot(\chi/2) = 4\pi\epsilon_0 mbv^2/qQ \tag{2.4}$$

This result is easily generalised (see Footnote 3, p. 19) to treat the case when the two particles both have finite masses m and m' moving with a centre of mass velocity $\mathbf{V} = (\mathbf{v}+\mathbf{v}')/2$ and relative velocity $\mathbf{u} = \mathbf{v}-\mathbf{v}'$. Since the field is central, the interaction may be described in the centre of mass frame as the motion of a particle of mass $\overline{m} = mm'/(m+m')$, known as the reduced mass, at a distance r equal to the separation of the particles in the field of a massive particle located at the centre of mass. The individual particle distances from the centre of mass are, therefore, $m'\mathbf{r}/(m+m')$ and $-m\mathbf{r}/(m+m')$ for particles m and m', respectively; the relation, Eq. (2.4), between the impact parameter b and the angle of scatter χ of the relative velocity in the centre of mass frame remains unchanged.

The differential Rutherford cross section $\mathrm{d}\sigma$ for scattering by a central Coulomb field into a small solid angle $\mathrm{d}\Omega = \sin\chi\,\mathrm{d}\chi\,\mathrm{d}\psi$ in the centre of mass frame is given by $b\mathrm{d}b$

$$\mathrm{d}\sigma = \frac{1}{4}\left(\frac{qQ}{(4\pi\epsilon_0)\overline{m}u^2}\right)^2\csc^4(\chi/2)\mathrm{d}\Omega \tag{2.5}$$

We may note that this value for the cross section is valid quite generally whether the scattering is quantum mechanical or classical (Landau and Lifshitz 1959, §112).

This result immediately raises a serious problem for as $\chi \to 0$, $\cot(\chi/2) \to \infty$ and b becomes large. The cross section area, i.e. the probability, for small angle scattering becomes very large. For atoms and molecules, where the law of force falls off more rapidly than inverse square law long-range collisions do not give rise to difficulties. However, the problems with Coulomb collisions may be accounted for as exemplified in Eq. (2.7), which exhibits a logarithmic singularity which is treated by introducing a cut-off based on physical intuition.

2.1.2 Momentum Transfer Cross Section

An important term which occurs frequently in calculations of the collisional behaviour of transport effects, such as conductivity, or in ion–electron energy exchange is the term expressing the slowing down of particles – particularly electrons – the *momentum transfer* or *transport* cross section.

The momentum loss in the direction of the incoming velocity by an electron in a collision with a heavy ion is $mv(1 - \cos\chi)$, where χ is the angle of scatter. Electrons with impact parameters b to $(b + db)$ will be scattered through χ to $(\chi + d\chi)$. Therefore, the momentum transfer cross section reflecting the mean fraction momentum loss by electrons in collision

$$\sigma_d = 2\pi \int_0^{b_{\max}} b\,db(1 - \cos\chi) = 4\pi \int_0^{b_{\max}} b\,db\ \sin^2(\chi/2) = 4\pi \int_0^{b_{\max}} \frac{b\,db}{[1 + \cot^2(\chi/2)]} \tag{2.6}$$

where we have introduced an upper limit $b_{\max}$ to limit long-range small angle scattering. Using the Rutherford scattering formula, Eq. (2.4) and introducing the Landau length $b_0 = qQ/4\pi\epsilon_0\overline{m}u^2$ corresponding to 90° scattering, we obtain

$$\sigma_d = 2\pi {b_0}^2 \int_0^{(b_{\max}/b_0)^2} \frac{dx}{(1+x)} = 2\pi {b_0}^2 \ln\left[1 + \left(\frac{b_{\max}}{b_0}\right)^2\right] \approx \frac{4\pi q^2 Q^2 \ln\Lambda}{(4\pi\epsilon_0)^2 \overline{m}^2 u^4} \tag{2.7}$$

where $\ln\Lambda = \ln(b_{\max}/b_0)$ is the Spitzer or Coulomb logarithm.[1]

The need for the introduction of a cut-off to the integral over the impact parameter at small scattering angles is clearly seen from the logarithmic divergence of the appropriate integral, which occurs generally with Coulomb scattering. However, as we have seen in Section 1.3, the field is screened at distances greater than the Debye length by background charges. It is, therefore, appropriate to replace the cut-off $b_{\max} \to \lambda_d$ in the above equation.

$$\ln\Lambda = \ln\frac{\lambda_d}{b_0} = \ln\left[\frac{\sqrt{\epsilon_0 \kappa T_e/n_e e^2}}{qQ/4\pi\epsilon_0 m u^2}\right] \tag{2.8}$$

It is easy to see that $\Lambda \sim 4\pi n_e {\lambda_d}^3$, is identical to the plasma parameter defined in Eq. (1.3.2), i.e. the number of particles in the Debye sphere. From this, we may draw two important conclusions

- The theory is only valid for dilute plasmas when $\Lambda \gg 1$.
- Many particles interact simultaneously with the test particle, but weakly.

1 We have assumed that the electron motion can be described by classical mechanics to derive the above cross section. However, if the mean particle velocity is high, the collision must be described quantum mechanically, i.e. when the electron wavelength is larger than the Landau parameter $\lambda_e = \hbar/\overline{m}u \gtrsim b_0 = qQ/(4\pi\epsilon_0\overline{m}u^2)$. Using the Born approximation for the cross section, this corresponds to the replacement of the Landau length b_0 by the electron wavelength λ_e in the Coulomb logarithm.

2.1.2.1 Dynamical Friction and Diffusion

Two important quantities follow the momentum change following collision. In the centre of mass frame of reference, the averages of the velocity change $\langle\Delta\mathbf{v}\rangle$ and the outer product[2] of the velocity change with itself $\langle\Delta\mathbf{v}\,\Delta\mathbf{v}\rangle$ in the direction of the incoming velocity is given by an integral over the outgoing direction of the scattered particle

$$\langle\Delta\mathbf{v}\rangle = \int \Delta\mathbf{v}\,\mathrm{d}\Omega \qquad \text{and} \qquad \langle\Delta\mathbf{v}\,\Delta\mathbf{v}\rangle = \int \Delta\mathbf{v}\,\Delta\mathbf{v}\,\mathrm{d}\Omega \tag{2.9}$$

known as the *coefficient of dynamical friction* and the *coefficient of diffusion*, respectively.

Taking components parallel and perpendicular to the direction of the incoming particles in the centre of mass frame and remembering that the velocity of the scattering particle is $m'u/(m+m')$ and that the centre of mass velocity is unchanged on collision

$$\begin{aligned}\langle\Delta v_{\parallel}\rangle &= -\int \frac{1}{4}\left(\frac{qQ}{(4\pi\epsilon_0)\overline{m}u^2}\right)^2 u\,\frac{m'u}{(m+m')}(1-\cos\chi)\,\csc^4(\chi/2)\,\mathrm{d}\Omega \\ &= -\frac{4\pi q^2Q^2N}{(4\pi\epsilon_0)^2m^2u^2}\,\frac{(m+m')}{m'}\ln\Lambda \end{aligned}\tag{2.10a}$$

$$\langle\Delta v_{\perp}\rangle = \int \frac{1}{4}\left(\frac{qQ}{(4\pi\epsilon_0)\overline{m}u^2}\right)^2 u\,\frac{m'u}{(m+m')}\sin\chi\cos\phi\,\csc^4(\chi/2)\,\mathrm{d}\Omega = 0 \tag{2.10b}$$

$$\begin{aligned}\langle\Delta v_{\perp}{}^2\rangle &= \int \frac{1}{4}\left(\frac{qQ}{(4\pi\epsilon_0)\overline{m}u^2}\right)^2 u\left[\frac{m'u}{(m+m')}\right]^2 \sin^2\chi\,\cos^2\phi\,\csc^4(\chi/2)\,\mathrm{d}\Omega \\ &= \frac{4\pi q^2Q^2N}{(4\pi\epsilon_0)^2m^2u}\left(2\ln\Lambda - \frac{\Lambda^2}{(1+\Lambda^2)}\right)\end{aligned}\tag{2.10c}$$

$$\begin{aligned}\langle\Delta v_{\parallel}{}^2\rangle &= \int \frac{1}{4}\left(\frac{qQ}{(4\pi\epsilon_0)\overline{m}u^2}\right)^2 u\left[\frac{m'u}{(m+m')}\right]^2 (1-\cos\chi)^2\csc^4(\chi/2)\,\mathrm{d}\Omega \\ &= \frac{4\pi q^2Q^2N}{(4\pi\epsilon_0)^2\,m^2\,u}\,\frac{\Lambda^2}{(1+\Lambda^2)}\end{aligned}\tag{2.10d}$$

It is normally assumed that $\ln\Lambda \gg 1$ and that the terms ~ 1 can be neglected.

The factor $(m+m')/m'$ in the expression for $\langle\Delta v_{\parallel}\rangle$ is due to the recoil of the scattering particle of finite mass. This term can often be neglected for heavy scattering particles, e.g. electrons scattering from ions.

Since these rates are specified in directions relative to the inter-particle velocity in the centre of mass frame for the specific pair of particles, it is necessary to transform them into some fixed laboratory frame and integrate over the probability distribution of scatterers. The transformation can be taken to either spherical polar co-ordinates (Cohen et al. 1950; Spitzer and Härm 1953) or more commonly into Cartesian co-ordinates (MacDonald et al. 1957; Shkarofsky et al. 1966). We will defer this procedure until Chapter 7.

2.1.3 Many Body Collisions – Impulse Approximation

Many body collisions involve the test particle interacting simultaneously with many separate scatterers: nearly all of which are at distances $\gg b_0$ and therefore weak, producing only a small change

2 The term **AB** is the dyadic product of the vector **A** with **B** which may be written as an outer product of two column matrices $\mathbf{AB} \equiv \mathbf{A} \otimes \mathbf{B} \equiv \mathbf{A}\mathbf{B}^T \equiv \mathrm{A_i B_j}$. The dyadic product of two vectors is a tensor of order 2. We shall use the dyadic product form throughout this text.

in the particle's path. However, the electrostatic field is cut-off at distances of the order of the Debye length, λ_D, as we have argued earlier (Section 1.3). The range of impact parameters is, therefore, $b_0 < b < \lambda_D$. An individual collision is completed in a time $\lambda_D/\overline{v}_e = \Pi_e^{-1}$. If short-range collisions ($b \sim b_0$) occur during a prescribed time interval, their duration is very short, but for a very brief time they dominate the others, which continue in the background.

Since each individual interaction is weak, it has only a very small effect on the test particle, i.e. the particle path is almost a straight line through the background of the perturbers, but with continuous random small impulses on it. During each interaction, the deflection of the path can be neglected, and we can calculate the effects of many interactions by summing the impulses given by each random interaction independent of the others. The characteristic time for a particle to be deflected from its straight-line path is determined by the electron–ion collision frequency ν_{ei}. At any instant, the test particle is interacting with a large number of particles in a Debye sphere $N \sim n\lambda_D{}^3$. The particle's trajectory is, therefore, determined statistically by averaging over the forces due to a large number of randomly distributed perturbers. Within this *impulse approximation*, we treat each interaction as separate and independent, and calculate the total effect by the vector sum of the impulses.

Provided the test particle is well separated from the scatterers, each impulse can be treated as a perturbation on the motion. To determine the deflection of the test particle velocity, we calculate the total impulse on the particle normal to its path in a short time interval Δt, during which the particle interacts with scattering particles located at many randomly distributed locations. In a dilute plasma, the plasma parameter $\Lambda \gg 1$ is large, and we may identify a time step $\Pi_e^{-1} \gg \Delta t \gg \nu_{ei}^{-1}$ such that collisions are completed within the time step whilst the particle path is little disturbed. Since the path is nearly straight, symmetry requires that each completed interaction leads to an impulse normal to the path. The mean deflection of the path over a finite time is obtained from the average over the scatterer particle distribution by summing over many small time steps.

Consider the scattering of a test particle by randomly distributed stationary perturbers of mean density n. We treat the situation along an interval of the path of the test particle of length $\Delta z = v\,\Delta t$ which is sufficiently short that the path is nearly a straight line, while the test particle samples the field from a representative group of scatterers. In the time interval Δt, it receives an impulse $\Delta\mathbf{p}_n$ from scatterer n. Since the scatterers are randomly distributed, individual impulses are in a direction along the normal from the scatterer to the test particle path. Since the scatterers are randomly distributed, the average net impulse on the test particle in Δt is zero justifying on a statistical basis the straight-line approximation. In the direction x_i,

$$\langle \Delta p_i \rangle = \sum_{n=1}^{N} \Delta p_{ni}/N = 0 \tag{2.11}$$

However, the mean squared impulse per interaction

$$\langle \Delta p_i{}^2 \rangle = \frac{1}{N}\left[\sum_{n=1}^{N} \Delta p_{ni}{}^2 + 2\cancel{\sum_{m=1}^{N}\sum_{\substack{n=1\\ n\neq m}}^{N} \Delta p_{mi}\,\Delta p_{ni}}\right] \quad \neq \quad 0 \tag{2.12}$$

is non-zero, where N is the total number of scatterers interacting with the test particle in time Δt (Figure 2.2).

On average in the time interval Δt, the particle experiences a representative sample of every type of impact since the scatterers are uniformly distributed, i.e. with every permitted impact parameter normal to the direction of motion z and at every distance along the path. In time Δt, the particle experiences $n\,\mathrm{d}z\,b\,\mathrm{d}b\,\mathrm{d}\theta$ encounters with scatterers at impact parameters b to $b + \mathrm{d}b$ at

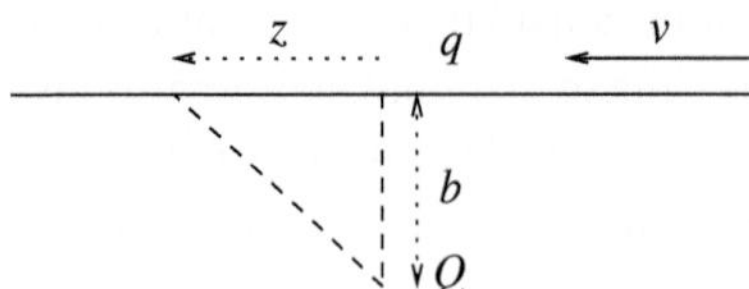

Figure 2.2 The weak collision trajectory of a light ion q of velocity v passing a stationary heavy ion Q at impact parameter b.

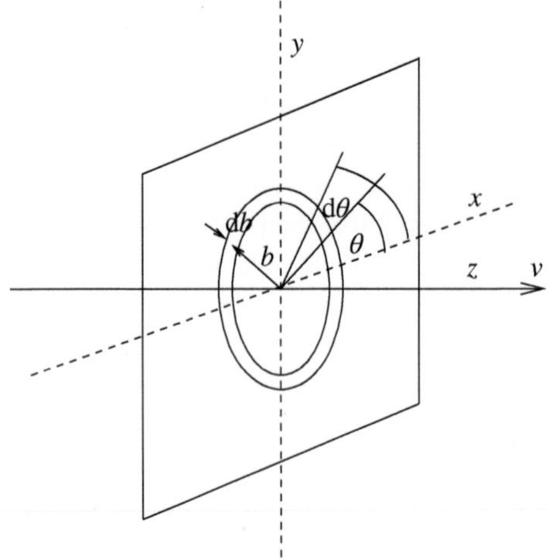

Figure 2.3 The weak collision trajectory of two repulsive ions at impact parameter b giving a scattering angle χ.

an azimuthal angle θ to $\theta + \mathrm{d}\theta$ along an interval of path $\mathrm{d}z$ (Figure 2.3). The contribution to the impulse from a scatterer at impact radius b along a line $z = -\infty$ to $z = \infty$ normal to the line of motion is easily calculated. Consider a test particle of mass m, charge q, and velocity v moving in the z direction interacting with a stationary massive particle of charge Q along a straight line with impact parameter b. The net impulse in the direction perpendicular to the test particle path from the force centre is

$$\Delta p(b) = \frac{qQ}{4\pi\epsilon_0 v}\int_{-\infty}^{\infty}\frac{b\,\mathrm{d}z}{(b^2+z^2)^{3/2}} = \frac{2qQ}{4\pi\epsilon_0 b v} \tag{2.13}$$

The mean squared impulse in the transverse direction x in time Δt is

$$\langle \Delta p_x{}^2\rangle = \iint \{\Delta p(b)\cos\theta\}^2\, n v b\, db\, d\theta\, \Delta t \tag{2.14}$$

where θ is the azimuthal angle with the x direction and b the impact parameter.

Since the impulses are random in both magnitude and direction, we may invoke the central limit theorem to predict that on averaging their distribution is normal with variance $\langle \Delta p_i{}^2\rangle$ in direction x_i, which increases linearly with the number of terms. As the particle moves along its path, the total mean squared impulse is a sum over the total number of scatterers encountered by the test particle along an element of path Δz. The variance increases linearly with path length or equivalently time. Thus, if the test particles initially form a collimated beam with small divergence, the beam will spread out to have a normal (Gaussian) distribution in angle with variance given by $\langle \Delta p_x{}^2\rangle$ after moving a distance Δz.

$$\begin{aligned}\langle \Delta p_x{}^2\rangle &= \int_{b_{\min}}^{b_{\max}} b\,db \int_0^{2\pi} d\theta \left[\frac{2qQ\cos\theta}{4\pi\epsilon_0 b m v^2}\right]^2 n v \Delta t \\ &= \frac{4\pi q^2 Q^2 n}{(4\pi\epsilon_0)^2 v}\int_{b_{\min}}^{b_{\max}}\frac{db}{b}\,\Delta t\end{aligned} \tag{2.15}$$

in time Δt and we have introduced cut-offs to avoid the logarithmic singularity at large and small impact parameters. When b is large, the scatterer potential at distances $b \gtrsim \lambda_D$ is screened, giving $b_{\max} \sim \lambda_D$.[3] When b is small, the deflection is large, the interaction is a two particle collision described by the Rutherford formula, Eq. (2.4) which is equivalent to a cut-off at $b_{\min} = b_0$ (the Landau parameter/distance) and corresponds to 90° scattering[4]

$$\langle \Delta p_x{}^2 \rangle = \frac{4\pi q^2 Q^2 n}{(4\pi\epsilon_0)^2 m^2 v^3} \ln\left[\frac{\lambda_D}{b_{\min}}\right] \Delta t = \frac{4\pi q^2 Q^2 n}{(4\pi\epsilon_0)^2 v} \ln \Lambda \, \Delta t \tag{2.16}$$

where $\ln \Lambda = \ln(\lambda_D / b_{\min})$ is the Spitzer or Coulomb logarithm introduced earlier, Eq. (2.8), whose value is typically (~5–10).

The mean squared deflection angle in the direction x is given by $\langle \Delta\chi_x^2 \rangle = \langle \Delta p_x^2 \rangle / (mv)^2$ since (mv) is the momentum of the particle. Noting that $\langle \Delta\chi_x^2 \rangle = \langle \Delta\chi_y^2 \rangle = \frac{1}{2}\langle \Delta\chi^2 \rangle$, it follows from Eq. (2.16) that the average time taken for the particle path to be deflected by an angle χ in any direction is

$$\tau \approx \left[\frac{8\pi q^2 Q^2 n \ln \Lambda}{(4\pi\epsilon_0)^2 m^2 v^3 \chi^2}\right]^{-1} \tag{2.17}$$

In contrast, the average time taken for a deflection χ in a single two-body collision is

$$\tau \approx \left[\frac{\pi q^2 Q^2 n}{(4\pi\epsilon_0)^2 m^2 v^3 \tan^2(\chi/2)}\right]^{-1} \approx \left[\frac{4\pi q^2 Q^2 n}{(4\pi\epsilon_0)^2 m^2 v^3 \chi^2}\right]^{-1} \tag{2.18}$$

The physical significance of the Coulomb logarithm (or the plasma parameter) $\ln \Lambda$ in this context is that it represents the importance of small-angle, large-impact collisions to large-angle, small-impact collisions. Thus, provided $\ln \Lambda$ is large, the scattering is well described by the accumulation of weak collisions added cumulatively.[5]

It is useful to examine the average rate of change of the velocity of electrons in the course of weak collisions with stationary ions per unit time. Thus, we have

$$\langle \Delta v_x \rangle = \langle \Delta_y \rangle \approx 0 \quad \langle \Delta v_x^2 \rangle = \langle \Delta v_y^2 \rangle \not\approx 0 \quad \langle \Delta v_x \Delta v_y \rangle \approx 0 \tag{2.19}$$

where x and y are perpendicular to the electron motion. Writing $\perp$ for the total perpendicular velocity component and $\|$ for the parallel, it follows from

$$\Delta v_\perp = v \sin\chi \approx v\chi \quad \Delta v_\| = -v(1 - \cos\chi) \approx -\frac{1}{2} v\chi^2 \tag{2.20}$$

that the rate of changes of velocity components are

$$\langle \Delta v_\perp \rangle = 0 \tag{2.21a}$$

$$\langle \Delta v_\| \rangle = -\frac{1}{2}\frac{8\pi q^2 Q^2 N \ln \Lambda}{(4\pi\epsilon_0)^2 m^2 v^2} = -n\sigma_d v^2 \tag{2.21b}$$

$$\langle \Delta v_\perp{}^2 \rangle = \frac{8\pi q^2 Q^2 N \ln \Lambda}{(4\pi\epsilon_0)^2 m^2 v} \tag{2.21c}$$

3 In early work (for example, Jeans 1961; Chandrasekhar 1960), it was assumed that the cut-off was at the inter-particle separation distance ($n^{-1/3}$). Persico (1926) seems to have been the first to draw attention to the importance of Debye shielding in this context. Spitzer (Cohen et al. 1950) gives a detailed argument for the use of the Debye length.

4 See Footnote 1, page 15.

5 The stochastic nature of long-range weak collisions and their dominance over close encounter strong collisions was first noted by Jeans (1961, p. 318).

Finally, since each collision is elastic, the energy change

$$\Delta E = \frac{1}{2}m[(v+\Delta v_{\parallel})^2 + \Delta v_{\perp}{}^2] - \frac{1}{2}mv^2 = m\left[v\Delta v_{\parallel} + \frac{1}{2}\Delta v_{\parallel}{}^2 + \frac{1}{2}\Delta v_{\perp}{}^2\right] = 0$$

it follows on averaging that

$$\langle \Delta v_{\parallel}{}^2 \rangle = 0 \tag{2.22}$$

It is perhaps surprising that there is equivalence of the two particle collision rate for heavy scattering particles ($m' \gg m$) as given by Eqs. (2.10) with the multiple scatterer results exemplified by Eq. (2.21). The two results are obtained by physical behaviour which involve different statistical procedures. In two particle collisions, one interaction is completed before the next, whereas in multiple particle scattering, many interactions occur together. As a result although the averages are equal, the angular distribution of the test particles cannot be guaranteed to be identical in the two cases. The momentum scattering cross section, Eq. (2.7) is derived using the assumption of well separated binary collisions for electron–ion collisions. The averaging over the impact parameter arises from the averaging in time involving the summation of many separate independent impacts. In contrast in simultaneous multiple particle scattering, the averaging arises from the multiple forces from distributed scatterers acting at an instant of time. The integration in time is accomplished by the summation of many independent, but continuous short time intervals. Because the dominant interactions are weak, so that most test particle paths remain almost straight, the two averaging procedures are nearly equivalent. In both cases, small impulses add independently and linearly as vectors. Consequently, the averaging processes involved in each calculation commute, and their order is immaterial. It should be noted that this is a particular property of the Coulomb interaction applying to plasmas, but not to gases where the interactions are short range. Plasma collisional effects should be calculated using multiple collision theory, whereas binary collision theory is appropriate for gases. The equivalence confirms that correctness of the inner cut-off applied to Eq. (2.16).

2.1.4 Relaxation Times

Equation (2.17) determines the time taken for one species of particle, typically the electrons, to be deflected by second much heavier ones (ions), which are therefore almost stationary. It therefore identifies the time taken for an electron distribution to be randomised by collisions with ions, or more specifically for any directed motion to be lost. In calculating the transport of current or heat, electrons dominate due to their larger velocity, the ions provide the resistive friction. The drag force on the directed drift motion of the electrons is therefore determined by the momentum loss due to electron/ion collisions, time averaged over the particle distribution. Averaging the collision frequency, $N\sigma_d v$, given by the momentum transfer cross section, Eq. (2.7), over a Maxwell distribution the collision time for electrons τ_e, the following value is conventionally taken for this relaxation time (Braginskii 1965, p. 215) (see Eq. (2.A.11))

$$\tau_{ei} = \frac{3(4\pi\epsilon_0)^2 m^{1/2}(\κT_e)^{3/2}}{4\sqrt{2\pi}\,Z^2 e^4 N \ln\Lambda} \tag{2.23}$$

where T_e is the temperature of the electron's Maxwellian distribution.

We may also calculate a momentum transfer collision time for like particles, for example for ions with ions, taking into account the relative motion of both particles (Jeans 1954, §33) (Section 2.4.3)

$$\tau_{ii} = \frac{3(4\pi\epsilon_0)^2 M^{1/2}(\κT_i)^{3/2}}{4\sqrt{\pi}\,Z^4 e^4 N \ln\Lambda} \tag{2.24}$$

where T_i is the ion temperature. This term accounts for the drag force on ions when this is important, for example in determining viscosity.

Equation (2.17) has a particularly important property in that the relaxation times for particles of differing velocities are larger for faster particles. Consequently, fast particles suffer less resistance and contribute more strongly to collisional transport behaviour.

In addition to playing the dominant role limiting collisional transport effects, through the particle drag force, there are several other related relaxation times Eq. (2.A.8) associated with momentum transfer between particles

- Electron–electron collisions determine the time taken for the electron distribution to relax to a Maxwellian – τ_{ee} (see Section 7.5).
- Electron–ion collisions are the dominant terms determining electrical conductivity and other transport coefficients – τ_{ei}.
- Ion–ion collisions determine the time taken for the ion distribution to relax to a Maxwellian – τ_{ii}.
- Ion–electron collisions are relatively inefficient at transferring momentum for the ions to the electrons due to the large mass ratio but play a role in magnetised plasma – τ_{ie}.
- Electron–ion energy exchange time. The time taken for an electron and ion distribution at different temperatures to reach thermal equilibrium – $\tau_{eq} \approx \frac{1}{2}\tau_{ie}$ (Eq. (7.59)).

Because of the large mass difference between ions and electrons, these terms all have markedly different values in approximate ratios

$$\tau_{ee} : \tau_{ei} : \tau_{ii} : \tau_{ie} : \tau_{eq} :: 1 : \frac{1}{\sqrt{2}Z^2} : \frac{1}{Z^4}\sqrt{\frac{M}{m}} : \frac{1}{\sqrt{2}Z^2}\frac{M}{m} : \frac{1}{2\sqrt{2}Z^2}\frac{M}{m} \tag{2.25}$$

the latter result following as the energy transfer from electrons to ions per collision is only $\sim m/M$. However, as the average fractional energy exchange between particles of like mass is about $1/2$, it follows that $\tau_{ee} \sim \tau_e$ and $\tau_{ii} \sim \tau_i$.

2.2 Collisional Transport Effects

Many processes in plasma involve the transport of particles for example electric current, heat conduction, diffusion, etc. Collisions play a dominant role limiting the speed at which the transport can take place. The rate at which these transports processes occur is described by a linear relation between the flux, namely the quantity flowing per unit area per unit time and the force driving the motion, typical cases are

- *Ohm's law*: The linear relationship between electric current density $\mathbf{j}$ and the driving electric field $\mathbf{E}$ expressed in terms of the conductivity σ

  $$\mathbf{j} = \sigma\,\mathbf{E} = -\sigma\,V \tag{2.26}$$

  An alternative statement is obtained by relating the electric current to the mobility flux of charge carriers defining the mobility $\mu = \sigma/nq$.
- *Fick's law*: The linear relationship between the diffusive drift $\boldsymbol{\Gamma}$ and the density gradient ∇n giving rise to the drift expressed in terms of the diffusivity D

  $$\boldsymbol{\Gamma} = -D\nabla n \tag{2.27}$$

- *Fourier's law*: The linear relationship between the heat flux q and the temperature gradient ∇T giving rise to a thermal flux expressed in terms of the thermal conductivity κ

$$\mathbf{q} = -\kappa \nabla T \tag{2.28}$$

All are linear relationships, provided the variation of the driving force X over a mean free path is small, i.e. $\lambda |\nabla X|/X \ll 1$. If the medium is anisotropic, these coefficients become tensor quantities.

It is useful at this point to examine relatively simple models, which although developed for other media, give physical insight and order of magnitude scalings for the relevant transport coefficients. These models were originally developed to treat transport phenomena in gases and are extremely valuable clearly identifying the underlying physical behaviour of many physical phenomena. As a result, although the results in plasma cannot be treated as exact, principally because the collisional terms such as the mean free path and the collision frequency, cannot be evaluated by simple means. The approach is very flexible and often used for more complex problems.

2.2.1 Random Walk Model for Transport Effects

Diffusion, viscosity, and thermal conduction all represent the transfer of a physical quantity $Q(X)$, namely, respectively, particles, momentum, and energy, by individual particles in the gradient of a field X, respectively, density, velocity, and temperature. In each case, the mechanism is similar; as a result, we may consider the three together by a simple model representing the transfer of molecules by a random walk across a plane perpendicular to the gradient of the field. The resultant coefficients, although not accurate, are very useful for order of magnitude estimates.

Provided the number of individual walks is large, the random walk mechanism may be clearly identified as an application of the central limit theorem, Section 2.B. Consider a group of particles of speed v starting at the origin suffering a series of collisions[6] with mean free path $\lambda(v)$, the probability of a path with projection $(x - x + \mathrm{d}x)$ in the x direction is

$$p_x \,\mathrm{d}x = \left\{ \int_0^\infty \int_0^\pi \exp(-r/\lambda)\delta(x - r\cos\theta) \sin\theta \,\mathrm{d}r\,\mathrm{d}\theta \right\} \Bigg/ \left\{ \int_0^\infty \int_0^\pi \exp(-r/\lambda) \sin\theta \,\mathrm{d}r\,\mathrm{d}\theta \right\} \mathrm{d}x \tag{2.29}$$

where $\delta(x)$ is the familiar Dirac delta function. The mean and the variance of the path in the x direction between collisions each with free path r and mean free path λ are therefore,

$$\langle x \rangle = \int_0^\infty \int_0^\pi \exp(-r/\lambda) r \cos\theta \sin\theta \,\mathrm{d}r\,\mathrm{d}\theta \Bigg/ \int_0^\infty \int_0^\pi \exp(-r/\lambda) \sin\theta \,\mathrm{d}r\,\mathrm{d}\theta = 0 \tag{2.30}$$

$$\langle x^2 \rangle = \int_0^\infty \int_0^\pi \exp\left(-r \big/ \lambda\right) r^2 \cos^2\theta \sin\theta \,\mathrm{d}r\,\mathrm{d}\theta \Big/ \int_0^\infty \int_0^\pi \exp(-r/\lambda) \sin\theta \,\mathrm{d}r\,\mathrm{d}\theta = 2\lambda^2/3 \tag{2.31}$$

After N steps, each lasting on average a time $\tau = \lambda/v$ the variance will be $N\langle x^2 \rangle$, the particles will be distributed in the x direction with a normal profile

$$p_x \,\mathrm{d}x = \frac{1}{\sqrt{2\pi N \langle x^2 \rangle}} \exp\left\{ -\frac{x^2}{2N\langle x^2 \rangle} \right\} \mathrm{d}x \tag{2.32}$$

with similar values for the distribution in the y and z directions.

---

6 Although unrealistic, we assume the scattering is isotropic.

Comparing this result (2.32) with the standard expression for diffusion from a point source in one direction only

$$p_x\,\mathrm{d}x = \frac{1}{\sqrt{4\pi Dt}}\exp\left\{-\frac{x^2}{4Dt}\right\}\mathrm{d}x \tag{2.33}$$

Hence, we obtain the diffusion coefficient

$$D = \frac{1}{2}\frac{N}{t}\langle x^2\rangle = \frac{1}{3}v\lambda \tag{2.34}$$

Since the sets of position co-ordinates (x, y, z) or (r, θ, ϕ) are each statistically independent, we may write the particle distribution in either Cartesian or polar co-ordinates

$$p_x(x,y,z)p_y(x,y,z)p_z(x,y,z)\,\mathrm{d}x\,\mathrm{d}y\,\mathrm{d}z = p_r(r,\theta,\phi)p_\theta(r,\theta,\phi)p_\phi(r,\theta,\phi)\,\mathrm{d}r\,\mathrm{d}\theta\,\mathrm{d}\phi \tag{2.35}$$

Hence, since the diffusion from a point source is isotropic, we obtain the probability of finding a particle in the shell $(r, \mathrm{d}r)$ (see Problem 1)

$$p_r\,\mathrm{d}r = \frac{4\pi}{(4\pi\lambda vt/3)^{3/2}}r^2\exp\left[-\frac{r^2}{4\lambda vt/3}\right]\mathrm{d}r = \frac{4\pi r^2}{(4\pi Dt)^{3/2}}\exp\left[-\frac{r^2}{4Dt}\right]\mathrm{d}r \tag{2.36}$$

That diffusion, viscosity, and thermal conduction are all *random walk* phenomena, may be demonstrated as follows: Assume that particles travel a distance ℓ between collisions over a time τ, the collision time. At each collision, the quantity $Q(X)$ is reset with the value appropriate to the value of $X(x)$ at its position x. Suppose, for simplicity, particles only move parallel to the x direction, which is parallel to the gradient in Q. At a collision, the particle's value of Q is reset to the local value. After the collision, $\frac{1}{2}$ the particles move forward and $\frac{1}{2}$ backwards. Taking the origin $x = 0$ at the surface, where the value of $Q = Q_0$, those coming from the left after a collision at $x < 0$ carry a value $Q(x) = Q_0 - x\,\mathrm{d}Q/\mathrm{d}X \cdot \mathrm{d}X/\mathrm{d}x$, and those from the right $x\rangle 0$ carry $Q(x) = Q_0 + x\,\mathrm{d}Q/\mathrm{d}X \cdot \mathrm{d}X/\mathrm{d}x$. There is therefore a net flux of $Q(X)$ through the surface $x = 0$ of approximately

$$\frac{1}{2}\frac{1}{\tau}n\left\{\int_{-\ell}^{0}x\,\mathrm{d}x - \int_{0}^{\ell}x\,\mathrm{d}x\right\}\frac{\mathrm{d}Q}{\mathrm{d}X}\cdot\left.\frac{\mathrm{d}X}{\mathrm{d}x}\right|_0 = -\frac{1}{2}n\frac{\ell^2}{\tau}\frac{\mathrm{d}Q}{\mathrm{d}X}\left.\frac{\mathrm{d}X}{\mathrm{d}x}\right|_0 \tag{2.37}$$

In the usual case, ℓ is the mean free path and τ the collision interval $\ell = \lambda = v\tau$, so that apart from a numerical factor this expression is identical to Eq. (2.43). The general problem of random walks in three dimensions is discussed by Chandrasekhar (1943, §3). The Maxwell's mean free path model, described below, is the asymptotic case when the number of walks is very large.

2.2.2 Maxwell's Mean Free Path Model of Transport Phenomena

This relatively simple model generalises the random walk model by following particle drift both diffusion and current flow, and the transport of heat (thermal conduction) and momentum (viscosity). The model requires that the particle distribution is only weakly perturbed by the driving force and the gas remains in approximately thermal equilibrium. This condition is obeyed if the driven flux is much smaller than the background random flux, which averages to zero (see Section 2.2.2.1). The model was originally derived by Maxwell (1867) in his classic study of kinetic theory (Figure 2.4).

Particles are allowed to travel in any direction and with a velocity determined by the appropriate distribution as originally treated by Maxwell (1867, p. 73). A quantity $Q(X)$ per molecule is transported across a plane x_0 normal to the gradient of a quantity X, namely $\mathrm{d}X/\mathrm{d}x$. As before, we assume that molecules when suffering a collision acquire the value of Q appropriate to the position

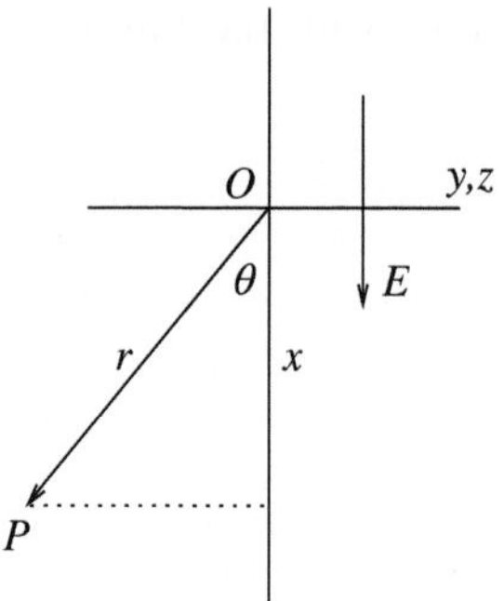

Figure 2.4 Drift collision geometry.

of the collision x in the gradient of X. The small surface element ΔS normal to the axis defined by the gradient is placed at O the origin of a set of spherical polar co-ordinates (r, θ, ϕ). Consider the number of molecules, which suffer a collision at P in the volume element $(\mathrm{d}r, \mathrm{d}\theta, \mathrm{d}\phi)$ at (r, θ, ϕ) passing through ΔS without suffering another collision. The probability of a collision in a path length $\mathrm{d}r$ is $\mathrm{d}r/\lambda$ and, therefore, the probability of a free path greater than r is $\exp(-r/\lambda)$, where λ is the mean free path. The solid angle subtended by ΔS at the element is $\Delta S \cos\theta/4\pi r^2$. Assuming the distribution is isotropic and varies only weakly with the gradient, the number of particles having suffered a collision in the element, and which pass through the surface element ΔS per unit time is given by

$$\int \mathrm{d}\mathbf{v} f(\mathbf{v})\, v[\exp(-r/\lambda(v))][\Delta S\cos\theta/(4\pi r^2)]\, r^2 \sin\theta\, \mathrm{d}\phi\, \mathrm{d}\theta\, \mathrm{d}r/\lambda(v)$$
$$= \frac{1}{4\pi}\int \mathrm{d}\mathbf{v} f(\mathbf{v})\, v[\exp(-r/\lambda(v))/\lambda(v)] \sin\theta \cos\theta\, \mathrm{d}r\, \mathrm{d}\theta\, \mathrm{d}\phi\, \Delta S \tag{2.38}$$

where the *distribution function* $f(\mathbf{r}, \mathbf{v})$ is the number of particles in the phase space volume $(\mathbf{r}, \mathrm{d}r, \mathbf{v}, \mathrm{d}v)$ (see Chapter 4).

To describe mobility and diffusion, it is useful to treat the particle potential in the electric field and the density n explicitly by introducing the *probability density distribution function* $\tilde{f}(\mathbf{r}, \mathbf{v}) = f(\mathbf{r}, \mathbf{v})/n$; namely, the probability of finding a particle in element of phase space $(\mathrm{d}\mathbf{r}, \mathrm{d}\mathbf{v})$. Assuming the plasma is in thermal equilibrium,[7] the density distribution at P is related to that at O by $\tilde{f}_P = \tilde{f}_O e^{-q\Delta V/\Bbbk T}$ where $-\nabla V = \mathbf{E}$ is the electric field. The net number of particles passing through the area ΔS is therefore

$$-\frac{1}{4\pi}\int_0^{2\pi}\int_0^{\pi}\int_0^{\infty}\int \mathrm{d}\mathbf{v}\tilde{f}(\mathbf{v})\, v \left[\frac{\exp(-r/\lambda(v))}{\lambda(v)}\right]\left[n_0 + \left(\frac{\mathrm{d}n}{\mathrm{d}x} + \frac{nq}{\Bbbk T}\frac{\mathrm{d}V}{\mathrm{d}x}\right) r\cos\theta\right]$$
$$\times \sin\theta\cos\theta\, \mathrm{d}r\, \mathrm{d}\theta\, \mathrm{d}\phi\, \Delta S \tag{2.39}$$
$$= -\frac{1}{3}\int \mathrm{d}\mathbf{v}\, \tilde{f}(\mathbf{v})\; v\; \lambda(v) \left(\frac{\mathrm{d}n}{\mathrm{d}x} - \frac{nq}{\Bbbk T}E\right)$$

This equation is known as the *general mobility equation.*

Since $\int \tilde{f}(\mathbf{v}\mathrm{d}\mathbf{v}) = 1$ the diffusivity is

$$D = \frac{1}{3}\int \mathrm{d}\mathbf{v}\tilde{f}(\mathbf{v})\, v\, \lambda(v) = \frac{1}{3}\langle v\, \lambda(v)\rangle = \frac{1}{3}\langle v^2\, \tau\rangle = \frac{\tau \Bbbk T}{m} \tag{2.40}$$

where $\tau = v\,\lambda$ is the collision time, assumed constant, and $m\langle v^2\rangle = 3\Bbbk T$.

7 Thermal equilibrium is maintained between O and P provided the total mobility flux associated with the perturbation $\tilde{f}_1$ is small compared to forward or backward fluxes associated with $\tilde{f}_0$.

The mobility for charged particles q follows immediately in conformity with the Einstein–Smoluchowski relation (1.27a).[8]

$$\mu = \frac{q}{\mathrm{k}T}D = \frac{1}{3}\frac{q}{\mathrm{k}T}\langle v\,\lambda(v)\rangle = \frac{q\tau}{m} \tag{2.42}$$

The quantity X may be written as $X_0 + x\mathrm{d}X/\mathrm{d}x$ and, since the mean free path $\lambda \ll X/(\mathrm{d}X/\mathrm{d}x)$, $Q = Q_0 + x\mathrm{d}Q/\mathrm{d}X \cdot \mathrm{d}X/\mathrm{d}x$. Hence, the quantity of Q flowing through ΔS per unit time is

$$-\frac{1}{4\pi}\int_0^{2\pi}\int_0^{\pi}\int_0^{\infty}\int \mathrm{d}\mathbf{v} f(\mathbf{v})\,v\left[\frac{\exp(-r/\lambda(v))}{\lambda(v)}\right]\left(Q_0 + \frac{\mathrm{d}Q}{\mathrm{d}X}\frac{\mathrm{d}X}{\mathrm{d}x}r\cos\theta\right)\sin\theta\cos\theta\,\mathrm{d}r\,\mathrm{d}\theta\,\mathrm{d}\phi\,\Delta S$$
$$= -\frac{1}{3}\int \mathrm{d}\mathbf{v} f(\mathbf{v})\,v\,\lambda(v)\frac{\mathrm{d}Q}{\mathrm{d}X}\frac{\mathrm{d}X}{\mathrm{d}x} \tag{2.43}$$

To treat thermal conduction, X is the temperature, and $Q = c_v T$, c_v being the specific heat per molecule. The thermal conductivity is therefore,

$$\kappa = \frac{1}{3}\int \mathrm{d}\mathbf{v} f(\mathbf{v})\,v\,\lambda(v)c_v(v) = \frac{1}{3}n\langle v\,\lambda(v)c_v(v)\rangle \tag{2.44}$$

the average being taken over the velocity distribution.

To treat viscosity, Q is the momentum transfer, and X is the flow velocity. The momentum transfer per unit time per unit area corresponds to stress at the plane $x = 0$ normal to the axis, the viscosity is

$$\eta = \frac{1}{3}nm\langle v\,\lambda(v)\rangle \approx p\tau \tag{2.45}$$

since the free path $\lambda(v) = v\tau(v)$ and the average $\langle mv^2\rangle = 3\,\mathrm{k}T$ and the pressure $p = n\,\mathrm{k}T$ and where τ is the average collision time. Both the thermal conductivity and viscosity may be written in terms of the *diffusivity* $\chi = \frac{1}{3}\langle v\,\lambda(v)\rangle$, or as approximate averages of the free path λ and the collision time τ

$$\kappa = nc_v\chi \approx \frac{1}{3}nc_v\frac{\lambda^2}{\tau} \qquad \text{and} \qquad \eta = nm\chi \approx \frac{1}{3}nm\frac{\lambda^2}{\tau} \tag{2.46}$$

2.2.2.1 Flux Limitation

It is immediately apparent that these models are only justified if the gradient of the quantity X is sufficiently small that the change in the value of X over a mean free path is small, i.e. $\lambda\,|\nabla X| \ll X$. If this condition is not satisfied, the simple mean free path law over-estimates the transport. It is obvious that there must be an upper bound on the transport of the quantity $Q(X)$ established by the random motion of the particles across unit surface namely

$$\text{Flux} \le \frac{1}{4}n\langle v\rangle\frac{\mathrm{d}Q}{\mathrm{d}X} \tag{2.47}$$

---

8 The mobility may be obtained directly from the model used here by noting that in travelling a distance r at speed v, the charged particle acquires velocity qEr/mv in the direction of the field. Noting that the backward velocity of particles from points with $0 < \theta < \pi/2$ is decreased and forward speed of those with $\pi/2 < \theta < \pi$ increased, the mean velocity of particles passing through the element ΔS is

$$\frac{1}{4\pi}\int_0^{2\pi}\int_0^{\pi}\int_0^{\infty}\int \mathrm{d}\mathbf{v}\tilde{f}(\mathbf{v})\frac{qEr}{mv}v\left[\frac{\exp(-r/\lambda(v))}{\lambda(v)}\right]\sin\theta\,|\cos\theta|\,\mathrm{d}r\,\mathrm{d}\theta\,\mathrm{d}\phi\,\Delta S = \left\langle\frac{q\lambda}{mv}\right\rangle E \approx \frac{1}{3}\frac{q}{\mathrm{k}T}\langle\lambda v\rangle E \tag{2.41}$$

Taking heat flow as an example, the actual flow of heat, when limited, is much reduced below this value. As a result, an empirical relationship

$$\text{Flux} = -\text{Min}(\kappa\,|\nabla T|, f\,n\,\langle v\rangle c_v\,T)\text{parallel to} - \nabla\,T \tag{2.48}$$

is often used in calculation with the empirical constant $f \sim 0.1$ estimated from experiment. Alternatively, a harmonic mean form of the thermal conductivity is sometimes convenient for routine calculation

$$\kappa = \kappa_0/[1 + \lambda\,|\nabla T|/T] \tag{2.49}$$

where $\kappa_0\,T/\lambda = f\,n\,k\!\!\!\!^{-}T\,(k\!\!\!\!^{-}T/m)^{1/2}$ and κ_0 is the thermal conduction in the absence of flux limitation.

Further problems with the simple flux limitation model to treat the breakdown of the mean free path model will be immediately apparent. The gradient scale length is much smaller than the mean free path. Consequently, particles are drawn from regions where the driving term (e.g. temperature) differs markedly from that at the point of measurement. The transport is therefore no longer local, as is assumed by the simple flux limiter of Eq. (2.48). We will return to this problem later and briefly discuss how it can be dealt with.

This model gives a clear physical picture of the underlying physics of transport phenomena. However, it is clearly lacking in several aspects regarding the nature of the collision. In particular, it is assumed that the angular distribution of the particles after scattering is isotropic (*hard sphere model*), and that their distribution corresponds to that at the point of collision. The expressions derived from the model are therefore limited by the inherent inaccuracy of the approximation. We will examine methods without these limitations later. Nevertheless, these models provide simple approximations suitable for many purposes, particularly for gases rather than plasma, where the forces are short range and the hard sphere more closely approximates. The clear association of the coefficients found by the model is reflected, for example by the nearly constant Prandtl number (the ratio of the thermal diffusivity to kinematic viscosity) of gases found experimentally in fluid mechanics.

2.2.3 Drude Model of Electrical Conductivity

A simple, but clear, physical picture of the conductivity of plasma by electrons is provided by the *Drude model* introduced by Drude (1900a).[9] This work was extended by Lorentz (2004, p. 63) and given a more formal derivation (Section 8.A.1), in which the electrons gain directed motion from the electric field, and loose it in collisions. In a time interval dt, electrons are accelerated by the electric field gaining momentum $-e\mathbf{E}\,\mathrm{d}t$ in the direction of the field. On average, an electron will suffer $\nu_e\,\mathrm{d}t$ collisions in the same time, where $\nu_e = 1/\tau_e$ is the electron–ion momentum loss collisional relaxation frequency (Eq. (2.23)).[10] It is assumed that in each collision, the directed electron motion is randomly distributed, so that, on average, after collisions the electron drift velocity v in the direction of the field is zero relative to the ions. Taking the time average over the electrons, the momentum balance is

$$m_e\,\frac{\mathrm{d}\langle\mathbf{v}\rangle}{\mathrm{d}t} = -e\mathbf{E} - \nu_e m_e\,\langle\mathbf{v}\rangle \tag{2.50}$$

9 Drude introduced his model to account for the conductivity of metals and extended it to include Hall effects and thermal conduction (Drude 1900b). In his original paper, Drude (1900a) estimated the electrical conductivity to be only half that in Eq. (2.51) due to a failure to take into account the random variation in the free path, identifying the collision time simply as the time between successive collisions (see McDaniel (1964, p. 429)), rather than as a relaxation time.

10 It is assumed that the collisional relaxation time is constant, independent of the electron velocity. This is the case for the Maxwell molecule, where the force of interaction between the particles varies as r^{-5}.

where $\langle \mathbf{v} \rangle = \mathbf{u}_e - \mathbf{u} = -\mathbf{j} / n_e e$ is the mean drift velocity relative to the ions which determines the current. Hence, in the steady state, we obtain a simple value for the conductivity based on the underlying physical picture

$$\mathbf{j} = \frac{n_e e^2 \tau_e}{m_e} \mathbf{E} = \epsilon_0 \Pi_e{}^2 \tau_e \mathbf{E} = \sigma_0 \mathbf{E} \tag{2.51}$$

where $\tau_e = 1/\nu_e$ is the electron–ion relaxation time given by Eq. (2.23) and Π_e the electron plasma frequency, Eq. (1.3).

The current carried by the electrons is simply due to the electron drift in the electric field defined by the electron mobility. Hence, the conductivity and mobility are related by $\sigma = n(-e)\,\mu_e$; consistent with Eq. (2.42). Drude's model is therefore consistent with Maxwell's model.

2.2.3.1 Alternating Electric Field, No Magnetic Field

We may solve this Eq. (2.50) for the general case when the field is sinusoidal $\mathbf{E} = \mathbf{E}_0\, e^{-\imath\omega t}$ namely

$$\langle \mathbf{v} \rangle = \frac{-e\,\mathbf{E}_0}{m_e(\nu_e - \imath\omega)} \exp(-\imath\omega t) \tag{2.52}$$

neglecting the initial decaying transient. Since the current is $\mathbf{j} = -n_e e\langle \mathbf{v} \rangle$, the ac conductivity of a plasma may, therefore, be expressed as $\mathbf{j} = \mathbf{j}_0\, e^{-\imath\omega t}$ where

$$\mathbf{j}_0 = \frac{n_e e^2}{m_e(\nu_e - \imath\omega)} \mathbf{E_0} = \sigma_0 \frac{\nu_e(\nu_e + \imath\omega)}{(\upsilon_e{}^2 + \omega^2)} \mathbf{E}_0 \tag{2.53}$$

having both in-phase (resistive) and out-of-phase (reactive) components. If the oscillation frequency is very large $\omega \gg \nu_e$, the reactive component is due to the electron velocity $eE_0/m_e\omega$, which is the familiar *quiver velocity* of free electrons. The resistive component is associated with *collisional* or *inverse bremsstrahlung* absorption in laser fields.

2.2.3.2 Steady Electric Field, Finite Magnetic Field

In the presence of a magnetic field, the electron motion is no longer parallel to the electric fields, components parallel and perpendicular to the magnetic field being markedly different (Section 3.2.2). Adding the Lorentz force to the equation of motion (2.50), and Eq. (2.51) becomes

$$m_e \frac{\mathrm{d}\langle \mathbf{v} \rangle}{\mathrm{d}t} = -e\mathbf{E} - e\langle \mathbf{v} \rangle \wedge \mathbf{B} - \nu_e m_e \langle \mathbf{v} \rangle \tag{2.54}$$

$$\Longrightarrow \quad \mathbf{E} = \eta_0 \mathbf{j} + \frac{\mathbf{j} \wedge \mathbf{B}}{n_e e} = \eta_0 \mathbf{j} + \rho_H \mathbf{j} \wedge \hat{\mathbf{b}} \tag{2.55}$$

defining the linear coefficients due to the resistivity $\eta_0 = 1/\sigma_0 = m_e/n_e e^2 \tau_e$ and the Hall term $\eta_H = B/n_e e$; $\hat{\mathbf{b}}$ is the unit vector parallel to the magnetic field $\mathbf{B}$. The equation is clearly linear in j and E so that we may write *Ohm's law* in terms of a linear resistivity tensor which explicitly includes the Hall term[11]

$$E_i = \eta_{ij} j_j \tag{2.56}$$

Taking co-ordinates with $\mathbf{B}$ parallel to the z axis, $\mathbf{E}$ lying in the plane x, z, x denoted by sub-script $\perp$, being perpendicular to z and y perpendicular to the plane x, z denoted by sub-script $\wedge$

$$\eta_{ij} = \begin{pmatrix} \eta_\perp & \eta_\wedge & 0 \\ -\eta_\wedge & \eta_\perp & 0 \\ 0 & 0 & \eta_0 \end{pmatrix} \tag{2.57}$$

11 It is easily seen that if the plasma is collisionless ($\tau \to \infty$), the Hall current is just the $\mathbf{E} \wedge \mathbf{B}$ drift current of the electrons (3.2.2).

where $\eta_\perp = \eta_0$ and $\eta_\wedge = \Omega_e \tau_e \rho_0$, where $\Omega_e = eB/m_e$ is the electron cyclotron frequency (Eq. (3.3)). The simultaneous Eqs. (2.56) and (2.57) may be written as

$$E_x = \eta_\perp j_x + \eta_\wedge j_y \qquad E_y = -\eta_\wedge j_x + \eta_\perp j_y \qquad E_z = \eta_0 j_z \tag{2.58}$$

The inverse of the resistivity tensor $\sigma_{ij} = \eta_{ij}^{-1}$ is easily found by solving the simultaneous equations and yields the conductivity tensor

$$j_i = \sigma_{ij} E_j \tag{2.59}$$

Considering only the terms perpendicular to **B**, we have

$$j_x = \sigma_\perp E_x - \sigma_\wedge E_y \qquad j_y = \sigma_\wedge E_x + \sigma_\perp E_y \tag{2.60}$$

which are found by solving Eqs. (2.58)

$$\sigma_\perp = \frac{\eta_\perp}{\eta_\perp^2 + \eta_\wedge^2} = \frac{1}{(1 + {\Omega_e}^2 {\tau_e}^2)}\sigma_0 \qquad \sigma_\wedge = \frac{\eta_\wedge}{\eta_\perp^2 + \eta_\wedge^2} = \frac{\Omega_e \tau_e}{(1 + {\Omega_e}^2 {\tau_e}^2)}\sigma_0 \tag{2.61}$$

where the conductivity tensor becomes

$$\sigma_{ij} = \begin{pmatrix} \sigma_\perp & -\sigma_\wedge & 0 \\ \sigma_\wedge & \sigma_\perp & 0 \\ 0 & 0 & \sigma_\parallel \end{pmatrix} = \sigma_0 \begin{pmatrix} \dfrac{1}{(1 + {\Omega_e}^2 {\tau_e}^2)} & -\dfrac{\Omega_e \tau_e}{(1 + {\Omega_e}^2 {\tau_e}^2)} & 0 \\ \dfrac{\Omega_e \tau_e}{(1 + {\Omega_e}^2 {\tau_e}^2)} & \dfrac{1}{(1 + {\Omega_e}^2 \tau_e^2)} & 0 \\ 0 & 0 & 1 \end{pmatrix} \tag{2.62}$$

where $\sigma_0 = 1/\rho_0$ and $\Omega_e = eB/m_e$ is the electron cyclotron frequency, Eq. (3.3).

When the magnetic field becomes very small $B \to 0$, the conductivity reduces to the field free value $\sigma_\perp \to \sigma_0$ and $\sigma_\wedge \to 0$. Similarly, if the collisions are weak $\tau_e \to \infty$, the conductivity reduces to that due to the Hall field $E_\perp = j_\wedge B/n_e e$ alone. We note that $\rho_H/\rho_0 = \Omega_e \tau_e$, where $\Omega_e = eB/m_e$ is the cyclotron frequency.

As the magnetic field becomes very large $\Omega_e \tau_e \gg 1$, $\sigma_\wedge \sim \sigma_0/\Omega_e \tau_e$, which corresponds to familiar $\mathbf{E} \wedge \mathbf{B}$ drift, Section 3.2.2.

2.2.3.3 Oscillatory Electric Field, Finite Magnetic Field

The final case involves both an oscillatory electric field varying as $E_0 \exp(-\imath\omega t)$ and is therefore a generalisation of the previous case (Section 2.2.3.2). The basic equation of motion becomes

$$(\nu_e - \imath\omega) m_e \langle \mathbf{v} \rangle = -e\mathbf{E} - e\langle \mathbf{v} \rangle \wedge \mathbf{B} \tag{2.63}$$

which is identical to that for the dc case with the replacement $\nu_e \to \nu_e - \imath\omega$. Alternatively, we may consider the frequency $\omega \to \omega + \imath\nu_e$ to be complex, the imaginary part representing damping of the wave. We may, therefore, use the results of the previous section directly provided we make this substitution. In particular taking the magnetic field parallel to z axis, the conductivity tensor becomes

$$\sigma_{ij} = \frac{n_e e^2}{m_e} \begin{pmatrix} \dfrac{(\nu_e - \imath\omega)}{[(\nu_e - \imath\omega)^2 + \Omega_e^2]} & -\dfrac{\Omega_e}{[(\nu_e - \imath\omega)^2 + \Omega_e^2]} & 0 \\ \dfrac{\Omega_e}{[(\nu_e - \imath\omega)^2 + \Omega^2]} & \dfrac{(\nu_e - \imath\omega)}{[(\nu_e - \imath\omega)^2 + \Omega_e^2]} & 0 \\ 0 & 0 & \dfrac{1}{(\nu_e - \imath\omega)} \end{pmatrix} \tag{2.64}$$

Although the simple Drude model is defective in several respects:

- No account is taken of the variation of the collision frequency with electron velocity.
- No account is taken of the memory of the velocity after collision.
- No account is taken of the possible currents driven by temperature gradients (Nernst and related effects).
- No account is taken of electron–electron collisions.

It is widely used as a simple approximation avoiding the complexity of accurate calculations. Despite its elementary nature, the approximation reflects the response of the plasma to both an oscillatory electric field and the presence of an ambient magnetic field. The appropriate collision time τ_e to use is generally that given by momentum loss cross section, Eq. (2.23).

Transport effects are dominated by electron–ion collisions. However, electron–electron collisions play an important role because of the variation of the scattering rates with velocity. This is significant due to the distortion of the Maxwellian equilibrium distribution by particle heating, which results from the electric field and ion collisions. To overcome these limitations, a model based on the kinetic description of the particles is necessary. We present a simplified account of the theory of the transport coefficients. We initially treat the effects of ions alone (Lorentzian plasma) and then add the electron effects. The analysis is performed with stationary ions, i.e. in the rest frame of reference of the plasma and assuming that the ion temperature is less than or of order of the electron ($T_i \lesssim T_e$). A more formal account is given by Shkarofsky et al. (1966) to which reference should be made.

2.2.4 Diffusivity and Mobility in a Uniform Magnetic Field

As was noted in Sections 1.4 and (2.42), the diffusion of a charged particle is closely related to its mobility and therefore conductivity, by the Einstein–Smoluchowski relation. In hot plasma, the medium consists of stripped ions and electrons and is normally described by a bulk motion with current flow accounting for the flow differentials. However, magnetically confined plasma is trapped within a closed magnetic surface moving along the magnetic field lines as discussed later in more detail in Chapter 15. Motion across the field lines leading to plasma loss is often described as a diffusive process – *cross field magnetic diffusion*.

It follows from the Einstein–Smoluchowski relation that the diffusivity in a steady magnetic field must show a similar tensor dependence on the conductivity. Therefore, we conclude that in a magnetic field, the diffusivity is a tensor with differing components parallel, perpendicular and cross to the magnetic field:

$$D_{ij} = D_0 \begin{pmatrix} \dfrac{1}{1+\Omega^2\tau^2} & -\dfrac{\Omega\tau}{1+\Omega^2\tau^2} & 0 \\ \dfrac{\Omega\tau}{1+\Omega^2\tau^2} & \dfrac{1}{1+\Omega^2\tau^2} & 0 \\ 0 & 0 & 1 \end{pmatrix} \tag{2.65}$$

where from Eq. (2.40) $D_0 = \kappa T\tau/m$.

Early work (see Chen 1983) found experimentally that cross-field diffusion was characterised by the so-called *Bohm diffusion* coefficient

$$D_{\mathrm{Bohm}} = \frac{1}{16}\frac{\kappa T}{eB} \tag{2.66}$$

The factor $1/16$ is empirical based on experimental data, and therefore not exact, uncertain to within a factor of 2 or 3. A number of causes of Bohm diffusion have been proposed, and it is probably associated with micro-turbulence in the plasma. The large rate of Bohm diffusion was thought to prevent the operation of magnetically confined fusion plasmas. Fortunately, newer designs, principally tokamaks discussed later in Chapter 15 were found to be able to operate outside the Bohm regime and therefore without this limitation. As a result, Bohm diffusion is now principally of historic interest only.

2.3 Plasma Permittivity

In the standard theory of electro-magnetism, a non-conducting dielectric material in a steady electric field responds to a small displacement of the positive and negative charges leading to the creation of an assembly of dipoles, whose moments are proportional to the applied field. The net dipole moment per unit volume defines the polarisation $\mathbf{P} = \chi \mathbf{E}$ in terms of the susceptibility χ, and leads to a net volume charge of density $-\nabla P$ in addition to any free charge per unit volume ρ_0 in the dielectric. Thus, from Poisson's equation

$$\epsilon_0 \nabla \cdot \mathbf{E} = (\rho_0 - \nabla \cdot \mathbf{P}) \quad \text{or} \quad \nabla \cdot \mathbf{D} = \nabla \cdot (\epsilon_0 \mathbf{E} + \mathbf{P}) = \rho_0 \tag{2.67}$$

where $\mathbf{D} = \epsilon_0 \mathbf{E} + \mathbf{P} = \epsilon \mathbf{E}$ where ϵ is the permittivity.

Maxwell's equations for the electro-magnetic field involve the (normally) linear relations between the dielectric displacement and the current with the electric field, and the magnetic intensity with the magnetic induction. We may approach the identification of these terms in two ways (see Brandstatter (1963, §8) for a concise exposition)

1. *Macroscopic–Maxwell* (1865): The medium is treated as a continuum, which is polarised by an applied electric field. The resulting displacement of charges gives rise to a current, which is additional to the normal conduction current. The polarisation, currents and magnetisation of the medium are treated implicitly by a set of phenomenological bulk parameters, namely permittivity ϵ, conductivity σ and permeability μ.
2. *Microscopic–Lorentz* (2004): The medium is treated as free space in which the motion of discrete charges, both bound and free, is treated explicitly thereby accounting for the internal fields. The permittivity and permeability are treated explicitly through the action of the current, which is calculated from the dynamics of both the bound and free charged particles.

The permittivity and permeability are useful concepts in the treatment of electro-magnetic waves from Maxwell's equation. In a plasma, where the bound charge is often nearly zero and currents due to the free charges dominate, it is still extremely useful to regard the medium as a dielectric with permittivity and permeability. Since the magnetic effects are small, the permeability is that of free space μ_0, and using the Lorentz picture the permittivity may be defined from the conductivity. In plasma, the dc permittivity has no meaning, we therefore consider the Maxwell equation:

$$\nabla \wedge \mathbf{H} = \mathbf{j} + \frac{\partial \mathbf{D}}{\partial t} = \sigma \mathbf{E} + \epsilon_0 \frac{\partial \mathbf{E}}{\partial t} \tag{2.68}$$

using both the Maxwell and Lorentz formulations, respectively. Assuming there is no externally generated current $\mathbf{j} = 0$, consider the time variation subject to a harmonic oscillation $\exp(-\imath\omega t)$ with frequency ω it follows that

$$\epsilon_{ij} = \epsilon_0 \delta_{ij} + \imath \frac{1}{\omega} \sigma_{ij}$$

$$= \epsilon_0 \begin{pmatrix} 1 - \dfrac{\Pi_e{}^2(\omega + \imath\nu_e)}{\{\omega[(\omega+\imath\nu_e)^2 - \Omega_e{}^2]\}} & \imath\dfrac{\Pi_e{}^2\Omega_e}{\{\omega[(\omega+\imath\nu_e)^2 - \Omega_e{}^2]\}} & 0 \\ -\imath\dfrac{\Pi_e{}^2\Omega_e}{[\omega[(\omega+\imath\nu_e)^2 - \Omega_e{}^2]\}} & 1 - \dfrac{\Pi_e{}^2(\omega + \imath\nu_e)}{\{\omega[(\omega+\imath\nu_e)^2 - \Omega_e{}^2]\}} & 0 \\ 0 & 0 & 1 - \dfrac{\Pi_e{}^2}{[\omega(\omega+\imath\nu_e)]} \end{pmatrix} \tag{2.69}$$

taking into account the tensor nature of the conductivity in inhomogeneous plasma. We note that the permittivity is complex, reflecting the propagation (real part) and damping (imaginary part) of electro-magnetic plasma waves.

Equation (2.69) for the permittivity due to the electrons is easily generalised to include other charged particles when the damping is negligible, see Section 11.2. The collisionless spectrum of cold plasma waves generated as a result form the content of Chapter 11.

2.3.1 Poynting's Theorem – Energy Balance in an Electro-magnetic Field

In a plasma, work is done by the electric field to generate the electro-magnetic field, even in the collisionless case where the Ohmic effects are negligible. The Maxwell equations

$$\frac{\partial \mathbf{B}}{\partial t} = -\nabla \wedge \mathbf{E} \qquad \text{and} \qquad \frac{\partial \mathbf{D}}{\partial t} = \nabla \wedge \mathbf{H} - \mathbf{j} \tag{2.70}$$

in terms of the magnetic flux density $\mathbf{B}$, the electric field $\mathbf{E}$, the magnetic intensity $\mathbf{H} = \mathbf{B}/\mu$, the dielectric displacement $\mathbf{D} = \epsilon\mathbf{E}$, and the current density $\mathbf{j}$. In the simplest case, when is dispersion is absent, the permeability μ and permittivity ϵ are both constant. Take the scalar product of the first equation with $\mathbf{B}/\mu$ and the second with $\mathbf{E}$ and add

$$\mathbf{B} \cdot \frac{\partial \mathbf{B}}{\partial t}/\mu + \epsilon\mathbf{E} \cdot \frac{\partial \mathbf{E}}{\partial t} = -\mathbf{B} \cdot \nabla \wedge \mathbf{E}/\mu + \mathbf{E} \cdot \nabla \wedge \mathbf{B}/\mu - \mathbf{E} \cdot \mathbf{j} = 0 \tag{2.71}$$

Collecting terms together and using the vector identity $\nabla(\mathbf{A} \wedge \mathbf{B}) = \mathbf{A} \cdot \nabla \wedge \mathbf{B} - \mathbf{B} \cdot \nabla \wedge \mathbf{A}$ we obtain

$$\mathbf{E} \cdot \mathbf{j} + \frac{1}{\mu}\nabla \cdot (\mathbf{E} \wedge \mathbf{B}) + \epsilon\frac{\partial(E^2/2)}{\partial t} + \frac{1}{\mu}\frac{\partial(B^2/2)}{\partial t} = 0 \tag{2.72}$$

This equation is readily interpreted in terms of the electro-magnetic energy stored in the field as an energy conservation law where the energy stored per unit volume is balanced against the outgoing flux and work done:

1. $W_E = \frac{1}{2}\epsilon E^2$ is the energy stored in the electric field per unit volume.
2. $W_B = \frac{1}{2}B^2/\mu$ is the energy stored in the magnetic field per unit volume.
3. $\mathbf{S} = \mathbf{E} \wedge \mathbf{B}/\mu$ is the flux of energy per unit area transported by the field perpendicular to both the electric and magnetic fields. This term is known as the *Poynting vector*.
4. $\mathbf{E} \cdot \mathbf{j}$ the rate per unit volume at which the fields do work on external objects.

Consider a plane monochromatic electro-magnetic wave of wavenumber $\mathbf{k}$ and frequency ω with fields $\mathbf{E} = \mathbf{E}_0 \cos(\mathbf{k} \cdot \mathbf{r} - \omega t)$ and $\mathbf{B} = \mathbf{B}_0 \cos(\mathbf{k} \cdot \mathbf{r} - \omega t)$. From Maxwell's equation $\omega\mathbf{B}_0 = \mathbf{k} \wedge \mathbf{E}_0$ and $\epsilon\omega\mathbf{E}_0 = -\mathbf{k} \wedge \mathbf{B}_0/\mu$.

Hence, we obtain that the wave phase velocity is $c = \dfrac{\omega}{k} = \sqrt{\dfrac{1}{\epsilon\mu}} = c_0\sqrt{\dfrac{\epsilon_0\mu_0}{\epsilon\mu}}$. Since the permeability of plasma is that of free space μ_0, there being no intrinsic magnetic effects in plasma, the refractive index $n = c/c_0 = 1/\sqrt{\epsilon\mu/\epsilon_0\mu_0}$, where c_0 is the velocity of light in free space.

Finally, we obtain the *irradiance*[12]

$$I = \frac{1}{2\mu}E_0B_0 = \frac{1}{2\mu c}{E_0}^2 = \frac{1}{2}\frac{\epsilon {E_0}^2}{\epsilon \mu c_0} = \frac{1}{2}n c_0 \epsilon {E_0}^2 \tag{2.73}$$

As we shall show in Section 5.5, the energy transport speed is equal to the *group velocity* – the speed at which a pulse propagates in the medium

$$\text{Energy transport speed} = \frac{\langle\text{Flux}\rangle}{\langle\text{Energy density}\rangle} = \frac{\frac{1}{2}E_0 B_0}{\frac{1}{4}(\epsilon {E_0}^2 + {B_0}^2/\mu)} = \frac{c_0}{n} \tag{2.74}$$

where $\langle\cdots\rangle$ represents averaging over a cycle. This result is clearly paradoxical in plasma, where $n < 1$ (in free space $n = 1$). The issue is clarified in Section 5.5, where we consider the effects of dispersion and the finite spectral width of the radiation.

2.4 Plasma as a Fluid – Two Fluid Model

As we shall discuss in Chapter 9, plasma may be treated as a fluid medium with two (or more) components (electrons and ions) but with additional forces due to the particle charge and electric and magnetic fields. As a result, we may follow the standard methods of fluid mechanics to derive equations for the individual particles based on the conservation of particles, momentum and energy, and treating the ions and electron as separate fluids interacting through the forces generated by their mutual charges; the plasma necessarily subject to the condition that the Debye length is small, i.e. the characteristic length in the plasma $\ell \gg \lambda_D$ is much greater than the Debye length. The plasma fluid is normally quasi-neutral with approximately equal charge densities $\sum q_s = Z_s n_s e \approx 0$ and flow velocities differ only by the electron electrical current.[13] Considering conservation of the general particle s

$$\frac{\partial n_s}{\partial t} + \nabla \cdot (n_s \mathbf{u}) = 0 \tag{2.75}$$

and its momentum conservation

$$\frac{\partial}{\partial t}(n_s m_s \mathbf{u}_s) + \frac{\partial}{\partial x_j}(n_s m_s u_{si} u_{sj} + p_s) - q_s n_s (E_i + \epsilon_{ijk} u_{sj} B_k) = \mathcal{P}_s \tag{2.76}$$

where p_s is the partial pressure of particle s and $\mathcal{P}_s$ the inter-particle collisional momentum exchange term, which determines the electrical conductivity Section 8.4. For the present, we will consider collisionless plasma $\mathcal{P} = 0$, and neglect these dissipative terms. In which case, the energy equation is replaced by an adiabatic equation of state

$$p_s = n_s \mathcal{k} T_s = \text{const}\, n_s^{\gamma_s} = \text{const} \tag{2.77}$$

where T_s is the temperature of particles s, $c_s = \sqrt{\gamma_s p_s / m_s n_s}$ their sound speed, m_s their mass and γ_s their polytropic index, which usually, but not always, takes the value $\gamma_s = 5/3$ for both electrons and ions.

---

12 The irradiance is the energy transport speed in general media with dispersion, Poynting's vector is the particular value in a dispersion-free medium. It is often incorrectly called the intensity in dispersionless media, where the permittivity is constant (usually free space) as the time averaged Poynting vector.

13 If the particle collisional mean free path is shorter than the characteristic length. $\ell \gg \lambda$, the plasma behaves as a conventional single fluid with common velocities, etc. (see Ginzburg 1970, p. 91).

The fluid equations are supplemented by Maxwell's equations for the fields

$$\nabla \wedge \mathbf{E} = -\frac{\partial \mathbf{B}}{\partial t} \qquad \nabla \mathbf{D} = \sum q_s$$
$$\nabla \wedge \mathbf{H} = \mathbf{j} + \frac{\partial \mathbf{D}}{\partial t} \qquad \nabla \mathbf{B} = 0 \tag{2.78}$$

where the magnetic induction and intensity are related by $\mathbf{B} = \mu_0 \mathbf{H}$ and the current density $\mathbf{j} = \sum q_s n_s \mathbf{u}_s$.

2.4.1 Waves in Plasma

One of the most characteristic features of plasma is the complexity of waves which are supported by the interaction of the particle motions with the self-generated electro-magnetic fields. In this section, we briefly introduce these waves within the fluid approximation in the absence of a background magnetic field. In this case, the characteristic length is given by the wavelength $1/k$, where k is the wavenumber. The two fluid approximation is therefore valid if $k\lambda_D \ll 1$ and the single fluid approximation if $k\lambda_D \gg 1$. The case $k\lambda_D \sim 1$ requires a more complete treatment using kinetic models; in terms of the phase velocity, this implies a resonance condition $\omega/k \sim v_t \sim \sqrt{kT/m_e}$ discussed in Chapter 12. More detailed analyses of the waves using kinetic and fluid models are given later in Chapters 10–12.

Following the elementary approach, we consider a wave propagating with wave number $\mathbf{k}$ and frequency ω as a perturbation in a uniform plasma of density n_{s0} with pressure p_{s0} at rest $\mathbf{u}_{s0} = 0$ in a field free environment $\mathbf{E}_0 = 0$ and $\mathbf{B}_0 = 0$. The properties of the medium through which the wave propagates impose a condition on the wave, namely the *dispersion relation* through a solution of the dispersion relation

$$D(\mathbf{k}, \omega) = 0 \tag{2.79}$$

in the medium with dispersion coefficient $D(\mathbf{k}, \omega)$. Consequently, the frequency and wavelength are not independent. We shall find that in more general cases, the dispersion plays an important role in understanding plasma behaviour. Linearising the dynamic equations remembering that the perturbed quantities are small, we obtain

$$\frac{\partial n_{s1}}{\partial t} + n_{s0} \nabla \cdot \mathbf{u}_{s1} = 0 \tag{2.80a}$$

$$m_s n_{s0} \frac{\partial \mathbf{u}_{s1}}{\partial t} + \nabla p_{s1} = q_s n_{s0} (\mathbf{E}_1 + \cancel{\mathbf{u}_{s1} \wedge \mathbf{B}_0}) \tag{2.80b}$$

$$\frac{p_{s1}}{p_{s0}} = \gamma_s \frac{n_{s1}}{n_{s0}} \tag{2.80c}$$

taken together with the Maxwell's equations for the fields

$$\nabla \wedge \mathbf{E}_1 = -\frac{\partial \mathbf{B}_1}{\partial t} \qquad \epsilon_0 \nabla \cdot \mathbf{E}_1 = \sum q_s n_{s1} \tag{2.81a}$$

$$\frac{1}{\mu_0} \nabla \wedge \mathbf{B}_1 = q_s n_{s0} \mathbf{u}_{s1} + \epsilon_0 \frac{\partial \mathbf{E}_1}{\partial t} \qquad \nabla \cdot \mathbf{B}_1 = 0 \tag{2.81b}$$

The wave takes the form of a perturbation $a = A \exp[\imath(\mathbf{k} \cdot \mathbf{r} - \omega t)]$ propagating in the direction of $\mathbf{k}$ with frequency ω. Clearly, the operators

$$\nabla \Longrightarrow \imath\mathbf{k} \qquad \text{and} \qquad \frac{\partial}{\partial t} \Longrightarrow -\imath\omega \tag{2.82}$$

which transform the dynamic equations to

$$\omega \left(\frac{n_{s1}}{n_{s0}} \right) = \mathbf{k} \cdot \mathbf{u}_{s1} \tag{2.83a}$$

$$\omega \mathbf{u}_{s1} - c_{s0}{}^2 \left(\frac{n_{s1}}{n_{s0}} \right) \mathbf{k} = \imath \frac{q_s}{m_s} (\mathbf{E}_1 + \mathbf{u}_{s1} \wedge \mathbf{B}_0) \tag{2.83b}$$

and introducing the permittivity from Eq. (2.69), Maxwell's equations become

$$\mathbf{k} \wedge \mathbf{E}_1 = \omega \mathbf{B}_1 \qquad \epsilon \mathbf{k} \cdot \mathbf{E}_1 = 0 \tag{2.84a}$$

$$\frac{1}{\mu_0} \mathbf{k} \wedge \mathbf{B}_1 = -\epsilon \omega \mathbf{E}_1 \qquad \mathbf{k} \cdot \mathbf{B}_1 = 0 \tag{2.84b}$$

It is easy to see that the governing Eqs. (2.83), and (2.84) separate into two independent groups of transverse Ⓣ and longitudinal waves Ⓛ (Figure 2.5).

- *Transverse waves*: Consider the case where $\nabla \cdot \mathbf{E}_1 = \nabla \cdot \mathbf{B}_1 = 0$. It follows that $\mathbf{E}_1$ and $\mathbf{B}_1$ are perpendicular to $\mathbf{k}$. Consequently, if both the electron and ion velocities also lie in the plane perpendicular to $\mathbf{k}$, it follows that the density n_{s1} and pressure p_{s1} perturbations are zero. The transverse wave is therefore an electro-magnetic wave and the plasma behaves as a classical dielectric with permittivity given by Section 2.3. We normally neglect the effect of the ions as their mass is large, and setting the magnetic field and the collision frequency to zero in Eq. (2.69), the permittivity $\epsilon = \epsilon_0[1 - (\Pi_e/\omega)^2]$ and the dispersion relation is therefore

$$k^2 = (\omega^2 + \Pi_e{}^2)/c^2 \tag{2.85}$$

where $c = 1/\sqrt{\epsilon_0 \mu_0}$ is the velocity of light in vacuo. The refractive index of the plasma is therefore

$$n = kc/\omega = \sqrt{1 - \Pi_e{}^2/\omega^2} \tag{2.86}$$

When the wave frequency $\omega < \Pi_e$ the refractive index is negative and the wave evanescent. This consequences of this behaviour are considered in Section 12.2.
Finally, we note there are two independent transverse wave modes with the electric field vector perpendicular to each other corresponding to differing polarisations of the waves.
- *Longitudinal waves*: Consider the case where the electric field is parallel to the wave vector $\mathbf{E} \parallel \mathbf{k}$. It follows from Eq. (2.84a) that the magnetic field perturbation is zero $\mathbf{B}_1 = 0$. The velocity perturbation is parallel to the wave vector $\mathbf{u}_{s1} \parallel \mathbf{k}$. The problem therefore reduces to a one-dimensional one. Since both ions and electrons can play a significant role, it is convenient to separate their contribution explicitly. Using Eq. (2.81b) and remembering that the condition of quasi-neutrality requires that $q_i = Z n_{i0} e = -q_e = -e n_{e0}$

$$E = \imath \frac{e n_{e0}}{\epsilon_0 \omega} (u_{i1} - u_{e1}) \tag{2.87}$$

and hence from Eq. (2.80b)

$$u_{i1} = -\imath \frac{e}{m_i \omega} E + c_i{}^2 \left(\frac{k^2}{\omega^2} \right) u_{i1} \qquad \text{and} \qquad u_{e1} = \imath \frac{e}{m_e \omega} E + c_e{}^2 \left(\frac{k^2}{\omega^2} \right) u_{e1} \tag{2.88}$$

Substituting for E, we obtain

$$u_{i1}(\omega^2 - \Pi_i{}^2 - c_i{}^2 k^2) + \Pi_1{}^2 u_{e1} \qquad \text{and} \qquad u_{e1}(\omega^2 - \Pi_e{}^2 - c_e{}^2 k^2) + \Pi_e{}^2 u_{i1} \tag{2.89}$$

where $\Pi_i - \sqrt{n_{i0}(Ze)^2/\epsilon_0 m_i}$ is the ion plasma frequency.

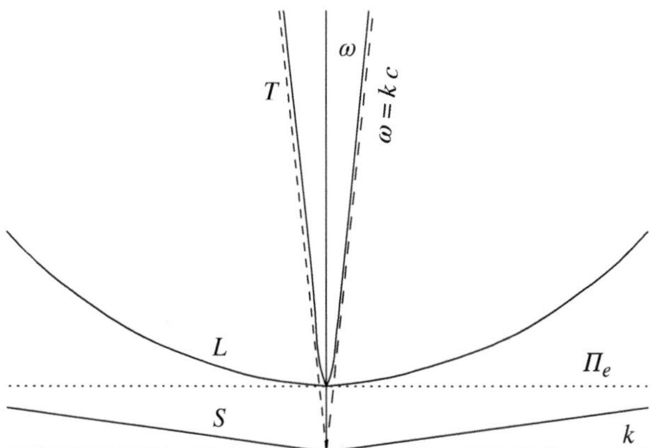

Figure 2.5 Sketch of the dispersion relations for plasma waves in the fluid approximation. T is the transverse wave, L the longitudinal plasma wave, and S the sound wave.

As is well known, this pair of homogeneous simultaneous equations only have solutions if the determinant formed by the coefficients is zero, i.e. if

$$(\omega^2 - \Pi_i{}^2 - c_i{}^2 k^2)(\omega^2 - \Pi_e{}^2 - c_e{}^2 k^2) - \Pi_i{}^2 \Pi_e{}^2 = 0 \tag{2.90}$$

to obtain a quartic in ω as a function k which is easily solved

$$\omega^4 - [\Pi_e{}^2 + \Pi_i{}^2 + k^2(c_e{}^2 + c_i{}^2)]\omega^2 + [k^2(\Pi_e{}^2 c_i{}^2 + \Pi_i{}^2 c_e{}^2) + k^4 c_e{}^2 c_i{}^2] = 0 \tag{2.91}$$

1. *Plasma wave*: However since $m_e \ll m_i$, $\Pi_e \gg \Pi_i$ and normally $c_e \sim c_e$, one approximate solution is

$$\omega^2 \approx \Pi_e{}^2 + c_e{}^2 k^2 = \Pi_e{}^2 + \gamma_e \mathcal{k} T_e k^2 \tag{2.92}$$

which we recognise as the generalisation of the plasma frequency to a warm plasma. Comparing this result with Eq. (Bohm–Gross frequency) for the Bohm–Gross frequency, we see that the value of $\gamma_e = 3$ is appropriate to this case, reflecting the one dimensional motion limiting the number of degrees of freedom.

2. *Ion sound wave*: Assuming that k is small, the second solution is

$$\begin{aligned}\omega^2 &\approx \frac{(\Pi_e{}^2 c_i{}^2 + \Pi_i{}^2 c_e{}^2)}{(\Pi_e{}^2 + \Pi_i{}^2)} k^2 \\ &= \frac{\gamma_i \mathcal{k} T_i + \gamma_e Z \mathcal{k} T_e}{m_i + Z m_e} k^2 = \frac{3 \mathcal{k} T_i + Z \mathcal{k} T_e}{m_i + Z m_e} k^2\end{aligned} \tag{2.93}$$

where $\gamma_i = 3$ since the ion motion is one dimensional and $\gamma_e = 1$, the electrons being isothermal ($p_e \propto n_e$) due to the short wavelength. The oscillation is readily seen to be a sound wave *ion sound wave*, where $u_{i1} \approx u_{e0}$ and $E_1 \approx 0$. The dynamics are determine the particles total pressure and density. Since the quasi-neutrality condition $k \lambda_D \ll 1$ is necessary, the full description of these waves using kinetic theory yields the same dispersion relation, thereby confirming that the electrons behave as an isothermal fluid. These waves are strongly damped unless $T_i \ll T_e$ (see Section 12.7.2.2).

2.4.2 Beam Instabilities

One of the most characteristic features of plasma is their lack of stability in which the waves instead of being simply sustained become unstable. Many of these involve either kinetic (thermal) interactions and/or bulk motion, which we treat later in Section 12.8 and Chapter 16, respectively.

2.4.2.1 Plasma Bunching

Typical of this behaviour is the interaction between two counter-propagating beams of charged particles – *beam instabilities*. The instability results from the phenomenon known as *bunching*. A velocity modulation is imposed on an electron stream by an applied longitudinal periodic field. Assuming the electron stream is uniform, electrons arriving in the trough of the field oscillation are decelerated, and those near the peak accelerated leading to bunching in the electron density near the mid-point of the cycle. This mechanism is responsible for the amplification and oscillation in microwave devices such as klystrons, magnetrons, and travelling wave devices. In a group of plasma beams, bunching, which leads to local decreases and increases in the density, gives rise to additional charge separation and resultant electric fields and may lead to unstable growth of the wave fields.

2.4.2.2 Two Stream Instability

We consider the simplest case of two interacting beams: charge q_1, mass m_1, and velocity v_1 and charge q_2, mass m_2, and velocity v_2 propagating in the same direction with no magnetic field. The dynamics of the interaction as a result of the perturbation of the particle densities n_j and velocities v_j through the space charge electric field E can be treated by the fluid equations:

$$\frac{\partial n_j}{\partial t} + \frac{\partial}{\partial x}(n_j v_j) = 0 \qquad \frac{\partial v_j}{\partial t} + v_j \frac{\partial v_j}{\partial x} + \frac{q_j}{m_j} E = 0 \tag{2.94}$$

and Poisson's equation:

$$\epsilon_0 \frac{\partial E}{\partial x} = \sum q_j n_j \tag{2.95}$$

We consider the system perturbed by a wave $\exp[\imath(kx - \omega t)]$, instability occurring for a constant value of the frequency when k is complex, or *vice versa*. Linearising to obtain the perturbed values[14]

$$\begin{aligned} -\imath\omega n_j + \imath k v_{j_0} n_j + \imath k n_{j_0} v_j &= 0 \\ -\imath\omega v_j + \imath k v_{j_0} v_j + \frac{q_j}{m_j} E &= 0 \\ \imath k E - \sum \frac{1}{\epsilon_0} q_j n_j &= 0 \end{aligned} \tag{2.96}$$

Solving for the perturbed density and velocity n_j and v_j

$$n_j = \imath \frac{q_j n_{i0} k E}{m_j(\omega - k v_{j_0})^2} \qquad v_j = \imath \frac{q_j E}{m_j(\omega - k v_{j_0})} \tag{2.97}$$

clearly demonstrates the bunching behaviour described earlier. Substituting in Poisson's equation, we get the dispersion relation

$$1 = \sum \frac{n_{j_0} q_j^2}{\epsilon_0 m_j(\omega - k v_{j_0})^2} = \sum \left\{ \frac{\Pi_j}{(\omega - k v_{j_0})} \right\}^2 \tag{2.98}$$

where $\Pi_j = n_{j_0} q_j^2 / \epsilon_0 m_j$ is the plasma frequency for particle j.

---

14 These equations also describe bunching in slow travelling longitudinal electro-static fields as used, for example in travelling wave tubes, if the corresponding charge q is set to zero.

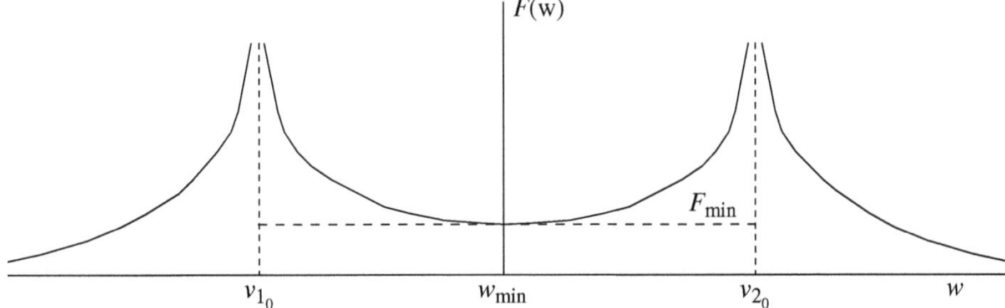

Figure 2.6 Sketch of the function $F(\mathrm{w})$.

We identify the form of the solution by introducing the terms $\mathrm{w} = \omega/k$

$$F(\mathrm{w}) = \frac{\Pi_1{}^2}{(\mathrm{w} - v_{10})^2} + \frac{\Pi_2{}^2}{(\mathrm{w} - v_{20})^2} = k^2 \tag{2.99}$$

It is easily shown that a sketch of the function $F(\mathrm{w})$ takes the form shown in Figure 2.6. Since Eq. (2.99) yields a quartic equation with real coefficients in w, the solutions are either four real, or two real, and two conjugate complex. Inspecting Figure 2.6, it can be seen that for values of $k = \sqrt{F(\mathrm{w})}$ less than the minimum two complex solutions are found. Differentiating $F(\mathrm{w})$, and taking the cube root, it is easily shown that the value $\mathrm{w}_{\mathrm{min}} = (\Pi_1{}^{2/3} v_{02} + \Pi_2{}^{2/3} v_{01})/(\Pi_1{}^{2/3} + \Pi_2{}^{2/3})$ and the corresponding value of $k_{\mathrm{min}} = \{(\Pi_1{}^{2/3} + \Pi_2{}^{2/3})^3/(v_{10} - v_{20})^2)\}^{1/2}$. Two complex solutions are obtained for $k < k_{\mathrm{min}}$ one of which corresponds to unstable growth corresponding to instability.

A simple analytic solution is found if the two beams have equal plasma frequencies $\Pi_1 = \Pi_2$, the quartic equation for the dispersion relation reduces to a simpler biquadratic:

$$(\omega - kU)^4 - (\Pi^2 + 2k^2V^2)(\omega - kU)^2 + k^2V^2(k^2V^2 - \Pi^2) = 0 \tag{2.100}$$

where $U = \frac{1}{2}(v_{10} + v_{20})$ and $V = \frac{1}{2}(v_{10} - v_{20})$ are the mean and relative velocities of the two beams, and $\Pi^2 = \sqrt{\Pi_1{}^2 + \Pi_2{}^2}$ the 'effective plasma frequency'. We note that $(\omega - kV)$ is the frequency Doppler shifted to the mean velocity of the waves. Defining the 'phase velocity' ratio $X = kV/\Pi$

$$(\omega - kU)^2 = \Pi^2 \left[\frac{1 + 2X^2 \pm (1 + 8X^2)^{1/2}}{2}\right] \tag{2.101}$$

which is unstable when $|X| < 1$ when the value of the right-hand side is negative and $(w - kU)^2$ imaginary.

Since $X = kV/\Pi$, it is clear that the beams remain stable for short wavelengths, k large; the oscillation leading to instability requiring a reasonably large distance to evolve.

2.4.3 Kinematics of Growing Waves

When more than one wave is present in the plasma, the dispersion function $D(\mathbf{k}, \omega)$ becomes more complex, often of polynomial form. In the absence of collisions, the coefficients of the dispersion parameters are all real. In consequence, the solutions are either real or complex conjugate pairs. Furthermore, there may be multiple solutions of the wave number k for a given value of the frequency ω and *vice versa*. Evanescence is identified with decay or growth of waves (or more clearly temporal wave packets) in space due to the imaginary part of the wave number. Instability on the

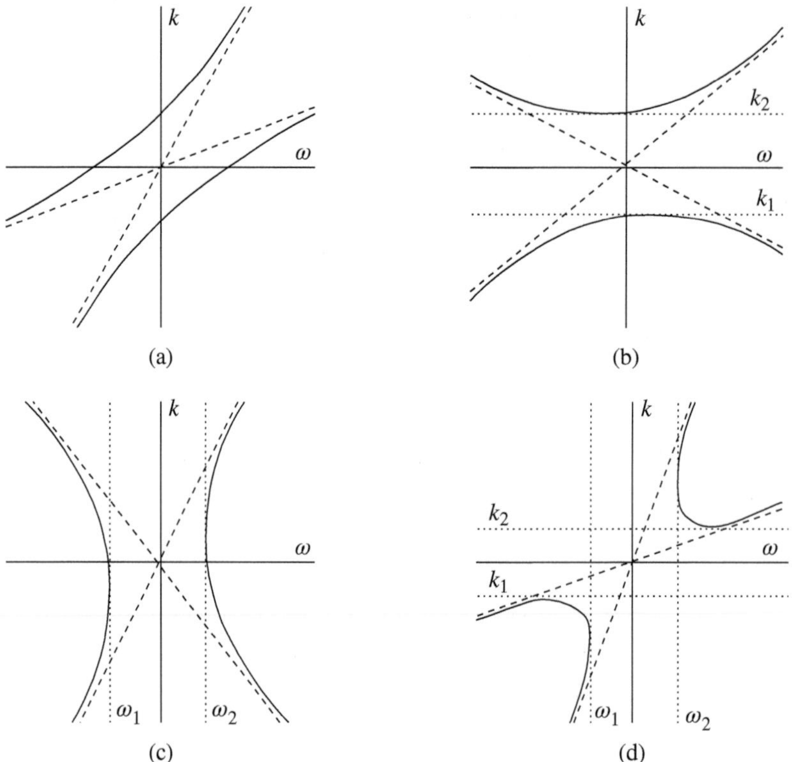

Figure 2.7 Plots of the real solutions of the dispersion equation for k as a function of ω: (a) real solutions exist throughout, (b) no solution $k_1 < k < k_2$, (c) no solution $\omega_1 < \omega < \omega_2$, (d) no solution $k_1 < k < k_2$ and $\omega_1 < \omega < \omega_2$. The dashed lines represent the underlying non-interacting waves.

other hand is generally identified with the temporal growth (or decay) of spatial wave packets associated with the imaginary part of the frequency.

Figure 2.7 illustrates the different solutions that may be found for a simple, but generic form of dispersion relation. Note there are four distinctive forms of the solution. In the regions where no real solution exists, complex values of k or ω occur leading to growth or decay of the way in space or time, respectively. The formal analysis of the behaviour in these regions is due to Sturrock (1958), Briggs (1964, §2.7.1), and Clemmow and Dougherty (1969) whose key results are summarised below:

1. Real solution throughout the range. Regular wave propagation. No growth or attenuation.
2. Complex solution for k in the range $k_1 < k < k_2$ for real ω, oscillatory in time, but decaying (or growing) in space *evanescent condition* (or *amplifying wave*).
3. Complex solution for ω in the range $\omega_1 < \omega < \omega_2$ for real k. Wave packet grows (or decaying) in time, but is stationary in space: *absolute instability*, Figure 2.8.
4. Complex solution in the range $k_1 < k < k_2$ and $\omega_1 < \omega < \omega_2$. Growth (or decay) of wave packet as it propagates: *convected instability*, Figure 2.8.

As an exemplar, we consider a simple generic model of two simple weakly interacting modes in the neighbourhood of their phase matched intersection at (k_0, ω_0) when both frequency and wave

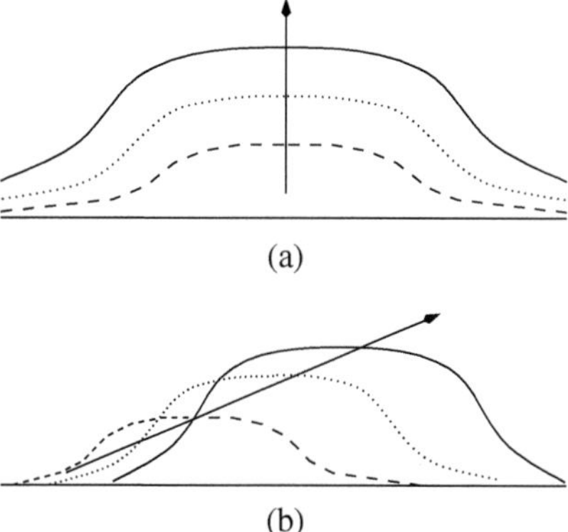

Figure 2.8 Sketch to illustrate the difference in growth of (a) absolute and (b) convective instabilities.

number of each wave are the same. The overall dispersion relation can be written as

$$D(k,\omega) = [\omega - \omega_0 - (k-k_0)v_1] \times [\omega - \omega_0 - (k-k_0)v_2] - \varepsilon = 0 \tag{2.102}$$

where v_1 and v_2 are the respective group velocities of the two modes, and ε is the weak mode coupling interaction. This simple approximation is valid for many mode coupling phenomena in the neighbourhood of the phase matching condition. Solving for $\omega(k)$ and $k(\omega)$ in Eq. (2.102)

$$\omega(k) - \omega_0 = \frac{1}{2}\left\{(k-k_0)(v_1+v_2) \pm \sqrt{[(k-k_0)^2(v_1-v_2)^2 + 4\varepsilon]}\right\} \tag{2.103:a}$$

$$k(\omega) - k_0 = \frac{1}{2v_1 v_2}\left\{(\omega-\omega_0)(v_1+v_2) \pm \sqrt{[(\omega-\omega_0)^2(v_1-v_2)^2 + 4\varepsilon v_1 v_2]}\right\} \tag{2.103a}$$

and the plots of Figure 2.7 correspond to the cases

Plot	Coupling coefficient	Velocity product	Plot	Coupling coefficient	Velocity product
(a)	$\varepsilon > 0$	$v_1 v_2 > 0$	(b)	$\varepsilon > 0$	$v_1 v_2 < 0$
(c)	$\varepsilon < 0$	$v_1 v_2 < 0$	(d)	$\varepsilon < 0$	$v_1 v_2 > 0$

Real solutions being found outside the ranges defined by

$$(\omega-\omega_0)^2 < |\varepsilon v_1 v_2|/(v_2 - v_1)^2 \qquad \text{and} \qquad (k-k_0)^2 < |\varepsilon|/(v_2-v_1)^2 \tag{2.104}$$

Appendix 2.A Momentum Transfer Collision Rate

Consider the interaction between a test particle from a set of mass m_1 with a background of stationary particles of mass m_2. The individual particles have velocities $\mathbf{v}_1$ and $\mathbf{v}_2$ respectively, with average values across each set of $\overline{\mathbf{v}}_1$ and $\overline{\mathbf{v}}_2$ in their mutual centre of mass frame. The two sets therefore drift towards or away from each other with a net velocity $\overline{\mathbf{v}} = \overline{\mathbf{v}}_1 - \overline{\mathbf{v}}_2$.

As is well known, the dynamics of the interaction of a test particle of type ① with a particle of type ② can be described as that between particles with the reduced mass $\overline{m} = m_1 m_2/(m_1 + m_2)$ about their individual centre of mass (see Footnote 3 on p. 19). The velocity difference between particle ① and particle ② is $\mathbf{u} = \mathbf{v}_1 - \mathbf{v}_2$. The velocity of particle ① is $\mathbf{u}_1 = m_2/(m_1 + m_2)\mathbf{u}$ with respect to the centre of mass, and of particle ② $\mathbf{u}_2 = -m_1/(m_1 + m_2)\mathbf{u}$. The velocity of the centre of mass $\mathbf{V} = (m_1\mathbf{v}_1 + m_2\mathbf{v}_2)/(m_1 + m_2)$. The momentum loss in the x direction by particle ① in a collision with a particle of type ② with scattering angle χ in the centre of mass frame is, therefore,

$$\Delta p_x = \frac{m_2}{m_1 + m_2} p_x(1 - \cos\chi) \approx \frac{1}{2}\frac{m_2}{(m_1 + m_2)} p_x \chi^2 \tag{2.A.1}$$

where p_x is the component of momentum in the centre of mass frame.

Following through the analysis in the centre of mass frame leading to Eqs. (2.21), each particle of type ① loses momentum to particles of type ② at the momentum collision frequency given by the rate of change of the velocity in the parallel direction. Averaging over multiple collisions with different impact parameters yields, the momentum loss per unit time. From Eq. (2.7) generalised to include finite mass particles in the centre of mass frame

$$\left\langle \frac{\mathrm{d}\Delta p_1}{\mathrm{d}t} \right\rangle = \left[\frac{4\pi {q_1}^2 {q_2}^2 \ln\Lambda}{(4\pi\epsilon_0)^2 [m_1 m_2(m_1 + m_2)]^2 u^4} \right] \cdot n_2 \cdot \frac{m_2}{(m_1 + m_2)} u^2 \tag{2.A.2}$$

where the velocity of particle ξ in the centre of mass frame is $u_1 = m_2 u/(m_1 + m_2)$. Hence, the averaged collision frequency:

$$\nu_{12} = \frac{1}{p_1}\frac{\mathrm{d}p_1}{\mathrm{d}t} = \frac{4\pi {q_1}^2 {q_2}^2 n_2 \ln\Lambda}{(4\pi\epsilon_0)^2 \overline{m} m_1 u^3} \tag{2.21b$'$}$$

The average rate of momentum transfer from particle ① to ② is a summation over the particle probability distribution $f_1(\mathbf{v})$ taking into account the distribution of particle ② velocities

$$\frac{\mathrm{d}\mathbf{p}}{\mathrm{d}t} = m \int \nu_{12}(v)\mathbf{v} f(\mathbf{v})\mathrm{d}\mathbf{v} \tag{2.A.3}$$

We now consider the two different groups of particles moving within the joint centre of mass frame of reference of the ensemble. Assume that the test particle ① is a member of a set with a thermal equilibrium Maxwell distribution with temperature T drifting with mean speed $\overline{\mathbf{v}}_1$ in the joint centre of mass frame, relative to the particles ②, which are also in equilibrium with the same temperature T and drifting with mean velocity $\overline{\mathbf{v}}_2$ in the same frame. Therefore, the distribution of particles ⓙ in the overall centre of mass frame is

$$f_j(\mathbf{v}_j) = n_j \left(\frac{m_j}{2\pi kT} \right)^{3/2} \exp\left[-\frac{m_j(\mathbf{v}_j - \overline{\mathbf{v}}_j)^2}{2kT} \right] \tag{2.A.4}$$

where $m_1\overline{\mathbf{v}}_1 = -m_2\overline{\mathbf{v}}_2$. Since the drift speeds are assumed to be much less than the thermal speed $\overline{v} \ll v_j$, the product of the distribution functions may be expressed in the centre of mass frame in terms of centre of mass velocity $\mathbf{V} = (m_1\mathbf{v}_1 + m_2\mathbf{v}_2)/(m_1 + m_2)$ and the approach velocity $\mathbf{u} = \mathbf{v}_1 - \mathbf{v}_2$ of the pair of particles

$$\begin{aligned} f_1 f_2 \approx{}& n_1 n_2 \left(\frac{m_1}{2\pi kT} \right)^{3/2} \left(\frac{m_2}{2\pi kT} \right)^{3/2} \exp\left[-\left(\frac{m_1 {v_1}^2}{2kT} + \frac{m_2 {v_2}^2}{2kT} \right) \right] \\ &\times \left[1 + \frac{m_1 \mathbf{v}_1 \cdot \overline{\mathbf{v}}_1}{kT} + \frac{m_2 \mathbf{v}_2 \cdot \overline{\mathbf{v}}_2}{kT} \right] \\ ={}& n_1 n_2 \left(\frac{M}{2\pi kT} \right)^{3/2} \left(\frac{\overline{m}}{2\pi kT} \right)^{3/2} \exp\left[-\left(\frac{MV^2}{2kT} + \frac{\overline{m}u^2}{2kT} \right) \right] \\ &\times \left(1 + \frac{\overline{m}\overline{\mathbf{v}} \cdot \mathbf{u}}{kT} \right) \end{aligned} \tag{2.A.5}$$

where $M = m_1 + m_2$ is the mass of the pair of particles.

We require the integral of Eq. (2.21b) taken over the distributions $\mathbf{v}_1$ and $\mathbf{v}_2$, which perform by transforming to integrals over $\mathbf{u}$ and $\mathbf{V}$. It can be shown that the volume element $\mathrm{d}\mathbf{v}_1\,\mathrm{d}\mathbf{v}_2 = \mathrm{d}\mathbf{V}\,\mathrm{d}\mathbf{u}$ as follows. Taking Cartesian co-ordinates

$$\frac{\partial(\mathbf{V},\mathbf{u})}{\partial(\mathbf{v}_1,\mathbf{v}_2)} = \frac{1}{(m_1+m_2)^3}\begin{vmatrix} 1 & 0 & 0 & -1 & 0 & 0 \\ 0 & 1 & 0 & 0 & -1 & 0 \\ 0 & 0 & 1 & 0 & 0 & -1 \\ m_1 & 0 & 0 & m_2 & 0 & 0 \\ 0 & m_1 & 0 & 0 & m_2 & 0 \\ 0 & 0 & m_1 & 0 & 0 & m_2 \end{vmatrix} = \frac{{m_1}^3 + 3{m_1}^2 m_2 + 3m_1{m_2}^2 + {m_2}^3}{(m_1+m_2)^3}$$
$$= 1 \tag{2.A.6}$$

Substituting for ${u_1}^{-3}$ in Eq. (2.21b), we require an integral for $\mathbf{u}$ taken over the unshifted Maxwellian $f_0(v)$ of the form

$$\int \frac{{v_x}^2}{v^3} f_0(\mathbf{v})\,\mathrm{d}\mathbf{v} = \frac{1}{3}\int \frac{{v_x}^2+{v_y}^2+{v_z}^2}{v^3} f_0(\mathbf{v})\,\mathrm{d}\mathbf{v} = \frac{1}{3}\int \frac{v^2}{v^3} f_0(\mathbf{v})\,\mathrm{d}\mathbf{v}$$
$$= \frac{1}{3}4\pi\int_0^\infty v f_0(v)\,\mathrm{d}v = \frac{2}{3}\left(\frac{m}{2\pi \mathit{k}T}\right)^{1/2} \tag{2.A.7}$$

since the distribution is isotropic.

Whence the momentum loss collision frequency of particles ① drifting with velocity $\mathbf{u}$ interacting with particles ②

$$\nu_{12} = \frac{4\pi q_1^2 q_2^2 n_2 \ln\Lambda}{(4\pi\epsilon_0)^2 \overline{m} m_1}\cdot\frac{2\pi}{3}\left(\frac{\overline{m}}{2\pi \mathit{k}T}\right)^{3/2} = \frac{4\sqrt{2\pi}{q_1}^2{q_2}^2\sqrt{\overline{m}}\,n_2\ln\Lambda}{3(4\pi\epsilon_0)^2 m_1(\mathit{k}T)^{3/2}} \tag{2.A.8}$$

Equations (2.23) and (2.24) follow immediately. The former for the case $m_2 \to \infty$ since the ion mass is much larger that the electron $M \gg m$. The latter when $m_1 = m_2$, the test particle being scattered off particles of the same type.

Appendix 2.B The Central Limit Theorem

The central limit theorem plays a key role in the interpretation of data, both from experiment and simulation. It finds wide application in many areas of plasma physics, particularly those associated with multiple collisions, although it is often implicitly assumed. Its classical form, which we use here, may be stated as follows:

> *Given a set of a large number N of independently and identically distributed random variables with a probability distribution having well defined mean $\overline{x}$ and variance σ^2, the mean of the sum is $N\overline{x}$ and the variance $N\sigma^2$. The distribution of the sum about its mean is normal.*

The central limit theorem is proved in many statistics texts, for example Whittaker and Robinson (1944, §85, 86). However, a simple non-rigorous justification follows, which demonstrates the essential asymptotic nature of the result, but avoids the use of Fourier transforms. The theorem is valid asymptotically as $N \to \infty$.

Suppose we have a series of independent measurements of a random variable y_i to a well-defined differential probability $p_1(y)$ which satisfies the above conditions for the central limit theorem to apply. Consider the sum formed by adding a set of n values of $x_n = \sum_0^n y_i$. Assuming n is large and y is small, so that after the nth addition, $x_n = x_{n-1} + y_n$ the probability $p_n(x)$ of a value $x_n = \sum_{i=1}^n y_i$ is given by the convolution:

$$p_n(x) = \int p_{n-1}(x-y)p_1(y)\,\mathrm{d}y \approx \int \left(p_{n-1} - y\left[\mathrm{d}p_{n-1}(x)\Big/\mathrm{d}x\right] + \frac{1}{2}y^2\left[\mathrm{d}^2p_{n-1}(x)\Big/\mathrm{d}x^2\right]\right) p_1(y)\,\mathrm{d}y$$
$$= p_{n-1}(x) - \frac{\mathrm{d}}{\mathrm{d}x}p_{n-1}(x)\langle y\rangle + \frac{1}{2}\frac{\mathrm{d}^2}{\mathrm{d}x^2}p_{n-1}(x)\langle y^2\rangle \tag{2.B.1}$$

where the probability $p_1(y)$ is normalised to unit integral, and $\langle y\rangle$ and $\langle y^2\rangle$ are the mean and variance taken over the same probability.

For a large number of trials, we may treat n as a continuous variable, and Eq. (2.B.1) becomes

$$\frac{\partial p}{\partial n} + \langle y\rangle\frac{\partial p}{\partial x} - \frac{1}{2}\langle y^2\rangle\frac{\partial^2 p}{\partial x^2} \approx 0 \tag{2.B.2}$$

which we recognise as the diffusion equation whose solution may be written as follows:

$$p(x) = \frac{1}{\sqrt{2\pi n\langle y^2\rangle}}\exp\left[-\frac{(x - n\langle y\rangle)^2}{2n\langle y^2\rangle}\right] \tag{2.B.3}$$

The final distribution of the sum $x = \sum y_i$ is therefore normal, displaced to the mean $n\langle x\rangle$ and variance $n\langle x^2\rangle$. Two important applications follow

1. In applications such as random walks, the total path length is a randomly oriented set of independent path of constant length. In the asymptotic limit of many walks, the central limit theorem may be used to estimate the statistical behaviour of the displacement in a given direction, as in Section 2.2.1.
2. The central limit theorem is widely assumed in data analysis where a set of N independent measurements are used to improve the accuracy of a measured quantity. It follows that the average of the results divided by the number of trials $\overline{x} = N\langle y\rangle/N$ gives the best value for the value x and the variance divided by the number of trials $\sigma^2 = N\langle y^2\rangle/N$ the mean square deviation. It follows that the error measured as the r.m.s. deviation decreases as $1/\sqrt{N}$.[15]

15 The corresponding improvement of the result towards the central value (average) gives rise to the name of the theorem.

3

Single Particle Motion – Guiding Centre Model

3.1 Introduction

In Chapter 1, we studied behaviour introduced by the interactions of electrostatic fields with charged particles. However, charged particles interact not only with electric fields but also with magnetic. In contrast to electric fields which are often applied by the particles themselves, magnetic fields are commonly generated by external field coils. It is important to understand the motion of individual charged particles in external fields, i.e. with no inter-particle interaction. Such motions are conveniently described by the *guiding centre model*, provided the magnetic field is not too large or the density too high.

The motion of a charged particle in a static, uniform magnetic field with zero electric field is a circular rotation about a line of induction, with the particle on a surface which extends helically along an axis along the line of induction. The centre of the rotation, known as the guiding centre, moves steadily along the field line. Although such a simple motion is not of interest, the radius, ρ, and period, $2\pi/\Omega$, of the rotation are generally very small compared to the characteristic scale of spatial and temporal variations in the field. These field variations induce perturbations in the rotation, but because the scale parameters of the variations are very small, the motion nearly closes over each cycle. As a result it is possible to average over the cycle to determine the successive positions of the guiding centre, avoiding the complication of the detailed movement of the particle. Averaging over the cycle is fundamental in the guiding centre model (Bogoliubov and Mitropolsky, 1961; Northrop, 1963; Morozov and Solev'ev, 1966). The model resolves the particle only to within the radius and period of the rotation. The guiding centre is used to describe the general motion; the position of the particle around its loop is defined as a consequence of the averaging.[1]

The guiding centre model is a relatively simple approach, useful when particles move in well-defined electric and magnetic fields usually applied externally. Collisions and self-generated fields, unless known, are not included. The method is mainly applicable to low-density plasma such as the magnetosphere and magnetically confined fusion devices. However, in many cases, it has now been superseded by 'fluid' models, which can handle large particle numbers. Nonetheless, the behaviour demonstrated by particles using this model continues to provide essential qualitative insights, even if its quantitative predictions are limited. In particular, the identification

1 The identification of a current with the motion of the guiding centre may be erroneous. The discrepancy arises from the form of averaging implicit in the model. The drift velocity (or current) is obtained from an instantaneous average over a small region of space containing many particles. On the other hand, the guiding centre motion is an average over time of a single particle (see Rose and Clark, 1961, p. 305).

Foundations of Plasma Physics for Physicists and Mathematicians, First Edition. Geoffrey J. Pert.

of particle drifts in crossed and gradient fields is a recurrent theme in plasma physics. The second key element identified by the guiding centre model is the three constants of motion known as adiabatic invariants.

3.2 Motion in Stationary and Uniform Fields

3.2.1 Static Uniform Magnetic Field – Cyclotron Motion

A particle of charge q moving in a constant and uniform magnetic field of induction $\mathbf{B}$ experiences a Lorentz force due to its velocity $\mathbf{v}$

$$\mathbf{F} = q\,\mathbf{v} \wedge \mathbf{B} \tag{3.1}$$

Separating the components of the force parallel to $\mathbf{F}_{\|}$ and perpendicular $\mathbf{F}_{\perp}$ to the induction we have

$$\mathbf{F}_{\|} = 0 \qquad \text{and} \qquad \mathbf{F}_{\perp} = q\,\mathbf{v}_{\perp} \wedge \mathbf{B} \tag{3.2}$$

Clearly, the velocity parallel to the magnetic field $\mathbf{v}_{\|}$ is constant.

Since the force perpendicular to the field $\mathbf{F}_{\perp} \cdot \mathbf{v}_{\perp} = 0$, the perpendicular speed $\mathbf{v}_{\perp}$ is constant, and further since $\mathbf{F}_{\perp} \perp \mathbf{v}_{\perp}$ the motion in the perpendicular plane is a circle of radius ρ called the *Larmor radius* or *gyroradius* determined by the provision of the necessary centripetal force by the Lorentz force

$$\frac{m{v_{\perp}}^2}{\rho^2}\,\rho = -q\,\mathbf{v} \wedge \mathbf{B}$$

The motion is easily shown to be anti-clockwise about the line of induction if q is positive. Therefore, the Larmor radius is $\rho = mv_{\perp}/qB$ and the angular frequency – known as the *cyclotron frequency* or *gyrofrequency*

$$\boldsymbol{\Omega} = -\frac{q}{m}\,\mathbf{B} \tag{3.3}$$

in vector form.

The particle therefore executes a helical motion around the line of induction. It is convenient to identify the particle path by the motion of the centre of the cyclotron orbit known as the *guiding centre*. Clearly, in this simple case, the guiding centre follows the line of induction.

The electron velocity is therefore made up of a periodic, rapidly rotating, component $\mathbf{u}$ in the plane normal to the induction and a slower secular (i.e. long-lasting, non-periodic) motion component $\mathbf{U}$ associated with the motion of the guiding centre.

$$\mathbf{v} = \mathbf{U} + \mathbf{u} \tag{3.4}$$

It is frequently convenient to separate the guiding centre motion explicitly so that the particle location is the sum of the guiding centre $\mathbf{R}$ and the orbital ρ position vectors

$$\mathbf{r} = \mathbf{R} + \rho = \mathbf{R} + \frac{m}{q\,B^2}\,\mathbf{B} \wedge \mathbf{u} \tag{3.5}$$

where the orbital position follows from the balance of the centripetal and Lorentz forces

$$\rho = \frac{m}{q\,B^2}\,\mathbf{B} \wedge \mathbf{u} \tag{3.6}$$

It is important to distinguish the guiding centre motion from the actual particle motion. The particle current is the instantaneous average over many particles in a small volume of space regardless

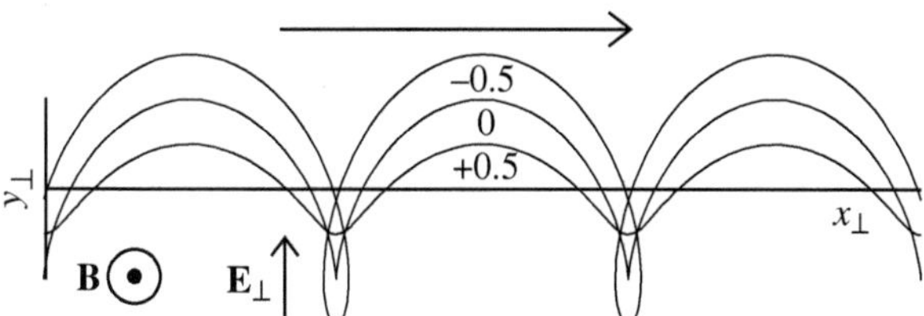

Figure 3.1 Plots of the charged particle trajectory in crossed electric and magnetic fields. The lines represent the particle initially at rest (0) and with velocity $\pm\frac{1}{2}v_d$ the drift velocity.

of the position of their guiding centres. The 'drift' velocity of the guiding centre, on the other hand, represents the average over time of the motion of a single particle. The two are not necessarily the same (see Section 3.9). Rose and Clark (1961, p. 205) discuss a well-known example where the guiding centre picture is inadequate to describe the behaviour.

3.2.2 Uniform Static Electric and Magnetic Fields

We consider next the case of a static uniform magnetic and electric fields. As before, we separate the fields and motions into components parallel (||) and perpendicular (⊥) the magnetic field. Thus,

$$\mathbf{F}_{\parallel} = q\,\mathbf{E}_{\parallel} \qquad \text{and} \qquad \mathbf{F}_{\perp} = q\,\mathbf{E}_{\perp} + q\,\mathbf{u}_{\perp} \wedge \mathbf{B} \tag{3.7}$$

Writing

$$\mathbf{u}_{\perp} = \mathbf{u}'_{\perp} + \frac{1}{B^2}\,\mathbf{E}_{\perp} \wedge \mathbf{B} \tag{3.8}$$

it follows that $\mathbf{F}_{\perp} = q\,\mathbf{u}'_{\perp} \wedge \mathbf{B}$ and that the velocity $\mathbf{u}'_{\perp}$ describes a cyclotron motion.

The complete particle motion is therefore a cyclotron rotation $\mathbf{u}'_{\perp}$ and a drift perpendicular to both the electric and magnetic fields with velocity $\mathbf{U}_E = \mathbf{E} \wedge \mathbf{B}/B^2$. We note the following points:

a. The particle is not accelerated in the transverse plane – in particular, the guiding centre does not move parallel to $\mathbf{E}_{\perp}$.
b. The guiding centre moves with constant velocity $\mathbf{E} \wedge \mathbf{B}/B^2$ in the transverse plane perpendicular to both the electric and magnetic fields across the magnetic field lines.
c. The particle is accelerated along the magnetic field direction by the parallel electric field $E_{\parallel}$.

This behaviour in the transverse plane is easily visualised. Suppose the particle is initially at rest. The electric field accelerates it in the direction of the field. Once moving, the path is rotated by the magnetic field until it moves perpendicular to the electric field. Further rotation brings the particle parallel to the electric field, but moving against it until it is brought to rest, and the cycle is repeated. The motion is clearly that of a circle rolling along a line, i.e. a cycloid (Figure 3.1). If the particle is initially at rest, the path is a true cycloid.

However, if the particle has an initial velocity, the motion is that of a point either on the inside or outside of the rim rolling along the line depending on the sign of the initial velocity.

3.3 The Guiding Centre Approximation

We now turn the situation where the fields vary slowly in space and time referred to earlier (Section 3.1). The motion of the particle consists of the rapid cyclotron rotation superimposed on a slow secular motion associated with the variations of the fields in space and time. In view of the gross differences between these two motions, we average over the rapid cyclotron rotation to identify the slow variation associated with the changes in the fields over distances L and times T, such that $\rho/L \ll 1$ and $\Omega\, T \gg 1$. To characterise the scale of the variation, we introduce a small parameter ϵ and develop the motion as a perturbation expansion in terms of ϵ whilst averaging over the cyclotron rotation; the secular motion being the result of differences in the field terms as the particle orbit varies. This behaviour consisting of a rapid rotation superimposed on a slow secular motion may be described by a general averaging procedure (Bogoliubov and Mitropolsky, 1961; Northrop, 1963; Morozov and Solev'ev, 1966). The scaling parameter ϵ is used to explicitly identify the scales of the individual terms in the expansion; although it need not be explicitly defined, its value $\epsilon \sim \max(\rho/L,\ 1/\ \Omega\ T) \ll 1$. The averaging procedure provides a general method for treating problems where a slow secular development is the average over a rapid oscillation, which occurs in many problems. Although well defined physically, such problems give rise to mathematical difficulty (see (Bogoliubov and Mitropolsky, 1961)). The treatment described here is based on the general method of averaging for the asymptotic expansion for non-linear oscillations developed by Bogoliubov and Mitropolsky (1961, Ch.5) and adapted to this problem by Morozov and Solev'ev (1966). The simplified presentation appropriate for guiding centre motion given here closely follows that due to Fitzpatrick (2015) based on Morozov and Solev'ev (1966). Northrop (1963) uses an equivalent, but less general, approach to the averaging, which is briefly outlined in Appendix 3.A and leads to identical results.

3.3.1 The Method of Averaging

Consider the development of a parameter $\mathbf{z}(t)$ of a particle, for example its position vector $\mathbf{r}$, which satisfies an evolution described by a differential equation:

$$\frac{\mathrm{d}\mathbf{z}}{\mathrm{d}t} = \mathbf{f}(\mathbf{z}, t) \tag{3.9}$$

where the driving 'force' $\mathbf{f}$ contains an oscillating term with period which is much shorter than that of the mean drift motion. The scaling of the rotational time to the drift time suggests the introduction of a rotational time variable, typically the phase of the rotation

$$\tau = t/\epsilon \tag{3.10}$$

with period $\mathcal{T}$, which is scaled to bring the values to a common magnitude. It is to be expected that the averaging over the period of the rotation will introduce two distinct motions: a slow secular drift and a rapid gyro-motion. Since, on averaging, the latter will generate nearly zero contributions to $\mathbf{z}$, we may split the functional dependence of $\mathbf{f}$ on time into two parts, one associated with the slow secular motion t and one with the rapid rotation τ

$$\mathbf{f}(\mathbf{z}, t) \to \mathbf{f}(\mathbf{z}, t, \tau) \tag{3.11}$$

In order to clearly identify the ordering of terms as the calculation proceeds, it is helpful to overtly introduce the parameter ϵ into the expressions for the various terms to bring their values to a common size, and thereby introduce a convenient method to identify the ordering of the terms in the asymptotic expansion. We can easily return to values in the laboratory scale by setting $\epsilon = 1$, although this is generally unnecessary.

The general procedure sets up an expansion of the quantities involved in the motion in terms of the expansion parameter, ϵ. The expansion terms depend on both the characteristic parameters of the guiding centre and the rotation. At each level of expansion, the equations developed are treated in two parts:

a. The 'condition of solubility' is obtained by averaging over the rotation. In this way, the rotational terms are eliminated and a relationship between terms depending only on the guiding centre motion is obtained.
b. The 'condition of solubility' is used to simplify the overall equation to yield an equation for the guiding centre motion.

We now treat t and τ as independent variables to generate solutions $\mathbf{z}(t, \tau)$ which are periodic in τ. In general, this procedure is indeterminate unless an additional equation is available to define the variable τ. However, as we are only interested in the slow development of the problem in time, this requirement is circumvented by averaging the rotational terms over the oscillation period $\mathcal{T}$. The secular drift motion is associated with the variable t and the rotation with τ. We may crudely envisage the procedure as equivalent to treating $\mathbf{z}(t)$ as a piecewise continuous function of t, sampled at intervals of the rotation period $\mathcal{T}$, which are very small. Since $\epsilon \ll 1$, the function $\mathbf{f}$ is nearly regular.
Taking the scaling (3.10) into account, the governing differential equation becomes

$$\frac{\partial \mathbf{z}}{\partial t} + \frac{1}{\epsilon}\frac{\partial \mathbf{z}}{\partial \tau} = \mathbf{f}(\mathbf{z}, t, \tau) \tag{3.12}$$

The solution $\mathbf{z}$ contains both the secular and rotation motions, which we separate, by the different scalings of the two terms into the guiding centre term $\mathbf{Z}(t)$ and the rotation $\boldsymbol{\zeta}(\mathbf{Z}, t, \tau)$, the latter being of order ϵ

$$\mathbf{z}(t, \tau) = \mathbf{Z}(t) + \epsilon\ \boldsymbol{\zeta}(\mathbf{Z}, t, \tau) \tag{3.13}$$

where $\boldsymbol{\zeta}(\mathbf{Z}, t, \tau)$ is a periodic function with mean zero.

$$\langle \boldsymbol{\zeta}(\mathbf{Z}, t, \tau)\rangle = \frac{1}{\mathcal{T}}\int_0^{\mathcal{T}} \boldsymbol{\zeta}(\mathbf{Z}, t, \tau)\ \mathrm{d}\tau = 0 \tag{3.14}$$

To proceed, we develop $\boldsymbol{\zeta}$ and $\mathbf{F}$ as perturbation expansions in ϵ:

$$\boldsymbol{\zeta} = \boldsymbol{\zeta}_0(\mathbf{Z}, t, \tau) + \epsilon\ \boldsymbol{\zeta}_1(\mathbf{Z}, t, \tau) + \epsilon^2\ \boldsymbol{\zeta}_2(\mathbf{Z}, t, \tau) + \cdots \tag{3.15}$$

$$\frac{\mathrm{d}\mathbf{Z}}{\mathrm{d}t} = \mathbf{F}_0(\mathbf{Z}, t) + \epsilon\ \mathbf{F}_1(\mathbf{Z}, t) + \epsilon^2\ \mathbf{F}_2(\mathbf{Z}, t) + \cdots \tag{3.16}$$

and substitute in Eqs. (3.12) and (3.13) to obtain at lowest order in ϵ:

$$\mathbf{F}_0(\mathbf{Z}, t) + \frac{\partial \boldsymbol{\zeta}_0}{\partial \tau} = \mathbf{f}(\mathbf{Z}, t, \tau) \tag{3.17}$$

Averaging over the periodic motion, the term in $\boldsymbol{\zeta}$ vanishes. Hence, setting the zero-order term in $\mathbf{F}$ to the cycle averaged value of $\mathbf{f}$ yields the appropriate 'condition of solubility':

$$\mathbf{F}_0(\mathbf{Z}, t) = \langle \mathbf{f}(\mathbf{Z}, t, \tau)\rangle \tag{3.18}$$

and integrating the Eq. (3.17) recovers the zero order oscillation:

$$\boldsymbol{\zeta}_0(\mathbf{Z}, t, \tau) = \int_0^{\tau} (\mathbf{f} - \langle \mathbf{f}\rangle)\mathrm{d}\tau' \tag{3.19}$$

Proceeding to first order, noting that

$$\frac{\partial \boldsymbol{\zeta}}{\partial t} = \left.\frac{\partial \boldsymbol{\zeta}}{\partial t}\right|_{\mathbf{Z}} + \left(\frac{\mathrm{d}\mathbf{Z}}{\mathrm{d}t}\cdot\frac{\partial}{\partial \mathbf{Z}}\right)\boldsymbol{\zeta} \qquad \text{and} \qquad \mathbf{f}(\mathbf{z}, t, \tau) \approx \langle \mathbf{f}\rangle(\mathbf{Z}, t) + \epsilon\ \left(\boldsymbol{\zeta}\cdot\frac{\partial}{\partial \mathbf{Z}}\right)\mathbf{f}(\mathbf{Z}, t, \tau)$$

we obtain

$$\mathbf{F}_1 + \frac{\partial \zeta_0}{\partial t} + (\mathbf{F}_0 \cdot \nabla)\zeta_0 + \frac{\partial \zeta_1}{\partial \tau} = (\zeta_0 \cdot \nabla)\mathbf{f} \tag{3.20}$$

Setting the first-order term in **F** to the cycle average eliminates the terms in ζ giving the first-order 'condition of solubility':

$$\mathbf{F}_1 = \langle(\zeta_0 \cdot \nabla)\mathbf{f}\rangle \tag{3.21}$$

and substituting Eqs. (3.18) and (3.21) in Eq. (3.16), we obtain

$$\frac{\mathrm{d}\mathbf{Z}}{\mathrm{d}t} = \langle\mathbf{f}\rangle + \epsilon\langle(\zeta_0 \cdot \nabla)\mathbf{f}\rangle + O(\epsilon^2) \tag{3.22}$$

The first-order rotational term ζ_1 is rarely needed, but may be obtained by integrating Eq. (3.20) having substituted for terms in ζ_0 from Eq. (3.19).

This is the governing equation of guiding centre motion. The particle dynamics, Eq. (3.9) involving the complete state of particle motion are replaced by an approximation for the guiding centre. This procedure averages over the small-scale variations associated with cyclotron rotation and provides a description based entirely on the secular motion, Eq. (3.22) obtained by averaging over the 'fluctuations'. The driving term therefore contains only the 'force' at the guiding centre and the correlation due to variation in position and the spatial or temporal gradient of the field.

3.3.2 The Guiding Centre Model for Charged Particles

In the case that we have the two conditions that the rotational motion is nearly constant as a result of temporal and spatial variations:

$$\frac{1}{B}\frac{\partial B}{\partial t} \ll \frac{1}{\Omega} \qquad \text{and} \qquad \frac{1}{B}\,|\nabla \mathbf{B}| \ll \rho$$

Including the gravitational acceleration **g**, the motion is described by

$$\frac{\mathrm{d}\mathbf{r}}{\mathrm{d}t} = \mathbf{v} \qquad \text{and} \qquad \frac{\mathrm{d}\mathbf{v}}{\mathrm{d}t} = \frac{q}{\epsilon\, m}(\mathbf{E} + \mathbf{v} \wedge \mathbf{B}) + \mathbf{g} \tag{3.23}$$

In this case, the gyrofrequency scales as $\Omega \sim \epsilon^{-1}$. It follows from Eqs. (3.23) that since $\Omega = qB/m$, the magnetic field scales as Ω and, therefore, $\mathbf{B},\ \mathbf{E},\ \mathbf{g} \sim \epsilon^{-1}$. The gyroradius scales as $\rho \sim \epsilon$ and therefore the rotational velocity as $u \sim \epsilon^0$. To overtly introduce the scaling, we change

$$\begin{aligned}(\Omega,\ \mathbf{B},\ \mathbf{E},\ \mathbf{g}) &\rightarrow (\epsilon^{-1}\Omega,\ \epsilon^{-1}\mathbf{B},\ \epsilon^{-1}\mathbf{E}, \epsilon^{-1}\mathbf{g})\\ (\mathbf{r},\ \mathbf{v}) &\rightarrow (\mathbf{r},\ \mathbf{v})\\ \rho &\rightarrow \epsilon\, \rho\end{aligned} \tag{3.24}$$

The rotational motion is separated from the secular by the introduction of the guiding centre position **R** and velocity **U**, and the values associated with the rotation: position ρ and velocity **u**

$$\mathbf{r} = \mathbf{R} + \epsilon\rho\,(\mathbf{R}, \mathbf{U}, t, \theta) \qquad \text{and} \qquad \mathbf{v} = \mathbf{U} + \mathbf{u}\,(\mathbf{R}, \mathbf{U}, t, \theta) \tag{3.25}$$

where θ is the phase angle of the rotation. Clearly, the averages over the rotation:

$$\langle\rho\rangle = 0 \qquad \text{and} \qquad \langle\mathbf{u}\rangle = 0 \tag{3.26}$$

Introducing the expansions:

$$\rho = \rho_0\,(\mathbf{R}, \mathbf{U}, t, \theta) + \epsilon\ \rho_1\,(\mathbf{R}, \mathbf{U}, t, \theta) \quad \text{and} \quad \mathbf{u} = \mathbf{u}_0\,(\mathbf{R}, \mathbf{U}, t, \theta) + \epsilon\ \mathbf{u}_1\,(\mathbf{R}, \mathbf{U}, t, \theta) \tag{3.27}$$

together with the corresponding one for the phase

$$\frac{d\theta}{dt} = \epsilon^{-1}\omega_{-1}(\mathbf{R}, \mathbf{U}, t) + \omega_0(\mathbf{R}, \mathbf{U}, t) \tag{3.28}$$

At each level of approximation, the evolution of the guiding centre is controlled by equations such as (3.18) and (3.21) relating the guiding centre position **R** and velocity **u** to the cycle averaged velocity. Since the equation for the position is linear

$$\frac{d\mathbf{R}}{dt} = \langle u(\mathbf{R}, \mathbf{U}, t, \theta)\rangle = \mathbf{U} \tag{3.29}$$

is obeyed at all orders of expansion.

The average of the phase equation (3.28) taken at the lowest order yields $\omega_{-1} = \Omega$. Turning now to the momentum equation, at lowest order $O(\epsilon^{-1})$, and noting that $\mathbf{v} = \mathbf{u} + \mathbf{U}$ and $\Omega = qB/m$, and that at this level of approximation $d/dt \to \omega\, \partial/\partial\theta$, we obtain

$$\omega\frac{\partial \mathbf{u}}{\partial \theta} - \Omega\mathbf{u} \wedge \hat{\mathbf{b}} = \frac{q}{m}(\mathbf{E} + \mathbf{U}_0 \wedge \mathbf{B}) + \mathbf{g} \tag{3.30}$$

where all unnecessary subscripts have been omitted. $\hat{\mathbf{b}} = \mathbf{B}/B$ is the unit vector parallel to the magnetic induction.

Taking the phase average

$$0 = \mathbf{E} + \mathbf{U}_0 \wedge \mathbf{B} + \frac{m}{q}\mathbf{g} \tag{3.31}$$

Clearly, the components of **E** and **g** parallel to the magnetic field, $E_{\|}$ and $g_{\|}$ are zero at this level of approximation. They must therefore be terms of higher-order $E_{\|}, g_{\|} \sim \epsilon^0$ than assumed in Eq. (3.31). Solving for $\mathbf{U}_0$ by taking the vector product with $\hat{\mathbf{b}}$, we obtain

$$\mathbf{U}_0 = U_{\|}\,\hat{\mathbf{b}} + \frac{\mathbf{E} \wedge \mathbf{B}}{B^2} + \frac{m}{q}\frac{\mathbf{g} \wedge \mathbf{B}}{B^2} = U_{\|}\,\hat{\mathbf{b}} + \mathbf{U}_E + \mathbf{U}_g \tag{3.32}$$

The parallel velocity component being of higher order is undefined at this stage – a term of the next order, Eq. (3.41) being required to establish its value. The perpendicular guiding centre velocity at the lowest order is just the $\mathbf{E} \wedge \mathbf{B}$ drift velocity $\mathbf{U}_E$ found in static fields.

Returning to Eq. (3.30), a simple integration yields

$$\mathbf{u} = \mathbf{c} + u_\perp\,[\hat{\mathbf{i}}_1 \sin(\Omega\,\theta/\omega) + \hat{\mathbf{i}}_2 \cos(\Omega\,\theta/\omega)] \tag{3.33}$$

where $\hat{\mathbf{i}}_1$ and $\hat{\mathbf{i}}_2$ form a right handed orthogonal set of unit vectors with $\hat{\mathbf{b}} = \hat{\mathbf{i}}_3$. It follows from the constraints induced by the cycle averaging that $\mathbf{c} = 0$ and $\omega = \Omega$. Therefore,

$$\mathbf{u} = u_\perp\,(\hat{\mathbf{i}}_1 \sin\theta + \hat{\mathbf{i}}_2 \cos\theta) \tag{3.34}$$

The perpendicular speed $u_\perp$ is undefined by this approximation. The phase is given by Eq. (3.28) at this level of approximation:

$$\theta = \theta_0 + \Omega\,t \tag{3.35}$$

and from the zero order kinematic equation, we obtain that

$$\Omega\frac{d\rho}{d\theta} = \mathbf{u} \longrightarrow \rho = \rho(-\hat{\mathbf{i}}_1 \cos\theta + \hat{\mathbf{i}}_2 \sin\theta) \tag{3.36}$$

where $\rho = u_\perp/\Omega$ and $\mathbf{u} = \Omega\,\rho \wedge \hat{\mathbf{b}}$.

Thus far the equations have simply reproduced the values for constant electric and magnetic fields, albeit with small gradients. To obtain the motion is spatially and temporally varying fields, we must consider the motion at the next level of approximation. Taking the phase average of the momentum equation at the order $O\epsilon$, we obtain

$$\frac{\mathrm{d}\mathbf{U}_0}{\mathrm{d}t} = \frac{q}{m}[E_{\parallel}\hat{\mathbf{b}} + \mathbf{U}_1 \wedge \mathbf{B} + \langle \mathbf{u} \wedge (\boldsymbol{\rho} \cdot \nabla)\mathbf{B}\rangle] + \mathbf{g} \tag{3.37}$$

Substituting for $\mathbf{u}$ and expanding the vector triple product yields a term:

$$\langle(\boldsymbol{\rho} \wedge \hat{\mathbf{b}}) \wedge (\boldsymbol{\rho} \cdot \nabla)\mathbf{B}\rangle = \hat{\mathbf{b}}\langle \boldsymbol{\rho} \cdot (\boldsymbol{\rho} \cdot \nabla)\mathbf{B}\rangle - \langle \boldsymbol{\rho}(\boldsymbol{\rho} \cdot \nabla B)\rangle = \left\langle \hat{b}_i \rho_j \rho_k \frac{\partial B_j}{\partial x_k} \right\rangle - \left\langle \rho_i \rho_j \frac{\partial B}{\partial x_j} \right\rangle$$

$$= \frac{1}{2}\rho^2 \left(\hat{b}_i \frac{\partial B_j}{\partial x_j} - \hat{b}_i \hat{b}_j \hat{b}_k \frac{\partial B_j}{\partial x_k} - \frac{\partial B}{\partial x_i} + \hat{b}_i \hat{b}_j \frac{\partial B}{\partial x_j} \right) = -\frac{1}{2}\,\rho^2\,\nabla B$$

since $\nabla \cdot \mathbf{B} = \partial B_i/\partial x_i = 0$ and $\hat{b}_i \equiv \hat{\mathbf{b}} = \hat{\mathbf{i}}_3$. The average

$$\langle \rho_i \rho_j \rangle = \frac{1}{2}\rho^2(\delta_{ij} - \hat{b}_i \hat{b}_j) \tag{3.38}$$

The rotating charge in the cyclotron motion constitutes a current loop of current $I = q\,u/2\,\pi\,\rho$ and area $\pi\,\rho^2$. The loop therefore constitutes a dipole of moment

$$\mu = \pi\,\rho^2\,I = \frac{m\,u^2}{2\,B} \tag{3.39}$$

The term $\mu = m\,u^2/2\,B = w_{\perp}/B$ is the magnetic moment and plays an important role in guiding centre motion.[2] As we shall show later in Section 3.6, the magnetic moment is conserved in a varying magnetic field. Equation (3.37) therefore becomes

$$\frac{\mathrm{d}\mathbf{U}_0}{\mathrm{d}t} = \frac{q}{m}[E_{\parallel}\hat{\mathbf{b}} + \mathbf{U}_1 \wedge \mathbf{B}] - \frac{\mu}{m}\nabla B + \mathbf{g} \tag{3.40}$$

The motion of the guiding centre along the lines of force therefore satisfies Eq. (3.32)

$$\frac{\mathrm{d}U_{0\parallel}}{\mathrm{d}t} = \frac{q}{m}E_{\parallel} - \frac{\mu}{m}\frac{\partial B}{\partial s} - \hat{\mathbf{b}} \cdot \frac{\mathrm{d}\mathbf{U}_E}{\mathrm{d}t} + g_{\parallel} \tag{3.41}$$

where s is the distance along the field line and making use of Eq. (3.31).

The motion perpendicular to the field lines is given by the vector product of Eq. (3.40) with $\hat{\mathbf{b}}$, namely,

$$\mathbf{U}_{1\perp} = \frac{1}{\Omega}\left\{ \hat{\mathbf{b}} \wedge \left(\frac{\mathrm{d}\mathbf{U}_0}{\mathrm{d}t} + \frac{\mu}{m}\nabla B \right) - \hat{\mathbf{b}} \wedge \mathbf{g} \right\}$$

$$= \frac{1}{\Omega}\left\{ \underbrace{\frac{\mu}{m}\,\hat{\mathbf{b}} \wedge \nabla B}_{\text{Grad B drift}} + \underbrace{U_{0\parallel}\,\hat{\mathbf{b}} \wedge \frac{\mathrm{d}\hat{\mathbf{b}}}{\mathrm{d}t}}_{\text{Inertial drift}} + \underbrace{\hat{\mathbf{b}} \wedge \frac{\mathrm{d}\mathbf{U}_E}{\mathrm{d}t}}_{\text{Polarisation drift}} - \hat{\mathbf{b}} \wedge \mathbf{g} \right\} \tag{3.42}$$

making use of Eq. (3.40).

---

2 The direction of the magnetic moment is independent of the charge. Due to the anti-clockwise rotation of a positively charged particle, it is anti-parallel to the induction

$$\boldsymbol{\mu} = -\frac{w_{\perp}}{B^2}\,\mathbf{B} \tag{3.39:a}$$

Plasma is therefore diamagnetic; the field generated by the magnetic dipole opposing that which generates the rotation of the plasma particle.

Since

$$\frac{\mathrm{d}\hat{\mathbf{b}}}{\mathrm{d}t} = \frac{\partial \hat{\mathbf{b}}}{\partial t} + (\mathbf{U}_E \cdot \nabla)\hat{\mathbf{b}} + U_{0\parallel}\,(\hat{\mathbf{b}} \cdot \nabla)\hat{\mathbf{b}} \tag{3.43}$$

the guiding centre convective derivative in Eq. (3.42) is a term of order 0. Substituting this value for $\mathrm{d}\hat{\mathbf{b}}/\mathrm{d}t$ the inertial drift term may be expanded to

$$U_{\text{Inert}} = \frac{U_{0\parallel}}{\Omega}\,\hat{\mathbf{b}} \wedge \left\{ \frac{\partial \hat{\mathbf{b}}}{\partial t} + (\mathbf{U}_E \cdot \nabla)\hat{\mathbf{b}} \right\} + \underbrace{\frac{{U_{0\parallel}}^2}{\Omega}\hat{\mathbf{b}} \wedge (\hat{\mathbf{b}} \cdot \nabla)\hat{\mathbf{b}}}_{\substack{\text{Curvature} \\ \text{drift}}} \tag{3.44}$$

3.4 Particle Kinetic Energy

The kinetic energy of the particle

$$\frac{1}{2}mv^2 = \frac{1}{2}m\,(U^2 + 2\mathbf{U} \cdot \mathbf{u} + u^2) \tag{3.45}$$

is given by

$$w = \frac{1}{2}m(\overbrace{{U_\parallel}^2}^{w_\parallel} + \overbrace{{U_E}^2}^{w_E} + \overbrace{u^2}^{w_\perp}) = w_\parallel + w_E + w_\perp \tag{3.46}$$

retaining only terms of first order after averaging. The term $w_\parallel$ is the kinetic energy due to the motion along the field lines, w_E that due to the $\mathbf{E} \wedge \mathbf{B}$ drift and $w_\perp$ that due to the cyclotron rotation around the field line. The temporal derivative is taken along the guiding centre trajectory $\mathrm{d}/\mathrm{d}t = \partial/\partial t + (\mathbf{U} \cdot \nabla)$. The increase in kinetic energy is due to the work done on the particle by the electromagnetic forces:

$$\frac{\mathrm{d}w}{\mathrm{d}t} = \left\langle \mathbf{v} \cdot \left[\frac{q}{m}(\mathbf{E} + \mathbf{v} \wedge \mathbf{B}) + \mathbf{g} \right] \right\rangle \tag{3.47}$$

Neglecting the gravitational force as it is easily included with the electric field, the kinetic energy averaged over the rotational cycle taking the expansion to first order, w, is, therefore,

$$\frac{\mathrm{d}w}{\mathrm{d}t} = \frac{m}{2}\frac{\mathrm{d}}{\mathrm{d}t}({\mathbf{U}_0}^2 + {\mathbf{U}_E}^2 + {\mathbf{u}_\perp}^2) = q\,U_{0\parallel}\,E_\parallel + q\,\mathbf{U}_1 \cdot \mathbf{E} + q\langle \mathbf{u} \cdot (\boldsymbol{\rho} \cdot \nabla)\mathbf{E} \rangle \tag{3.48}$$

Noting that the only non-zero averages are

$$\langle u_1\,\rho_2 \rangle = -\langle u_2\,\rho_1 \rangle = \frac{1}{2}\rho^2\,\Omega \tag{3.49}$$

and that

$$\langle \mathbf{u} \cdot (\boldsymbol{\rho} \cdot \nabla)\mathbf{E} \rangle = \frac{1}{2}\,\rho^2\,\Omega\,\left(\frac{\partial E_1}{\partial x_2} - \frac{\partial E_2}{\partial x_1} \right) = -\frac{1}{2}\,\rho^2\,\Omega\,(\nabla \wedge \mathbf{E}\,)_\parallel \tag{3.50}$$

we obtain

$$\frac{\mathrm{d}w}{\mathrm{d}t} = q\,\mathbf{U} \cdot \mathbf{E} + \boldsymbol{\mu} \cdot \nabla \wedge \mathbf{E} = q\,U_{0\parallel}\,E_\parallel + q\,\mathbf{U_1} \cdot \mathbf{E} + \mu\,\frac{\partial B}{\partial t} \tag{3.51}$$

from Eq. (3.32).

It follows that kinetic energy changes are associated either with an electric field directed along the path of the guiding centre, or with *betatron* action due to the rising magnetic field, which generates an e.m.f. around the cyclotron loop. Clearly, in the absence of an electric field, $\mathbf{E} = 0$, and the magnetic field held constant, the energy of the particle is conserved.

The rate of change of the energy associated with the zero-order velocity is given by Eq. (3.40):

$$\frac{\mathrm{d}(w_{\|}+w_E)}{\mathrm{d}t} = m\,\mathbf{U}_0 \cdot \frac{\mathrm{d}\mathbf{U}_0}{\mathrm{d}t} = q\,[U_{\|}E_{\|} + \mathbf{U}_0 \cdot (\mathbf{U}_1 \wedge \mathbf{B})] - \mathbf{U}_0 \cdot \mu\,\nabla B \tag{3.52}$$

The rate of change of the magnetic moment is related to that of the perpendicular energy associated with the rotation, which is given by

$$\begin{aligned}\frac{\mathrm{d}w_{\perp}}{\mathrm{d}t} &= \frac{\mathrm{d}w}{\mathrm{d}t} - \frac{\mathrm{d}w_{\|}}{\mathrm{d}t} - \frac{\mathrm{d}w_E}{\mathrm{d}t}\\ &= \cancel{q\,\mathbf{U}_0 \cdot E_{\|}\hat{\mathbf{b}}} + \cancel{q\mathbf{U}_1 \cdot \mathbf{E}} + \mu\frac{\partial \mathbf{B}}{\partial t} - \cancel{qE_{\|}U_{0\|}} - \cancel{q\mathbf{U}_0 \cdot \mathbf{U}_1 \wedge \mathbf{B}} + \mathbf{U}_0 \cdot \mu\,\nabla B\\ &= \mu\frac{\mathrm{d}B}{\mathrm{d}t}\end{aligned} \tag{3.53}$$

using Eqs. (3.31), (3.40) and (3.42). Since $\mu = w_{\perp}/B$ it follows that

$$\frac{\mathrm{d}\mu}{\mathrm{d}t} = \frac{1}{B}\frac{\mathrm{d}w_{\perp}}{\mathrm{d}t} - \frac{w_{\perp}}{B^2}\frac{\mathrm{d}B}{\mathrm{d}t} = 0 \tag{3.54}$$

and the magnetic moment is constant along the guiding trajectory at least as a first-order approximation. In fact, it can be shown to be an exact invariant of the motion (Kruskal, 1962).

3.5 Motion in a Static Inhomogeneous Magnetic Field

If the electric and magnetic fields are constant in time, but varying in space, Eqs. (3.41) and (3.42) show that only the speed along the field line changes. In a static magnetic field alone, where the induction varies slowly in space, the particle energy is constant since the force is normal to the velocity. If L is the spatial length associated with the gradient, we assume as before that the gradients are small. The gradient of an inhomogeneous field contains nine spatial derivatives of the induction of which only eight are independent since $\nabla \cdot \mathbf{B} = 0$,

$$\begin{pmatrix} \dfrac{\partial B_1}{\partial x_1} & \dfrac{\partial B_1}{\partial x_2} & \dfrac{\partial B_1}{\partial x_3}\\ \dfrac{\partial B_2}{\partial x_1} & \dfrac{\partial B_2}{\partial x_2} & \dfrac{\partial B_2}{\partial x_3}\\ \dfrac{\partial B_3}{\partial x_1} & \dfrac{\partial B_3}{\partial x_2} & \dfrac{\partial B_3}{\partial x_3} \end{pmatrix}$$

where x_3 is the direction of the magnetic induction, $\hat{\mathbf{b}}$, and x_1 and x_2 are orthogonal directions perpendicular to the field.

In a current-free environment $\nabla \wedge \mathbf{B} = 0$, off-diagonal terms such as $\partial B_x/\partial y = \partial B_y/\partial x$ so that $\nabla\mathbf{B}$ is represented by a traceless, symmetric matrix. The number of independent components is therefore reduced to five.

Although it is straightforward to analyse the guiding centre motion of the particle in the field quite generally from Eqs. (3.42) and (3.41) (Northrop, 1963), this formal approach loses sight of the underlying physics. It is easier to understand their behaviour if they are taken separately emphasising their basic physical behaviour. Since the effects of the varying induction are hypothesised to be small, it is satisfactory to assume that each may be treated as a perturbation on the underlying cyclotron motion, and that each is independent of the others.

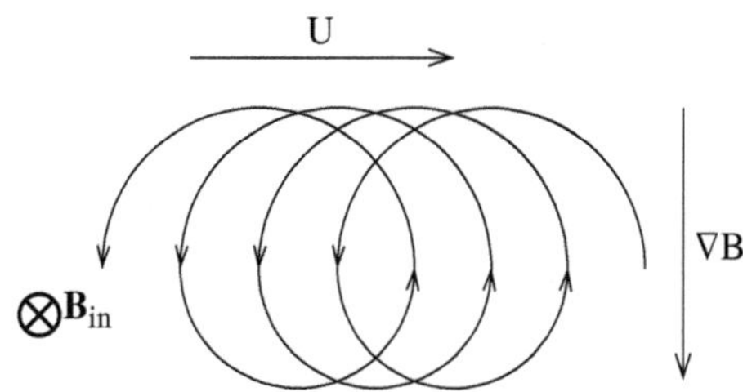

Figure 3.2 Sketch of the motion of a positively charged particle in a magnetic field directed out of the plane and decreasing downwards.

3.5.1 Field Gradient Drift

We consider first the terms $\partial B_3/\partial x_1$ and $\partial B_3/\partial x_2$ associated with transverse gradients in the magnetic field. The direction of the field is constant but varies in magnitude transversely to the field lines; the electric field **E** and all other gradients terms are zero. In the absence of the electric field, we may write the guiding centre velocity directly from Eq. (3.6):

$$\mathbf{U} = \mathbf{v} - \frac{m}{qB^2}\left(\mathbf{B} \wedge \frac{\mathrm{d}\mathbf{u}}{\mathrm{d}t}\right) - \frac{m}{q}\left[\frac{\mathrm{d}}{\mathrm{d}t}\left(\frac{\mathbf{B}}{B^2}\right) \wedge \mathbf{u}\right] \tag{3.55}$$

where $\mathrm{d}/\mathrm{d}t = \partial/\partial t + (\mathbf{v} \cdot \nabla)$ indicates differentiation with respect to time along the particle path.

Since the direction of the field is constant, $\mathbf{B}/B$ is constant along the particle path. Hence, making use of the results that

$$\frac{\mathrm{d}\mathbf{u}}{\mathrm{d}t} = \frac{q}{m}\mathbf{u} \wedge \mathbf{B} \qquad \text{and} \qquad \frac{m}{q} \cdot \frac{q}{m}\frac{\mathbf{B} \wedge (\mathbf{u} \wedge \mathbf{B})}{B^2} = u_\perp \tag{3.56}$$

we obtain

$$\mathbf{U} = \mathbf{v}_\parallel + \frac{q}{mB^2}\frac{\mathrm{d}B}{\mathrm{d}t}(\mathbf{B} \wedge \mathbf{u}) \tag{3.57}$$

Since the magnetic field is static, and the averages $\langle u_1 \cdot u_1 \rangle = \langle u_2 \cdot u_2 \rangle = \frac{1}{2}u^2$ and $\langle u_1 \cdot u_2 = 0 \rangle$, we may average over a particle orbit to obtain

$$\left\langle \frac{\mathrm{d}B}{\mathrm{d}t}(\mathbf{B} \wedge \mathbf{u}) \right\rangle = \langle [(\mathbf{u} \cdot \nabla)B](\mathbf{B} \wedge \mathbf{u}) \rangle = \frac{1}{2}\, u^2\, [\mathbf{B} \wedge \nabla(B)] \tag{3.58}$$

where $u = \rho\, \Omega$ is the speed of the cyclotron rotation.

The drift velocity associated with the gradient of the magnetic induction is, therefore,

$$\mathbf{U}_{\text{grad}} = \frac{1}{4}\frac{m\,u^2}{q\,B^4}\mathbf{B} \wedge \nabla(B^2) = \frac{\mu}{m\,\Omega}\hat{\mathbf{b}} \wedge \nabla B \tag{3.59}$$

which is identical to the corresponding term of Eq. (3.42).

The origin of gradient drift is easily seen from Figure 3.2, where it can be seen the 'radius' of the Larmor orbit in the upper part of the diagram, where the field is smaller, is greater than in the lower where it is larger. Consequently, the motion in the upper and lower halves are unbalanced, leading to the drift: Problem 2 gives an approximate method of calculating the drift using this picture.

3.5.2 Curvature Drift

When a field line of constant induction is curved, either $\partial B_1/\partial x_3 \neq 0$ or $\partial B_2/\partial x_3 \neq 0$, and the field line has a radius of curvature perpendicular to the field, Figure 3.3:

$$R \approx \frac{\Delta x_3}{\Delta\theta} \approx \frac{B\,\Delta x_3}{\Delta B} \tag{3.60}$$

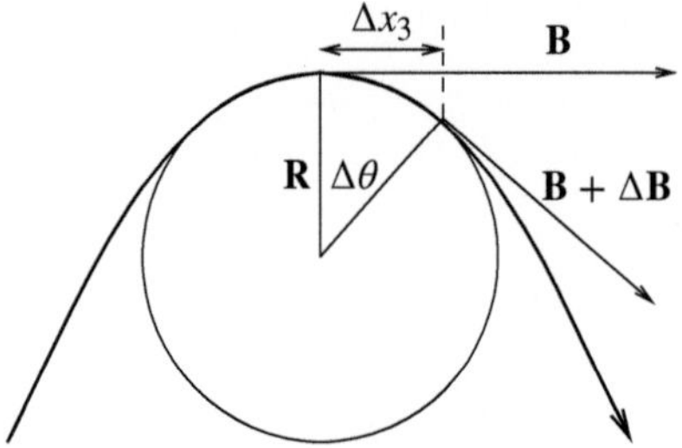

Figure 3.3 Sketch of curved field line showing the radius of curvature.

where $\Delta\theta$ is the angular deflection of the field line corresponding to a change ΔB in the induction over a distance Δx_3. Assigning a direction to the radius of curvature $\mathbf{R}$ from the centre of rotation, i.e. anti-parallel to $\Delta\mathbf{B}$, we obtain

$$\frac{\mathbf{R}}{R^2} = -\frac{(\mathbf{B}\cdot\nabla)\mathbf{B}}{B^2} \tag{3.61}$$

As the guiding centre of the particle attempts to follow the curvature of the field line, there is a centrifugal force:

$$\mathbf{F} = \frac{m{v_\parallel}^2\mathbf{R}}{R^2} = -m{v_\parallel}^2\frac{(\mathbf{B}\cdot\nabla)\mathbf{B}}{B^2} \tag{3.62}$$

which imparts a cross-field drift

$$\mathbf{U}_{\mathrm{R_c}} = -\frac{m\,{v_\parallel}^2\mathbf{R}\wedge\mathbf{B}}{q\,R^2\,B^2} = \frac{m\,{v_\parallel}^2\mathbf{B}\wedge(\mathbf{B}\cdot\nabla)\mathbf{B}}{q\,B^4} = \frac{{U_\parallel}^2\hat{\mathbf{b}}\wedge(\hat{\mathbf{b}}\cdot\nabla)\hat{\mathbf{b}}}{\Omega} \tag{3.63}$$

consistent with the appropriate term across the field lines from the inertial drift in Eq. (3.44).

There is a further term associated with the field curvature. It follows from Maxwell's equations that there is a field gradient along the radius of curvature associated with curvature. In the absence of a current density at the point of measurement $\mathbf{j} = \nabla\wedge\mathbf{B} = 0$ and therefore making use of the vector identity $\frac{1}{2}\nabla B^2 = (\mathbf{B}\cdot\nabla)\mathbf{B} + \mathbf{B}\wedge(\nabla\wedge\mathbf{B})$,[3] the corresponding drift speed is

$$\mathbf{U}_{\mathrm{R_g}} = \frac{1}{2}\frac{m\,{v_\perp}^2}{q\,B^4}\mathbf{B}\wedge(\mathbf{B}\cdot\nabla)\mathbf{B} = \frac{1}{2}\frac{m\,{v_\perp}^2\mathbf{R}\wedge\mathbf{B}}{q\,R^2\,B^2} \tag{3.64}$$

Summing these two terms, we obtain for the drift associated with field line curvature in the absence of an external current:

$$\mathbf{U}_{\mathrm{R}} = \frac{m}{q\,B^4}\left({v_\parallel}^2 + \frac{1}{2}\,{v_\perp}^2\right)\mathbf{B}\wedge(\mathbf{B}\cdot\nabla)\mathbf{B} \tag{3.65}$$

An alternative form for plasma with no current flow where $(\mathbf{B}\cdot\nabla)\mathbf{B} = B\,\nabla B$ may be useful for studying magnetically confined fusion devices with low β such as tokamaks and stellarators. Using the vector identity,

$$\nabla\wedge\left(v_\parallel\,\frac{\mathbf{B}}{B}\right) = \nabla v_\parallel\wedge\frac{\mathbf{B}}{B} - v_\parallel\,\frac{1}{B^2}\,\nabla B\wedge\mathbf{B}$$

Since the magnetic moment μ and the total kinetic energy $E = \frac{1}{2}mv^2$ are both constants of motion, we obtain

$$\nabla v_\parallel = \nabla\sqrt{\frac{2}{m}(E-\mu B)} = -\frac{\mu}{m v_\parallel}\nabla B$$

---

3 Alternatively, consider the local osculating plane of the field line with the origin taken at the centre of curvature. Assuming that the current density is zero $\nabla\wedge\mathbf{B} = 0$ at the field line $r = R$, implies that locally $B_\theta = B\,R/r$ and therefore $\nabla_\perp B = -B/R^2\,\mathbf{R}$ is the local field gradient.

Noting that $\mu\ B = m\ v_\perp^2/2$ and substituting. we obtain a form written in terms of $v_\parallel$

$$\mathbf{U}_\mathrm{R} = \frac{m}{q\ B^3}\left({v_\parallel}^2 + \frac{1}{2}\ {v_\perp}^2\right)\ \mathbf{B} \wedge \nabla B = \frac{m\ v_\parallel}{q\ B}\ \nabla \wedge \left(v_\parallel\ \frac{\mathbf{B}}{B}\right) \tag{3.66}$$

3.5.3 Divergent Field Lines

It follows from the basic equation $\nabla \cdot \mathbf{B} = 0$ that if $\partial B_1/\partial x_1$ and $\partial B_2/\partial x_2$ do not vanish, the lines of force have an angular divergence. As a consequence. the particle experiences a force along the field line due to the Lorentz force resulting from the cyclotron rotation and the transverse field components resulting from the gradients. On averaging using Eqs. (3.34) and (3.36), only terms of second order in the rotational motion are non-zero:

$$\langle u_1\ \rho_2\rangle = -\langle u_1\ \rho_2\rangle = \frac{1}{2}\ \rho\ u \tag{3.67}$$

to give

$$\begin{aligned} F_\parallel &= q\left\langle u_1\ \frac{\partial B_2}{\partial x_2}\delta x_2 - u_2\frac{\partial B_1}{\partial x_1}\ \delta x_1\right\rangle = \frac{1}{2}\ q\ \rho\ u\left\{\frac{\partial B_1}{\partial x_1} + \frac{\partial B_2}{\partial x_2}\right\} \\ &= -\mu\frac{\partial B_3}{\partial x_3} \approx \boldsymbol{\mu}\ \nabla B \end{aligned} \tag{3.68}$$

where δx_1 and δx_2 are the particle displacements from the guiding centre, and $B_1, B_2 \ll B_3$. This result is consistent with magnetostatics, provided the sign of $\boldsymbol{\mu}$ is adjusted to be consistent with the particle rotation (see Footnote 2, p. 60).

Letting s be the length of arc of a line of induction $U_\parallel = \mathrm{d}s/\mathrm{d}t$ and

$$m\ \frac{\mathrm{d}s}{\mathrm{d}t}\ \frac{\mathrm{d}U_\parallel}{\mathrm{d}s} = m\ U_\parallel\ \frac{\mathrm{d}U_\parallel}{\mathrm{d}s} = \frac{\mathrm{d}w_\parallel}{\mathrm{d}s} = -\frac{w_\perp}{B}\frac{\mathrm{d}B}{\mathrm{d}s} \tag{3.69}$$

where $w_\parallel$ and $w_\perp$ are the kinetic energies of the parallel $\frac{1}{2}m{v_\parallel}^2$ and perpendicular $\frac{1}{2}m{v_\perp}^2$ components, respectively. But since the force is purely magnetic, no work is done on the motion so that any increase in the parallel kinetic energy must be at the expense of the perpendicular. The total kinetic energy $w_\parallel + w_\perp = \text{const}$ and

$$\frac{\mathrm{d}w_\perp}{w_\perp} = \frac{\mathrm{d}B}{B} \qquad \therefore \quad \frac{w_\perp}{B} = \text{const} \qquad \text{and} \qquad \mu = \text{const} \tag{3.70}$$

This value of the transverse kinetic energy $w_\perp$ is obtained directly via betatron acceleration in Section 3.6. Importantly, we note that the magnetic moment is an approximate constant of the motion.

For future reference, we note that the decrease in the parallel component of the kinetic energy resulting from an increase ΔB of the induction is

$$\Delta w_\parallel = -\Delta w_\perp = -\mu\Delta B \tag{3.71}$$

3.5.4 Twisted Field Lines

The terms $\partial B_1/\partial x_2$ and $\partial B_2/\partial x_1$ cause the field line to twist as it progresses in x_3 direction. The orbit is slightly changed, but the guiding centre essentially follows the field line. No drift is introduced.

3.6 Motion in a Time Varying Magnetic Field

In the absence of an applied electric field, when the magnetic field is varying in time, an electric field is generated by induction

$$\nabla \wedge \mathbf{E} = -\frac{\partial \mathbf{B}}{\partial t} \tag{3.72}$$

Any electric field component parallel to the magnetic induction $\mathbf{E}_{\|}$ accelerates the particle along the line of induction in the usual way. However, the transverse field will generate an e.m.f. around the cyclotron loop thereby increasing the perpendicular velocity $\mathbf{v}_{\perp}$.[4] Since the flux through the loop is $\pi\rho^2 B$, the e.m.f. generated around a single rotation is $-\pi\ \rho^2\ \partial B/\partial t$. Assuming the field is only slowly varying and $\Delta T = 2\ \pi/\Omega$ is the period of the rotation, the increase in the kinetic energy perpendicular to the magnetic field per rotation, anti-clockwise is

$$\Delta w_{\perp} = q\ \pi\ \rho^2\ \frac{\partial B}{\partial t} = \mu\ T\ \frac{\partial B}{\partial t} = \mu\ \Delta B \tag{3.73}$$

Since the magnetic moment $\mu = w_{\perp}/B$, it follows that $\Delta w_{\perp} = w_{\perp}\ \Delta B/B$ and therefore that

$$\mu = w_{\perp}/B = \text{const} \tag{3.74}$$

i.e. the magnetic moment is invariant.

If the particle is moving in a divergent field, it experiences an additional time varying magnetic induction due to its motion:

$$\frac{\mathrm{d}B}{\mathrm{d}t} = \frac{\partial B}{\partial t} + (\mathbf{U} \cdot \nabla)B \tag{3.75}$$

which changes the flux through the loop and generates the accelerating e.m.f. around the cyclotron loop as above. The increase in the transverse kinetic energy is therefore given by Eq. (3.73). This contains a term due to the convective induction variation whose value is in agreement with that previously obtained from the conservation of energy, Eq. (3.70).

We note the important difference between the convective and temporal field effects. Convective changes involve no net increase in the total energy of the particle, longitudinal energy being transferred to transverse. In contrast, the temporal change of induction increases the kinetic energy, thereby heating the particles.

3.7 Motion in a Time Varying Electric Field

Consider the motion of a particle in a static uniform magnetic field under the influence of a time varying uniform electric field. Since the consequence of the electric field component parallel to magnetic lines of induction $E_{\|}$ is simply acceleration along the field lines, we may consider only the case where the electric field is perpendicular to the magnetic field. Writing $\mathbf{v} = \mathbf{u} + \mathbf{U}_E$ and using $\mathrm{d}\mathbf{v}/\mathrm{d}t = q/m(\mathbf{E} + \mathbf{v} \wedge \mathbf{B})$, we have

$$\frac{\mathrm{d}\mathbf{u}}{\mathrm{d}t} + \frac{\mathrm{d}\mathbf{U}_E}{\mathrm{d}t} = \frac{q}{m}\ \mathbf{u} \wedge \mathbf{B} \tag{3.76}$$

Introducing the polarisation drift velocity consistent with Eq. (3.42)

$$\mathbf{U}_P = \frac{m}{qB^2}\ \mathbf{B} \wedge \frac{\mathrm{d}\mathbf{U}_E}{\mathrm{d}t} = \frac{m}{qB^2}\frac{\partial \mathbf{E}}{\partial t} = \frac{1}{\Omega B}\frac{\partial E}{\partial t} \tag{3.77}$$

4 This is the familiar *betatron* action.

and writing $\mathbf{u} = \mathbf{U}_P + \mathbf{u}'$ yields

$$\frac{d\mathbf{u}'}{dt} + \frac{d\mathbf{U}_P}{dt} = \frac{q}{m}\,\mathbf{u}' \wedge \mathbf{B} \tag{3.78}$$

It is easy to show that U_P is a term of high order in this equation

$$\frac{m|d\mathbf{u}'/dt|}{q|\mathbf{u}' \wedge \mathbf{B}|} = \left(\frac{m}{qB}\right)^2 \frac{q}{u'\,B}|\ddot{\mathbf{E}}\,| \sim \frac{E}{u'\,B}\,\frac{1}{(\Omega\;T)^2} = \frac{U_E}{u'}\,\frac{1}{(\Omega\;T)^2} \tag{3.79}$$

and by hypothesis, the time constant $T \gg 1/\Omega$. The velocity $\mathbf{u}'$ therefore reduces to the familiar cyclotron rotation $\mathbf{u}' \approx \mathbf{u}$. The electric field effects are therefore superimposed on the rotation as an additional drift of the guiding centre

$$\mathbf{v} \approx \mathbf{u} + \mathbf{U}_E + \mathbf{U}_P \tag{3.80}$$

$\mathbf{U}_P$ is known as the *polarisation drift* whose value agrees with that given by the full analysis (Eq. (3.44)).

The direction of the polarisation drift velocity depends on the sign of the charge. In a neutral plasma with equal positive and negative charge density, the velocity is independent of the charge and no current results in a static field. In contrast when the field is time varying, the positive and negative charges move in opposite directions, generating a current. A classical polarised medium results with the plasma acting as a dielectric. The total current

$$\mathbf{J}_P = \sum n_i\; q_i\; \mathbf{U}_{Pi} = \frac{\rho}{B^2}\frac{\partial E}{\partial t} \tag{3.81}$$

where n_i is the number density of particles of charge q_i and $\rho = \sum N_i\; m_i$ the mass density. Classically, we write the dielectric displacement $\mathbf{D} = \mathbf{P} + \epsilon_0\mathbf{E} = \epsilon\mathbf{E}$. Since $\mathbf{J}_p = d\mathbf{P}/dt$, we obtain the permittivity

$$\epsilon = \epsilon_0(1 + \rho/\epsilon_0 B^2) \tag{3.82}$$

The plasma therefore is acting as a dielectric to an oscillating field. Typically, the susceptibility ϵ/ϵ_0 may be large.[5]

The origin of the term is easily seen by considering the behaviour when an electric field is switched. Referring to Figure 3.1 at time zero when the field is switched on, the particle is at the minimum corresponding to rest. Initially, the particle is accelerated along the field direction, before the velocity increases sufficiently for the magnetic field to re-direct the motion. Over roughly a quarter cycle, the particle is accelerated to reach the cycloidal motion with the guiding centre acquiring velocity $\mathbf{U}_E$. The particle displacement due to a field increase ΔE over an interval $\Delta t \sim \Omega_1$ is therefore,

$$d \sim U_P \Delta t \sim \frac{q\;\Delta E}{m}\Delta t^2 \sim \frac{\Delta E}{\Omega\;B} \sim \frac{1}{\Omega B}\frac{\partial E}{\partial t}\Delta t \tag{3.83}$$

More generally, if the electric field is changed, inertia of the particles gives rise to a delay of about a quarter cycle before the cycloidal motion is re-established with the values appropriate to the new field.

The polarisation shift is parallel to the electric field. Therefore, work is done by a small change in the field $\Delta\mathbf{E}$ in time Δt, giving rise to a shift of the guiding centre $\Delta\mathbf{r} = \mathbf{U}_P\Delta t$. The work done appears an increase in the energy of the electron:

$$\Delta w = q\;\mathbf{E} \cdot \mathbf{U}_P\;\Delta t = \frac{m}{B^2}\;\mathbf{E} \cdot \frac{\partial \mathbf{E}}{\partial t}\;\Delta t \approx \Delta\left(\frac{1}{2}\;m\;\frac{E^2}{B^2}\right) = \Delta\left(\frac{1}{2}\;m\;{U_E}^2\right) = \Delta w_E \tag{3.84}$$

5 Note that due to their larger mass, the ions contribute more strongly to polarisation current and susceptibility.

The energy change is due to the change in the kinetic energy associated with the $\mathbf{E} \wedge \mathbf{B}$ drift.

Alternatively, we may write the energy per unit volume as the sum of the vacuum field and the polarisation energy:

$$\Delta \mathcal{E} = \frac{1}{2}\, \epsilon_0\, \Delta(E^2) + \sum \frac{1}{2}\, n_i\, m_i\, \Delta \left(\frac{E^2}{B^2} \right) = \frac{1}{2}\, \epsilon\, \Delta(E^2) \tag{3.85}$$

i.e. the value for a polarised dielectric showing that the plasma behaves as a classical dielectric.

3.8 Collisional Drift

There is one further source of drift due to collisions, see Section 15.8. Assume collisions are instantaneous so that the position of the particle is unchanged, although its velocity is changed by $\Delta\mathbf{v}$. Then the guiding centre is shifted to adjust to this new velocity:

$$\mathbf{r} = \mathbf{r}_g + \frac{m}{q}\frac{\mathbf{B} \wedge \mathbf{v}}{B^2} = \mathbf{r}_g + \Delta\mathbf{r}_g + \frac{m}{q}\frac{\mathbf{B} \wedge (\mathbf{v} + \Delta\mathbf{v})}{B^2} \tag{3.86}$$

The shift of the guiding centre is, therefore,

$$\Delta\mathbf{r}_g = \frac{m}{q}\frac{\Delta\mathbf{v} \wedge \mathbf{B}}{B^2} \tag{3.87}$$

3.9 Plasma Diamagnetism

In the Larmor motion about the magnetic field, each charged particle constitutes a circular current loop in the context of classical magnetism, and forms an elementary dipole of moment $\boldsymbol{\mu} = -\dfrac{m{v_\perp}^2}{2\,B^2}\,\mathbf{B} = -\dfrac{w_\perp}{B^2}\,\mathbf{B}$ anti-parallel to the magnetic field, Eq. (3.39:a).

The magnetisation $\mathbf{M}$ defined as the dipole moment per unit volume is, therefore,

$$\mathbf{M} = n\langle \boldsymbol{\mu} \rangle = -\frac{1}{2}\, n\, m\, \langle {v_\perp}^2 \rangle\, \frac{1}{B^2}\mathbf{B} = -\frac{p_\perp}{B^2}\mathbf{B} \tag{3.88}$$

Since $v_\perp$ has two degree of freedom, each having energy $\frac{1}{2}\mathcal{k}T$ in thermal equilibrium,[6] $\frac{1}{2}\, m\langle v_\perp^2 \rangle = \mathcal{k}T$, the transverse pressure $p_\perp = mn\langle v_\perp^2 \rangle = n\langle w_\perp \rangle$, the averages being taken over the particle distribution.

It follows that the current associated with the plasma diamagnetism due to gradients in the particle density and temperature is, therefore,

$$\mathbf{j}_M = \nabla \wedge \mathbf{M} = -\frac{\nabla p_\perp \wedge \mathbf{B}}{B^2} - p_\perp \nabla \wedge \left(\frac{\mathbf{B}}{B^2} \right) \tag{3.90}$$

The physical origin of the diamagnetic current is easily understood in terms of the cyclotron orbits of particles whose guiding centres lie in regions of varying density or temperature. In the absence of a temperature gradients, more particle orbits with centres on the higher density side pass through a reference point than those on the lower density side (Tonks, 1955). Since the orbits

6 Alternatively for a two-dimensional Maxwellian distribution:

$$\langle v_\perp^2 \rangle = \int v_\perp^2\, f(v_\perp)\, v_\perp\, \mathrm{d}v_\perp = \frac{m}{\mathcal{k}T} \int v_\perp^2\, \exp(-mv_\perp^2/2m\, \mathcal{k}T)\, v_\perp\, \mathrm{d}v_\perp = 2\, \mathcal{k}T/m \tag{3.89}$$

must all pass through the measurement point, their centres must lie approximately one Larmor radius above or below this point, the current is

$$j_M \approx q\, v_\perp \frac{mv_\perp}{qB}|\nabla n| \tag{3.91}$$

A similar pattern arises in the presence of a temperature gradient due to the higher particle velocities on the high temperature side. The guiding centre model of the diamagnetic current is treated fully in Problem 3. It is notable that the current is generated by the orbital motion of the charged particles, without the guiding centre contributing to the motion (see Footnote 1 on page 43).

The diamagnetic current implies a motion of the charged elements corresponding to a *diamagnetic drift*. This motion is perhaps surprising in that it is not accompanied by a corresponding guiding centre drift but solely with a motion of the individual charges, as discussed earlier (Footnote 1 on page 43).

The second term in Eq. (3.90), which has been cancelled, is due to a gradient in the magnetic field strength. This leads to current due to the guiding centre drift which is equal and opposite to that due to the orbital motion,[7] as derived in Problem 4 – a result originally due to Tonks (1955). In contrast to the previous case of a pressure gradient, no current flows despite the guiding centre *grad B* drift being present.

The magnetisation current induces a magnetic field $\mathbf{B}'$ in addition to the applied field $\mathbf{B}$. Since it follows from Ampère's equation that $\mu_0 \nabla \wedge \mathbf{B}' = \mathbf{j}_M = \nabla \wedge \mathbf{M}$, the relative magnitude of the induced field

$$\frac{B'}{B} \sim \frac{p_\perp}{B^2/2\mu_0} = \beta \tag{3.93}$$

where β, the plasma β, is defined later (Eq. (9.26)) as the ratio between the gas kinetic and the magnetic pressures.

3.10 Particle Trapping and Magnetic Mirrors

When the field varies slowly in space and time, the magnetic moment of the particle is conserved,

$$\mu = \frac{1}{2}m{v_\perp}^2/B = \text{const} = \frac{1}{2}m{v_{\perp 0}}^2/B_0$$

and in a static field the total energy is constant

$$E = \frac{1}{2}m({v_\parallel}^2 + {v_\perp}^2) = \text{const} = \frac{1}{2}m{v_0}^2$$

Hence,

$$v_\parallel^2 = {v_0}^2 - {v_\perp}^2 = {v_0}^2\left(1 - \frac{B}{B_0}\frac{{v_{\perp 0}}^2}{{v_0}^2}\right) \tag{3.94}$$

---

7 Taking axis such that $\mathbf{B} = B(x)\,\hat{\mathbf{k}}$ and the field gradient $\dot{\mathbf{B}} = \dot{B}(x)\,\hat{\mathbf{i}}$, then the cancelled term

$$\mathbf{j}_M = \cdots - p_\perp \nabla \wedge \left(\frac{\mathbf{B}}{B^2}\right) = \cdots - p_\perp \frac{1}{B^3}\frac{\mathrm{d}B}{\mathrm{d}x}\,\hat{\mathbf{j}} = \cdots - \frac{1}{2}\frac{n\,m\,{v_\perp}^2}{B^4}\mathbf{B} \wedge \nabla_\perp \mathbf{B} \tag{3.92}$$

which is equal to the current due to the orbital motion alone, to which must be added the equal and opposite guiding centre motion, Eq. (3.59).

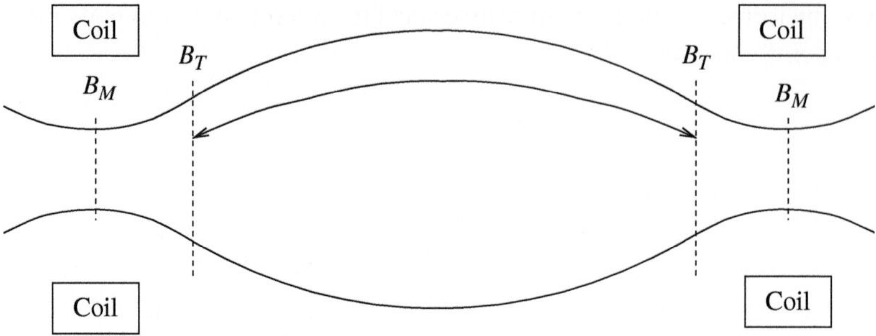

Figure 3.4 Sketch of the layout of a magnetic bottle with mirrors at each end. Particles drift between the mirror points B_T.

where the values B_0 and v_0 are taken at some convenient point, for example at the weakest field point. As the field increases, $v_\parallel$ decreases until when

$$B = B_T = \left(\frac{{v_0}^2}{{v_{\perp 0}}^2} \right) B_0 \tag{3.95}$$

the parallel velocity along the field line $v_\parallel =$ is reduced to zero. At which point, the particle is reflected and returned along its inward path. This is the phenomenon of *magnetic trapping* where the particles are held in a magnetic bottle formed by magnetic mirrors (Figure 3.4). If B_M is the maximum field, all particles with $B_T < B_M$ are trapped. However, particles with a large initial ratio of parallel to perpendicular velocities will be lost

$$\frac{{v_{\parallel 0}}^2}{{v_{\perp 0}}^2} \geq \frac{B_M - B_0}{B_0} \tag{3.96}$$

This region of loss in velocity space is known as the loss cone. Particles are transferred into it by collisions and departures from invariance of the magnetic moment (Figure 3.5).

The simple magnetic bottle arrangement formed the basis of several particle trapping configurations. A number of early fusion reactor proposals involved magnetic mirrors with deuterium and tritium ions injected in some way into the bottle (Rose and Clark, 1961, Ch.15). The ions initially had an approximately uniform velocity distribution. However, collisions resulting from multi-particle distant interactions limited any potential applications in a simple arrangement both by the Maxwellianisation of the distribution and moving particles into the loss cone. The plasma within the mirror also suffered from instability. However, modern fusion devices such as tokamaks

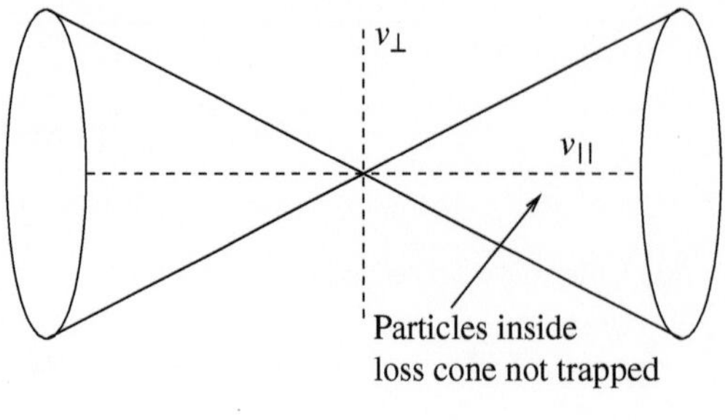

Figure 3.5 All particle whose velocities lie inside the lost cone are *not* reflected at the mirror.

make use of mirroring in the torus. Magnetic mirrors are critical to structure of the trapped radiation in the earth's magnetosphere, where particles bounce back and forth between the stronger fields at the poles.

3.10.1 Fermi Acceleration

If the lines of induction are moving, charged particles moving towards the field are reflected with an increased velocity. Fermi (1949) suggested that this mechanism could be responsible for the generation of high-energy cosmic ray particles through repeated collisions with magnetic fields. The variants are proposed (Northrop, 1963)

a. Interactions with shock wave magnetic inhomogeneities.
b. Interactions with random magnetic fields contained in gas clouds.

3.11 Adiabatic Invariance

Adiabatic invariance is a very useful concept which is applied in several fields of physics (Ter Haar, 2006) whereby a certain quantity is nearly constant whilst very slow changes occur in the external parameters. It is a result of Hamiltonian mechanics in many cases (Landau and Lifshitz, 1976, §49–51) amongst which the motion of charged particles in electric and magnetic fields is of particular interest here. The historical importance of adiabatic invariance stems from a question posed by Lorentz at the 1911 Solvay conference: 'How does a simple pendulum behave when the length of the string is gradually shortened?' The answer was given by Einstein 'that if the suspending thread is shortened infinitely slowly, then the energy E will increase proportionately to the "instantaneous" frequency'; in other words, E/ν is an adiabatic invariant, the quantum of energy therefore remaining the same quantum of energy. The analogy of the pendulum with a quantum mechanical oscillator where the adiabatic invariant is Planck's constant $h = E/\nu$. The importance of this result to the quantum theory of that time is evident and led to the Bohr–Sommerfeld quantisation rule.

The concept of adiabatic invariance follows from classical mechanics as periodic motion is slowly perturbed. Consider periodic motion in the simplest case of one dimension, the generalisation to further degrees of freedom is straightforward. The *action integral* of a classical orbit is defined by the area enclosed in phase space by the cycle

$$J = \int_0^T p(t)\, \dot{q}(t)\, \mathrm{d}t \tag{3.97}$$

and is a function of the energy alone since the integral is taken over the period T. When the Hamiltonian is constant in time, J is constant in time and the energy is constant. Using Hamilton's equation, the variable θ, canonically conjugate to J, increases at a steady rate

$$\frac{\mathrm{d}\theta}{\mathrm{d}t} = \frac{\partial H}{\partial J} = \nu(J) \tag{3.98}$$

Since the energy is constant, the variable θ increases steadily with time and represents an increasing rotational angle. The constant ν therefore defines the angular rate. Differentiating Eq. (3.97)

$$\frac{\mathrm{d}J}{\mathrm{d}J} = \int_0^T \left(\frac{\partial p}{\partial J}\, \dot{q} + p\, \frac{\partial \dot{q}}{\partial J} \right) \mathrm{d}t = \nu \int_0^T \left(\frac{\partial p}{\partial J} \frac{\partial q}{\partial \theta} - \frac{\partial p}{\partial \theta} \frac{\partial q}{\partial J} \right) \mathrm{d}t = \nu \int_0^T \{p,\, q\} \mathrm{d}T = 1 \tag{3.99}$$

where $\{p, q\}$ is the Poisson bracket of the canonical variables p and q. Since the Poisson bracket of conjugate variables is 1 (Landau and Lifshitz, 1976, §45)[8] we obtain

$$1 = \nu \int_0^T \{p,\ q\}\mathrm{d}T = \nu T \tag{3.103}$$

so that ν, is the frequency, i.e. $\omega/2\,\pi$, where ω is the angular frequency. Thus, a rotation is expressed by a constant value of the action integral J associated with a constant value of the Hamiltonian H and

$$\theta = \omega t = 2\,\pi\,\nu\,t \tag{3.104}$$

We now consider the situation when the Hamiltonian is slowly varying in time. The variation in the action J is

$$\frac{\mathrm{d}J}{\mathrm{d}t} = \int_0^T \left(\frac{\mathrm{d}p}{\mathrm{d}t}\frac{\mathrm{d}q}{\mathrm{d}t} + p\,\frac{\mathrm{d}}{\mathrm{d}t}\frac{\mathrm{d}q}{\mathrm{d}t}\right)\mathrm{d}t = \omega \int_0^{2\,\pi} \left(\frac{\partial p}{\partial\theta}\frac{\partial q}{\partial\theta} + p\,\frac{\partial}{\partial\theta}\frac{\partial q}{\partial\theta}\right)\mathrm{d}\theta \tag{3.105}$$

where the time derivatives have been replaced by angular from Eq. (3.104). Provided the variables J and θ do not change appreciably over a period from their values when the Hamiltonian is constant, i.e. the variation of the Hamiltonian in time is sufficiently slow and $|\mathrm{d}\omega/\mathrm{d}t| \ll \omega^2$, we may integrate by parts to obtain

$$\frac{\mathrm{d}J}{\mathrm{d}t} = 0 + O\left(\frac{1}{\omega^2}\left|\frac{\mathrm{d}\omega}{\mathrm{d}t}\right|\right) \tag{3.106}$$

and J is therefore approximately constant.

In principle, there may be as many periodic motions as degrees of freedom. However, not all may exist in the particular experimental situation. Generally, the adiabatic invariants are the action variables J.

For a harmonic oscillator, the motion in phase space is an ellipse of constant energy E

$$E = \frac{p^2}{2\,m} + \frac{m\,\omega^2\,x^2}{2} \tag{3.107}$$

with axes $\sqrt{2\,m\,E}$ and $\sqrt{2\,E/m\,\omega^2}$. The phase space area $J = 2\,\pi\,E/\omega$ is an adiabatic invariant, in agreement with Einstein's response to Lorentz' question. The behaviour of the adiabatic invariant of the harmonic oscillator is discussed in some detail by Chandrasekhar (1958) with particular relevance to the first invariant of charged particle motions.

Adiabatic invariants are approximate integrals of Hamiltonian mechanics. The invariant property implies that the perturbation parameter ϵ on the 'orbiting motion' is small. In many

8 This result is easily obtained. Since the Hamiltonian is constant, the canonical transformation from the set (p, q) to (P, Q) must satisfy

$$\left.\frac{\partial Q_m}{\partial p_n}\right|_{p,q} = -\left.\frac{\partial q_n}{\partial P_m}\right|_{P,Q} \quad \text{and} \quad \left.\frac{\partial Q_m}{\partial q_n}\right|_{p,q} = \left.\frac{\partial p_n}{\partial P_m}\right|_{P,Q} \tag{3.100}$$

$$\left.\frac{\partial P_m}{\partial p_n}\right|_{p,q} = \left.\frac{\partial q_n}{\partial Q_m}\right|_{P,Q} \quad \text{and} \quad \left.\frac{\partial P_m}{\partial q_n}\right|_{p,q} = -\left.\frac{\partial p_n}{\partial Q_m}\right|_{P,Q} \tag{3.101}$$

Hence, since $P(p, q)$ and $Q(p, q)$ and using the chain rule, we obtain

$$\{P, Q\} = \sum_n \frac{\partial P_m}{\partial p_n}\frac{\partial Q_m}{\partial q_n} - \frac{\partial Q_m}{\partial p_n}\frac{\partial P_m}{\partial q_n} = \sum_n \frac{\partial P_m}{\partial p_n}\frac{\partial p_n}{\partial P_m} + \frac{\partial P_m}{\partial q_n}\frac{\partial q_n}{\partial P_m} = \frac{\partial P_m}{\partial P_m} = 1 \tag{3.102}$$

cases, the result of Eq. (3.106) can be shown to be the first term in an asymptotic expansion in the parameter ϵ

$$J(\epsilon) = J_0 + J_1\ \epsilon + J_2\ \epsilon^2 + \cdots \tag{3.108}$$

is constant.

3.12 Adiabatic Invariants of Charged Particle Motions

A single particle has three degrees of freedom.[9] It therefore has a maximum of three adiabatic invariants in its motion. These are[10]

- *First invariant – magnetic moment*: The first invariant is associated with the rotation of the particle around the magnetic field lines. The action variable J for the particle spiralling about the field easily calculated if the fields are constant

$$\begin{aligned} \mathcal{I} &= \oint p_i\ q_i = 2\ \pi\ r_L\ v_\perp - \pi\ {r_L}^2\ B \\ &= \frac{\pi\ {v_\perp}^2}{\omega_L} = \frac{2\ \pi\ w_\perp}{qB} = \frac{2\ \pi\ m}{q}\mu \end{aligned} \tag{3.109}$$

  since the canonical momentum $p_i = mv_i + qA_i$ and $\mathbf{A}$ is the vector potential of the field.[11] As we have seen, the magnetic moment is approximately constant if the fields are slowly changing in space and time. It can be shown (Northrop, 1963) that the magnetic moment is the leading term in an asymptotic expansion in the adiabatic parameter $\epsilon = r_L/L$ such that the series

$$\mu\epsilon = \mu_0 + \mu_1\epsilon + \mu_2\epsilon^2 + \cdots \tag{3.110}$$

  is constant to all orders in n (Kruskal, 1962). This means that

$$\lim_{\epsilon\to 0}[\{\mu(\epsilon) - \text{const}\}/\epsilon^N] = 0 \qquad \text{for any } N \tag{3.111}$$

  It implies that the deviation from a constant value goes to zero faster than any power of the parameter as the latter tends to zero. The development of the terms $\mu_1,\ \mu_2, \ldots$ is discussed by Hastie (1981).
- *Second or longitudinal invariant*: In a converging magnetic mirror field (Figure 3.4), a particle oscillates between the two mirror points along the field lines. It follows that the resulting action variable $\int p_\parallel dq_\parallel$ calculated along the field lines is constant.

$$\mathcal{J} = \oint p_\parallel\ d\ell_\parallel = \oint \{m\ v_\parallel + q\ A_\parallel\} d\ell_\parallel = 2\ m \int_1^2 v_\parallel\ d\ell_\parallel \tag{3.112}$$

  where ① and ② are the mirror points. The flux term clearly integrates to zero over the full path. The longitudinal invariant is, therefore,

$$\mathcal{J} = \int_1^2 v_\parallel\ d\ell_\parallel \tag{3.113}$$

9 The number of degrees of freedom is the total number of variables which are needed to uniquely specify the position of the system, i.e. N particles specify a system of 3 N degrees of freedom.

10 The adiabatic invariants are most easily identified through Hamilton's equations. *Ab initio* methods as derived by Northrop (1963) are more complex, but allow the higher order terms to be identified.

11 Note the sign of magnetic field term is negative to take account of the anti-clockwise rotation of the particle (see Footnote (2) in Section 3.3.2).

In a practical situation, the invariant is only approximately constant due to the changes that occur in the field, for example the curvature of the lines of induction that induce cross-field drift. The motion therefore moves from one field line to a neighbour from cycle to cycle. As a result, the invariant is an adiabatic invariant subject to small parameters dependent on the field gradients (ϵ). As with the first invariant (magnetic moment), we may calculate corrections based on the parameters for the perturbed magnetic surfaces over which the drift occurs where $J = J_0 + J_1\,\epsilon + \cdots \approx \text{const.}$

- *Third or flux invariant*: Due to cross-field drifts induced by the field gradients, the particles circle slowly around closed loops perpendicular to the induction, provided such exist and allow a closed periodic motion. The corresponding invariant is obtained by integrating over a loop over which the longitudinal drift occurs. To specify the integration loop we define a centre, *bound centre*, of the motion between the two mirror points. As time proceeds, the bound centre moves around the field lines eventually returning to its starting point. Assuming the field lines remain unchanged during the cycle, the corresponding adiabatic invariant should remain constant:

$$\mathcal{K} = \oint \mathbf{p}_\perp \cdot d\ell_\perp = \oint (m\mathbf{v}_\perp + q\mathbf{A}_\perp) \cdot \mathrm{d}\ell_\perp \approx q\Phi \tag{3.114}$$

where Φ is the flux enclosed by the orbit of the bounce points. The term in $v_\perp$ may be neglected as the drift is very slow. The condition for adiabatic invariance is that the magnetic fields vary over times much longer than the drift period.

In most practical systems, the variation of the magnetic field is too rapid to allow all three invariants to exist. However, the motion of particles in the magnetosphere exhibits the effects of all three as the particles precess around the earth in the long-lived Van Allen belts. These are

a. Gyration around the geomagnetic field lines.
b. North-south oscillation.
c. Precession around the earth.

Appendix 3.A Northrop's Expansion Procedure

As we have seen, the motion of a charged particle in a magnetic field is a circular cyclotron rotation in the plane normal to the line of induction and a motion along the line. If the gradients in the electric and magnetic fields are small so that the Larmor radius changes slowly over the path, we may adapt this qualitative picture to a quantitative model of charged particle motion in varying fields – *guiding centre model*. We identify the motion of the particle, position vector $\mathbf{r}$, in terms of the cyclotron rotation $\mathbf{R}$ and the position of the guiding centre ρ

$$\mathbf{r} = \mathbf{R} + \rho \tag{3.5'}$$

The position of the guiding centre defined by Eq. (3.5') is not fixed during the rotation due to the varying fields. However, when averaged over the cyclotron motion, the steady drift of the centre, representing the mean motion of the particle, is obtained.

In the presence of an electric field, the guiding centre drifts across the lines of induction with velocity $\mathbf{v}_D = \mathbf{E} \wedge \mathbf{B}/B^2$ independent of the sign of charge, and with an acceleration due to the parallel field component $\mathbf{E}_\parallel/m$ along the lines of induction. Equation (3.6) for the location of the guiding centre **R** with respect to the particle position **r** is modified to

$$\mathbf{R} = \mathbf{r} - \rho = \mathbf{r} - \frac{m}{qB^2}\mathbf{B} \wedge \left(\mathbf{v} - \frac{\mathbf{E} \wedge \mathbf{B}}{B^2}\right) \tag{3.6'}$$

to take account of the drift on the particle velocity.

The motion of the particle is given by Newton's second law of motion

$$m\,\dot{\mathbf{v}} = q\,(\mathbf{E} + \mathbf{v} \wedge \mathbf{B}) + m\,\mathbf{g} \tag{3.A.1}$$

The electric (**E**) and magnetic (**B**) fields are both measured at the particle. We assume the gradients are small compared to the Larmor radius, the expansion parameter $\epsilon = r_L/L = mv/eB \ll 1$,[12] where L is the characteristic distance over which the fields change. The variation in the fields over the cyclotron orbit are small, so that we may use Taylor's expansion to write the fields in terms of those at the guiding centre retaining only terms of order ϵ:

$$\mathbf{E}(\mathbf{r}) = \mathbf{E}(\mathbf{R}) + (\rho \cdot \nabla)\mathbf{E}\,|_{\mathbf{R}} \qquad \text{and} \qquad \mathbf{B}(\mathbf{r}) = \mathbf{B}(\mathbf{R}) + (\rho \cdot \nabla)\mathbf{B}\,\Big|_{\mathbf{R}} \tag{3.A.2}$$

Differentiating Eq. (3.6') and using (3.A.2) in (3.A.1), we obtain

$$\dot{\mathbf{U}} + \dot{\mathbf{u}} = \mathbf{g} + (q/m)\{[\mathbf{E}(\mathbf{R}) + (\rho \cdot \nabla)\,\mathbf{E}(\mathbf{R})] + (\mathbf{U} + \mathbf{u}) \wedge [\mathbf{B}(\mathbf{R}) + (\rho \cdot \nabla)\,\mathbf{B}(\mathbf{R})]\} \tag{3.A.3}$$

The guiding centre motion is obtained by averaging Eq. (3.A.3) over a cycle of the cyclotron rotation. Introducing the orthogonal right-handed co-ordinate set $(x_1\,\hat{\imath}_1,\ x_2\,\hat{\imath}_2,\ x_3\,\hat{\imath}_3)$ with $\hat{\imath}_3$ parallel to the magnetic field **B** and $\hat{\imath}_1$ and $\hat{\imath}_2$ in the perpendicular plane, we write $\rho = \rho(\hat{\imath}_1\ \sin\theta + \hat{\imath}_2\ \cos\theta)$ where $\theta = \omega_c\,t$. Averaging over a cycle, noting that $\langle\cos\theta\rangle = \langle\sin\theta\rangle = \langle\cos\theta\ \sin\theta\rangle = 0$ and that $\langle\cos^2\theta\rangle = \langle\sin^2\theta\rangle = \frac{1}{2}$, we obtain $\langle\rho\rangle = \langle\mathbf{u}\rangle = \langle\dot{\mathbf{u}}\rangle = 0$. The only non-zero average term involving ρ is

$$\begin{aligned}
\mathbf{u} \wedge (\rho \cdot \nabla)\mathbf{B} &= \frac{1}{2}\,\rho^2\{(\cos\theta\,\hat{\imath}_1 - \sin\theta\,\hat{\imath}_2) \wedge \\
&\quad \left(\sin\theta\frac{\partial}{\partial x_1} + \cos\theta\,\frac{\partial}{\partial x_2}\right)(B_1\hat{\imath}_1 + B_2\hat{\imath}_2 + B_3\hat{\imath}_3)\} \\
&= \frac{1}{2}\,r_c^{\,2}\left\{-\frac{\partial B_3}{\partial x_1}\hat{\imath}_1 - \frac{\partial B_3}{\partial x_2}\hat{\imath}_2 - \frac{\partial B_3}{\partial x_3}\hat{\imath}_3 + \frac{\partial B_1}{\partial x_1}\hat{\imath}_3 + \frac{\partial B_2}{\partial x_2}\hat{\imath}_3 + \frac{\partial B_3}{\partial x_3}\hat{\imath}_3\right\} \\
&= -\,\frac{1}{2}\,\rho^2\,\nabla B
\end{aligned} \tag{3.A.4}$$

since $\nabla \cdot \mathbf{B} = 0$ and the field is aligned along the direction $\hat{\imath}_3$.

The average guiding centre acceleration is therefore

$$\dot{\mathbf{U}} = \mathbf{g} + \left(\frac{q}{m}\right)\left\{[\mathbf{E}(\mathbf{R}) + \mathbf{u} \wedge \mathbf{B}(\mathbf{R})] - \frac{1}{2}\rho^2\,\nabla B\right\} \tag{3.A.5}$$

which is the basic equation for the guiding centre motion, which we recognise as (3.40).

3.A.1 Drift Velocity and Longitudinal Motion along the Field Lines

To derive the drift velocity across the field lines, we take the vector product of Eq. (3.A.8) with the magnetic induction **B** to obtain

$$\mathbf{U}_\perp = \underbrace{\frac{\mathbf{E} \wedge \mathbf{B}}{B^2}}_{\text{ⓐ}} + \underbrace{\frac{\mu}{q}\frac{\mathbf{B} \wedge \nabla B}{B^2}}_{\text{ⓑ}} + \underbrace{\frac{m}{q}\frac{(\mathbf{g} - \dot{\mathbf{U}}) \wedge \mathbf{B}}{B^2}}_{\text{ⓒ}} \tag{3.A.6}$$

12 The development of the guiding centre equations in terms of the order of ϵ is considered in detail by Northrop (1963).

The different terms in this equation are easily identified:

a. The $\mathbf{E} \wedge \mathbf{B}$ drift, Section 3.2.2.
b. The gradient B drift due to variations in the field strength across the field lines.
c. The acceleration drift associated with forces along the line of induction.

Taking the scalar product of Eq. (3.A.5) with the magnetic induction **B**, we obtain

$$\dot{\mathbf{U}}_{\|} = \mathbf{g}_{\|} + \frac{q}{m}\mathbf{E}_{\|} - \frac{\mu}{q}\frac{\partial B}{\partial s} \tag{3.A.7}$$

where s is the arc length of the line of induction and $g_{\|}$ and $E_{\|}$ the components of **g** and **E** parallel to the magnetic field.

It is easily seen that these two equations are identical to (3.42) and (3.41), respectively.

4
Kinetic Theory of Gases

4.1 Introduction

The foundation of the theory of many particle systems is kinetic theory. Although the single particle theory of charged particle motion is sufficient when inter-particle interactions can be neglected, it is clearly inadequate to handle the behaviour when the particles interact amongst themselves. In a plasma, the dominant interactions are between charged particles via the long-range Coulomb force. This behaviour is in contrast to gases, where short-range forces, for example van der Waals forces, between the molecules, are important. As we shall see this leads to major differences in the kinetic behaviour of plasmas and gases. Nonetheless, it is easier to develop kinetic theory in terms of systems interacting via short-range forces and generalise to include plasmas later. Initially, we will examine the fundamentals of kinetic theory in terms of a dilute gas of single species of identical molecules interacting via short-range forces.

Kinetic theory develops the behaviour of a system of interacting particles via their statistical properties. In view of the very large number of particles in a laboratory volume, it would be impracticable to follow the dynamics of individual particles in time – although such calculations form the basis of molecular dynamics simulation. Individual particle motions are defined by Hamilton's equations, and as a consequence the particle trajectories are reversible, i.e. if we reverse the direction of time and the velocities, the particles retrace their paths, as for the single particle model. The system is therefore thermodynamically reversible. However, we know from classical thermodynamics that gas systems are subject to dissipation from diffusion, viscosity, and thermal conduction, and in consequence irreversible. If we take into account the statistical nature and exceptionally large number of particles, it becomes apparent that only after exceedingly long times can the system return to its original state. In general, over practicable observation times, the system tends to its state of maximum probability, maximum entropy consistent with the second law of thermodynamics. In most cases, this is the Maxwell–Boltzmann velocity distribution from which only infinitesimally small fluctuations are found.

Kinetic theory was established in the second half of the nineteenth century when it was realised that a gas consisted of freely moving molecules subject to irregular occasional collisions between the particles. Maxwell found the fundamental particle velocity distribution law and derived expressions for the collisional transport coefficients, such as viscosity and thermal conduction, for a gas. The theory was placed on a sound foundation by Boltzmann, who provided a rigorous theory of the distribution law (Maxwell–Boltzmann equation) via the H-theorem. This led to the famous relation between entropy and probability $S = k \log W$. Boltzmann was also foremost in refuting the many 'paradoxes' propounded against kinetic theory.

Foundations of Plasma Physics for Physicists and Mathematicians, First Edition. Geoffrey J. Pert.

The major limiting condition in Boltzmann's theory is the restriction to two particle collisional interactions, i.e. the limitation of the kinetic theory to dilute gases. In contrast, in both plasmas and liquids many particles interact simultaneously. As a result multi-particle correlations become important, and kinetic theory must be developed to encompass these effects. Since the theory of dilute gases is comparatively straightforward, we develop it first before moving on to consider the more complex case of plasma. This is a consequence of the fact that in a gas, the force of interaction between neutral molecules is short range ($a \ll n^{-1/3}$), where a is the range of the interaction and n the particle density. In contrast, in plasma, the particles are charged and interact through the long-range Coulomb force.

4.2 Phase Space

In a simple dynamic system such as a plasma, the state of each particle is specified by its position and momentum alone. The configuration of a dynamic system of N particles is defined by their generalised conjugate co-ordinates p_i, q_i in the phase space of $\mathcal{N} = 6N$ dimensions mapped by p_i, q_i (see Landau and Lifshitz, 1980, for a full discussion of generalised co-ordinates). Such systems satisfy Hamiltonian's equations of motion in a conservative system or one modified to include electromagnetic fields. In this case, the Hamiltonian function H is simply the total energy and

$$\dot{p}_i = -\frac{\partial H}{\partial q_i} \quad \text{and} \quad \dot{q}_i = \frac{\partial H}{\partial p_i} \tag{4.1}$$

the dot indicating differentiation with respect to time.

It is often convenient to identify the generalised momentum p with the particle velocity $\mathbf{v}$ provided magnetic fields are absent. Similarly, the generalised co-ordinate q can be identified with the position vector $\mathbf{r}$.

Phase space is used to specify the particle distribution in configuration space and velocity space. This may be achieved in two dissimilar ways.

4.2.1 Γ Phase Space

A general description of the system as a whole is given by Γ phase space in which each particle i identifies a set of six co-ordinates by its position q_i and momentum p_i. The complete state of the system is described by a single point in the Γ phase space of $\mathcal{N} = 6N$ dimensions spanning the complete set of variables $p_1, q_1, \ldots, p_{\mathcal{N}}, q_{\mathcal{N}}$ in the entire Hamiltonian system

$$X = p_1,\ q_1,\ p_2,\ q_2, \ldots, p_{\mathcal{N}},\ q_{\mathcal{N}}$$

We may define a distribution function in Γ space known as the *Liouville distribution function*, which defines the probability of finding a particle in each of the cells $(\mathrm{d}p_1,\ \mathrm{d}q_1) \cdots (\mathrm{d}p_{\mathcal{N}},\ q_{\mathcal{N}})$ at time t, namely,

$$\mathcal{F}(p_1, q_1, \ldots, p_{\mathcal{N}},\ q_{\mathcal{N}},\ t)\ \mathrm{d}p_1\ \mathrm{d}q_1\ \cdots \mathrm{d}p_{\mathcal{N}}\ \mathrm{d}q_{\mathcal{N}}$$

If we know the position and velocity of each particle $p_i(t)$, $q_i(t)$ at time t, we may immediately write down the Liouville distribution function:

$$\mathcal{F}(p_1, q_1, \ldots, p_{\mathcal{N}},\ q_{\mathcal{N}},\ t) = \prod_{i=1}^{\mathcal{N}} \delta(p - p_i(t))\ \delta(q - q_i(t)) \tag{4.2}$$

The integral of the probability overall Γ space is, therefore,

$$\int \cdots \int \mathcal{F}(p_1, q_1, \ldots, p_{\mathcal{N}}, q_{\mathcal{N}})\, \mathrm{d}p_1\, \mathrm{d}q_1 \cdots \mathrm{d}p_{\mathcal{N}}\, \mathrm{d}q_{\mathcal{N}} = 1 \tag{4.3}$$

namely, unity, since the system exists in some configuration.

4.2.1.1 Liouville's Equation

The Liouville probability function satisfies an important relation known as *Liouville's equation*:

$$\frac{\partial \mathcal{F}}{\partial t} + \sum_{i=1}^{\mathcal{N}} \left[\frac{\partial \mathcal{F}}{\partial p_i} \dot{p}_i + \frac{\partial \mathcal{F}}{\partial q_i} \dot{q}_i \right] = 0 \tag{4.4}$$

Introducing the Poisson bracket and making use of Hamilton's equation:

$$(\mathcal{F}, H) = \sum \frac{\partial H}{\partial p_i} \frac{\partial \mathcal{F}}{\partial q_i} - \frac{\partial H}{\partial q_i} \frac{\partial \mathcal{F}}{\partial p_i} = \sum \dot{q}_i \frac{\partial \mathcal{F}}{\partial q_i} + \dot{p}_i \frac{\partial \mathcal{F}}{\partial p_i} \tag{4.5}$$

in terms of which Liouville's equation takes the form:

$$\frac{\partial \mathcal{F}}{\partial t} + (\mathcal{F}, H) = 0 \tag{4.6}$$

The probability $\mathcal{F}$ is constant moving along the trajectory through phase space followed by the system.

Since it is clear that the total probability is a conserved quantity, we may apply a generalisation of the continuity arguments used with regard to the conservation laws in fluid mechanics. Thus, the rate of change of the probability in a volume V in phase space must be balanced by the loss of probability due to movement through its complete bounding surface S, given by

$$\begin{aligned} \int_V \frac{\partial \mathcal{F}}{\partial t}\, \mathrm{d}V &= -\int_S \sum [\mathcal{F}(\dot{p}_i)\, \mathrm{d}S_i^p + \mathcal{F}(\dot{q}_i)\, \mathrm{d}S_i^q] \\ &= -\int_V \sum \left[\frac{\partial(\mathcal{F}\dot{p}_i)}{\partial p_i} + \frac{\partial(\mathcal{F}\dot{q}_i)}{\partial q_i} \right] \mathrm{d}V \end{aligned}$$

where $\mathrm{d}S_i^p$ and $\mathrm{d}S_i^q$ are the surface area elements normal to p_i and q_i, respectively, and we have used the generalisation of Gauss' theorem to phase space. Since the volume V is arbitrary, we may equate the integrands in this relation. Hence, differentiating the products and making use of Hamilton's equations (4.1), we obtain Liouville's equation (4.4).

An alternative proof of Liouville's theorem considers a phase space volume element ΔV which moves with the system through phase space, changing from ΔV at time t to $\Delta V'$ at time t'. Clearly, the total probability within the volume $\mathcal{F}\, \Delta V$ must remain constant. Therefore, if the volume of the element remains constant in time, i.e. $\Delta V = \Delta V'$, the probability $\mathcal{F}$ along the trajectory is constant as required. Consider the Jacobian for the transformation from p_i, q_i to p_i', q_i' by treating the move from p_i, q_i to p_i', q_i' over time t to t' in two steps:

a. p_i, q_i to p_i', q_i, i.e. q_i constant

$$p_i = p_i' - \int_t^{t'} \dot{p}_i\, \mathrm{d}t = p_i' + \int_t^{t'} \frac{\partial H}{\partial q_i}\, \mathrm{d}t$$

b. p_i', q_i to p_i', q_i', i.e. p_i constant

$$q_i' = q_i + \int_t^{t'} \dot{q}_i\, \mathrm{d}t = q_i + \int_t^{t'} \frac{\partial H}{\partial p_i}\, \mathrm{d}t$$

The Jacobian for the transformation from the system at time t to t' may be written as follows:

$$\frac{\Delta V'}{\Delta V} = \frac{\partial(q_1', q_2', \ldots, q_{\mathcal{N}}')}{\partial(q_1, q_2, \ldots, q_{\mathcal{N}})} \Big/ \frac{\partial(p_1, p_2, \ldots, p_{\mathcal{N}})}{\partial(p_1', p_2', \ldots, p_{\mathcal{N}}')} \tag{4.7}$$

But

$$\frac{\partial q_i}{\partial q_j} = \delta_{ij} + \int_t^{t'} \frac{\partial^2 H}{\partial \mathbf{p_i}\, \partial q_j}\, \mathrm{d}t$$

$$\frac{\partial p_i}{\partial p_j} = \delta_{ij} + \int_t^{t'} \frac{\partial^2 H}{\partial q_i\, \partial p_j}\, \mathrm{d}t$$

Therefore, each Jacobian in Eq. (4.7) is the transpose of the other, and the respective determinants are equal. Consequently,

$$\frac{\mathcal{F}'}{\mathcal{F}} = \frac{\Delta V}{\Delta V'} = 1 \tag{4.8}$$

The probability is, therefore, constant along the trajectory thereby re-establishing the Liouville result.

We note that the Liouville distribution expresses the probability of a particular particle distribution within the ensemble.

In practice, the state of the system is not accurately known. Rather we have an ensemble of possible configurations each with a known probability amongst the possible systems, for example with constant energy. Clearly, Liouville's theorem applies to each such system, and therefore, to the complete ensemble with the probability function $\mathcal{F}$ now specifying the probability of each individual distribution of the particles.

4.2.2 μ Space

An alternative, rather simpler formulation of kinetic theory defines the statistical behaviour of the system in terms of the six-dimensional phase space specified by the particle position $\mathbf{r}$ and velocity $\mathbf{v}$. Henceforward, we replace p_i and q_i by the particle velocity $\mathbf{v}_i$ and position $\mathbf{r}_i$, respectively. The number of particles in a cell $(\mathbf{dr},\ \mathbf{dv})$ at time t is

$$f(\mathbf{r},\ \mathbf{v},\ t)\, \mathbf{dr}\, \mathbf{dv}$$

In contrast to Γ phase space, the distribution function in μ space expresses the combination of two different probabilities:

a. The particle distribution within a particular statistical arrangement.
b. The probability of that particular arrangement within the ensemble.

If the positions and velocities of all N particles are known at time t, the distribution function has the simple form:

$$f(\mathbf{r},\ \mathbf{v},\ t) = \sum_{n=1}^{N} [\delta(\mathbf{r} - \mathbf{r}_n(t))\, \delta(\mathbf{v} - \mathbf{v}_n(t))] \tag{4.9}$$

The integral of the distribution function over μ space

$$\int f(\mathbf{r},\ \mathbf{v},\ t)\, \mathbf{dr}\, \mathbf{dv} = N \tag{4.10}$$

is equal to the total number of particles N.

This form of phase space is most satisfactory for dealing with *dilute systems*, where the interaction between the particles is weak. The particle motions are nearly independent of one another; in a gas, each particle only interacting with another through infrequent binary collisions. The distribution function defines only the position of single independent particles and is, therefore, strictly called the *single particle distribution function* denoted $f^{(1)}(\mathbf{r}, \mathbf{v}, t)$.

To take account of possible correlations whereby one particle is at $(\mathbf{r}, \mathbf{v})$, simultaneously, with another at $(\mathbf{r}', \mathbf{v}')$, we define the *two particle distribution function* $f^{(2)}(\mathbf{r}, \mathbf{v}, \mathbf{r}', \mathbf{v}', t)$. The probability of finding a particle at $(\mathbf{r}_1, \mathbf{v}_1)$ in $(d\mathbf{r}_1, d\mathbf{v}_1)$ and a second in $(d\mathbf{r}_2, d\mathbf{v}_2)$ at time t is

$$f^{(2)}(\mathbf{r}_1, \mathbf{v}_1, \mathbf{r}_2, \mathbf{v}_2, t)\, d\mathbf{r}_1\, d\mathbf{v}_1\, d\mathbf{r}_2\, d\mathbf{v}_2 \tag{4.11}$$

If the particles are independently distributed:

$$f^{(2)}(\mathbf{r}_1, \mathbf{v}_1, \mathbf{r}_2, \mathbf{v}_2, t) = f^{(1)}(\mathbf{r}_1, \mathbf{v}_1, t)\, f^{(1)}(\mathbf{r}_2, \mathbf{v}_2, t) \tag{4.12}$$

We may generalise the multi-particle distribution to define the q particle distribution, where the probability of simultaneously finding a particle in each of the cells $d\mathbf{r}_1, d\mathbf{v}_1, \ldots, d\mathbf{r}_q, d\mathbf{v}_q$ at time t

$$f^{(q)}(\mathbf{r}_1, \mathbf{v}_1, \ldots, \mathbf{r}_q, \mathbf{v}_q, t)\, d\mathbf{r}_1\, d\mathbf{v}_1 \cdots d\mathbf{r}_q\, d\mathbf{v}_q \tag{4.13}$$

The complete set of multi-particle distribution functions $f^{(q)}$, $1 \le q \le N$, in μ space are equivalent to the distribution $\mathcal{F}$ in Γ space. Each contains a full description of the correlation within the system.

4.3 Relationship Between Γ Space and μ Space

If we integrate the Γ space distribution function over $(N - 1)$ particles, we generate the probability of finding a particle in the remaining phase space cell, i.e. the one-particle distribution:

$$f^{(1)}(\mathbf{r}_1, \mathbf{v}_1, t) = N \int \cdots \int \mathcal{F}(\mathbf{r}_1, \mathbf{v}_1, \mathbf{r}_2, \mathbf{v}_2, \ldots, \mathbf{r}_N, \mathbf{v}_N, t)\, d\mathbf{r}_2\, d\mathbf{v}_2 \cdots d\mathbf{r}_N\, d\mathbf{v}_N \tag{4.14}$$

The factor N arises because the particles in μ space are indistinguishable. It may be checked by further integrating over μ space and noting that the integral of $\mathcal{F}$ is 1, whereas that of f is N, the number of particles. In a similar fashion, integrating $\mathcal{F}$ over $(N - 2)$ particles yield the two-particle distribution:

$$\begin{aligned} &f^{(2)}(\mathbf{r}_1, \mathbf{v}_1, \mathbf{r}_2, \mathbf{v}_2, t) \\ &\quad = N\,(N-1) \int \cdots \int \mathcal{F}(\mathbf{r}_1, \mathbf{v}_1, \mathbf{r}_2, \mathbf{v}_2, \mathbf{r}_3, \mathbf{v}_3, \ldots, \mathbf{r}_N, \mathbf{v}_N, t)\, d\mathbf{r}_3\, d\mathbf{v}_3 \cdots d\mathbf{r}_N\, d\mathbf{v}_N \end{aligned} \tag{4.15}$$

In Γ space, the particles are treated as distinguishable, the number of equivalent ways in which the integration can be performed is, therefore, the permutation of two objects from N, namely $N!/(N-2)! = N(N-1)$. This result is easily generalised to q particles:

$$\begin{aligned} &f^{(q)}(\mathbf{r}_1, \mathbf{v}_1, \ldots, \mathbf{r}_q, \mathbf{v}_q, t) \\ &= \frac{N!}{(N-q)!} \int \cdots \int \mathcal{F}(\mathbf{r}_1, \mathbf{v}_1, \ldots, \mathbf{r}_q, \mathbf{v}_q, \mathbf{r}_{(q+1)}, \mathbf{v}_{(q+1)}, \ldots, \mathbf{r}_N, \mathbf{v}_N, t) \\ &\times d\mathbf{r}_{(q+1)}\, d\mathbf{v}_{(q+1)} \cdots d\mathbf{r}_N\, d\mathbf{v}_N \end{aligned} \tag{4.16}$$

4.3.1 Integrals of the Liouville Equation

The forces on a particle may be considered to be of two different types:

- **External forces** $\mathbf{F}_i$ on particle i due to applied gravitational, and electric and magnetic fields.
- **Internal forces** $\mathbf{F}_{ij}(|\mathbf{r_i} - \mathbf{r}_j|)$ representing the interaction between two particles i and j depending on their spatial separation.

Including the force terms explicitly, Liouville's equation becomes

$$\frac{\partial \mathcal{F}}{\partial t} + \sum \mathbf{v}_i \cdot \frac{\partial \mathcal{F}}{\partial \mathbf{r}_i} + \sum \dot{\mathbf{v}}_i \cdot \frac{\partial \mathcal{F}}{\partial \mathbf{v}_i} = -\sum_{i=1}^{N} \sum_{\substack{j=1 \\ j \neq i}}^{N} \frac{\mathbf{F}_{ij}}{m} \cdot \frac{\partial \mathcal{F}}{\partial \mathbf{v}_i} \tag{4.17}$$

where m is the particle mass.

In the following, we will have several integrals of the following type, which may all be evaluated by the use of Gauss' theorem:

$$\int \mathbf{v}_i \cdot \frac{\partial \mathcal{F}}{\partial \mathbf{r}_i} \, \mathrm{d}\mathbf{r}_i = \left\{ \oint \mathrm{d}\mathrm{S}_i \cdot \mathbf{v}_i \, \mathcal{F} - \int \mathrm{d}\mathbf{r}_i \, \mathcal{F} \frac{\partial \mathbf{v}_i}{\partial \mathbf{r}_i} \right\} = 0 \tag{4.18}$$

where the surface integral is over a surface at infinity, where $\mathcal{F} \to 0$, and since $\mathbf{v}_i$ and $\mathbf{r}_i$ are independent variables $\partial \mathbf{v}_i / \partial \mathbf{r}_i = 0$.

Multiply each term by $N!/(N-n)!$ before integrating term by term. Integrating the time derivative over particles $n+1$ to N, we obtain

$$\frac{N!}{(N-n)!} \int \frac{\partial \mathcal{F}}{\partial t} \, \mathbf{dr}_{n+1} \, \mathbf{dv}_{n+1} \cdots \mathbf{dr}_N \, \mathbf{dv}_N = \frac{\partial f^{(n)}}{\partial t}(\mathbf{r}_1 \, \mathbf{v}_1 \cdots \mathbf{r}_n \, \mathbf{v}_n, \, t) \tag{4.19}$$

In configuration space, we obtain integrals of the type (4.18) for $i > n$ and obtain

$$\frac{N!}{(N-n)!} \int \mathbf{v}_i \cdot \frac{\partial \mathcal{F}}{\partial \mathbf{r}_i} \, \mathrm{d}\mathbf{r}_{n+1} \, \mathrm{d}\mathbf{v}_{n+1} \dots \mathrm{d}\mathbf{r}_N \, \mathrm{d}\mathbf{v}_N = \begin{cases} \mathbf{v}_i \cdot \dfrac{\partial f^{(n)}}{\partial \mathbf{r}_i}(\mathbf{r}_1, \mathbf{v}_1, \dots, \mathbf{r}_N, \mathbf{v}_N) & \text{if } i \leq n \\ 0 & \text{if } i > n \end{cases} \tag{4.20}$$

In a gas, the external forces are conservative. However, in a plasma system, the Lorentz force due to an external magnetic field, $\mathbf{v} \wedge \mathbf{B}$, is perpendicular to the velocity, and consequently, the term $\partial(\mathbf{v} \wedge \mathbf{B})/\partial \mathbf{v} = 0$. Hence,

$$\begin{aligned} &\frac{N!}{(N-n)!} \frac{1}{m} \int \mathbf{F}_i \cdot \frac{\partial \mathcal{F}}{\partial \mathbf{v}_i} \, \mathrm{d}\mathbf{r}_{n+1} \mathrm{d}\mathbf{v}_{n+1} \cdots \mathrm{d}\mathbf{r}_N \mathrm{d}\mathbf{v}_N \\ &= \begin{cases} \dfrac{1}{m} \mathbf{F}_i \cdot \dfrac{\partial f^{(n)}}{\partial \mathbf{v}_i}(\mathbf{r}_1, \mathbf{v}_1, \dots, \mathbf{r}_N, \mathbf{v}_N) & \text{if } i \leq n \\ 0 & \text{if } i > n \end{cases} \end{aligned} \tag{4.21}$$

Finally, we treat the mutual interaction between the particles by splitting the sum into three parts:

$$\sum_{i=1}^{N} \sum_{\substack{j=1 \\ j \neq i}}^{N} = \sum_{i=1}^{n} \sum_{\substack{j=1 \\ j \neq i}}^{n} + \sum_{i=1}^{n} \sum_{\substack{j=n+1 \\ j \neq i}}^{N} + \sum_{i=n+1}^{N} \sum_{\substack{j=1 \\ j \neq i}}^{N} \tag{4.22}$$

When $i \leq n$ and $j < n$

$$\frac{N!}{(N-n)!}\int \frac{\mathbf{F}_{ij}(|\mathbf{r}_i - \mathbf{r}_j)|}{m}\cdot\frac{\partial \mathcal{F}}{\partial \mathbf{v}_i}\,\mathrm{d}\mathbf{r}_{n+1}\,\mathrm{d}\mathbf{v}_{n+1}\cdots\,\mathrm{d}\mathbf{r}_N\,\mathrm{d}\mathbf{v}_N = \frac{\mathbf{F}_{ij}(\mathbf{r}_i - \mathbf{r}_j)}{m}\cdot\frac{\partial f^{(n)}}{\partial \mathbf{v}_i} \tag{4.23}$$

when $i \leq n$ and $j > n$

$$\frac{N!}{(N-n)!}\int \frac{\mathbf{F}_{ij}(|\mathbf{r}_i - \mathbf{r}_j|)}{m}\cdot\frac{\partial \mathcal{F}}{\partial \mathbf{v}_i}\mathrm{d}\mathbf{r}_{n+1}\,\mathrm{d}\mathbf{v}_{n+1}\cdots\,\mathrm{d}\mathbf{r}_N\,\mathrm{d}\mathbf{v}_N = \frac{1}{(N-n)}\int \frac{\mathbf{F}_{ij}(\mathbf{r}_i - \mathbf{r}_j)}{m}\cdot\frac{\partial f^{(n+1)}}{\partial \mathbf{v}_i}\mathrm{d}\mathbf{r}_j\,\mathrm{d}\mathbf{v}_j \tag{4.24}$$

however, this term is multiplied by $(N-n)$ in the sum since the term j can be selected in $(N-n)$ different ways. Finally, when $i > n$

$$\frac{N!}{(N-n)!}\int \frac{\mathbf{F}_{ij}(|\mathbf{r}_i - \mathbf{r}_j|)}{m}\cdot\frac{\partial \mathcal{F}}{\partial \mathbf{v}_i}\,\mathrm{d}\mathbf{r}_{n+1}\,\mathrm{d}\mathbf{v}_{n+1}\cdots\,\mathrm{d}\mathbf{r}_N\,\mathrm{d}\mathbf{v}_N = 0 \tag{4.25}$$

The total contribution from the interactive term gives

$$\frac{N!}{(N-n)!}\int \frac{\mathbf{F}_{ij}(\mathbf{r}_i - \mathbf{r}_j)}{m}\frac{\partial \mathcal{F}}{\partial \mathbf{v}_i}\,\mathrm{d}\mathbf{r}_{n+1}\,\mathrm{d}\mathbf{v}_{n+1}\cdots\mathrm{d}\mathbf{r}_N\,\mathrm{d}\mathbf{v}_N = \begin{cases} \dfrac{\mathbf{F}_{ij}(\mathbf{r}_i - \mathbf{r}_j)}{m}\cdot\dfrac{\partial f^{(n)}}{\partial \mathbf{v}_i} & \text{if } i \leq n \text{ and } j \leq n \\ \displaystyle\int \frac{\mathbf{F}_{ij}(\mathbf{r}_i - \mathbf{r}_j)}{m}\cdot\frac{\partial f^{(n+1)}}{\partial \mathbf{v}_i}\mathrm{d}\mathbf{r}_j\,\mathrm{d}\mathbf{v}_j & \text{if } i \leq n \text{ and } j > n \end{cases} \tag{4.26}$$

4.4 The BBGKY (Bogoliubov–Born–Green–Kirkwood–Yvon) Hierarchy

The complete behaviour of the dynamic system is contained in the Γ space distribution function $\mathcal{F}$, although in a form which is too complex to use for calculation. In contrast, the distribution function $f^{(1)}$ in μ space is sufficiently simple to provide a basis for classical and quantum statistical mechanics where macroscopic quantities are related to microscopic. Although as we shall see, $f^{(1)}$ is adequate to describe dilute systems, where the particles are nearly independent, the complete description in the general case can only be provided by the full set of multiple particle distributions $f^{(1)}, \dots, f^{(n)}, \dots, f^{(N)}$. Thus, in the most general case, no improvement in the tractability of the problem is obtained. However, in many physical situations, it is possible to introduce closure after a limited number of terms and thereby reduce the complexity. Since the Liouville equation is an expression of the Hamilton equation, which are reversible, the full set of the multiple particle distributions $f^{(n)}$, themselves expressions $\mathcal{F}$, must be reversible. Therefore, the essential irreversibility of physical systems required by the second law of thermodynamics must be a consequence of closure and the resultant truncation.

Substituting the integrals (4.19)–(4.26) into the Liouville equation, we obtain

$$\frac{\partial f^{(n)}}{\partial t} + \sum_{i=1}^{n} \mathbf{v}_i \cdot \frac{\partial f^{(n)}}{\partial \mathbf{r}_i} + \sum_{i=1}^{n} \frac{\mathbf{F}_i}{m} \cdot \frac{\partial f^{(n)}}{\partial \mathbf{v}_i} + \sum_{i=1}^{n} \sum_{\substack{j=i \\ j\neq i}}^{n} \frac{\mathbf{F}_{ij}}{m} \cdot \frac{\partial f^{(n)}}{\partial \mathbf{v}_i}$$
$$= -\sum_{i=1}^{n} \sum_{j=n+1}^{N} \int d\mathbf{r}_j \, d\mathbf{v}_j \, \frac{\mathbf{F}_{ij}(\mathbf{r}_i - \mathbf{r}_j)}{m} \cdot \frac{\partial f^{(n+1)}}{\partial \mathbf{v}_i} \tag{4.27}$$

This set of equations constitute the BBGKY hierarchy for the calculation of the distribution function. Unfortunately, to determine $f^{(n)}$ requires a knowledge of the next higher distribution function $f^{(n+1)}$, and so on. Hence, a complete solution requires the evaluation of the full set $f^{(1)} \cdots f^{(N)}$. Progress can only be made if it is possible to introduce additional physical constraints which allow the progression to be closed at a value of n typically 2 or 3. Closure[1] is normally introduced at the second term as demonstrated in Section 4.5.

If we consider the first term of the hierarchy, $n = 1$, we obtain

$$\frac{\partial f^{(1)}}{\partial t} + \mathbf{v}_1 \cdot \frac{\partial f^{(1)}}{\partial \mathbf{r}_1} + \frac{\mathbf{F}_1}{m} \cdot \frac{\partial f^{(1)}}{\partial \mathbf{v}_1} = \underbrace{-\frac{1}{m} \int F_{12}(\mathbf{r}_1 - \mathbf{r}_2) \cdot \frac{\partial}{\partial \mathbf{v}_1} \{f^{(2)}(\mathbf{r}_1, \mathbf{v}_1, \mathbf{r}_2, \mathbf{v}_2)\} \, d\mathbf{r}_2 \, d\mathbf{v}_2}$$
$$\tag{4.28}$$

the underlined term represents the two body inter-particle interaction term, which can normally be identified as collisions, provided the interaction is local and rapid.

Proceeding to the next term, we obtain

$$\frac{\partial f^{(2)}}{\partial t} + \mathbf{v}_1 \cdot \frac{\partial f^{(2)}}{\partial \mathbf{r}_1} + \mathbf{v}_2 \cdot \frac{\partial f^{(2)}}{\partial \mathbf{r}_2} + \frac{\mathbf{F}_1}{m} \cdot \frac{\partial f^{(2)}}{\partial \mathbf{v}_1} + \frac{\mathbf{F}_2}{m} \cdot \frac{\partial f^{(2)}}{\partial \mathbf{v}_2}$$
$$+ \frac{1}{m} \left\{ \mathbf{F}_{12}(\mathbf{r}_1 - \mathbf{r}_2) \cdot \frac{\partial f^{(2)}}{\partial \mathbf{v}_1} + \mathbf{F}_{21}(\mathbf{r}_2 - \mathbf{r}_1) \cdot \frac{\partial f^{(2)}}{\partial \mathbf{v}_2} \right\}$$
$$= -\int \left[\mathbf{F}_{13}(\mathbf{r}_1 - \mathbf{r}_3) \cdot \frac{\partial f^{(3)}}{\partial \mathbf{v}_1} + \mathbf{F}_{23}(\mathbf{r}_2 - \mathbf{r}_3) \cdot \frac{\partial f^{(3)}}{\partial \mathbf{v}_2} \right] d\mathbf{r}_3 \, d\mathbf{v}_3 \tag{4.29}$$

which introduces the third-order correlation term $f^{(3)}$.

4.5 Bogoliubov's Hypothesis for Dilute Gases

At this stage, the set of equations deriving directly from the Liouville equation are still reversible. To obtain irreversibility, consistent with the second law of thermodynamics, we must introduce an additional assumption. To illustrate this requirement, we will consider the short-range elastic

---

1 Many problems in the theory of fluids suffer from a lack of closure whereby the relation determining the nth term in the sequence $f^{(n)}$ depends on the subsequent one $f^{(n+1)}$. The sequence is closed either by

- *Truncation schemes* in which the sequence is closed by arbitrarily assuming all higher terms than n are zero, or equivalently that a relationship is imposed on physical grounds which assigns a value for $f^{(n+1)}$ in terms of the earlier members of the set $f^{(s)}$. It will be seen that the assumption of *molecular chaos* will provide such a relationship.
- *Asymptotic schemes* in which a small parameter can be used to develop an expansion, which can be truncated when the parameter is small. Such an approach is characteristic of the Chapman–Enskog method discussed in Section 4.11.

interaction between two identical particles, a condition typical of a neutral gas of a single species. This is the situation considered by the classic *Boltzmann's ansatz*, more usually in the manner derived later in Section 4.7.

We consider a low-density gas such that multiple interactions between the particles are rare and only occur infrequently. The distance over which particles interact (interaction length) is assumed to be very small compared to the inter-particle separation distance and to the mean free path. Collisions can consequently assumed to be binary and well separated in space and time: a situation characteristic of a dilute gas, where the inter-particle forces are typically van der Waals interactions, but as we shall see is not appropriate to a plasma. Under these conditions, we identify three well separated time scales:

a. *Collision time, i.e. the duration of a typical collision*: $\tau_0 \sim a/v$, where a is the range of intermolecular force or interaction length.
b. *Collision interval, i.e. the time between collisions*: $t_0 \sim \lambda/v$ where λ is the mean free path.
c. *Macroscopic time scale over which experimental phenomena take place*: $\theta_0 \sim L/v_s$, where L is the typical length of a laboratory experiment, and v_s the speed of sound.

The term

$$\frac{Df^{(1)}}{Dt} \equiv \frac{\partial f^{(1)}}{\partial t} + \mathbf{v}_1 \cdot \frac{\partial f^{(1)}}{\partial \mathbf{r}_1} + \frac{\mathbf{F}_1}{m} \cdot \frac{\partial f^{(1)}}{\partial \mathbf{v}_1}$$

is identified as the streaming derivative for the one particle distribution in phase space. In the fluid dynamic analogy, it is the Lagrangian time derivative for the 'fluid' flowing in phase space and represents changes in the distribution function as the fluid moves through phase space under its velocity and external forces. We note that this term does not contain a term dependent on the particle interactions. The remaining term in Eq. (4.28) due to the interaction forces is determined by the two particle distribution and represents the effect of collisions, which occur infrequently at low density at intervals of the mean free time t_0. The characteristic time scale for changes in $f^{(1)}$ is, therefore, t_0. If at time $t = 0$ the system is prepared in a state far from equilibrium, rapid changes in the multi-particle distribution functions $f^{(s)}$, $s \neq 1$ occur over times τ_0 due to the inter-particle forces, affecting all terms in the expansion except the first. The latter only includes this force through the collision term (Eq. (4.28)), and as a result it is dominated by its 'stream' term $Df^{(1)}/Dt$ which changes slowly over time scales of the order t_0.

In a dilute gas $a \ll \lambda \ll L$. Since $v \approx v_s$, $\tau_0 \ll t_0 \ll \theta_0$ and we see that the three time scales are indeed well-separated. Furthermore, since $a \ll n^{-1/3}$, where n is the particle density, the likelihood of more than two particles interacting simultaneously is extremely small. At low density, we therefore, consider only binary, well-separated collisions.

In practice, we are not interested in the detailed dynamics of the system over short times $\sim \tau_0$, but in the statistical behaviour over times $\sim t_0$, i.e. we average out the detailed fluctuations over times $\tau_0 \lesssim t \lesssim t_0$. More formally since $\tau_0 \ll t_0$, f_2 is represented by its asymptotic state over time scales $\sim t_0$. As a result for times $t \gtrsim t_0$, it may be assumed that we may treat the terms $f^{(s)}$ for $s \geq 2$ as depending on time only through a functional dependence on $f^{(1)}$, thus

$$f^{(s)}(\mathbf{r}_1, \mathbf{v}_1, \dots, \mathbf{r}_s, \mathbf{v}_s, t) \equiv f^{(s)}(\mathbf{r}_1, \mathbf{v}_1, \dots, \mathbf{r}_s, \mathbf{v}_s \mid f^{(1)}(t))$$

This important condition, known as *Bogoliubov's hypothesis*, ensures closure of the set in a manner that allows the first equation to take the form

$$\frac{Df^{(1)}}{Dt} \equiv \frac{\partial f^{(1)}}{\partial t} + \mathbf{v}_1 \cdot \frac{\partial f^{(1)}}{\partial \mathbf{r}_1} + \frac{\mathbf{F}_1}{m} \cdot \frac{\partial f^{(1)}}{\partial \mathbf{v}_1} = A(\mathbf{r}, \mathbf{v} \mid f^{(1)}) = \left.\frac{\partial f^{(1)}}{\partial t}\right|_{\text{coll}} \tag{4.30}$$

frequently designated as the Boltzmann equation in terms of an evaluable collision operator.

4.6 Derivation of the Boltzmann Collision Integral from the BBGKY Hierarchy

Since we have argued that in dilute gases collisions are binary and completed over very short timescales, we may express these conditions in terms of the following assumptions:

(i) Three-body (and higher) terms are negligible.
(ii) Time variation of the distribution during collision is negligible.
(iii) Spatial variation of the distribution over the collision range (a) is negligible.
(iv) Change in velocity due to external forces *during* collisions is negligible.

Consider the second term of the BBGKY hierarchy (4.29). Making use of the above assumptions:

$$\overbrace{\cancel{\frac{\partial f^{(2)}}{\partial t}}}^{\text{(ii)}} + \mathbf{v}\cdot\frac{\partial f^{(2)}}{\partial \mathbf{r}} + \mathbf{v}'\cdot\frac{\partial f^{(2)}}{\partial \mathbf{r}'} + \overbrace{\cancel{\frac{\mathbf{F}}{m}\cdot\frac{\partial f^{(2)}}{\partial \mathbf{v}}}}^{\text{(iv)}} + \overbrace{\cancel{\frac{\mathbf{F}'}{m}\cdot\frac{\partial f^{(2)}}{\partial \mathbf{v}'}}}^{\text{(iv)}}$$
$$+ \frac{\mathbf{F}(\mathbf{r}-\mathbf{r}')}{m}\cdot\frac{\partial f^{(2)}}{\partial \mathbf{v}} + \frac{\mathbf{F}(\mathbf{r}-\mathbf{r}')}{m}\cdot\frac{\partial f^{(2)}}{\partial \mathbf{v}'}$$
$$= -\frac{1}{m}\int\left[\overbrace{\cancel{\mathbf{F}(\mathbf{r}-\mathbf{r}'')\cdot\frac{\partial f^{(3)}}{\partial \mathbf{r}}}}^{\text{(i)}} + \overbrace{\cancel{\mathbf{F}(\mathbf{r}'-\mathbf{r}'')\cdot\frac{\partial f^{(3)}}{\partial \mathbf{r}'}}}^{\text{(i)}}\right]\mathrm{d}\mathbf{r}''\,\mathrm{d}\mathbf{v}'' \tag{4.31}$$

where the labels indicate the assumption allowing the term's cancellation.

Since there is no spatial inhomogeneity (assumption (iii)), a small displacement ρ leaves the two particle distribution unchanged.

$$f^{(2)}(\mathbf{r},\mathbf{v},\mathbf{r}',\mathbf{v}',t) = f^{(2)}((\mathbf{r}+\rho),\mathbf{v},(\mathbf{r}'+\rho),\mathbf{v}',t) \tag{4.32}$$

We can now differentiate with respect to ρ setting $\rho = 0$:

$$\frac{\partial f^{(2)}}{\partial \mathbf{r}} + \frac{\partial f^{(2)}}{\partial \mathbf{r}'} = 0 \tag{4.33}$$

and simplify Eq. (4.31) to:

$$-(\mathbf{v}-\mathbf{v}')\,\frac{\partial f^{(2)}}{\partial \mathbf{r}'} + \frac{\mathbf{F}(\mathbf{r}-\mathbf{r}')}{m}\cdot\frac{\partial f^{(2)}}{\partial \mathbf{v}} + \frac{\mathbf{F}(\mathbf{r}'-\mathbf{r})}{m}\cdot\frac{\partial f^{(2)}}{\partial \mathbf{v}'} = 0 \tag{4.34}$$

The first-order BBGKY equation (4.28) is

$$\frac{\partial f^{(1)}}{\partial t} + \mathbf{v}\cdot\frac{\partial f^{(1)}}{\partial \mathbf{r}} + \frac{\mathbf{F}}{m}\frac{\partial f^{(1)}}{\partial \mathbf{v}} = \underbrace{-\int\int\left(\frac{\mathbf{F}(\mathbf{r}'-\mathbf{r})}{m}\cdot\frac{\partial f^{(2)}}{\partial \mathbf{v}}\right)\mathrm{d}\mathbf{r}'\,\mathrm{d}\mathbf{v}'}_{J} \tag{4.35}$$

Denoting the integral by J, the integrand can be substituted from (4.34) to give

$$J = -\int\int\left[(\mathbf{v}-\mathbf{v}')\cdot\frac{\partial f^{(2)}}{\partial \mathbf{r}'} - \frac{\mathbf{F}(\mathbf{r}'-\mathbf{r})}{m}\cdot\frac{\partial f^{(2)}}{\partial \mathbf{v}'}\right]\mathrm{d}\mathbf{r}'\,\mathrm{d}\mathbf{v}' \tag{4.36}$$

Integrating the second term over $\mathbf{v}'$, using Gauss' theorem in velocity space:

$$\int\frac{\mathbf{F}(\mathbf{r}'-\mathbf{r})}{m}\cdot\frac{\partial f^{(2)}}{\partial \mathbf{v}'}\,\mathrm{d}\mathbf{v}' = 0 \tag{4.37}$$

To integrate the first term over $\mathbf{r}'$, we again use Gauss' theorem in configuration space to give

$$J = -\int \mathrm{d}\mathbf{v}' \oint \mathbf{dS}' \cdot (\mathbf{v} - \mathbf{v}')\, f^{(2)} - \int \mathrm{d}\mathbf{r}' \cdot \frac{\partial}{\partial \mathbf{r}'}(\mathbf{v} - \mathbf{v}')^{0}\, f^{(2)} \tag{4.38}$$

The second term on the right-hand side is cancelled as $\mathbf{v}$ and $\mathbf{v}'$ are not functions of $\mathbf{r}'$. The surface integral is over the collision volume, i.e. the region where $f^{(2)}$ changes due to the interaction, bounded by $|\mathbf{r} - \mathbf{r}'| \gg a$ and $F(\mathbf{r}' - \mathbf{r}) \approx 0$. Thus, we consider the surface of integration S' to be a sphere-centred on the interaction centre in the centre of mass frame. We divide the surface of the sphere into two parts the upper $\hat{r}$ and the lower $\check{r}$ by the equatorial plane perpendicular to the relative velocity of the incoming particles $\mathbf{V} = \mathbf{v} - \mathbf{v}'$ (Figure 4.1). Since the sphere is taken in the centre of mass frame and the particles have equal mass, the path of each particle is the reflection of the other in the equatorial plane. The particles enter through the surface area element $\mathbf{dS}$ whose projection on the equatorial plane is $\mathrm{d}\sigma$ with impact parameter b yielding a scattering angle χ. $\mathrm{d}\sigma$ is, therefore, the cross section for scattering through χ at a solid angle $\mathrm{d}\Omega$ of the incoming particles:

$$\mathrm{d}\sigma = \sigma(V, \chi)\, \mathrm{d}\Omega \tag{4.39}$$

To calculate the collision integral J, we need to know the two particle distribution functions before and after the collision, i.e. on the incoming and outgoing surfaces of the sphere. This is determined by the final assumption of *molecular chaos.*

(v) Since the interval between collisions between any two specified particles is very long during which each particle has suffered separately many randomising collisions, their distributions are independent *before* collision, i.e. $f^{(2)}(\mathbf{v}_1, \mathbf{v}_2) = f^{(1)}(\mathbf{v}_1)\, f^{(1)}(\mathbf{v}_2)$.

However, the very nature of the collision in which the velocity of one particle is changed in response of the other requires that correlation is introduced between the two particles *after* the collision and that $f^{(2)}(\mathbf{v}'_1, \mathbf{v}'_2) \neq f^{(1)}(\mathbf{v}'_1) f^{(1)}(\mathbf{v}'_2)$, although none existed before their interaction. There is, therefore, a major difference between *before* and *after* giving a clear direction to the 'arrow of time'. In particular, we have introduced irreversibility into the motion, reflecting the essential dissipation in collisions through the assumption of molecular chaos (assumption (v)).

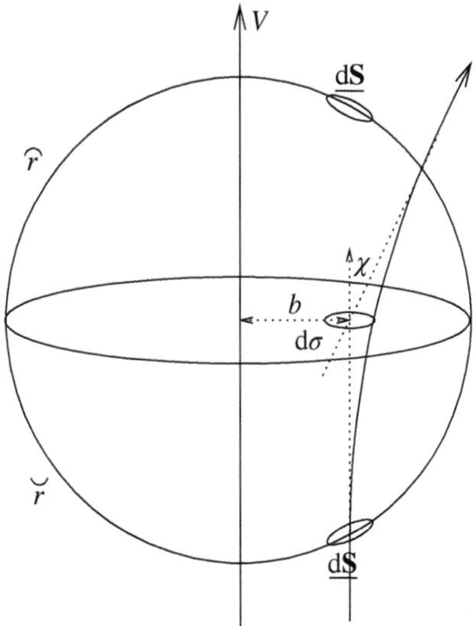

Figure 4.1 The interaction sphere showing the incoming particles in the lower hemisphere ($\check{r}$) and the outgoing in the upper ($\hat{r}$) $\mathrm{d}\sigma$ is the projection of the incoming area element on to equatorial plane, with impact distance b.

Since the collision involves only two particles, the two body distribution function after collision may be obtained from Liouville's theorem applied to the motion: namely, the two particle distribution function is unchanged by the collision. Therefore, if we calculate the initial velocities of two particles $\bar{\mathbf{v}}_1$ and $\bar{\mathbf{v}}_2$ whose final velocities are $\mathbf{v}_1$ and $\mathbf{v}_1$, we obtain the required two particle distribution function $f^{(2)}(\mathbf{v}_1, \mathbf{v}_2) = f^{(1)}(\bar{\mathbf{v}}_1)\, f^{(1)}(\bar{\mathbf{v}}_2)$ after collision. More completely, we have

$$
\begin{aligned}
f^{(2)}(\mathbf{r}, \mathbf{v}, \mathbf{r}', \mathbf{v}', \hat{t}) &= f^{(2)}(\bar{\mathbf{r}}, \bar{\mathbf{v}}, \bar{\mathbf{r}}', \bar{\mathbf{v}}', t) \\
&= f^{(1)}(\bar{\mathbf{r}}, \bar{\mathbf{v}}, t)\, f^{(1)}(\bar{\mathbf{r}}', \bar{\mathbf{v}}', t)
\end{aligned}
$$

Making use of the earlier assumptions (ii) and (iii):

$$
\begin{aligned}
f^{(1)}(\check{\mathbf{r}}', \mathbf{v}', t) &\approx f^{(1)}(\mathbf{r}, \mathbf{v}, t) \\
f^{(1)}(\bar{\mathbf{r}}, \bar{\mathbf{v}}, \bar{t}) &\approx f^{(1)}(\mathbf{r}, \bar{\mathbf{v}}, t) \\
f^{(1)}(\bar{\mathbf{r}}', \bar{\mathbf{v}}', \bar{t}) &\approx f^{(1)}(\mathbf{r}, \bar{\mathbf{v}}', t)
\end{aligned}
$$

Hence, we obtain by integration:

$$
\begin{aligned}
J &= \int \mathrm{d}\mathbf{v}' \int \mathrm{d}\sigma\; V\, \{f^{(1)}(\mathbf{r}, \bar{\mathbf{v}}, t)\, f^{(1)}(\mathbf{r}, \bar{\mathbf{v}}', t) - f^{(1)}(\mathbf{r}, \mathbf{v}, t)\, f^{(1)}(\mathbf{r}, \mathbf{v}', t)\} \\
&= \int \mathrm{d}\mathbf{v}' \int \mathrm{d}\Omega\; V\, \sigma(V, \chi)\, \{f^{(1)}(\mathbf{r}, \bar{\mathbf{v}}, t)\, f^{(1)}(\mathbf{r}, \bar{\mathbf{v}}', t) - f^{(1)}(\mathbf{r}, \mathbf{v}, t)\, f^{(1)}(\mathbf{r}, \mathbf{v}', t)\}
\end{aligned} \tag{4.40}
$$

which is Boltzmann's collision integral.

4.7 Boltzmann Collision Operator

The Boltzmann collision term for a dilute gases is not normally derived through the BBGKY hierarchy, but via a probability argument applied directly to the single-particle distribution function $f(\mathbf{r}, \mathbf{v}, t) \equiv f^{(1)}(\mathbf{r}, \mathbf{v}, t)$. As we have argued, in a dilute system, the distribution on the collisional time scale $\sim t_0$ is averaged over individual binary and higher interaction terms. By evaluating the probability of a particular change in the distribution function and averaging over all possible interactions, the Boltzmann equation (4.30) is derived as a simple physical picture in terms of a 'fluid' equation of continuity for the distribution in μ space, where the collisional term represents loss/gain term between individual phase cells.

$$
\frac{\partial f}{\partial t} + \mathbf{v} \cdot \frac{\partial f}{\partial \mathbf{r}} + \frac{\mathbf{F}}{m} \cdot \frac{\partial f}{\partial \mathbf{v}} = \left.\frac{\partial f}{\partial t}\right|_{\mathrm{coll}} \tag{4.41}
$$

where $f \equiv f^{(1)}$ and $J \equiv \partial f / \partial t|_{\mathrm{coll}}$.[2] This key equation originally derived by Boltzmann (1896, §17) underlies the development of kinetic theory provided a suitable approximation to the collision term is known. In this representation, it is easy to calculate the Boltzmann collision integral appropriate to the short-range interactions in gases using the definition of detailed balance:

Definition 4.1 ***Detailed balance*** The rate of change of f due to collisions into the phase space volume $\mathrm{d}\mathbf{r}\, \mathrm{d}\mathbf{v}$ = the rate of particles entering – rate of particles exiting

2 Henceforward, we omit distribution function superscripts unless necessary.

The number of collisions exiting $\mathbf{dr}\,\mathbf{dv}$ per unit time is determined by the density of the two particles undergoing collisions at their relative speed V and the cross section $\sigma(V)$ namely

$$f(\mathbf{r},\mathbf{v})\ \mathrm{d}\mathbf{r}\ \mathrm{d}\mathbf{v}\int f(\mathbf{r}',\mathbf{v}')\ \sigma\ V\ \mathrm{d}\mathbf{r}'\ \mathrm{d}\mathbf{v}'$$

To calculate the number of collisions entering the volume element, we consider all the collisions whose initial velocities lead to those of the particles in the cell after collision. $\overline{\mathbf{r}},\overline{\mathbf{v}},\ \overline{\mathbf{r}}',\overline{\mathbf{v}}' \Rightarrow \mathbf{r},\ \mathbf{v},\ \mathbf{r}',\ \mathbf{v}'$. Noting that since the cross section depends only on the relative speed V of the colliding particles, which is unchanged in an elastic collision, the cross sections for the forwards and backwards collisions are the same

$$\int f(\overline{\mathbf{r}},\overline{\mathbf{v}})f(\overline{\mathbf{r}}',\overline{\mathbf{v}}')\ \sigma\ V\ \mathrm{d}\overline{\mathbf{r}}\ \mathrm{d}\overline{\mathbf{v}}\ \mathrm{d}\overline{\mathbf{r}}'\ \mathrm{d}\overline{\mathbf{v}}'$$

The volume elements for the phase space before and after collision are equal as can be easily seen by the transformation of each in the centre of mass frame, where one is transformed to the other by a simple rotation since the collision is elastic. Hence,

$$\mathrm{d}\overline{\mathbf{r}}\ \mathrm{d}\overline{\mathbf{v}}\ \mathrm{d}\overline{\mathbf{r}}'\ \mathrm{d}\overline{\mathbf{v}}' = \mathrm{d}\mathbf{r}\ \mathrm{d}\mathbf{v}\ \mathrm{d}\mathbf{r}'\ \mathrm{d}\mathbf{v}'$$

Calculating the net gain of particles in the cell $\mathbf{dr}\,\mathbf{dv}$, we obtain the same result as before (4.40)

$$\left.\frac{\partial f}{\partial t}\right|_{\mathrm{coll}} = \int [f(\mathbf{r},\overline{\mathbf{v}})f(\mathbf{r},\overline{\mathbf{v}}') - f(\mathbf{r},\mathbf{v})f(\mathbf{r},\mathbf{v}')]\ \sigma\ V\ \mathrm{d}\mathbf{v}'\ \mathrm{d}\Omega \tag{4.42}$$

Since the collision volume is small, we ignore any variation in the spatial position $\mathbf{r}$.

4.7.1 Summation Invariants

In a binary collision between two simple molecules, where the energy is entirely kinetic, we may identify five quantities, which are unchanged in value before and after the collision. These are known as *summation invariants* and are easily seen to be

$$\psi_1 = m \qquad \boldsymbol{\psi}_2 = m\mathbf{v} \qquad \psi_3 = \frac{1}{2}mv^2 \tag{4.43}$$

so that

$$\psi_i + \psi_i' = \overline{\psi}_i + \overline{\psi}_i' \tag{4.44}$$

It can be shown (Jeans, 1954, §25; Chapman and Cowling, 1952, §3.2) that for simple molecules, these are a complete set of dynamical summation invariants, and that any others can therefore, be expressed as a sum of these.

4.8 Boltzmann's H Theorem

Define the function

$$H = \int f(\mathbf{v},t)\ \ln f(\mathbf{v},t)\ \mathrm{d}\mathbf{v} \tag{4.45}$$

Differentiating with respect to time, t, and making use of Boltzmann's collision integral (4.42), we obtain

$$\frac{\mathrm{d}H}{\mathrm{d}t} = \int \cdots \int [1 + \ln f(\mathbf{v},t)][f(\overline{\mathbf{v}},t)\ f(\overline{\mathbf{v}}',t) - f(\mathbf{v},t)\ f(\mathbf{v}',t)]\ \ \sigma\ V\ \mathrm{d}\mathbf{v}\ \mathrm{d}\mathbf{v}'\ \mathrm{d}\Omega \tag{4.46a}$$

$$= \int \cdots \int [1 + \ln f(\mathbf{v}', t)][f(\overline{\mathbf{v}}, t)\, f(\overline{\mathbf{v}}', t) - f(\mathbf{v}, t)\, f(\mathbf{v}', t)] \;\; \sigma\, V\, \mathbf{dv}\, \mathbf{dv}'\, \mathrm{d}\Omega \tag{4.46b}$$

$$= -\int \cdots \int [1 + \ln f(\overline{\mathbf{v}}, t)][f(\overline{\mathbf{v}}, t)\, f(\overline{\mathbf{v}}', t) - f(\mathbf{v}, t)\, f(\mathbf{v}', t)] \;\; \sigma\, V\, \mathbf{dv}\, \mathbf{dv}'\, \mathrm{d}\Omega \tag{4.46c}$$

$$= -\int \cdots \int [1 + \ln f(\overline{\mathbf{v}}', t)][f(\overline{\mathbf{v}}, t)\, f(\overline{\mathbf{v}}', t) - f(\mathbf{v}, t)\, f(\mathbf{v}', t)]\; \sigma\, V\, \mathbf{dv}\, \mathbf{dv}'\, \mathrm{d}\Omega \tag{4.46d}$$

$$\frac{\mathrm{d}H}{\mathrm{d}t} = \frac{1}{4} \int \cdots \int \ln \left[\frac{f(\mathbf{v}, t)\, f(\mathbf{v}', t)}{f(\overline{\mathbf{v}}, t)\, f(\overline{\mathbf{v}}', t)} \right] [f(\overline{\mathbf{v}}, t)\, f(\overline{\mathbf{v}}', t) - f(\mathbf{v}, t)\, f(\mathbf{v}', t)]\; \sigma\, V\, \mathbf{dv}\, \mathbf{dv}'\, \mathrm{d}\Omega \tag{4.47}$$

Since the integrand is always ≤ 0, it follows that $\mathrm{d}H/\mathrm{d}t \leq 0$ and the H function always decreases to reach a constant value when $f(\mathbf{v})\, f(\mathbf{v}') = f(\overline{\mathbf{v}})\, f(\overline{\mathbf{v}}')$, i.e. the distribution has a Maxwellian equilibrium form. Since H has a finite lower bound (Chapman and Cowling, 1952, §4.1), it follows that H decreases until this bound is reached, when $\mathrm{d}H/\mathrm{d}t = 0$ which, as we show in the next section, corresponds to the well-known Maxwellian form.

4.9 The Equilibrium Maxwell–Boltzmann Distribution

It follows from Eqs. (4.42 and (4.47) that

$$f(\mathbf{r}, \overline{\mathbf{v}})\, f(\mathbf{r}, \overline{\mathbf{v}}') = f(\mathbf{r}, \mathbf{v})\, f(\mathbf{r}, \mathbf{v}') \tag{4.48}$$

for all values of $\mathbf{v}$ and $\mathbf{v}'$ is both a necessary and sufficient condition that the distribution has reached a steady state or equilibrium form.

Taking the logarithm of Eq. (4.48), we obtain

$$\ln f + \ln f' - \ln \overline{f} - \ln \overline{f}' = 0 \tag{4.49}$$

and conclude that $\ln f$ is also a summation invariant. It can, therefore, be written as a linear sum of the dynamical set (4.44)

$$\begin{aligned} \ln f &= \alpha_1\, \psi_1 + \alpha_2\, \psi_2 - \alpha_3\, \psi_3 \\ &= m \left\{ 1 + \alpha_2\, \mathbf{v} - \frac{1}{2}\, \alpha_3\, v^2 \right\} \\ &= \tilde{\alpha} - \frac{1}{2}\, \alpha_3\, m\, [\mathbf{v} - \alpha_2/\alpha_3]^2 \end{aligned} \tag{4.50}$$

Since

$$n = \int f(\mathbf{v})\, \mathbf{dv} \qquad \mathbf{u} = \int f(\mathbf{v})\, \mathbf{v}\, \mathbf{dv}$$

$$\frac{3}{2} kT = \frac{1}{2} m \overline{w^2} = \frac{1}{m} \int f(\mathbf{v})\, (\mathbf{v} - \mathbf{u})^2\, \mathbf{dv} \tag{4.51}$$

The constants α_1, α_2, and α_3 can be determined in terms of the number density n, the mean velocity, $\mathbf{u}$, and the random or thermal velocity $\mathbf{w}$ or temperature T. Thus, $\mathbf{w} = \mathbf{v} - \mathbf{u}$. Finally, we obtain the familiar Maxwell–Boltzmann velocity distribution function in a uniform steady state:

$$f_0(\mathbf{v}) = n \left(\frac{m}{2\pi kT} \right)^{3/2} \exp\left[-\frac{m(\mathbf{v} - \mathbf{u})^2}{2kT} \right] \tag{4.52}$$

Since the distribution is isotropic, this is usually written in terms of the thermal speed w as

$$f_0(w)\,\mathrm{d}w = n\,\frac{4\pi w^2}{(2\pi \mathcal{k}T/m)^{3/2}}\exp\left\{-\frac{mw^2}{2\mathcal{k}T}\right\}\,\mathrm{d}w \tag{4.53}$$

where $\mathcal{k}$ is Boltzmann's constant.

4.9.1 Entropy and the *H* function

It is easily shown by direct integration that in the equilibrium state, the value of the H function,

$$H_0 = \int \ln(f_0)\,f_0\,\mathbf{dv} = n\left[\ln(n) + \frac{3}{2}\ln\left(\frac{m}{2\pi\mathcal{k}T}\right) - \frac{3}{2}\right] \tag{4.54}$$

The entropy of a mass of n particles of an gas is given by classical thermodynamics[3]

$$S = n\mathcal{k}\ln\left(\frac{T^{3/2}}{n}\right) + \text{const} \tag{4.56}$$

Therefore, apart from the undefined constant H_0 is directly related to the entropy S, namely,

$$S = -\mathcal{k}\,H_0 \tag{4.57}$$

For non-uniform or non-equilibrium states, entropy is not thermo-dynamically defined. However, if the above result is generalised, Boltzmann's H theorem becomes a statement of the second law of thermo-dynamics. In particular, it establishes the equivalence between microscopic statistical thermodynamics introduced by Boltzmann and macroscopic classical thermodynamics of Clausius.

From the H theorem, we derive three important results, concerning the fluid as consequences of binary elastic collisions amongst the particles.

1. These exists an equilibrium configuration described by a Maxwell–Boltzmann distribution.
2. Collisions drive the fluid towards the equilibrium state.
3. The entropy of the system increases due to collisions consistent with the second law of thermo-dynamics.

4.10 Hydrodynamic Limit – Method of Moments

At this stage, our description of the system is entirely expressed in terms of the distribution function $f(\mathbf{r}, \mathbf{v}, t)$, which is an expression of the microscopic state. This contains information in much greater detail than is needed for macroscopic laboratory experiments, which typically require the values of quantities averaged over the particle distribution. Evaluating the quantities per unit volume:

Mass	$\rho(\mathbf{r}, t) = m\int f(\mathbf{r}, \mathbf{v}, t)\,\mathbf{dv}$
Momentum	$\mathrm{p}(\mathbf{r}, t) = m\int f(\mathbf{r}, \mathbf{v}, t)\,\mathbf{v}\,\mathbf{dv}$
Energy	$E(\mathbf{r}, t) = \frac{1}{2}m\int f(\mathbf{r}, \mathbf{v}, t)\,v^2\,\mathbf{dv}$

3 Making use of the first law of thermo-dynamics $T\,\mathrm{d}S = \mathrm{d}U + p\,\mathrm{d}V$ for a mass of an ideal monatomic gas, where the internal energy $U = C_v\,T = \frac{3}{2}\,\mathcal{R}T$, and the pressure $p = \mathcal{R}T/V$

$$S = \int T^{-1}\left\{\frac{\partial S}{\partial T}\mathrm{d}T + \frac{\partial S}{\partial V}\mathrm{d}V\right\} = \left\{\int \frac{3}{2}\mathcal{R}\mathrm{d}T/T + \int \mathcal{R}\mathrm{d}V/V\right\} = \ln(V\,T^{3/2}) + \text{const} \tag{4.55}$$

to give the mass density $\rho(\mathbf{r}, t)$, momentum density $p(\mathbf{r}, t)$ and energy density $E(\mathbf{r}, t)$ within the fluid. Each of these quantities is the integral of a power of the velocity over the distribution, and may be expressed as a general term $Q(\mathbf{v})$. The average value of $Q(\mathbf{v})$, known as the *moment of* Q, is, therefore,

$$\overline{Q}(\mathbf{r}, t) = \int Q(\mathbf{v})\, f(\mathbf{r}, \mathbf{v}, t)\, \mathrm{d}\mathbf{v} \Big/ \int f(\mathbf{r}, \mathbf{v}, t)\, \mathrm{d}\mathbf{v} \tag{4.58}$$

In order to proceed, we form the moments of Q with respect to various terms in the Boltzmann equation (4.30).

The external force on the particles is due to the gravity alone:

$$\mathbf{F} = -m\nabla\phi \tag{4.59}$$

where ϕ is the gravitational potential.

Forming the moments of $Q(\mathbf{v})$ about the terms in the Boltzmann equation:

$$\int Q\, \frac{\partial f}{\partial t} \mathrm{d}\mathbf{v} = \frac{\partial}{\partial t} \int Q\, f \mathrm{d}\mathbf{v} = \frac{\partial(n\, \overline{Q})}{\partial t} \tag{4.60a}$$

$$\int Q\, \mathbf{v} \cdot \frac{\partial f}{\partial \mathbf{r}}\, \mathrm{d}\mathbf{v} = \frac{\partial}{\partial \mathbf{r}} \int Q\, \mathbf{v}\, f(\mathbf{v})\, \mathrm{d}\mathbf{v} = \frac{\partial}{\partial \mathbf{r}} (n \overline{\mathbf{v}\, Q}) \tag{4.60b}$$

$$\int Q\, \mathbf{a} \cdot \frac{\partial f}{\partial \mathbf{v}} \mathrm{d}\mathbf{v} = \int Q\, a_i\, \frac{\partial f}{\partial v_i}\, \mathrm{d}\mathbf{v} = -\int f a_i\, \frac{\partial Q}{\partial v_i}\, \mathrm{d}\mathbf{v} = -n\, \mathbf{a}\, \overline{\frac{\partial Q}{\partial \mathbf{v}}} \tag{4.60c}$$

after integration by parts, noting that $\mathbf{a}$ is independent of $\mathbf{v}$ and that at ∞ the boundary values $f \to 0$ as $\mathbf{v} \to \infty$. The particle density (number of particles per unit volume) is $n(\mathbf{r}.t) = \int f(\mathbf{r}, \mathbf{v}.t)\, \mathrm{d}v$.

The collision term

$$\int Q \frac{\partial f}{\partial t}\bigg|_{\text{coll}} \mathrm{d}\mathbf{v} = \frac{\partial}{\partial t} (n\overline{Q})\bigg|_{\text{coll}} \tag{4.61}$$

For a simple fluid mass, momentum and energy are conserved in collisions. The five terms, given by $Q(\mathbf{v}) = 1,\ m\, \mathbf{v},\ \frac{1}{2} m v^2$, are the set of summation invariants.

Inserting these terms into the Boltzmann equation and integrating yields

$$\frac{\partial}{\partial t}(n\, \overline{Q}) + \frac{\partial}{\partial \mathbf{r}}(n\, \overline{Q\, \mathbf{v}}) - n\, \mathbf{a} \cdot \overline{\frac{\partial Q}{\partial \mathbf{v}}} = \frac{\partial}{\partial t}(n\, \overline{Q})\bigg|_{\text{coll}} \tag{4.62}$$

Equation (4.62) may be regarded as generating the nth order moment of the velocity $\mathbf{v}$ in terms of the collisional moment. However, the equation also contains the $(n+1)$th moment through the term $\overline{Q\mathbf{v}}$. The set of equations thus form an infinite set similar to the BBGKY set. For practical applications, this set must be closed at a finite number of terms. This is achieved by using a condition similar to the Bogoliubov hypothesis (Section 4.5) and leads to the well-known equations of fluid dynamics (Section 4.11).

At this stage, it is convenient to separate the mean velocity of the particles $\mathbf{u}$ from the 'random' thermal part $\mathbf{w}$

$$\mathbf{v} = \mathbf{u} + \mathbf{w} \quad \mathbf{u} = \overline{\mathbf{v}} \quad \overline{\mathbf{w}} = 0 \tag{4.63}$$

and therefore, the kinetic energy of the particles is also divided into the mean kinetic energy of flow and the internal (thermal) energy $\epsilon = \frac{1}{2}\, \overline{w^2}$ of the particles per unit mass

$$\frac{1}{2}\, m\, \overline{v^2} = \frac{1}{2}\, m\, u^2 + m\, \overline{\mathbf{u} \cdot \mathbf{w}} + \frac{1}{2}\, m\, \overline{w^2} \tag{4.64}$$

4.10.1 Conservation of Mass

If $Q(\mathbf{v}) = 1$, we obtain an equation for the mass conservation

$$\frac{\partial n}{\partial t} + \frac{\partial}{\partial \mathbf{r}}(n\mathbf{v}) = 0 \tag{4.65}$$

This is normally written in terms of the mass density $\rho = n\ m$

$$\frac{\partial \rho}{\partial t} + \nabla \cdot (\rho\ \mathbf{v}) = 0 \tag{4.66}$$

to yield the *equation of continuity* of fluid mechanics.

4.10.2 Conservation of Momentum

If $Q(\mathbf{v}) = m\mathbf{v}$, we obtain the equation for the momentum conservation in the fluid:

$$\frac{\partial}{\partial t}(n\ m\ v_i) + \frac{\partial}{\partial r_j}(nm\overline{v_i v_j}) - \overline{\frac{n\,F_i}{m}\frac{\partial}{\partial v_j}(mv_j)} = 0 \tag{4.67}$$

in Cartesian tensor notation. $\mathbf{F}$ is the total force on each particle, m its mass. The tensor containing the two velocity components can be simplified:

$$\overline{v_i\ v_j} = u_i\ u_j + \cancel{\overline{u_i w_j}} + \cancel{\overline{w_i u_j}} + \overline{w_i\ w_j} \tag{4.68}$$

These terms are easily recognised from fluid mechanics. nmu_iu_j is the momentum flux, the convected transfer of the ith component of the bulk fluid momentum per unit time across unit surface in the jth direction. The stress tensor

$$\mathbf{P}_{ij} = nm\overline{w_i w_j} \tag{4.69}$$

is the momentum transfer due to the random motion and represents the force per unit area in the direction i exerted on a surface with area vector in the j direction.

If the fluid is isotropic, described by the equilibrium (Maxwellian) distribution, the stress tensor is solely due to the pressure, which acts normal inwards to the surface and has the same value in all directions (isotropic). The stress tensor then takes the simpler form:

$$\mathbf{P}_{ij} = n\ m\ \overline{w_i\ w_j} = \frac{1}{3}\ n\ m\overline{w^2}\delta_{ij} = -p\ \delta_{ij} \tag{4.70}$$

Generalising this result, the hydrostatic pressure is defined as the negative of $\frac{1}{3}$ the trace of the stress tensor $p = -\frac{1}{3}\ \mathrm{tr}(\mathbf{P})$. The traceless part of the stress tensor is the viscosity tensor $\mathbf{T} = \boldsymbol{p} - \frac{1}{3}\ \mathrm{tr}(\mathbf{P}) = \boldsymbol{p} + p\ \mathbf{I}$. The identity tensor $\mathbf{I}$ and the Kronecker delta δ_{ij} are defined:

$$\mathbf{I} \equiv \delta_{ij} = \begin{cases} 1 & \text{if} \quad (i = j) \\ 0 & \text{otherwise} \end{cases}$$

The *equation of momentum conservation* is, therefore,

$$\frac{\partial}{\partial t}(\rho\ u_i) + \frac{\partial\,\Gamma_{ij}}{\partial x_j} = \frac{\partial}{\partial t}(\rho\ u_i) + \frac{\partial}{\partial x_j}(\rho\ u_i\ u_j + \mathbf{P}_{ij}) = n\ F_i \tag{4.71}$$

where Γ_{ij} is the *momentum flux tensor*.

Making use of the equation of continuity (4.65), we obtain the more familiar equation from fluid mechanics for isotropic dissipationless fluids, namely *Euler's equation*

$$\frac{\partial \mathbf{u}}{\partial t} + (\mathbf{u} \cdot \nabla)\mathbf{u} + \frac{1}{\rho}\ \nabla p = \frac{1}{m}\mathbf{F} \tag{4.72}$$

4.10.3 Conservation of Energy

If $Q(\mathbf{v}) = \frac{1}{2}mv^2$, we obtain the equation for the conservation of energy by the fluid:

$$\frac{\partial}{\partial t}\left(\frac{1}{2}\, n\, m\, \overline{v^2}\right) + \frac{\partial}{\partial \mathbf{r}}\left(\frac{1}{2} n\, m\, \overline{v^2 \mathbf{v}}\right) - n\overline{\mathbf{v}} \cdot \mathbf{F} = 0 \tag{4.73}$$

The fluid energy may also be considered as two terms, the kinetic energy of the flow and the internal energy associated with the kinetic energy of the thermal motion; the total energy per unit mass being their sum

$$E = \epsilon + \frac{1}{2}\, u^2 \tag{4.74}$$

where the thermal energy per unit mass is given by

$$\epsilon = \frac{1}{2}\, \overline{w^2} = \int \frac{1}{2}\, w^2\, f(\mathbf{v})\mathrm{d}\mathbf{v} = \frac{3}{2}\frac{kT}{m} \tag{4.75}$$

for simple monatomic molecules with a Maxwellian distribution, k being Boltzmann's constant and T the temperature.

The energy flux term expands to

$$\frac{1}{2}\, n\, m\, \overline{v^2 \mathbf{v}} = \frac{1}{2}\, n\, m\, (u^2 + \overline{w^2})\, \mathbf{u} + \frac{1}{2}\, n\, m\, \overline{w^2\, \mathbf{w}} \tag{4.76}$$

where the thermal heat flux $q = \frac{1}{2}\, n\, m\, \overline{w^2 \mathbf{w}}$ is zero if the distribution is isotropic.

Hence, since the gravitational potential ϕ is constant in time

$$\frac{\partial}{\partial t}\left[\rho\left(\epsilon + \frac{1}{2}\, u^2 + \phi\right)\right] + \frac{\partial}{\partial\, r_i}\left[\rho\left(\epsilon + \frac{1}{2}\, u^2 + \phi\right)\, u_i + \mathsf{P}_{ij}\, u_j + q_i\right] = 0 \tag{4.77}$$

Equation (4.77) can be written in an alternative form by using Eqs. (4.65) and (4.71)

$$\frac{\partial \epsilon}{\partial t} + u_i\, \frac{\partial \epsilon}{\partial r_i} + \frac{1}{\rho}\, \mathsf{P}_{ij}\, \dot{e}_{ij} + \frac{1}{\rho}\, \frac{\partial q_i}{\partial r_i} = g_i\, u_i \tag{4.78}$$

where $\dot{e}_{ij} = \frac{1}{2}(\partial u_i/\partial r_j + \partial u_j/\partial r_i)$ is the rate of strain tensor.

For an isotropic fluid, we introduce the enthalpy $h = \epsilon + p/\rho$, and making use of Eq. (4.59) obtain the standard *equation of energy conservation* in an inviscid fluid:

$$\frac{\partial}{\partial t}\left[\rho\left(\epsilon + \frac{1}{2}u^2 + \phi\right)\right] + \nabla \cdot \left[\rho\left(h + \frac{1}{2}u^2 + \phi\right)\, \mathbf{u} + q\right] = 0 \tag{4.79}$$

4.11 The Departure from Steady Homogeneous Flow: The Chapman–Enskog Approximation

We have now obtained the classic equations of hydrodynamics in 'ideal' dissipationless fluids. By using the collision invariants as the moments, we have ensured that the collisional terms are zero after integration over the Boltzmann equation. By including only the equilibrium (Maxwellian) distribution, no collisions and consequently, no irreversibility are introduced into the equations.

In performing the integrations leading to the moment equations, we have eliminated the variation over the velocity, but at a price. The set of equations is not closed in that the momentum (first) moment equation (momentum) introduces a second-order moment (stress). Similarly, the energy (second) moment introduces a third-order term (heat flux). Each moment introduces a next order moment. The equations thus reduce from a set of equations in seven-dimensional space $(\mathbf{r},\ \mathbf{v},\ t)$

to an infinite set in a four-dimensional space ($\mathbf{r}$, t). As with the BBGKY set, closure can only be obtained by introducing appropriate physical constraints.

In most practical situations, the system is neither homogeneous nor steady state. As a result, the distribution is perturbed from the equilibrium Maxwellian form. The consequence is the introduction of additional terms in the stress and heat flux. In addition the H function and, therefore, entropy change in time, the flow becoming dissipational through the introduction of viscosity and thermal conduction into the pressure tensor and heat flux vector. In general, this is a difficult problem, and requires numerical simulation. However, in most cases, the variations are small, so that gradients introduce small changes over the mean free path $\lambda|\nabla T|/T \ll 1$, and we may use a perturbation expansion to solve the Boltzmann equation (4.30) when the gradient parameter is small, thereby reducing it to a manageable problem.

$$f(\mathbf{r},\mathbf{v},t) = f_0(\mathbf{r},\mathbf{v},t) + \xi\, f_1(\mathbf{r},\mathbf{v},t) + \xi^2\, f_2(\mathbf{r},\mathbf{v},t) + \cdots \tag{4.80}$$

where ξ is a small parameter expressing the variation from steady equilibrium conditions; ξ is chosen so that the changes in the macroscopic quantities are small over a mean free path. Therefore, it is chosen to be proportional to the gradient, for example an appropriate form is $\xi \sim \lambda\ |\nabla T|/T \ll 1$.[4] Once equilibrium is established, we can introduce the slowly varying values of the hydrodynamic variables:

$$\begin{aligned} n &= \int f(\mathbf{r},\mathbf{v},t)\ \mathrm{d}\mathbf{v} \\ n\,\mathbf{u} &= \int \mathbf{v} f(\mathbf{r},\mathbf{v},t)\ \mathrm{d}\mathbf{v} \\ n\,\frac{3}{2}\,k\,T &= \int \frac{1}{2}\,mw^2 f(\mathbf{r},\mathbf{v},t)\ \mathrm{d}\mathbf{v} \end{aligned} \tag{4.81}$$

where $\mathbf{w} = \mathbf{v} - \mathbf{u}$ is the random thermal velocity as before.

As noted above, the moment equations have a closure problem. The general approach has similarity to Bogoliubov's reduction of the BBGKY hierarchy of equations (Section 4.4) in that we can identify two distinct times associated with the equilibrium distribution and the perturbation:

1. Collisional interval: the time between collisions $t_0 \sim \lambda/v$
2. Macroscopic time scale: $\theta_0 \sim L/v_s$

where L is the gradient scale length. If the gas is sufficiently dense that $\lambda \ll L$, the two time scales are well separated. The perturbation varies on time scales $\sim t_0$, whereas the equilibrium distribution varies much more slowly as $\sim \theta_0$. In hydrodynamics, we are interested in the behaviour over time scales $\sim \theta_0$, but not in the detailed behaviour over the collision interval, and can, therefore, average over times $\sim t_0$. Consequently, we may treat the functional dependence of the perturbation distribution as depending on time only as the slowly varying values of the equilibrium distribution or alternatively of the slowly varying hydrodynamic variables n, $\mathbf{u}$ and T

$$f(\mathbf{r},\mathbf{v},t) \rightarrow f(\mathbf{r},\mathbf{v}|n,\ \mathbf{u},\ T) \tag{4.82}$$

To proceed, we insert Eq. (4.80) into the Boltzmann equation (4.30) with the Boltzmann collision operator (4.42) retaining terms of the same order in ξ. Since the left-hand side of the Boltzmann equation contains terms of the order of the gradient, it is clear that the collisional term on the

4 Since ξ is simply an ordering parameter, its exact value is not used in the calculation and an estimate of its scale is sufficient.

right-hand side is of higher order. Writing the Boltzmann collision integral (4.42) as[5]

$$J(f,\ f') = \iiint [f(\mathbf{r},\overline{\mathbf{v}})f(\mathbf{r},\overline{\mathbf{v}}') - f(\mathbf{r},\mathbf{v})f(\mathbf{r},\mathbf{v}')]\ \sigma\ V\ \mathrm{d}\mathbf{v}'\ \mathrm{d}\Omega \tag{4.83}$$

we may formally take this into account by writing the Boltzmann equation (4.30)

$$\frac{\partial f}{\partial t} + \mathbf{v}\cdot\frac{\partial f}{\partial \mathbf{r}} + \frac{\mathbf{F}}{m}\cdot\frac{\partial f}{\partial \mathbf{v}} = \frac{1}{\xi}J(f,\ f') \tag{4.84}$$

The zero-order equation is immediately seen to be $J(f,\ f') = 0$ namely, the equation for the equilibrium (Maxwellian) distribution. Therefore, to zero order $f(\mathbf{r},\mathbf{v}) = f_0(\mathbf{r},\mathbf{v})$. The Enskog series is, therefore,

$$\begin{aligned} 0 &= J(f_0,\ f_0{}') \\ \frac{\partial f_0}{\partial t} + \mathbf{v}\cdot\frac{\partial f_0}{\partial \mathbf{r}} + \frac{\mathbf{F}}{m}\cdot\frac{\partial f_0}{\partial \mathbf{v}} &= [J(f_0,\ f_1{}') + J(f_1,\ f_0{}')] \\ \frac{\partial f_1}{\partial t} + \mathbf{v}\cdot\frac{\partial f_1}{\partial \mathbf{r}} + \frac{\mathbf{F}}{m}\cdot\frac{\partial f_1}{\partial \mathbf{v}} &= [J(f_0,\ f_2{}') + 2J(f_1,\ f_1{}') + J(f_2,\ f_0{}')] \end{aligned} \tag{4.85}$$

The zero order solution is

$$\begin{aligned} &f_0 = n\left(\frac{m}{2\,\pi\,k\,T}\right)^{3/2}\exp\left(-\frac{mw^2}{2kT}\right) \\ &\mathbf{P}_{ij}^{(0)} = p\ \delta_{ij} = n\ k\ T\ \delta_{ij} \qquad p \equiv \text{hydrostatic pressure} \\ &\frac{\mathrm{d}n}{\mathrm{d}t} = -n\nabla\cdot\mathbf{u} \qquad \frac{\mathrm{d}\mathbf{u}}{\mathrm{d}t} = \frac{\mathbf{F}}{m} - \frac{1}{\rho}\nabla p \\ &\frac{\mathrm{d}T}{\mathrm{d}t} = -\frac{2T}{3} \qquad \frac{\mathrm{d}}{\mathrm{d}t}(nT^{-3/2}) = 0 \end{aligned} \tag{4.86}$$

where $\mathrm{d}/\mathrm{d}T = \partial/\partial t + (\mathbf{u}\cdot\nabla)$ is the standard convective (Lagrangian) time derivative of fluid mechanics.

Turning now to the first-order terms in the Enskog expansion (4.85), we obtain after making use of Eqs. (4.86)

$$\begin{aligned} \frac{\mathrm{d}f_0}{\mathrm{d}t} &= \left[\left(\frac{me^2}{2kT} - \frac{5}{2}\right)\mathbf{w}\cdot\frac{\partial \ln T}{\partial \mathbf{r}} + \frac{m}{kT}\left(w_i\ w_j - \frac{1}{3}w^2\ \delta_{ij}\right)\ \dot{e}_{ij}\right] f_0 \\ &= \int f_0(\mathbf{v})\ f_0(\mathbf{v}')(\varphi(\overline{\mathbf{v}}) + \varphi(\overline{\mathbf{v}'}) - \varphi(\mathbf{v}) - \varphi(\mathbf{v}'))\ \sigma\ V\mathrm{d}\mathbf{v}'\ \mathrm{d}\Omega \end{aligned} \tag{4.87}$$

where

$$f_1(\mathbf{v}) = f_0(\mathbf{v})\ \varphi(\mathbf{v}) \tag{4.88}$$

Equation (4.87) is an inhomogeneous integral equation. The solution is the sum of the complementary function and the particular integral. The complementary function is the solution of the corresponding homogeneous equation and must be orthogonal to the inhomogeneous part if the solution exists (Courant and Hilbert, 1953, pp. 115–121).

---

5 Note the essential bilinearity of the collision operator $J(f,f')$ which is necessary to allow the expansion (4.85) and therefore, the Chapman–Enskog closure scheme.

Since the homogeneous equation is a reduced form of Boltzmann's equation for binary collisions, its solution is a sum of the summation invariants. Since

$$\int \frac{\mathrm{d}}{\mathrm{d}t}\begin{Bmatrix} 1 \\ m\mathbf{w} \\ \frac{1}{2}mw^2 \end{Bmatrix} f_0 \, \mathrm{d}\mathbf{w} = 0$$

as a consequence of the conservation laws in binary collisions. The solution, therefore, exists.

The form of the particular integral follows from the form of the term on the right-hand side of Eq. (4.87). Thus, we obtain a solution:

$$\varphi(\mathbf{w}) = a + \mathbf{b}\cdot\mathbf{w} + c\,w^2 - \frac{1}{n}\left(\frac{2kT}{m}\right)^{-1/2}\mathbf{A}\cdot\frac{\partial \ln T}{\partial \mathbf{r}} - \frac{1}{n}\mathbf{B}_{ij}\,\dot{e}_{ij} \tag{4.89}$$

The three constants a, b and c are determined from the conditions that

$$\int f_1 \,\mathrm{d}\mathbf{v} = \int f_1\,\mathbf{v}\,\mathrm{d}\mathbf{v} = \int f_1\,v^2\,\mathrm{d}\mathbf{v} = 0 \tag{4.90}$$

which follow from the normalisation of f. It follows (Chapman and Cowling, 1952, §7.41) that $a = c = 0$, and that $\mathbf{b} \propto \partial T/\partial \mathbf{r}$ and may therefore, be absorbed into the corresponding term in Eq. (4.89). The terms $\mathbf{A}$ and $\mathbf{B}_{ij}$ can only be formed from the variables $\mathbf{w}$, and n and T. Therefore, since the only vector quantity is $\mathbf{w}$

$$\mathbf{A} = A(w)\,\mathbf{w} \qquad \text{and} \qquad \mathbf{B}_{ij} = \left(w_i\,w_j - \frac{1}{3}w^2\,\delta_{ij}\right)B(w) \tag{4.91}$$

A and B being calculated from the solution to the integral equation. They determine the heat flux and the pressure tensor in weakly inhomogeneous fluids to first order (Chapman and Cowling, 1952, §7.4; Lifshitz and Pitaevskii, 1981, §7,8)

$$q^{(1)} = \frac{1}{2}m\int f_1 w^2\mathbf{w}\,\mathrm{d}\mathbf{w} = -\lambda\nabla T \tag{4.92}$$

$$\mathbf{P}^{(1)}_{ij} = m\int f_1 w_i\,w_j\,\mathrm{d}\mathbf{w} = -2\eta\left(\dot{e}_{ij} - \frac{1}{3}\dot{e}_{kk}\,\delta_{ij}\right) \tag{4.93}$$

where λ is the thermal conductivity and η the viscosity. Substituting for f_1, we obtain

$$\lambda = \frac{k}{3\,n}\int f_0\,A(n,\mathbf{w},T)\left(\frac{mw^2}{2kT} - \frac{5}{2}\right)w^2\,\mathrm{d}\mathbf{w}$$

$$\eta = \frac{m}{10\,n}\int f_0\,B(n,\mathbf{w},T)\left(w_i\,w_j - \frac{1}{3}w^2\,\delta_{ij}\right)\mathrm{d}\mathbf{w} \tag{4.94}$$

since $\int f_1\,w^2\,\mathrm{d}\mathbf{w} = 0$. Because only monatomic gases are considered here the first-order pressure tensor contains solely a term depending on the divergence/trace free rate of strain, i.e. the rate of strain deviator or rate of distortion. Consequently, there is no viscous term depending on the rate of dilation ($\nabla\cdot\mathbf{u}$), namely, the second viscosity. This term is introduced by the inclusion of additional modes of energy, for example rotation, in polyatomic molecules (Lifshitz and Pitaevskii, 1981).

The parameters A and B are calculated from the solution of the integral equation by an expansion in Sonine polynomials and the values obtained either algebraically (Chapman and Cowling, 1952, §7.5) or by a variational method (Hirschfelder et al., 1964, §7.3) for appropriate forms of the intermolecular force.

The generalisation of this approach to consider polyatomic molecules and gas mixtures can be found in the standard texts by Chapman and Cowling (1952), Hirschfelder et al. (1964), and Lifshitz and Pitaevskii (1981).

5

Wave Propagation in Inhomogeneous, Dispersive Media

5.1 Introduction

One of the most characteristic features of plasma dynamics is the ability of the plasma to support a wide variety of wave motions. Therefore, before describing the many possible plasma wave motions, it is useful to review some of the basic general properties of waves, which impact on plasma waves. Many important effects can be described simply by assuming an isotropic medium in which there is a linear dispersion relation, namely, the relationship between the frequency of the wave (ω) and the wave number (k). In an isotropic medium, the constant $n = k\, c/\omega$, the *refractive index*, is independent of the direction in which the wave propagates, where c is the velocity of the wave in vacuo. In general, the medium is *dispersive*, and the refractive index is a function of the wavenumber $\omega(k)$. The medium is assumed to be dissipationless, except where some damping is necessary at discontinuities, and is introduced through a small arbitrary imaginary addition to the refractive index (Ginzburg, 1970). The introduction of anisotropy, which is normally a straightforward extension of isotropic behaviour, plays an important role in plasma in an ambient magnetic field, but is deferred until later. In many cases, plasma waves are electro-magnetic waves involving both the vector electric and magnetic fields, again, these cases will be discussed later. When specific field effects are not important, a simplifying condition may be used – *physical optics* – and the propagation described by the scalar wave equation:

$$\nabla^2 U - \frac{1}{V^2}\frac{\partial^2 U}{\partial t^2} = 0 \tag{5.1}$$

where U is the wave amplitude and $V = c/n$ the phase velocity. For a monochromatic wave of frequency ω, the amplitude $U \sim U_0 \exp(-\imath\, \omega\, t)$, this simplifies to the *Helmholtz equation*:

$$\nabla^2 U + k^2\, U = 0 \tag{5.2}$$

where $k = \omega/c$ is the wave number. The intensity of the wave, namely, the energy flowing through unit area in unit time, is given by $|U|^2$.

Our approach is that taken in the development of optics (Born and Wolf, 1965). The relationship with electro-magnetic theory is straightforward, the amplitude U represents either the electric or magnetic field provided $E \sim H$, and the intensity $|U|^2$ the Poynting vector $\mathbf{E} \wedge \mathbf{H}$. Provided the wavelength is short, a further simplification is possible, leading to *geometrical optics*, and to the *WKB approximation*.

Foundations of Plasma Physics for Physicists and Mathematicians, First Edition. Geoffrey J. Pert.

5.2 Basic Concepts of Wave Propagation – The Geometrical Optics Approximation

When the wavelength of the radiation is short compared to the characteristic lengths in the system, $\lambda \to 0$, the full description using the complete set of Maxwell equations may be simplified by invoking the geometrical optics approximation (Born and Wolf, 1965; Sommerfeld, 1964). During the propagation of plane electro-magnetic waves, a surface of constant phase occurs in the plane defined by the electric **E** and magnetic field **B** vectors: the Poynting vector $\mathbf{E} \wedge \mathbf{B}$ specifying that the propagation of energy is normal to this plane. In the limit $\lambda \to 0$, the fields vary relatively slowly and locally form a plane wave. In consequence, we may specify a surface of constant phase, *wave surface*, which is locally plane. Energy propagates normal to these wave surfaces. Consequently, small wave-packets propagate without diffraction along well-defined paths known as *rays* normal to the surfaces of constant phase. Since energy may be directly associated with a group of these wave-packets through the use of the cycle averaged Poynting vector, this provides a simple method of calculating the intensity in the wave field, namely, the energy flux $I = \mathbf{E} \wedge \mathbf{B}/\mu_0$ in free space.

Within this approximation, any quantity describing the wave field (either **E** or **B**) is given by a form:

$$\Phi = \Phi_0 \exp\{\imath\, \tilde{S}\} \tag{5.3}$$

where the amplitude Φ_0 is a slowly varying function of $\mathbf{r}$ and t. The phase $\tilde{S}(\mathbf{r}, t)$, known as the *eikonal* is a large quantity which varies rapidly; it plays an important role in geometrical optics. Assuming the wave is monochromatic, the time derivative of the eikonal is the frequency $\omega = -\partial\tilde{S}/\partial t$ and the gradient is the wave number $\mathbf{k} = \nabla\tilde{S}$. If the wave is steady, the frequency is constant, and we may write

$$\tilde{S}(\mathbf{r}, t) = -\omega\, t + \frac{\omega}{c}\, S(\mathbf{r}) \tag{5.4}$$

where

$$\nabla S = \mathbf{n} \tag{5.5}$$

where $\mathbf{n} = c\,\mathbf{k}/\omega$ is the refractive index vector. The phase velocity of the wave $\mathbf{V}_p = \hat{\mathbf{k}}\,\omega/k = \hat{\mathbf{n}}\,c/n$ is the velocity with which the phase moves locally in the medium, $\hat{\mathbf{k}}$ and $\hat{\mathbf{n}}$ are unit vectors parallel to **k** and **n**, respectively.

The eikonal is usually expressed as the solution of the equation

$$|\nabla S|^2 = \left(\frac{\partial S}{\partial x}\right)^2 + \left(\frac{\partial S}{\partial y}\right)^2 + \left(\frac{\partial S}{\partial z}\right)^2 = n^2 \tag{5.6}$$

in terms of the refractive index n. Knowing the distribution of the refractive index, this equation formally allows the direct calculation of the eikonal, and thus the solution of the wave equation within the limits of the geometrical optics approximation.

Rays are normal to the surface of constant phase. The direction of the ray is, therefore, given by the unit vector:

$$\hat{\mathbf{s}} = \frac{\nabla S}{|\nabla S|} = \frac{\nabla S}{n} \tag{5.7}$$

It follows from Eq. (5.5) that the optical path difference between two points ① and ② defines the phase difference between them

$$\int_1^2 n\, \mathrm{d}s = \int_1^2 \nabla S \cdot \mathrm{d}\mathbf{s} = S_2 - S_1 \tag{5.8}$$

where s is the distance measured along the ray.

Consider the osculating plane of a curved ray,[1] and calculate the principal curvature. By definition $|\hat{\mathbf{s}}| = 1$, consequently $|\mathrm{d}\hat{\mathbf{s}}| = \mathrm{d}\alpha$ directed along the normal, where α is the angle the tangent makes with a fixed axis. We, therefore, define the curvature vector directed along the principal normal, whose magnitude is the reciprocal of the radius of curvature

$$\mathbf{K} = \frac{\mathrm{d}\hat{\mathbf{s}}}{\mathrm{d}s} = (\hat{\mathbf{s}} \cdot \nabla)\hat{\mathbf{s}} \tag{5.10}$$

since in Cartesian co-ordinates $\dfrac{\partial \hat{s}_i}{\partial x_j}\dfrac{\partial x_j}{\partial s} = \hat{s}_j\dfrac{\partial \hat{s}_i}{\partial x_j} = (\hat{\mathbf{s}} \cdot \nabla)\hat{\mathbf{s}}$.

But since $\hat{\mathbf{s}}$ is a unit vector $\nabla|\hat{\mathbf{s}}|^2 = 0$, it follows from the vector identity $(\mathbf{v} \cdot \nabla)\mathbf{v} = \frac{1}{2}\nabla|\mathbf{v}|^2 - \mathbf{v} \wedge (\nabla \wedge \mathbf{v})$ that

$$\mathbf{K} = (\nabla \wedge \hat{\mathbf{s}}) \wedge \hat{\mathbf{s}} \tag{5.11}$$

From Eq. (5.7), we obtain that $\nabla \wedge (n\ \hat{\mathbf{s}}) = 0$ and hence,

$$n\ \nabla \wedge \hat{\mathbf{s}} = \hat{\mathbf{s}} \wedge \nabla n \tag{5.12}$$

Therefore,

$$\mathbf{K} = \frac{1}{n}\ (\hat{\mathbf{s}} \wedge \nabla n) \wedge \hat{\mathbf{s}} = \frac{1}{n}[\nabla n - (\hat{\mathbf{s}} \cdot \boldsymbol{\nabla}\mathbf{n})\hat{\mathbf{s}}] \tag{5.13}$$

which is the fractional component of the refractive index gradient perpendicular to the path in the osculating plane. The principal radius of curvature, therefore, lies in the osculating plane perpendicular to the path. The curvature is, therefore, the magnitude of the vector **K**, namely,

$$K = |(\nabla \wedge \hat{\mathbf{s}}) \wedge \hat{\mathbf{s}}| = \frac{1}{n}|\hat{\mathbf{s}} \wedge \nabla \mathbf{n}| = \frac{1}{n}|\nabla n|\ \sin\alpha \tag{5.14}$$

where α is the angle between the ray and the refractive index gradient. From this result, we easily deduce several important results of elementary geometrical optics:

1. *Straight-line propagation*: In a uniform medium, the refractive index n is constant. Therefore, $K = 0$, i.e. the curvature of ray is zero, and the ray path a straight line.
2. *Snell's law*: If the medium is layered with parallel strata, e.g. a plasma with continuously varying refractive index, then defining a length r parallel to ∇n, so that $\mathrm{d}s = \mathrm{d}r/\cos\alpha$ and remembering $K = \mathrm{d}\alpha/\mathrm{d}s$, we obtain

$$\cos\alpha\ \frac{\mathrm{d}\alpha}{\mathrm{d}r} = \frac{1}{n}\frac{\mathrm{d}n}{\mathrm{d}r}\ \sin\alpha \tag{5.15}$$

Integrating we obtain the desired result.

$$n\ \sin\alpha = \text{const} = n_0\ \sin\alpha_0 \tag{5.16}$$

where n_0 and α_0 the refractive index and angle of incidence of the incoming ray.

---

1 The *osculating plane* to a curve at a specified point is the plane which touches the curve at that point. It is defined by three limiting points on the curve in the neighbourhood of the given point. The *principal normal* is the normal to the curve lying in the osculating plane.
The *curvature* of a curve is the rate of change of the angle of the tangent with respect to the distance along the curve, i.e. $K = \mathrm{d}\alpha/\mathrm{d}s$, where α is the angle the tangent makes with a fixed axis and s the distance along the arc. The reciprocal of the curvature is the *principal radius of curvature* directed along the principal normal. The principal radius of curvature is the radius of the *osculating circle*, namely the circle lying in the osculating plane which touches the curve at the specified point.
Formally, if $\hat{\mathbf{t}}$ is a unit vector along the tangent, then

$$\frac{\mathrm{d}\hat{\mathbf{t}}}{\mathrm{d}s} = K\ \hat{\mathbf{n}} \tag{5.9}$$

where $\hat{\mathbf{n}}$ is the unit vector along the normal and K the principal curvature (see for example Lowry and Hayden (1957, p. 207)). The plane of $\hat{\mathbf{t}}$ and $\hat{\mathbf{n}}$ is the osculating plane.

3. *Corollary to Snell's law*: This result has the useful corollary. The wave vector **k** may be resolved into components $k_{\|} = k \cos\alpha$ and $k_{\perp} = k \sin\alpha$ with respect to the density stata. Since the frequency of the wave ω is constant, it follows that $k_{\perp} = k_{\perp 0}$, where $\mathbf{k}_0$ is the wave vector of the incoming ray.
4. *Total internal reflection*: In particular, if the incoming ray is normal to the refractive index gradient $\alpha = 0$, then the ray is reflected at the surface, where the refractive index is zero $n = 0$. This case treated in more detail in Section 5.4.1.
5. *'Parabolic path'*: Generally, if the ray is incident obliquely at an angle α_0, then the ray is bent towards or away from the normal. In particular, if the refractive index is decreasing the ray will follow a concave ('parabolic') path, and is reflected when $\alpha = \pi/2$, i.e. $k_{\|} = \sqrt{k^2 - {k_{\perp}}^2} = 0$ or $n = n_0 \,|\sin\alpha_0|$.

The speed at which the eikonal surface propagates is known as the *phase velocity* $V_p = \omega/k$: the direction of propagation being along the wave vector, normal to the eikonal surfaces. The velocity of energy propagation is given by the *group velocity* (Section 5.5).

5.3 The WKB Approximation

We may extend the geometrical optics approximation to investigate the propagation of a wave using the WKB (Wentzel–Kramers-Brillouin) approximation. The WKB method is usually described as an approximation to the wave equation, but equally can be regarded as a development of the geometrical optics limit. For simplicity, we consider the (Helmholtz) scalar wave equation, which is used to study steady-state propagation in an isotropic, but not necessarily homogeneous, medium within physical optics:

$$\nabla^2 U + k^2 U = 0 \tag{5.17}$$

where U is the wave amplitude function and $k = n\,k_0$ the wave number in the medium ($k_0 = c/\omega$ being the free space wavenumber). Limiting ourselves to one dimension (z) and to waves propagating in the z direction only, and as a trial solution, we may try a modified form of Eq. (5.3) $U = U_0 \exp\{\pm\imath\, k_0\, n\, z\}$ allowing forward and backward waves. Treating U_0 as a constant and substituting in the Helmholtz equation (5.2):

$$\left(\frac{\mathrm{d}S}{\mathrm{d}z}\right)^2 = {k_0}^2\, n^2 + \imath \frac{\mathrm{d}^2 S}{\mathrm{d}z^2} \tag{5.18}$$

If n is a slowly varying function of z, then it follows from the assumption $S \approx k_0\, n\, z$ that the last term is small. As a result, we obtain

$$\frac{\mathrm{d}S}{\mathrm{d}z} \approx \pm k_0\, n \qquad \text{and} \qquad \frac{\mathrm{d}^2 S}{\mathrm{d}z^2} \approx \pm k_0 \frac{\mathrm{d}n}{\mathrm{d}z} \tag{5.19}$$

consistent with Eq. (5.5). Clearly, the condition of slow variation is $|\mathrm{d}n/\mathrm{d}z|/(k_0\, n)^2 \ll 1$, i.e. the changes in refractive index take place on a scale length much greater than the wavelength: similarly to the geometrical optics approximation.

To obtain the next level of approximation, we use the approximate value of $\mathrm{d}^2S/\mathrm{d}z^2$ in Eq. (5.18) to obtain

$$\frac{\mathrm{d}S}{\mathrm{d}z} \approx \pm\left({k_0}^2\, n^2 \pm \imath\, k_0 \frac{\mathrm{d}n}{\mathrm{d}z}\right)^{1/2} \approx \pm k_0\, n + \frac{\imath}{2n}\frac{\mathrm{d}n}{\mathrm{d}z} \tag{5.20}$$

which is easily integrated to give

$$S \approx \pm k_0 \int^z n \; dz + \imath \; \log(n^{1/2}) \tag{5.21}$$

and hence, obtain the *WKB approximation* for the wave function:

$$U \approx U_0 \; n^{-1/2} \; \exp\left(\pm \imath \; k_0 \int^z n \; dz\right) \tag{5.22}$$

Substituting in the Helmholtz equation it is easily seen that this solution is valid if

$$\left(\frac{1}{k_0}\right)^2 \left|\frac{3}{4}\left(\frac{1}{n}\frac{dn}{dz}\right)^2 - \frac{1}{2n}\frac{d^2n}{dz^2}\right| \ll 1 \tag{5.23}$$

It is obvious that the WKB approximation always fails near a reflection, where $n \approx 0$, however, small the refractive index gradient.

5.3.1 Oblique Incidence

Thus far we have applied the WKB approximation to the most simple case of wave propagation parallel to the refractive index gradient. However, we may make use of the corollary of Snell's law, namely, that the component of the wavevector $\mathbf{k}$ perpendicular to the refractive index gradient, namely, $k_\perp = k_0 \; \sin\alpha_0$, is invariant. As a result since the phase term contains the optical path length increment:

$$\mathbf{k} \cdot \mathbf{ds} = k_z \; dz + k_x \; dx = \sqrt{k^2 - {k_\perp}^2} \; dz + k_\perp \; dx \tag{5.24}$$

where x is the distance along the ray in the plane of incidence perpendicular to the density gradient. Since $k_\perp$ is constant, the solution is separable into a term varying with z and a phase term along x, namely, $U(x, z) = U'(z) \; \exp(\imath \; k_\perp \; x)$ where the term $U'(z)$ satisfies

$$\frac{d^2U'(z)}{dz^2} + (k(z)^2 - {k_\perp}^2) \; U'(z) = \frac{d^2U'(z)}{dz^2} + {k_0}^2\zeta(z)^2 \; U'(z) = 0 \tag{5.25}$$

where $\zeta^2 = (n^2 - \sin^2\alpha_0)$. This is the one-dimensional Helmholtz equation whose WKB solution we have already obtained but with the replacement $k(z) \to k_0 \; \zeta(z)$. The WKB approximation in this case is, therefore,

$$U \approx U_0 \left[\frac{\zeta(0)}{\zeta(z)}\right]^{1/2} \exp\left[\pm \imath \left(k_0 \int^z \zeta(z') \, dz'\right) + \imath \; k_\perp \; x\right] \tag{5.26}$$

subject to the condition on the incoming ray at entry $U = U_0$ when $n = n_0$ and $x = z = 0$.

The WKB solution for oblique incidence is valid if

$$\frac{1}{({k_0}^2\zeta(z)^2)} \left|\frac{3}{4}\left(\frac{1}{\zeta(z)}\frac{d\zeta(z)}{dz}\right)^2 - \frac{1}{2\;\zeta(z)}\frac{d^2\zeta(z)}{dz^2}\right| \ll 1 \tag{5.27}$$

Since the wave is reflected when $\zeta(z) = 0$, it can again be seen that the WKB approximation fails near the surface of reflection.

5.4 Singularities in Waves

Two important singularities of the WKB solutions occur when the refractive index becomes either very small or very large. The case $n \to 0$ corresponds to a plasma wave resonance with a plasma oscillation, e.g. $\omega = \Pi_e$, and to total internal reflection at a solid interface. The case $n \to \infty$ occurs in plasma with magnetic fields, for example at the electron cyclotron resonance. Examples of these cases in magnetised plasma are discussed in Section 11.2.1.

- When $n \to 0$ the phase velocity becomes very large and the simple solution based on the propagation of the waves fronts using geometrical optics becomes un-physical. In this case, the wave is prevented from further propagation. The refractive index becomes imaginary and the wave evanescent. The singular point is known as a *cut-off* or *turning point*. The wave is reflected at the turning point. This phenomenon, familiar from optics, is simply *total internal reflection* at the boundary of an optically dense medium with a less dense one.
- When $n \to \pm\infty$, the phase velocity becomes zero. In this case, the frequency matches the natural frequency of one of the modes of oscillation in the medium. The optical analogue in this case is the absorption and emission of light by electronic transitions in the medium.

We will examine both phenomena as solutions of the Helmholtz equation by considering an incoming plane parallel wave incident from a region where the refractive index is unity. Far from the singularity, the propagating wave can be determined by the WKB method. In the neighbourhood of the singularity, the direct solution of the Helmholtz equation must be obtained, and using asymptotic expansions matched to incoming and outgoing waves in the region of propagation $n^2 > 0$, and subject to the condition that the wave is well behaved in evanescent zone $n^2 < 0$.

5.4.1 Cut-off or Turning Point

It is clear that both the geometrical optics limit and the WKB approximation fail as $n \to 0$ and the wave length becomes large. We consider the behaviour of a wave travelling parallel to the gradient in the refractive index from $z = -\infty$, where the refractive index is unity to a turning point at $z = 0$. Away from the turning point, the wave and its reflection are described by the WKB approximation:

$$U = U_0\, n^{-1/2} \left[\exp\left(\imath\, k_0 \int_0^z n\, \mathrm{d}z\right) + R\, \exp\left(-\imath\, k_0 \int_0^z n\, \mathrm{d}z\right)\right] \tag{5.28}$$

Thus, we have an incoming wave travelling parallel to z and an outgoing wave travelling anti-parallel to z with complex amplitude reflectivity at R at $z = 0$.

Let us suppose that in the neighbourhood of the cut-off point, where $n \to 0$ the refractive index can be written as

$$n^2 = -a\, z + O(z^2) \tag{5.29}$$

where $a > 0$ is a positive linear coefficient. Clearly, if $z < 0$, the medium allows the wave to propagate, but if $z > 0$, n is imaginary and the wave becomes evanescent. The WKB approximation with this refractive index function takes the form:

$$U = U_0\, (k_0/a)^{1/6}\, \zeta^{-1/4} \left[\exp\left(\imath\, \frac{2}{3}\, \zeta^{3/2}\right) + R\, \exp\left(-\imath\, \frac{2}{3}\, \zeta^{3/2}\right)\right] \tag{5.30}$$

where $\zeta = ({k_0}^2\, a)^{1/3} z$ and R is the reflection coefficient. Clearly, this solution fails near the turning points when $\zeta \to 0$.

To understand the behaviour of waves in the neighbourhood of the turning point at $z = 0$, we must return to the basic wave equation (5.2). Substituting for z and n^2, we obtain

$$\frac{\mathrm{d}^2 U}{\mathrm{d}\zeta^2} - \zeta\, U = 0 \tag{5.31}$$

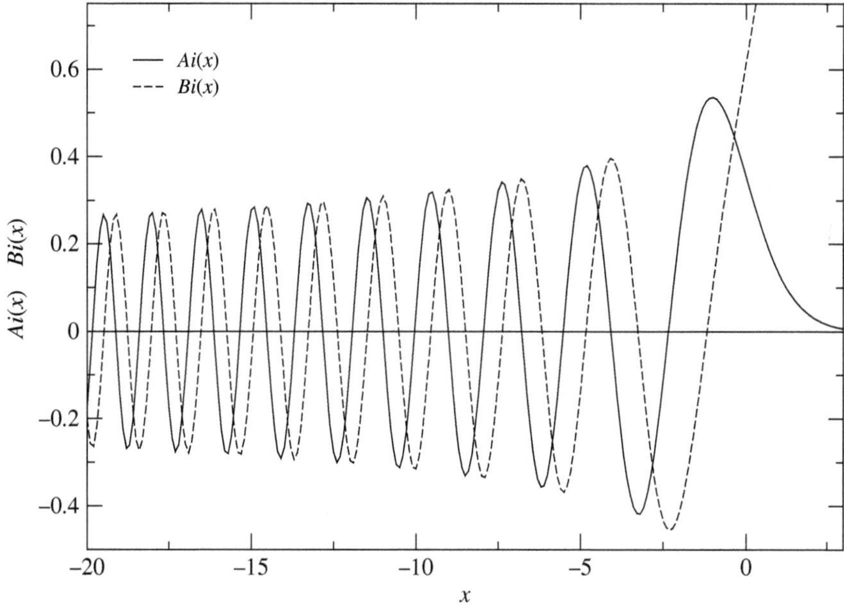

Figure 5.1 Plots of the Airy functions $Ai(x)$ and $Bi(x)$.

which is a standard form known as Airy's equation possessing two independent solutions $Ai(\zeta)$ and $Bi(\zeta)$, shown plotted in Figure 5.1. The Airy functions are given by the series expansions (Abramowitz and Stegun, 1965, p. 446):

$$Ai(z) = c_1\, f(z) - c_2\, g(z) \qquad \text{and} \qquad Bi(z) = \sqrt{3}\,[c_1\, f(z) + c_2\, g(z)] \tag{5.32}$$

where

$$f(x) = \frac{1}{3!}x^3 + \frac{1\cdot 4}{6!}x^6 + \frac{1\cdot 4\cdot 7}{9!}x^9 \cdots \qquad \text{and} \qquad g(x) = \frac{2}{4!}x^4 + \frac{2\cdot 5}{7!}x^7 + \frac{2\cdot 5\cdot 8}{10!}x^{10} \cdots \tag{5.33}$$

where $c_1 = 3^{2/3}/\Gamma(2/3) = 0.35503$ and $c_2 = 3^{-1/3}/\Gamma(1/3) = 0.25519$.

More useful for our purposes are the asymptotic expansions for large positive and negative x. For $x \to \infty$:

$$Ai(x) \approx \frac{1}{2}\,\pi^{-1/2}\,x^{-1/4}\,\exp\left(-\frac{2}{3}x^{3/2}\right) \quad \text{and} \quad Bi(x) \approx \pi^{-1/2}\,x^{-1/4}\,\exp\left(\frac{2}{3}x^{3/2}\right) \tag{5.34}$$

and for $x \to -\infty$

$$Ai(-x) \approx \pi^{-1/2}\,x^{-1/4}\,\sin\left(\frac{2}{3}x^{3/2} + \frac{\pi}{4}\right) \quad \text{and} \quad Bi(-x) \approx \pi^{-1/2}\,x^{-1/4}\,\cos\left(\frac{2}{3}x^{3/2} + \frac{\pi}{4}\right) \tag{5.35}$$

In terms of the Airy functions, the solution of Eq. (5.31) is

$$U = U_1\, Ai(\zeta) + U_2\, Bi(\zeta) \tag{5.36}$$

subject to boundary conditions imposed the input wave (5.30). These are simply expressed by the incoming and outgoing waves from $z = -\infty$, and the condition $U = 0$ at $z = \infty$ of the evanescent

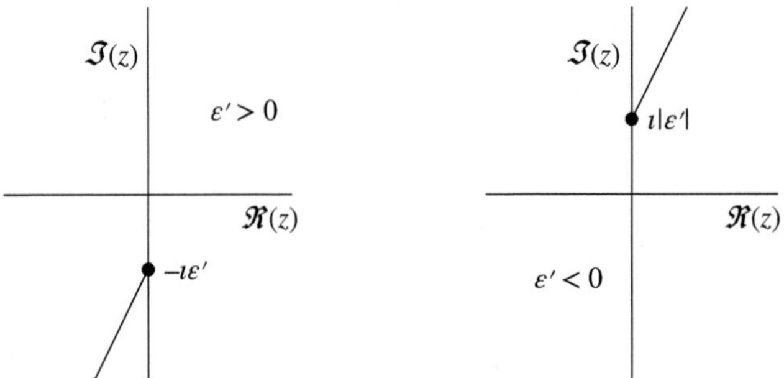

Figure 5.2 Sketch of the branch cuts (thick lines) needed to ensure a well-behaved solution at the resonance point.

wave. From the asymptotic forms of the Airy function for large positive values (5.34), $Bi(\zeta) \to \infty$ as $\zeta \to \infty$ and it is clear that $U_2 = 0$. Furthermore, using the asymptotic form of $Ai(\zeta)$ for $\zeta > 0$ (5.35):

$$U \approx \frac{U_1}{2}\sqrt{\frac{\iota}{\pi}}\left[\zeta^{1/4}\exp\left(\iota\,\frac{2}{3}\zeta^{3/2}\right) - \iota\,\zeta^{1/4}\exp\left(-\iota\,\frac{2}{3}\zeta^{3/2}\right)\right] \tag{5.37}$$

Hence, $U_0 = -\frac{1}{2}\sqrt{\iota/\pi}(a/k_0)^{1/6}\,U_1$ and $R = -\iota$. The wave is, therefore, totally reflected with no change in amplitude, but with a phase change of $(\pi/2)$.

5.4.2 Resonance Point

At a resonance point, the refractive index becomes infinite. We simulate this condition by considering a medium, where $n^2 = -b/(z + \iota\epsilon) + O(1)$, where $b > 0$ and ϵ is a small quantity, which is included to avoid the singularity on the real axis of z, and will be allowed to approach 0. In physical terms, the imaginary part of the refractive index corresponds to either attenuation or amplification of the wave. We shall find that ϵ represents either damping or gain of the incoming wave depending on the sign of ϵ. With this form of the refractive index, the Helmholtz equation becomes

$$\frac{d^2U}{dz^2} + \frac{k_0^2\, b}{z + \iota\,\epsilon}\, U = 0 \tag{5.38}$$

which we simplify to

$$\frac{d^2U}{d\zeta^2} + \frac{U}{\zeta + \iota\,\epsilon'} = 0 \tag{5.38:a}$$

where $\zeta = -k_0{}^2\, b\, z$ and $\epsilon' = -k_0{}^2\, b\, \epsilon$.[2]

Equation (5.38:a) has a pole (or more correctly a branch point) at $\zeta = -\iota\,\epsilon'$. Consequently, the solution is multi-valued and not an entire function.[3] This may be avoided by a branch cut from the pole to a point at infinity. Since we require the solution on the real axis $\Im(z) = 0$ to be continuous, the branch cut must be chosen to avoid the real axis as illustrated in Figure 5.2. Taking the argument of z on the real axis for positive $\Re(z) > 0$ to be 0, and avoiding the branch cut for

2 The change of sign, corresponding to a reversal of the direction of propagation, makes the calculation becomes simpler.

3 A short introduction to the theory of complex functions is given in the supplementary material Section M.3 to which reference should be made.

negative $\Re(z) < 0$, we proceed along the real axis with $\Im(z) > 0$ if $\epsilon' > 0$ and *vice versa* if $\epsilon' < 0$ (see Figure 5.2). Therefore,

$$\arg z = \begin{cases} +\pi & \text{for } \epsilon' > 0 \\ -\pi & \text{for } \epsilon' < 0 \end{cases}$$

To proceed define

$$\eta = 2\sqrt{\zeta} \qquad \text{and} \qquad U = \eta\ \Psi(\eta) \tag{5.39}$$

In the limit, $\epsilon' \to 0$ the Helmholtz equation becomes

$$\frac{d^2\Psi}{d\eta^2} + \frac{1}{\eta}\frac{d\Psi}{d\eta} + \left(1 - \frac{1}{\eta^2}\right)\Psi = 0, \tag{5.40}$$

which is Bessel's equation of order one, and whose solution can be written in terms of the two independent solutions $J_1(\eta)$ and $Y_1(\eta)$ (Abramowitz and Stegun, 1965, p. 358), namely,

$$U = U_1\ \sqrt{(-\zeta)}\, J_1(2\sqrt{\zeta}) + U_1'\ Y_1(2\sqrt{\zeta}) \tag{5.41}$$

where U_1 and U_1' are two constants to be determined by the boundary conditions. Since $\sqrt{\zeta}$ is only real when z is real and negative, this represents the solution using Bessel's equation for real arguments only if $\Re(z) < 0$.

For positive values of $\Re(z)$, the argument of the Bessel functions becomes imaginary. However, since Bessel functions are well defined over the entire complex plane this presents no problem, apart from dealing with the branch cut. Introducing the variables,

$$\eta' = 2\sqrt{\alpha\,\zeta} \qquad \text{and} \qquad U = \eta'\ \Psi'(\eta') \tag{5.42}$$

where $\alpha = \exp[-\imath\ \pi\ \mathrm{sgn}(\epsilon')]$, so that for negative real values of ζ the argument of η is zero. As $\epsilon' \to 0$ Helmholtz' equation transforms to

$$\frac{d^2\Psi'}{d\eta'^2} + \frac{1}{\eta'}\ \frac{d\Psi'}{d\eta'} - \left(1 + \frac{1}{\eta'^2}\right)\Psi' = 0 \tag{5.43}$$

whose solution we require for real η. This is the modified Bessel's equation with two independent solutions $I_1(\eta)$ and $K_1(\eta)$ (Abramowitz and Stegun, 1965, p. 374). These are the solutions to Bessel's equation for imaginary arguments referred to earlier. Clearly, since these are also entire functions, the solution in terms of I_1 and K_1 must be completely equivalent to that in terms of J_1 and K_1:

$$U = U_2\ \sqrt{\alpha\,\zeta}\ I_1(2\sqrt{\alpha\,\zeta}) + U_2'\ \sqrt{\alpha\,\zeta}\ K_1(2\sqrt{\alpha\,\zeta}) \tag{5.44}$$

To find the relationship between the two solutions, we may use the first terms in the power series expansions for J_1 and Y_1 and I_1 and K_1 (Abramowitz and Stegun, 1965, pp. 360 & 375), namely,

$$\sqrt{\zeta}\ J_1(2\ \sqrt{\zeta}) = \zeta + O(z^2) \tag{5.45a}$$

$$\sqrt{\alpha\,\zeta}\ I_1(2\ \sqrt{\alpha\,\zeta}) = -\zeta + O(\zeta^2) \tag{5.45b}$$

$$\sqrt{\zeta}\ Y_1(2\ \sqrt{\zeta}) = -\frac{1}{\pi}\ \{1 - [\ln(\zeta) + 2\gamma - 1]\zeta\} + O(\zeta^2) \tag{5.45c}$$

$$\sqrt{\alpha\,\zeta}\ K_1(2\ \sqrt{\alpha\,\zeta}) = \frac{1}{2}\ \{1 - [\ln(|\zeta|) + 2\gamma - 1]\ \zeta - \imath\ \arg(\alpha)\ \zeta\} + O(\zeta^2), \tag{5.45d}$$

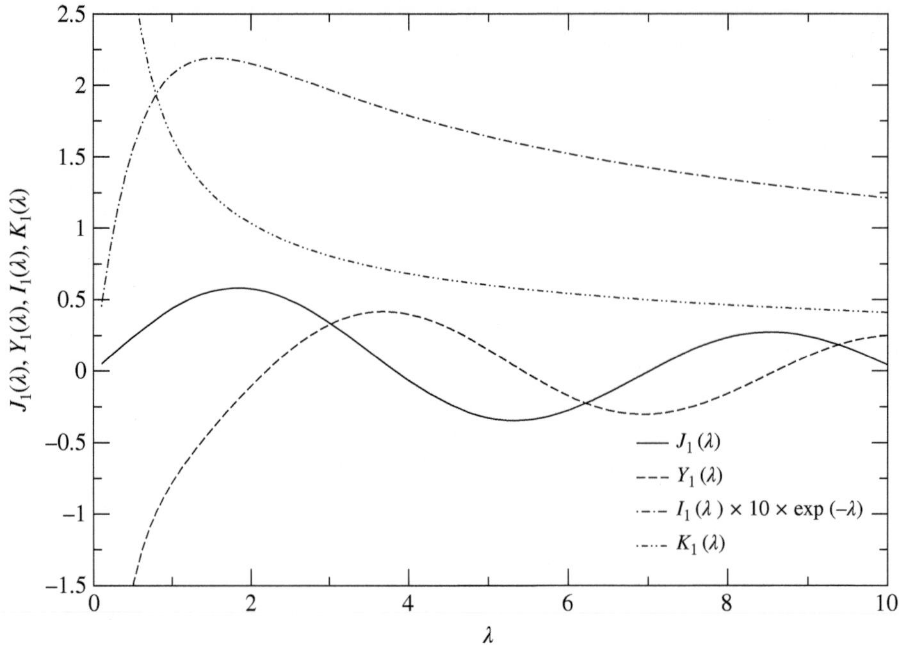

Figure 5.3 Plots of the Bessel functions $J_1(x)$, $Y_1(x)$, $I_1(x)$ and $K_1(x)$ for real arguments x.

where $\gamma = 0.577216$ is Euler's constant and $\zeta > 0$ is real. Retaining only terms of order ζ, we obtain a consistent solution:

$$U_2 = -U_1 + \imath\, \frac{\pi}{2}\, \mathrm{sgn}(\epsilon')\, U_2' = -U_1 - \imath\, \mathrm{sgn}(\epsilon')\, U_1' \tag{5.46a}$$

$$U_2' = -\frac{2}{\pi}\, U_1', \tag{5.46b}$$

ensuring the Eqs. (5.41) and (5.44) are equivalent (Figure 5.3).

This rather cumbersome solution of the Helmholtz equation near a resonance has the merit that we only require the asymptotic forms for real arguments to match the boundary conditions imposed by the incoming wave $\zeta > 0$ and the vanishing solution in the evanescent region $\zeta < 0$. For large negative values of z (positive values of ζ) corresponding to the propagating waves (Abramowitz and Stegun, 1965, p. 364):

$$\begin{aligned} \sqrt{\zeta}\, J_1(2\sqrt{\zeta}) &\approx \frac{\zeta^{1/4}}{\sqrt{\pi}} \cos\left(2\sqrt{\zeta} - \tfrac{3}{4}\,\pi\right) \\ \sqrt{\zeta}\, Y_1(2\sqrt{\zeta}) &\approx \frac{\zeta^{1/4}}{\sqrt{\pi}} \sin\left(2\sqrt{\zeta} - \tfrac{3}{4}\,\pi\right) \end{aligned} \tag{5.47}$$

and for large positive values of z (Abramowitz and Stegun, 1965, p. 377):

$$\begin{aligned} \sqrt{\alpha\,\zeta}\, I_1(2\sqrt{\alpha\,\zeta}) &\approx \frac{|\zeta|^{1/4}}{2\sqrt{\pi}} \exp\left(2\sqrt{|\zeta|}\right) \\ \sqrt{\alpha\,\zeta}\, K_1(2\sqrt{\alpha\,\zeta}) &\approx \frac{\sqrt{\pi}\,|\zeta|^{1/4}}{2} \exp\left(-2\sqrt{|\zeta|}\right) \end{aligned} \tag{5.48}$$

Since the amplitude U of the wave must remain finite as $z \to \infty$, it is clear that $U_2 = 0$ and that in consequence $U_1 = -\imath\, \mathrm{sgn}(\epsilon) U_2'$. Hence, we obtain the final general solution for the propagating waves $z < 0$:

$$U(\zeta) = U_0 \left\{ [\mathrm{sgn}(\epsilon) + 1]\zeta^{1/4} \exp\left[+\imath\left(2\sqrt{\zeta} - \frac{3}{4}\pi\right)\right] \right.$$
$$\left. + [\mathrm{sgn}(\epsilon) - 1]\zeta^{1/4} \exp\left[-\imath\left(2\sqrt{\zeta} + \frac{3}{4}\pi\right)\right] \right\} \tag{5.49}$$

It is easily seen that since the refractive index $n = \sqrt{b/(-z)}$ and $k_0 \int n \, dz = 2\sqrt{\zeta}$ that the WKB solution can be written as

$$U = (k_0\, b)^{-1/2}[U_0' \, \zeta^{-1/4} \exp(-\imath\, 2\, \sqrt{\zeta}) + U_0'' \, \zeta^{-1/4} \exp(+\imath\, 2\, \sqrt{\zeta})] \tag{5.50}$$

These two terms are, respectively, the incoming and outgoing waves. In particular, if $\epsilon < 0$, there is no return wave and the incoming wave is totally absorbed by the resonance. On the other hand, if $\epsilon > 0$, there is no incoming wave, but an outgoing one corresponding to spontaneous emission. In both these cases, the wave is responding to a natural oscillation in the medium.

It is noteworthy that in the case of absorption, there is no account of the mechanism (e.g. collisions) by which the medium can disperse the energy out of the wave into other modes. Consequently, the medium accepts all the energy delivered to it by the incoming wave independent of the ability of resonant wave to dissipate it. As a result, the intensity of the latter may become sufficiently large that non-linear behaviour sets in and causes breakup of the resonance – for example wave breaking in cold plasma discussed in Section 1.2.1.

5.4.3 Resonance Layer and Collisional Damping

In the neighbourhood of the resonance, we have assumed that the refractive index includes an imaginary term through $n^2 = -b/(z + \imath\, \epsilon)$. Anticipating some of the subsequent discussion of the refractive index of electro-magnetic waves in a dielectric medium, Section 13.1, we consider a plane transverse sinusoidal wave travelling in the z direction with electric field in the y direction and magnetic field in x. The energy absorbed per unit area per unit time from the wave is given by the rate of decrease of the time averaged Poynting vector P_z along the propagation direction z (see Footnote 4, page 323)

$$P_z = \frac{(E_y B_x^* + E_y^*\, B_x)}{4\, \mu_0} \tag{5.51}$$

where $\mathrm{d}E_y/\mathrm{d}z = -\imath\, \omega\, B_x$ from the Maxwell induction equation. For electro-magnetic waves, the refractive index n^2 is equal to the relative permittivity and consequently,

$$\frac{\mathrm{d}^2 E_y}{\mathrm{d}z^2} + \frac{{k_0}^2\, b}{z + \imath\, \epsilon} E_y = 0 \tag{5.52}$$

Substituting for $\mathrm{d}^2 E_y/\mathrm{d}z^2$ we obtain

$$W = -\frac{\imath}{4\, \mu_0\, \omega} \left(\frac{\mathrm{d}^2 E_y}{\mathrm{d}z^2} E_y^* - \frac{\mathrm{d}^2 E_y^*}{\mathrm{d}z^2} E_y \right) = -\frac{{k_0}^2\, b}{2\, \mu_0\, \omega} \frac{\epsilon}{(z^2 + \epsilon^2)} |E_y|^2 \tag{5.53}$$

where $|E_y|^2/4\, \mu_0$ is the averaged energy density of the wave. The resonance, therefore, has a finite spatial width ϵ over which the absorption takes place.

In the neighbourhood of a resonance, the frequency of the wave ω matches one of the characteristic frequencies of the medium Ω. Implicit in the assumption that the refractive index varies as $n^2 = -b/(z + \imath\, \epsilon)$ near the resonance is the condition that the resonance frequency itself varies along the ray path, i.e. $\Omega = \omega + \Omega' z$. From Eqs. (13.20), (13.15) and (2.51), the refractive index typically has the form:

$$n^2 = 1 + \frac{\Pi^2}{\omega\,(\omega - \Omega - \imath\,\nu)} \tag{5.54}$$

where ν is the collision frequency. Hence, $b = \Pi^2/\omega\,\Omega'$ and $\epsilon = -\nu/\Omega'$. The width of the resonance is consequently,

$$|\epsilon| \approx \nu/\Omega' \tag{5.55}$$

which determines the frequency width of the resonance $\Omega'\,|\epsilon| \approx \nu$, but not the fraction of the energy absorbed, which in this model is 100%.

5.5 The Propagation of Energy

The behaviour of waves in anisotropic media is extensively studied in the optics of birefringent materials (Born and Wolf, 1965; Landau and Lifshitz, 1984; Sommerfeld, 1964). However, the behaviour described is much more general, applicable to mechanical systems involving fluids and gases as well as plasma. The subject was originally treated by Rayleigh (1896) and, subsequently, for many types of waves in fluids by Lighthill (1978), to which reference may be made for a wide general discussion.

The propagation of the maximum signal amplitude of an electro-magnetic wave in anisotropic media is examined from two different, but equivalent, viewpoints, either by considering a wave group, *group velocity* or as an envelope of the surfaces of constant phase known as the *ray surface*. In each case, the peak amplitude arises due to the coherent interference of different wave elements. The equivalence of the group velocity with the energy transport velocity is demonstrated by two methods: the first due to Brillouin (1960) treats electro-magnetic waves in an isotropic medium, and the second following Lighthill (1960) considers the anisotropic mechanical behaviour of the waves in a very general fluid medium.

5.5.1 Group Velocity of Waves in Dispersive Media

In a transparent medium, the propagation of a quasi-harmonic wave, i.e. one which is a pulse of radiation and whose frequency spread $\Delta\omega$ is small ($\delta\omega \ll \omega$) is calculated from the wave equations and the properties of the medium. However, in practice, this is circumvented by the use of the dispersion relation, which by separately calculating the frequency response of the medium at the particular wave number, automatically treats the details of the interaction of each element of the wave with the medium, provided the medium is linear, i.e. the amplitude of the wave is small. As a result, the general wave may be written as a Fourier term:

$$\int \mathbf{A}(\mathbf{k}, \omega)\; e^{[\imath(\mathbf{k}\cdot\mathbf{r} - \omega(\mathbf{k})t)]}\; d\mathbf{k} \tag{5.56}$$

where $\omega(\mathbf{k})$ is the frequency of the wave with wave number $\mathbf{k}$ permitted by the dispersion relation.

Consider a pulse formed by a group of waves within a narrow band in wave number and frequency about the central values $\mathbf{k}_0$ and $\omega_0(\mathbf{k}_0)$ so that $\mathbf{k} = \mathbf{k}_0 + \mathbf{k}'$ and $\omega = \omega_0 + \omega'$ and $\mathbf{A}(\mathbf{k}, \omega)$ is narrow peaked about $\mathbf{k}_0, \omega_0$. If the pulse has only a narrow frequency spread

$$\omega' \approx \frac{\partial \omega}{\partial \mathbf{k}} \cdot \mathbf{k}' \tag{5.57}$$

The general wave becomes

$$e^{[\imath(\mathbf{k}_0\cdot\mathbf{r}-\omega_0 t)]}\int \mathbf{A}(\mathbf{k},\omega)\, e^{[\imath(\mathbf{k}'\cdot\mathbf{r}-\omega'(\mathbf{k}')t)]}\,\mathrm{d}\mathbf{k}' = e^{[\imath(\mathbf{k}_0\cdot\mathbf{r}-\omega_0 t)]}\int \mathbf{A}(\mathbf{k},\omega)\, e^{[\imath\mathbf{k}'\cdot(\mathbf{r}-\partial\omega/\partial\mathbf{k}\, t)]}\,\mathrm{d}\mathbf{k}' \tag{5.58}$$

which represents a wave of wave-number $\mathbf{k}_0$ frequency ω_0 with varying amplitude. A reference point on the pulse waveform (for example the peak amplitude) moves with the steady velocity:

$$\mathbf{V}_g = \frac{\partial\omega}{\partial\mathbf{k}} \tag{5.59}$$

known as the *group velocity*.

5.5.2 Waves in Dispersive Isotropic Media

We consider the behaviour of electro-magnetic waves in an isotropic homogeneous medium, where the permittivity, and therefore, refractive index, vary as functions of the frequency $\epsilon \equiv \epsilon(\omega)$ of the radiation – the situation already discussed in Section 5.5.1. We consider the modification to Poynting's theorem on the transfer of energy as a result of dispersion and show that the energy transfer velocity defined as the ratio of the energy flux to the energy density is identical to the signal (or group velocity).

As in Section 5.5.1, we consider the response to a pulse of radiation within a narrow band of frequencies ($\Delta\omega \ll \omega$). Let the electric field in the pulse be described by

$$E(\omega) = E_0\, g(\omega)\, e^{\imath\,\omega\,t} + c.c. \tag{5.60}$$

where *c.c.* is the complex conjugate and $g(\omega)$ specifies the frequency profile of the radiation.

Referring back to Eq. (2.72), we see that the electric energy W_E is determined by the term:

$$\frac{\partial W_E}{\partial t} = \mathbf{E}\cdot\partial\mathbf{D}/\partial t \tag{5.61}$$

Averaging over times large compared to the period (see Footnote 4, page 286), we require the term $\mathbf{E}^*\mathbf{D}$. Integrating over the frequency profile, we require terms:

$$\int \mathbf{E}^*(\omega')\cdot\frac{\partial}{\partial t}\mathbf{D}(\omega)\,\mathrm{d}\omega\,\mathrm{d}\omega' = \int \mathbf{E}^*(\omega')\cdot\frac{\partial}{\partial t}[\epsilon(\omega)\,\mathbf{E}(\omega)]\,\mathrm{d}\omega\,\mathrm{d}\omega'$$
$$= \int \imath\,\omega\,\epsilon(\omega')\,\mathbf{E}(\omega')\mathbf{E}^*(\omega)\,\mathrm{d}\omega\,\mathrm{d}\omega' \tag{5.62}$$

Noting that the frequency variation about the central value (ω_0) is small, we write $\omega = \omega_0 + \Omega$, so that

$$\omega\,\epsilon(\omega) \approx (\omega_0+\Omega)\,(\epsilon|_{\omega_0}0 + \mathrm{d}\epsilon/\mathrm{d}\omega|_{\omega_0}\Omega) \approx \omega_0\,\epsilon|_{\omega_0} + \Omega\,(\epsilon|_{\omega_0} + \mathrm{d}\epsilon/\mathrm{d}\omega|_{\omega_0}) \tag{5.63}$$

Since the field initially is assumed to be zero, we may integrate over time and average over the wave period to obtain the mean electrical energy density:

$$W_E = \frac{1}{2}\mathbf{E}_0\,\mathbf{E}_0^*\,(\epsilon|_{\omega_0} + \mathrm{d}\epsilon/\mathrm{d}\omega|_{\omega_0})\int_0^t \mathrm{d}t\int_{-\infty}^{\infty}\mathrm{d}\Omega\int_{-\infty}^{\infty}\mathrm{d}\Omega' g(\Omega)\,g^*(\Omega')\exp[\imath\,(\Omega-\Omega')\,t] + c.c. \tag{5.64}$$

Fortunately, we do not need to evaluate these integrals, as if the permittivity is constant $\partial\epsilon/\partial\omega|_{\omega_0} = 0$, and we revert to a dissipationless medium, where the average energy density $\frac{1}{4}\,\epsilon|_{\omega_0}\,{E_0}^2$. Therefore, the mean energy density in an isotropic, dispersive medium without absorption is given by Brillouin (1960)

$$W_E = \frac{1}{4}\,\left(\epsilon|_{\omega_0} + \omega_0\frac{\partial\epsilon}{\partial\omega}\bigg|_{\omega_0}\right)\,{E_0}^2 \tag{5.65}$$

The derivative term does not appear in the dispersionless result (Section 2.3.1, see also Footnote 12 on page 32). It is interpreted as associated with the kinetic energy of charged particles in the Lorentz picture, which is a result of the field and whose energy varies with frequency (see Bers in Allis et al., 1963, pt 2). The averaged magnetic energy density we found earlier (see Section 2.3.1 item 2 on page 31) $W_B = \frac{1}{4}B^2/\mu_0$. The total average energy density:

$$W = W_E + W_B = \frac{1}{4}\left\{\frac{\partial(\omega\epsilon)}{\partial\omega}|_{\omega_0} + \epsilon|_{\omega_0}\right\} {E_0}^2 = \frac{1}{2}\, n\, \left\{n + \omega_0 \frac{\partial n}{\partial\omega}\right\} E_0/2 \tag{5.66}$$

where $n = \sqrt{\epsilon c\, k/\omega}$ is the refractive index.

The average energy flux is given by Poynting's vector (Section 2.3.1 item 3 on page 31) $S = \frac{1}{2}\epsilon {E_0}^2 = \frac{1}{2}\, n^2\, {E_0}^2$. Hence, the energy transport velocity

$$V_E = S/W = \frac{n}{\partial(\omega\, n)/\partial\omega} = \frac{\partial\omega}{\partial k} = V_g \tag{5.67}$$

Therefore, as expected, the energy transport velocity is equal to the group velocity V_g.

5.6 Group Velocity of Waves in Anisotropic Dispersive Media

This general expression for the group velocity encompasses the general case of anisotropic media, where the group velocity depends on the direction in which the ray propagates. Let us suppose that the system has a single axis (such as the ambient magnetic field) and that the propagation vector $\mathbf{k}$ is inclined at an angle θ to the axis. Noting that $\omega = kV_p(\theta)$ and writing the derivative in components we obtain

$$\mathbf{V}_g = \left.\frac{\partial\omega}{\partial k}\right|_\theta \hat{\mathbf{k}} + \frac{1}{k}\left.\frac{\partial\omega}{\partial\theta}\right|_k \hat{\theta} = \mathbf{V}_p + k\left.\frac{\partial V_p}{\partial k}\right|_\theta \hat{\mathbf{k}} + \left.\frac{\partial V_p}{\partial\theta}\right|_k \hat{\theta} \tag{5.68}$$

where $k = |k|$, $\hat{\mathbf{k}}$ is the unit vector parallel to the wave vector $\mathbf{k}$ and, therefore, also to the phase velocity V_p and $\hat{\theta}$ the unit vector orthogonal to $\hat{\mathbf{k}}$ in the plane of the axis and the wave vector $\mathbf{k}$. In this case of asymmetry with a single axis, as in magnetised plasma, it follows from Eq. (5.68) that the group velocity vector is co-planar with the axis and the propagation vector.

- The first term in this expression equals the phase velocity $\mathbf{V}_p$ and is parallel to the incident wave vector $\mathbf{k}$.
- The second term is due to constructive interference of successive waves of differing wave number overtaking their neighbours, and gives the familiar group velocity of an isotropic *dispersive* medium, when the phase velocity depends on the wave number k and is also parallel to the wave vector $\mathbf{k}$.
- The third term is due to the constructive interference of differing spatial elements on the wave front in an *anisotropic* dispersion-less medium, where the phase velocity depends on the direction θ of the propagation vector k, but *not on its magnitude* with respect to the symmetry axis.

Since the group velocity represents the velocity of the amplitude of a group of waves, it represents propagation of the energy by the wave due to the time averaged Poynting flux (Brillouin, 1960; Born and Wolf, 1965; Landau and Lifshitz, 1984; Sommerfeld, 1964). In geometrical optics, the propagation of light takes place along rays, whose path is described by the group velocity. The velocity of transport of the electro-magnetic energy flux of plane quasi-monochromatic waves, Eq. (2.73), namely, the averaged Poynting flux divided by the energy density, can be shown to be equal to the group velocity, using either Maxwell's description of the dispersion as a phenomenological property of the medium (Landau and Lifshitz, 1984; Ginzburg, 1970), or Lorentz's explicit mechanistic form (see Bers in Allis et al., 1963, pt 2). The polar plot of the group velocity is known as the *wave front* in plasma physics or the *ray surface* in optics. It represents the output wave from a point source (Figure 10.2) at unit time. The phase is, therefore, constant over the entire ray surface.

If the medium is anisotropic, the direction of propagation of the pulse is not collinear with the phase, i.e. $\mathbf{V}_{\text{group}} \nparallel \mathbf{V}_{\text{phase}}$. The group velocity defines the speed of propagation of the amplitude of the wave at a particular value of the wave vector ($\mathbf{k}$), but not its phase. The latter value is determined by the projection of the group velocity on to the wave vector $\mathbf{k}$, and its speed is equal to the phase velocity $\mathbf{V}_p(\mathbf{k}) = \omega/k\,\hat{\mathbf{k}}$ in the direction of the wave vector $\hat{\mathbf{k}}$ and whose value varies with $\hat{\mathbf{k}}$. Remembering that the phase of the ray is determined by the plane normal to the wave vector on which the phase is constant, we define the *ray vector* $\mathbf{s}$ whose direction is the group velocity and whose magnitude is defined by the phase relation $\mathbf{s} \cdot \mathbf{n} = 1$, where $\mathbf{n} = c\,\mathbf{k}/\omega$ is the *refractive index vector*. The vector $\mathbf{s} = \mathbf{V}_g/c$, then determines the group velocity. The polar plot of the refractive index vector $\mathbf{n}$ in $\mathbf{k}$ space defines the *wave normal surface*, and similarly the polar plot of the ray vectors $\mathbf{s}$ the *ray surface* in optics.

We consider the geometrical relationship of the ray surface with the wave normal surface on a polar plot, Figure 5.4. The phase of the wave is given by the phase change along a ray path of length ℓ

$$S = \int^{\ell} \mathbf{n} \cdot \mathbf{d}\boldsymbol{\ell} = \int^{\ell} \mathbf{n} \cdot \mathbf{s}\, \mathrm{d}\ell/s = \int^{\ell} \mathrm{d}\ell/s \tag{5.69}$$

Therefore, a set of rays of length a constant multiple of s or more simply s itself form a set of planar surfaces of constant phase, *phase planes*, normal to the refractive index vector $\mathbf{n}$ at the wave normal surface:

As with our previous calculation, it follows from the method of stationary phase that the peak of a pulse, and therefore, the ray in geometrical optics, occurs when multiple elements interfere constructively. In this context, neighbouring phase planes meet generating constructive interference at the ray. The ray surface is, therefore, the envelope of the phase planes and the wave normal

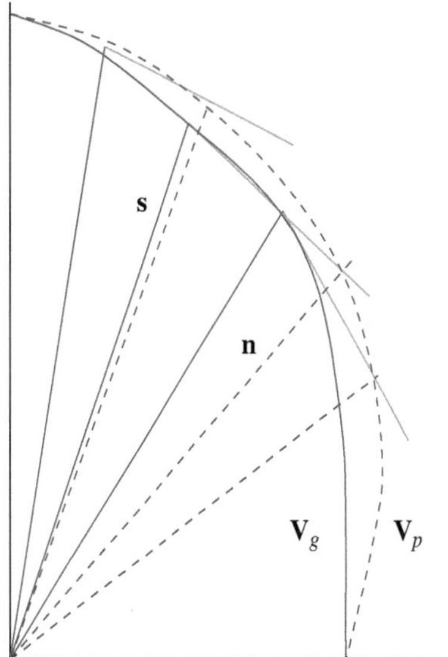

Figure 5.4 Sketch to illustrate the relationship between the normal surface (blue dashed) and the ray surface (red full). The normal planes at the phase points (green full) (**n**) being tangent to the ray surface at the ray points (**s**).

surface its evolute.[4] We now show that this is indeed the case consistent with Eq. (5.59), and that the ray surface is, therefore, the envelope of planes of constant phase, or equivalently that the normal surface is the pedal surface to the ray surface. Let $\mathbf{n}(\mathbf{k})$ be a point on the normal surface and $\mathbf{s}(\mathbf{k})$ the dual point on the ray surface. If $f(\mathbf{k}, \omega) = 0$ is the wave normal surface, we may write the group velocity:

$$\mathbf{V}_g = \frac{\partial \omega}{\partial \mathbf{k}} = -\frac{\partial f/\partial \mathbf{k}}{\partial f/\partial \omega} \tag{5.70}$$

If the phase velocity depends on the direction of propagation as in many cases in plasma physics, for example magneto-sonic waves (Section 10.2), or optical waves in axial crystals (Born and Wolf, 1965; Landau and Lifshitz, 1984; Sommerfeld, 1964), the group velocity $\mathbf{V}_g$ and. therefore, the ray vector $\mathbf{s}$ are parallel to the value of $\partial f/\partial \mathbf{k}$ calculated at $\mathbf{n}$. The latter term is the gradient of the function $f(\mathbf{k}, \omega)$ in $\mathbf{k}$ space at $\mathbf{n}$. It is the normal to the vector $\mathbf{k}$ on the wave normal surface $f(\mathbf{k}, \omega) = 0$ at that point.[5] The ray point $\mathbf{s}$ and the phase point $\mathbf{n}$, therefore, share a correspondence with each other, in that the ray vector $\mathbf{s}$ is parallel to the normal of the wave normal surface at the phase point $\mathbf{n}$.

Consider a small element $\delta\mathbf{n}$ which lies in the wave normal surface;it follows from the preceding result that $\mathbf{s} \cdot \delta\mathbf{n} = 0$. Since the phase on the ray surface at $\mathbf{s}$ is given by $\mathbf{k} \cdot \mathbf{s}$, the projection $\mathbf{s} \cdot \mathbf{n} =$ const over the ray surface. The small element $\delta\mathbf{n}$ corresponds to a small change in the ray vector $\delta\mathbf{s}$ lying in the ray surface, which therefore, satisfies $\mathbf{s} \cdot \delta\mathbf{n} = -\mathbf{n} \cdot \delta\mathbf{s} = 0$. Since $\delta\mathbf{s}$ is tangential to the ray surface, $\mathbf{n}$ is perpendicular to the ray surface. Consequently, the normal to the ray surface at $\mathbf{s}$ is parallel to the wave vector at $\mathbf{n}$, or equivalently, the tangent to the wave vector at $\mathbf{s}$ is perpendicular to the wave vector at $\mathbf{n}$, the latter being the plane of constant phase (Figure 5.4). Moving over the normal surface, we see that the ray surface is the envelope of the planes of constant phase. It can be shown (Born and Wolf, 1965, §14.2.1) that the ray defined in this way with the corresponding velocity represents the averaged Poynting vector of the corresponding electro-magnetic wave[6].

To illustrate this behaviour, we consider a simple case in which the value of the group velocity is easily calculated in a system with a single axis of symmetry, when the ray propagation vector lies in the plane of the axis and the wave vector. Using Cartesian co-ordinates $x,\ y,\ z$, where the z axis is parallel to the symmetry axis and the plane x, z contains the direction of propagation $\mathbf{k}$. The

---

4 Note Fermat's principle of least time (Born and Wolf, 1965) immediately follows from this construction on application of the method of stationary phase.

5 Note that the function $f(\mathbf{k}, \omega)$ is *not* the phase speed V_p, but typically $f(\mathbf{k}, \omega) = \mathbf{k}\ V_p(\mathbf{k})/\omega$.

6 An alternative direct calculation for a system with a single axis is as follows. The envelope of the family of lines in parametric form $f(x, y, t) = 0$ is given by the intersection of $f(x, y, t) = 0$ with $\partial f(x, y, t)/\partial t = 0$ (Lowry and Hayden, 1957, p. 43). The envelope of the lines $\mathbf{s} \cdot \mathbf{n}(\theta)$ is therefore, given by

$$\mathbf{s} \cdot \mathbf{n} = 1 \qquad \text{and} \qquad \mathbf{s} \cdot \frac{d\mathbf{n}}{d\theta} = 0 \tag{5.71}$$

where $\mathbf{s} = \mathbf{V}_g/c$ and $\mathbf{n} = c\ \mathbf{k}/\omega = c/V_p(\theta)\ \hat{\mathbf{k}}$, when θ is the angle between the phase velocity and the anisotropy axis. In Cartesian co-ordinate form parallel and perpendicular to the axis

$$s_\parallel\ \cos\theta + s_\perp\ \sin\theta = \frac{V_p}{c} \qquad \text{and} \qquad -s_\parallel\ \sin\theta + s_\perp\ \cos\theta = \frac{1}{c}\frac{dV_p}{d\theta} \tag{5.72}$$

Solving for the group velocity $V_{g\parallel}$ and $V_{g\perp}$, we obtain

$$V_{g\parallel} = c\, s_\parallel = V_p\ \cos\theta - \frac{dV_p}{d\theta}\ \sin\theta \qquad \text{and} \qquad V_{g\perp} = c\, s_\perp = V_p\ \sin\theta + \frac{dV_p}{d\theta}\ \cos\theta \tag{5.73}$$

in accord with Eq. (5.75).

phase velocity is assumed to depend on the angle θ between the symmetry axis and the propagation vector $\mathbf{k}$ only. Since $\omega = k\ V_p(\theta)$, where V_p is the phase velocity, and

$$\frac{\partial k}{\partial k_z} = \cos\theta \qquad \text{and} \qquad \frac{\partial k}{\partial k_x} = \sin\theta$$

$$\frac{\partial \theta}{\partial k_z} = -\frac{1}{k}\ \sin\theta \qquad \text{and} \qquad \frac{\partial \theta}{\partial k_x} = \frac{1}{k}\ \cos\theta$$

it follows that the group velocity $\partial\omega/\partial\mathbf{k}$ components parallel and perpendicular to the axis are

$$V_g|_{\parallel} = V_p\ \cos\theta - \frac{\mathrm{d}V_p}{\mathrm{d}\theta}\ \sin\theta \qquad \text{and} \qquad V_g|_{\perp} = V_p\ \sin\theta + \frac{\mathrm{d}V_p}{\mathrm{d}\theta}\ \cos\theta \tag{5.74}$$

It is easily shown that there is a simple interpretation of this result. On a polar plot, the group velocity lies in the wave normal plane containing the axis and the phase velocity (fig 5.5). It follows from fig 5.5 that the group (ray) velocity is always greater than the phase (wave) velocity. The peak amplitude of the pulse clearly relates to the situation where the different frequency components are in phase. As a result, waves of different frequency interfere coherently resulting in a large local amplitude and therefore, energy. The propagation of a pulse is, therefore, accomplished along a ray defined by the group (ray) velocity. The wave (phase) velocity in contrast refers to the speed with which the phase of an infinite plane parallel wave changes as it is propagates.

The angular variation in the phase speed may make a large contribution to the ray surface. As a result, rays may have only a narrow angle of propagation as exemplified in Figures 10.1 and 10.2. A graphic illustration of this behaviour is provided by Kelvin's wedge, namely, the form of the wake generated by a ship moving on the surface (Pert, 2013, §4.I.i). Gravity waves spread out from the bow of the ship propagating along the group velocity giving the wake its distinctive V profile, which is a clear visualisation of the ray surface.

In dispersive media, the phase velocity depends directly on the frequency (or wave number) as well as the propagation direction $V_p(k, \theta)$. As a result, the group velocity includes an additional term to take the consequent temporal dispersion into account. Since

$$\mathbf{V}_g = \frac{\partial V_p}{\partial k}\ \tilde{\mathbf{k}} + \frac{1}{k}\frac{\partial V_p}{\partial \theta}\ \tilde{\theta}$$

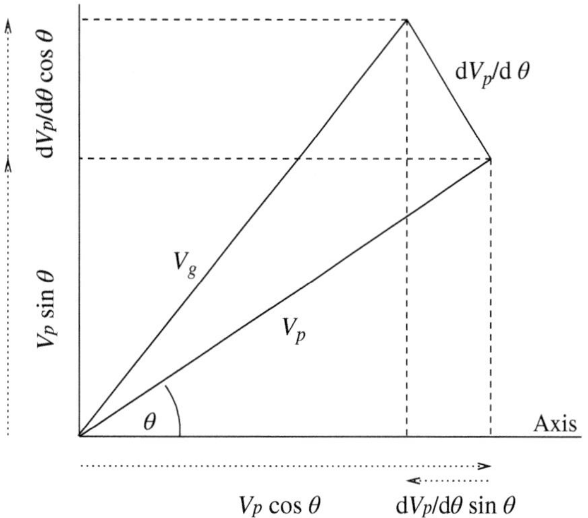

Figure 5.5 Geometrical interpretation of Eqs. (5.74) showing the relationship between the group (V_g) and phase (V_p) velocities.

where $\tilde{\mathbf{k}}$ is unit vector in the $\mathbf{k}$ direction and $\tilde{\theta}$ the unit vector normal to $\tilde{\mathbf{k}}$ in the plane of incidence, namely, the axis and $\tilde{\mathbf{k}}$. It is immediately apparent that the general result for dispersive media is

$$\begin{aligned} V_g\Big|_{\parallel} &= \left(V_p + k\,\frac{\partial V_p}{\partial k}\right)\cos\theta - \frac{\partial V_p}{\partial \theta}\sin\theta \\ V_g\Big|_{\perp} &= \left(V_p + k\,\frac{\partial V_p}{\partial k}\right)\sin\theta + \frac{\partial V_p}{\partial \theta}\cos\theta \end{aligned} \tag{5.75}$$

As a check, we note that if the dispersion is isotropic, i.e. $\partial V_p/\partial\theta = 0$, this reduces to the familiar result $V_g = \partial\omega/\partial k$.

5.6.1 Equivalence of Energy Transport Velocity and Group Velocity

We may demonstrate the equivalence of the group velocity with the energy transport velocity generically for the propagation of a wave of general properties in an arbitrary medium following the approach due to Lighthill (1960). Consider an arbitrary plane wave propagating in a homogeneous conservative medium with the wave vector $\mathbf{k}$ namely,

$$\mathbf{u} = \mathbf{a}\exp\left[\imath\,(\mathbf{k}\cdot\mathbf{r} - \omega t)\right] \tag{5.76}$$

where the wave vector $\mathbf{k}$ and frequency ω are related by a dispersion relation

$$G(\mathbf{k},\omega) = 0 \tag{5.77}$$

The derivatives of ω with respect to the wave vector $\mathbf{k}$ are given by the set

$$\nabla_{\mathbf{k}}(\omega) \equiv \frac{\partial\omega}{\partial k_i} = -\frac{\partial G}{\partial k_i}\Big/\frac{\partial G}{\partial\omega} \tag{5.78}$$

Since the medium is non-dissipative $G(\mathbf{k},\omega)$ is real.

To determine the velocity of energy transmission, we calculate

$$\mathbf{U} = \frac{\text{Flux of energy crossing unit area}}{\text{Energy per unit volume}} = \frac{\mathbf{I}}{E} \tag{5.79}$$

Quite generally, the energy E per unit volume is a fixed multiple of the square of the wave amplitude, thus $E = E_0 a^2$.

To calculate the energy flux, imagine that dispersion relation contains a small imaginary term, so that the wave is slightly damped as it propagates by a non-conservative force term $\mathbf{k} \to \mathbf{k} + \frac{1}{2}\,\imath\,\epsilon$, where $\epsilon \to 0$ so that the wave becomes described by

$$\mathbf{u} = \mathbf{a}\exp\left[\imath\,(\mathbf{k}\cdot\mathbf{r} - \omega t) - \frac{1}{2}\epsilon\cdot\mathbf{r}\right] \tag{5.80}$$

The wave frequency must also be modified consistently by the non-conservative force thereby introducing an equivalent damping term to produce a motion consonant with the equations of motion.

$$\omega \to \omega - \frac{1}{2}\,\imath\,\epsilon\cdot\nabla_{\mathbf{k}}(\omega) \tag{5.81}$$

The minus sign being chosen appropriate to a stationary set of decaying waves travelling in the $\mathbf{k}$ direction. The energy per unit volume at $\mathbf{r}$ is damped to $E_0\,a^2\exp(-\epsilon\cdot\mathbf{r})$ and the total energy per unit area to the right of the plane $\hat{\mathbf{n}}$ normal to $\mathbf{I}$ is, therefore,

$$H = E_0\,a^2/(\epsilon\cdot\hat{\mathbf{n}}) \tag{5.82}$$

The wave is also damped in time so that the energy loss per unit area per unit time to the right of the zero plane $\hat{n}$ is $\nabla_{\mathbf{k}}(\omega) \cdot \epsilon\ H = \nabla_{\mathbf{k}}(\omega) \cdot \epsilon/(\hat{\mathbf{n}} \cdot \epsilon)\ E_0\ a^2$ must be compensated by a flux of energy in through the zero plane **I**. In a steady system, this energy is the required input energy flux **I**.[7]

Hence, since ϵ is arbitrary, the energy transport velocity becomes

$$\mathbf{U} = \frac{\nabla_{\mathbf{k}}(\omega) \cdot \epsilon/(\hat{\mathbf{n}} \cdot \epsilon)\ E_0\ a^2}{E_0\ a^2} = \nabla_{\mathbf{k}}\omega = \left.\frac{\partial \omega}{\partial \mathbf{k}}\right|_{G=0} \tag{5.85}$$

as we let $\epsilon \rightarrow 0$.

Comparing this result with that from Eq. (5.59), we see that the energy transport speed is equal to the group velocity in conservative (i.e. dissipationless) media.

Appendix 5.A Waves in Anisotropic Inhomogeneous Media

Waves are formed as a consequence of the successive interactions of two or more coupled dynamical effects. For example the simplest cases such as sound waves or waves on a string involve two such terms: acceleration by stresses which arise from the displacement of the particles, and particle displacement by the velocity established by the acceleration. Similarly, electro-magnetic waves involve the electro-magnetic fields **E** and **B** coupled by Maxwell's equations. In a general medium, we may identify the wave as perturbation produced by the interaction between terms expressed by a set of vector quantities **X** containing N independent terms,[8] which are coupled by a relationship of the form known as the *dispersion relation* which expresses the nature of the coupling:

$$\mathbb{M}[(-\imath\ \nabla),\ (\imath\ \partial/\partial t), \mathbf{r}, t]\mathbf{\Psi}(\mathbf{r}, t) = 0 \tag{5.A.1}$$

where $\mathbf{\Psi}(\mathbf{r}, t)$ is the general vector wave function describing the perturbation in the N dimensional space spanned by **X**.

Within the 'geometrical optics' approximation, we consider a general monochromatic travelling wave in the direction **k**, whose wave elements vary as $\exp\left[\imath\ (\mathbf{k} \cdot \mathbf{r} - -\omega t)\right]$. Applying the WKB approximation to the wave-function, the changes of the amplitude of the wave function are slowly varying and can be neglected in comparison with those associated with the phase. The tensor operator $\mathbb{M}$, therefore, transforms to the algebraic form $\mathbb{M}\ (\mathbf{k}, \omega, \mathbf{r}, t)$. If we now assume that the wave is weak so that the disturbance is linear, being only a small perturbation on the ambient background state, then $\mathbb{M}$ becomes a simple matrix linearly connecting the elements $\mathbf{\Psi}$, namely,

$$\mathbb{M}_{ij}\ \Psi_j = 0 \qquad \text{for all } i \tag{5.A.2}$$

7 Alternatively, following Lighthill (1960) Eq. (5.81) implies that the inertial force (minus the mass times the acceleration) would change from $m\omega^2\mathbf{r}$ to

$$m\left(\omega - \frac{1}{2}\ \imath\ \epsilon \cdot \nabla_{\mathbf{k}}\right)^2\ \mathbf{r} \approx m\omega^2\mathbf{r} - (\epsilon \cdot \nabla_{\mathbf{k}})\ \imath\ m\ \omega\ \mathbf{r} \tag{5.83}$$

This would be the same as if an additional force $-(\epsilon \cdot \nabla_{\mathbf{k}})$ times its momentum were applied to every particle (Lighthill, 1960). The sum of the rate at which work is done by this inertial force and the energy transmission across the zero plane **I** must be zero:

$$\mathbf{I} = (\epsilon \cdot \nabla_{\mathbf{k}})m\dot{\mathbf{r}} \cdot \dot{\mathbf{r}}\ \hat{\mathbf{n}} = (\epsilon \cdot \nabla_{\mathbf{k}})2\ T\ \hat{\mathbf{n}} = (\epsilon \cdot \nabla_{\mathbf{k}})H\ \hat{\mathbf{n}} \tag{5.84}$$

since in a progressive wave, the kinetic and potential energies are equal.

8 In many cases, the number of elements in the set N will be infinite, for example in a continuous medium such as plasma. Note also that the set **X** will involve not only different physical terms, for example not just the fields **E** and **B**, but also their separate components in different directions.

which is a set of homogeneous linear equations and can only have solutions if

$$\det \mathbb{M}(\mathbf{k}, \omega, \mathbf{r}, t) = \mathcal{M}(\mathbf{k}, \omega, \mathbf{r}, t) = 0 \tag{5.A.3}$$

which is satisfied by

$$\omega = \Omega(\mathbf{k}, \mathbf{r}, t) \tag{5.A.4}$$

This relation, which specifies the possible wave vectors for a given frequency, or *vice versa* is known as the *dispersion relation*. We shall meet it in several forms in our studies of waves in plasma (for example Section 13.2).

Since the phase of the wave changes as $\mathrm{d}S = (\mathbf{k} \cdot \mathrm{d}\mathbf{r} - \omega\, \mathrm{d}t)$ along the ray $\mathbf{r}(t)$, it follows that, provided they are slowly varying,

$$\mathbf{k} = \nabla S \qquad \text{and} \qquad \omega = -\partial S/\partial t \tag{5.A.5}$$

give the local value of the wavenumber and frequency, even if either are varying due to inhomogeneity as the wave travels. The velocity of phase propagation is given by the lines $S =$ const or the phase velocity

$$V_p = \omega/k \tag{5.A.6}$$

in the direction of $\mathbf{k}$.

The refractive index (if appropriate) is defined by the ratio of the phase velocity in vacuo c to that in the medium

$$n = c/V_p \tag{5.A.7}$$

An arbitrary pulse can be written as

$$\Psi(\mathbf{r}, t) = \int \mathbf{A}(\mathbf{k}) \exp(\imath\, S)\, \mathrm{d}\mathbf{k} \tag{5.A.8}$$

Applying the method of stationary phase to this integral, we can see that the major contribution comes from the region near $\partial S/\partial \mathbf{k} = \mathrm{d}\mathbf{r} - (\partial\Omega/\partial \mathbf{k})\, \mathrm{d}t \approx 0$. Therefore, the peak of the pulse moves with velocity

$$\mathbf{V}_g = \lim_{\mathrm{d}t \to 0} \frac{\mathrm{d}\mathbf{r}}{\mathrm{d}t} = \frac{\partial \Omega}{\partial \mathbf{k}} \tag{5.A.9}$$

Namely, the *group velocity* as we found before (Eq. (5.59)).

It follows from the definitions of $\mathbf{k}$ and ω as derivatives of the phase S that

$$\frac{\partial \mathbf{k}}{\partial t} + \nabla \Omega = 0 \qquad \text{and} \qquad \nabla \wedge \mathbf{k} = 0 \tag{5.A.10}$$

Therefore, noting that the frequency $\omega(\mathbf{r}, t)$ is a function of space and time only, whereas the dispersion term $\Omega(\mathbf{k}, \mathbf{r}, t)$ is also a function of the wavenumber $\mathbf{k}$, we obtain

$$\frac{\partial k_i}{\partial t} = -\frac{\partial \omega}{\partial x_j} = -\frac{\partial \Omega}{\partial x_i} - \frac{\partial \Omega}{\partial k_j}\frac{\partial k_j}{\partial x_i} = -\frac{\partial \Omega}{\partial x_i} - \frac{\partial \Omega}{\partial k_j}\frac{\partial k_i}{\partial x_j} \tag{5.A.11}$$

it follows that the time derivative moving with the pulse

$$\frac{\mathrm{d}\mathbf{k}}{\mathrm{d}t} = \frac{\partial \mathbf{k}}{\partial t} + (\mathbf{V}_g \cdot \nabla)\mathbf{k} = -\nabla \Omega \tag{5.A.12}$$

Similarly,

$$\frac{\partial \omega}{\partial t} = \frac{\partial \Omega}{\partial k_j}\frac{\partial k_j}{\partial t} + \frac{\partial \Omega}{\partial t} = -\frac{\partial \Omega}{\partial k_j}\frac{\partial \omega}{\partial x_j} + \frac{\partial \Omega}{\partial t} \tag{5.A.13}$$

and thus

$$\frac{\mathrm{d}\omega}{\mathrm{d}t} = \frac{\partial \omega}{\partial t} + (\mathbf{V}_g \cdot \nabla)\omega = \frac{\partial \Omega}{\partial t} \tag{5.A.14}$$

Hence, the frequency variation of the wave pulse as it moves along its path is given by the temporal variation of the solution of the dispersion relation alone.

Once the dispersion relation $\Omega(\mathbf{k}, \mathbf{r}, t)$ in a spatially and temporally varying medium is known, we may directly write down the equation of pulses through the ray tracing equations, which may be expressed in terms of the determinant of the dispersion matrix $\mathcal{M}(\mathbf{k}, \omega, \mathbf{r}, t)$ using the total differential

$$\mathrm{d}\mathcal{M} = \frac{\partial \mathcal{M}}{\partial \mathbf{k}} \cdot \mathrm{d}\mathbf{k} + \frac{\partial \mathcal{M}}{\partial \boldsymbol{\omega}}\mathrm{d}\omega + \frac{\partial \mathcal{M}}{\partial \mathbf{r}} \cdot \mathrm{d}\mathbf{r} + \frac{\partial \mathcal{M}}{\partial t}\mathrm{d}t = 0 \tag{5.A.15}$$

to form the appropriate temporal derivatives moving with the rays directly in terms of the dispersion matrix without necessarily solving it explicitly.

$$\frac{\mathrm{d}\mathbf{r}}{\mathrm{d}t} = \frac{\partial \Omega}{\partial \mathbf{k}} = -\frac{\partial \mathcal{M}/\partial \mathbf{k}}{\partial \mathcal{M}/\partial \omega} \tag{5.A.16}$$

$$\frac{\mathrm{d}\mathbf{k}}{\mathrm{d}t} = -\frac{\partial \Omega}{\partial \mathbf{r}} = \frac{\partial \mathcal{M}/\partial \mathbf{r}}{\partial \mathcal{M}/\partial \omega} \tag{5.A.17}$$

$$\frac{\mathrm{d}\omega}{\mathrm{d}t} = \frac{\partial \Omega}{\partial t} = -\frac{\partial \mathcal{M}/\partial t}{\partial \mathcal{M}/\partial \omega} \tag{5.A.18}$$

6
Kinetic Theory of Plasmas – Collisionless Models

6.1 Introduction

Unlike the interaction force between uncharged molecules, the force between charged particles in a plasma is not negligible at large distances. Although the force decreases as an inverse square law of distance, the number of particles involved increases commensurately, so that the net force is nearly independent of the separation, provided the contribution from neighbouring particles adds in a scalar fashion; behaviour observed in the cold plasma wave (Section 1.2). As we found in Section 1.3, the fields due to individual particles can only be observed over distances shorter than the Debye length. At distances less than the Debye length, the particles undergo individual interactions, albeit with many other individual particles simultaneously (Section 2.1) – short range collisional behaviour. The plasma, therefore, exhibits behaviour similar to that found in collisions between molecules in gases. However, the long-range nature of the Coulomb force introduces another form of interaction in which the motions of a large number of particles are tightly co-ordinate – namely collective effects – frequently giving rise to collective oscillation such as plasma waves discussed in Section 1.2.

In dilute plasma such as we are investigating in this study, the plasma is sufficiently low density so that the collisional frequency is small and such effects can be neglected. In this case, the motion of the plasma is dominated by collective effects. A typical example of this behaviour is demonstrated by the wall sheath in free fall limit (Section 1.5). The particle dynamics in the collision-free limit are essentially those of their single particle motion as described in Chapter 3. However, in contrast to our earlier study, the electro-magnetic fields are usually determined by the charge density and current resulting from the particle motion. Therefore, rather than studying the behaviour of individual charged particles, it becomes convenient for the study of these problems to use the averages given by the distribution function modified by the electro-magnetic fields calculated by Maxwell's equations. The most important application of the resulting equations will be to investigate waves in plasma, which will be deferred to Chapter 11.

6.2 Vlasov Equation

When the interaction is over distances ($\ell < \lambda_D$) less than the Debye length the particles act independently in a manner similar to the collisions experienced by gas molecules. The system acts more like a fluid and any significant local space charge differences due to ion/electron separation are screened out. As a result we lose the 'graininess' associated with discrete particles. This limit gives rise to the Vlasov equation describing the one particle distribution function $f^{(1)}$ (Vlasov, 1938).

Foundations of Plasma Physics for Physicists and Mathematicians, First Edition. Geoffrey J. Pert.

To formally establish this condition, we imagine that the plasma particles are subdivided in such a way that the ratio e/m is kept constant, where $-e$ is the particle (electron) charge and m the mass. Macroscopic phenomena associated with electric and magnetic fields are unchanged. We allow the particle density $n \to \infty$ and the particle charge and mass $e, m \to 0$ so that the charge and mass densities $q = n\,e$, $\rho = n\,m$ and the charge to mass ratio e/m are finite and remain held constant. The static electric and magnetic fields, $\mathbf{E}^{(0)}$ and $\mathbf{B}^{(0)}$, respectively, generate a constant acceleration:

$$\mathbf{a}_i^{(0)} = -\frac{e}{m}[\mathbf{E}^{(0)} + \mathbf{v}_i \times \mathbf{B}^{(0)}] \tag{6.1}$$

since e/m is constant.

Now consider the interaction between two particles, i and j, due to their electric charges and which we shall show is always small ($\sim e^2/m$) is due to two terms. Electrostatic interaction between the particles namely the Coulomb force on particle i due to particle j is

$$a_i^{(j)} = -\frac{e^2}{4\pi\epsilon_0 m}\frac{(\mathbf{r}_i - \mathbf{r}_j)}{|\mathbf{r}_i - \mathbf{r}_j|^3} \sim \frac{e^2}{m} \to 0 \tag{6.2}$$

Magnetostatic interaction due to the charge carried by particle j moving with velocity v_j which constitutes a current element and produces a magnetic field given by the Biot–Savart equation

$$B_i^j = -\frac{e\mu_0}{4\pi}\frac{\mathbf{v}_j \wedge (\mathbf{r}_i - \mathbf{r}_j)}{|\mathbf{r}_i - \mathbf{r}_j|^3}$$

and exerts a force on the moving particle i giving an acceleration[1]

$$a_i^{(j)} = -\frac{e^2}{4\pi m}\frac{\mathbf{v}_i \wedge (\mathbf{v}_j \wedge (\mathbf{r}_i - \mathbf{r}_j))}{|\mathbf{r}_i - \mathbf{r}_j|^3} \sim \frac{e^2}{m} \to 0.$$

It is easily seen that the relative strength of the electric and magnetic forces is as $1 : v^2/c^2$. Provided the plasma is nonrelativistic, the magnetic force can generally be neglected. Both interaction forces scale as $e^2/m \to 0$. The interaction is therefore of higher order than forces generated by the externally applied fields. Retaining the interactions terms as ones of lower order in the expansion, we may consider the particles to be uncorrelated so that the two particle distribution function $f^{(2)}$ may written as the product of the single particle terms for particles ① and ②:

$$f^{(2)}(\mathbf{r}, \mathbf{v}, \mathbf{r}', \mathbf{v}') = f^{(1)}(\mathbf{r}, \mathbf{v})\, f^{(1)}(\mathbf{r}', \mathbf{v}') \tag{6.3}$$

Substituting these terms into the leading term of the BBGKY expansion:

$$\frac{\partial f^{(1)}}{\partial t} + \mathbf{v}_1 \cdot \frac{\partial f^{(1)}}{\partial \mathbf{r}} + \left[\mathbf{a}^{(0)} + \int \mathbf{a}^{(2)}\, f^{(1)}(\mathbf{r}', \mathbf{v}', t)\, d\mathbf{r}'\, d\mathbf{v}'\right] \cdot \frac{\partial f^{(1)}}{\partial \mathbf{v}} = 0 \tag{6.4}$$

where $\mathbf{a}^{(0)}$ is the static field term, and the integral is over the self-consistent field term.

Substituting for $\mathbf{a}^{(2)}$, we introduce the self-consistent electric and magnetic fields:

$$\begin{aligned} \mathbf{E}^{\text{self}}(\mathbf{r}, t) &= -\frac{1}{4\pi\epsilon_0}\, e \int \frac{\mathbf{r} - \mathbf{r}'}{|\mathbf{r} - \mathbf{r}'|^3}\, f^{(1)}(\mathbf{r}', \mathbf{v}', t)\, d\mathbf{r}' d\mathbf{v}' \\ \mathbf{B}^{\text{self}}(\mathbf{r}, t) &= -\frac{\mu_0}{4\pi}\, e \int \frac{\mathbf{v}' \wedge (\mathbf{r} - \mathbf{r}')}{|\mathbf{r} - \mathbf{r}'|^3}\, f^{(1)}(\mathbf{r}', \mathbf{v}', t)\, d\mathbf{r}' d\mathbf{v}'. \end{aligned} \tag{6.5}$$

---

1 It is well known that the magnetic interaction between two current elements does not satisfy Newton's third law of motion (see Page and Adams, 1945). This need not concern us here as averaging will replace the elements by bulk current loops.

Thus, we finally obtain the Vlasov or collisionless Boltzmann equation:

$$\frac{\partial f}{\partial t} + \mathbf{v} \cdot \frac{\partial f}{\partial \mathbf{r}} + \mathbf{a} \cdot \frac{\partial f}{\partial \mathbf{v}} = 0 \tag{6.6}$$

omitting the (unnecessary) superscript on the distribution function, and

$$\mathbf{a} = -\frac{e}{m}[\mathbf{E}^0 + \mathbf{E}^{\text{self}} + \mathbf{v} \times (\mathbf{B}^0 + \mathbf{B}^{\text{self}})] \tag{6.7}$$

and $\mathbf{E}^{\text{self}}$ and $\mathbf{B}^{\text{self}}$ are found by integrating over the local charge density and current density throughout the plasma. Generalising to a plasma the space charge density q and current density $\mathbf{j}$ are obtained from the distribution functions for the electrons f_e and the ions f_i

$$q = e \int \{Z f_i(\mathbf{v}) - f_e(\mathbf{v})\} d\mathbf{v} \qquad \text{and} \qquad \mathbf{j} = e \int \mathbf{v}\{Z f_i(\mathbf{v}) - f_e(\mathbf{v})\}\, d\mathbf{v} \tag{6.8}$$

where Z is the ion charge number.[2] The self-consistent fields are obtained then from Maxwell's equations:

$$\frac{1}{\mu_0} \nabla \wedge \mathbf{B}^{\text{self}} = \mathbf{j} + \epsilon_0 \frac{\partial \mathbf{E}^{\text{self}}}{\partial t} \qquad \nabla \cdot \mathbf{B}^{\text{self}} = 0 \qquad \nabla \wedge \mathbf{E}^{\text{self}} = -\frac{\partial \mathbf{B}^{\text{self}}}{\partial t} \qquad \nabla \cdot \mathbf{E}^{\text{self}} = \frac{1}{\epsilon_0} q \tag{6.9}$$

The Vlasov equation describes collision free, fluid-like long-range interactions. The interaction term is collective, not collisional, the interaction between two particles simultaneously involves a very large number in a coherent response induced by the self-generated fields. The fields are macroscopic in scale and therefore don not lead to an approach to thermal equilibrium, in contrast to short-range collisions. The process is therefore random and cannot lead to an increase in entropy, or in other words is thermodynamically reversible. The interactions between the particles are not considered to be interactions between individual particles, but rather with the fields established by a continuum fluid. In this sense, it is 'collisional free' and treats only co-operative or collective effects. It is assumed that the collective effect of distant particles is much greater than any systematic effect arising from the interaction of one, or a small number of particles. That is we average over the discrete nature of the particles by sub-division. Clearly, the validity of this approach requires $n\lambda_D^3 \sim n^{-1/2}e^{-3} \sim e^{-5/2} \gg 1$, i.e. the number of particles in the Debye sphere is large, and we consider only problems whose scale lengths $\ell \gg \lambda_D$.[3]

The Vlasov equation is a relatively simple form of the BBGKY hierarchy set of equations in the particular case that the individual collisions are negligible and that collective (macroscopic) interactions dominate.[4] Introducing time reversal into the equations

$$\begin{array}{lll} \mathbf{r} \to \mathbf{r} & t \to -t & \mathbf{v} \to -\mathbf{v} \\ \mathbf{E} \to \mathbf{E} & \mathbf{B} \to -\mathbf{B} & \mathbf{j} \to -\mathbf{j} \end{array}$$

Any solutions of the equation, therefore, are reversible. 'Time's arrow' – a direction of time – must be introduced either within the temporal boundary conditions imposed on the solution or by an

2 Since the distribution functions $f(\mathbf{v})$ scale as n, it is clear that $\mathbf{E}^{\text{self}}$ and $\mathbf{B}^{\text{self}}$ are constants in response to the limit scaling discussed earlier.

3 Vlasov (1938) specified the cut-off at the inter-particle separation distance $\sim n_i^{-1/3}$ rather than the Debye length λ_D assuming only binary collisions contributed: the effect of multi-particle collisions at distances less than the Debye length not being recognised.

4 Closure of the BBGKY set is introduced into the Vlasov equation by the assumption that particles are uncorrelated due to the weakness of the interaction force between specified particles. Therefore, $f^{(2)}$ is simply given by Eq. (6.3) and does not involve higher-order distributions.

ad hoc inclusion of irreversibility through collisions. For these purposes, it is often convenient to empirically add a collision term to the Vlasov equation:

$$\frac{\partial f}{\partial t} + \mathbf{v} \cdot \frac{\partial f}{\partial \mathbf{r}} + \mathbf{a} \cdot \frac{\partial f}{\partial \mathbf{v}} = \left.\frac{\partial f}{\partial t}\right|_{\text{coll}} \tag{6.10}$$

The Vlasov equation in this form including a collisional term is the Boltzmann equation (4.30) taking into account the self-generated fields. It is sometimes known as the *collisional Vlasov equation*.

The collision term is often expressed in a very simple form known as the BGK (*Bhatnagar–Gross–Krook*) collision integral, which expresses the role of collisions in driving the distribution to an equilibrium Maxwellian form (Section 4.8):

$$\left.\frac{\partial f}{\partial t}\right|_{\text{coll}} = -\frac{(f - f_0)}{\tau} \tag{6.11}$$

where f_0 is the equilibrium distribution and τ is the appropriate particle relaxation time (e.g. τ_{ee} for electrons) as discussed in Section 2.1.4.

Although useful for qualitative or order of magnitude scaling arguments, the BGK collision term is not accurate and in consequence either a Boltzmann, Eq. (4.42) or preferably a Fokker–Planck, Eq. (7.3) should be used for quantitative studies.

In this chapter, we investigate some of the behaviour of collisionless plasma through solutions of the Vlasov equation. Many of the solutions of the collision-free Vlasov equation represent waves. Collisionless damping of the wave occurs as a result of causality, introduced either as the initial (switch-on) condition or through a vanishingly small collisional term.

6.3 Particle Trapping by a Potential Well

Consider first the steady-state interaction of electrons having a Maxwellian velocity distribution function with a planar potential barrier $\phi(x)$. Since there is no force in the y, z directions we may consider the behaviour of the x component of velocity alone. In the steady state, the velocity distribution function following the motion of the particles in phase space is given by

$$\frac{\mathrm{d}f}{\mathrm{d}t} = 0 \tag{6.12}$$

The velocity distribution function is therefore a function of the invariants of the motion.

When the electrons are free, one invariant is the total energy $\varepsilon = \frac{1}{2}mu^2 + U$, where $U = -e\,\phi$ is the electron potential energy; the second invariant is a Lagrangian particle identifier of no interest here. Suppose the electrons are incident from both sides at infinity with density N_0 on the potential barrier $\phi(x)$, which is uniform in the y, z plane. Considering only the one-dimensional motion and assuming the distribution function far from the barrier at $\pm\infty$ is a one-dimensional thermal distribution:

$$f(\pm\infty, u) = N_0\sqrt{\frac{m}{2\pi kT}}\exp\left(-\frac{mu^2}{2kT}\right).$$

If the initial particle velocity is such that its energy exceeds that of the barrier $u_0{}^2 > -2\,U_0/m$, the electron travels over the barrier U_0, otherwise, it is reflected

$$f(x,u) = N_0\sqrt{\frac{m}{2\pi kT}}\begin{cases}\exp(-\varepsilon(x,u)/kT) & \text{if } \varepsilon(x,u) > 0\\ 0 & \text{otherwise}\end{cases} \tag{6.13}$$

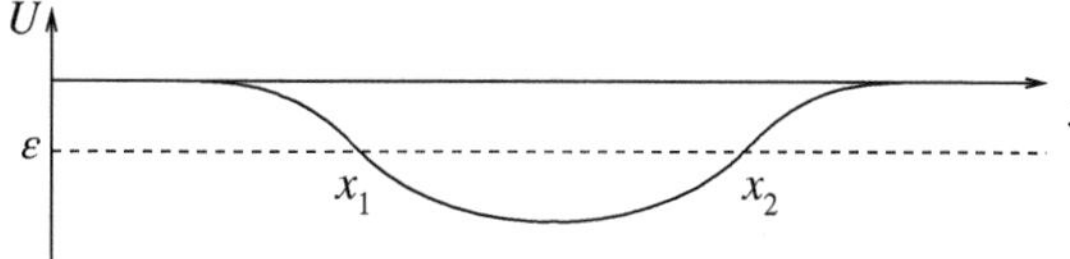

Figure 6.1 A simple symmetric potential well with base line potential $U = 0$. Electrons with energy $\varepsilon < 0$ are trapped by the minimum between the reflection points at x_1 and x_2, where $U(x) = \varepsilon$.

and the density at the point x

$$N(x) = \int_{\sqrt{2U(x)/m}}^{\infty} f(x,u)\;\mathrm{d}u = N_0 \exp\left(-U(x)/\kappa T\right) \tag{6.14}$$

This simple model used here for the case of particles incident from both sides on the barrier is easily generalised to cover the cases of particles incident from one side only (Section §1.5) and/or a non-zero incident mean particle velocity.

Turning now to the case of particles trapped by the potential well, we consider the basic picture of how trapped particles arise in a simple well (Figure 6.1). In the steady state, there is no mechanism to enable particles to enter the potential well in the absence of collisions. Trapping, therefore, takes place as the depth of the well is increased. We can imagine a particle with very small relative velocity enters the region of the well. Whilst travelling slowly through the well, the depth of the latter increases sufficiently for the particle to be reflected when it meets the first barrier. Returning the particle now finds the second barrier is also too large to allow it to escape and it becomes trapped between the two walls.

If the well potential were stationary, the energy of the particles would be constant, and the distribution function would be a function of the particle energy as above. However, it is not possible to easily identify the invariant as the system is no longer steady state, and the particles being trapped in the well are not connected with the external environment. The state of a particle is, therefore, determined by its value when originally trapped and can only be determined if the history of the trapping is known.

If the time constant for changes in the potential $\tau \gg L/\overline{v}_e$, electrons are trapped adiabatically by a slowly increasing potential well of depth $-e\phi$ and width L. However, the energy inside the well is not stationary as the depth of the well slowly increases. The preceding analysis must therefore be modified. Provided the changes in the well are slow, there is an adiabatic invariant (see Section 3.12) determined by the motion in the well, namely,

$$J = \int_0^T u\;\mathrm{d}x = 2\int_{x_1}^{x_2} \sqrt{2\left(\varepsilon - U(t,x)\right)/m}\;\mathrm{d}x \tag{6.15}$$

where x_1 and x_2 are the turning points at which the electron velocity is zero. The trapped particle distribution function is therefore given by $f_{tr} = f_{tr}(J(t,\varepsilon))$, and equal to that of the incoming particles with zero energy.

Consider a simple well (Figure 6.1) with the potential energy $U(x)$ having the same value on either side. The distribution function for the trapped particles is determined by the fact that it must be continuous across the boundary between trapped and untrapped particles, $\varepsilon = 0$. Hence, for the particles with adiabatic invariant $J, f_t r(J) = f(0)$, i.e. the distribution function is a constant. Within the well, the distribution function is, therefore,

$$f(u) = \begin{cases} f(\varepsilon) & \text{if } \varepsilon > 0 \\ f(0) & \text{otherwise} \end{cases} \tag{6.16}$$

The number of particles per unit length at x is, therefore,

$$:N = 2\int_{u_1}^{\infty} f(\varepsilon)\;\mathrm{d}u + 2\int_0^{u_1} f(0)\;\mathrm{d}u \tag{6.17}$$

where $u_1 = -\sqrt{2\,U(x)/m}$ is the minimum velocity of untrapped particles in the well. The factors of 2 take account of particles moving in both directions. In general, if $|U|$ is small, the density is dominated by the distribution of the untrapped particles, but if $|U|$ large, the density varies proportionately to square root of the well depth.

To illustrate this behaviour, suppose that the distribution is Maxwellian at infinity, the distribution function is

$$f(u) = N_0\sqrt{\frac{m}{2\pi \text{k}T}}\begin{cases}\exp(-\varepsilon(u)/\text{k}T) & \text{if } \varepsilon > 0\\ 1 & \text{otherwise}\end{cases} \tag{6.18}$$

and the density is

$$N(t,x) = N_0\,\{\exp(|U|/\text{k}\,T)[1-\Phi(\sqrt{|U|/\pi\ \text{k}T})] + 2\sqrt{|U|/\text{k}T}\} \tag{6.19}$$

where $\Phi(x) = (2/\sqrt{\pi})\ \int_0^x e^{-t^2}\ dt$ is the error function. Making use of the small and large argument expansions of the error function (Abramowitz and Stegun, 1965), we obtain

$$N \approx \begin{cases} N_0\left[1+\dfrac{|U|}{\text{k}T}-\dfrac{4}{3\sqrt{\pi}}\left(\dfrac{|U|}{\text{k}T}\right)^{3/2}\right] & |U| \ll \text{k}T \\ 2N_0\left(\dfrac{|U|}{\pi\text{k}T}\right)^{1/2} & |U| \gg \text{k}T \end{cases} \tag{6.20}$$

As can be seen the number of particles for small well depths starts to depart from the Maxwellian in the term in $(|U|/\text{k}T)^{3/2}$ due to the trapping. At large well depths, the number of particles increases much more slowly than predicted by a Maxwellian distribution overall.

Note on particle trapping in static fields In principle, charged particles may be trapped in either electro-static fields via their charge as we have seen above (Section 6.3) or by magnetic fields in magnetic mirrors through their dipole moment (Section 3.10). However, it is a result of Earnshaw's theorem that such trapping cannot be three dimensional. The underlying reason is very simple and due to the fact that in either static field case the trapped particle must have a stable position at a turning point of the field in order to be reflected at the boundary. Since both static electric and magnetic fields in vacuo satisfy Laplace's equation $\nabla^2\phi = 0$, it follows that any turning point must be at least a saddle point (see Pert, 2013, §2.6), thereby opening a clear escape path. We conclude that particles can only be trapped in one or two dimensions by a static electro-field. In fact, charged particles can be trapped in an alternating field of differing strengths in separate directions, namely quadrupole fields. In a quadrupole trap[5] the particles are held by an r.f. field applied between hyperboloid electrodes in a hyperboloidal enclosure; the period of the field being smaller than the escape time of particles from the enclosure.

It should be noted that Earnshaw's theorem only applies in simply connected spaces. Thus, in a multiply connected space, it becomes possible to trap particles on closed paths which are not reducible. This result opens the door to the possibility of equilibrium magnetic configurations such as tori and infinite linear systems (Section 15.2).

Simple trapped configurations are commonly found throughout plasma physics, often embedded in more complex systems such as trapped particle in non-linear Landau damping (Section 12.5.1) and banana orbits in tokamaks (Section 15.9).

5 Also known as a Paul trap or r.f trap. W. Paul was awarded the 1989 Nobel prize in Physics for the development of these systems (Paul, 1990), which had a major impact in several fields for their ability to reduce the effects of Döppler broadening (see Case Study 6.1).

Case Study 6.1

Quadrupole Ion Trap

We follow the presentation of the analysis of these devices by Paul (1990) in his Nobel Prize lecture. Consider a cylindrical system shown in Figure 6.2 with (x, y) normal to the axis (z). Laplace's equation in a charge or current free environment takes the form $\nabla^2\phi = 0$ for the potential of the electric or magnetic fields. Taking account of the cylindrical symmetry:

$$\frac{\partial^2\phi}{\partial x^2} + \frac{\partial^2\phi}{\partial y^2} + \frac{\partial^2\phi}{\partial z^2} = 0 \tag{6.21}$$

integrating and assuming the potentials applied to the end-caps with respect to the ring are equal,

$$\phi = A(r^2 - 2z^2) + C \tag{6.22}$$

where $r^2 = x^2 + y^2$. Constructing the end-caps to be hyperbolae and the ring a hyperboloid centred on the centre of the field automatically satisfies the field boundary conditions since the potential along the surface of a conductor is a constant. Letting $(0, \pm z_0)$ be the centre of the end-caps and $(r_0, 0)$ that of the ring, the potential difference between the end-caps and the ring electrodes is $\phi_0 = A({r_0}^2 + 2{z_0}^2)$. The equipotentials form a set of nested hyperbolae. Referring to Figure 6.2, the field generated by a pair of hyperbolic end-caps and a cylindrical hyperboloidal ring, respectively,

$$\frac{r^2}{2\,{z_0}^2} - \frac{z^2}{{z_0}^2} = 1 \qquad \text{and} \qquad \frac{r^2}{{r_0}^2} - 2\,\frac{z^2}{{r_0}^2} = 1 \tag{6.23}$$

is given by Eq. (6.22). Referring the zero of potential to a point on the end-cap, we obtain $C = 2\,\phi_0\,{z_0}^2/({r_0}^2 + 2\,{z_0}^2)$, and the general expression for the potential is

$$\phi(r, z) = \frac{\phi_0}{({r_0}^2 + 2\,{z_0}^2)}(r^2 - 2z^2) + \frac{2\,\phi_0\,{z_0}^2}{({r_0}^2 + 2\,{z_0}^2)} \tag{6.24}$$

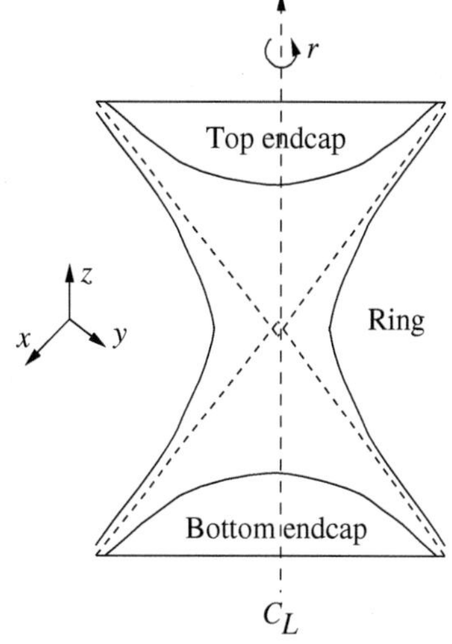

Figure 6.2 Schematic of the electrode arrangement in an electrostatic quadrupole trap.

(Continued)

Case Study 6.1 (Continued)

If $\phi_0 > 0$, the potential at the centre has a minimum on the plane $z = 0$ and a maximum along z at the centre, and *vice versa* if $\phi_0 < 0$. If we calculate the motion of a charged particle $q > 0$ in this field, we find it is trapped in but escapes along the z direction. To avoid this happening, we oscillate the field between the ring and the end-caps sufficiently rapidly that the anti-trapping motion is cancelled by successive impulses. Introducing an oscillatory component to the field, we let the potential $\phi_0 = U + V\cos\Omega t$ between the ring and the end-caps. This gives the net potential:

$$\phi(r,z) = \frac{U + V\cos\Omega t}{{r_0}^2 + 2{z_0}^2}(r^2 - 2z^2) + 2\left(\frac{U + V\cos\Omega t}{{r_0}^2 + 2{z_0}^2}\right){z_0}^2 \tag{6.25}$$

Taking the gradient of the potential yields the force on the particle $-Q\nabla\phi$, and hence the accelerations in the component direction r and z:

$$\frac{d^2r}{dt^2} = -\frac{2\,Q}{M}\left(\frac{U + V\cos\Omega t}{{r_0}^2 + 2{z_0}^2}\right)r \quad \text{and} \quad \frac{d^2z}{dt^2} = \frac{4\,Q}{M}\left(\frac{U + V\cos\Omega t}{{r_0}^2 + 2{z_0}^2}\right)z \tag{6.26}$$

where Q and M are the charge and mass of the particle, respectively.

Substituting

$$a_r = \frac{8QU}{M\,({r_0}^2 + 2{z_0}^2)\Omega^2} \quad \text{and} \quad q_r = -\frac{4QV}{M\,({r_0}^2 + 2\,{z_0}^2)\Omega^2} \tag{6.27}$$

$$a_z = -\frac{16QU}{M\,({r_0}^2 + 2{z_0}^2)\Omega^2} \quad \text{and} \quad q_r = \frac{8QV}{M\,({r_0}^2 + 2\,{z_0}^2)\Omega^2} \tag{6.28}$$

reduces the equations of motion to Mathieu's form

$$\frac{d^2u}{dt^2} + [a_u - 2q_u\cos(2\xi)]u = 0 \tag{6.29}$$

The solutions of the Mathieu equation describe two alternative scenarios depending on their Lyapunov stability:

1. *Stable (bounded)*: The particles oscillate in the (x, z) plane with limited amplitude. They pass through the quadrupole fields without striking the electrodes.
2. *Unstable (unbounded)*: The amplitudes of the oscillations grow exponentially and the particles are lost.

Clearly, the trap operates in the region where the solutions $r(t)$ and $z(t)$ are both stable. These solutions show that the motion of the ions can be described by a slow (secular) oscillation with fundamental frequencies $\omega_{r,z} = \beta_{r,z}\,\Omega$ modulated with a micromotion which is a much faster oscillation. The general form of the motion is

$$u(t) = R_u\cos(\omega_u t)\left(1 + \frac{q_u}{2}\cos(\Omega t)\right) \tag{6.30}$$

Stable solutions are given by $0 < \beta < 1$. The value of β is given by a pair of continued fractions

$$\beta^2 = a + \cfrac{q^2}{(\beta+2)^2 - a - \cfrac{q^2}{(\beta+4)^2 - a - \cfrac{q^2}{(\beta+6)^2 - a - \cdots}}} + \cfrac{q^2}{(\beta-2)^2 - a - \cfrac{q^2}{(\beta-4)^2 - a - \cfrac{q^2}{(\beta-6)^2 - a - \cdots}}} \tag{6.31}$$

A simple approximation for $q_u < 0.4$ and $a_u \ll q_u$ is $\beta_u = \sqrt{a_u + q_u^2/2}$, but is only accurate within the stated limits.

Since $a_z = -2a_r$ and $q_z = -2q_r$ it is clear that the stability conditions in the two directions are linked. Therefore we construct a plot in the space (a_z, q_z) within which the motion is stable $0 < \beta_r < 1, 0 < \beta_z < 1$. Figure 6.3 shows the basic form of such a plot.[6] The trap is stable within the enclosing quadrilateral formed by arcs of the lines $\beta_r \in 0.1$ and $\beta_z \in 0, 1$.

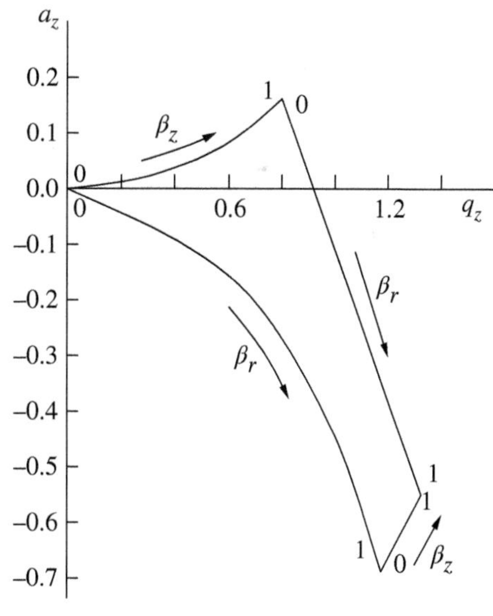

Figure 6.3 Sketch of the characteristic stability for the quadrupole trap.

6 The plot in Figure 6.3 is only approximate and should not be used for calculation. Trap stability plots have been widely constructed and accurate values may easily be found, e.g. Paul (1990) or on the Internet.

7
Kinetic Theory of Plasmas

7.1 Introduction

Thus far, we have considered only collisionless plasma where the mutual interactions are due to long-range collective forces generated by the electric and magnetic fields consequent on bulk charge differences and currents. In this context, we imposed the condition that the only forces were associated with particle separations greater than the Debye length. At these separation distances, the screening of the plasma is sufficiently strong that the field due to individual ions is lost and only the collective macroscopic field remains. Such interactions are reversible and do not lead to thermodynamic equilibrium. They are therefore not associated with an entropy increase.

At distances less than the Debye length, we saw in Section 1.3 that the field due to individual charges exists in the plasma and can lead to scattering of particular electrons or ions. These are typical of collisions as exemplified by those between gas molecules in Section 4.7. As noted in Section 2.1.3, at large impact parameters b, many particles interact simultaneously, but not collectively, provided $b \lesssim \lambda_D$. However, if the impact parameter is less than the inter-particle separation $b < n^{-1/3}$, the collision will be two body and strong[1] :

- Particle interactions with $b_0 \lesssim b \lesssim n^{-1/3}$ are strong binary collisions.
- Particle interactions with $n^{-1/3} \lesssim b \lesssim \lambda_D$ are weak multi-particle collisions.
- Particle interactions with $\lambda_D \lesssim b$ are collective.

Clearly, weak collisions dominate if $n\lambda_D{}^3 \gg 1$, i.e. in the normal condition that the plasma is dilute and weakly coupled (Section 2.1.3).[2]

Under these conditions, the field seen by a test particle is simultaneously due to a large number of randomly distributed perturbers and is rapidly varying in both time and space. As we argued in Section 2.1.3, the particle motion is determined by the statistical average of the independent impulses of the consequent forces due to the perturbers. Since the interactions are all elastic, collisions with stationary perturbers simply rotate the particle's velocity vector inducing a random velocity in the plane transverse to the original motion and a corresponding reduction longitudinally, the variance of the transverse velocity distribution increasing linearly with time. This behaviour is characteristic of a stochastic process and was first analysed as such in an astrophysical context (Jeans, 1961, p. 318). Arguing within a stochastic framework (Chandrasekhar 1960, 1943)

1 Only binary collisions with $b \sim b_0$ are strong, where b_0 is the Landau length, the impact parameter for 90° scattering. If $n^{-1/3} \gg b_0$, a large fraction of binary collisions will be weak.
2 This condition is strictly the weaker one $b_0 \ll \lambda_D$.

Foundations of Plasma Physics for Physicists and Mathematicians, First Edition. Geoffrey J. Pert.

showed that the dynamics of galaxies governed by forces determined by the inverse square law of gravity could be described by a statistical model based on the Fokker–Planck equation originally obtained for Brownian motion. Since plasmas are also governed by similar inverse square law electrostatic forces, it was quickly realised that this approach offered a more satisfactory description than one based on the Boltzmann equation (Cohen et al., 1950; Spitzer, 1956; Allis, 1956).

In a gas, the interactions are short range so that the collision time is short, typically less than the fluctuation time of the molecules. In contrast in plasma, the long range of the Coulomb force ensures that the collision time is long compared to the fluctuation time. This important difference leads to the distinction between the well-defined two body form of the Boltzmann collision operator for gases and the stochastic Fokker–Planck term for plasma. In plasma, therefore, the Fokker–Planck collision term, therefore, replaces the Boltzmann collision operator in the dynamic Boltzmann equation (4.41).

In addition to being derived from stochastic principles, the Fokker–Planck collision term can be derived directly from the dynamics of multiple small angle collisions (Rosenbluth et al., 1957). It is, therefore, possible to obtain the Fokker–Planck operator for Coulomb collisions as the weak collision limit of the Boltzmann term (Landau, 1936) when third-order correlations in the BBGKY hierarchy can be neglected (Section 4.6). In the case of simple gases, this condition is achieved by limiting the derivation to binary collisions, where the probability of a three particle collision is very small. In contrast, in plasma interactions are many body, but weak. Consequently, the third-order correlation term is weak. Thus, we may anticipate that the Boltzmann equation, with the proviso that the interactions are weak, may be applied in this case also, but with the clear understanding of the limitations of the physical picture introduced thereby. The mathematical relationship indicates the derivation of the Fokker–Planck collision term from the BBGKY hierarchy when third- and higher-order correlations are neglected due to the long-range, weak interaction nature of the Coulomb field in plasma.

The direct derivation of the Fokker–Planck equation for plasmas from the BBGKY hierarchy is achieved using the globular expansion. Imposing Bogoliubov's condition to plasmas leads to the Balescu–Lenard equation (Balescu, 1960; Lenard, 1960), which can be reduced to the standard form of the Fokker–Planck equation.

7.2 The Fokker–Planck Equation – The Stochastic Approach

The Fokker–Planck equation as applied to plasma is most satisfactorily introduced through a similar probability argument to that used earlier dealing with the Boltzmann collision integral (Section 4.7). The plasma is assumed to be sufficiently tenuous that its dynamical state may be described by a single particle distribution function. Although a particle is simultaneously subjected to forces from many neighbouring particles, the effect of each is small and the result is that of many independent impacts (Section 2.1.3), not involving the simultaneous interaction from a cluster of correlated particles, i.e. the plasma is weakly coupled (Section 1.3.1). As shown in Section 2.1, in this limit collisional changes are dominated by weak, multi-particle interactions by a factor $\ln\Lambda$, approximately equal to the logarithm of the necessarily large plasma parameter. In principle infrequent strong, close impacts can be accounted for approximately by including a suitable cut-off for small impacts equal to the Landau parameter b_0 in the weak scattering formula.

Multiple small interactions change the distribution. We, therefore, investigate the rate of change in the distribution function that an electron with velocity $\mathbf{v}$ undergoes through a series of collisions that change its velocity by a small amount $\mathbf{\Delta v}$ in the range $\mathrm{d}\mathbf{\Delta v}$ with a probability $\psi(\mathbf{v}, \mathbf{\Delta v})\,\mathrm{d}\mathbf{\Delta v}$ in time Δt. It is assumed that the collisions are between random particles and the collisions are Markovian, i.e. the probability of a given transition depends only on the instantaneous values, not on those during the earlier history (see Chandrasekhar, 1943) – a requirement equivalent to the condition of molecular chaos (item (v), p. 90).The distribution function at time t is therefore the modified value of that at the earlier time $t - \Delta t$ resulting from these changes:

$$f(\mathbf{r}, \mathbf{v}, t) = \int \lfloor f(\mathbf{r}, (\mathbf{v} - \mathbf{\Delta v}), (t - \Delta t t))\rfloor\, \psi(\mathbf{v}, \mathbf{\Delta v})\,\mathrm{d}\mathbf{\Delta v} \tag{7.1}$$

where $\psi(\mathbf{v}, \mathbf{\Delta v})\,\mathrm{d}\mathbf{\Delta v}$ is the probability that the velocity changes from $\mathbf{v} - \mathbf{\Delta v}$ in the range $\mathrm{d}\mathbf{\Delta v}$ to $\mathbf{v}$ in time Δt.

If Δt and $\mathbf{\Delta v}$ are small, we can expand the terms in a Taylor series to give

$$\begin{aligned} f(\mathbf{r}, \mathbf{v}, t) &= \int d(\Delta v)\, f(\mathbf{r}, \mathbf{v}, (t - \Delta t))\, \psi(\mathbf{v}, \mathbf{\Delta v}) \\ &\quad - \mathbf{\Delta v} \cdot \left[\frac{\partial f}{\partial \mathbf{v}}\, \psi + f\, \frac{\partial \psi}{\partial \mathbf{v}}\right] + \frac{1}{2}\, \Delta v_i\, \Delta v_j \left[\psi\, \frac{\partial^2 f}{\partial v_i\, \partial v_j} + 2 \frac{\partial f}{\partial v_i}\, \frac{\partial \psi}{\partial v_j} + f\, \frac{\partial^2 \psi}{\partial v_i\, \partial v_j}\right] \end{aligned}$$

Since the total probability of all possible events must be unity:

$$\int \psi(\mathbf{v}, \mathbf{\Delta v})\,\mathrm{d}\mathbf{\Delta v} = 1 \tag{7.2}$$

Hence, the rate of change of the distribution function due to collisions is

$$\left.\frac{\partial f}{\partial t}\right|_{\mathrm{coll}} = -\frac{\partial}{\partial v_i}[f\langle \Delta v_i \rangle] + \frac{1}{2}\,\frac{\partial^2}{\partial v_i\, \partial v_j}[f\langle \Delta v_i \Delta v_j \rangle] \tag{7.3}$$

The averages of the velocity and mean squared velocity changes per unit time are given by the probability distribution: $\psi(\mathbf{v})$.

$$\left\{ \begin{array}{c} \langle \Delta v_i \rangle \\ \langle \Delta v_i \Delta v_j \rangle \end{array} \right\} = \frac{1}{\Delta t} \int (\mathbf{\Delta v}) \left\{ \begin{array}{c} \Delta v_i \\ \Delta v_i \Delta v_j \end{array} \right\} \psi(\mathbf{v}),\, \mathrm{d}\mathbf{\Delta v}.$$

The upper line refers to the mean change in the velocity and lower to the mean change in the product of two velocity components.

As a simple example of a set of Fokker–Planck coefficients, we may identify Eqs. (2.21) for the elementary case of electrons interacting with stationary ions. As this does not involve changes in the electron energy, it cannot treat the relaxation of the electron velocity to a Maxwellian thermal distribution, only the isotropisation of a beam of particles. More general sets of coefficients allowing for energy transfer and consequently thermal relaxation are derived later.

In elastic scattering, the relative velocity vector of a particle is simply rotated, but its magnitude unchanged. Therefore, the only non-zero average terms are the change in velocity parallel to the incoming motion $\langle \Delta v_{\parallel} \rangle < 0$ and the mean squared velocity transverse to it $\langle \Delta v_{\perp}{}^2 \rangle > 0$, Eqs. (2.21).

The Fokker–Planck equation represents a flux of particles in velocity space driven by a Markovian probability $\psi(\mathbf{v}, \mathbf{\Delta v})$

$$\begin{aligned} \mathcal{J}_i &= \langle \Delta v_i \rangle f - \frac{1}{2}\frac{\partial}{\partial v_j}(\langle \Delta v_i\, \Delta v_j \rangle\, f) \\ &= \underbrace{\left[\langle \Delta v_i \rangle - \frac{1}{2}\frac{\partial}{\partial v_j}(\langle \Delta v_i\, \Delta v_j \rangle)\right]}_{A_i} f - \underbrace{\frac{1}{2}(\langle \Delta v_i\, \Delta v_j \rangle)}_{B_{ij}}\, \frac{\partial f}{\partial v_j} \end{aligned} \tag{7.4}$$

such that

$$\left.\frac{\partial f}{\partial t}\right|_{\text{coll}} + \frac{\partial \mathcal{J}_i}{\partial v_i} = 0 \qquad (7.5)$$

The first term A_i is known as the *coefficient of dynamical friction* and the second B_{ij} is the *coefficient of diffusion* (Chandrasekhar, 1960). The origin of these names is easily seen from Eqs. (2.21): respectively the term, $\langle \Delta v_\parallel \rangle$ represents the slowing down of a beam of particles and the term $\langle \Delta {v_\perp}^2 \rangle$ the transverse spreading of the beam. *Dynamical friction* draws the velocity of individual particles towards the 'mean', whilst *diffusion in velocity* spreads the particle velocity distribution towards a 'bell' shaped curve by diffusion in velocity space. In equilibrium, these two effects balance against one another giving rise to a Maxwellian distribution.

The Fokker–Planck equation has a very general form describing the changes in the distribution function associated with small random impulses, applicable to many stochastic processes in physics besides collisions. It is easily seen that the general form takes the structure of a diffusion equation in an external field $\mathbf{F}$:

$$\frac{\partial n}{\partial t} = -\nabla \cdot [\mu\, \mathbf{F}\, n - D\, \nabla n] \qquad (7.6)$$

where μ is the mobility, D the diffusivity, and n the density. When the force is conservative $\mathbf{F} = -\nabla \phi$, the mobility and diffusivity are connected by Einstein's relation $D = \mu\, \mathcal{k}T$ since in thermal equilibrium $n = n_0 \exp(-\phi/\mathcal{k}T)$ and $\partial n/\partial t = 0$. The mobility is equivalent to the *coefficient of dynamical friction* and the diffusivity to the *coefficient of diffusion*. In a similar fashion to Einstein's relation, we infer that in thermal equilibrium, the flux $j_i = 0$ is zero and the distribution function has the Boltzmann form $f = \text{const}\ \exp(-mv^2/2\mathcal{k}T)$ and, therefore, the terms A_i and B_{ij} are not independent but satisfy

$$A_i\, \mathcal{k}T + B_{ij}\, m\, v_j = 0 \qquad (7.7)$$

where T is the temperature when equilibrium is established.

7.2.1 The Scattering Integral for Coulomb Collisions

The detailed calculation of the collision terms following Coulomb scattering is straightforward, but complicated (Rosenbluth et al., 1957). It is predicated on the assumption that it is satisfactory to perform the calculation using binary collisions in a regime, where the interactions are dominantly multi-particle but weak, as discussed in Section 2.1.3.

We consider the case of two particles of charges Z and Z' with masses m and m', respectively, moving with velocities $\mathbf{v}$ and $\mathbf{v}'$. As is well known, an elastic collision occurs with scattering about the centre of mass involving a rotation of the relative velocity vector $\mathbf{u} = \mathbf{v} - \mathbf{v}'$. The dynamics of the interaction are described by the motion of a particle of mass equal to the reduced mass $\overline{m} = mm'/(m + m')$ about the centre of mass with velocity equal to the relative velocity u.[3]

3 Two particles of mass m and m' are moving with velocities $\mathbf{v}$ and $\mathbf{v}'$ and position vectors $\mathbf{r}$ and $\mathbf{r}'$, respectively. The interaction forces between the particles are $\mathbf{F}(\rho)$ and $-\mathbf{F}(\rho)$, respectively, where $\rho = \mathbf{r} - \mathbf{r}'$ is their separation. The centre of mass velocity $\mathbf{V} = (m'\mathbf{v} + m\mathbf{v}')/(m + m')$. Relative to the centre of mass the position vectors are $m'\rho/(m + m')$ and $-m\rho/(m + m')$ and the velocities $m'\mathbf{u}/(m + m')$ and $-m\mathbf{u}/(m + m')$ respectively. Hence,

$$m\frac{\mathrm{d}^2\mathbf{r}}{\mathrm{d}t^2} = \mathbf{F}(\rho) = -m'\frac{\mathrm{d}^2\mathbf{r}'}{\mathrm{d}t^2}.$$

Therefore, the centre of mass moves with constant velocity, and the separation:

$$\frac{\mathrm{d}^2\rho}{\mathrm{d}t^2} = \left(\frac{1}{m} + \frac{1}{m'}\right)\mathbf{F}(\rho)$$

The velocities of the individual particles in terms of the centre of mass velocity **V** and the relative velocity **u** are

$$\mathbf{v} = \mathbf{V} + \frac{m'}{m+m'}\mathbf{u}$$
$$\mathbf{v}' = \mathbf{V} - \frac{m}{m+m'}\mathbf{u}.$$

Hence, the change in the velocity **v** as a result of the collision:

$$\Delta \mathbf{v} = \frac{m'}{m+m'}\,\Delta \mathbf{u}.$$

The scattering gives rise to velocity changes in the perpendicular plane, $\perp$, and parallel, $\|$, to the initial relative velocity **u** as above. The cross section for scatter through an angle ϕ is modified by the inclusion of the reduced mass $\overline{m}$ but is otherwise unchanged. Therefore, we may use the averages we have already calculated, Eqs. (2.21) but with the mass replaced by the reduced mass.

The changes in direction of the relative velocity **u** and the centre of mass velocity **V** must be removed before we can relate the particle velocities to their values in the laboratory frame. To do this, we identify a set of directions with respect to **u**, namely, $\hat{\lambda}$ parallel to **u** and $\hat{\mu}$ and $\hat{\nu}$ perpendicular to **u** so that

$$\begin{aligned}
\Delta u_\lambda &= -u(1-\cos\phi) = 2u\sin^2(\phi/2)\\
\Delta u_\mu &= u\sin\phi\,\cos\chi = 2u\sin(\phi/2)\,\cos(\phi/2)\,\cos\chi\\
\Delta u_\nu &= u\sin\phi\,\sin\chi = 2u\sin(\phi/2)\,\cos(\phi/2)\,\sin\chi
\end{aligned} \tag{7.8}$$

χ being the rotation angle in the plane $\hat{\mu}$, $\hat{\nu}$ perpendicular to **u**.

Using either the small angle approach as in Eq. (2.13) or directly from the full Rutherford cross section, retaining the dominant ($\ln\Lambda$) term Eq. (2.5) for particles of charge $-e$ and Ze. The resultant averages are carried out as described in Section 2.1, but in the centre of mass frame

$$\begin{aligned}
\langle \Delta u_\lambda\rangle &= \langle \Delta u_\|\rangle = -\frac{4\pi Z^2 Z'^2 e^4}{(4\,\pi\epsilon_0)^2\,\overline{m}^2 u^2}\ln\Lambda\\
\langle \Delta u_\mu\rangle &= \langle \Delta u_\nu\rangle = \langle \Delta u_\perp\rangle = 0\\
\langle \Delta u_\lambda{}^2\rangle &= \langle \Delta u_\|{}^2\rangle = 0\\
\langle \Delta u_\mu{}^2\rangle &= \langle \Delta u_\nu{}^2\rangle = \left\langle \frac{1}{2}\Delta u_\perp{}^2\right\rangle = \frac{4\pi Z^2 Z'^2 e^4}{(4\,\pi\epsilon_0)^2\,\overline{m}^2 u}\ln\Lambda.
\end{aligned} \tag{7.9}$$

The unit vectors in the directions $\hat{\lambda}$, $\hat{\mu}$ and $\hat{\nu}$ are

$$\hat{\lambda} = \frac{\mathbf{u}}{u} \qquad \hat{\mu} = \frac{\hat{\mathbf{k}}\wedge\mathbf{u}}{u_x{}^2 + u_y{}^2} \qquad \text{and} \qquad \hat{\nu} = \hat{\lambda}\wedge\hat{\mu},$$

where ($\hat{\imath}$, $\hat{\mathbf{j}}$, $\hat{\mathbf{k}}$) are the unit vectors in the (x, y, z) directions.

Thus, $\boldsymbol{\Delta}\mathbf{u}$ in the laboratory set of Cartesian co-ordinates (x, y, z) may written

$$\Delta u_x = (\hat{\imath}\cdot\hat{\lambda})\Delta u_\lambda + (\hat{\imath}\cdot\hat{\mu})\Delta u_\mu + (\hat{\imath}\cdot\hat{\nu})\Delta u_\nu,$$

with similar expressions for Δu_y and Δu_z. Taking the average

$$\begin{aligned}
\langle \Delta u_x\rangle &= (\hat{\imath}\cdot\hat{\lambda})\langle \Delta u_\lambda\rangle + (\hat{\imath}\cdot\hat{\mu})\cancelto{0}{\langle \Delta u_\mu\rangle} + (\hat{\imath}\cdot\hat{\nu})\cancelto{0}{\langle \Delta u_\nu\rangle}\\
&= \frac{u_x}{u}\langle \Delta u_\|\rangle
\end{aligned} \tag{7.10}$$

---

satisfies the dynamics of a particle of mass $m\,m'/(m+m')$ (the reduced mass) moving under the action of the force $\mathbf{F}(\rho)$ at a distance ρ about the centre of mass.

with similar expressions for $\langle u_y\rangle$ and $\langle u_z\rangle$.

The product terms $\langle\Delta u_x\,\Delta u_y\rangle$, etc., are obtained in a similar manner, noting that several terms cancel to zero:

$$\begin{aligned}\langle\Delta u_x\,\Delta u_y\rangle &= \Big\langle[(\hat{\imath}\cdot\hat{\lambda})\Delta u_\lambda+(\hat{\imath}\cdot\hat{\mu})\Delta u_\mu+(\hat{\imath}\cdot\hat{\nu})\Delta u_\nu][(\hat{\mathbf{j}}\cdot\hat{\lambda})\Delta u_\lambda+(\hat{\mathbf{j}}\cdot\hat{\mu})\Delta u_\mu+(\hat{\mathbf{j}}\cdot\hat{\nu})\Delta u_\nu]\Big\rangle\\
&= \Big\langle[(\hat{\imath}\cdot\hat{\lambda})(\hat{\mathbf{j}}\cdot\hat{\lambda})]\cancel{\Delta u_\lambda^2}+[(\hat{\imath}\cdot\hat{\lambda})(\hat{\mathbf{j}}\cdot\hat{\mu})+(\hat{\imath}\cdot\hat{\mu})(\hat{\mathbf{j}}\cdot\hat{\lambda})]\cancel{\Delta u_\lambda\,\Delta u_\mu}\\
&\quad+[(\hat{\imath}\cdot\hat{\mu})(\hat{\mathbf{j}}\cdot\hat{\mu})]\Delta u_\mu{}^2+[(\hat{\imath}\cdot\hat{\nu})(\hat{\mathbf{j}}\cdot\hat{\lambda})+(\hat{\imath}\cdot\hat{\lambda})(\hat{\mathbf{j}}\cdot\hat{\nu})]\cancel{\Delta u_\lambda\,\Delta u_\nu}\\
&\quad+[(\hat{\imath}\cdot\hat{\nu})(\hat{\mathbf{j}}\cdot\hat{\nu})]\Delta u_\nu{}^2+[(\hat{\imath}\cdot\hat{\mu})(\hat{\mathbf{j}}\cdot\hat{\nu})+(\hat{\imath}\cdot\hat{\nu})(\hat{\mathbf{j}}\cdot\hat{\mu})]\cancel{\Delta u_\mu\,\Delta u_\nu}\Big\rangle\\
&= [(\hat{\imath}\cdot\hat{\mu})(\hat{\mathbf{j}}\cdot\hat{\mu})+(\hat{\imath}\cdot\hat{\nu})(\hat{\mathbf{j}}\cdot\hat{\nu})]\,\frac{1}{2}\,\langle\Delta u_\perp{}^2\rangle\,.\end{aligned}$$

Hence, generalising

$$\begin{aligned}\langle\Delta u_i\,\Delta u_j\rangle &= [\hat{\mu}_i\,\hat{\mu}_j+\hat{\nu}_i\,\hat{\nu}_j]\,\frac{1}{2}\,\langle\Delta u_\perp{}^2\rangle\\
&= [\delta_{ij}-\hat{\lambda}_i\,\hat{\lambda}_j]\,\frac{1}{2}\,\langle\Delta u_\perp{}^2\rangle\end{aligned}\tag{7.11}$$

since the components of the unit vectors $\hat{\lambda},\ \hat{\mu},\ \hat{\nu}$ on the orthogonal set of unit vectors $\hat{\imath},\ \hat{\mathbf{j}},\ \hat{\mathbf{k}}$, namely, $(\hat{\lambda}_i,\ \hat{\lambda}_j,\ \hat{\lambda}_k)$, $(\hat{\mu}_i,\ \hat{\mu}_j,\ \hat{\mu}_k)$ and $(\hat{\nu}_i,\ \hat{\nu}_j,\ \hat{\nu}_k)$, respectively, satisfy the condition:

$$\hat{\lambda}_i\hat{\lambda}_j+\hat{\mu}_i\hat{\mu}_j+\hat{\nu}_i\hat{\nu}_j=\delta_{ij}.$$

The projection matrix $\delta_{ij}-u_iu_j/u^2$ projects the product of the components $\Delta u_i\Delta u_j$ on to the transverse plane.

Returning to the laboratory frame, the change in the particle averages are given by

$$\begin{aligned}\langle\Delta\,v_i\rangle &= \frac{m'}{m+m'}\,\langle\Delta\,u_i\rangle\\
&= -\frac{4\,\pi\,Z^2\,Z'^2\,e^4\,\ln\Lambda}{(4\,\pi\,\epsilon_0)^2\,\overline{m}\,m}\,\frac{u_i}{u^3}\\
&= \frac{4\,\pi\,Z^2\,Z'^2\,e^4\,\ln\Lambda}{(4\,\pi\,\epsilon_0)^2\,\overline{m}\,m}\,\frac{\partial}{\partial v_i}\left(\frac{1}{u}\right)\end{aligned}\tag{7.12}$$

since $u=\sqrt{(\mathbf{v}-\mathbf{v}')\cdot(\mathbf{v}-\mathbf{v}')}$ and

$$\begin{aligned}\langle\Delta v_i\,\Delta v_j\rangle &= \frac{4\,\pi\,Z^2\,Z'^2\,e^4\,\ln\Lambda}{(4\,\pi\,\epsilon_0)^2\,m^2\,u}\left[\delta_{ij}-\frac{u_i\,u_j}{u^2}\right]\\
&= \frac{4\,\pi\,Z^2\,Z'^2\,e^4\,\ln\Lambda}{(4\,\pi\,\epsilon_0)^2\,m^2}\,\frac{\partial^2 u}{\partial v_i\,\partial v_j}.\end{aligned}\tag{7.13}$$

These terms are written in terms of the functions known as the Rosenbluth potentials

$$\mathcal{G}(\mathbf{v})=\sum_{\text{pert}}\int \mathrm{d}\mathbf{v}' f'(\mathbf{v}')|\mathbf{v}-\mathbf{v}'|\tag{7.14a}$$

$$\mathcal{H}(\mathbf{v})=\sum_{\text{pert}}\frac{m+m'}{m'}\int \mathrm{d}\mathbf{v}' f'(\mathbf{v}')|\mathbf{v}-\mathbf{v}'|^{-1},\tag{7.14b}$$

the sums being taken over all the primed perturbing species. Noting the derivatives $\partial u/\partial u_i=u_i/u$ and $\partial u_j/\partial u_i=\delta_{ij}$, and defining the modified projection matrix

$$\omega_{ij}=\frac{1}{u}\left[\delta_{ij}-\frac{u_i\,u_j}{u^2}\right]\tag{7.15}$$

which has derivatives

$$\frac{\partial \omega_{ij}}{\partial u_j} = \frac{\partial \omega_{ij}}{\partial v_j} = -\frac{\partial \omega_{ij}}{\partial v'_j} = -\frac{2u_i}{u^3} \tag{7.16}$$

and trace $\omega_{ii} = 2/u$.

The derivatives of $\mathcal{G}(\mathbf{v})$ are

$$\frac{\partial \mathcal{G}}{\partial v_i} = \sum_{\text{pert}} \int \mathrm{d}\mathbf{v}' f'(\mathbf{v}') \frac{u_i}{u}$$

$$\frac{\partial^2 \mathcal{G}}{\partial v_i \, \partial v_j} = \sum_{\text{pert}} \int \mathrm{d}\mathbf{v}' f'(\mathbf{v}') \frac{1}{u} \left[\delta_{ij} - \frac{u_i u_j}{u^2}\right] = \sum_{\text{pert}} \int \mathrm{d}\mathbf{v}' f'(\mathbf{v}') \, \omega_{ij} \tag{7.17}$$

$$\frac{\partial^2 \mathcal{G}}{\partial v_i \, \partial v_i} = 2 \sum_{\text{pert}} \int \mathrm{d}\mathbf{v}' \, \frac{1}{u} f'(\mathbf{v}'), \tag{7.18}$$

and those of $\mathcal{H}(\mathbf{v})$

$$\begin{aligned} \frac{\partial \mathcal{H}}{\partial v_i} &= \sum_{\text{pert}} \frac{(m+m')}{m'} \int \mathrm{d}\mathbf{v}' f'(\mathbf{v}') \frac{\partial}{\partial v_i}\left(\frac{1}{u}\right) = -\sum_{\text{pert}} \frac{(m+m')}{m'} \int \mathrm{d}\mathbf{v}' f'(\mathbf{v}') \frac{u_i}{u^3} \\ &= \sum_{\text{pert}} \frac{(m+m')}{2\,m'} \int \mathrm{d}\mathbf{v}' f'(\mathbf{v}') \frac{\partial \omega_{ij}}{\partial v'_j} = -\sum_{\text{pert}} \frac{(m+m')}{2\,m'} \int \mathrm{d}\mathbf{v}' \frac{\partial f'}{\partial v'_j} \omega_{ij} \end{aligned} \tag{7.19}$$

$$\begin{aligned} \frac{\partial^2 \mathcal{H}}{\partial v_i \, \partial v_i} &= -\sum_{\text{pert}} \frac{(m+m')}{m'} \int \mathrm{d}v' f'(\mathbf{v}') \frac{\partial^2}{\partial v_i \, \partial v_i}\left(\frac{1}{u}\right) \\ &= -\sum_{\text{pert}} 4\,\pi \, \frac{(m+m')}{m'} f'(\mathbf{v}). \end{aligned} \tag{7.20}$$

These results follow directly from the definitions, Eqs. (7.14), and the familiar integrals used in potential theory $\nabla^2 r = 2/r$, and $\nabla^2(1/r) = -4\,\pi\,\delta(r)$. For a single perturber, these results simplify to

$$\frac{\partial^2 \mathcal{G}}{\partial v_i \, \partial v_i} = \nabla_{\mathbf{v}}{}^2 \, \mathcal{G}(\mathbf{v}) = \frac{2\,m'}{(m+m')} \mathcal{H}(\mathbf{v}) \tag{7.21}$$

$$\frac{\partial^2 \mathcal{H}}{\partial v_i \, \partial v_i} = \nabla_{\mathbf{v}}{}^2 \, \mathcal{H}(\mathbf{v}) = -4\,\pi \, \frac{(m+m')}{m'} f'(\mathbf{v}), \tag{7.22}$$

which we note take the form of Poisson's equation in velocity space.

The higher derivatives of $\mathcal{G}(\mathbf{v})$ follow directly

$$\frac{\partial^3 \mathcal{G}}{\partial v_i \, \partial v_j \, \partial v_j} = \frac{2m'}{(m+m')} \frac{\partial \mathcal{H}}{\partial v_i} \tag{7.23}$$

$$\frac{\partial^4 \mathcal{G}}{\partial v_i \, \partial v_i \, \partial v_j \, \partial v_j} = \frac{2m'}{(m+m')} \frac{\partial^2 \mathcal{H}}{\partial v_i \, \partial v_i} = -8\,\pi\, f'(\mathbf{v}). \tag{7.24}$$

Substituting, we obtain the Rosenbluth et al. (1957) form of the Fokker–Planck collision terms

$$\left\langle \Delta v_i \right\rangle = \frac{4\,\pi\, Z^2\, Z'^2\, e^4\, \ln \Lambda}{(4\,\pi\,\epsilon_0)^2\, m^2} \frac{\partial \mathcal{H}}{\partial v_i} \tag{7.25}$$

$$\left\langle \Delta v_i \, \Delta v_j \right\rangle = \frac{4\,\pi\, Z^2\, Z'^2\, e^4\, \ln \Lambda}{(4\,\pi\,\epsilon_0)^2\, m^2} \frac{\partial^2 \mathcal{G}}{\partial v_i \, \partial v_j}. \tag{7.26}$$

Hence,

$$\left.\frac{\partial f}{\partial t}\right|_{coll} = \frac{4\,\pi\,Z^2\,Z'^2\,e^4}{(4\,\pi\,\epsilon_0)^2\,m^2}\ln\Lambda\left\{-\frac{\partial}{\partial v_i}\left[f(\mathbf{v})\underbrace{\frac{\partial \mathcal{H}}{\partial v_i}}_{\langle \Delta v_i\rangle}\right] + \frac{1}{2}\frac{\partial^2}{\partial v_i\,\partial v_j}\left[f(\mathbf{v})\underbrace{\frac{\partial^2 \mathcal{G}}{\partial v_i\,\partial v_j}}_{\langle \Delta v_i\,\Delta v_j\rangle}\right]\right\} \tag{7.27}$$

Rosenbluth et al. (1957) generalise this expression to an equation in general curvilinear co-ordinates. This allows the development in terms of cylindrical and other co-ordinate systems. In polar co-ordinates, without symmetry, the analysis gives basically a similar, but more complicated, result. In particular, Eq. (7.27) can be written in terms of vectors and dyads of the velocity gradient operator $\nabla_\mathbf{v} \equiv \partial/\partial v_i$ as follows:

$$\left.\frac{\partial f}{\partial t}\right|_{coll} = \frac{4\,\pi\,Z^2\,Z'^2\,e^4}{(4\,\pi\,\epsilon_0)^2\,m^2}\ln\Lambda\left\{-\nabla_\mathbf{v}\cdot[f(\mathbf{v})\nabla_\mathbf{v}\mathcal{H}(\mathbf{v})] + \frac{1}{2}\nabla_\mathbf{v}\nabla_\mathbf{v} : [f(\mathbf{v})\nabla_\mathbf{v}\nabla_\mathbf{v}\mathcal{G}(\mathbf{v})]\right\} \tag{7.27:a}$$

which allows the form for polar co-ordinates to be easily identified.

The probability flux in velocity space from Eq. (7.4) is easily expressed as

$$\begin{aligned}\mathbf{j} &= \frac{4\,\pi\,Z^2\,Z'^2\,e^4}{(4\,\pi\,\epsilon_0)^2\,m^2}\ln\Lambda\left\{\left[\nabla_\mathbf{v}\mathcal{H}(\mathbf{v}) - \frac{1}{2}\nabla_\mathbf{v}\cdot\nabla_\mathbf{v}\nabla_\mathbf{v}\mathcal{G}(\mathbf{v})\right]f(\mathbf{v}) - \frac{1}{2}\nabla_\mathbf{v}\nabla_\mathbf{v}\mathcal{G}(\mathbf{v})\cdot\nabla_\mathbf{v}f(\mathbf{v})\right\}\\ &= \frac{4\,\pi\,Z^2\,Z'^2\,e^4}{(4\,\pi\,\epsilon_0)^2\,m^2}\ln\Lambda\left\{\frac{m}{m+m'}\nabla_\mathbf{v}\mathcal{H}(\mathbf{v}) - \frac{1}{2}\nabla_\mathbf{v}\nabla_\mathbf{v}\mathcal{G}(\mathbf{v})\cdot\nabla_\mathbf{v}f(\mathbf{v})\right\}\end{aligned} \tag{7.28}$$

We may cast the resulting Fokker–Planck equation in a number of equivalent forms. For simplicity, consider a single perturbing particle and transform the principal term using Eqs. (7.18), (7.23) and (7.19):

$$\begin{aligned}&\frac{\partial f}{\partial v_j}\frac{\partial^2\mathcal{G}}{\partial v_i\,\partial v_j} + f(\mathbf{v})\frac{\partial^3\mathcal{G}}{\partial v_i\,\partial v_j\,\partial v_j} - 2f(\mathbf{v})\frac{\partial\mathcal{H}}{\partial v_i}\\ &= \int d\mathbf{v}'\left[f'(\mathbf{v}')\frac{\partial f}{\partial v_j} + f(\mathbf{v})\frac{\partial f'}{\partial v'_j}\right]\omega_{ij} + \frac{(m+m')}{m'}\int d\mathbf{v}'\,f(\mathbf{v})f'(\mathbf{v}')\frac{\partial\omega_{ij}}{\partial v'_j}\\ &= \int d\mathbf{v}'\left[f'(\mathbf{v}')\frac{\partial f}{\partial v_j} - \frac{m}{m'}f(\mathbf{v})\frac{\partial f'}{\partial v'_j}\right]\omega_{ij}\end{aligned}$$

to obtain an alternative form of the collision operator

$$\left.\frac{\partial f}{\partial t}\right|_{coll} = \frac{2\,\pi\,Z^2\,e^4}{(4\,\pi\,\epsilon_0)^2\,m^2}\ln\Lambda\frac{\partial}{\partial v_i}\left\{\int d\mathbf{v}'\left[f'(\mathbf{v}')\frac{\partial f}{\partial v_j} - \frac{m}{m'}f(\mathbf{v})\frac{\partial f'}{\partial v'_j}\right]\omega_{ij}\right\} \tag{7.29}$$

7.3 The Fokker–Planck Equation – The Landau Equation

The Fokker–Planck equation accounts for simultaneous many body interactions. As we argued in Section 2.1.3, when individual interactions are weak, each delivers a small impulse to the test particle and only slightly perturbs its motion. Each corresponding velocity increment is, therefore, added independently to the others (see Chandrasekhar (1960, §2.3.iii) for a discussion of this assumption). The resultant is a linear vectoral sum of a series of transverse impulses to the particle velocity. The test particle path is, therefore, subject to a series of random transverse increments described by a stochastic analysis, the particle velocity describing a random walk in velocity space.

The large number of small simultaneous binary collisions, thus, constitutes a many body collision, but three or more body correlations, which might be expected to be introduced by the many body behaviour, are small due to low density, dilute medium. It is, therefore, reasonable to expect that the collision operator in Boltzmann's equation (4.41) can be written in a Fokker–Planck form provided the collisions are weak (Landau, 1936). The consequent equivalence of these two forms enables the transition probability ψ to be calculated and allows the Fokker–Planck method to be used to calculate the collision operator in dilute plasmas.

We consider a similar approach to that of the Boltzmann collision operator (Section 4.7), but with multiple weak collisions only, i.e.

The rate of change of f due to collisions into the phase space volume $d\mathbf{r}\, d\mathbf{v}$ =
the rate of particles entering – rate of particles exiting.

For simplicity, we assume collisions between a test particle m with distribution $f(\mathbf{v})$ and scatterers of mass m' and distribution $f'(\mathbf{v}')$. The test particle ① with velocity $\mathbf{v}$ scatters from a particle ② with velocity $\mathbf{v}'$, the relative velocity being $\mathbf{u} = \mathbf{v} - \mathbf{v}'$ and centre of mass velocity $\mathbf{v} = (m\mathbf{v} + m'\mathbf{v}')/(m + m')$. After collision, the velocity of particle ① is increased by $\Delta\mathbf{v}$ and that of ② decreased by $\Delta\mathbf{v}' = (m/m')\Delta\mathbf{v}$ by an equal momentum change. The differential cross section for a collision between two simple particles, e.g. electrons and/or ions, is a function of the relative speed u and the angle of deflection ϕ only, namely $\sigma(u, \phi)d\Omega$. In order to extend the discussion to multiple collisions, the cross section must be replaced by the equivalent term, the velocity transition probability accounting for many particle simultaneous transitions:

$$u\,\sigma(u,\phi)\, d\Omega \Rightarrow \psi\left((\mathbf{v} + \Delta\mathbf{v}/2), (\mathbf{v}' - \Delta\mathbf{v}'/2), \Delta\mathbf{v}\right) d\Delta\mathbf{v} \tag{7.30}$$

calculated at the mid-point of the transition (i.e. the half-sum of the velocities of the initial and final states). It follows from either the above or the principle of detailed balance that the probability is unchanged if the transition is reversed, namely, a transition from $[\mathbf{v}, \mathbf{v}']$ to $[(\mathbf{v} + \Delta\mathbf{v}), (\mathbf{v}' - \Delta\mathbf{v}')]$, with velocity change $\Delta\mathbf{v}$ and from $[(\mathbf{v} + \Delta\mathbf{v}), (\mathbf{v}' - \Delta\mathbf{v}')]$ to $[\mathbf{v}, \mathbf{v}']$ with velocity change $-\Delta\mathbf{v}$. Therefore,

$$\psi((\mathbf{v} + \Delta\mathbf{v}/2), (\mathbf{v}' - \Delta\mathbf{v}'/2), \Delta\mathbf{v}) = \psi((\mathbf{v} + \Delta\mathbf{v}/2), (\mathbf{v}' - \Delta\mathbf{v}'/2), -\Delta\mathbf{v}) \tag{7.31}$$

and that the probability $\psi(\mathbf{v}, \mathbf{v}', \Delta\mathbf{v})$ is an even function of $\Delta\mathbf{v}$.

The Boltzmann collision integral can be written in terms of the transition probability

$$\begin{aligned}\left.\frac{\partial f}{\partial t}\right|_{\text{coll}} = \iint & \left[f(\mathbf{v} + \Delta\mathbf{v})f(\mathbf{v}' - \Delta\mathbf{v}') - f(\mathbf{v})f(\mathbf{v}')\right] \\ & \times \psi((\mathbf{v} + \Delta\mathbf{v}/2), (\mathbf{v}' - \Delta\mathbf{v}'/2), \Delta\mathbf{v})\, d\mathbf{v}'\, d\Delta\mathbf{v}\end{aligned} \tag{7.32}$$

Expanding the terms in $\Delta\mathbf{v}$, the zero-order terms cancel to zero. The first-order terms

$$\int d\Delta\mathbf{v}\, \psi(\mathbf{v}, \mathbf{v}', \Delta\mathbf{v}) \left(f(\mathbf{v}')\, \frac{\partial f}{\partial \mathbf{v}} \cdot \Delta\mathbf{v} - f(\mathbf{v})\, \frac{\partial f}{\partial \mathbf{v}'} \cdot \Delta\mathbf{v}'\right) = 0,$$

since the probability $\psi(\mathbf{v}, \mathbf{v}', \Delta\mathbf{v})$ is an even function of $\Delta\mathbf{v}$.

Turning now to the second-order terms, we have

$$\begin{aligned}&\iint d\mathbf{v}'\, d\Delta\mathbf{v}\, \psi \left[\frac{\Delta v_i \Delta v_j}{2} f'(\mathbf{v}') \frac{\partial^2 f}{\partial v_i\, \partial v_j} - \Delta v_i\, \Delta v_j' \frac{\partial f}{\partial v_i} \frac{\partial f'}{\partial v_j'} + \frac{\Delta v_i' \Delta v_j'}{2} f(\mathbf{v}) \frac{\partial^2 f'}{\partial v_i'\, \partial v_j'}\right] \\ &+ \iint d\mathbf{v}'\, d\Delta\mathbf{v} \tfrac{1}{2} \left(\frac{\partial \psi}{\partial v_i} \Delta v_i - \frac{\partial \psi}{\partial v_i'} \Delta v_i'\right) \left(f'(\mathbf{v}')\, \tfrac{\partial f}{\partial v_j} \Delta v_j - f(\mathbf{v})\, \tfrac{\partial f'}{\partial v_j'} \Delta v_j'\right).\end{aligned}$$

Integrating with respect to $\mathbf{v}'$ and neglecting terms at the surface at infinity, where $f'(\mathbf{v}') = 0$, we obtain

$$-\frac{1}{2}\iint \mathrm{d}\mathbf{v}'\,\mathrm{d}\boldsymbol{\Delta}\mathbf{v}\frac{\partial\psi}{\partial v_i'}f'(\mathbf{v}')\frac{\partial f}{\partial v_j}\,\Delta v_i'\,\Delta v_j$$

$$=\frac{1}{2}\iint \mathrm{d}\mathbf{v}'\,\mathrm{d}\boldsymbol{\Delta}\mathbf{v}\,\psi(\mathbf{v},\mathbf{v}',\boldsymbol{\Delta}\mathbf{v})\,\frac{\partial f'}{\partial v_i'}\,\frac{\partial f}{\partial v_j}\,\Delta v_i'\,\Delta v_j$$

$$\frac{1}{2}\iint \mathrm{d}\mathbf{v}'\,\mathrm{d}\boldsymbol{\Delta}\mathbf{v}\frac{\partial\psi}{\partial v_i'}\,\frac{\partial f'}{\partial v_j'}\,f(\mathbf{v})\,\Delta v_i'\,\Delta v_j'$$

$$=-\frac{1}{2}\iint \mathrm{d}\mathbf{v}'\,\mathrm{d}\boldsymbol{\Delta}\mathbf{v}\,\psi(\mathbf{v},\mathbf{v}',\boldsymbol{\Delta}\mathbf{v})\,\frac{\partial^2 f'}{\partial v_i'\,\partial v_j'}\,f(\mathbf{v})\,\Delta v_i'\,\Delta v_j'$$

Substituting and collecting terms, we obtain

$$\left.\frac{\partial f}{\partial t}\right|_{\mathrm{coll}} = \frac{\partial}{\partial v_i}\iint \mathrm{d}\mathbf{v}'\,\mathrm{d}\boldsymbol{\Delta}\mathbf{v}\,\psi(\mathbf{v},\mathbf{v}',\boldsymbol{\Delta}\mathbf{v})\left(\frac{\Delta v_i\,\Delta v_j}{2}f'(\mathbf{v}')\frac{\partial f}{\partial v_j} - \frac{\Delta v_i\,\Delta v_j'}{2}f(\mathbf{v})\frac{\partial f'}{\partial v_j'}\right)$$

$$= \frac{\partial}{\partial v_i}\iint \mathrm{d}\mathbf{v}'\left(\frac{\langle \Delta v_i\,\Delta v_j\rangle}{2}f'(\mathbf{v}')\frac{\partial f}{\partial v_j} - \frac{\langle \Delta v_i\,\Delta v_j'\rangle}{2}f(\mathbf{v})\frac{\partial f'}{\partial v_j'}\right), \tag{7.33}$$

which is clearly of the Fokker–Planck form.

Since momentum is conserved in collisions $m\,\boldsymbol{\Delta}\mathbf{v} = m'\,\boldsymbol{\Delta}\mathbf{v}'$, it follows that $\partial f/\partial t\big|_{\mathrm{coll}} = 0$ if both $f(\mathbf{v})$ and $f'(\mathbf{v}')$ are Maxwellian with the same temperature $T = T'$. A Maxwellian distribution is therefore a steady solution.

7.3.1 Application to Collisions between Charged Particles

Momentum is conserved in binary collisions $m\boldsymbol{\Delta}\mathbf{v} = m'\boldsymbol{\Delta}\mathbf{v}'$. Since the Coulomb field is central, the calculation of the dynamics can be reduced to that of particle of reduced mass $\overline{m} = m\,m'/(m+m')$ about the centre of mass (see Footnote 3, Section 7.2.1). The term $\int \mathrm{d}\boldsymbol{\Delta}\mathbf{v}\,\psi(\mathbf{v},\mathbf{v}',\boldsymbol{\Delta}\mathbf{v})\,\Delta v_i\,\Delta v_j$ is the average taken over all possible collisions of the change in the relative velocity $\mathbf{u}$ of the particles, which we have already calculated for a Coulomb field in Eq. (2.21c). Taking into account the reduced mass $\boldsymbol{\Delta}\mathbf{v} = [m'/(m+m')]\boldsymbol{\Delta}\mathbf{u}$ and $\boldsymbol{\Delta}\mathbf{v}' = [m/(m+m')]\boldsymbol{\Delta}\mathbf{u}$, we obtain the average on the plane perpendicular to $\mathbf{u}$

$$\left\langle \Delta u_\perp{}^2\right\rangle = \frac{8\pi(Ze)^2(Z'e)^2\ln\Lambda}{(4\pi\epsilon_0)^2\overline{m}^2 u} \tag{7.34}$$

Since we need the values in a set of arbitrary Cartesian directions, this result must be projected into the appropriate directions i, j using the projection matrix equation (7.11) and the general terms associated with directions i, j become

$$\left\langle \Delta u_i\,\Delta u_j\right\rangle = \frac{4\pi Z^2 Z'^2 e^4\ln\Lambda}{(4\pi\epsilon_0)^2\overline{m}^2 u}\left(\delta_{ij} - \frac{u_i\,u_j}{u^2}\right) \tag{7.35}$$

Substituting in Eq. (7.33), we obtain

$$\left.\frac{\partial f}{\partial t}\right|_{\mathrm{coll}} = 2\pi\left(\frac{Z\,Z'\,e^2}{(4\pi\epsilon_0)\,m}\right)^2\ln\Lambda\,\frac{\partial}{\partial v_i}\left\{\int \mathrm{d}\mathbf{v}'\left[f'(\mathbf{v}')\frac{\partial f}{\partial v_j} - \frac{m}{m'}f(\mathbf{v})\frac{\partial f'}{\partial v_j'}\right]\omega_{ij}\right\} \tag{7.36}$$

This equation, known as the Landau equation, is identical with that obtained earlier (Eq. (7.29)).

This may be cast into the standard form for the Fokker–Planck equation by the transformation of the term in square brackets as follows:

$$
\begin{aligned}
&\frac{\partial}{\partial v_j}\int \mathrm{d}\mathbf{v}' \left\{\left[f'(\mathbf{v}')\frac{\partial f}{\partial v_i} - \frac{m}{m'}f(\mathbf{v})\frac{\partial f'}{\partial v_i'}\right]\omega_{ij}\right\}\\
&= -\frac{m}{m'}\frac{\partial}{\partial v_j}\left\{f(\mathbf{v})\left[\int \mathrm{d}\mathbf{v}'\frac{\partial f'}{\partial v_i'}\right]\omega_{ij}\right\} + \frac{\partial^2}{\partial v_j\,\partial v_i}\left\{f(\mathbf{v})\left[\int \mathrm{d}\mathbf{v}' f'(\mathbf{v}')\omega_{ij}\right]\right\}\\
&\quad -\frac{\partial}{\partial v_j}\left\{f(\mathbf{v})\int \mathrm{d}\mathbf{v}' f'(\mathbf{v}')\frac{\partial \omega_{ij}}{\partial v_i}\right\}\\
&= -\frac{(m+m')}{m'}\frac{\partial}{\partial v_j}\left\{f(v_i)\underbrace{\left[\int \mathrm{d}\mathbf{v}'\frac{\partial f'}{\partial v_i'}\omega_{ij}\right]}_{\langle \Delta v_i\rangle}\right\} + \frac{\partial^2}{\partial v_i\,\partial v_j}\left\{f(\mathbf{v})\underbrace{\left[\int \mathrm{d}\mathbf{v}' f'(\mathbf{v}')\,\omega_{ij}\right]}_{\langle \Delta v_i\,\Delta v_j\rangle}\right\}.
\end{aligned}
\tag{7.37}
$$

The mathematical equivalence of the Boltzmann and Fokker–Planck collision operators as limit forms allows us to infer that many of the properties of the Boltzmann term (Section 4.8) will hold for the Fokker–Planck; full proofs are given by Hinton (1983):

- The collision operator conserves
  - Particle density
  - Total momentum
  - Total energy
- The H-theorem holds, and consequently, the distribution relaxes to an equilibrium state.
- The final equilibrium state is an isotropic Maxwellian, even in the presence of a magnetic field.

Additional properties of the Fokker–Planck term may be deduced from the nature of the diffusion equation since the matrix B_{ij} is positive definite

- If the distribution is initially non-negative, it will remain so. Since $\lim_{\mathbf{v}\to\infty} f(\mathbf{v}) \to 0$, it follows that if the distribution function becomes negative, a minimum must be formed, where $\nabla_{\mathbf{v}} f = 0$, and $B_{ij} \sim \langle \Delta v_\perp{}^2\rangle > 0$. Therefore at a minimum $\partial f/\partial t > 0$ and cannot form.
- Since the eigenvalues of the matrix $B_{ij} > 0$ the Fokker–Planck is stable, i.e. solutions decay to a steady state or remain constant.

7.4 The Fokker–Planck Equation – The Cluster Expansion

Our discussion of the Fokker–Planck equation thus far has been intuitive. First, we have assumed that a stochastic picture satisfactorily described the random nature of the multi-particle collisions. Second, we have assumed that a form of the ergodic hypothesis[4] allows us to average the simultaneous effects of many interacting particles as separate events in time. In fact, both assumptions are well upheld under most conditions. However, a more satisfactory approach is through the

4 In statistical mechanics, the assertion that the average values over time of the physical quantities that characterise a system are equal to the statistical average values of the quantities.

BBGKY hierarchy of for gases. Some progress towards this result is obtained from Landau's modification of the Boltzmann collision term, but it still relies on the summation of individual collisions, an argument accounted for physically (Section 2.1.3). A more satisfactory approach is to directly account for the total field at the particle and to calculate the response of the particle to it through the BBGKY equations. This approach leads to the Balescu–Lenard equation, which plays the same role for plasma as the Boltzmann collision integral for gases, providing a bridge from the complexity of the BBGKY foundation of kinetic theory to practical equations, which are valid for dilute systems.

7.4.1 The Balescu–Lenard Equation

We may re-write the two, three, … particle distribution in a form which brings out the particle correlation of a particular order by removing at each stage the lower-order correlations:

$$
\begin{aligned}
f^{(1)} &= f(1) \\
f^{(2)} &= f(1)f(2) + \mathcal{G}(2,1) \\
f^{(3)} &= f(1)f(2)f(3) + \mathcal{G}(1,2)f(3) + \mathcal{G}(2,3)f(1) + \mathcal{G}(3,1)f(2) + \mathcal{H}(1,2,3)
\end{aligned}
\tag{7.38}
$$

$\mathcal{G}$ and $\mathcal{H}$ being known as the second- and third-order correlation coefficients. Higher-order correlation coefficients are treated similarly.

Writing the electrostatic field between particles with charges $-e$ and Ze as a potential

$$\mathbf{F}(\mathbf{r}_1 - \mathbf{r}_2) = \frac{\partial \phi_{12}}{\partial \mathbf{r}_1} = \frac{\partial \phi_{12}}{\partial \mathbf{r}_2} \tag{7.39}$$

where the potential

$$\phi_{12} = -\frac{ZZ'e^2}{4\pi\epsilon_0 |\mathbf{r}_1 - \mathbf{r}_2|} \tag{7.40}$$

Substituting in Eq. (4.28), we obtain for the first-order term:

$$
\begin{aligned}
&\frac{\partial f(1)}{\partial t} + \mathbf{v}_1 \cdot \frac{\partial f(1)}{\partial \mathbf{r}_1} + \frac{\mathbf{F}_1^{\mathrm{ext}}}{m} \cdot \frac{\partial f(1)}{\partial \mathbf{v}_1} - \frac{1}{m}\frac{\partial f(1)}{\partial \mathbf{v}_1} \cdot \int \frac{\partial \phi_{12}}{\partial \mathbf{r}_1} f(2)\, \mathrm{d}\mathbf{r}_2\, \mathrm{d}\mathbf{v}_2 \\
&= \frac{1}{m} \int \frac{\partial \phi_{12}}{\partial \mathbf{r}_1} \cdot \frac{\partial \mathcal{G}(1,2)}{\partial \mathbf{v}_1}\, \mathrm{d}\mathbf{r}_2\, \mathrm{d}\mathbf{v}_2.
\end{aligned}
\tag{7.41}
$$

We note that the left-hand side of this equation, i.e. if $\mathcal{G}(1,2) = 0$ is just the Vlasov equation (6.6).

Turning now to the second-order equation (4.29), and substituting

$$
\begin{aligned}
&\frac{\partial \mathcal{G}(1,2)}{\partial t} + \mathbf{v}_1 \cdot \frac{\partial \mathcal{G}(1,2)}{\partial \mathbf{r}_1} + \mathbf{v}_2 \cdot \frac{\partial \mathcal{G}(1,2)}{\partial \mathbf{r}_2} + \frac{\mathbf{F}_1^{\mathrm{ext}}}{m} \cdot \frac{\partial \mathcal{G}(1,2)}{\partial \mathbf{v}_1} + \frac{\mathbf{F}_2^{\mathrm{ext}}}{m} \cdot \frac{\partial \mathcal{G}(1,2)}{\partial \mathbf{v}_2} \\
&\quad - \frac{1}{m}\frac{\partial \phi_{12}}{\partial \mathbf{r}_1} \cdot \left(\frac{\partial f(1)}{\partial \mathbf{v}_1} f(2) - \frac{\partial f(2)}{\partial \mathbf{v}_2} f(1) + \cancel{\frac{\partial \mathcal{G}(1,2)}{\partial \mathbf{v}_1}} - \cancel{\frac{\partial \mathcal{G}(1,2)}{\partial \mathbf{v}_2}} \right) \\
&= \frac{1}{m} \int \left\{ \frac{\partial \phi_{13}}{\partial \mathbf{r}_1} \cdot \frac{\partial}{\partial \mathbf{v}_1} [f(1)\mathcal{G}(2,3) + f(3)\mathcal{G}(1,2) + \cancel{\mathcal{H}(1,2,3)}] \right. \\
&\quad \left. + \frac{\partial \phi_{23}}{\partial \mathbf{r}_2} \cdot \frac{\partial}{\partial \mathbf{v}_2} [f(2)\mathcal{G}(1,3) + f(3)\mathcal{G}(1,2) + \cancel{\mathcal{H}(1,2,3)}] \right\} \mathrm{d}\mathbf{r}_3\, \mathrm{d}\mathbf{v}_3.
\end{aligned}
\tag{7.42}
$$

In most situations. the correlation terms satisfy the condition $|\mathcal{H}| \ll |\mathcal{G}|\, f \ll f\,f\,f$, known as the *weak coupling approximation*, is obeyed and allows the cancellation of terms as indicated, thereby truncating the expansion at second-order correlations.

Thus, we obtain a pair of simultaneous equations to solve for the functions f and $\mathcal{G}$. However, the nature of the governing equations makes their solution a very complicated problem. To resolve this difficulty, we invoke Bogoliubov's hypothesis. The characteristic time for the fields, which govern the changes in the two-particle correlation function, $\mathcal{G}$, is the plasma frequency $\tau_0 = \Pi_e^{-1}$: on the other hand, for the single-particle distribution function f, it is t_0, the collision interval. In the plasma case $t_0 \sim \tau$, the time calculated is as earlier in Eq. (2.17). For all plasmas studied, $t_0 \gg \Pi_e^{-1}$ and the time scales of the changes in f and $\mathcal{G}$ are much different. Considering the problem over times $\sim t_0$, it is clear that we may use $\mathcal{G}$ in its asymptotic form $\mathcal{G}(\infty)$ in the equation for f, similarly to the conclusion obtained in Section 4.5 for gases. The application of Bogoliubov's hypothesis is equivalent to assumption of Markovian processes in the stochastic picture (Section 7.2). Up to this point, the system is reversible, but clearly the application of Bogoliubov's hypothesis introduces time's arrow and thereby irreversibility.

The inclusion of Bogoliubov's hypothesis makes the solution of this pair of equations tractable, but complex. A number of solutions have been given by Balescu (1960); Lenard (1960); and Dupree (1961). The result is a modified form of Landau equation, known as the *Balescu–Lenard equation*. The change to Landau's equation (7.36) is given in Eq. (7.43),

$$\omega_{ij} \ln \Lambda \Rightarrow \frac{1}{\pi} \int \mathrm{d}\mathbf{k} \, \frac{k_i\, k_j\, \delta\,(\mathbf{k}\cdot\mathbf{v} - \mathbf{k}\cdot\mathbf{v}')}{k^4\, |\epsilon\,(\mathbf{k}\cdot\mathbf{v},\, k)|^2} \tag{7.43}$$

where $\epsilon(\omega, \mathbf{k})$ is the longitudinal dielectric function equation (12.29). A clear account of the mathematics leading to this result is given by Montgomery and Tidman (1964, §6.2) and Clemmow and Dougherty (1969, §12.5).

A clearer picture of the physics underlying this modification is given by Hubbard (1961a) and Lifshitz and Pitaevskii (1981, §46) (see also Shkarofsky et al. 1966) using the stochastic picture and the impulse approximation. The derivation of the Landau equation suffers from the need to impose cut-offs at the upper- and lower-impact parameters. Its accuracy is therefore no better than logarithmic. The Balescu–Lenard equation by treating the collective behaviour of the scatterers includes the Debye length cut-off, but the problem of strong scattering at large angles remains. We have seen in the previous section that the behaviour at small-impact parameters can be accounted for by treating the Coulomb scattering accurately. The two approximations for strong and weak scattering may be matched at intermediate scattering angles (Hubbard 1961b). The careful analysis of the interaction between particles through the Coulomb field must take account of a more sophisticated model of shielding. A moving electron has a distorted shielding cloud, indeed in the limit a rapidly moving charge radiates Langmuir waves by Cerenkov radiation. The displacement $\mathbf{D}$ of a charge Q moving with velocity $\mathbf{v}'$ is given by

$$\nabla \cdot \mathbf{D} = \frac{Q}{\epsilon_0}\, \delta(\mathbf{r} - \mathbf{v}'\, t) \tag{7.44}$$

Assuming a linear relationship between the displacement $\mathbf{D}$ and the intensity $\mathbf{E}$ as in Section 13.1, the Fourier transform of the potential

$$\tilde{\phi}(\mathbf{k}) = \frac{Q}{(2\,\pi)^{3/2}\, k^2\, \epsilon(\mathbf{k}\cdot\mathbf{v}',\, k)} \exp(-\imath\, \mathbf{k}\cdot\mathbf{v}'\, t) \tag{7.45}$$

where $\epsilon(\omega, k)$ is the longitudinal dielectric constant for an effective frequency ω (Eq. (12.29)). Assuming the scattered particle of charge q follows a straight-line path with velocity $\mathbf{v}$, as in Section 2.1.3, the momentum transfer is given by

$$\Delta\mathbf{p} = -q \int \frac{\partial \phi}{\partial \mathbf{r}} \mathrm{d}t = -\imath\, q\, Q \int \frac{\mathrm{d}\mathbf{k}}{(2\,\pi)^3} \left\{ \frac{\mathbf{k} e^{\imath\, \mathbf{k}\cdot\mathbf{r}}}{k^2\, \epsilon(\omega,\, k)} \int_{-\infty}^{\infty} e^{-\imath\, \mathbf{k}\cdot(\mathbf{v}'-\mathbf{v})\, t}\, \mathrm{d}t \right\} \tag{7.46}$$

Noting that $2\,\pi\,\delta(\alpha) = \int \exp(\iota\,\alpha\,x)\,dx$, the integral over the time t may be used to reduce the integral of the three-dimensional vector $\mathbf{k}$ to one over the two-dimensional one $\mathbf{k}_\perp$ in the plane perpendicular to velocity difference $\mathbf{u} = \mathbf{v} - \mathbf{v}'$ so that

$$\begin{aligned}\langle \Delta v_i\, \Delta v_j \rangle &= -\left(\frac{q\,Q}{(2\,\pi)^2\,m\,u}\right)^2 u \int d\mathbf{r}_\perp \int d\mathbf{k}_\perp \frac{k_i}{k^2\epsilon(\omega,k)} \int d\mathbf{k}'_\perp \frac{k'_j}{{k'}^2\epsilon(\omega',k')} e^{\iota\,(\mathbf{k}_\perp+\mathbf{k}'_\perp)\cdot\mathbf{r}_\perp} \\ &= \left(\frac{q\,Q}{2\,\pi\,m}\right)^2 \frac{1}{u} \int d\mathbf{k}_\perp \frac{k_i k_j}{k^4|\epsilon(\omega,k)|^2} \end{aligned} \tag{7.47}$$

since $\epsilon(-\omega,k) = \epsilon^*(\omega,k)$.

The choice of the value of the frequency ω is as yet undefined. The centre of mass velocity $\mathbf{V} = \frac{1}{2}(\mathbf{v}+\mathbf{v}')$ and consequently, $\mathbf{v}$ or $\mathbf{v}' = \mathbf{V} \pm \frac{1}{2}\mathbf{u}$ respectively. Since $\mathbf{k}_\perp \cdot \mathbf{u} = 0$, it follows that $\mathbf{k}_\perp \cdot \mathbf{V} = \mathbf{k}_\perp \cdot \mathbf{v} = \mathbf{k}_\perp \cdot \mathbf{v}'$, and therefore, that $\omega = \mathbf{k}_\perp \cdot \mathbf{V}$ is an appropriate choice. It is easy to show that this result is identical to Eq. (7.43).

The integral over wavenumbers is equivalent to that over impact parameters which leads to the Coulomb logarithm $\ln\Lambda$ in the Landau (1936) or Rosenbluth et al. (1957) theories. In this case, as $k \to 0$, $\epsilon \to k^{-2}$ and the integral is well behaved, and no cut-off for large-impact parameters (λ_D) is required. The model however cannot take into account the large angle scattering at small-impact parameters $\sim b_0$, as it is based on the straight-line path approximation. A cut-off is therefore required at large values of $k \sim 1/b_0$. If the particle velocity is not much larger than the average particle velocity, the dielectric constant takes the simple form corresponding to zero frequency $\epsilon(0,k) \approx 1 + 1/k^2\,{\lambda_D}^2$, and the Landau form is recovered using a cut-off at large k: the Debye length however appears naturally within the calculation.

The singularity at small-impact parameters is treated properly in an exact calculation of the Landau equation, which includes the accurate Rutherford cross section. Since $\epsilon(0,k) \to 1$ in the limit $k \to \infty$, we may expect that the calculation is correctly treated for large k. The two calculations may, therefore, be matched at an intermediate value of k to give a correction to higher-order (Hubbard, 1961b; Lifshitz and Pitaevskii, 1981).

If the scattering is quantum mechanical, the scattering cross section may be calculated using the Born approximation with the inter-particle electric field modified to include collective effects using the dielectric coefficient (Lifshitz and Pitaevskii, 1981). Substituting in the Boltzmann collision operator (4.42) gives the modified quantum mechanical form of Eq. (7.A.1).

$$\left.\frac{\partial f}{\partial t}\right|_{\text{coll}} = \int \{f(\mathbf{p}+\mathbf{q})\,f(\mathbf{p}'-\mathbf{q}) - f(\mathbf{p})f(\mathbf{p}')\}\,\frac{\mathbf{u}\,\sigma_R u}{\epsilon(\omega,\mathbf{k})}\,d\mathbf{p}'\,d\Omega \tag{7.48}$$

where the distribution function is written in terms of momentum $\mathbf{p} = m\mathbf{v}$, and $\hbar\,\omega$ and $\hbar\,\mathbf{k} = \mathbf{q}$ are the energy and momentum exchanged in the collision, respectively, $\sigma_R u$ is the Rutherford cross section (2.5). The resulting Eq. (7.48) is free from divergence at both large- and small-scattering angles (Lifshitz and Pitaevskii, 1981).

Applying Landau's approximation to this equation, we obtain the Balescu–Lenard equation (7.43), but the parameter $\mathbf{k}$ is defined by the momentum exchanged by the particles in the collision. The Balescu–Lenard equation generally offers little practical improvement over the Landau equation (Dolinsky, 1965), but provides the formal justification for the latter result. Fortunately, the correction is generally small, and the approximation of a dominant Coulomb logarithm is accurate, and therefore, normally used for the calculation of transport coefficients.

7.5 Relaxation of a Distribution to the Equilibrium Form

7.5.1 Isotropic Distribution

If the distribution is spherically symmetric, i.e. the distribution function $f(\mathbf{v})$ depends on v alone, the number of particles with velocity vector given by the speed range $(v,\ \mathrm{d}v)$ in the angle intervals $(\theta,\ \mathrm{d}\theta)$ and $(\phi,\ \mathrm{d}\phi)$ is $f(v)\,v^2\,\mathrm{d}v\ \sin\theta\,\mathrm{d}\theta\,\mathrm{d}\phi$. The distribution is normalised by

$$4\,\pi\int f(v)\,v^2\,\mathrm{d}v = n$$

the collision integral for collisions between similar particles of charge $Z\,e$ and mass m takes a simple form (Problem 10)

$$\left.\frac{\partial f}{\partial t}\right|_{\mathrm{coll}} = \frac{(4\,\pi)^2\,Z^4\,e^4}{(4\,\pi\,\epsilon_0)^2\,m^2}\ln\Lambda\frac{1}{v^2}\frac{\partial}{\partial v}\left\{\alpha f + \beta\frac{\partial f}{\partial v}\right\} \tag{7.49}$$

where

$$\alpha = \int_0^v f(v')v'^2\mathrm{d}v' \qquad \text{and} \qquad \beta = \frac{1}{3}\left[\frac{1}{v}\int_0^v f(v')v'^4\mathrm{d}v' + v^2\int_v^\infty f(v')v'\mathrm{d}v'\right] \tag{7.50}$$

This problem is conveniently cast (MacDonald et al., 1957) into a similar form by introducing the constant parameters, namely, the total number of particles n and the average particle energy E, the latter being used to identify a characteristic velocity v_0. Thus, we may write the distribution function in dimensionless form as

$$f(v,t) = \frac{A}{4\,\pi\,{v_0}^3}\,\varphi(\xi,\tau) \tag{7.51}$$

where

$$v = v_0\,\xi \qquad \text{and} \qquad t = \frac{(4\,\pi\,\epsilon_0)^2\,m^2\,{v_0}^3}{4\,\pi\,A\,Z^4\,e^4\,\ln\Lambda}\,\tau \tag{7.52}$$

where A and v_0 are defined by the invariant relations

$$n = A\,I_2 \qquad \text{and} \qquad E = \frac{1}{2}\,m\,{v_0}^2\frac{I_4}{I_2} \tag{7.53}$$

in terms of the constant integrals, which are defined by the initial distribution $\varphi(\xi, 0)$

$$I_n = \int_0^\infty \xi^n\,\varphi(\xi,\tau)\,\mathrm{d}\xi \tag{7.54}$$

for $n = 2$ and $n = 4$. For a Maxwellian distribution $I_2 = \sqrt{\pi}/4$ and $I_4 = 3\sqrt{\pi}/8$, $A = 4n/\sqrt{\pi}$ and $v_0 = \sqrt{4\,E/3\,m} = \sqrt{2\,k T/m}$, where T is the equilibrium temperature. In this case, the characteristic relaxation time for electron–electron collisions is therefore consistent with Eq. (2.23) and the ion–ion term a factor of $\sqrt{M/m}$ longer as discussed earlier.

The equilibration history is determined by the reduced Fokker–Planck equation, which takes the simple form

$$\frac{\partial\varphi}{\partial\tau} = \frac{1}{\xi^2}\frac{\mathrm{d}}{\mathrm{d}\xi}\left\{\alpha\,\varphi + \beta\,\frac{\partial\varphi}{\partial\xi}\right\} \tag{7.55}$$

where

$$\alpha = \int_0^\xi \varphi(\xi')\,\xi'^2\,\mathrm{d}\xi' \qquad \text{and} \qquad \beta = \frac{1}{3}\left[\frac{1}{\xi}\int_0^\xi \varphi(\xi')\,\xi'^4\,\mathrm{d}\xi' + \xi^2\int_\xi^\infty \varphi(\xi')\,\xi'\,\mathrm{d}\xi'\right] \tag{7.56}$$

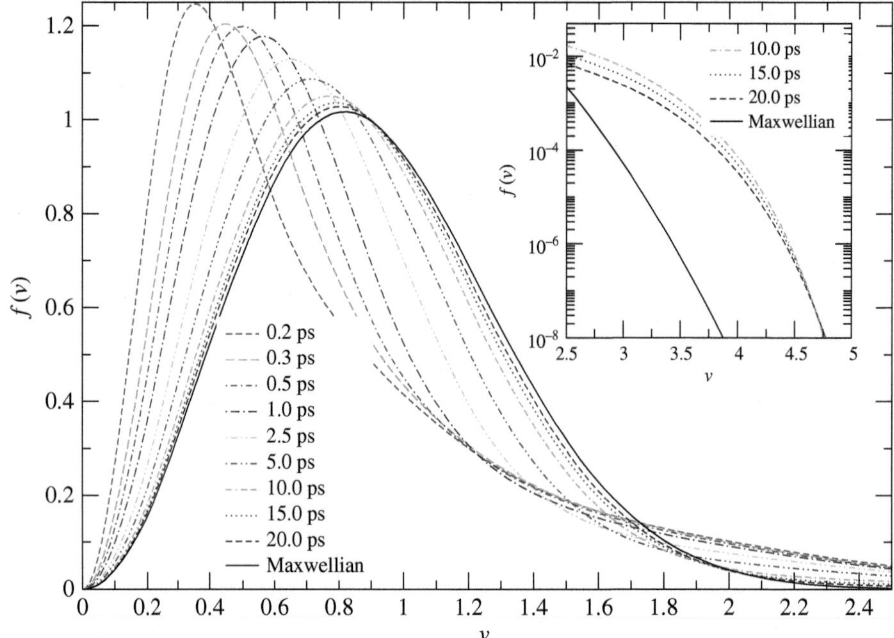

Figure 7.1 Relaxation of the distribution to Maxwellian following above threshold ionisation of an helium atom with ion density 10^{18} cm^{-3} by linearly polarised laser radiation at wavelength 0.6164 μm and intensity 10^{16} W cm^{-2}. The inset diagram shows the slow relaxation at high velocities. Velocities are measured as a fraction of the root mean square speed $\sqrt{\overline{v^2}} = 3.143 \times 10^8$cm s^{-1}.

This result clearly indicates the non-linear form of the Fokker–Planck equation for this problem. The integration of the equation is easily performed numerically using a simple finite difference algorithm. A straightforward iteration can be used to overcome this non-linearity (Pert, 2001). The generalisation of the algorithm to treat ion–ion collisions is straightforward. However, since the energy exchange in ion–electron collisions is very small, their contribution to relaxation of the distribution to the Maxwellian can usually be ignored.

Figure 7.1 illustrates the relaxation of helium at an ion density 10^{18} cm^{-3} having been fully ionised by linearly polarised light of 0.6164 μm wavelength and 10^{16} W cm^{-2} intensity, calculated using the finite difference algorithm described in Appendix 7.B. The initial distribution at the conclusion of the laser pulse is strongly non-equilibrium reflecting the different intensities required to ionise the first and second electrons. As can be seen, the bulk of the distribution is thermalised over a time of approximately 10 ps. Since the final equilibrium temperature is 19.09 eV, the electron–electron relaxation time from Eq. (2.24) is approximately 8 ps consistent with the plots. Following MacDonald et al. (1957), it can be seen that although the bulk of the distribution attains the Maxwellian form in the electron–electron relaxation time, the fast electrons are much slower as their collision times increase as v^3. Correspondingly, there tends to be a higher population of faster or slower electrons at later times than predicted by the Maxwell–Boltzmann distribution. Several further examples of the relaxation of the electron distribution following high-field ionisation are given in Pert (2001).

7.5.2 Anisotropic Distribution

In general, an initially non-thermal distribution is also anisotropic when formed, for example following above threshold ionisation (ATI) by a high irradiance laser. The approach to equilibrium, therefore, includes both isotropisation and equilibration. Both effects are treated together by the general three-dimensional Fokker–Planck equation. However, the solution of such a general form is extremely complex, although Shkarofsky et al. (1966) treat the case of a two temperature plasma distribution when the temperature difference is small and derive an analytic result. Numerical investigations of the relaxation of anisotropic Maxwellian distributions were performed by Jorna and Wood (1987b) and McGowan and Sanderson (1992).

Some general results can be deduced by applying the same reduction to a scaling form as in the isotropic case, Eqs. (7.51), (7.52), and (7.53). It is clear that the same relaxation time governs the general approach to equilibrium including both isotropisation and equilibration. However, in a plasma, due to the large mass difference, ion–electron collisions change the direction of the electron motion without any corresponding energy shift. Isotropisation an electron distribution, therefore, takes place at slightly faster rate than the energy re-distribution required for equilibration. For most purposes, the relaxation of the ion or electron distributions can be satisfactorily treated by considering only the isotropic Fokker–Planck equation as above.

The standard method of treating anisotropic relaxation involves expanding the distribution function in an expansion in spherical harmonics (Allis, 1956). Since many problems possess axial symmetry, these can usually be reduced to an expansion in Legendre polynomials. The appropriate forms of the Rosenbluth potentials in this case are given by Rosenbluth et al. (1957) and a method of numerical integration by Jorna and Wood (1987a).

A general treatment of particle dynamics requires the full complexity of a many particle molecular dynamics approach (David and Hooker, 2003). The general approach involves integration over a complex multi-dimensional space. This is the type of problem for which MonteCarlo numerical methods are very suitable, particularly if a high level of accuracy is not required and the resulting stochastic noise can be tolerated. The statistical nature of the scattering problem allows variance reduction methodology to be introduced to the treatment of the overall many body dynamics by using multi-body scattering. The algorithm is described in Appendix 7.C.

ATI ionisation by a linearly polarised laser beam generates an electron distribution directed along the field axis. Figure 7.2 illustrates the equilibration of a typical example following such an ionisation, the electric field of the laser beam is along the z direction. Fast electrons are generated during the ionisation with their velocities along the field z; in contrast the x and y components of the velocity are small, Figure 7.2a. It can be seen that the isotropisation and equilibration to a Maxwell–Boltzmann distribution take place together: an isotropic distribution is obtained after about 5 ps (Figure 7.2a), whereas the equilibrium Maxwellian requires longer, about 10 ps to be achieved (Figure 7.2b). Although these calculations were performed with 500 000 electrons, the statistical noise on the outputs clearly shows the limitations of the Monte Carlo approach.

Figures 7.1 and 7.2 are both calculated under the same physical conditions. Comparing Figure 7.1 with Figure 7.2b it can be seen that there is good general agreement, but that minor differences are apparent particularly at early times. These reflect the differences introduced by the angular averages inherent in the finite difference calculation compared with the full velocity profile used in the Monte Carlo. The calculations are both performed with collisional heating from absorption of the laser beam included. The finite difference calculation gives a final energy per electron of 28.08 eV,

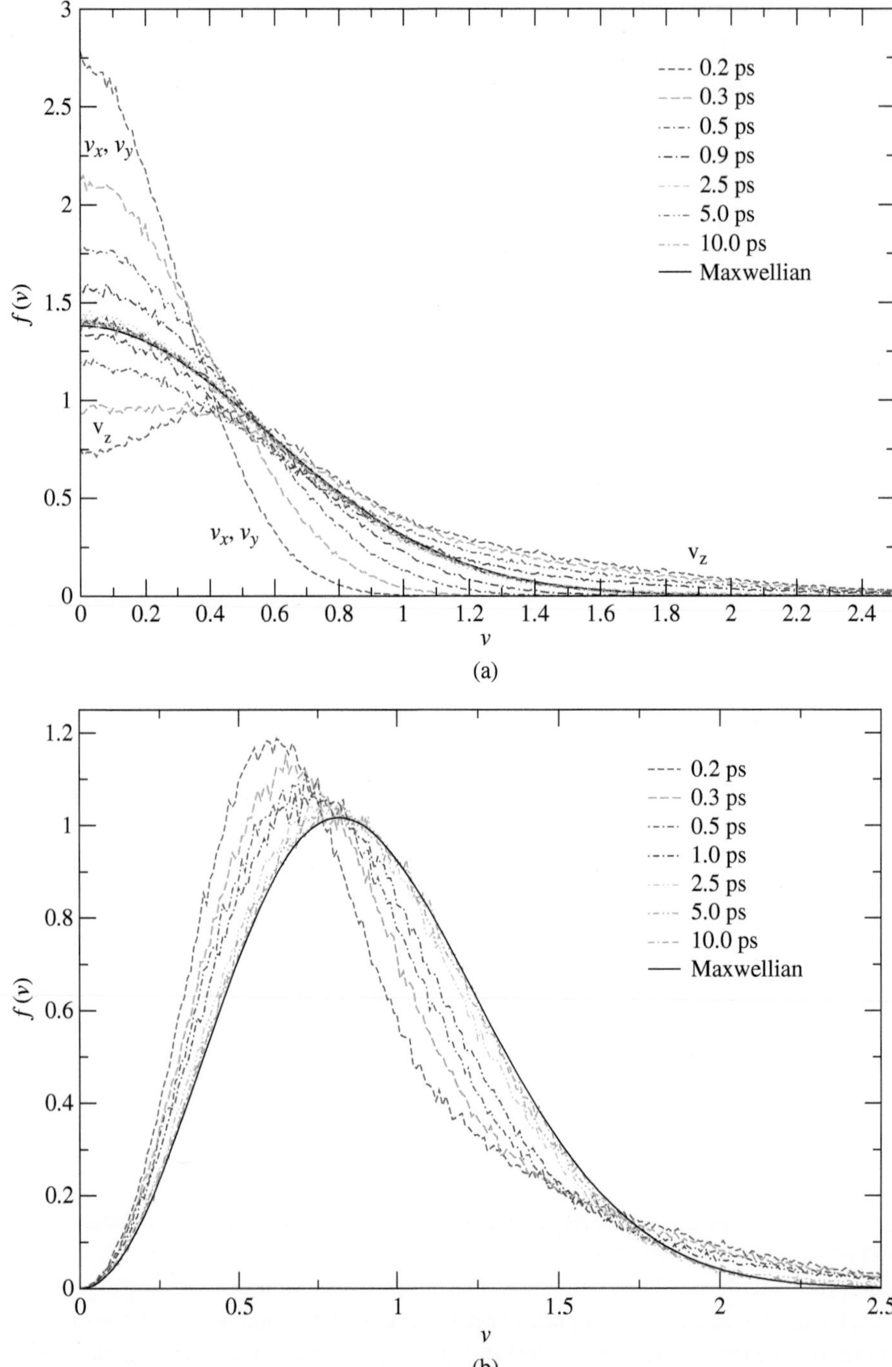

Figure 7.2 Relaxation of the anisotropic velocity distribution to Maxwellian following above threshold ionisation of an helium atom with ion density 10^{18} cm^{-3} by linearly polarised laser radiation at wavelength 0.6164 μm and intensity 10^{16} W cm^{-2}. The thermal relaxation time is approximately 10 ps. The laser beam has electric field in the *z* direction. Velocities are measured as a fraction of the root mean square speed $\sqrt{\overline{v^2}} = 3.174 \times 10^8$cm s^{-1}. (a) Relaxation of the anisotropic velocity components following ionisation. (b) Relaxation of the total velocity to an equilibrium form.

the Monte Carlo calculation gives a slightly larger value of 28.64 eV. However, since the differences are small, it is satisfactory to use the simpler one-dimensional calculation for most problems involving relaxation to a Maxwellian.

7.6 Ion–Electron Thermal Equilibration by Coulomb Collisions

Thus far we have investigated the equilibration of a single species (electrons or ions) with particles of the same type. As noted earlier in Section 2.1.4, there is large disparity between the electron–electron and ion–electron collision relaxation rates, and that between electrons and ions. The large mass ratio of ions to electrons $M \gg m$ ensures that only a small fraction of the energy of either particle is transferred to the other in an elastic collision. As a result, the electron and ion distributions may be assumed to be isotropic Maxwellians, but with differing temperatures T_e and T_i, respectively, which become equal once thermal equilibrium is achieved.

The calculation of the rate of equilibration was derived independently by Landau (1936) and Spitzer (1940). Since the electron and ion distributions are isotropic Maxwellian, the equilibration rate is obtained from the isotropic Fokker–Planck equation for dissimilar particles:

$$\begin{aligned}\left.\frac{\partial f}{\partial t}\right|_{\text{coll}} &= \frac{(4\,\pi)^2\,Z^2\,Z'^2\,e^4}{(4\,\pi\,\epsilon_0)^2\,m^2}\ln\Lambda \\ &\times \frac{4\,\pi}{v^2}\frac{\mathrm{d}}{\mathrm{d}v}\left\{\frac{m}{m'}\left[\int_0^v v'^2\,f'(v')\,\mathrm{d}v'\right]f(v)\right. \\ &\left.+\frac{1}{3}v^2\left[\int_0^v \frac{v'^4}{v^3}\,f'(v')\,\mathrm{d}v' + \int_v^\infty v'\,f'(v')\,\mathrm{d}v'\right]\frac{\mathrm{d}f}{\mathrm{d}v}\right\}\end{aligned} \tag{7.57}$$

The energy gain by the unprimed particles per unit volume per unit time is therefore,

$$\begin{aligned}\frac{\mathrm{d}E}{\mathrm{d}t} &= (4\,\pi)\int \frac{1}{2}\,m\,v^2\left.\frac{\partial f}{\partial t}\right|_{\text{coll}} v^2\,\mathrm{d}v = \frac{1}{2}\,(4\,\pi)\,m\,\frac{(4\,\pi)^2\,Z^2\,Z'^2\,e^4}{(4\,\pi\,\epsilon_0)^2\,m^2}\ln\Lambda \\ &\times \int v^4\,\mathrm{d}v\,\frac{1}{v^2}\frac{\mathrm{d}}{\mathrm{d}v}\left\{\frac{m}{m'}\left[\int_0^v v'^2\,f'(v')\,\mathrm{d}v'\right]f(v)\right. \\ &\left.+\frac{1}{3}v^2\left[\int_0^v \frac{v'^4}{v^3}\,f'(v')\,\mathrm{d}v' + \int_v^\infty v'\,f'(v')\,\mathrm{d}v'\right]\frac{\mathrm{d}f}{\mathrm{d}v}\right\}\end{aligned}$$

Integrating by parts and neglecting the surface terms at infinity

$$\begin{aligned}\frac{\mathrm{d}E}{\mathrm{d}t} &= -(4\,\pi)\,m\,\frac{(4\,\pi)^2\,Z^2\,Z'^2\,e^4}{(4\,\pi\,\epsilon_0)^2\,m^2}\ln\Lambda \\ &\times \int \mathrm{d}v\,v\left\{\left[\frac{m}{m'}\int_0^v v'^2\,f'(v')\,\mathrm{d}v'\right]f(v)\right. \\ &\left.+\frac{1}{3}\left[\frac{1}{v}\int_0^v v'^4\,f'(v')\,\mathrm{d}v' + v^2\int_v^\infty v'\,f'(v')\,\mathrm{d}v'\right]\frac{\mathrm{d}f}{\mathrm{d}v}\right\}\end{aligned}$$

Integrating a second times by parts

$$\begin{aligned}\frac{\mathrm{d}E}{\mathrm{d}t} &= -4\,\pi\,m\,\frac{(4\,\pi)^2\,Z^2\,Z'^2\,e^4}{(4\,\pi\,\epsilon_0)^2\,m^2}\ln\Lambda \\ &\times \int \mathrm{d}v\left\{\left[\frac{m}{m'}\,v\int_0^v v'^2\,f'(v')\,\mathrm{d}v'\right]f(v) - v^2\int_v^\infty v'\,f'(v')\,\mathrm{d}v'\,f(v)\right\}\end{aligned}$$

Finally, reversing the order of integration

$$\frac{\mathrm{d}E}{\mathrm{d}t} = -4\,\pi\,m\,\frac{(4\,\pi)^2\,Z^2\,Z'^2\,e^4}{(4\,\pi\,\epsilon_0)^2\,m^2}\,\ln\Lambda$$
$$\times\left\{\frac{m}{m'}\int_0^\infty v'^2 f'(v')\,\mathrm{d}v'\int_{v'}^\infty v\,f(v)\,\mathrm{d}v - \int_0^\infty v^2 f(v)\,\mathrm{d}v\int_v^\infty v'\,f'(v')\,\mathrm{d}v'\right\}$$

Interchanging the primed and unprimed terms, it is clear that $\mathrm{d}E/\mathrm{d}t = -\mathrm{d}E'/\mathrm{d}t$, and therefore that, as required, energy is conserved in the collisions.

The equilibrium Maxwellian distribution is

$$f(v) = n\left(\frac{m}{2\,\pi\,kT}\right)^{3/2}\exp\left(-\frac{m\,v^2}{2kT}\right)$$

Hence, the integral

$$\int_0^\infty v'^2 f'(v')\,\mathrm{d}v'\int_{v'}^\infty v\,f(v)\,\mathrm{d}v$$
$$= n\,n'\left\{\frac{m\,m'}{(2\,\pi kT)\,(2\,\pi kT')}\right\}^{3/2}\frac{kT}{m}\int_0^\infty v'^2\exp\left\{-\left(\frac{m}{2kT}+\frac{m'}{2kT'}\right)v'^2\right\}$$
$$= \frac{\sqrt{\pi}}{4}\,n\,n'\left\{\frac{m\,m'}{(2\,\pi kT)\,(2\,\pi kT')}\right\}^{3/2}\Big/\left(\frac{m}{2\,kT}+\frac{m'}{2\,kT'}\right\}^{3/2}\frac{kT}{m} \tag{7.58}$$

Substituting

$$\frac{\mathrm{d}E}{\mathrm{d}t} = -\frac{\mathrm{d}E'}{\mathrm{d}t} - \frac{4\sqrt{2\,\pi}n\,n'\,Z^2\,Z'^2\,e^4}{(4\,\pi\,\epsilon_0)^2\,m\,m'}\ln\Lambda\,(kT - kT')\Big/\left\{\frac{kT}{m}+\frac{kT'}{m'}\right\}^{3/2} \tag{7.59}$$

Since the energy per unit volume $E = \frac{3}{2}\,n\,kT$, Spitzer's result (Spitzer, 1940, 1956) for the equilibration time between two ions Z, m and Z', m' due to Coulomb collisions follows:

$$\tau_e q = \frac{T' - T}{\mathrm{d}T/\mathrm{d}t} = \frac{3\,(4\,\pi\epsilon_0)^2\,m\,m'}{8\sqrt{2\,\pi}\,n'\,Z^2\,Z'^2\,e^4\,\ln\Lambda}\left\{\frac{kT}{m}+\frac{kT'}{m'}\right\}^{3/2} \tag{7.60}$$

Implicit in this result is the condition that the distributions of both particles are thermalised much faster than their mutual equilibration. As we have seen, this implies that there is a large mass imbalance between the two species, e.g. ions and electrons.

7.7 Dynamical Friction

The interaction of a particle with those in its environment gives rise to a force, whose nature is a drag if the particle velocity is larger than the 'mean' of those in its neighbourhood. Thus, consider a test particle with velocity $\mathbf{v}$ from the distribution $f(\mathbf{v})$ interacting with particles from the distribution $f'(\mathbf{v})$. The Boltzmann equation for the test particle distribution is obtained by using the Fokker–Planck collision term in the form of Eq. (7.27a)

$$\frac{\partial f}{\partial t} + (\mathbf{v}\cdot\nabla)\cdot f(\mathbf{v}) + \left[\frac{q}{m}(\mathbf{E}+\mathbf{v}\wedge\mathbf{B})\cdot\nabla_{\mathbf{v}}\right]f(\mathbf{v})$$
$$= \frac{4\,\pi\,Z^2\,Z'^2\,e^4}{(4\,\pi\,\epsilon_0)^2\,m^2}\ln\Lambda\,\nabla_{\mathbf{v}}\cdot\left\{-[f(\mathbf{v})\nabla_{\mathbf{v}}\mathcal{H}(\mathbf{v})] + \frac{1}{2}\nabla_{\mathbf{v}}\cdot[f(\mathbf{v})\nabla_{\mathbf{v}}\nabla_{\mathbf{v}}\mathcal{G}(\mathbf{v})]\right\} \tag{7.61}$$

We may include the collision term into the total force on the particles since it is obvious that $\Delta \mathbf{v} \cdot (\mathbf{E} + \mathbf{v} \wedge \mathbf{B}) = 0$. We may therefore re-write the Boltzmann equation as

$$\frac{\partial f}{\partial t} + (\mathbf{v} \cdot \nabla) f(\mathbf{v}) + \nabla_{\mathbf{v}} \cdot \left\{ \frac{1}{m} [q(\mathbf{E} + \mathbf{v} \wedge \mathbf{B}) + \mathcal{J}(\mathbf{v})] f(\mathbf{v}) \right\} \tag{7.62}$$

where the dynamic friction drag force due to the collisions is equivalent to the momentum flux

$$\begin{aligned} \mathcal{J}(\mathbf{v}) &= \frac{4\pi Z^2 Z'^2 e^4}{(4\pi\epsilon_0)^2 m^2} \ln\Lambda \left\{ [f(\mathbf{v}) \nabla_{\mathbf{v}} \mathcal{H}(\mathbf{v})] - \frac{1}{2} \nabla_{\mathbf{v}} \cdot [f(\mathbf{v}) \nabla_{\mathbf{v}} \nabla_{\mathbf{v}} \mathcal{G}(\mathbf{v})] \right\} \\ &= \frac{4\pi Z^2 Z'^2 e^4}{(4\pi\epsilon_0)^2 m^2} \ln\Lambda \left\{ \left[\nabla_{\mathbf{v}} \mathcal{H}(\mathbf{v}) - \frac{1}{2} \nabla_{\mathbf{v}} \cdot \nabla_{\mathbf{v}} \nabla_{\mathbf{v}} \mathcal{G}(\mathbf{v}) \right] f(\mathbf{v}) - \frac{1}{2} [\nabla_{\mathbf{v}} \nabla_{\mathbf{v}} \mathcal{G}(\mathbf{v})] \nabla_{\mathbf{v}} \cdot f(\mathbf{v}) \right\} \end{aligned} \tag{7.63}$$

Let us consider the dynamical friction induced on the scattered particles by the primed background, which we assume have an isotropic Maxwellian distribution of temperature T':

$$f'(\mathbf{v}) = n' \left(\frac{m'}{2\pi kT'} \right)^{3/2} \exp\left(-\frac{m' v^2}{2kT'} \right) \tag{7.64}$$

For an isotropic distribution making use of Eq. (7.22)

$$\nabla_{\mathbf{v}}^2 \mathcal{H}(\mathbf{v}) = \frac{1}{v^2} \frac{\mathrm{d}}{\mathrm{d}v} \left(v^2 \frac{\mathrm{d}\mathcal{H}(v)}{\mathrm{d}v} \right) = \frac{1}{v} \frac{\mathrm{d}^2}{\mathrm{d}v^2} (v\mathcal{H}) = \frac{2}{\sqrt{\pi}} \frac{(m+m')}{m'} n' \left(\frac{m'}{kT'} \right)^{3/2} \exp\left(-\frac{m' v^2}{2kT'} \right) \tag{7.65}$$

which, provided $\mathcal{H}(0)$ is finite, may be integrated twice to give

$$\mathcal{H}(\zeta) = \frac{n'}{v_T'} \frac{(m+m')}{m'} \frac{\mathrm{erf}(\zeta)}{\zeta} \tag{7.66}$$

where $\zeta = v/v_T'$ with $v_T' = \sqrt{2kT'/m'}$, and $\mathrm{erf}(\zeta) = (2/\sqrt{\pi}) \int_0^{\zeta} \exp(-\zeta^2)\, \mathrm{d}\zeta$ is the standard error function.

Similarly, making use of Eq. (7.21), we obtain an equivalent differential equation for $\mathcal{G}(\zeta)$

$$\frac{\mathrm{d}^2}{\mathrm{d}\zeta^2} (\zeta\, \mathcal{G}) = 2\, n'\, v_T'\, \mathrm{erf}(\zeta) \tag{7.67}$$

which in turn may be integrated, if $\mathcal{G}(0)$ is finite, to give

$$\mathcal{G}(\zeta) = \frac{n'\, v_T'}{2\zeta} \left[\zeta \frac{\mathrm{d}\, \mathrm{erf}(\zeta)}{\mathrm{d}\zeta} + (1 + 2\zeta^2)\, \mathrm{erf}(\zeta) \right] \tag{7.68}$$

Differentiating these terms, we obtain

$$\nabla_{\mathbf{v}} \mathcal{H}(\zeta) = -n'[(m+m')/m'][\mathrm{erf}(\zeta) - \zeta\, \dot{\mathrm{erf}}(\zeta)]\, \mathbf{v}/v^3 \tag{7.69}$$

$$\nabla_{\mathbf{v}} \cdot \nabla_{\mathbf{v}} \nabla_{\mathbf{v}} \mathcal{G}(\zeta) = 2\, m'/(m+m')\, \nabla_{\mathbf{v}} \mathcal{H}(\zeta) = -2\, n'\, [\mathrm{erf}(\zeta) - \zeta\, \dot{\mathrm{erf}}(\zeta)]\, \mathbf{v}/v^3 \tag{7.70}$$

$$\begin{aligned} \nabla_{\mathbf{v}} \nabla_{\mathbf{v}} \mathcal{G}(\zeta) = \frac{n'\, v_T'^{\,2}}{2\, v^3} \{ & [-(1 - 2\zeta^2)\, \mathrm{erf}(\zeta) - \zeta\, \dot{\mathrm{erf}}(\zeta)] \mathbf{I} \\ & + 3[(1 - 2/3\zeta^2)\, \mathrm{erf}(\zeta) - \zeta\, \dot{\mathrm{erf}}(\zeta)]\, \mathbf{v}\mathbf{v}/v^2 \} \end{aligned} \tag{7.71}$$

and hence the probability flux

$$\begin{aligned} \mathcal{J} = \frac{2\pi Z^2 Z'^2 e^4 n'}{(4\pi\epsilon_0)^2 m'} \ln\Lambda & \left\{ -2\, [\mathrm{erf}(\zeta) - \zeta\, \dot{\mathrm{erf}}(\zeta)] \frac{\mathbf{v}}{v^3} f(\mathbf{v}) - \frac{m'}{m} \frac{v_T'^{\,2}}{2\, v^3} \right. \\ & \left. \times \left\{ [-(1 - 2\zeta^2)\, \mathrm{erf}(\zeta) - \zeta\, \dot{\mathrm{erf}}(\zeta)] \mathbf{I} + 3\, [(1 - 2/3\zeta^2)\, \mathrm{erf}(\zeta) - \zeta\, \dot{\mathrm{erf}}(\zeta)] \frac{\mathbf{v}\mathbf{v}}{v^2} \right\} \right\} \cdot \nabla_{\mathbf{v}} f(\mathbf{v}) \end{aligned} \tag{7.72}$$

This expression is simplified if the distribution $f(\mathbf{v}) = f(v)$ is isotropic when $\nabla_{\mathbf{v}} f(\mathbf{v}) = \dot{f}(v)\,\mathbf{v}/v$ and therefore,

$$\mathcal{J} = -\frac{4\,\pi\,Z^2\,Z'^2\,e^4\,n'}{(4\,\pi\,\epsilon_0)^2\,m'} \ln\Lambda\,[\mathrm{erf}(\zeta) - \zeta\,\dot{\mathrm{erf}}(\zeta)] \left(f(v) + \frac{m'\,v_T'^2}{2\,m\,v}\frac{\mathrm{d}f(v)}{\mathrm{d}v} \right) \frac{\mathbf{v}}{v^3} \tag{7.73}$$

Appendix 7.A Reduction of the Boltzmann Equation to Fokker–Planck Form in the Weak Collision Limit

We may address the problem of calculating the Fokker–Planck coefficients for Coulomb collisions directly from the Boltzmann equation. Consider the binary Coulomb collision of two particles of mass m and m' and charge $-e$ and Ze, with velocities $\mathbf{v}$ and $\mathbf{v}'$. The relative velocity $\mathbf{u} = \mathbf{v} - \mathbf{v}'$, and centre of mass velocity $\mathbf{V} = (\mathbf{v} + \mathbf{v}')/2$.[5] The cross-section is given by the Rutherford cross section (Eq. (2.5)), in terms of $\overline{m} = mm'/(m+m')$ the reduced mass of the particles. The Boltzmann collision integral takes the form

$$\left.\frac{\partial f}{\partial t}\right|_{\mathrm{coll}} = \left(\frac{Z\,Z'\,e^2}{2\,(4\pi\epsilon_0)\,\overline{m}}\right)^2 \int \mathrm{d}\mathbf{v}' \int \sin\theta\,\mathrm{d}\theta\,\mathrm{d}\phi \frac{u}{[u\,\sin(\theta/2)]^4}$$
$$\times \left\{ f\left(\mathbf{v} + \frac{m'}{m+m'}\mathbf{\Delta u}\right) f\left(\mathbf{v}' - \frac{m}{m+m'}\mathbf{\Delta u}\right) - f(\mathbf{v})f(\mathbf{v}') \right\} \tag{7.A.1}$$

where $\mathbf{\Delta u}$ is the change in the relative velocity after the collision. Since energy is conserved, $\mathbf{u}$ is simply rotated by the collision through an angle ϕ, the scattering angle. We form a set of orthonormal unit vectors $(\hat{\lambda}, \hat{\mu}, \hat{\nu})$, $\hat{\lambda}$ is parallel to the incoming relative velocity $\mathbf{u}$, $\hat{\mu}$ and $\hat{\nu}$ are in the plane normal to $\mathbf{u}$. The change in the relative velocity $\mathbf{\Delta v}$ can be expressed in terms of these unit vectors as

$$\mathbf{\Delta u} = 2\,u \sin(\phi/2)\{-\hat{\lambda}\sin(\phi/2) + \hat{\mu}\cos(\phi/2)\cos(\psi) + \hat{\nu}\cos(\phi/2)\sin(\psi)\} \tag{7.A.2}$$

the magnitude of the vector $\mathbf{u}$ being left unchanged by elastic scattering.

If the scattering is weak, $\mathbf{\Delta u}$ is small, and we can expand the terms in the Boltzmann integral, Eq. (7.A.1) using Taylor series:

$$f\left(\mathbf{v} + \frac{m'}{(m+m')}\mathbf{\Delta\,u}\right) f\left(\mathbf{v}' - \frac{m}{(m+m')}\mathbf{\Delta u}\right) - f(\mathbf{v})\,f(\mathbf{v}')$$
$$= \cancel{f(\mathbf{v})f(\mathbf{v}')} - \cancel{f(\mathbf{v})f(\mathbf{v}')} + \left[\frac{m'}{(m+m')} f'(\mathbf{v}') \frac{\partial f}{\partial \mathbf{v}} - \frac{m}{(m+m')} f(\mathbf{v}) \frac{\partial f'}{\partial \mathbf{v}'}\right] \mathbf{\Delta u} \tag{7.A.3}$$
$$+ \frac{1}{2}\left[\frac{m'^2}{(m+m')^2} f'(\mathbf{v}') \frac{\partial^2 f}{\partial v_i\,\partial v_j} - 2\frac{m\,m'}{(m+m')^2}\frac{\partial f}{\partial v_i}\frac{\partial f'}{\partial v_j'} + \frac{m^2}{(m+m')^2} f(\mathbf{v}) \frac{\partial^2 f'}{\partial v_i'\,\partial v_j'}\right] \Delta u_i\,\Delta u_j$$

Noting that integration over the azimuthal angle ψ eliminates terms in $\cos\psi\,\sin\psi$, $\cos\psi$ and $\sin\psi$ since

$$\int_0^{2\pi} \mathrm{d}\psi\,\cos\psi = \int_0^{2\pi} \mathrm{d}\psi\,\sin\psi = \int_0^{2\pi} \mathrm{d}\psi\,\sin\psi\,\cos\psi = 0$$

and

$$\int_0^{2\pi} \mathrm{d}\psi\,\cos^2\psi = \int_0^{2\pi} \mathrm{d}\psi\,\sin^2\psi = \frac{1}{2}$$

---

5 See Footnote 3, Section 7.2.1.

Substituting in Eq. (7.A.1), we obtain for the collision integral $\partial f/\partial t|_{\rm coll}$

$$\frac{\partial f}{\partial t}\bigg|_{\rm coll} = 4\,\pi\left(\frac{Z\,Z'\,e^2}{(4\,\pi\epsilon_0)\,\overline{m}}\right)^2 \int \mathrm{d}\mathbf{v}' \int_0^{\pi} \mathrm{d}\left(\frac{\phi}{2}\right)\cos\left(\frac{\phi}{2}\right)$$
$$\left\{\frac{-\hat{\lambda}_i}{u^2\sin(\phi/2)}\left[\frac{m'}{(m+m')}f'(\mathbf{v}')\frac{\partial f}{\partial v_i} - \frac{m}{(m+m')}f(\mathbf{v})\frac{\partial f'}{\partial v_i'}\right]\right.$$
$$+\left[\frac{m'^2}{(m+m')^2}f'(\mathbf{v}')\frac{\partial^2 f}{\partial v_i\,\partial v_j} - 2\frac{m\,m'}{(m+m')^2}\frac{\partial f}{\partial v_i}\frac{\partial f'}{\partial v_j'} + \frac{m^2}{(m+m')^2}f(\mathbf{v})\frac{\partial^2 f'}{\partial v_i'\,\partial v_j'}\right]$$
$$\left.\times\frac{1}{u}\left(\hat{\lambda}_i\,\hat{\lambda}_j\sin\left(\frac{\phi}{2}\right) + \frac{1}{2}(\hat{\mu}_i\,\hat{\mu}_j + \hat{v}_i\,\hat{v}_j)\left(\frac{1}{\sin(\phi/2)} - \sin\left(\frac{\phi}{2}\right)\right)\right)\right]\Bigg\} \tag{7.A.4}$$

The integral $\int_0^{\pi} \mathrm{d}(\phi/2)\cos(\phi/2)/\sin(\phi/2)$ diverges for small ϕ. Physically, this is due to the neglect of screening at small- and large-impact parameters, small deflections $\phi \approx 0$. As before, when we treated the similar divergence in Eq. (2.16), we cut-off the integral for large impact parameters at the Debye length corresponding to

$$\phi_{\rm min} \approx \frac{b_{\rm min}}{\lambda_D} \tag{7.A.5}$$

from Eq. (2.4) and introduce the Spitzer or Coulomb logarithm

$$\int_0^{\pi} \mathrm{d}(\phi/2)\frac{\cos(\phi/2)}{\sin(\phi/2)} \approx \ln\left[\frac{\sin(\phi_{\rm max}/2)}{\sin(\phi_{\rm min}/2)}\right] = \ln\Lambda \tag{7.A.6}$$

$$\int_0^{\pi} \mathrm{d}(\phi/2)\cos(\phi/2)\sin(\phi/2) = 1/2 \tag{7.A.7}$$

In dilute plasma, the Coulomb logarithm $\ln\Lambda$ is large, and we retain only the terms containing $\ln\Lambda$.[6] The tensor $\hat{\mu}_i\hat{\mu}_j + \hat{v}_i\hat{v}_j$ is a projection matrix which projects a vector, $\mathbf{a}$, on to the plane perpendicular to $\mathbf{u}$.

$$[\hat{\mu}_i\hat{\mu}_j + \hat{v}_i\hat{v}_j]a_j = (\hat{\boldsymbol{\mu}}\cdot\mathbf{a})\hat{\boldsymbol{\mu}} + (\hat{\mathbf{v}}\cdot\mathbf{a})\hat{\mathbf{v}} = \mathbf{a} - (\hat{\boldsymbol{\lambda}}\cdot\mathbf{a})\hat{\boldsymbol{\lambda}} = (\delta_{ij} - \hat{\lambda}_i\hat{\lambda}_j)a_j \tag{7.A.8}$$

Noting that $\hat{\lambda}$ is parallel to $\mathbf{u}$, we again introduce the modified projection matrix ω_{ij} Eq. (7.15) which has derivatives

$$\frac{\partial\omega_{ij}}{\partial u_i} = \frac{\partial\omega_{ij}}{\partial v_j} = -\frac{\partial\omega_{ij}}{\partial v_i'} = \frac{\partial\omega_{ij}}{\partial v_i} = \frac{\partial\omega_{ij}}{\partial v_j'} \tag{7.A.9}$$

Retaining only the terms in $\ln\Lambda$, we obtain the collision integral

$$\frac{\partial f}{\partial t}\bigg|_{\rm coll} = 2\,\pi\left(\frac{Z\,Z'\,e^2}{(4\,\pi\,\epsilon_0)\,\overline{m}}\right)^2 \ln\Lambda \int \mathrm{d}\mathbf{v}'\left\{\left[\frac{m'}{(m+m')}f'(\mathbf{v}')\,\frac{\partial f}{\partial v_i} - \frac{m}{(m+m')}f(\mathbf{v})\,\frac{\partial f'}{\partial v_i'}\right]\frac{\partial\omega_{ij}}{\partial u_i}\right.$$
$$+\frac{\partial}{\partial v_j}\left[\frac{m'^2}{(m+m')^2}f'(\mathbf{v}')\,\frac{\partial f}{\partial v_i} - \frac{m\,m'}{(m+m')^2}f(\mathbf{v})\,\frac{\partial f'}{\partial v_i'}\right]\omega_{ij}$$
$$\left.+\left[\frac{m^2}{(m+m')^2}f(\mathbf{v})\,\frac{\partial^2 f'}{\partial v_i'\,\partial v_j'} - \frac{m\,m'}{(m+m')^2}\,\frac{\partial f}{\partial v_i}\,\frac{\partial f'}{\partial v_j'}\right]\omega_{ij}\right\} \tag{7.A.10}$$

6 The terms not containing $\ln\Lambda$ are associated with the deviations from a straight line path associated with binary collisions (see Section 2.1.2.1). As we argued in Section 2.1.3, these deviation are reduced by the random nature of the deflection in many body collisions. This magnitude of this approximation is discussed by Chandrasekhar (1960, p. 64) and Spitzer (1956, p. 74). Numerical checks by Dolinsky (1965) comparing the solution of the Fokker–Planck equation with those of the Balescu–Lenard form (Eq. (7.43)), where the outer cut-off is automatically included, found that the differences were small.

Noting the derivatives of ω_{ij} (7.A.9) and integrating by parts as necessary, to obtain Landau's equation:

$$\left.\frac{\partial f}{\partial t}\right|_{\text{coll}} = 2\pi\left(\frac{Z Z' e^2}{(4\pi\epsilon_0)\overline{m}}\right)^2 \ln\Lambda \int \mathrm{d}\mathbf{v}' \left\{\left[\frac{m'}{(m+m')} f'(\mathbf{v}')\frac{\partial f}{\partial v_i} - \frac{m}{(m+m')} f(\mathbf{v})\frac{\partial f'}{\partial v_i'}\right]\frac{\partial \omega_{ij}}{\partial u_j}\right.$$
$$+\frac{\partial}{\partial v_j}\left[\frac{m'^2}{(m+m')^2} f'(\mathbf{v}')\frac{\partial f}{\partial v_i} - \frac{m\,m'}{(m+m')^2} f(\mathbf{v})\frac{\partial f'}{\partial v_i'}\right]\omega_{ij}$$
$$-\left[\frac{m^2}{(m+m')^2} f(\mathbf{v})\frac{\partial f'}{\partial v_i'} - \frac{m\,m'}{(m+m')^2}\frac{\partial f}{\partial v_i} f'(\mathbf{v}')\right]\frac{\partial \omega_{ij}}{\partial v_j'} \tag{7.A.11a}$$

$$= 2\pi\left(\frac{Z Z' e^2}{(4\pi\epsilon_0)\overline{m}}\right)^2 \ln\Lambda \int \mathrm{d}\mathbf{v}' \left\{\left[\frac{m'^2}{(m+m')^2} f'(\mathbf{v}')\frac{\partial f}{\partial v_i} - \frac{mm'}{(m+m')^2} f(\mathbf{v})\frac{\partial f'}{\partial v_i'}\right]\frac{\partial \omega_{ij}}{\partial v_j}\right.$$
$$\left.+\frac{\partial}{\partial v_j}\left[\frac{m'^2}{(m+m')^2} f'(\mathbf{v}')\frac{\partial f}{\partial v_i} - \frac{m\,m'}{(m+m')^2} f(\mathbf{v})\frac{\partial f'}{\partial v_i'}\right]\omega_{ij}\right\} \tag{7.A.11b}$$

$$= 2\pi\left(\frac{Z Z' e^2}{(4\pi\epsilon_0)\overline{m}}\right)^2 \ln\Lambda \int \mathrm{d}\mathbf{v}' \left\{\frac{m'}{(m+m')^2}\left[m' f'(\mathbf{v}')\frac{\partial f}{\partial v_i} - m f(\mathbf{v})\frac{\partial f'}{\partial v_i'}\right]\frac{\partial \omega_{ij}}{\partial v_j}\right.$$
$$\left.+\frac{\partial}{\partial v_j}\frac{m'}{(m+m')^2}\left[m' f'(\mathbf{v}')\frac{\partial f}{\partial v_i} - m f(\mathbf{v})\frac{\partial f'}{\partial v_i'}\right]\omega_{ij}\right\} \tag{7.A.11c}$$

$$= 2\pi\left(\frac{Z Z' e^2}{(4\pi\epsilon_0)\, m}\right)^2 \ln\Lambda \frac{\partial}{\partial v_j}\int \mathrm{d}\mathbf{v}' \left\{\left[f'(\mathbf{v}')\frac{\partial f}{\partial v_i} - \frac{m}{m'} f(\mathbf{v})\frac{\partial f'}{\partial v_i'}\right]\omega_{ij}\right\} \tag{7.A.11d}$$

which is consistent with Eqs. (7.29) and (7.36).

Appendix 7.B Finite Difference Algorithm for Integrating the Isotropic Fokker–Planck Equation

Establish a one-dimensional mesh in velocity space with cells centred at v_j and boundaries $v_{(j-1/2)}$ and $v_{(j+1/2)}$ symmetrically distributed in velocity space so that $v_{1/2} = 0$. The finite difference of Eq. (7.49) takes the centred difference form in both space and time

$$\frac{f_j^{n+1/2} - f_j^{n-1/2}}{Dt} = \frac{1}{2}\frac{1}{v_j^2\,\Delta v_j} h\left\{\left(g_{(j+1/2)}^{(n+1)} - g_{(j-1/2)}^{(n+1)}\right) + \left(g_{(j+1/2)}^{n} - g_{(j-1/2)}^{n}\right)\right\} \tag{7.B.1}$$

where Dt is the time-step, $\Delta v_j = v_{(j+1/2)} - v_{(j-1/2)}$ is the cell width, $h = (4\pi)^2 e^4 \ln(\Lambda)/m^2$ and $g_{j+1/2}^n$ is the flux across the face $(j+1/2)$ at time $n\,Dt$

$$g_{(j+1/2)} = \frac{1}{2} p_{(j+1/2)}\,(f_{(j+1)} + f_j) + q_{(j+1/2)}\,(f_{(j+1)} - f_j)/\Delta v_{(j+1/2)} \tag{7.B.2}$$

where

$$p_{(j+1/2)} = p_{(j-1/2)} + f_j\, v_j^2\,\Delta v_j \tag{7.B.3a}$$

$$q_{(j+1/2)} = q_{(j-1/2)} + \frac{1}{3}\left[\frac{1}{v_{(j+1/2)}}\Phi_{(j+1/2)} + v_{(j+1/2)}^2\,\Psi_{(j+1/2)}\right] \tag{7.B.3b}$$

$$\Phi_{(j+1/2)} = \Phi_{(j-1/2)} f_j\, v_j^4\,\Delta v_j \tag{7.B.3c}$$

$$\Psi_{(j-1/2)} = \Psi_{(j+1/2)} - f_j\, v_j\, \Delta v_j. \tag{7.B.3d}$$

The terms $\phi_{(j+1/2)}$ and $\Psi_{(j+1/2)}$ are calculated recursively. The solution of Eq. (7.B.1) is easily calculated by the standard tri-diagonal matrix solver provided the fluxes $g^{(n+1)}_{(j+1/2)}$ are known at the later time $(n+1)\,Dt$. This requires an iteration of Eqs. (7.B.2) and (7.B.3). In most cases, a single iteration is sufficient, leading to a predictor–corrector form.

The centred difference form of the algorithm ensures that it is unconditionally stable. However, to maintain reasonable accuracy, it is advisable that the time step be limited to a reasonable fraction ($\lesssim 1/2$) of the relaxation time $(4\,\pi\,\epsilon_0)^2\,m^2\,v_0{}^3/(4\,\pi)^2\,n\,e^4\,\ln(\Lambda)$. A full discussion of the method is given by Pert (2001). The application for electron–electron and ion–ion scattering is obvious.

Appendix 7.C Monte Carlo Algorithm for Integrating the Fokker–Planck Equation

The inherently stochastic nature of the scattering process makes it natural that the development of the distribution function due to collisions can be treated by Monte Carlo simulation.[7] Instead of treating each two-body collision independently, the multi-particle scattering inherent in plasma allows a considerable simplification of the method as the angular distribution of the scattered particle can be expressed as a normal (Gaussian) distribution whose variance increases linearly with time (Section 2.1.3) encompassing many separate interactions.

Provided the scattering is weak and the collisions are perturbative on the particle motion, they change the momentum vectorally normal to the path. As a result, the particle suffers a series of random impulses in each of the two directions in the plane normal to its path. Since each impulse is one of a set of independent and identically distributed random variables with well-defined mean and variance in two perpendicular directions, we may apply the central limit theorem (Appendix 2.B) to their sum over multiple interactions, and conclude that overall distribution of momentum transfer must be normal (Gaussian) in each direction with mean zero and variance determined by the momentum transfer cross section (2.16). Thus, if θ_1 and θ_2 are the deflection angles in the two directions in the plane perpendicular to the particle velocity, and θ the total angle of scatter, the variance of the angle of scatter over the interval Δt is

$$\begin{aligned}\left\langle \sin^2\left(\tfrac{1}{2}\,\theta_1\right)\right\rangle &= \left\langle \sin^2\left(\tfrac{1}{2}\,\theta_2\right)\right\rangle = \tfrac{1}{2}\left\langle \sin^2\left(\tfrac{1}{2}\,\theta\right)\right\rangle \\ &= \frac{1}{4}\,n\,u\int (1-\cos\theta)\,\sigma(\theta)\,d\theta\,\Delta t = \frac{1}{4}\,n\,u\,\sigma_d\,\Delta t\end{aligned} \tag{7.C.1}$$

where u is the relative velocity of the scattering particles. In the context of plasma particle scattering, the variance of the angle of scatter becomes

$$\langle\theta^2\rangle \approx 8\,\pi\,b_0{}^2\,n\,u\,\ln\Lambda = \frac{8\,\pi\,Z_1{}^2\,Z_2{}^2\,e^4\,\ln\Lambda}{\overline{m}^2\,u^3}\,\Delta t \tag{7.C.2}$$

consistent with Eq. (2.16).

---

7 Stochastic methods have a long history (Hammersley and Handscomb, 1964), but were not practicable for general use until the introduction of electronic computers made extensive numerical calculation rapid and efficient. Modern methods stem from work on neutron scattering developed at Los Alamos by von Neumann, Ulam, and co-workers (Metropolis, 1989; Eckhardt, 1989). A reasonably detailed introduction to Monte Carlo methods is given by MacKeown (1997).

As argued above, the distribution is normal in both the orthogonal x and y directions so that the probability of a deflection $\sin(\frac{1}{2}\theta) = \xi \approx \sqrt{\xi_1{}^2 + \xi_2{}^2}$ is

$$p(\xi)\,\mathrm{d}\xi = \frac{2}{\langle \xi^2 \rangle} \exp\left(-\frac{\xi^2}{\langle \xi^2 \rangle}\right)\,\xi\,\mathrm{d}\xi \tag{7.C.3}$$

for which deviates are easily generated by the transformation method (MacKeown, 1997). In addition, we require a rotation in the plane x, y, which is simply a uniform deviate ϕ in the range $(0,\ 2\pi)$. Occasionally, large values of $\xi > 1$ will be generated, which correspond to strong collisions and for which the weak collision is not valid. Such values must either be rejected or the calculation repeated with the distribution corresponding to a single binary collision using the Rutherford formula.

The algorithm is completed by randomly selecting two particles from either the same distribution (for identical particles) or from each (for unlike). This is most simply done by a sweep through the each distribution for one particle and randomly selecting the second from the appropriate set, in which case, each particle is counted twice and the mean squared deviation is calculated with half the step $Dt/2$. The scattering is treated as a two body collision with scattering angles 2 arcsin ξ and ϕ thereby ensuring that energy and momentum are conserved. Defining a set of co-ordinates $(\hat{\imath},\ \hat{\boldsymbol{j}},\ \hat{\boldsymbol{k}})$ with $\hat{\boldsymbol{k}}$ *not* parallel to $\mathbf{u}$, we obtain a new value of the relative velocity:

$$u'_x = u_x\ \cos\theta - \frac{u_x\,u_z}{u_r} \sin\theta\ \cos\phi + \frac{u\,u_z}{u_r}\ \sin\theta\ \sin\phi \tag{7.C.4a}$$

$$u'_y = u_y\ \cos\theta - \frac{u_y\,u_z}{u_r} \sin\theta\ \cos\phi - \frac{u\,u_z}{u_r}\ \sin\theta\ \sin\phi \tag{7.C.4b}$$

$$u'_z = u_z\ \cos\theta + u_r \sin\theta\ \cos\phi \tag{7.C.4c}$$

where $u_r = \sqrt{u_x{}^2 + u_y{}^2}$ and $u = \sqrt{u_r{}^2 + u_z{}^2}$, and $\sin\theta = 2\xi\,\sqrt{1-\xi^2}$ and $\cos\theta = 1 - 2\,\xi^2$.

It is easily seen that the Monte Carlo model is equivalent to the Fokker–Planck description of collisions provided the time step Dt is sufficiently small that the values of ξ generated are small, that the collisions are always weak. The multi-particle nature of the collision process is ensured by generating a typical statistical distribution with the appropriate variance for the events. The global conservation of momentum and energy are ensured by the two-body representation of the collision process. The representation in terms of a summation of weak binary collisions is equivalent to the Rosenbluth et al. (1957) approach, the averages being replaced by random sampling instead of algebraic integrals. Each simulation particle represents a large number of real particles and each event a random sample from a large number of simultaneous multi-particle interactions.

The MonteCarlo method replaces the two quasi-analytic integrals of the Fokker–Planck equation, namely over impact parameter and distribution function, by two equivalent 'random' integrations, but in the reverse order. The outer (Monte-Carlo) integration is accomplished over the particle distribution by summing randomly selected pairs of interacting particles. The inner integration is generated by a well-defined pair of random impact parameters selected from the appropriate distributions for each pair of particles; thereby generating the resultant scattering. The ordering is important for the choice of the value of the Spitzer parameter $\ln\Lambda$, which should therefore correspond to the interacting pairs of particles, not to the average value over the particle distribution. In David and Hooker (2003, Figure 5) is shown a comparison between *ab initio* molecular dynamics, Monte-Carlo and Fokker–Planck models. In can be seen that the use of the Spitzer logarithm for individual pairs of particles gives markedly better results than the distribution averaged value.[8] There is good agreement between the calculations of the MonteCarlo using the averaged value of $\ln \Lambda$ and the Fokker–Planck models, as is expected since both have used the same distribution averaged value.

A full discussion of the method is given in Pert (1999).

Appendix 7.D Landau's Calculation of the Electron–Ion Equilibration Rate

Landau (1936) calculated the equilibration rate directly calculated from his form of the Fokker–Planck equation (7.36). Noting that the net energy transfer between like species is zero, the energy transfer rate from species m' to species m per unit volume is

$$\begin{aligned}
\frac{dE}{dt} &= \int \frac{1}{2} m\, v^2\, f(\mathbf{v})\, d\mathbf{v} \\
&= \frac{2\pi Z^2 e^4 \ln \Lambda}{(4\pi\epsilon_0)^2 m^2} \int d\mathbf{v}\, \frac{1}{2} m\, v^2 \frac{\partial}{\partial v_i} \left\{ \int d\mathbf{v}' \left(f'(\mathbf{v}') \frac{\partial f}{\partial v_j} - \frac{m}{m'} f(\mathbf{v}) \frac{\partial f'}{\partial v'_j} \right) \omega_{ij} \right\} \\
&= -\frac{2\pi Z^2 e^4 \ln \Lambda}{(4\pi\epsilon_0)^2 m^2}\, m \int d\mathbf{v}\, v_i \left\{ \int d\mathbf{v}' \left(f'(\mathbf{v}') \frac{\partial f}{\partial v_j} - \frac{m}{m'} f(\mathbf{v}) \frac{\partial f'}{\partial v'_j} \right) \omega_{ij} \right\} \\
&= \frac{2\pi Z^2 e^4 \ln \Lambda}{(4\pi\epsilon_0)^2 m^2}\, m \int d\mathbf{v} \int d\mathbf{v}'\, \omega_{ij}\, v_i \left(\frac{m\, v_j}{kT} - \frac{m}{m'} \frac{m'\, v'_j}{kT'} \right) f(\mathbf{v}) f'(\mathbf{v}')
\end{aligned} \tag{7.D.1}$$

after integrating by parts, neglecting the surface integral at infinity and substituting for a Maxwellian velocity distribution. Since $\mathbf{u} = \mathbf{v} - \mathbf{v}'$, it follows that

$$\omega_{ij}\,(v_j - v'_j) = \omega_{ij} u_j = u_i - u_i \frac{u_j u_j}{u^2} = 0 \tag{7.D.2}$$

reflecting the condition that the matrix product $\omega_{ij}\, a_j$ projects the components of the vector $\mathbf{a}$ on to the plane perpendicular to $\mathbf{u}$. Hence, we obtain

$$\frac{dE}{dt} = \frac{2\,\pi\, n\, Z^2\, e^4\, \ln \Lambda}{(4\,\pi\,\epsilon_0)^2} \left(\frac{1}{kT} - \frac{1}{kT'} \right) \int d\mathbf{v}\, v_i\, v_j\, f(\mathbf{v}) \int d\mathbf{v}'\, \omega_{ij}\, f'(\mathbf{v}') \tag{7.D.3}$$

Interchanging the primed and unprimed quantities, it is clear that $dE'/dt = -dE/dt$ and that energy is, therefore, conserved in the collisions.

Although this result applies generally, integration over the relative velocity is awkward (see Appendix 7.B). Landau (1936) sidestepped the problem by specifically specifying the ions and electrons. Identifying the heavier ions as the unprimed quantities and the lighter electron as the primed, $m \gg m'$; therefore, the electron velocity is much larger than the ion $v \ll v'$, $\mathbf{u} \approx \mathbf{v}'$ and $\omega_{ij} \approx [\delta_{ij} - v'_i\, v'_j / v'^2]/v'$. In consequence, we may separate the integrals over $\mathbf{v}$ and $\mathbf{v}'$. Furthermore,

$$\int v_i\, v_j\, f(\mathbf{v})\, d\mathbf{v} = \frac{1}{3}\, \delta_{ij}\, n\, \overline{v^2} = \delta_{ij}\, n\, \frac{kT}{m} \qquad \text{and} \qquad \int \frac{1}{v'} f'(\mathbf{v}') d\mathbf{v}' = n' \sqrt{\frac{2\, m'}{\pi\, kT'}} \tag{7.D.4}$$

where n and m are the ion density and mass, respectively, and n' and m' those of the electrons.

8 Note unfortunately the caption has the symbols for these two calculations interchanged.

Substituting in Eq. (7.D.3), we obtain energy transfer rate from electrons to ions per unit volume and from ions to electrons, respectively,

$$\frac{\mathrm{d}E}{\mathrm{d}t} = -\frac{\mathrm{d}E'}{\mathrm{d}t} \approx -\frac{4\sqrt{2\pi}\, n\, n'\, Z^2\, e^4\, \sqrt{m'}\, \ln\Lambda}{(4\,\pi\,\epsilon_0)^2\, m(kT')^{3/2}}(kT - kT') \tag{7.D.5}$$

which is identical to Spitzer's result (7.59) if $m \gg m'$.

8

The Hydrodynamic Limit for Plasma

8.1 Introduction – Individual Particle Fluid Equations

As for fluids (Section 4.10), it is usually inconvenient to describe a plasma in terms of its microscopic state. For many practical purposes, we require only the averaged macroscopic quantities, which are measured directly in the laboratory. However, the macroscopic situation in plasma is more complex than that in gases due to the fact that there are at least two distinct species of particles (ions and electrons) both of which are electrically charged and which may have different states (e.g. different temperatures). For simplicity, we will consider only a two component plasma of ions i of charge Ze and electrons e with charge $-e$. The generalisation to include additional species is straightforward. As with gases, we identify five moments, but separately for each of the ions and electrons[1]:

Number	$n_i(\mathbf{r},t) = \int f_i(\mathbf{r},\mathbf{v}_i,t)\,\mathrm{d}\mathbf{v}_i$	$n_e(\mathbf{r},t) = \int f_e(\mathbf{r},\mathbf{v}_e,t)\,\mathrm{d}\mathbf{v}_e$
Momentum	$\mathbf{p}_i(\mathbf{r},t) = m_i \int f_i(\mathbf{r},\mathbf{v}_i,t) q\,\mathbf{v}_i\,\mathrm{d}\mathbf{v}_i$	$\mathbf{p}_e(\mathbf{r},t) = m_e \int f_e(\mathbf{r},\mathbf{v}_e,t)\,\mathbf{v}_e\,\mathrm{d}\mathbf{v}_e$
Energy	$E_i(\mathbf{r},t) = \frac{1}{2} m_i \int f_i(\mathbf{r},\mathbf{v}_i,t)\,v_i^2\,\mathrm{d}\mathbf{v}_i$	$E_e(\mathbf{r},t) = \frac{1}{2} m_e \int f_e(\mathbf{r},\mathbf{v}_e,t)\,v_e^2\,\mathrm{d}\mathbf{v}_e$

Following the methodology of Section 4.10, we introduce the general velocity dependent term $Q(\mathbf{v})$ which we integrate over the Boltzmann equation to form the moments of Q. The external force on the particles is

$$\mathbf{f}_i = Ze(\mathbf{E} + \mathbf{v}_i \wedge \mathbf{B}) - m_i \nabla\phi \qquad \mathbf{f}_e = -e(\mathbf{E} + \mathbf{v}_e \wedge \mathbf{B}) - m_e \nabla\phi \tag{8.1}$$

where $\mathbf{E}$ and $\mathbf{B}$ are the total electric and magnetic fields including the self-generated fields and ϕ the gravitational potential. The integration over the terms in the Boltzmann equation is accomplished as in Eq. (4.60) except for the term (4.60c), which involves the derivative $\partial a_i/\partial v_i$. However, since $\mathbf{v} \wedge \mathbf{B}$ is perpendicular to $\mathbf{v}$, the acceleration a_i contains no term dependent on the component v_i and the derivative is zero, $\partial a_i/\partial v_i = 0$.

Hence, finally, we obtain the same general equation as before, which may be applied to each species:

$$\frac{\partial}{\partial t}(n\,\overline{Q}) + \frac{\partial}{\partial \mathbf{r}}(n\,\overline{Q\,\mathbf{v}}) - n\,\mathbf{a}\cdot\overline{\frac{\partial Q}{\partial \mathbf{v}}} = \left.\frac{\partial}{\partial t}(n\,\overline{Q})\right|_{\text{coll}} \tag{4.62}$$

However, in contrast to the fluid made up of a single species, the collisional terms are not zero to allow for the exchange of momentum and energy between the ions and electrons.

1 Quantities referring to particular species are identified by a superscript i or e for ions or electrons, respectively.

Foundations of Plasma Physics for Physicists and Mathematicians, First Edition. Geoffrey J. Pert.

Proceeding as before we set

- *Electron/ion number conservation* $Q(\mathbf{v}) = 1$:

$$\frac{\partial n_i}{\partial t} + \nabla \cdot (n_i \mathbf{u}_i) = 0 \qquad \frac{\partial n_e}{\partial t} + \nabla \cdot (n_e \mathbf{u}_e) = 0 \tag{8.2}$$

- *Momentum conservation* $Q(\mathbf{v}) = m\mathbf{v}$:

$$\begin{aligned} &\frac{\partial}{\partial t}(n_i\, m_i\, u_{i_i}) + \frac{\partial}{\partial x_j}(n_i\, m_i\, u_{i_i}\, u_{i_j} + \mathsf{P}_{i_{ij}}) - Z\, e\, n_i\, (E_i + \epsilon_{ijk}\, u_{i_j}\, B_k) - n_i m_i \frac{\partial \phi}{\partial x_i} = \mathcal{P}^{ie}_i \\ &\frac{\partial}{\partial t}(n_e\, m_e\, u_{ei}) + \frac{\partial}{\partial x_j}(n_e\, m_e\, u_{e_i}\, u_{e_j} + \mathsf{P}_{e_{ij}}) + e\, n_e\, (E_i + \epsilon_{ijk}\, u_{e_j}\, B_k) - n_e m_e \frac{\partial \phi}{\partial x_i} = -\mathcal{P}^{ie}_i \end{aligned} \tag{8.3}$$

where $\mathbf{P}$ is the stress tensor and $\mathcal{P}^{ie}$ is the momentum exchange rate from electrons to ions known as the *ion–electron drag force*.[2]

- *Energy conservation* $Q(\mathbf{v}) = \frac{1}{2}mv^2$:

$$\begin{aligned} \frac{\partial}{\partial t}\left[n_i\, m_i\, \left(\epsilon_i + \frac{1}{2}\, u_i^{\,2} + \phi\right)\right] + \frac{\partial}{\partial\, x_i}\left[n_i\, m_i\, \left(\epsilon_i + \frac{1}{2}\, u_i^{\,2} + \phi\right)\, u_{ii} + \mathbf{P}_{i_{ij}}\, u_{i_j} + q_{i_i}\right] \\ = Z\, e\, n_i\, E_i\, u_{ii} + Q^{ie} \\ \frac{\partial}{\partial t}\left[n_e\, m_e\, \left(\epsilon_e + \frac{1}{2}\, u_e^{\,2} + \phi\right)\right] + \frac{\partial}{\partial\, x_i}\left[n_e\, m_e\, \left(\epsilon_e + \frac{1}{2}\, u_e^{\,2} + \phi\right)\, u_{e_i} + \mathbf{P}_{e_{ij}}\, u_{ej} + q_{e_i}\right] \\ = -e\, n_e E_i u_{ei} - Q^{ie} \end{aligned} \tag{8.4}$$

where ϵ_i and ϵ_e are the individual specific energies of the ion and electrons, i.e. the energy per unit mass of either ion or electron as appropriate. $\mathbf{q}_i$ and $\mathbf{q}_e$ are the ion and electron heat fluxes, respectively. Q^{ie} is the *collisional energy transfer rate* from electrons to ions, which we treated earlier in Section 7.6.

8.2 The Departure from Steady, Homogeneous Flow: The Transport Coefficients

Unfortunately, at this stage, we are left with the same problem that we found for gases (Section 4.11), namely, that the equations for momentum and energy each involve a higher, unknown moment. As before these are associated with gradients of velocity and temperature, which gave rise to viscosity and thermal conduction. Closure is again obtained through the Chapman–Enskog approximation, Eq. (4.82). In this case, we must identify two differing possibilities depending on the relative strengths of collisions and the magnetic field. Although individual collisions are weak, we may conveniently identify a scale length associated with collisions from the time taken for a particle to be scattered through 90° typically given by Eq. (2.23) for electrons and (2.24) for ions. The corresponding mean free paths, λ are readily obtained. The presence of a background magnetic field introduces further plasma scale lengths, namely, the Larmor radii, ρ.

In general, in plasma, we may identify one of two conditions to be applicable:

1. *Collision dominated plasma*: $\lambda \ll L$ and $\lambda \ll \rho$ when $\xi = \lambda/L$,
2. *Collisionless plasma* $\rho \ll L$ and $\rho \ll \lambda$ when $\xi = \rho/L$,

where L is the characteristic length of the gradients and ξ the expansion parameter.

2 The first subscripts i and e represent the particle, ions i, and electrons e, respectively, subsequent subscripts i, j, etc., represent directional components, e.g. u_{i_i} is the ith component of the ion mean velocity $\mathbf{u}_i$.

Provided one of these conditions is met, may use the Enskog approximation (Eq. (4.85)) applied to the appropriate kinetic equation. Due to the long-range nature of the force, we must modify the Boltzmann equation by using the Fokker–Planck collision operator to include the required gradients. Since the Fokker–Planck collision operator is bilinear in the distribution function in either the Rosenbluth, MacDonald, and Judd form or the Landau form, we may use the Enskog expansion (4.80) to obtain closure in the same way as in the Chapman–Enskog method for gases.

If neither condition is met, the particles are essentially free and their behaviour may be treated as in Chapter 3.

8.3 Magneto-hydrodynamic Equations

In many applications, it is convenient to treat the plasma as a single fluid with appropriate quantities:

$$\begin{array}{llll} \text{Mass density} & \rho = n_i\, m_i + n_e\, m_e & \text{Flow velocity} & \mathbf{u} = (n_i\, m_i\, \mathbf{u}_i + n_e\, m_e\, \mathbf{u}_e)/\rho \\ \text{Charge density} & q = Z\, e\, n_i - e\, n_e & \text{Current density} & \mathbf{j} = Z\, e\, n_i\, \mathbf{u}_i - e\, n_e\, \mathbf{u}_e \\ \text{Specific internal energy} & \multicolumn{3}{l}{\epsilon = (n_i\, m_i\, \epsilon_i + n_e\, m_e\, \epsilon_e)/(n_i\, m_i + n_e\, m_e)} \\ \text{Total stress tensor} & \multicolumn{3}{l}{\mathsf{P} = \mathsf{P}_i + \mathsf{P}_e} \end{array} \tag{8.5}$$

solving for the individual particle variable in terms of the fluid terms. These equations form a generalisation of the standard equations of fluid mechanics to include the electro-magnetic field. As such they can be expressed in terms of the familiar conservation laws of hydro-dynamics (Pert, 2013) with additional terms and Maxwell's equations. We shall consider only non-relativistic plasma, where the characteristic length is short $L \ll cT$. c being the velocity of light and T the characteristic time.

Unfortunately, the moment equations cannot be expressed simply in the single fluid variable set (8.5). It is therefore necessary to simplify them by taking advantage of the large difference in mass between the ions and electrons $m_i \gg m_e$. Furthermore, since a small charge imbalance leads to a large neutralising electric field, the space charge q is approximately zero (*quasi-neutral approximation*) and $Z\, n_i \approx n_e$. These two approximations allow us to re-write to express the individual quantities in terms of the set (8.5) and reduce them to a simpler form

$$\begin{aligned} n_i &= \frac{e\,\rho - m_e\, q}{e\,(m_i + Z\, m_e)} && \approx \frac{\rho}{m_i} \\ n_e &= \frac{Z\, e\,\rho - m_i\, q}{e\,(m_i + Z\, m_e)} && \approx \frac{Z\,\rho}{n_i} - \frac{q}{e} \quad \approx \frac{Z\,\rho}{n_i} \\ \mathbf{u}_i &= \frac{e\,\rho\,\mathbf{u} + m_e \mathbf{j}}{e\rho + m_e q} && \approx \mathbf{u} \\ \mathbf{u}_e &= \frac{Z\, e\,\rho \bar{\mathbf{u}} - m_i \mathbf{j}}{Z\, e\,\rho - m_i\, q} && \approx \mathbf{u} - \frac{m_i}{Z\, e\,\rho}\mathbf{j} \quad \approx \mathbf{u} - \frac{\mathbf{j}}{e\, n_e} \end{aligned} \tag{8.6}$$

8.3.1 Equation of Mass Conservation

Adding the two individual particle conservation equation, we obtain the *equation of mass continuity*:

$$\frac{\partial \rho}{\partial t} + \nabla \cdot (\rho \mathbf{u}) = 0 \tag{8.7}$$

and the *equation of charge conservation*

$$\frac{\partial q}{\partial t} + \nabla \cdot \mathbf{j} = 0 \qquad \longrightarrow \qquad \nabla \cdot \mathbf{j} \approx 0 \tag{8.8}$$

The latter equation implies that the current paths form either closed loops in the plasma or start and end at infinity.

8.3.2 Equation of Momentum Conservation

Assuming the normal condition for charge neutrality, i.e. that the Debye length is much smaller than the characteristic lengths in the flow ($\lambda_D \ll L$), we may treat the combination of electrons and ions as a single fluid, by adding the individual momentum equations. The fluid as a bulk moves with the centre of mass velocity. The differential velocity of the electrons with respect to the ions is treated separately as an electric current.

Since[3]

$$(\mathbf{A} \cdot \nabla)\mathbf{A} = A_j \frac{\partial A_i}{\partial x_j} = \frac{\partial (A_i A_j)}{\partial x_j} - A_i \frac{\partial A_j}{\partial x_j} = \nabla \cdot (\mathbf{AA}) - \mathbf{A}\ \nabla \cdot \mathbf{A} \tag{8.9}$$

the identity takes the form

$$(\nabla \wedge \mathbf{A}) \wedge \mathbf{A} = \frac{1}{2}\nabla A^2 - (\mathbf{A} \cdot \nabla)\mathbf{A} = \frac{1}{2}\nabla A^2 - \nabla \cdot (\mathbf{AA}) + \mathbf{A}\ \nabla \cdot \mathbf{A} \tag{8.10}$$

From Maxwell's equations,

$$\begin{aligned} \nabla \wedge \mathbf{E} &= -\frac{\partial \mathbf{B}}{\partial t} \qquad \text{and} \qquad \nabla \wedge \mathbf{B} = \mu_0 \left(\mathbf{j} + \epsilon_0 \frac{\partial \mathbf{E}}{\partial t}\right) \\ \nabla \cdot \mathbf{E} &= \frac{q}{\epsilon_0} \qquad \text{and} \qquad \nabla \cdot \mathbf{B} = 0 \end{aligned} \tag{8.11}$$

we obtain

$$\begin{aligned} q\mathbf{E} + \mathbf{j} \wedge \mathbf{B} &= \epsilon_0\ \mathbf{E}\ \nabla \cdot \mathbf{E} - \epsilon_0 (\nabla \wedge \mathbf{E}) \wedge \mathbf{E} - \tfrac{1}{\mu_0}(\nabla \wedge \mathbf{B}) \wedge \mathbf{B} - \epsilon_0\ \frac{\partial (\mathbf{E} \wedge \mathbf{B})}{\partial t} \\ &= \nabla \cdot \left(\epsilon_0\ \mathbf{EE} + \tfrac{1}{\mu_0}\ \mathbf{BB}\right) - \tfrac{1}{2}\nabla\left(\epsilon_0\ E^2 + \tfrac{1}{\mu_0} B^2\right) - \epsilon_0\ \frac{\partial (\mathbf{E} \wedge \mathbf{B})}{\partial t} \end{aligned} \tag{8.12}$$

Neglecting the small momentum of the electrons due to the current in view of the small electron mass

$$\begin{aligned} &\frac{\partial (\rho\ \mathbf{u})}{\partial t} + \nabla \cdot [(n_i m_i \mathbf{u}_i\ \mathbf{u}_i + n_e m_e \mathbf{u}_e\ \mathbf{u}_e) + \mathsf{P}] - q\ \mathbf{E} - \mathbf{j} \wedge \mathbf{B} - \rho\ \nabla\phi = 0 \\ &\quad \approx \frac{\partial (\rho\ \mathbf{u})}{\partial t} + \nabla \cdot [(\rho\ \mathbf{u}\ \mathbf{u}) + \mathsf{P}] - \mathbf{j} \wedge \mathbf{B} - \rho\ \nabla\phi \\ &\quad = \frac{\partial}{\partial t}[\rho\ \mathbf{u} + \epsilon_0\ \mathbf{E} \wedge \mathbf{B}] \\ &\quad + \nabla \cdot \left[(\rho\ \mathbf{u}\ \mathbf{u}) + \mathsf{P} + \frac{1}{2}\left(\epsilon_0 E^2 + \frac{1}{\mu_0} B^2\right)\mathsf{I} - \left(\epsilon_0 \mathbf{EE} + \frac{1}{\mu_0}\mathbf{BB}\right)\right] - \rho\ \nabla\phi \approx 0 \end{aligned} \tag{8.13}$$

where $\mathsf{I} \equiv \delta_{ij}$ is the identity dyadic.

Excluding the gravitational term, which cannot be cast into this form, this form of the equation of momentum balance equation is expressed in general conservation law form in terms of the

3 It is convenient to use dyadic forms denoted by bold sans-serif characters rather than Cartesian tensors for these equations. Thus, $\mathsf{P} \equiv \mathsf{P}_{ij}$. Inner product operations $\mathbf{a} \cdot \mathsf{P} \equiv a_i \mathsf{P}_{ij} \neq \mathsf{P}_{ij} a_j \equiv \mathsf{P} \cdot \mathbf{a}$ (dyadics are NOT commutative). Two vectors $\mathbf{A}$ and $\mathbf{B}$ form a dyadic $\mathbf{A}\ \mathbf{B} \neq \mathbf{B}\ \mathbf{A}$.

momentum density vector and the total momentum stress (Pert, 2013). It represents the generalisation of the standard equation of fluid mechanics to magneto-hydrodynamics. The term $\boldsymbol{\Phi} = \left[\frac{1}{2}(\mathbf{E}\cdot\mathbf{D}+\mathbf{B}\cdot\mathbf{H})\,\mathsf{I} - (\mathbf{ED}+\mathbf{BH})\right]$ will be recognised as the Maxwell stress tensor,[4] and the term $\mathbf{G} = \mathbf{D}\wedge\mathbf{B}$ the momentum density of the electro-magnetic field.[5]

In analogy with classical fluid mechanics, this equation can written in Lagrangian form as a generalisation of the Navier–Stokes equation by making use of the equation of continuity (8.7):

$$\frac{\mathrm{d}\mathbf{u}}{\mathrm{d}t} = \frac{\partial u_i}{\partial t} + \left(u_j\frac{\partial}{\partial x_j}\right)u_i + \frac{\epsilon_0}{\rho}\frac{\partial(\epsilon_{ijk}E_jB_k)}{\partial t} = -\frac{1}{\rho}\frac{\partial p}{\partial x_i} + \frac{1}{\rho}\frac{\partial \sigma_{ij}}{\partial x_j} - \frac{1}{\rho}\left\{\frac{1}{2}\frac{\partial}{\partial x_i}\left(\epsilon_0 E^2 + \frac{1}{\mu_0}B^2\right) - \left[\epsilon_0\left(E_j\frac{\partial}{\partial x_j}\right)E_i + \frac{1}{\mu_0}\left(B_j\frac{\partial}{\partial x_j}\right)B_i\right]\right\} + \nabla\phi \tag{8.15}$$

where σ_{ij} is the viscous stress tensor and ϵ_{ijk} the permutation symbol and where

$$\frac{\mathrm{d}}{\mathrm{d}t} = \frac{\partial}{\partial t} + \mathbf{u}\cdot\nabla \tag{8.16}$$

is the Lagrangian time derivative taken following a fluid point moving with the plasma. This equation may be written more compactly using the total stress tensor:

$$\mathsf{T} = \mathsf{P} + \boldsymbol{\Phi} = \mathsf{P} + \left[\frac{1}{2}(\mathbf{E}\cdot\mathbf{D}+\mathbf{B}\cdot\mathbf{H})\,\mathsf{I} - (\mathbf{ED}+\mathbf{BH})\right] \tag{8.17}$$

and the electro-magnetic momentum density

$$\mathbf{G} = \mathbf{D}\wedge\mathbf{B} \tag{8.18}$$

to give

$$\rho\,\frac{\mathrm{d}\mathbf{u}}{\mathrm{d}t} + \frac{\partial\mathbf{G}}{\partial t} = -\nabla\cdot\mathsf{T} + \rho\,\nabla\phi \tag{8.15'}$$

In the absence of a gravitational field, the conservation of momentum in an isolated body of gas is immediately established by integrating $\int \rho\,\nabla\phi\,\mathrm{d}V$ over the total volume. In an astrophysical context, an isolated body forms a self-interacting gravitational mass (such as a star), where element $\mathrm{d}V$ of mass $\rho(V)\,\mathrm{d}V$ at position $\mathbf{r}$ attracts another $\mathrm{d}V'$ of mass $\rho(V')\,\mathrm{d}V'$ at $\mathbf{r}'$ with force $G\,\rho(V)\,\mathrm{d}V\,\rho(V')\,\mathrm{d}V'(\mathbf{r}'-\mathbf{r})|\mathbf{r}'-\mathbf{r}|^3$, where G is the gravitational constant. The rate of change of the total momentum due to internal gravitation forces on elements V due to elements V' is, therefore,

$$\int \rho\,\nabla\phi\,\mathrm{d}V = \int \rho(V)\,\mathrm{d}V \int \rho(V')\,\mathrm{d}V'\frac{G}{|\mathbf{r}'-\mathbf{r}|^3}(\mathbf{r}'-\mathbf{r}) = 0 \tag{8.19}$$

as may be seen by interchanging the primed and un-primed quantities. Hence, overall momentum conservation is established due to internal gravitational forces.

4 The Maxwell stress tensor is more commonly written as a Cartesian tensor in matrix form:

$$\Phi_{ij} = \left\{\begin{matrix} \begin{pmatrix}\frac{1}{2}(\mathbf{E}\cdot\mathbf{D}+\mathbf{B}\cdot\mathbf{H}) \\ -(E_x\,D_x+B_x\,H_x)\end{pmatrix} & -(E_x\,D_y+B_x\,H_y) & -(E_x\,D_z+B_x\,H_z) \\ -(E_y\,D_x+B_y\,H_x) & \begin{pmatrix}\frac{1}{2}(\mathbf{E}\cdot\mathbf{D}+\mathbf{B}\cdot\mathbf{H}) \\ -(E_yD_y+B_yH_y)\end{pmatrix} & -(E_yD_z+B_yD_z) \\ -(E_z\,D_x+B_z\,H_x) & -(E_z\,D_y+B_z\,H_y) & \begin{pmatrix}\frac{1}{2}(\mathbf{E}\cdot\mathbf{D}+\mathbf{B}\cdot\mathbf{H}) \\ -(E_zD_z+B_zH_z)\end{pmatrix}\end{matrix}\right. \tag{8.14}$$

5 Since all particles are treated explicitly, $\mathbf{D} = \epsilon_0\mathbf{E}$ is the dielectric displacement and $\mathbf{H} = \mathbf{B}/\mu_0$ the magnetic intensity.

8.3.3 Virial Theorem

Consider a finite mass of plasma occupying a volume V, which varies in time. The virial theorem follows from Eq. (8.15') by taking its scalar product with distance $\mathbf{r}$ and integrating over the volume of the fluid.

We define the scalar moment of inertia about the origin $I = \int \rho r^2\, \mathrm{d}V$. Noting that since $\rho\, \delta V$ is the mass of the fluid element δV, which moves with the fluid, it is therefore constant, and the Lagrangian time derivative $\mathrm{d}\,(\rho\, \delta V)/\mathrm{d}t = 0$. Hence,

$$\frac{\mathrm{d}I}{\mathrm{d}t} = 2\int \rho\, \mathbf{r}\cdot\mathbf{u}\, \mathrm{d}V \qquad \text{and} \qquad \frac{\mathrm{d}^2 I}{\mathrm{d}t^2} = 2\int \rho\left(\mathbf{r}\cdot\frac{\mathrm{d}\mathbf{u}}{\mathrm{d}t} + u^2\right)\mathrm{d}V \tag{8.20}$$

Carrying out the scalar products and performing the appropriate integrals:

$$\frac{1}{2}\frac{\mathrm{d}^2 I}{\mathrm{d}t} + \int_V \mathbf{r}\cdot\frac{\partial \mathbf{G}}{\partial t}\,\mathrm{d}V = 2K + \int_V \mathrm{tr}(\mathbf{T})\,\mathrm{d}V - \int_S \mathbf{r}\cdot\mathbf{T}\cdot\mathbf{ds} + \int_V \rho\mathbf{r}\cdot\nabla\phi\,\mathrm{d}V \tag{8.21}$$

where $K = \frac{1}{2}\int \rho\, u^2\, \mathrm{d}V$ is the total kinetic energy, and we have used the identity $\mathbf{r}\cdot(\nabla\cdot\mathbf{T}) = \nabla\cdot(\mathbf{r}\cdot\mathbf{T}) - \mathrm{tr}(\mathbf{T})$; S is the bounding surface of the fluid volume V. Neglecting the viscous stress $\mathrm{tr}(\mathbf{T}) = 3\,p + W$, where p is the kinetic pressure (assumed isotropic) and $W = \frac{1}{2}\,(\mathbf{E}\cdot\mathbf{D} + \mathbf{B}\cdot\mathbf{H})$ is the electro-magnetic energy density.

In laboratory plasma physics, gravity normally plays an insignificant role. However, in astrophysics gravitational forces within an isolated massive body play an important role. The virial of the gravitational term for an isolated mass is easily calculated by considering the summation of pairs of elemental masses:

$$\int \rho(V)\,\mathrm{d}V \int \rho(V')\,\mathrm{d}V'\,(\mathbf{r}-\mathbf{r}')\cdot\frac{G}{|\mathbf{r}'-\mathbf{r}|^3}\,(\mathbf{r}'-\mathbf{r}) = -\int \mathrm{d}V \int \mathrm{d}V' \frac{G\,\rho(V)\,\rho(V')}{|\mathbf{r}-\mathbf{r}'|} = 2\,U \tag{8.22}$$

The factor of two appears because the integral over the volume is performed twice. U is the total gravitational potential energy of particles in the self-generated gravitational field.[6]

Including the gravitational term the complete virial equation including gravity is

$$\frac{1}{2}\frac{\mathrm{d}^2 I}{\mathrm{d}t} + \int_V \mathbf{r}\cdot\frac{\partial \mathbf{G}}{\partial t}\,\mathrm{d}V = 2K + \int \mathrm{tr}(\mathbf{T})\,\mathrm{d}V - \int_S \mathbf{r}\cdot\mathbf{T}\cdot\mathbf{ds} + U \tag{8.21$'$}$$

8.3.4 Equation of Current Flow

The current flow is obtained from the difference between the electron and ion momentum equations (8.3):

$$\begin{aligned}
&\frac{m_e}{n_e\, e^2}\left[\frac{\partial \mathbf{j}}{\partial t} + \nabla\cdot\left(\mathbf{u}\,\mathbf{j} + \mathbf{j}\,\mathbf{u} - \frac{m_i}{Z\,e\rho}\mathbf{j}\,\mathbf{j}\right)\right] \\
&\quad = \frac{m_e}{n_e e^2}\left\{\left[\frac{\partial}{\partial t}(Z\,e\,n_i\mathbf{u}_i) + \nabla\cdot(Z\,e\,n_i\mathbf{u}_i\mathbf{u}_i)\right] - \left[\frac{\partial}{\partial t}(e\,n_e\mathbf{u}_e) + \nabla\cdot(Z\,e\,n_i\mathbf{u}_i\mathbf{u}_i)\right]\right\} \\
&\quad \approx \mathbf{E} + \overline{\mathbf{u}}\wedge\mathbf{B} - \frac{\mathbf{j}\wedge\mathbf{B}}{n_e\,e} + \frac{\nabla\cdot\mathbf{P}_e}{n_e\,e} - \frac{\boldsymbol{p}^{ie}}{n_e\,e}.
\end{aligned} \tag{8.23}$$

6 Since the body is bound by the gravitational forces, its total potential energy is negative. In astrophysics, the binding energy (the negative of the total potential energy) of a uniform sphere of radius R, and mass M, namely $3\,G\,M^2/5\,R$, is often used as the virial term.

The convective and time-dependent terms on the left-hand side of this equation can normally be neglected to give the generalised form of Ohm's law:

$$\mathbf{E} + \overline{\mathbf{u}} \wedge \mathbf{B} - \frac{\mathbf{j} \wedge \mathbf{B}}{n_e\, e} + \frac{\nabla \cdot \mathbf{P}_e}{n_e\, e} - \frac{\mathcal{P}^{ie}}{n_e\, e} \approx 0 \tag{8.24}$$

The interpretation of this equation is easily seen by considering the momentum exchange $\mathcal{P}^{ie}$ from the electrons to the ions as a result of collisions. The electrons are accelerated by the generalised driving force $\mathbf{F} = -e(\mathbf{E} + \overline{\mathbf{u}} \wedge \mathbf{B}) + \mathbf{j} \wedge \mathbf{B}/n_e - \nabla \cdot \mathbf{P}_e/n_e$. In a collision with an ion, the electrons transfer momentum to the ions at a rate determined by $\langle \Delta v_{\|} \rangle$ (Eq. (2.21b′)). It may therefore be expected that the momentum exchange $\mathcal{P}^{ie}$ is parallel to the electron velocity relative to ions; a result consistent with Eq. (2.21a) (since $\langle \Delta v_{\perp} \rangle = 0$), i.e. $\mathcal{P}^{ie} \propto \mathbf{j}$. This physical behaviour is simply illustrated by the Drude model discussed in Section 2.2.3. However, a systematic analysis requires a full description of the velocity dependence of the collision frequency using the Fokker–Planck equation.

8.3.5 Equation of Energy Conservation

The general equation for the total conservation of energy summed over both ions and electrons is rarely used due to the slow equilibration rate between the particles. As a consequence, the electron and ion temperatures are generally unequal, and the equilibration between them must be treated explicitly as part of the time development of each. Nonetheless, the general equation, which is valid if the system time development is sufficiently slow, is of interest and, therefore, we develop it here.

Summing the equivalent equations for the electrons and ions, we obtain

$$\frac{\partial}{\partial t}\left\{\begin{array}{l} \quad n_i\, m_i\, \left(\epsilon_i + \frac{1}{2}\, u_i^{\,2}\right) \\ +\; n_e\, m_e\, \left(\epsilon_e + \frac{1}{2}\, u_e^{\,2}\right) \end{array}\right\} + \frac{\partial}{\partial \mathbf{r}} \cdot \left\{\begin{array}{ll} \quad n_i\, m_i\, \mathbf{u}_i\, \left(\epsilon_i + \frac{1}{2}\, u_i^{\,2}\right) & + \mathbf{p}_i \cdot \mathbf{u}_i + \mathbf{q}_i \\ +\; n_e\, m_e\, \mathbf{u}_e\, \left(\epsilon_e + \frac{1}{2}\, u_e^{\,2}\right) & + \mathbf{p}_e \cdot \mathbf{u}_e + \mathbf{q}_e \end{array}\right\} - \mathbf{E} \cdot \mathbf{j} = 0 \tag{8.25}$$

The current density

$$\mathbf{j} = n_i\, q_i \mathbf{u}_i + n_e\, q_e\, \mathbf{u}_e = \nabla \wedge \mathbf{H} - \frac{\partial \mathbf{D}}{\partial t} = \frac{1}{\mu_0}\, \nabla \wedge \mathbf{B} - \epsilon_0\, \frac{\partial \mathbf{E}}{\partial t} \tag{8.26}$$

and the Ohmic work term:

$$\begin{aligned} \mathbf{E} \cdot \mathbf{j} &= \frac{1}{\mu_0}\, \mathbf{E} \cdot \nabla \wedge \mathbf{B} - \epsilon_0\, \mathbf{E} \cdot \frac{\partial \mathbf{E}}{\partial t} \\ &= -\frac{1}{\mu_0}\, \nabla \cdot (\mathbf{E} \wedge \mathbf{B}) + \frac{1}{\mu_0}\, \mathbf{B} \cdot \nabla \wedge \mathbf{E} - \epsilon_0\, \mathbf{E} \cdot \frac{\partial \mathbf{E}}{\partial t} \\ &= -\frac{1}{\mu_0}\, \nabla \cdot (\mathbf{E} \wedge \mathbf{B}) - \frac{1}{2\mu_0}\, \frac{\partial (B^2)}{\partial t} - \frac{\epsilon_0}{2}\, \frac{\partial (E^2)}{\partial t}. \end{aligned} \tag{8.27}$$

Hence, replacing the Ohmic work term, we finally obtain the equation in conservation law form:

$$\frac{\partial}{\partial t}\left\{\begin{array}{l} n_i\, m_i\, \left(\epsilon_i + \frac{1}{2}\, u_i^{\,2}\right) + \\ n_e\, m_e\, \left(\epsilon_e + \frac{1}{2}\, u_e^{\,2}\right) + \\ \epsilon_0\, E^2/2 + B^2/2\, \mu_0 \end{array}\right\} + \frac{\partial}{\partial \mathbf{r}} \cdot \left\{\begin{array}{l} n_i\, m_i\, \mathbf{u}_i\, \left(\epsilon_i + \frac{1}{2}\, u_i^{\,2}\right) + \mathbf{P}_i \cdot \mathbf{u}_i + \mathbf{q}_i + \\ n_e\, m_e\, \mathbf{u}_e\, \left(\epsilon_e + \frac{1}{2}\, u_e^{\,2}\right) + \mathbf{P}_e \cdot \mathbf{u}_e + \mathbf{q}_e + \\ \mathbf{E} \wedge \mathbf{B}/\mu_0 \end{array}\right\} = 0 \tag{8.28}$$

Note the introduction of the electro-magnetic energy flux determined by the Poynting vector $\mathbf{E} \wedge \mathbf{B}/\mu_0$. This equation can be cast into the more familiar fluid dynamics form by replacing the total stress $\mathbf{P}$ separating the pressure p from the 'viscous' stress $\mathbf{T}$, namely, $\mathbf{P} = p\,\mathbf{I} - \mathbf{T}$, where $\mathbf{I}$ is the identity tensor, and introducing the enthalpies $h = \epsilon + p/n\,m$.

$$\frac{\partial}{\partial t}\begin{Bmatrix} n_i\, m_i\, \left(\epsilon_i + \frac{1}{2}\, u_i^{\,2}\right) + \\ n_e\, m_e\, \left(\epsilon_e + \frac{1}{2}\, u_e^{\,2}\right) + \\ \epsilon_0\, E^2/2 + B^2/2\,\mu_0 \end{Bmatrix} + \frac{\partial}{\partial \mathbf{r}} \cdot \begin{Bmatrix} n_i\, m_i\, \mathbf{u}_i\, \left(h_i + \frac{1}{2}\, u_i^{\,2}\right) - \mathbf{T}_i \cdot \mathbf{u}_i + \mathbf{q}_i + \\ n_e\, m_e\, \mathbf{u}_e\, \left(h_e + \frac{1}{2}\, u_e^{\,2}\right) - \mathbf{T}_e \cdot \mathbf{u}_e + \mathbf{q}_e + \\ \mathbf{E} \wedge \mathbf{B}/\mu_0 \end{Bmatrix} = 0. \qquad (8.29)$$

When the ion and electron temperatures are equal, this result may be cast into simpler form which is the generalisation of the classical fluid mechanics energy conservation equation. Including the convective heat flow associated with the current in the total electron heat flow and writing

$$\rho\,\epsilon = n_i m_i \epsilon_i + n_e m_e \epsilon_e = \frac{3}{2}\, n_i(1+Z)\, kT \qquad \mathbf{T} = \mathbf{T}_i + \mathbf{T}_e \qquad p = p_i + p_e = n_i(1+Z)\, kT \qquad (8.30)$$

The specific enthalpy $h = \epsilon + p/\rho$. The heat flux must be modified to include a further term to explicitly account for the heat transported by the current carrying electrons:

$$\mathbf{q} = \mathbf{q}_i + \mathbf{q}_e + n_e\, m_e\, h_e(\mathbf{u}_e - \mathbf{u}) = \mathbf{q}_i + \mathbf{q}_e - \frac{5}{2}\, (kT)\,\mathbf{j}/e \qquad (8.31)$$

we obtain

$$\frac{\partial}{\partial t}\left[\rho\, \left(\epsilon + \frac{1}{2}u^2\right) + \frac{1}{2}\left(\epsilon_0\, E^2 + \frac{1}{\mu_0}\, B^2\right)\right] + \frac{\partial}{\partial \mathbf{r}}\left[\rho\mathbf{u}\left(h + \frac{1}{2}u^2\right) - \mathbf{T} + \mathbf{q} + \frac{1}{\mu_0}\, \mathbf{E} \wedge \mathbf{B}\right] = 0 \qquad (8.32)$$

where the first term $\left[\rho\, \left(\epsilon + \frac{1}{2}u^2\right) + \frac{1}{2}\left(\epsilon_0\, E^2 + \frac{1}{\mu_0}\, B^2\right)\right]$ accounts for the time dependence of the energy density and the second $\left[\rho\mathbf{u}\left(h + \frac{1}{2}u^2\right) - \mathbf{T} + \mathbf{q} + \frac{1}{\mu_0}\, \mathbf{E} \wedge \mathbf{B}\right]$ the divergence of the energy flux.

This equation is the generalisation of the standard energy conservation law of fluid mechanics to magneto-hydrodynamics. It includes the additional terms, the energy density associated with the electro-magnetic field, namely $\frac{1}{2}(\mathbf{E} \cdot \mathbf{D} + \mathbf{B} \cdot \mathbf{H})$ and the (Poynting) energy flux $\mathbf{E} \wedge \mathbf{H}$. The heat flux $\mathbf{q}$ is that due to thermal conduction and thermo-electric effects including the current convected energy.

This equation of energy transport can also be conveniently written in Lagrangian form in analogy with classical fluid mechanics:

$$\frac{\mathrm{d}\epsilon}{\mathrm{d}t} = \frac{\partial \epsilon}{\partial t} + (\mathbf{u} \cdot \nabla)\epsilon = -\frac{1}{\rho}[p\, \nabla \cdot \mathbf{u} + \sigma_{ij}\dot{e}_{ij} + \nabla \cdot \mathbf{q} - \mathbf{E} \cdot \mathbf{j}] \qquad (8.33)$$

8.4 Transport Equations

As in the case of classical fluid mechanics, the moment equations have each introduced quantities of higher moment than that being evaluated, the first-order moment (momentum) of the Boltzmann equation has introduced a second-order moment (stress); and the second-order moment (energy) equation has introduced a third-order moment (heat flux). To obtain closure, we use similar arguments to those for gases, Section 4.10, with the additional complication that there are now two small-scale lengths[7] which may be used to ensure the convergence of the approximation, namely, the mean free path λ for 90° scattering given by Eq. (2.23) or (2.24), and the particle Larmor radius ρ Eq. (3.3) for an appropriate average particle velocity. There are correspondingly two distinct cases:

1. *Collision-dominated plasma* characterised by the mean free path, λ, of the particles less than the Larmor radius, ρ.
2. *Collision-free plasma* characterised by the Larmor radius ρ shorter than the mean free path λ.

The ordering parameter is, therefore, typically $\min(\rho, \lambda)/L$.

Important roles are played by the collisional exchange terms $\mathcal{P}^{ie}$ and Q^{ie} known as the friction or drag terms. Momentum drag is associated with the resistance to current flow as described by the elementary Drude model (Section 2.2.3). From this model, it is immediately apparent that, due to the large mass ratio, the current is dominated by electrons.

Similarly, it follows from the elementary picture of thermal conduction (Section 2.2.2) that electron thermal conduction dominates that of the ions, and that it is also controlled by momentum drag.

However, the viscosity depends on the particle mass (Eq. (2.45)), so that ion momentum drag determines the viscosity. Viscosity is normally weak in plasmas and generally plays a minor role in controlling the dynamics. However, its treatment involving a linear relationship between two tensors of second order, presents an interesting technical problem.

The energy drag term Q^{ie} is the controlling factor in ion–electron equilibration discussed earlier in Section 7.6.

8.4.1 Collision Times

The times for electron–electron, ion–ion momentum transfer collisions, and electron–ion energy relaxation scale as $(\tau_{ee} \sim \tau_e : \tau_{ii} \sim \tau_i : \tau_{ei}) :: (1 : \sqrt{m_i/m_e} : (m_i/m_e))$, Eq. (2.25). It is clear that both electrons and ions remain in approximate Maxwellian distributions, but as the rate of energy transfer is much slower than the individual relaxation rates, the electrons and ions may have separate temperatures. The momentum transfer rate from the electrons to the ions, which determines the drag force, is of the same order of magnitude as the electron–electron collision rate, although ion–electron momentum transfer and energy transfer are greatly reduced by the large mass ratio of the ions. Since most transport terms are determined by momentum transfer from electrons to ions, the appropriate collision rate for most transport terms is given by the electron–ion collision time τ_e given by Eq. (2.23). Similar terms, which are a consequence of ion–ion collisions such as ion thermal conductivity or viscosity are determined by the ion–ion momentum collision time τ_i given by Eq. (2.24).

In a magnetic field B, the electron and ion cyclotron frequencies which are given by the standard expressions, Eq. (3.3):

$$\omega_e = \frac{e\,B}{m_e} \qquad \text{and} \qquad \omega_i = \frac{Z\,e\,B}{m_i} \tag{8.34}$$

determine the characteristic particle times.

7 There is, of course, a further scaling length, namely the Debye length, λ_D which must be shorter than all others to allow the fluid approximations to be valid. The Debye length provides the essential Coulomb field cut-off to the cross sections as discussed earlier.

8.4.2 Symmetry of the Transport Equations

It is a familiar experimental result that provided the departure from thermal equilibrium is small, the transport terms may be expressed as a linear relation between a characteristic *flux* and the *force* driving it. In the case of heat and current flow, there are two characteristic fluxes, namely, the heat flux and the electrical current. It follows that the generation of entropy by these two coupled irreversible terms may be expressed by linear relations involving pairs of thermodynamically conjugate functions (J_n, F_n)[8] namely in this case, the interacting force-flux pairs $[\mathbf{j},\ (1/T)\nabla\mu]$ and $[\mathbf{q},\ \nabla(1/T)]$, where μ is the generalised chemical potential including the electric potential.[9] These relationships are expressed in the general form of the Onsager expansion (see for example Callen (1948), de Groot and Mazur (1984, Ch.XIII), Landau and Lifshitz (1980, Ch.XII) or Landau and Lifshitz (1984, §27)):

$$J_m = \sum_n L_{mn}\, F_n \tag{8.37}$$

In the absence of a magnetic field, the familiar symmetry relation (Onsager relation) $L_{mn} = L_{nm}$ holds. In a magnetic field, the relationship is generalised to

$$L_{mn}(\mathbf{B}) = L_{nm}(-\mathbf{B}) \tag{8.38}$$

Let us write the linear relationship identified later in Eqs. (8.51) as

$$j_i = a_{ij}\left(-\frac{1}{\cancel{k}T}\right)\frac{\partial \mu}{\partial x_j} + b_{ij}\frac{\partial}{\partial x_j}\left(\frac{1}{\cancel{k}T}\right) \tag{8.39a}$$

$$q_i = c_{ij}\left(-\frac{1}{\cancel{k}T}\right)\frac{\partial \mu}{\partial x_j} + d_{ij}\frac{\partial}{\partial x_j}\left(\frac{1}{\cancel{k}T}\right) \tag{8.39b}$$

In a magnetic field, the system has axial symmetry about the magnetic field, which separates the charged particle motion, and therefore, the transport coefficients, into two distinct sets:

1. In the direction along the electron axis, the forces associated with the field gradient, and consequently, the resulting velocity are parallel to the field. The flux due to the magnetic field is, therefore, zero.
2. In a plane normal to the magnetic field, the magnetic field induces velocities parallel and perpendicular to the field gradient force lying in the plane.

Consider the behaviour when the magnetic field is reversed, the diagonal terms, which correspond to flux parallel to the force, are independent of the field reversal, whereas the off-diagonal terms,

8 Thermodynamically, conjugate pairs of quantities involve an intensive function analogous to a potential F_n, whose gradient is known as the thermodynamic force, and an extensive function known as the generalised displacement or flux J_n. The conjugate nature of the two functions is expressed in terms of the rate of entropy production

$$\frac{\partial S}{\partial t} = \sum_n J_n \cdot F_n \tag{8.35}$$

(or alternatively, $J_n = \partial S/\partial F_n$) for a set of functions whose coupling is expressed by the phenomenological relationship:

$$J_m = \sum_n L_{mn}\ \nabla F_n \tag{8.36}$$

demonstrating the linear relationship between the forces and the fluxes. In a system exhibiting time inversion, Onsager's relationship $L_{mn} = L_{nm}$ follows (Landau and Lifshitz, 1980; de Groot and Mazur, 1984).

9 The heat flux $\mathbf{q}$ excludes the energy flow associated with the current, namely, $\mathbf{q} = \mathbf{Q} - \mathbf{j}\,\phi$, where $\mathbf{Q}$ is the total heat flux and $\phi = -\frac{5}{2}\cancel{k}T/e$ the individual electron energy per unit charge.

which are due to the Lorentz force $\mathbf{v} \wedge \mathbf{B}$, are reversed by the change of field direction. In particular, the flux due to force parallel to the magnetic field is unaffected by the field. It is clear that the diagonal terms a_{ii} must be even and the off-diagonal odd in the field **B**.

The transport coefficients may be split into symmetric and anti-symmetric terms depending on whether they are even or odd functions of the magnetic field. Each of the transport coefficients must take the form below (Eq. (8.40)) in a co-ordinate system, where the z direction is taken parallel to the magnetic field **B** and the directions x and y lie in the plane of the field and the force gradient forming a right-handed set about z in the plane perpendicular to **B**. The diagonal component parallel to the field $\sigma_\parallel$ is independent of field reversal. The diagonal terms in the x, y plane $\sigma_\perp$ are even and the off-diagonal $\sigma_\wedge$ odd, since the plasma is isotropic both diagonal and off-diagonal terms are symmetric and have, the same magnitude. The general form of the coefficient matrix is therefore[10]:

$$\sigma_{ij} = \begin{pmatrix} \sigma_\perp & \sigma_\wedge & 0 \\ -\sigma_\wedge & \sigma_\perp & 0 \\ 0 & 0 & \sigma_\parallel \end{pmatrix} \tag{8.40}$$

These results are consistent with Onsager's relations, which also introduce the relationship between the current and heat thermo-electric terms b_{ij} and c_{ij}:

$$\begin{aligned} &a_{ij}(\mathbf{B}) = a_{ji}(-\mathbf{B}) \qquad d_{ij}(\mathbf{B}) = d_{ji}(-\mathbf{B}) \\ &b_{ij}(\mathbf{B}) = b_{ji}(-\mathbf{B}) = c_{ij}(-\mathbf{B}) = c_{ji}(\mathbf{B}) \end{aligned} \tag{8.41}$$

It is often more convenient to write the electric field as a function of the current, i.e. using the pair [**j**, **E**], instead of [**E**, **j**]. Solving for the transport coefficients as used in Eq. (8.51), we obtain

$$\alpha_{ij}(\mathbf{B}) = \alpha_{ji}(-\mathbf{B}) \qquad \kappa_{ij}(\mathbf{B}) = \kappa_{ji}(-\mathbf{B}) \tag{8.42}$$

$$\beta'_{ij}(\mathbf{B}) = \beta'_{ji}(-\mathbf{B}) = (kT)\, \beta_{ij}\ (\mathbf{B}) = (kT)\ \beta_{ji}(-\mathbf{B}), \tag{8.43}$$

where β' is the value of β in Eq. (8.51b). We conclude that a total of nine separate coefficients are required to describe the simultaneous flow of heat and charge in a magnetic field.

In order to relate these results to general co-ordinate set, it is convenient to write the matrix coefficients in vector form for the flux resulting from a force **s** as

$$\sigma_{ij}s_j = \sigma_\perp\ [\mathbf{s} - (\hat{\mathbf{b}} \cdot \mathbf{s})\ \hat{\mathbf{b}}] + \sigma_\wedge\ \hat{\mathbf{b}} \wedge \mathbf{s} + \sigma_\parallel\ [\hat{\mathbf{b}} \cdot \mathbf{s}]\ \hat{\mathbf{b}} \tag{8.44}$$

where $\hat{\mathbf{b}}$ is the unit vector in the direction of the magnetic field **B**.

Viscosity is more complicated to treat. Unlike electrical and thermal conduction, where the relationship between flux and force is expressed by a simple linear form between two vectors, i.e. a tensor of rank 2. The viscosity is a relationship between two tensor quantities, stress **T**, and the rate of distortion defined as

$$\mathsf{U}_{ij} = \frac{1}{2}\left(\frac{\partial u_i}{\partial x_j} + \frac{\partial u_j}{\partial x_i}\right) - \frac{1}{3}\nabla \cdot \mathbf{u}\ \delta_{ij} \tag{8.45}$$

As a result, the linear coefficient is a tensor of rank 4 with 81 individual components $\mathsf{T}_{ij} = \mu_{ij}^{kl} \dot{e}_{kl}$. Onsager's relation and the essential symmetry of the viscous stress **T** and the rate of strain tensors $\dot{\boldsymbol{e}}$

$$\mu_i\, j^{kl}(\mathbf{B}) = \mu_{kl}{}^{ij}(-\mathbf{B}) \qquad \text{and} \qquad \mu_{ij}{}^{kl} = \mu_{ji}{}^{kl}$$

reduces this to 21 components (de Groot and Mazur, 1984, Ch.XII). Introducing the cylindrical symmetry associated with the magnetic field further reduces this to seven components of which

10 As exemplified by Eq. (2.56) in the simple case with zero temperature gradient.

we may omit a further two, which yield zero viscosity (Lifshitz and Pitaevskii, 1981, §13) reducing the number to five terms.

In simple isotropic fluids, this reduces to two components – the *first or kinematic (μ)* and *second or bulk (ζ) components of viscosity*,

$$\mathsf{T}_{ij} = \underbrace{\mu \left(\frac{\partial u_i}{\partial x_j} + \frac{\partial u_j}{\partial x_i} - \frac{1}{3} \frac{\partial u_k}{\partial x_k} \delta_{ij} \right)}_{\text{First coefficient}} + \underbrace{\zeta \frac{\partial u_k}{\partial x_k} \delta_{ij}}_{\text{Second coefficient}} \tag{8.46}$$

In monatomic gases ($\gamma = \frac{5}{3}$), the second coefficient ($\zeta = 0$) is explicitly zero (Lifshitz and Pitaevskii, 1981, §8). In plasma, similarly, there are two bulk viscosity terms both of which contribute zero viscosity.

Plasma is more complicated than simple fluid due to the presence of magnetic fields which destroy the isotropy. However, the number of components is reduced by cylindrical symmetry about the magnetic field direction to 5. Consider the gradient of velocity

$$\nabla \mathbf{u} = \frac{1}{3} \nabla \cdot \mathbf{u} + \mathbf{U} + \boldsymbol{\Omega} \tag{8.47}$$

where the first term represents the rate of volumetric compression expressed by the distortion tensor (traceless rate of strain tensor) $\mathsf{U}_{ij} = \frac{1}{2}\left(\frac{\partial u_i}{\partial x_j} + \frac{\partial u_j}{\partial x_i}\right) - \frac{1}{3}\nabla \cdot \mathbf{u}$,[11] the second the rate of distortion of the fluid and the third a solid body rotation $\Omega_{ij} = \frac{1}{2}\left(\frac{\partial u_i}{\partial x_j} - \frac{\partial u_j}{\partial x_i}\right)$. The first two terms give rise to the bulk viscosity and kinematic viscosity, respectively. The third involving no body change of shape does not contribute to the internal forces. As noted above, the bulk viscosity is zero in plasma.

Since $\mathbf{U}_{ii} = \nabla \cdot \mathbf{u} = 0$ and since both the stress and rate of strain tensors are symmetric, there are five independent components[12] of $\mathbf{U}$ which may be conveniently arranged in three groups (see Kaufman, 1960) depending on the deformation associated with them. Forming a local right-handed Cartesian co-ordinate system with z parallel to magnetic field $\mathbf{B}$, and x and y lying in the plane normal to $\mathbf{B}$, these are

1. *Parallel viscosity* $\mathsf{U}_{zz} = -(\mathsf{U}_{xx} + \mathsf{U}_{yy})$ representing compression (or dilation) along the field compensated by dilation (or compression) normal to it.
2. *Perpendicular viscosity* U_{xy} and $\frac{1}{2}(\mathsf{U}_{xx} - \mathsf{U}_{yy})$ are deformations in the x, y plane.
3. *Gyro-viscosity* U_{xz} and U_{yz} are deformations in a plane containing the magnetic field.

It is easily shown (Kaufman, 1960) that these terms have the requisite symmetry with respect to the magnetic field.

The stress is a linear function of the rate of strain terms, each type of motion having an appropriate constant of proportionality, thus,

$$\mathsf{T}_{ij} = -\sum_{n=0}^{4} \mu_n \, \mathsf{U}_{n_{ij}} \tag{8.48}$$

---

11 Braginskii (1965) omits the factor $\frac{1}{2}$, we follow Chapman and Cowling (1952), Marshall (1960) and Lifshitz and Pitaevskii (1981), and conform with normal fluid mechanics practice.

12 In fact there are seven components including two associated with volumetric changes. Since there are three spatial terms, e.g. (8.40) there are 21 elements as required.

where the terms $\mathbf{U}_n$ are formed from the sets above and μ_n are appropriate constants for each set. Solving for the sets U_n, we obtain stress

$$\left.\begin{aligned}
\mathsf{T}_{zz} &= 2\mu_0\, \mathsf{U}_{zz} \\
\mathsf{T}_{xx} &= -\mu_0\, \mathsf{U}_{xx} + \mu_1\, (\mathsf{U}_{xx} - U_{yy}) + 2\mu_3\, \mathsf{U}_{xy} \\
\mathsf{T}_{yy} &= -\mu_0\, \mathsf{U}_{yy} + \mu_1\, (\mathsf{U}_{xx} - \mathsf{U}_{yy}) - 2\mu_3\, \mathsf{U}_{xy} \\
\mathsf{T}_{xy} &= 2\, \mu_1 \mathsf{U}_{xy} - \mu_3\, (\mathsf{U}_{xx} - \mathsf{U}_{yy}) \\
\mathsf{T}_{xz} &= 2\, \mu_2 \mathsf{U}_{xz} + 2\mu_4\, \mathsf{U}_{yz} \\
\mathsf{T}_{yz} &= 2\, \mu_3 \mathsf{U}_{yz} - 2\mu_4\, \mathsf{U}_{xz}
\end{aligned}\right\} \tag{8.49}$$

The constants $\mu_0 \ldots \mu_4$ must be found by calculation from the Boltzmann equation (see Chapman and Cowling, 1952; Marshall, 1960; Braginskii, 1965). An approximate method of deriving these coefficients using the BGK collision model is given in Section 8.A.2. The behaviour of the coefficients in a magnetic field is demonstrated therein.

8.5 Two Fluid MHD Equations – Braginskii Equations

The full set of MHD equations including explicit forms of the transport coefficients were first formulated by Braginskii (1965), who also elucidated the underlying physics associated with these terms in plasma, and gave values for the coefficients. Braginskii developed the transport equations in terms of the resistivity, thermal conductivity, and corresponding thermo-electric terms. A set of *force* terms were introduced to take into account the collisional effects modifying Eqs. (8.2)–(8.4). The current Hall effect was taken into account explicitly modifying the electric field (Eq. (8.51a)) as

$$\frac{\partial n_e}{\partial t} + n_e \nabla \cdot \mathbf{u}_e = 0 \tag{8.50:e.1}$$

$$m_e n_e \left[\frac{\partial \mathbf{u}_e}{\partial t} + (\mathbf{u}_e \cdot \nabla)\, \mathbf{u}_e\right] + \nabla p_e + \nabla \cdot \mathbf{T}_e - e n_e\, (\mathbf{E} + \mathbf{u}_e \wedge \mathbf{B}) = \mathbf{F} \tag{8.50:e.2}$$

$$\frac{3}{2}\left[\frac{\partial p_e}{\partial t} + (\mathbf{u}_e \cdot \nabla)\, p_e\right] + \frac{5}{2} p_e \nabla \cdot \mathbf{u}_e + \mathsf{T}_e : \nabla \mathbf{u}_e + \nabla \cdot \mathbf{q}_e = W_e \tag{8.50:e.3}$$

and

$$\frac{\partial n_i}{\partial t} + n_i \nabla \cdot \mathbf{u}_i = 0 \tag{8.50:i.1}$$

$$m_i n_i \left[\frac{\partial \mathbf{u}_i}{\partial t} + (\mathbf{u}_i \cdot \nabla)\, \mathbf{u}_i\right] + \nabla p_i + \nabla \cdot \mathbf{T}_i + Zen_i\, (\mathbf{E} + \mathbf{u}_i \wedge \mathbf{B}) = -\mathbf{F} \tag{8.50:i.2}$$

$$\frac{3}{2}\left[\frac{\partial p_i}{\partial t} + (\mathbf{u}_i \cdot \nabla)\, p_i\right] + \frac{5}{2} p_i \nabla \cdot \mathbf{u}_i + \mathsf{T}_i : \nabla \mathbf{u}_i + \nabla \cdot \mathbf{q}_i = W_i \tag{8.50:i.3}$$

where the internal energy is replaced by the pressure $\epsilon = \dfrac{3}{2}\dfrac{p}{\rho}$.

Since the current is carried predominately by the electrons, the electron heat, the electron energy flux $\mathbf{q}_e$ contains terms from the convected enthalpy $h_e\, \mathbf{j}/e$ and from the electro-magnetic field energy flux $\mathbf{E} \wedge \mathbf{B}/\mu_0 \approx -\phi\, \mathbf{j}$, where ϕ is the electric potential, together with the frictional energy transport terms to which the Onsager analysis is applied.

In weak gradients, where the parameter is $\xi \ll 1$, the frictional terms may be expressed as linear coefficients of the driving forces:

$$en_e\mathbf{E} = -\nabla p + \mathbf{j} \wedge \mathbf{B} + e\, n_e\, \boldsymbol{\alpha} \cdot \mathbf{j} - n_e \boldsymbol{\beta} \cdot \nabla(\mathit{k}T) \tag{8.51a}$$

$$\mathbf{q} = -\boldsymbol{\kappa} \cdot \nabla(\mathit{k}T) - \boldsymbol{\beta} \cdot \mathbf{j}\, (\mathit{k}T)/e \tag{8.51b}$$

where the second-order tensors $\boldsymbol{\alpha}$, $\boldsymbol{\beta}$ and $\boldsymbol{\kappa}$ are the resistivity, thermo-electric coefficient, and thermal conductivity, respectively. The equality of the factor $\boldsymbol{\beta}$ in the thermo-electric (last) terms in Eqs. (8.51) follows from the Onsager relation, as derived above and follows the standard theory of thermo-electricity (Landau and Lifshitz, 1984, §27).

The two terms in Eqs. (8.51) comprise the direct effects of collisions on the electron dynamics to which the reciprocal relations of irreversible thermodynamics may be applied. Thus, the drag force does not include the reversible field effect of the magnetic field, namely, the Hall current. Similarly, the energy flux relation excludes the terms contained in W_e and W_i, namely electron–ion energy transfer given by Eq. (7.D.5), and the work done by the electric field per electron driving the current against the drag force $\mathbf{j} \cdot \mathbf{F}$, namely, the Ohmic heating

$$\rho_i = -\frac{3\, m_e\, n_e}{m_i}\frac{(\mathit{k}T_e - \mathit{k}T_i)}{\tau_{eq}} \qquad \text{and} \qquad W_e = -W_i + \frac{\mathbf{j} \cdot \mathbf{F}}{n_e e} \tag{8.52}$$

8.5.1 Magnetic Field Equations

Currents driven by the electric pressure in Eq. (8.50:e.2) generate magnetic fields, known as *self-generated magnetic fields*. Although their influence remains unclear, theoretical simulations have suggested they may play an important role by modifying the simple transport properties of the plasma. Neglecting viscosity, the first-order moment of the Boltzmann equation can be written in the general form:

$$\rho_i \frac{\partial \mathbf{v}}{\partial t} + (\mathbf{v}_i \cdot \nabla).\mathbf{v}_i + \nabla \mathbf{p}_i - n_i\, q_i\, (\mathbf{E} + \mathbf{v}_i \wedge \mathbf{B}) = \mathbf{F}_i \tag{8.53}$$

for the species i, of charge q_i mass density ρ_i number density n_i and velocity $\mathbf{v}_i$, $\mathbf{F}_i$ is the frictional resistance term and $\mathbf{E}$ and $\mathbf{B}$ are the electrostatic intensity and magnetic flux density, respectively.

Applying this equation to electrons alone, we may, since their mass is small, neglect the acceleration term to yield:

$$\nabla p_e + e\, n_e(\mathbf{E} + \mathbf{v}_e \wedge \mathbf{B}) = \mathbf{F}_e \tag{8.54}$$

where $q_e = -e$. The electron velocity $\mathbf{v}_e$ is comprised of two parts, namely, the bulk mass velocity $\mathbf{v}$ and that due to the current density $\mathbf{j}$:

$$\mathbf{j} = -e\, n_e\, \mathbf{v}_j = -e\, n_e\, (\mathbf{v}_e - \mathbf{v}) \tag{8.55}$$

The current density and electric field may both be expressed in terms of the magnetic flux by Maxwell's equations:

$$\nabla \wedge \mathbf{E} = -\frac{\partial \mathbf{B}}{\partial t} \qquad \nabla \wedge \mathbf{B} = \mu_0\, \mathbf{j} \tag{8.56}$$

neglecting the displacement current, as is usual in plasmas. The frictional term may be written as follows:

$$\mathbf{F}_e = \boldsymbol{\alpha} \cdot \mathbf{v}_j. - \boldsymbol{\varphi} \cdot \nabla(\mathit{k}T_e) = e\, n_e\, \boldsymbol{\rho} \cdot \mathbf{j} - n_e\, \boldsymbol{\beta}^0 \cdot \nabla(\mathit{k}T_e) \tag{8.57}$$

where k is Boltzmann's constant, T_e the electron temperature, $\boldsymbol{\alpha}$, and $\boldsymbol{\rho}$ are Ohmic resistance tensors and $\boldsymbol{\varphi}$ and $\boldsymbol{\beta}^0$ thermo-electric tensors. Substituting Eqs. (8.55)–(8.57) in Eq. (8.54), we obtain

$$\frac{\partial \mathbf{B}}{\partial t} - \nabla \wedge (\mathbf{v} \wedge \mathbf{B}) + \nabla \wedge [\boldsymbol{\eta} \cdot (\nabla \wedge \mathbf{B})] = \nabla \wedge \mathbf{S} \tag{8.58}$$

where the diffusivity tensor $\boldsymbol{\eta} = \boldsymbol{\rho}/\mu_0$ and source vector $\mathbf{S}$ may be written in two alternative, but equivalent, forms:

(i) The Hall term is included in the source as a magnetic stress:

$$\mathbf{S} = \frac{1}{e}\left\{\frac{1}{n_e}\left[\nabla(p_e) + \nabla\left(\frac{B^2}{2\mu_0}\right) - \frac{1}{\mu_0}(\mathbf{B}\cdot\nabla)\cdot\mathbf{B}\right] + \boldsymbol{\beta}^0 \cdot \nabla(kT_e)\right\} \tag{8.59}$$

and the magnetic diffusivity

$$\eta = \rho/\mu_0 \tag{8.60}$$

(ii) or alternatively, it is included as a contribution to the tensor resistivity

$$\mathbf{S} = \frac{1}{e}\left\{\frac{1}{n_e}\nabla(p_e) + \boldsymbol{\beta}^0 \cdot \nabla(kT_e)\right\} \tag{8.59b}$$

and

$$\eta = \rho'/\mu_0 \quad \rho_{\parallel}' = \rho_{\parallel} \quad \rho_{\perp}' = \rho_{\perp} \quad \rho_{\wedge}' = \rho_{\wedge} + B/e\, n_e\, c \tag{8.60b}$$

where the subscripts $\parallel$, $\perp$, and $\wedge$ refer to electric field components parallel and perpendicular to the magnetic field, and to the component perpendicular to both.

The source vector $\mathbf{S}$ can also be written in a number of equivalent ways, since $\nabla \wedge \nabla\phi = 0$, we may equivalently replace

$$\frac{1}{n_e}\nabla(p_e) \to kt_e\nabla(\ln n_e) \to -\ln n_e\ \nabla(kT_e) \tag{8.59c}$$

in Eqs. (8.59a) and (8.59b), since $p_e = n_e\ kT_e$. Therefore, we may also write Eq. (8.59b) as

$$\mathbf{S} = -\frac{1}{e}\beta \cdot \nabla(kT_e) \tag{8.59d}$$

where $\beta_{\parallel} = {\beta_{\parallel}}^0 - \ln n_e$, $\beta_{\perp} = {\beta_{\perp}}^0 - \ln n_e$, $\beta_{\wedge} = {\beta_{\wedge}}^0$, and a similar form in (8.59a).

If the collisional thermo-electric term β^0 is negligible, we may write the magnetic field source term $\nabla \wedge \mathbf{S}$ in the usual collision-free form:

$$\nabla \wedge \mathbf{S} = \frac{1}{e}\nabla(kT_e) \wedge \nabla(\ln n_e) \tag{8.61}$$

Self-generated fields are therefore a consequence of the electron pressure in non-parallel density and temperature gradients.

Returning to Eq. (8.58) and neglecting the source term $\mathbf{S}$, we consider the simple case where the resistivity is a scalar constant and the equation reduces to

$$\frac{\partial \mathbf{B}}{\partial t} + \nabla \wedge (\mathbf{v} \wedge \mathbf{B}) - \eta\ \nabla^2\ \mathbf{B} = 0 \tag{8.62}$$

which represents the diffusion of the magnetic field through the plasma as a result of the resistance of the plasma to the internal currents. As we shall see in Section 9.1.1, in the absence of diffusion ($\eta = 0$), the field is tied into the plasma motion by induction. Introducing resistivity, the field diffuses through the plasma. The relative importance of these two effects is identified by comparing their relative strengths:

$$R_M \sim v\,\ell/\eta \tag{8.63}$$

where v is the characteristic velocity and ℓ the characteristic length. R_M is known as the magnetic Reynolds number, and plays a similar role in magneto-hydrodynamics to the Reynolds number in viscous fluid flow. Thus, if $R_M \gg 1$ magnetic diffusion is negligible or if $R_M \lesssim 1$ magnetic fields must be taken into account.

8.5.1.1 Energy Balance

The second-order moment of the Boltzmann equation for the species i may be written as

$$\left(\frac{\partial}{\partial t} + \mathbf{v}_i \cdot \nabla\right) \epsilon_i + \epsilon_i \, \nabla \cdot \mathbf{v}_i + p_i \, \nabla \cdot \mathbf{v}_i + \nabla \cdot \mathbf{q}_i = -\mathbf{v}_i \cdot \mathbf{F}_i + H_i \tag{8.64}$$

where e_i is the energy density of species i, $\mathbf{q}_i$ its thermal conduction flux, and H_i the volume distribution of heat sources.

Substituting Eqs. (8.54)–(8.61) in Eq. (8.64), we obtain for the electrons

$$\frac{\partial e_e}{\partial t} + \nabla \cdot (\epsilon_e \, \mathbf{v_e}) + p_e \, \nabla \cdot \mathbf{v}_e + \nabla \cdot \mathbf{q}_e = H_e - G - J + \mathbf{j} \cdot \nabla(\frac{3}{2} k T_e/e) + \rho \mathbf{j}^2 - (l/c)\mathbf{S} \cdot \mathbf{j} \tag{8.65}$$

where $\mathbf{S} = (c/e)kT_e \nabla[\ln n_e] + \boldsymbol{\beta}^0 \, \nabla(kT_e)$. The electron heat flux is the sum of the normal thermal conduction flux q_t and the thermo-electric flux, q_j.

$$\mathbf{q}_t = -\kappa \cdot \nabla(T_e) \qquad\qquad \mathbf{q}_j = -\boldsymbol{\beta}^0 \cdot \mathbf{j}(kT_e/e) \tag{8.66}$$

where we have used the Onsager relation for the thermo-electric flux coefficient. Hence,

$$\nabla \cdot \mathbf{q}_j + \mathbf{S} \cdot \mathbf{j} = -\nabla \cdot (\boldsymbol{\beta} \cdot \mathbf{j})(kTe/e) \tag{8.67}$$

and thus

$$\partial \epsilon_e/\partial t + \nabla \cdot (\epsilon_e \mathbf{v}) + p_e \nabla \cdot \mathbf{v} + \nabla \cdot \mathbf{q}_t = H_e - G - J + j \cdot \nabla(\frac{3}{2} kT_e/e) + \rho \mathbf{j}^2 + (kT_e/e)\nabla \cdot (\boldsymbol{\beta} \cdot \mathbf{j}) \tag{8.68}$$

where G is the ion–electron equilibration energy transfer rate (Spitzer, 1956) and $J = n_e \, e(\rho \cdot \mathbf{j}) \cdot \mathbf{v}$. We may also express this result in a conservative form:

$$\begin{aligned} \frac{\partial}{\partial t}\left(\epsilon_e + \frac{B^2}{2\,\mu_0}\right) + \nabla \cdot \left[\left(\epsilon_e + \frac{B^2}{2\,\mu_0}\right) \cdot \mathbf{v}\right] + \left(p_e + \frac{B^2}{2\,\mu_0}\right) \nabla \cdot \mathbf{v} - \frac{1}{\mu_0}[(\mathbf{B} \cdot \nabla)\mathbf{B}] \cdot \mathbf{v} \\ + \nabla \cdot \mathbf{q}_t + \nabla \cdot (\epsilon_e \, \mathbf{v}_j) + \nabla \cdot [(\rho \cdot \mathbf{j} - S) \wedge \mathbf{B}] = H_e - G - J \end{aligned} \tag{8.69}$$

Several points are worthy of note in this equation. Thus, the work terms, representing work done by the electron pressure and magnetic stress, appear in a non-conservative form, since we have not included their contribution to the rate of increase of the total kinetic energy. The form of this equation is independent of the form of $\boldsymbol{\eta}$ and $\mathbf{S}$ provided a compatible set from Eqs. (8.60a)–(8.61) is used. We observe two additional energy flux terms associated with the Ohmic diffusion of the magnetic field, and with the creation of the field, namely, and

$$Q_\eta = (\rho \cdot \mathbf{j}) \wedge \mathbf{B} = \mu_0 \, (\boldsymbol{\eta} \cdot \mathbf{j}) \cdot \mathbf{B} \tag{8.70}$$

and

$$\mathbf{Q}_s = -\mathbf{S} \wedge \mathbf{B} \tag{8.71}$$

We may remark that $\mathbf{Q}_s$ is to some extent arbitrary since any value of $\mathbf{S}$ given by the transformations (8.59a) may be used consistent with Eq. (8.60b).

8.6 Transport Coefficients

Following Braginskii (1965), the collisional drag and energy transfer terms have relatively simple physical descriptions which are useful to identify the scaling of each term.

8.6.1 Collisional Dominated Plasma

Consider first the case of zero (or small magnetic field, $\Omega\,\tau \ll 1$). In this case, the tensors $\boldsymbol{\alpha}$ and $\boldsymbol{\beta}$ become simple scalars.

8.6.1.1 Force Terms F

The force terms represent the momentum exchange due to friction between the electrons and ions. Braginskii's formulation in terms of the force terms allows a particularly straightforward interpretation of the physics underlying each of the transport coefficients.

The simplest term is the resistivity, which is the frictional force exerted on the current flow driven by an electric field. The drag force due to the current easily estimated by considering the momentum loss by an electron in a collision with a heavy ion. A simple representation of this process is given by the Drude model (Section 2.2.3), which assumes a constant collision frequency independent of the electron velocity.

$$\mathbf{F}_j = -\frac{n_e\,e}{\sigma}\,\mathbf{j} \sim -\frac{m_e}{e\,\tau_{ei}}\mathbf{j} \tag{8.72}$$

where σ is the electron conductivity. Since the electron–ion collision time varies as $\tau_{ei} \sim v^3$ (Eq. (2.17)), it is clear that the current is mostly carried by fast electrons.[13]

The thermal force arises from the differential drag on the electron random drift arising from the temperature gradient. In the absence of an electric field, there is zero net drift of electrons. Suppose there exists a temperature gradient in the x direction. Consider the situation near the reference plane $x = x_0$. The random flux of electrons, namely $\sim \frac{1}{4}n_e\overline{v_e}$, from the left side ($x < x_0$) is balanced by that from the right ($x > x_0$). In the presence of the temperature gradient, electrons travel a distance $\sim\lambda_{ei}$, the mean free path, between thermalising collisions and therefore, those from the positive side $x > x_0$ have a temperature difference $\Delta T \sim \lambda_{ei}\ \mathrm{d}T/\mathrm{d}x$ greater than those from negative $x < x_0$ and consequently a different collision frequency:

$$\Delta\nu_{ei} \approx \frac{3}{2}\,\frac{\lambda_{ei}}{T}\,\frac{\mathrm{d}T}{\mathrm{d}x}\,\nu_{ei} \approx \frac{3}{2}\,\frac{\overline{v}}{T}\,\frac{\mathrm{d}T}{\mathrm{d}x} \tag{8.73}$$

Since the mean free path $\lambda \sim \tau_{ei}\overline{v}$, the net drag force is

$$F_T \approx -m_e\,\Delta\nu_{ei}\cdot\frac{1}{4}n_e\,\overline{v_e} \sim -n_e\,\frac{\mathrm{d}T}{\mathrm{d}x} \tag{8.74}$$

8.6.1.2 Energy Flux Terms

The effects of collisions on the heat flux are due to two separate effects: Thermal conduction flux determined by the thermal conductivity $\kappa \sim \frac{1}{2}\,n_e\,\overline{v_e}\,\lambda$, given approximately by Eq. (2.44)

$$\mathbf{q}_e = -\kappa_e\,\nabla(kT) \qquad \text{where} \qquad \kappa_e \sim n_e\,(kT)\,\tau_{ei}/m_e \tag{8.75}$$

---

13 In most cases, the dominant contribution to the transport coefficients is due to fast electrons. The numerical coefficients determined by simple constant collision frequency models are therefore in error, although the scalings are correct.

The thermo-electric term $\mathbf{j} \cdot \mathbf{F}_T \sim -n_e \mathbf{j}$ is associated with the work done against the thermal force. This term is reversible and corresponds to the thermo-electric *Thomson effect* in metals. The corresponding heat flux

$$\mathbf{q}_T \sim n_e\, (kT)\, \mathbf{j} \tag{8.76}$$

8.6.1.3 Viscosity

In the absence of the magnetic field, the viscosity is well described by the simple kinetic theory model equation (2.45) so that

$$\mathsf{T} = \mu_0\, \mathsf{U} \tag{8.77}$$

From Eq. (2.45), it follows that the value of the field free viscosity is given approximately by $\mu_0 \approx p\tau$, where p is the ion pressure and τ an appropriate average of the ion–ion collision time. We note that since ion–ion collisions dominate in determining the ion viscosity by randomising the directed velocity (see Section 2.2.2), the appropriate collision term must be that between ions.

8.6.2 Field-Dominated Plasma

When the magnetic field is large ($\Omega\tau \gg 1$), the cyclotron rotation around the magnetic field lines dominates the motion of the particles under the transport forces. As a result the coefficients α and β both become tensors with components along, perpendicular to and across the field. The modification induced to the current by the Hall terms is not an irreversible collisional term, although it generates perpendicular and crossfield components (Section 2.2.3.2), is excluded from this discussion.

When the applied force is parallel to the magnetic field, the motion induced by the force is directly along the field lines, which consequently make no modification to the transport drift. The transport coefficients ($\sigma_{\parallel}$, $\kappa_{\parallel}$, $\beta_{\parallel}$) are therefore unchanged from their values in the absence of the magnetic field discussed in Section 8.5. There is no component perpendicular to the field induced by the collisional terms, although, of course, the field gives rise to the Hall terms. In some circumstances, it may be convenient to include the Hall terms in the conductivity.

8.6.2.1 Force Terms F

Consider the case where the perturbing force is applied perpendicularly to the magnetic field. The momentum exchange between the current carrying electrons and the ions gives rise to a drag force $F_{j\perp}$ parallel to the applied force as in the collisionless case, but with a different coefficient. The latter arises as a result of the anisotropy induced in the electron distribution function by the magnetic rotation. As a result, the electrical conductivity is reduced by a factor of approximately ¾ from that parallel to the field.

In the absence of a temperature gradient, symmetry ensures that there can be no current flow in the cross field direction, normal to both the magnetic and electric fields due to kinetic effects in addition to the collisionless Hall effect.

There is a drag force generated by the temperature gradient transverse to both fields. This arises from the motion of electrons from plasma elements at $x = \pm\rho$ which have a temperature difference of $\rho\ \mathrm{d}T/\mathrm{d}x$. Figure 8.1 illustrates this situation. Electrons from $x = -\rho$ and from $x = +\rho$ pass through the plane $x = 0$ in opposite directions. As in the absence of a magnetic field, those from $x > 0$ have a different temperature and therefore different collision frequency $\Delta\nu_{ei} \sim \frac{\rho}{T}\frac{\mathrm{d}T}{\mathrm{d}x}$. The net drag force at the element x_0 is the difference between the drag on the left and right-hand electron 'streams' as a resulting from their different collision frequencies. Hence, following the previous analysis the drag force

$$F_j \sim -m_e\, n_e\, \overline{v_e}\, \nu_{ei}\, \frac{\rho}{T}\frac{\mathrm{d}T}{\mathrm{d}x} \sim -\frac{1}{|\Omega_e|\tau_{ei}}\, n_e\, \hat{\mathbf{b}} \wedge \nabla T \tag{8.78}$$

where $\hat{\mathbf{b}}$ is the unit vector in the direction of the magnetic field. This is the classical thermo-electric *Nernst effect.*

8.6.2.2 Energy Flux Terms

When the temperature gradient is perpendicular to the magnetic field, the thermal conductivity is greatly reduced from its value along the field $\kappa_\perp \ll \kappa_\parallel$. The particle motion in a strong field is normally tied to the field lines and therefore the drift transverse to the field is small. To estimate the value of the thermal conductivity, it is convenient to use the random walk model (see Section 2.2.2). As discussed in Chapter 3, particles execute a rotation about the field lines with radius equal to the Larmor radius ρ unless disturbed by collisions which occur at an interval τ. In the context of the random walk model, the diffusion length in this case is approximately the Larmor radius $\ell \approx \rho$, whereas the collision time is unchanged. Compared with the thermal conductivity along the field lines where the length $\ell \approx \lambda = v\tau$ is the mean free path, it follows that the transverse thermal conductivity is reduced by a factor $\sim 1/(\Omega\ \tau)^2$ (see Eq. (2.46)) compared to the longitudinal conductivity. Hence,

$$\kappa_\parallel \sim \frac{n_e \tau_{ei}}{m_e} \qquad \text{and} \qquad \kappa_\perp \sim \frac{n_e \tau_{ei}}{m_e\ (\Omega_e \tau_e)^2} \tag{8.79}$$

The cross product term is due to cyclotron rotation. Consider the situation about the reference plane x_0 and referring to Figure 8.1 electrons with $x < x_0$ will have lower temperature than those with $x > x_0$. If the fluxes due to electrons from the left- and right-hand sides are balanced, there remains a net heat transfer due to their temperature difference. Referring to Figure 8.1, it can be seen that due to the rotation the flux is in the $-y$ direction, i.e. that of $-\hat{\mathbf{b}} \wedge \nabla T$.

$$\kappa_\wedge \sim -\frac{n_e\ T_e}{m_e|\Omega_e|} \tag{8.80}$$

A similar expression obtains for the ions, but the direction is reversed as the rotation is in the opposite sense. This term, which is closely related to the Hall effect, is known as the *Righi–Leduc effect*.

The remaining heat transfer terms due to the current, the terms dependent on β in Eq. (8.51b) whose values are obtained from the conjugate terms in the drag.

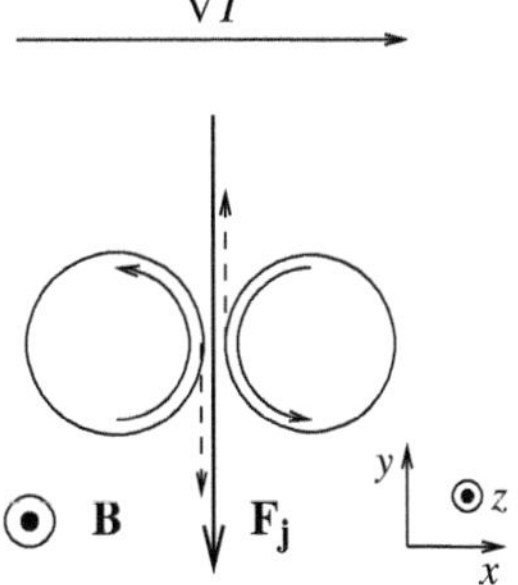

Figure 8.1 Sketch of the electron trajectories giving rise to thermal drag in a magnetic field **B** and temperature gradient ∇T.

8.6.2.3 Viscosity

It is useful to refer to Section 8.A.2 to elucidate the behaviour of the viscosity coefficients in a strong magnetic field. As noted earlier, the coefficients may be divided into three distinct types depending on the spatial relationship between the field and the velocity gradient. We also recall that plasma viscosity is dominated by the ions and that the appropriate relaxation time is therefore the ion–ion relaxation time and the magnetic field is expressed in terms of the ion Larmor frequency. The critical scaling parameter for viscosity is therefore $\Omega_i \tau_i \gg 1$, in strong fields.

When the velocity gradient only has a component U_{zz} associated with the field direction, the viscosity is unaffected by the magnetic field and the stress is therefore given be the field free value $\mathsf{T}_{zz} = -2\,\mu_0\,\mathsf{U}_{zz}$. This term is the *parallel viscosity* coefficient (Eq. (8.A.16:a)).

However, components in the other directions will be modified by the field. In particular, if the plasma has a velocity perpendicular to the field and a gradient in the same direction, the 'random walk' scattering length is approximately equal to the ion Larmor radius, the ion collision time is unchanged. The viscosity is therefore reduced by a factor $(\rho_i/\lambda_i)^2$ or

$$\mathsf{T}_{xx} \sim \frac{\mu_0}{(\Omega_i \tau_i)^2}\,\mathsf{U}_{xx} \quad \text{and} \quad \mathsf{T}_{yy} \sim \frac{\mu_0}{(\Omega_i \tau_i)^2}\,\mathsf{U}_{yy} \tag{8.81}$$

In a similar way, the stress associated with velocity gradients $(\partial u_x/\partial y)$ and $(\partial u_x/\partial y)$ must also have a random walk scattering length ρ_i and therefore also scale as

$$\mathsf{T}_{xy} \sim \frac{\mu_0}{(\Omega_i \tau_i)^2}\,\mathsf{U}_{xy} \quad \text{and} \quad \mathsf{T}_{yx} \sim \frac{\mu_0}{(\Omega_i \tau_i)^2}\,\mathsf{U}_{yx} \tag{8.82}$$

Terms scaling in this way constitute the *perpendicular viscosity* coefficients (see Eq. (8.A.16:b)).

There are several terms where the viscosity varies as $1/(\Omega_i \tau_i)$, for example

$$\mathsf{T}_{xx} \sim \frac{\mu_0}{\Omega_i\,\tau_i}\frac{\partial u_y}{\partial x} \sim \frac{nT}{\Omega_i}\frac{\partial u_y}{\partial x} \quad \text{and} \quad \mathsf{T}_{xz} \sim \frac{\mu_0}{\Omega_i\,\tau_i}\frac{\partial u_y}{\partial x} \sim \frac{nT}{\Omega_i}\frac{\partial u_y}{\partial x} \tag{8.83}$$

We note that the latter terms are independent of the ion collision time and are, therefore, collision-free; the viscosity being generated by the magnetic field even in the absence of collisions. These terms are known as *gyro-viscosity* coefficients (see Eq. (8.A.16:c)). Kaufman (1960) elucidates this behaviour by a more detailed analysis.

The most accurate values of the viscosity coefficients are due to Braginskii (1965) to which reference is made.

8.7 Calculation of the Transport Coefficients

The calculation of the transport coefficients follows along similar lines to that in gases using an expansion in terms of different orders in the small parameters ξ. The plasma is considered to be locally at rest in a state near to thermo-dynamic equilibrium, but subject to weak perturbations by forces such as those due to the electric or magnetic fields or spatial gradients of the temperature or pressure. Early calculations were carried out using the Boltzmann equation with a Boltzmann collision operator following the Chapman–Enskog prescription initially without a magnetic field (Landshoff, 1949, 1951); subsequently and more generally with a magnetic field (Marshall, 1960; Braginskii, 1965). Later workers used increasing numbers of terms in expansions in Laguerre (also known as Sonine) polynomials (see Epperlein and Haines, 1986). Following Shkarofsky (1961) subsequent calculations have been based on the Fokker–Planck collision operator with the distribution function expanded in Cartesian tensor terms.

If the system is at rest and in thermo-dynamic equilibrium so that the distribution is Maxwellian,

$$f_0(v) = \left(\frac{m}{2\,\pi\,\mathcal{k}T}\right)^{3/2} n\,\exp\left(-\frac{m\,v^2}{2\,\mathcal{k}T}\right) \tag{8.84}$$

the moment equations simply reduce to zero, and there is no closure problem.[14] However, if perturbations are introduced so that the distribution changes from point to point in space, additional terms are introduced associated with the perturbation. Assuming that the unperturbed distribution is isotropic, the perturbed distribution function become

$$f(\mathbf{r}, \mathbf{v}, t) = f_0(\mathbf{r}, \mathbf{v}) + f_1(\mathbf{r}, \mathbf{v}, t) \cdot \frac{\mathbf{v}}{v} + f_2(\mathbf{r}, \mathbf{v}, t) : \frac{\mathbf{vv}}{v^2} + \cdots \tag{8.85}$$

where the perturbation $f_1 \ll f_0$.[15] Taking the magnetic field, collision terms are taken to be zero order compared to the first-order perturbations due to the gradient terms, we obtain

$$\cancel{\frac{q}{m}\mathbf{v} \wedge \mathbf{B} \cdot \frac{\partial f_0(v)}{\partial \mathbf{v}}}^{\,0} = \left.\frac{\partial f_0}{\partial t}\right|_{\text{coll}} \tag{8.86a}$$

$$\frac{\partial f_0}{\partial t} + \mathbf{v} \cdot \frac{\partial f_0}{\partial \mathbf{r}} + \frac{q}{m}\mathbf{E} \cdot \frac{\partial f_0}{\partial \mathbf{v}} + \frac{q}{m}\mathbf{v} \wedge \mathbf{B} \cdot \frac{\partial f_1}{\partial \mathbf{v}} = \left.\frac{\partial f_1}{\partial t}\right|_{\text{coll}} \tag{8.86b}$$

Equation (8.86a) simply ensures that the zero-order (equilibrium) distribution f_0 is Maxwellian.[16] The perturbation distribution f_1 is related to the perturbations introduced by temporal, spatial, and external forces through Eq. (8.86b). The resulting fluxes, current, heat flow, etc., are calculated by taking the appropriate moment of f_1.

Electron transport results from electrical conduction and thermal gradients, which only slightly modify the particle distribution function, and can therefore be represented by an isotropic equilibrium (Maxwellian) distribution plus a small skewed perturbation induced by the fields[17]

$$f(\mathbf{v}) = f_0(v) + \frac{\mathbf{v}}{v} \cdot f_1(v) \tag{8.87}$$

In the absence of an electric field or a temperature gradient, the zero-order equilibrium distribution $f_0(v)$ when substituted in Fokker–Planck equation necessarily gives zero on both left- and right-hand sides.

The current $\mathbf{j}$ and total heat flux $\mathbf{Q}$ are given by integrals over the distribution[18]

$$\begin{aligned}\mathbf{j} &= -e \int f(\mathbf{v})\,\mathbf{v}\,\mathrm{d}\mathbf{v} = -e \int \left(\cancel{f_0(v)} + \frac{\mathbf{v}}{v} \cdot f_1(\mathbf{v})\right) \mathbf{v}\,\mathrm{d}\mathbf{v} \\ &= -2\pi e \iint f_1(v) v^3 \cos^2\theta \,\sin\theta \,\mathrm{d}\theta\,\mathrm{d}v = -\frac{4\pi}{3} e \int f_1(v)\, v^3 \,\mathrm{d}v\end{aligned} \tag{8.88a}$$

$$\begin{aligned}\mathbf{Q} &= \frac{1}{2} m \int f(\mathbf{v})\, v^2\, \mathbf{v}\,\mathrm{d}\mathbf{v} = \frac{1}{2} m \int \left(\cancel{f_0(v)} + \frac{\mathbf{v}}{v} \cdot f_1(v)\right) v^2\, \mathbf{v}\,\mathrm{d}\mathbf{v} \\ &= \pi m \iint f_1(v) v^5 \cos^2\theta \,\sin\theta \,\mathrm{d}\theta\,\mathrm{d}v = \frac{2\pi}{3} m \int f_1(v)\, v^5 \,\mathrm{d}v\end{aligned} \tag{8.88b}$$

Note that both the current and the heat flux are parallel to the vector component of the distribution function. The heat flux given by $\mathbf{Q}$ includes the enthalpy convected by the electron motion from

14 Bulk motion resulting from a mean plasma velocity $\mathbf{u}$ is included in the equilibrium distribution function. It is assumed to be eliminated by a local frame transformation the mean flow.

15 If the unperturbed distribution is non-isotropic, we must introduce an addition tensor perturbation term $\mathbf{f}_2 : \mathbf{vv}$.

16 In a magnetic field, the equilibrium distribution $f_0(v)$ is isotropic, and therefore the Lorentz force term $\mathbf{v} \wedge \mathbf{B} \cdot \partial f_0(v)/\partial \mathbf{v} = (\mathbf{v} \wedge \mathbf{B} \cdot \mathbf{v}/v)\, \mathrm{d}f_0(v)/\mathrm{d}v = 0$.

17 This Cartesian tensor form is the first-order term of an expansion in spherical harmonics (Johnston, 1960; Shkarofsky et al., 1966).

18 The integrals over $\mathbf{v}$ are performed by taking a polar set of co-ordinates with axis along f_1.

Eq. (8.4) and is the sum of the current and thermo-electric effects. The net heat flux due to the thermo-electric terms (in the electron rest frame) alone is, therefore,

$$\mathbf{q} = \mathbf{Q} + \frac{5}{2}(\mathcal{k}T)\,\mathbf{j}/e \tag{8.89}$$

Introducing the perturbing fields as first-order terms: a temperature gradient ∇T gives a non-zero first-order term due to $\partial f_0/\partial \mathbf{r}$. Since $f_0(v)$ is a function of v^2, $\partial f_0/\partial \mathbf{v} = (\mathbf{v}/v)\,\partial f_0/\partial v$ and the term $\mathbf{v} \wedge \mathbf{B} \cdot \partial f_0/\partial \mathbf{v} = 0$. The first-order magnetic field term is due to the perturbed distribution function f_1 alone. Noting again that $\partial f_0/\partial \mathbf{v} = (\mathbf{v}/v)\,\partial f_0/\partial v$ and retaining first-order terms in the Fokker–Planck equation, we obtain

$$\mathbf{v} \cdot \nabla f_0 - \frac{e}{m}\mathbf{E} \cdot \frac{\partial f_0}{\partial \mathbf{v}} + \mathbf{\Omega} \wedge f_1 \cdot \mathbf{v} = \frac{\partial f_1}{\partial t}\Big|_{\text{coll}} \tag{8.90}$$

The gradient terms

$$\begin{aligned} \frac{\partial f_0}{\partial \mathbf{r}} &= \frac{\mathrm{d}f_0(v,n)}{\mathrm{d}n}\nabla n + \frac{\mathrm{d}f_0(v,T)}{\mathrm{d}T}\nabla T = \left\{\frac{\nabla n}{n} + \frac{1}{2}\left(\frac{m_e\,v^2}{\mathcal{k}T} - 3\right)\frac{\nabla T}{T}\right\} f_0 \\ \frac{\partial f_0}{\partial \mathbf{v}} &= \frac{\mathrm{d}f_0(v,T)}{\mathrm{d}v}\frac{\mathbf{v}}{v} = -\frac{m\mathbf{v}}{\mathcal{k}T} f_0 \end{aligned} \tag{8.91}$$

represents the distortion of the distribution due to the temperature gradient, which gives rise to thermal conduction and thermo-electric effects.

The collisional relaxation term $\partial f_1/\partial t|_{\text{coll}}$ is made up of two terms:

- *Electron–ion collisions* are described by a relatively simple term. Since the mass of the ions is much larger than that of the electrons $m_i \gg m_e$, the ions are relatively slowly moving and may be considered to be stationary, the ion distribution in the centre of mass frame, therefore. takes the simple form $f_i(\mathbf{v}) \approx \delta(\mathbf{v})$. In addition due to the elastic nature of the collisions the electron speed is unchanged; the electron velocity vector being rotated, giving rise to momentum loss and hence 'drag'.
- *Electron–electron collisions*: In this case, the target particles are identical with the test, and cannot be regarded as stationary. The collision integral is consequently a complicated function of the perturbed and unperturbed electron distribution functions containing both first and second derivatives.

Once the collisional operator for the perturbed distribution is identified, it may be included with the transport equation (8.90) and solved for f_1 and, hence, to calculate the current and energy flux from Eqs. (8.88). These final expressions may be specified in two differing forms (see Eq. (8.39))

$$\mathbf{j} = \boldsymbol{\sigma} \cdot (\mathbf{E} + T\,\nabla n/n\,e) + \boldsymbol{\tau} \cdot \nabla T \tag{8.92a}$$

$$\mathbf{Q} = -\boldsymbol{\mu} \cdot (\mathbf{E} + T\,\nabla n/n\,e) - \mathbf{K} \cdot \nabla T \tag{8.92b}$$

or solving for the electric field (see Eq. (8.51))

$$\mathbf{E} = \boldsymbol{\alpha} \cdot \mathbf{j} + \boldsymbol{\beta} \cdot \nabla T \tag{8.93a}$$

$$\mathbf{q} = \boldsymbol{\beta} \cdot \mathbf{j}(\mathcal{k}T) - \mathbf{K} \cdot \nabla T. \tag{8.93b}$$

8.8 Lorentz Approximation

A satisfactory approximation for electron scattering from ions is provided by the Lorentz approximation in which the ions being very much heavier than the electrons ($m_e \ll m_i$) are assumed to be stationary with respect to the centre of mass, $\mathbf{v}_i \approx \overline{\mathbf{u}}$. Introducing a frame transformation to the mean speed $\overline{\mathbf{u}}$ brings the ions to rest, the ion distribution function is, therefore. $f_I(\mathbf{v}) = n_i\ \delta(\mathbf{v}_i)$, and the relative velocity of the electrons relative to the ions $\mathbf{u} = \mathbf{v}_e$.[19] The Fokker–Planck equation takes a simple form with the dynamic viscosity and diffusion coefficients given by the averages found earlier in Eqs. (2.21).

Incorporating the correct scattering term for stationary ions, given by Eq. (7.A.11d), the Fokker–Planck equation for the electrons takes the form

$$\frac{\partial f}{\partial t} + \mathbf{v} \cdot \frac{\partial f}{\partial \mathbf{r}} - \frac{e}{m}(\mathbf{E} + \mathbf{v} \wedge \mathbf{B}) \cdot \frac{\partial f}{\partial \mathbf{v}} = \mathscr{C}_{ei} + \mathscr{C}_{ee} \tag{8.94}$$

where $\mathscr{C}_{ei}$ and $\mathscr{C}_{ee}$ are the collision rates for electron–ion and electron–electron collisions, respectively, and $\omega_{ij} = \delta_{ij} - v_i\ v_j/v^2$. For the Lorentz model, $\mathscr{C}_{ee} = 0$, and the ion–electron collision term $\mathscr{C}_{ei} = \nu_{ei} \boldsymbol{f}_1 \cdot \mathbf{v}$ the above equation becomes

$$\frac{\partial f}{\partial t} + \mathbf{v} \cdot \frac{\partial f}{\partial \mathbf{r}} - \frac{e}{m}(\mathbf{E} + \mathbf{v} \wedge \mathbf{B}) \cdot \frac{\partial f}{\partial \mathbf{v}} = 2\,\pi \left(\frac{Ze^2}{(4\,\pi\,\epsilon_0)\,m}\right)^2 n_i\ \ln \Lambda\ \frac{\partial}{\partial v_i}\left(\omega_{ij} \frac{\partial f}{\partial v_j}\right) \tag{8.95}$$

After some algebraic effort,[20] we obtain

$$\frac{\partial f_0}{\partial \mathbf{r}} - \frac{e}{m}\ \mathbf{E}\ \frac{\partial f_0}{\partial v} + \boldsymbol{\Omega} \wedge \boldsymbol{f}_1 = -4\,\pi \frac{(Z\,e^2)^2\ n_i\ \ln \Lambda}{(4\,\pi\,\epsilon_0)^2\ m^2\ v^3} \boldsymbol{f}_1 = -\nu_{ei}\,\boldsymbol{f}_1 \tag{8.99}$$

since the velocity $\mathbf{v}$ is arbitrary. $\boldsymbol{\Omega} = e\ \mathbf{B}/m$ is the electron cyclotron frequency and $\nu_{ei} = n_i\ v\ \sigma_d = (Z\,e^2)^2\ n_i\ \ln \Lambda/(4\,\pi\,\epsilon_0)^2\ m^2\ v^3$, the effective collision frequency. In particular, the perturbation of the distribution function $\boldsymbol{f}_1(\mathbf{v_e})$ is found to be inversely proportional to the collision frequency:

$$\nu_{ei}(v) = [4\,\pi\ n_i\ (Z\ e^2/(4\ \pi\epsilon_0)\ m)^2 \ln \Lambda]/v^3 = \nu_0/v^3 \tag{8.100}$$

We note that there is no term resulting from the zero-order component of the distribution consequent of the assumption that the ion mass is large and that therefore no energy exchange from ions to electrons takes place on collision. The first-order term represents the rotation of the electron velocity vector eliminating any isotropy in the distribution.

As a consequence, the resulting equations take a simple linear form, which may be solved as a set of simultaneous equations. Detailed calculations of the transport coefficient in this limiting case has been given by (Epperlein, 1984). In practice, this case corresponds to the limit of very large ion charge $Z \to \infty$.

---

19 In this section, we omit the subscript e, it being redundant.

20 The first-order collision term is more simply obtained from the Boltzmann collision integral equation (4.42). Since the scattering is elastic, the particle speed v is unchanged, but the velocity turned through the scattering angle χ. The ions are stationary and the distribution of scatterers is therefore $f'(\mathbf{v}') = n_i\ \delta(\mathbf{v}')$. The collision integral:

$$\frac{\partial f}{\partial t}\Big|_{\text{coll}} = \int \sigma(V,\chi)V(f(\overline{\mathbf{v}})\,f(\overline{\mathbf{v}'}) - f(\mathbf{v})\,f(\mathbf{v}))\ \mathrm{d}\mathbf{v}'\ \mathrm{d}\Omega = \int \sigma(v,\chi)\,\boldsymbol{f}_1(v) \cdot (\overline{\mathbf{v}} - \mathbf{v})\ \mathrm{d}\Omega \tag{8.96}$$

Taking axes (x, y, z) with z along $\mathbf{v}$ and x normal to the plane containing $\mathbf{v}$ and $\boldsymbol{f}_1$ and defining the angle between $\mathbf{v}$ and $\boldsymbol{f}_1$ as θ. Expressing $\overline{\mathbf{v}}$ and $\boldsymbol{f}_1$ in component form and noting that the azimuthal term ϕ vanishes on integration over the solid angle

$$\frac{\partial f}{\partial t}\Big|_{\text{coll}} = n_i \int \sigma\ v\ \boldsymbol{f}_1(v)\ [(\cos\theta\ \cos\chi - \sin\theta\ \sin\chi\ \cos\phi) - \cos\theta]\ \mathrm{d}\Omega \tag{8.97}$$

$$= n_i \boldsymbol{f}_1 \cdot \mathbf{v} \int \sigma\ (\cos\chi - 1)\ \mathrm{d}\Omega = -n_i\ \sigma_d \boldsymbol{f}_1 \cdot \mathbf{v}, \tag{8.98}$$

where σ_d is the momentum transport cross section equation (2.7).

Introducing a co-ordinate system $(x,\ y,\ z)$ with x and y in the plane perpendicular to the magnetic field and z parallel to the field, including the Hall term explicitly, and noting that the resulting equations are valid for all velocities, Eq. (8.90) yields

$$\left\{\frac{1}{n}\frac{\partial n}{\partial x}+\frac{1}{2}\left(\frac{m\,v^2}{\kappa T}-3\right)\frac{1}{T}\frac{\partial T}{\partial x}\right\} f_0-\frac{e}{\kappa T}\,E_x\,f_0+\frac{1}{v}\,\Omega\,f_{1y} \qquad =-\nu_{ei}\,\frac{f_{1x}}{v} \tag{8.101a}$$

$$\left\{\frac{1}{n}\frac{\partial n}{\partial y}+\frac{1}{2}\left(\frac{m\,v^2}{\kappa T}-3\right)\frac{1}{T}\frac{\partial T}{\partial y}\right\} f_0-\frac{e}{\kappa T}\,E_y\,f_0-\frac{1}{v}\,\Omega\,f_{1x} \qquad =-\nu_{ei}\,\frac{f_{1y}}{v} \tag{8.101b}$$

$$\left\{\frac{1}{n}\frac{\partial n}{\partial z}+\frac{1}{2}\left(\frac{m\,v^2}{\kappa T}-3\right)\frac{1}{T}\frac{\partial T}{\partial z}\right\} f_0-\frac{e}{\kappa T}\,E_z\,f_0 \qquad =-\nu_{ei}\,\frac{f_{1z}}{v}, \tag{8.101c}$$

where Ω is the electron cyclotron frequency. From these equations, the components of f_1 are easily obtained.

Introducing the pressure $p = n\ \kappa T$ and writing $\mathbf{E}' = \mathbf{E} + \nabla p/n\,e$ in conformity with Eq. (8.24), and solving for the components of f_1 we obtain[21]

$$f_{1i}=-\frac{e\,v\,\tau_{ei}}{\kappa T}\,S_{ij}\,\left[E'_j+\frac{1}{2\,e}\left(\frac{m\,v^2}{\kappa T}-5\right)\frac{\partial(\kappa T)}{\partial x_j}\right]\,f_0 \tag{8.102}$$

where the tensor $\mathbf{S}$ is

$$\mathbf{S}=\begin{pmatrix}\dfrac{1}{(1+\Omega^2{\tau_{ei}}^2)} & \dfrac{-\Omega\tau_{ei}}{(1+\Omega^2{\tau_{ei}}^2)} & 0\\ \dfrac{\Omega\tau_{ei}}{(1+\Omega^2{\tau_{ei}}^2)} & \dfrac{1}{(1+\Omega^2{\tau_{ei}}^2)} & 0\\ 0 & 0 & 1\end{pmatrix} \tag{8.103}$$

where $\tau_{ei}(v) = 1/\nu_{ei}(v))$ is the collision interval for electrons of speed v. Integrating leads directly to the transport coefficients in the form used by Shkarofsky et al. (1963). First, the current

$$j_i=\sigma_{ij}\left(E_j+\frac{1}{n\,e}\,\kappa T\,\frac{\partial n}{\partial x_j}\right)+\tau_{ij}\frac{\partial(\kappa T)}{\partial x_j} \tag{8.104a}$$

$$=\frac{4\,\pi}{3}\frac{n\,e}{m}\frac{1}{\kappa T}\int_0^\infty \tau_{ei}\,S_{ij}\,\left\{e\,E'_j+\frac{\partial(\kappa T)}{\partial x_j}\left(\frac{m\,v^2}{2\,\kappa T}-\frac{5}{2}\right)\right\}\,v^4\,f_0(v)\,\mathrm{d}v \tag{8.104b}$$

and the heat flow rate

$$Q_i=-\mu_{ij}\left[E_j+\frac{1}{n\,e}T\,\frac{\partial n}{\partial x_j}\right] \tag{8.105a}$$

$$=-\frac{2\,\pi}{3}\,m\,\frac{1}{\kappa T}\int_0^\infty \tau_{ei}\,S_{ij}\,\left\{e\,E'_j+\frac{\partial(\kappa T)}{\partial x_j}\left(\frac{m\,v^2}{2\,\kappa T}-\frac{5}{2}\right)\right\}\,v^6\,f_0(v)\,\mathrm{d}v. \tag{8.105b}$$

Casting Eq. (8.104) into Cartesian co-ordinates and solving for the electric field $\mathbf{E}$ in terms of the current $\mathbf{j}$, and substituting the resulting expression for the field in Eq. (8.105) to yield the total heat flux $\mathbf{Q}$. From which the thermo-electric term and the thermal conductivity are obtained from Eq. (8.93) noting that these terms are obtained after subtracting the current convective heat flux, to obtain the thermal conduction term alone

$$\mathbf{q}=\mathbf{Q}-\frac{5}{2}\,\kappa T\,\mathbf{j} \tag{8.106}$$

As the calculations are somewhat messy, we refer the reader to the paper by Epperlein (1984).

21 This solution for f_1 applies generally to a wide range of transport processes when $\mathbf{E} = 0$ namely: diffusion $\nabla n \neq 0$, thermal conduction $\nabla T \neq 0$ and viscosity $\partial v_i/\partial x_j \neq 0$.

8.8.1 Electron–Electron Collisions

In order to treat the full plasma, we must add electron–electron collisions to the Lorentz model which only treats electron–ion collisions. The approximation of stationary ions does not introduce significant error due to the large ion/electron mass ratio, which ensures that the electrons move much more rapidly than the ions. The electron distribution is again given by Eq. (8.87) so that the electron–ion collision term is again given by Eq. (8.99). However, the electron–electron collision term is much more complex.

The calculation of the electron–electron collision term is most easily accomplished by the use of the Rosenbluth potentials, Eq. (7.14). Assuming that the distribution function of the scatterer can be expanded in spherical harmonics or equivalently in Cartesian tensor form (Johnston, 1960)

$$f'(\mathbf{v}) = \sum F'_{lmn} Y_{lmn}(\theta, \phi) = f'_0 + \mathbf{f}'_1 \cdot \frac{\mathbf{v}}{v} + \mathbf{f}'_2 : \frac{\mathbf{vv}}{v^2} + \cdots \tag{8.107}$$

After some tedious algebra treated in detail by Shkarofsky et al. (1966, §7.5):

$$\mathcal{G}(\mathbf{v}) = v \sum_{n=0} \left[\frac{1}{2\,n+3}(\mathbf{I}^n_{n+2} + \mathbf{J}^n_{-1-n}) - \frac{1}{2\,n-1}(\mathbf{I}^n_{n-1} + \mathbf{J}^n_{1-n}) \right] \underset{n}{\bullet} \frac{\mathbf{v}^n}{v^n} \tag{8.108}$$

$$\mathcal{H}(\mathbf{v}) = \frac{m+m'}{m'\,v} \sum_{n=0} \frac{1}{2\,n+1}(\mathbf{I}^n_n + \mathbf{J}^n_{-1-n}) \underset{n}{\bullet} \frac{\mathbf{v}^n}{v^n} \tag{8.109}$$

where $\underset{n}{\bullet}$ is an n-fold scalar product and

$$\mathbf{I}^i_i = \frac{4\,\pi}{v^j} \int_0^v \mathbf{f}'_j \, v'^{(j+2)} \, dv' \qquad \text{and} \qquad \mathbf{J}^i_j = \frac{4\,\pi}{v^j} \int_v^\infty \mathbf{f}'_j \, v'^{(j+2)} \, dv' \tag{8.110}$$

are the functions introduced by Allis (1956).

Retaining only second-order terms

$$\mathcal{G}(\mathbf{v}) = \mathcal{G}_0 + \mathcal{G}_1 \cdot \frac{\mathbf{v}}{v} + \mathsf{G}_2 : \frac{\mathbf{vv}}{v^2} \cdots \tag{8.111:a}$$

$$= v \left\{ \left[\frac{1}{3}(I^0_2 + J^0_{-1}) + I^0_0 + J^0_1 \right] + \left[\frac{1}{15}(\mathbf{I}^1_3 + \mathbf{J}^1_{-2}) - \frac{1}{3}(\mathbf{I}^1_1 + \mathbf{J}^1_0) \right] \cdot \frac{\mathbf{v}}{v} + \left[\frac{1}{35}(\mathbf{I}^2_4 + \mathbf{J}^2_{-3}) - \frac{1}{15}(\mathbf{I}^2_2 + \mathbf{J}^2_{-1}) \right] : \frac{\mathbf{vv}}{v^2} + \cdots \right\} \tag{8.111:b}$$

and

$$\mathcal{H}(\mathbf{v}) = \mathcal{H}_0 + \mathcal{H}_1 \cdot \frac{\mathbf{v}}{v} + \mathsf{H}_2 : \frac{\mathbf{vv}}{v^2} \cdots \tag{8.112:a}$$

$$= \frac{m+m'}{m'\,v} \left\{ I^0_0 + J^0_{-1} + \frac{1}{3} \left[(\mathbf{I}^1_1 + \mathbf{J}^1_{-2}) \cdot \frac{\mathbf{v}}{v} \right] + \frac{1}{5} \left[(\mathbf{I}^2_2 + \mathbf{J}^2_{-3}) : \frac{\mathbf{vv}}{v^2} \right] + \cdots \right\}. \tag{8.112:b}$$

The electron–electron collision terms $\mathscr{C}_{ee}$ are now calculated from the appropriate differentials of the Rosenbluth potentials (Shkarofsky et al., 1966, §7.6)

$$\mathscr{C}_{ee0} = Y \left\{ \frac{1}{3\,v^2} \frac{\partial}{\partial v} \left[\frac{3\,m}{m'} \mathbf{I}^0_0 f_0 + v\,(\mathbf{I}^0_2 + \mathbf{J}^0_{-1}) \frac{\partial f_0}{\partial v} \right] \right\} \tag{8.113:a}$$

$$\mathscr{C}_{ee1} = Y \left\{ \frac{1}{3\,v}(\mathbf{I}^1_3 + J^0_{-1}) \frac{\partial^2 f_1}{\partial v^2} + \frac{1}{3\,v^2} \left(\frac{3\,m}{m'} \mathbf{I}^0_0 - \mathbf{I}^0_2 + 2\mathbf{J}^0_{-1} \right) \frac{\partial f_1}{\partial v} + \frac{1}{3v^3}(-3\mathbf{I}^0_0 + \mathbf{I}^0_2 - 2\,\mathbf{J}^0_{-1}) f_1 + \frac{4\pi m}{m'}(f'_0 f_1 + f'_1 f_0) + \frac{1}{5}(\mathbf{I}^1_3 + \mathbf{J}^1_{-2}) \frac{\partial^2 f_0}{\partial v^2} + \frac{1}{15\,v^2} \left[-3\,\mathbf{I}^1_3 + \left(7 - \frac{5\,m}{m'} \right) \mathbf{J}^1_{-2} + \left(-5 + \frac{10\,m}{m'} \right) \mathbf{I}^1_1 \right] \frac{\partial f_0}{\partial v} \right\}, \tag{8.113:b}$$

where $Y = 4\,\pi(Z\,Z'\,e^2)^2/[(4\,\pi\,\epsilon_0)^2\,m^2]\,\ln\Lambda$, and for electron–electron scattering $Z = Z' = 1$ and $m = m'$. When the unperturbed distribution is Maxwellian (8.84), the integrals

$$\mathbf{I}_0^0 = n(\Phi(x) - x\,\dot{\Phi}(x)) \qquad \mathbf{J}_{-1}^0 = n\,x\,\dot{\Phi}(x) \qquad \mathbf{I}_0^2 = n\,[-x\,\dot{\Phi}(x) + (3/2x^2)(\Phi - x\,\dot{\Phi}(x))] \tag{8.114}$$

where $\Phi(x) = (2/\sqrt{\pi})\int_0^x exp(-x^2)\,\mathrm{d}x$ is the error function and $x = v\sqrt{m/2\,kT}$.

It is instructive to calculate the electron–ion collision rate in the Lorentz approximation using Eqs. (8.113), when $f'(\mathbf{v}) = \delta(\mathbf{v})$, $m/m' = 0$ and $I_j^i = J_j^i = 0$ unless $i = j = 0$, when $I_0^0 = n_i$. As a result $\mathscr{C}_{ei0} = 0$, and $\mathscr{C}_{ei1} = -\nu_{ei}\,f_1(\mathbf{v})$ in conformity with Eq. (8.100).

The total electron collision rate is therefore $\mathscr{C}_{e0} = \mathscr{C}_{ee0}$ and $\mathscr{C}_{e1} = \mathscr{C}_{ee1} + \mathscr{C}_{ei1}$.

Even in the absence of a magnetic field, the calculation is complicated; the first successful calculations of the transport terms were carried out by Spitzer and his co-workers (Cohen et al., 1950; Spitzer and Härm, 1953) using spherical polar geometry. The presence of a magnetic field leads to a complicated integro-differential equation requiring Cartesian co-ordinates, linear in f_1, but involving both integral and differential terms (see Shkarofsky et al., 1966). We shall not pursue this topic further and refer the reader to Epperlein and Haines (1986) and the references therein where various calculations of the transport coefficients are discussed and accurate tabulations of the electrical and thermal conductivity given.

8.8.2 Electron Runaway

In describing the electrical conductivity of plasma, it has been assumed that the energy gained by an electron between collisions is much smaller than the thermal energy. Summed over all the particles, this energy gain is simply Ohmic heating. It is assumed that the rate is small, and that the plasma can be considered to be Maxwellian with a steady temperature. Whilst this is a valid approximation for most particles in the plasma provided the electric field is weak, it may not be correct for electrons in the high velocity tail of the distribution.

We investigate the physics of this effect using a very simple picture. Let us consider the collisional interaction of a fast test particle, velocity $\mathbf{v} \gg 0$, with the plasma bulk having velocities near zero. Both scattering and scattered particles may be of the same species (e.g. electrons). In this elementary picture of the interaction of a fast test particle gives rise to a drag force, given approximately by $\langle v_\parallel \rangle$ from Eq. (2.10a). The rate of loss of momentum by the test particle is given by the momentum loss cross section in terms of the $\overline{m}$ and its velocity $\mathbf{v}$. In accordance with our previous discussion, the momentum loss is on the average in the direction of the incoming particle velocity namely $-\mathbf{v}$

$$\frac{\mathrm{d}}{\mathrm{d}t}(m\mathbf{v}) \approx -\sigma_d\,v\,f(0)\,\mathrm{d}\mathbf{v}\,\overline{m}\,\mathbf{v} = -\sum_{\text{particles}} 4\pi\left(\frac{q\,Q}{4\,\pi\overline{m}}\right)^2 \ln\Lambda\,f(0)\left(\frac{\overline{m}\,\mathbf{v}}{v^3}\right)\,\mathrm{d}\mathbf{v} \tag{8.115}$$

where the sum is taken over the constituent particles in the plasma, namely electrons and ions with fractional charge Z.

The momentum loss rate is equivalent to a force exerted on the test particle by the collisions in the form of a drag tending to reduce the particle speed to the mean velocity of the scatterers. In fact, there is a balance between the drag exemplified by the generalisation of the term $\langle \Delta v_\parallel \rangle$ and diffusion in velocity space $\langle \Delta {v_\perp}^2 \rangle$, as discussed in Chapter 7.

Inspection of Eq. (8.117) in velocity space shows mathematical similarity with the familiar expression in electrostatics for the intensity due to a charge distribution in configuration space. The consequent mathematical results, such as Gauss' law, are therefore replicated. In particular, given an isotropic distribution of scattering particles, we conclude that only those with velocity $u' < u$ contribute to the drag (Allis, 1956). Clearly, the dynamical friction drives particles towards the centre in velocity space.

It is apparent from Eq. (2.10a) that the drag on electrons decreases quite rapidly as their speed increases. In an electric field, electrons gain momentum from the field at a rate $-e\,\mathbf{E}$ and loose it in collisions at the rate $\langle \Delta v_{\|} \rangle$.

$$\frac{d(m\mathbf{v})}{dt} = m\langle \Delta v_{\|} \rangle \frac{\mathbf{v}}{v} - e\mathbf{E} = -\sum_{\text{particles}} \left\{ \frac{4\,\pi\,q^2\,Q^2\,N\,\ln\Lambda}{(4\pi\epsilon_0)^2 m v^3} \frac{(m+m')}{m'}\,\mathbf{v} \right\} - e\mathbf{E} \tag{8.116}$$

Provided the electron speed is not too high, the momentum loss compensates for the gain, and the electron remains within the normal (thermal) velocity distribution, contributing to the electric current (Section 2.2.3). However, fast electrons can no longer dissipate the momentum gained from the field by collisional loss, and became continually accelerated – a phenomenon known as *electron runaway*.

Since these electrons are much faster than those in the bulk, which may be treated approximately at rest. The relative collision velocity is approximately the speed of the runaway electron $u \approx v$. The onset velocity v_e above which the electrons runaway is given by balancing their dynamic friction loss to that gained from the field, i.e. when

$$\sum_{\text{particles}} eE > \sum_{\text{particles}} \frac{4\pi q^2 Q^2 N}{(4\pi\epsilon_0)^2 m v_e^2} \frac{(m+m')}{m'} \ln\Lambda = \frac{4\pi(2+Z)\,e^4 N}{(4\pi\epsilon_0)^2 m v_e^2} \ln\Lambda \tag{8.117}$$

Clearly, those with velocity along the electric field are most likely to exceed the escape velocity v_e at which runaway occurs. Considering a plasma comprising electrons and ions of charge Z, the escape electron speed

$$v_e = \left\{ \frac{4\pi e^4\,(1+Z/2)N}{(4\pi\epsilon_0)^2 m^2\,eE} \right\}^{1/2} = (1+Z/2)^{1/2} \frac{E_D}{E}\,v_{Th} \tag{8.118}$$

where $v_{Th} = \sqrt{\kappa T/m}$ is the thermal speed and

$$E_D = \frac{1}{2} \frac{m\,\nu\,v_{Th}}{e} = \frac{e^3\,N\,ln\Lambda}{4\,\pi\,\epsilon_0^2\,T} \tag{8.119}$$

is known as the Dreicer field, most simply expressed in terms of the collision frequency $\nu = 4\,\pi\,e^4\,N/(4\,\pi\,\epsilon_0\,m)^2\,{v_{Th}}^3$ so that particles moving parallel to the electric field with $v > v_e$ escape.

In deriving the Dreicer field, we have assumed the electrons to be non-relativistic. However, this may not always be valid and relativistic correction may be needed (Connor and Hastie, 1975). Since the electron velocity cannot exceed that of light, there is clearly a limiting critical electric field obtained by replacing the thermal speed v_{Th} by the speed of light c in Eq. (8.119)

$$E_c = m\,\nu\,c/2\,e \tag{8.120}$$

Therefore, for $E < E_c$, runaway electrons cannot be generated.

When the field in the plasma is equal to the Dreicer field $E = E_D$, the critical velocity is equal to the thermal speed $v_c = v_T$ and the distribution is far from equilibrium and a large fraction of particles runaway.

'Lines of force' in velocity space map the structure of the force field and consequently, the trajectory in velocity space of a particle under the action of the electric field and the dynamical friction equation (8.117). Normalising in terms of the runaway velocity $V = v/v_c$ the lines of force are given by

$$\frac{V\,d\theta}{dV} = \frac{V^2\,\sin\theta}{1 - V^2\,\cos\theta} \tag{8.121}$$

where θ is the angle the velocity makes with the electric field. Expanding the equation

$$V\,\sin^2\theta\,dV + V^2\,\sin\theta\,\cos\theta\,d\theta = V\,\sin\theta\,d(V\,\sin\theta) = \sin\theta\,d\theta \tag{8.122}$$

whose solution is

$$V^2 = 2/(1+\cos\theta) + K/\sin^2\theta \tag{8.123}$$

where K is the constant of integration, which identifies specific electron trajectories.

At the centre $V = 0$

$$2(1-\cos\theta) = -K \qquad \cos\theta = 1 + K/2 \tag{8.124}$$

Since $-1 \le \cos\theta \le 1$, it follows that only trajectories with $0 \ge K \ge -4$ can reach the origin. Trajectories with $K > 0$ therefore runaway (Figure 8.2). The value of the number K for a particular orbit is determined by the initial state of the electron. Provided the field is weak, i.e. $E \ll E_D$, the thermal electrons' distribution forms a cloud around the origin O well separated from the runaway limit orbit $K = 0$. Consequently, the number of runaways is small, but, being temporally continuous, may not be negligible in some situations.

Any attempt to use this simple model for quantitative modelling clearly has obvious deficiencies. It requires that the electron distribution be assumed to be Maxwellian at least up to the critical velocity. In practice, the electron distribution is no longer isotropic at the critical velocity being distorted by the electric field. An accurate velocity distribution, therefore, requires a calculation allowing for this distortion of the distribution. Provided the field is weak, the model may be assumed

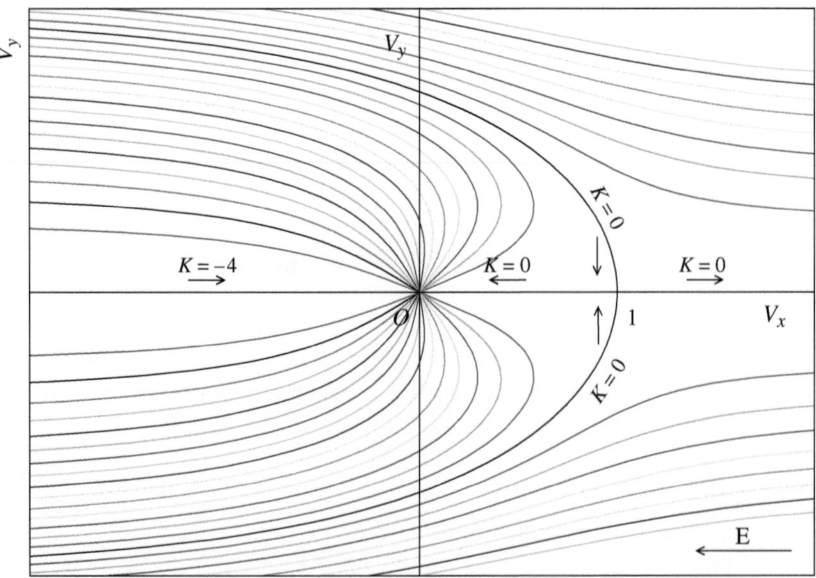

Figure 8.2 Lines of force in velocity space plotted at 0.25 intervals. The runaway region corresponds to $K > 0$ and the tied to $-4 \le K \le 0$. The electric field is parallel to the $-x$ direction. The direction of electron motion along the lines of force is indicated by the small arrows. Note the lines of force are cylindrically symmetric about the electric field axis (V_x).

steady state with a source of electrons at zero velocity to compensate for the loss. The analytic calculation of the loss rate due to runaway requires a complete solution of the electron distribution using the Fokker–Planck equation. This was first satisfactorily achieved by Kruskal (1962) using a set of approximations matched in five different zones out to the runaway limit to avoid the singularities (see also Cohen, 1976; Connor and Hastie, 1975). The final loss rate per unit time was found to be

$$S = C \; \nu \; v_{Th} \; N \; (E_D/E)^{3(Z+1)/16} \exp\{-E_D/4E - \sqrt{[(1+Z)E_D/E]}\} \tag{8.125}$$

where Z is the ion charge number. The constant C cannot be determined analytically. Comparison with model calculations carried out by Kulsrud et al. (1975) found that C is a weak function of the charge Z with value $C(1) = 0.32$ for the most important case of a hydrogen plasma. Relativistic effects can be included through an additional correction factor (Connor and Hastie, 1975).

In tokamak discharges, the heating and confinement are the result of a current driven by an electric field, which is itself generated by the toroidal induction resulting from the toroidal coils. The electric fields thus created naturally generate runaway electrons (Parail and Pogutse, 1986). Such runaway electrons can lead to instability and damage in toroidal discharges.

The runaway flux can be increased by avalanche in which thermal electrons are accelerated in collisions with previously accelerated electrons to velocities above the critical speed, and subsequently further accelerated by the electric field (Sokolov, 1979; Rosenbluth and Putvinski, 1997). This process can be easily understood when the electrons are relativistic $v \sim c$ as the recoil of a secondary electron in a close encounter. From Eq. (2.13), we obtain the transverse momentum of the secondary particle $p \approx e^2/2 \; \pi \; \epsilon_0 \; c \; b$ at impact parameter b. When $b < e^2/2 \; \pi \; \epsilon_0 \; c \; mv_e$, the secondary electrons runaway.

Runaway electrons can be extremely damaging in toroidal discharges as once the velocity exceeds the critical speed, they continue to accelerate until they leave the electric field. On interaction with the containment vessel, strong X-ray production and local heating take place. These problems have been long recognised, see for example Knoepfel and Spong (1979). As a result, runaway production needs to be avoided in large tokamaks such as ITER particularly during disruptions (Putvinski, 2011).

8.9 Deficiencies in the Spitzer/Braginskii Model of Transport Coefficients

The Spitzer–Braginskii model of conductivity, thermal conductivity, and viscosity is satisfactory and gives accurate results for most applications. However, it suffers from several defects if the perturbing force is not small. To some extent, the nature of the problem depends on the experimental configuration of application. Thus, we must distinguish between

i. Laser generated plasma, where the plasma is generated by a rapid, localised deposition of energy, Chapter 14,
ii. Magnetically confined plasma, Chapter 15.

In these two cases, the calculation of generalised transport coefficients introduces different methods.

- *Laser generated plasma*: In this case, the problems stem from the large field gradients, e.g. temperature, introduced by the nature of the driving force. As a consequence, the perturbation calculation used by the Spitzer/Braginskii theory is invalid. The core distribution can no longer be assumed to be Maxwellian. The solution is a direct attack on the solution of the Fokker–Planck equation using computational methods. This was first achieved for heat transport by Bell and his collaborators (Bell et al., 1981). This directly accounted for the two dominant problems in large gradient conduction, namely the free-streaming limit and non-locality. The former is dealt with Section 2.2.2.1. Non-locality obviously occurs if the mean free path is long compared to the scale length of the field gradient. The effect is emphasised by the reduction in cross section as the electron velocity increases as the heat flow is dominated by the faster electrons.

Appendix 8.A BGK Model for the Calculation of Transport Coefficients

8.A.1 BGK Conductivity Model

A simple picture of electron–ion interactions allowing the calculation of the conductivity and thermal conductivity is provided by the approximation of the electrons treated as Maxwell molecules, where the collision frequency (or time) is constant, i.e. independent of the electron velocity. Such an approximation, equivalent to the Drude model (Section 2.2.3) was considered in the pioneering work of Lorentz (2004, p. 267), treats the particle interaction as a central force $F \propto r^{-s}$ with $s = 5$ instead of $s = 2$ for Coulomb interaction. This modification allows us to demonstrate the basic method of calculation for a problem with a tractable analytic solution, but with obvious deficiencies. Since the collision frequency is assumed constant, independent of the electron velocity, it is clear from the preceding discussion of the underlying physical mechanism that the thermo-electric terms are zero in this case.

The BGK collision term, Eq. (6.11), for the electrons takes the simple form

$$\frac{\partial f}{\partial t}\Big|_{\text{coll}} = -\frac{\mathbf{v}\cdot \boldsymbol{f}_1}{v\,\tau} \tag{8.A.1}$$

where τ is the electron collision time, assumed to be independent of the electron speed.[22] Introducing a co-ordinate system $(x,\ y,\ z)$ with x and y in the plane perpendicular to the magnetic field and z parallel to the field, including the Hall term explicitly, and noting that the resulting equations are valid for all velocities, Eq. (8.90) yields Eq. (8.101) but with a constant collision interval τ

$$\left\{\frac{1}{n}\frac{\partial n}{\partial x} + \frac{1}{2}\left(\frac{m\,v^2}{k\,T} - 3\right)\frac{1}{T}\frac{\partial T}{\partial x}\right\} f_0 - \frac{e}{kT}\,E_x\,f_0 + \frac{1}{v}\,\Omega\,f_{1y} = -\frac{f_{1x}}{v\tau} \tag{8.A.2a}$$

$$\left\{\frac{1}{n}\frac{\partial n}{\partial y} + \frac{1}{2}\left(\frac{m\,v^2}{k\,T} - 3\right)\frac{1}{T}\frac{\partial T}{\partial y}\right\} f_0 - \frac{e}{kT}\,E_y\,f_0 - \frac{1}{v}\,\Omega\,f_{1x} = -\frac{f_{1y}}{v\tau} \tag{8.A.2b}$$

$$\left\{\frac{1}{n}\frac{\partial n}{\partial z} + \frac{1}{2}\left(\frac{m\,v^2}{k\,T} - 3\right)\frac{1}{T}\frac{\partial T}{\partial z}\right\} f_0 - \frac{e}{kT}\,E_z\,f_0 = -\frac{f_{1z}}{v\tau}, \tag{8.A.2c}$$

where Ω is the electron cyclotron frequency. From these equations, the components of f_1 are easily obtained.

---

22 We may remark that the BGK collision model is formally inconsistent with stationary ion–electron collisions, i.e. in the Lorentz approximation. The collision time τ represents the relaxation of the electrons to a Maxwellian. This can only be accomplished by energy transfer between the electrons, but only electron–ion collisions are allowed. However, for ions of infinite mass, collision dynamics prevents any energy transfer in collisions. For the calculation of transport coefficients, the relaxation time relates to momentum not energy transfer, i.e. to the isotropisation of the distribution. The Druyvesteyn distribution discussed in Appendix 8.C resolves this problem when dominant electron collisions are short-range interactions with heavy, but finite mass, atoms.

Introducing the pressure $p = n\,kT$ and writing $\mathbf{E}' = \mathbf{E} + \nabla p/n\,e$ in conformity with Eq. (8.24), and solving for the components of $\boldsymbol{f}_1$, we obtain[23]

$$f_{1i} = -\frac{e\,v\,\tau}{kT}\,S_{ij}\left[E_j' + \frac{1}{2\,e}\left(\frac{m\,v^2}{kT} - 5\right)\frac{\partial(kT)}{\partial x_j}\right] f_0 \tag{8.A.3}$$

where the tensor $\mathbf{S}$ is

$$\mathbf{S} = \begin{pmatrix} \dfrac{1}{(1+\Omega^2\tau^2)} & \dfrac{-\Omega\tau}{(1+\Omega^2\tau^2)} & 0 \\ \dfrac{\Omega\tau}{(1+\Omega^2\tau^2)} & \dfrac{1}{(1+\Omega^2\tau^2)} & 0 \\ 0 & 0 & 1 \end{pmatrix} \tag{8.A.4}$$

Substituting for $\boldsymbol{f}_1$ in Eq. (8.88a) and integrating for constant τ

$$j_i = \frac{4\,\pi}{3}\,\frac{n\,e}{m}\,\frac{1}{kT}\int_0^\infty \tau\,S_{ij}\left\{e\,E_j' + \frac{\partial(kT)}{\partial x_j}\left(\frac{m\,v^2}{2\,kT} - \frac{5}{2}\right)\right\} v^4\,f_0(v)\,\mathrm{d}v \tag{8.A.5a}$$

$$= \sigma_{ij}E_j' + \tau_{ij}\frac{\mathrm{d}T}{\mathrm{d}x_j} \tag{8.A.5b}$$

Treating the parameter $\Omega\,\tau$ as a constant as in the Drude model and including the Hall term implicitly, the conductivity

$$\sigma_{ij} = \begin{pmatrix} \sigma_\perp & -\sigma_\wedge & 0 \\ \sigma_\wedge & \sigma_\perp & 0 \\ 0 & 0 & \sigma_\| \end{pmatrix} = \frac{n\,e^2\tau}{m}\begin{pmatrix} \dfrac{1}{(1+\Omega^2\tau^2)} & \dfrac{-\Omega\tau}{(1+\Omega^2\tau^2)} & 0 \\ \dfrac{\Omega\tau}{(1+\Omega^2\tau^2)} & \dfrac{1}{(1+\Omega^2\tau^2)} & 0 \\ 0 & 0 & 1 \end{pmatrix} \tag{8.A.6}$$

Consistent with Eq. (2.62), where the physical basis of this result is described in more detail. The thermo-electric term $\tau_{ij} = 0$. Since

$$\int_0^\infty x^n\,e^{-x^2}\,\mathrm{d}x = \frac{1}{2}\int_0^\infty x^{(n-1)/2}\,e^{-x}\,\mathrm{d}x = \frac{1}{2}\,\Gamma((n+1)/2)$$

$$= \frac{1}{2}\begin{cases} \left[\frac{1\cdot 3\cdot 5\cdots(n-1)}{2^{n/2}}\right]\sqrt{\pi} & n \text{ even} \\ [(n-1)/2]! & n \text{ odd} \end{cases} \tag{8.A.7}$$

As noted earlier, the thermo-electric term τ_{ij} is zero because the collision frequency is constant. Similarly, calculating the heat flow rate

$$Q_i = -\frac{2\,\pi}{3}\,m\,\frac{1}{kT}\int_0^\infty \tau\,S_{ij}\left\{e\,E_j' + \frac{\partial(kT)}{\partial x_j}\left(\frac{m\,v^2}{2\,kT} - \frac{5}{2}\right)\right\} v^6\,f_0(v)\,\mathrm{d}v \tag{8.A.8a}$$

$$= -\frac{5}{2}\frac{n\,\tau\,(kT)}{m}S_{ij}\left[eE_j' + \frac{\partial(kT)}{\partial x_j}\right] \tag{8.A.8b}$$

Subtracting the current convective heat flux, we obtain the thermal conduction term alone

$$\mathbf{q} = -\frac{5}{2}\,\frac{n\,\tau\,(kT)}{m}\,\mathbf{S}\cdot\nabla(kT) \tag{8.A.9}$$

with zero thermo-electric term depending on the electric field consistent with the Onsager relation.

23 This solution for f_1 applies generally to a wide range of transport processes when $\mathbf{E} = 0$ namely: diffusion $\nabla n \neq 0$, thermal conduction $\nabla T \neq 0$ and viscosity $\partial v_i/\partial x_j \neq 0$.

8.A.2 BGK Viscosity Model

Kaufman (1960) has given a simplified calculation of the viscosity coefficients using the constant mean collision time (BGK) model, which we outline here.

We take the moment of $(v_i - u_i)(v_j - u_j)$ with respect to the Boltzmann equation (4.41) using the BGK collision term:

$$\frac{\partial f}{\partial t} + v_i \frac{\partial f}{\partial r_i} + \frac{F_i}{m} \frac{\partial f}{\partial v_i} = \frac{(f - f_0)}{\tau} \tag{8.A.10}$$

where $F_i = q(E_i + \epsilon_{ijk} v_j B_k)$ to obtain

$$\frac{\partial}{\partial t} \mathsf{P}_{ij} + \frac{\partial}{\partial x_k}(u_k \mathsf{P}_{ij} + \mathsf{Q}_{ijk}) + \left(\frac{\partial u_i}{\partial x_k} \mathsf{P}_{kj} + \frac{\partial u_j}{\partial x_k} \mathsf{P}_{ki} \right) - \Omega_k \left(\epsilon_{i\ell k} \mathsf{P}_{\ell j} + \epsilon_{j\ell k} \mathsf{P}_{\ell i} \right) = -\frac{1}{\tau} (\mathsf{P}_{ij} - p \delta_{ij}) \tag{8.A.11}$$

where

Mean velocity $\mathbf{u} = \int \mathbf{v}\, f(\mathbf{v})\, d\mathbf{v} / \int f(\mathbf{v}) d\mathbf{v}$ Stress tensor $\mathbf{P} = \int m\, (\mathbf{v} - \mathbf{u})(\mathbf{v} - \mathbf{u})\, f(\mathbf{v})\, d\mathbf{v}$
Heat flow tensor $\mathsf{Q}_{ijk} = \int m (v_i - u_i)(v_j - u_j)(v_k - u_k)\, f(\mathbf{v})\, d\mathbf{v}$: Larmor frequency $\boldsymbol{\Omega} = q\, \mathbf{B}/m$

If there are no spatial or temporal gradients, Eq. (8.A.11) has the solution $f = f_0$ the equilibrium distribution and $\mathbf{E} + \mathbf{v} \wedge \mathbf{B} = 0$, where the stress tensor $\mathsf{P}_{ij} = p\, \delta_{ij} = n(\not{k}T)\, \delta_{ij}$ and $\mathbf{Q} = 0$.

We look for a perturbation from the equilibrium solution based on linearising the derivatives and the departures from the zero order:

$$\left[\frac{\partial p}{\partial t} + \frac{\partial (u_k\, p)}{\partial x_k} \right] \delta_{ij} + p \left[\frac{\partial u_i}{\partial x_j} + \frac{\partial u_j}{\partial x_i} \right] - \Omega_k \left(\epsilon_{i\ell k} \mathsf{P}_{\ell j} + \epsilon_{j\ell k} \mathsf{P}_{\ell i} \right) = -\frac{1}{\tau} (\mathsf{P}_{ij} - p\, \delta_{ij}) \tag{8.A.12}$$

Taking the trace of Eq. (8.A.12), we obtain

$$\mathrm{tr} \left\{ \left[\frac{\partial p}{\partial t} + \nabla \cdot (p\, \mathbf{u}) \delta_{ij} \right] \right\} + 2\, p\, \nabla \cdot \mathbf{u} = 0 \tag{8.A.13}$$

Since $\mathrm{tr}(\delta_{ij}) = 3$, this equation becomes

$$\frac{1}{p} \frac{\mathrm{d}p}{\mathrm{d}t} = -\frac{5}{3} \nabla \cdot \mathbf{u} = \frac{1}{n} \frac{\mathrm{d}n}{\mathrm{d}t} \tag{8.A.14}$$

and we recover the familiar adiabatic expansion law.

If the magnetic field is taken along the z direction, $\Omega_i = \Omega \delta_{iz}$. Introducing the viscous stress tensor $\mathbf{T}$ as the stress deviator of the pressure tensor $\mathbf{P}$ and the rate of distortion $\mathbf{U}$ as the deviator of the rate of strain tensor $\nabla \mathbf{u}$, we obtain from Eq. (8.A.12) in component form

$$p\, \mathsf{U}_{zz} = -p\, (\mathsf{U}_{xx} + \mathsf{U}_{yy}) = -(1/2\, \tau)\, \mathsf{T}_{zz} = (1/2\tau)(\mathsf{T}_{xx} + \mathsf{T}_{yy}) \tag{8.A.15a}$$

$$\begin{cases} 2\, p\, \mathsf{U}_{xy} + \Omega\, (\mathsf{T}_{xx} - \mathsf{T}_{yy}) &= -(1/\tau)\, \mathsf{T}_{xy} \\ p\, (\mathsf{U}_{xx} - \mathsf{U}_{yy}) - 2\, \Omega\, \mathsf{T}_{xy} &= -(1/2\, \tau)\, (\mathsf{T}_{xx} - \mathsf{T}_{yy}) \end{cases} \tag{8.A.15b}$$

$$\begin{cases} 2\, p\, \mathsf{U}_{xz} - \Omega\, \mathsf{T}_{yz} &= -(1/\tau)\, \mathsf{T}_{xz} \\ 2\, p\, \mathsf{U}_{yz} + \Omega\, \mathsf{T}_{xz} &= -(1/\tau)\, \mathsf{T}_{yz} \end{cases} \tag{8.A.15c}$$

These equations form a group of three sets of simultaneous equations, which may be solved to give the components of the stress tensor:

$$\mathsf{T}_{zz} = -2\,\mu_0\,\mathsf{U}_{zz} \tag{8.A.16a}$$

$$\begin{cases} \mathsf{T}_{xx} = -\dfrac{2\,\mu_0}{(1+4\,\Omega^2\,\tau^2)}[\mathsf{U}_{xx} + 1/2\cdot 4\Omega^2\tau^2\,(\mathsf{U}_{xx}+\mathsf{U}_{yy}) + 4\,\Omega\,\tau\,\mathsf{U}_{xy}] \\ \mathsf{T}_{yy} = -\dfrac{2\,\mu_0}{(1+4\,\Omega^2\,\tau^2)}[\mathsf{U}_{yy} + 1/2\cdot 4\,\Omega^2\tau^2\,(\mathsf{U}_{xx}+\mathsf{U}_{yy}) - 4\,\Omega\,\tau\,\mathsf{U}_{xy}] \\ \mathsf{T}_{xy} = -\dfrac{2\,\mu_0}{(1+4\,\Omega^2\,\tau^2)}[\mathsf{U}_{xy} - 1/2\cdot 2\Omega\,\tau(\mathsf{U}_{xx}-\mathsf{U}_{yy})] \end{cases} \tag{8.A.15b}$$

$$\begin{cases} \mathsf{T}_{xz} = -\dfrac{2\,\mu_0}{(1+\Omega^2\,\tau^2)}(\mathsf{U}_{xz} - \Omega\,\tau\,\mathsf{U}_{yz}) \\ \mathsf{T}_{yz} = -\dfrac{2\,\mu_0}{(1+\Omega^2\,\tau^2)}(\mathsf{U}_{yz} - \Omega\,\tau\,\mathsf{U}_{xz}) \end{cases} \tag{8.A.16c}$$

where $\mu_0 = p\,\tau$. Apart from a slight difference in the definition of the collision time $\tau \Rightarrow (3/2)\,\tau$, these results are identical to the more accurate values quoted by Chapman and Cowling (1952, §18.44) and derived by Marshall (1960, Pt 3, §6) using the Chapman–Enskog approach. The value of the zero-field ion viscosity, given in these calculations is

$$\mu_0 = \frac{5\,(4\,\pi\epsilon_0)^2\,m_i^{1/2}(\mathcal{k}T)^{5/2}}{8\,\sqrt{\pi}\,Z^4\,e^4\,\ln\Lambda} \tag{8.A.17}$$

which yields a relaxation time approximately equal to ion–ion collision time (τ_i) given earlier in Eq. (2.24).

Appendix 8.B The Relationship Between the Flux Equations Given By Shkarofsky and Braginskii

The form of the flux equations given by Shkarofsky et al. (1963) are

$$\mathbf{j} = \boldsymbol{\sigma}\cdot(\mathbf{E} + \mathcal{k}T\,\nabla n_e/e\,n_e) + \boldsymbol{\tau}\cdot\nabla T \tag{8.B.1}$$

$$\mathbf{q}' = -\boldsymbol{\mu}\cdot(\mathbf{E} + \mathcal{k}T\,\nabla n_e/e\,n_e) - \mathbf{K}\cdot\nabla T \tag{8.B.2}$$

and those due to Braginskii (1965)

$$en_e\mathbf{E} = -\nabla p + \mathbf{j}\wedge\mathbf{B} + e\,n_e\,\boldsymbol{\alpha}\cdot\mathbf{j} - n_e\boldsymbol{\beta}\cdot\nabla(\mathcal{k}T) \tag{8.B.3}$$

$$\mathbf{q} = -\boldsymbol{\kappa}\cdot\nabla(\mathcal{k}T) - \boldsymbol{\beta}\cdot\mathbf{j}\,(\mathcal{k}T)/e \tag{8.B.4}$$

Adding and subtracting the term $\mathbf{j}\wedge\mathbf{B}$ in Shkarofsky's current equation (8.B.1) and writing the pressure $p = n_e(\mathcal{k}T)$

$$e\,n_e,\mathbf{j} + \boldsymbol{\sigma}\cdot(-\mathbf{j}\wedge\mathbf{B} + \nabla(\mathcal{k}T)) = \boldsymbol{\sigma}\cdot(e\,n_e\,\mathbf{E} - \mathbf{j}\wedge\mathbf{B} + \nabla p) + e\,n_e\,\boldsymbol{\tau}\cdot\nabla(\mathcal{k}T) \tag{8.B.5}$$

Hence, we obtain the electric field

$$e\,n_e\,\mathbf{E} = e\,n_e\,\boldsymbol{\sigma}^{-1}\cdot\mathbf{j} + \mathbf{j}\wedge\mathbf{B} - \nabla p - (e\,n_e\,\boldsymbol{\sigma}^{-1}\cdot\boldsymbol{\tau} - \mathbf{I})\cdot\nabla(\mathcal{k}T) \tag{8.B.6}$$

Using the $(x,\ y,\ z)$ co-ordinate system with z parallel to the field, the Hall term takes a simple form written as a matrix equation:

$$\mathbf{j}\wedge\mathbf{B} = \begin{pmatrix} 0 & -B & 0 \\ B & 0 & 0 \\ 0 & 0 & 0 \end{pmatrix}\cdot\mathbf{j}$$

we obtain Braginskii's form, Eq. (8.B.3)

$$\mathbf{E} - \mathbf{j} \wedge \mathbf{B} + \nabla p = e\, n_e\, \boldsymbol{\sigma}^{-1} \cdot \mathbf{j} - \mathbf{j} \wedge \mathbf{B} - (e\, n_e \boldsymbol{\sigma}^{-1} \cdot \boldsymbol{\tau} - \mathbf{I}) \cdot \nabla(\not{k}T)$$

Turning now to the heat transport, we substitute Eq. (8.B.1) into Eq. (8.B.2)

$$\mathbf{q} = \mathbf{q}' - \frac{3}{2}(\not{k}T) = -(\boldsymbol{\mu} \cdot \boldsymbol{\sigma}^{-1}) \cdot \mathbf{j} - (\mathbf{K} + \boldsymbol{\sigma}^{-1} \cdot \boldsymbol{\tau}) \cdot \nabla(\not{k}T) \tag{8.B.7}$$

which is Braginskii's energy transport equation (8.B.4). Thus, we obtain the equivalent coefficients

$$\boldsymbol{\alpha} = \begin{pmatrix} \sigma_\perp^{-1} & \sigma_\wedge^{-1} + B/e\, n_e & 0 \\ \sigma_\wedge^{-1} - B/e\, n_e & \sigma_\perp^{-1} & 0 \\ 0 & 0 & \sigma_\parallel^{-1} \end{pmatrix} \tag{8.B.8}$$

and

$$\boldsymbol{\beta} = e\, n_e\, \boldsymbol{\sigma}^{-1} \cdot \boldsymbol{\tau} - \mathbf{I} = \boldsymbol{\mu} \cdot \boldsymbol{\sigma}^{-1} \qquad \boldsymbol{\kappa} = \mathbf{K} + \boldsymbol{\sigma}^{-1} \cdot \boldsymbol{\tau} \tag{8.B.9}$$

For reference,

$$\boldsymbol{\sigma} = \begin{pmatrix} \sigma_\perp & -\sigma_\wedge & 0 \\ \sigma_\wedge & \sigma_\perp & 0 \\ 0 & 0 & \sigma_\parallel \end{pmatrix} \boldsymbol{\sigma}^{-1} = \begin{pmatrix} \dfrac{\sigma_\perp}{(\sigma_\perp{}^2 + \sigma_\wedge{}^2)} & \dfrac{\sigma_\wedge}{(\sigma_\perp{}^2 + \sigma_\wedge{}^2)} & 0 \\ -\dfrac{\sigma_\wedge}{(\sigma_\perp{}^2 + \sigma_\wedge{}^2)} & \dfrac{\sigma_\perp}{(\sigma_\perp{}^2 + \sigma_\wedge{}^2)} & 0 \\ 0 & 0 & \sigma_\parallel{}^{-1} \end{pmatrix}$$

and the product of two such matrices

$$\begin{pmatrix} \sigma_\perp & -\sigma_\wedge & 0 \\ \sigma_\wedge & \sigma_\perp & 0 \\ 0 & 0 & \sigma_\parallel \end{pmatrix} \cdot \begin{pmatrix} \mu_\perp & -\mu_\wedge & 0 \\ \mu_\wedge & \mu_\perp & 0 \\ 0 & 0 & \mu_\parallel \end{pmatrix} = \begin{pmatrix} (\sigma_\perp\mu_\perp - \sigma_\wedge\mu_\wedge) & -(\sigma_\wedge\mu_\perp + \sigma_\perp\mu_\wedge) & 0 \\ (\sigma_\wedge\mu_\perp + \sigma_\perp\mu_\wedge) & (\sigma_\perp\mu_\perp - \sigma_\wedge\mu_\wedge) & 0 \\ 0 & 0 & \sigma_\parallel\mu_\parallel \end{pmatrix}$$

Appendix 8.C Electrical Conductivity in a Weakly Ionised Gas and the Druyvesteyn Distribution

The calculation of the conductivity of the electrons in a weak electric field in a gas with a low degree of ionisation was first carried out by Lorentz (2004, p. 267) in 1905. It was assumed that the electrons had a Maxwellian distribution and interacted with cold stationary atoms via *hard sphere* interaction (i.e. constant mean free path). The electron density was sufficiently small that electron–electron collisions played no role. By neglecting electron–electron interactions, energy relaxation amongst electrons is not possible, and consequently, the electron velocity distribution cannot be expected to be Maxwellian.

In fact, electron energy is determined by a balance between the energy gain from the electric field and the loss in collisions with the cold background atoms. We follow Morse et al. (1935) (based on Lorentz (2004, p. 267)) in obtaining the solution to the problem by a direct application of the idea of detailed balance as defined in Section 4.7. The dominant electron collisions are assumed to be elastic with neutral un-ionised atoms, the ionisation being small. As is well-known in an elastic collision between a light particle of mass m and a stationary heavy one of mass M, the light particle energy and velocity are, respectively, reduced by

$$\frac{\Delta\epsilon}{\epsilon} = 2\frac{\Delta v}{v} = \frac{2\,m}{M}(1 - \cos\theta) \tag{8.C.1}$$

where θ and ϕ are the angles of scatter. (This term was neglected by Lorentz corresponding to the case of infinitely heavy atoms $M \to \infty$.)

Due to the presence of the electrical field, the distribution will no longer be isotropic but is aligned about the field axis x. Expanding the distribution function in a series of Legendre polynomials[24]

$$f(\mathbf{r},\mathbf{v}) = f_0(x,v) + P_1(\cos\chi)\, f_1(x,v) + P_2(\cos\chi)\, f_2(x,v) + \cdots \tag{8.C.2}$$

where ξ, η, ζ are the velocity components in the x, y, z directions, respectively, and $\chi = \arccos(\xi)/v$ is the angle the velocity makes with the axis.

Since we assume the field and, therefore, the perturbation to the distribution are weak, we may neglect all the terms in the series beyond the first. The rate of change in the distribution is given by Eq. (4.41), provided we can identify the collision rate $\partial f/\partial t|_{\text{coll}}$. In a homogeneous steady state, the number of electrons leaving the element of phase space dr dv per unit time reduces to

$$\left(\frac{\partial f}{\partial t} + \mathbf{v}\cdot\frac{\partial f}{\partial \mathbf{r}} - \frac{e\mathbf{E}}{m}\cdot\frac{\partial f}{\partial \mathbf{v}}\right) \mathrm{dr\ dv} = -\left(\frac{eE}{m}\frac{\partial f}{\partial \xi} + \xi\,\frac{\partial f}{\partial x}\right) \mathrm{dr\ dv} \tag{8.C.3}$$

Substituting for the distribution from Eq. (8.C.2) and retaining only the first-order term, it is clear that terms involving the square of $P_1(\cos\theta)$ appear in the expression for the loss. This is inconsistent with the level of approximation required and is avoided by replacing the squared terms by their average

$$\langle [P_1(\cos\theta)]^2\rangle = \langle \cos^2\theta\rangle = \frac{\int_0^\pi \cos^2\theta\ \sin\theta\ d\theta}{\int_0^\pi \sin\theta\ d\theta} = \frac{1}{3} \tag{8.C.4}$$

The number leaving the phase cell is given

$$c\ \mathrm{dr\ dv} = \left[-\frac{eE}{m}\left(\cos\chi\frac{\partial f_0}{\partial v} - \frac{1}{3}\frac{1}{v^2}\frac{\partial(v^2 f_1)}{\partial v}\right) + v\cos\chi\frac{\partial f_0}{\partial x} + \frac{v}{3}\frac{\partial f_1}{\partial x}\right] \tag{8.C.5}$$

To this must be added the number of electrons making collisions with atoms

$$a\ \mathrm{dr\ dv} = N\,v \int_0^\pi \sin\theta\ d\theta \int_0^{2\pi} d\phi\, f(v,x,\chi)\,\sigma(v,\theta)\ \mathrm{dr\ dv} \tag{8.C.6}$$

We must now calculate the number of electrons entering the phase cell. Before collision, these electrons had velocity $\mathbf{v}'$ in the velocity element $\mathbf{dv}'$, their initial speed was $v' = v + (m/M)\,v\,(1 - \cos\theta)$ after accounting for the energy loss in collision (Eq. (8.C.1)). The change in the phase cell volume corresponding to the velocity change $\mathbf{dv}' = (v'/v)^3$ dv. Hence, the total number of electrons entering the phase cell per unit time is given by Eq. (8.C.6) but with the primed values replaced by un-primed ones:

$$b\ \mathbf{dr}\ \mathbf{dv} = N\,v \int_0^\pi \sin\theta d\theta \int_0^{2\pi} d\phi\, f(v'.\chi')\sigma(v',\theta)\ (v'/v)^4\ \mathbf{dr}\ \mathbf{dv} \tag{8.C.7}$$

The difference between the number of particles entering and leaving the phase cell is, therefore,

$$(b-a)\ \mathbf{dr}\ \mathbf{dv} = N\,v^{-3} \int_0^\pi \sin\theta\ d\theta \int_0^{2\pi} d\phi\ [v'^4 f(v',\chi')\sigma(v',\theta) - v^4 f(v,\chi)\sigma(v,\theta)]\ \mathbf{dr}\ \mathbf{dv} \tag{8.C.8}$$

24 This procedure to first order is identical to that used in Section 8.A.1 which used Cartesian tensors instead of Legendre polynomials (Johnston, 1960).

Since the mass ratio $M/m \gg 1$ is large, the change in speed is small during the collision and also since $f_0(v) \gg f_1(v)$:

$$[v'^4\ f(v',\chi')\ \sigma(v',\theta) - v^4\ f(v,\chi')\sigma(v',\theta) - f(v,\chi)\ \sigma(v,\theta)]$$
$$\approx v^4\ [f(v,\chi') - f(v,\chi)]\ \ \sigma(v,\theta) + \Delta v\ \frac{\partial}{\partial v}[v^4\ f(v,\chi')]\ \sigma(v,\theta) \quad (8.C.9)$$
$$\approx v^4\ [f_1(v)\ (\cos\chi' - \cos\chi)]\ \ \sigma(v,\theta) + \Delta v\ \frac{\partial}{\partial v}[v^4\ f_0(v)]\ \sigma(v,\theta)$$

To calculate $\cos\chi'$, we take axes: x' along the velocity **v**, y' in the plane of **v** and the field **E** and z' normal to both. The unit vector along the field is $(\cos\chi, \sin\chi, 0)$ and that in the direction of the velocity **v′** is $(\cos\theta,\ \sin\theta\ \cos\phi,\ \sin\theta\ \sin\phi)$. Then

$$\cos\chi' = (\cos\chi, \sin\chi, 0)\cdot(\cos\theta,\ \sin\theta\ \cos\phi,\ \sin\theta\ \sin\phi) = \cos\theta\ \cos\chi + \sin\theta\ \sin\chi\ \cos\phi \quad (8.C.10)$$

Substituting into Eq. (8.C.8) and integrating over ϕ, the term in $\sin\chi$ becomes zero. Hence,

$$(b-a)\ \mathbf{dr}\ \mathbf{dv} = \left[-N\ v\ \sigma_D\ \cos\chi\ f_1(v) + \frac{m}{M}\frac{N}{v^2}\frac{\partial}{\partial v}(v^4\ \sigma_D\ f_0(v))\right]\ \mathbf{dr}\ \mathbf{dv} \quad (8.C.11)$$

where $\sigma_D = 2\ \pi\ \int_0^\pi (1-\cos\theta)\ \sin\theta\ d\theta$ is the momentum transfer cross section, Eq. (2.7).

The equation for a steady state is now obtained by equating the loss in the population of the phase cell due to the electric field from equation (8.C.5) to the gain due to collisions Eq. (8.C.11), i.e. $c = b - a$. Equating terms in powers of $\cos\chi$ gives two equations:

$$-\frac{e\ E}{2v}\ \frac{\partial}{\partial v}(v^2\ f_1) + \cancel{\frac{1}{2}\ m\ v^2\ \frac{\partial f_1}{\partial x}} = \frac{m^2}{M}\ \frac{3N}{2v}\ \frac{\partial}{\partial v}\ (v^4\ \sigma_D\ f_0) \quad (8.C.12a)$$

$$-e\ E\ \frac{\partial f_0}{\partial v} + \cancel{m\ v\ \frac{\partial f_0}{\partial x}} = -N\ \sigma_D\ m\ v\ f_1 \quad (8.C.12b)$$

(Lorentz (2004, p. 267) neglected the energy loss by electrons in heavy particle collisions consequently omitting Eq. (8.C.12a) and could not obtain a self-consistent form for f_0. He was forced to assume a Maxwellian zero-order distribution as a result.)

In homogeneous ionised gas, the cancelled terms are omitted, and Eq. (8.C.12b) can be directly integrated to give

$$-e\ E\ \epsilon\ f_1 = 6\ \frac{m}{M}\ N\ \sigma_D\ \epsilon^2\ f_0 - B \quad (8.C.13)$$

where B is a constant of integration.

The current carried by electrons in the energy range ϵ to $\epsilon + d\epsilon$ in the velocity cell $dv = (4\pi/m)\ v\ d\epsilon$ is $dJ = -\langle e\ \xi\ f\rangle dv = -e\ v\ f_1$ **dv** and, consequently, the energy extracted from the field per unit volume per unit time is $E\ dJ = -(8\ \pi\ e\ E/3m^2)\epsilon\ f_1\ d\epsilon$. Hence, the above Eq. (8.C.13) after multiplication by $(8\ \pi/3\ m^2)d\epsilon$ becomes

$$-\frac{8\ \pi\ e\ E}{3\ m^2}\epsilon\ f_1\ d\epsilon - \frac{16\ \pi\ N\ \sigma_d}{M^2}\ \frac{m}{M}\ \epsilon^2 f_0\ d\epsilon = -j\ d\epsilon \quad (8.C.14)$$

and represents the energy balance of the electrons. The first term is the energy gain from the field, the second the energy lost in collisions at an average rate $(2m/M)\ \epsilon$ per collision. We may argue that in equilibrium there must be equality between the energy gain from the field and that lost by collision. When these balance j and therefore B are zero. The perturbation distribution function f_1 is therefore given by

$$f_1 = -3\frac{m}{M}\frac{mv^2}{eE}N\sigma_D f_0 \quad (8.C.15)$$

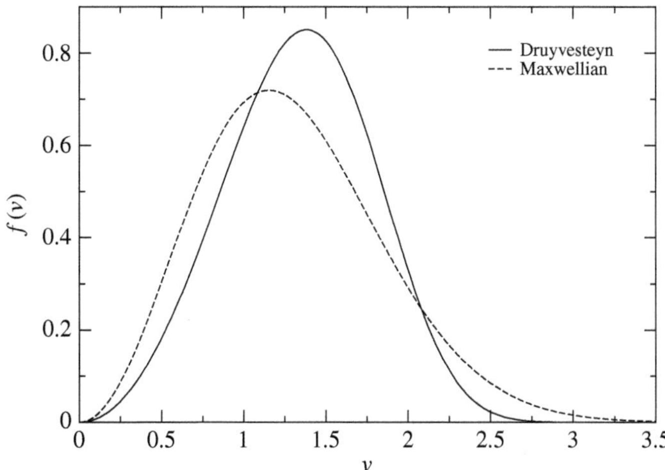

Figure 8.C.1 Comparison of the Druyvesteyn probability distribution with a Maxwellian.

Substituting for f_1 in Eq. (8.C.12b), we obtain

$$\frac{\partial f_0}{\partial v} = -\frac{3\,m}{M\,v}\left(\frac{mv^2}{eE}N\sigma_D\right)^2 f_0 \tag{8.C.16}$$

which, if the momentum cross section and therefore the mean free path are constant,[25] integrates directly to

$$f_0 = A\exp\left[-\left(\frac{3\,m}{M}\right)\left(\frac{mv^2}{2eE}N\sigma_D\right)^2\right] = A\ \exp(-h^4\ v^4) \tag{8.C.17}$$

where A is the normalisation constant and $h = \sqrt[4]{3\ m/M}\ \sqrt[2]{mN\sigma_D/2eE}$. The distribution function is normalised so that the integral over velocity space yields the electron density n

$$\int f_0\ \mathrm{dv} = 4\,\pi\,A\int \exp(-h^4v^4)\ v^2\ \mathrm{d}v = \pi\,A\int z^{-1/4}e^{-z}\mathrm{d}z/h^3 = \pi\Gamma(3/4)\ A/h^3 = 3.84976A/h^3 \tag{8.C.18}$$

Therefore, $A = n\ h^3/\pi\ \Gamma(3/4) = 0.25976\ n\ h^3$. The mean energy and drift velocity are easily calculated by similar integrals:

$$\begin{aligned}\bar{\epsilon} &= \frac{\Gamma(5/4)}{\Gamma(3/4)}\left(\frac{M}{3\,m}\right)^{1/2}\left(\frac{e\,E}{N\,\sigma_D}\right) = 0.4270\left(\frac{M}{m}\right)^{1/2}\left(\frac{e\,E}{N\,\sigma_D}\right)\\ v_d &= \frac{\pi^{1/2}}{3\,\Gamma(3/4)}\left(\frac{3\,m}{M}\right)^{1/4}\left(\frac{2\,e\,E}{m\,N\sigma_D}\right)^{1/2} = 0.6345\,\left(\frac{m}{M}\right)^{1/4}\left(\frac{2\,e\,E}{m\,N\sigma_D}\right)^{1/2}\end{aligned} \tag{8.C.19}$$

The energy and velocity clearly have characteristic values $\epsilon_0 = eE\lambda$ and $v_0 = \sqrt{2eE\lambda/m}$, corresponding to the energy delivered by the electric field to an electron moving through 1 mean free path $\lambda = 1/N\ \sigma_D$.

Figure 8.C.1 compares the Druyvesteyn distribution with the Maxwell both with the same energy per particle of one unit per particle. It is immediately apparent that the Maxwellian is significantly

25 Constant mean free path interactions are known as *hard sphere* collisions.

wider with a larger high velocity tail. This reflects the greater effectiveness of purely random processes in the transfer of energy to regions of low probability, in comparison with more deterministic ones. In the latter case, as here, the maximum energy transfer to an electron is limited to $eE\lambda$.

The distribution is usually written in terms of the electron energy.

This distribution is known as the *Druyvesteyn distribution* after it was first derived by Druyvestyn (1930); Druyvestyn and Penning (1940) and references therein. This work was based on the mobility equation derived by Hertz (1925) using the constant mean free path transport model in Section 2.2.2, adapted to include charged particles in an electric field.

Nowadays, the Druyvesteyn distribution tends to be of historical interest. It was originally developed to model electron behaviour in low ionisation gas discharges and in particular to enable calculation of the Townsend α coefficient for ionisation. It was found to give a reasonable description of the electron distribution, but was deficient in the ionisation rate, where inelastic collisions further depleted the relatively small population of electrons (Morse et al., 1935). However, a generalised form based on the Druyvesteyn distribution has been found useful in many cases

$$f(g,\epsilon)\,d\epsilon = \frac{1}{\Gamma(3/4g)}\left[\frac{\Gamma(5/4g)}{\Gamma(3/4g)}\right]^{3/2}\frac{\epsilon^{1/2}}{\overline{\epsilon}^{3/2}}\exp\left\{-\left[\frac{\Gamma(5/2g)}{\Gamma(3/2g)}\left(\frac{\epsilon}{\overline{\epsilon}}\right)^{g}\right]\right\}d\epsilon \tag{8.C.20}$$

where $1 \leq g \leq 2$ is an empirical constant found by experiment. When $g = 1$, we obtain the Maxwell distribution and $g = 2$ the Druyvesteyn. A more versatile distribution based on the same approach was developed by Margenau (1946).

When the electron density is reasonably large, the neglected electron–electron collisions dominate, driving the distribution towards the Maxwellian.

9

Ideal Magnetohydrodynamics

We return to the fluid model of plasma dynamics introduced in Section 8.3, namely magnetohydrodynamics. The resulting picture of plasma physics has many similarities with fluid mechanics. However, the presence of electrical conductivity and magnetic fields in the fluid greatly complicates the behaviour of its motion and, in consequence, its mathematical description. In this section, we will examine the generalisation of fluid mechanics introduced by magnetohydrodynamics and show that the behaviour of a simple fluid is markedly complicated by the presence an anisotropic magnetic field and fluid conductivity. The Debye length is assumed to be much smaller than all other characteristic lengths in the plasma. Consequently, the space charge fields are sufficiently strong that the ion and electron separation is small and the plasma is assumed to behave a single fluid.

The general equations are simplified by taking account of the quasi-neutrality, as described above, inherent in the single fluid description

$$q \approx 0 \tag{9.1}$$

and the non-relativistic nature of the flow, which implies that the displacement current

$$\left|\frac{\partial D}{\partial t}\right| \sim \epsilon_0 \frac{E}{T} \ll \frac{1}{\mu_0}\frac{B}{L} \sim |\nabla \wedge \mathbf{H}| \qquad \text{and} \qquad \left|\frac{\partial \mathbf{E}}{\partial t}\right| \sim \frac{E}{T} \sim \frac{B}{L} = |-\nabla \wedge \mathbf{B}| \tag{9.2}$$

is small since $L \ll cT$. Maxwell's equations become

$$\begin{aligned} \nabla \wedge \mathbf{E} &= -\frac{\partial \mathbf{B}}{\partial t} \qquad \text{and} \qquad \nabla \wedge \mathbf{B} = \mu_0 \mathbf{j} \\ \nabla \cdot \mathbf{E} &= 0 \qquad \text{and} \qquad \nabla \cdot \mathbf{B} = 0 \end{aligned} \tag{8.11:a}$$

As a result, the conservation equations simplify to the equation of mass conservation:

$$\frac{\partial \rho}{\partial t} + \nabla \cdot (\rho \mathbf{u}) = 0 \tag{8.7:a}$$

momentum conservation

$$\frac{\partial}{\partial t}[\rho\ \mathbf{u}] + \nabla \cdot \left[(\rho\ \mathbf{u}\ \mathbf{u}) + \mathsf{P} + \frac{1}{2}\frac{1}{\mu_0}B^2\mathsf{I} - \frac{1}{\mu_0}\mathbf{BB}\right] \approx 0 \tag{8.13:a}$$

and energy conservation

$$\frac{\partial}{\partial t}\left[\rho\ \left(\epsilon + \frac{1}{2}u^2\right) + \frac{1}{2}\left(\epsilon_0 E^2 + \frac{1}{\mu_0}\ B^2\right)\right] + \frac{\partial}{\partial \mathbf{r}}\left[\rho \mathbf{u}\left(h + \frac{1}{2}u^2\right) - \mathsf{T} + \mathbf{q} + \frac{1}{\mu_0}\ \mathbf{E} \wedge \mathbf{B}\right] \approx 0 \tag{8.32:a}$$

Foundations of Plasma Physics for Physicists and Mathematicians, First Edition. Geoffrey J. Pert.

If the magnetic field is weak, $\Omega\ \tau \lesssim 1$, then we assume that the density and the plasma transport coefficients, in particular the resistivity $\rho_0 = 1/\sigma_0$ the conductivity, are constants.[1] The electric field may be written as

$$\mathbf{E} = -\mathbf{v} \wedge \mathbf{B} + \frac{1}{n_e e}\ \nabla \mathbf{P}_e + \frac{\rho_0}{\mu_0}\ \nabla \wedge \mathbf{B} \tag{9.3}$$

Hence, from Maxwell's equations (8.11)

$$\frac{\partial \mathbf{B}}{\partial t} + (\mathbf{u} \cdot \nabla)\,\mathbf{B} = (\mathbf{B} \cdot \nabla)\,\mathbf{u} + \frac{\rho_0}{\mu_0} \nabla^2 \mathbf{B} \tag{9.4}$$

and Eq. (8.13) as

$$\frac{\mathrm{d}\mathbf{u}}{\mathrm{d}t} = \frac{\partial \mathbf{u}}{\partial t} + (\mathbf{u} \cdot \nabla)\,\mathbf{u} = \left\{ -\frac{1}{\rho} \nabla \left(p + \frac{1}{2} \frac{1}{\mu_0} B^2 \right) + \frac{1}{\mu_0\ \rho} (\mathbf{B} \cdot \nabla)\,\mathbf{B} \right\} + \frac{\mu}{\rho}\ \nabla^2 \mathbf{u} \tag{9.5}$$

where the stress tensor has been replaced by the pressure p and viscosity μ terms and $\mathrm{d}/\mathrm{d}t$ is the Lagrangian time derivative as before.

Comparing these two equations, we note their similarity in the inertial (first) and the dissipative (last) terms, the middle term being the driving 'force' in each case. In classical fluid mechanics, the dissipative effects are reflected in the ratio of the inertial to viscous terms, namely, $R = \rho\ u\ L/\mu$, namely, the Reynolds' number. When $R \gtrsim 1$, viscosity plays little role in the flow except near a solid surface (boundary layer). In contrast when $R \lesssim 1$, the flow is dominated by viscosity (Stokes or creeping flow). In magnetohydrodynamics, similarly the ratio of the inertial to the dissipative terms can be expressed by $R_M = \mu_0\ u\ L/\rho_0 = \mu_0\ \sigma_0\ u\ L$, known as the magnetic Reynold's number. When $R_M \gtrsim 1$ the resistivity plays little role in the development of the magnetic field, whereas if $R_M \lesssim 1$, it dominates the field. In each case, the solution is a balance between the driving (middle) term and either the inertial (first) or the dissipative (last) terms.

Examining Eq. (9.4), it is clear that the resistive term acts as a diffusive term for the magnetic field with diffusion coefficient ρ_0/μ_0.

9.1 Infinite Conductivity MHD Flow

We consider the case where the plasma is an ideal fluid in that no dissipation occurs. As for a classical fluid, the viscosity and thermal conduction are zero, on the other hand, the conductivity is infinite $\sigma_0 \to \infty$, so that Joule heating is negligible. This corresponds to the case where the magnetic Reynold's number is very large $R_M \gg 1$. Plasma approximating to this condition is frequently found, particularly in astrophysical and magnetospheric situations.

Since the current $\mathbf{j}$ must be finite, it follows from Eq. (9.3) that the electric field must satisfy the relation

$$\mathbf{E} + \mathbf{u} \wedge \mathbf{B} = 0 \tag{9.6}$$

neglecting any pressure gradient.

Infinite conductivity implies that the plasma is collision-free. This is identical to the condition imposed on the guiding centre model of plasma. The difference between the two models lies in the single particle picture used in the guiding centre model and the averaged particle description

1 If one transport coefficient is constant, the mean free time τ, must be a constant, i.e. the interaction is that for Maxwell molecules. All the transport coefficients are constants and the thermo-electric terms zero.

inherent in the use of the distribution function with bulk quantities such as density in the MHD picture. The direct equivalence of the two approaches can be demonstrated in simple cases (Tonks, 1955; Rose and Clark, 1961).

9.1.1 Frozen Field Condition

Taking the curl of this Eq. (9.6) and substituting we obtain

$$\frac{\partial \mathbf{B}}{\partial t} = \nabla \wedge (\mathbf{u} \wedge \mathbf{B}) \tag{9.7}$$

We now consider the magnetic flux Φ through an arbitrary loop S which is fixed in the fluid, i.e. made up of the same fluid particles for all time:

$$\Phi = \oint_S \mathbf{B} \cdot \mathbf{ds} \tag{9.8}$$

The elementary area vector $\mathbf{ds}$ associated with length element $\mathbf{dl}$ changes in time as the fluid moves:

$$\frac{\mathbf{ds}}{dt} = \mathbf{u} \wedge \mathbf{dl} \tag{9.9}$$

so that the rate of change of magnetic flux through the loop

$$\frac{\mathrm{d}\Phi}{\mathrm{d}t} = \underbrace{\oint_S \frac{\partial \mathbf{B}}{\partial t}}_{①} \cdot\ \mathbf{ds} + \underbrace{\oint_S \mathbf{B} \cdot \mathbf{u} \wedge \mathbf{dl}}_{②} \tag{9.10}$$

The first term ① is due to time variation in the magnetic field, the second ② due to flux swept out by the motion. Changing the order of terms in the scalar triple product and using Stokes theorem, we obtain:

$$\frac{\mathrm{d}\Phi}{\mathrm{d}t} = 0 \tag{9.11}$$

and the flux remains constant through the loop as the plasma moves and the loop follows the plasma particles.

To understand the physical consequence of this result, consider a tube of force whose walls are everywhere parallel to the lines of magnetic induction, i.e. parallel to $\mathbf{B}$. A loop lying in the wall of the tube has the flux $\Phi = 0$, since the normal component of $\mathbf{B}$ through the surface of the loop is zero. This value for the loop remains constant (= 0) as the plasma moves. However, the loop is arbitrary, and each loop is always made up from the same particles. Considering an ensemble of such loops, it is easy to see they all continue to lie on the displaced surface, which has zero flux through any element, i.e. is a tube of force. However, it also comprises the same particles and is therefore a tube of flow. Every flux tube is therefore tied to its corresponding flow tube. Allowing the tubes to become vanishingly thin, it follows that neighbouring particle trajectories always lie on the same field line, and *vice versa*. This is known as the *frozen field* condition.[2]

An alternative less-rigorous derivation is based on the equation for the time development of the quantity $\mathbf{B}/\rho$. Since

$$\nabla \wedge (\mathbf{u} \wedge \mathbf{B}) = (\mathbf{B} \cdot \nabla)\, \mathbf{u} - (\mathbf{u} \cdot \nabla)\, \mathbf{B} + \cancelto{0}{\mathbf{u}\, \nabla \cdot \mathbf{B}} - \mathbf{B}\, \nabla \cdot \mathbf{u} \tag{9.12}$$

2 The similarity of these relations with Helmholtz' theorem for vorticity in classical fluid mechanics (Pert, 2013) should be noted.

Hence, the development of the magnetic field:

$$\frac{\partial \mathbf{B}}{\partial t} + (\mathbf{u} \cdot \nabla)\mathbf{B} = (\mathbf{B} \cdot \nabla)\mathbf{u} - \mathbf{B}\, \nabla \cdot \mathbf{u} \tag{9.13}$$

which may be conflated with the equation of continuity

$$\frac{\partial \rho}{\partial t} + (\mathbf{u} \cdot \nabla)\rho = -\rho\, \nabla \cdot \mathbf{u} \tag{9.14}$$

to give

$$\frac{\mathrm{d}}{\mathrm{d}t}\left[\frac{\mathbf{B}}{\rho}\right] = \frac{\partial}{\partial t}\left[\frac{\mathbf{B}}{\rho}\right] + \mathbf{u} \cdot \nabla\left[\frac{\mathbf{B}}{\rho}\right] = \left(\frac{\mathbf{B}}{\rho} \cdot \nabla\right) \mathbf{u} \tag{9.15}$$

Consider the development of an infinitesimal element of length $\mathbf{\Delta l}$. The two ends of the element fixed in the plasma move with velocities $\mathbf{u}$ and $\mathbf{u} + (\mathbf{\Delta l} \cdot \nabla)\, \mathbf{u}$, the total rate of increase of the length $\mathbf{\Delta l}$ is, therefore,

$$\frac{\mathrm{d}}{\mathrm{d}t}[\mathbf{\Delta l}] = \frac{\partial}{\partial t}[\mathbf{\Delta l}] + \mathbf{u} \cdot \nabla[\mathbf{\Delta l}] = (\mathbf{\Delta l} \cdot \nabla)\mathbf{u} \tag{9.16}$$

Comparing Eqs. (9.15) and (9.16), it is clear that if the vectors $\mathbf{B}/\rho$ and $\mathbf{\Delta l}$ are initially parallel, then they remain so and that the ratio of their lengths is constant. Therefore, if the element $\mathbf{\Delta l}$ initially lies on a field line, it will continue to do so.

Since this is true for all infinitesimal elements along the elements comprising the line of induction, it must follow that the plasma flow overall follows the field lines. If the plasma is incompressible $\rho = \text{const}$, then the field varies as the lines of force are stretched or contracted. However, it is clear that the result holds if the plasma is compressible, and furthermore, that the only necessary condition is that the conductivity be infinite.

The origin of this result is easily understood from Faraday's law of induction and the fact that lines of induction form closed loops. Since the conductivity is infinite, the flux cannot change through any closed loop.

The similarity of this result to that for vorticity in ideal flow fluid mechanics should be noted. As has been seen, the resistivity of the plasma diffuses the magnetic field. As a result, the close relationship between the field lines and the particle is lost in real plasma. However in many cases, particularly where the density is low, for example magnetospheric and astrophysical plasmas, diffusion is sufficiently slow that for most practical purposes it can be neglected.

9.1.2 Adiabatic Equation of State

In the ideal MHD fluid limit, when all the dissipative terms are negligibly small, $\sigma \to \infty$ and $\mathbf{q} \approx 0$, there is no dissipation or entropy increase. The specific internal energy for a plasma with polytropic index[3] γ is $\epsilon = \frac{1}{(\gamma - 1)} \kappa T = \frac{1}{(\gamma - 1)} \frac{p}{\rho}$ Eq. (8.33) for the energy reduces to the simple form:

$$\frac{1}{(\gamma - 1)} \frac{\mathrm{d}p}{\mathrm{d}t} + \frac{\gamma}{(\gamma - 1)} \frac{p}{\rho} \frac{\mathrm{d}\rho}{dt} = 0 \tag{9.17}$$

which leads to the polytropic gas adiabatic equation of state

$$p/\rho^{\gamma} = \text{const} \tag{9.18}$$

3 For an ion–electron plasma γ takes the familiar value $5/3$ appropriate to particles with no internal degrees of freedom.

9.1.3 Pressure Balance

Consider the plasma at rest subject only to the forces due to the magnetic field and an isotropic pressure, the acceleration term can be rewritten as follows:

$$\rho \frac{\mathrm{d}\mathbf{u}}{\mathrm{d}t} = -\nabla p + \mathbf{j} \wedge \mathbf{B} = 0 \tag{9.19}$$

so that the pressure force balances the magnetic term due to the current

$$\nabla p = \mathbf{j} \wedge \mathbf{B} \tag{9.20}$$

Eliminating the current using Maxwell's equation $\mathbf{j} = (\nabla \wedge \mathbf{B})/\mu_0$

$$\nabla p = \frac{1}{\mu_0}(\nabla \wedge \mathbf{B}) \wedge \mathbf{B} \tag{9.21}$$

$$= \underbrace{-\frac{1}{2\mu_0}\nabla(B^2)}_{①} + \underbrace{\frac{1}{\mu_0}(\mathbf{B} \cdot \nabla)\mathbf{B}}_{②} \tag{9.22}$$

where the first term ① represents an isotropic pressure due to the magnetic field (*magnetic pressure*) and the second ② the longitudinal stress acting along the field lines, equivalent to a stretching tension along the field lines (*Maxwell stress*). The total stress tensor (*Maxwell stress tensor*) including both the pressure and the magnetic forces is

$$\mathbb{T} = p\,\mathbb{I} - \frac{1}{2\mu_0} B^2\,\mathbb{I} + \frac{1}{\mu_0}\,\mathbf{BB} = (p - p_M) + \frac{1}{\mu_0}\,\mathbf{BB} \tag{9.23}$$

where $\mathbb{I}$ is the identity dyadic. The equilibrium condition (9.20) becomes

$$\nabla \cdot \mathbb{T} \equiv \frac{\partial T_{ij}}{\partial r_j} = 0 \tag{9.24}$$

The Maxwell stress may be unimportant if we are interested in the ability of the magnetic field to support the plasma against its pressure and confine it within a 'magnetic bottle'. The equilibrium condition equation is then simply a pressure balance

$$p + \frac{1}{2\mu_0}B^2 = \text{const} \tag{9.25}$$

The plasma β is defined as follows:

$$\beta = \frac{p_{\max}}{(B_{\max}^2/2\mu_0)} \tag{9.26}$$

relates to the stability of the plasma containment for magnetic confinement devices. Clearly, for all confined plasmas, $\beta < 1$, and in practice should be significantly less than one.

Magnetic confinement devices for fusion power generation, such as *pinches* and *tokamaks*, are based on the concept of a 'stable' plasma confined within a magnetic bottle by the balance of the magnetic pressure against the kinetic pressure due to the temperature. Unfortunately, as we shall show plasma is extremely unstable and the major challenge of fusion physics has been to create an environment at a high temperature, which is sufficiently long lasting to allow fusion reaction to take place (see Chapter 15).

9.1.3.1 Virial Theorem

A useful result with important consequences for the design of magnetic confinement systems is contained in the *virial theorem*, whose derivation was given in Section 8.3.3. The application to a stationary equilibrium magnetic confinement system, which leads to an important result, is based on a simplified version of steady-state equation (8.21) in ideal MHD, where the electric fields may be neglected

$$\int_V \left\{ 3p + \frac{B^2}{2\,\mu_0} \right\} \mathrm{d}V = \int_S \left\{ \mathbf{r} \cdot \left[\left(p + \frac{B^2}{2\mu_0} \right) \mathsf{I} - \frac{1}{\mu_0} \mathbf{BB} \right] \right\} \cdot \mathbf{ds} \tag{9.27}$$

This equation must be satisfied if we wish to maintain a stable equilibrium plasma by means of magnetic fields. The left-hand side of this equation is positive definite and non-zero. However, at the plasma boundary surface $p \to 0$, and if the magnetic fields are due to currents inside the plasma alone, the magnetic field must decrease as a dipole field or faster, so that $B \lesssim 1/r^3$ and the surface integral tends to zero far from the plasma. Therefore, we conclude that the plasma must be supported by external currents, in addition to any currents flowing internally in the plasma.

9.2 Incompressible Approximation

It is instructive to see how well magnetohydrodynamics mirrors the classic relations of fluid mechanics. As with classical fluids a useful simplifying approximation is obtained by considering the plasma to be incompressible ρ = const. With this condition, the equation of motion simplifies to a generalisation of Euler's equation

$$\begin{aligned} \rho \left[\frac{\partial \mathbf{u}}{\partial t} + (\mathbf{u} \cdot \nabla)\, \mathbf{u} \right] &= \rho \left[\frac{\partial \mathbf{u}}{\partial t} + (\nabla \wedge \mathbf{u}) \wedge \mathbf{u} + \frac{1}{2} \nabla (u^2) \right] \\ &= -\nabla p + \frac{1}{\mu_0} (\nabla \wedge \mathbf{B}) \wedge \mathbf{B} = -\nabla \left(p + \frac{1}{2\mu_0} B^2 \right) + \frac{1}{\mu_0} (\mathbf{B} \cdot \nabla) \mathbf{B} \end{aligned} \tag{9.28}$$

If the Maxwell stress, namely the last term, is zero, the Lorentz force $\mathbf{j} \wedge \mathbf{B}$ can be written as the potential of the magnetic pressure, then the flow may be described in terms of the usual fluid mechanics relations with an additional pressure due to the magnetic field. More generally, we may write the stress term in terms of a component parallel to the field lines, and a second perpendicular to them. Let $\hat{\mathbf{b}} = \mathbf{B}/B$ be the unit vector in the field direction, and $\hat{\mathbf{n}}$ the unit vector along the principal normal to the field line, then

$$\frac{1}{\mu_0} \{ (\mathbf{B} \cdot \nabla) \mathbf{B} \} = \frac{1}{\mu_0} \left\{ \underbrace{\hat{\mathbf{b}} \cdot \nabla \left(\frac{1}{2} B^2 \right) \hat{\mathbf{b}}}_{\text{Parallel}} + \underbrace{\frac{B^2}{R}\, \hat{\mathbf{n}}}_{\text{Normal}} \right\} \tag{9.29}$$

where R is the radius of curvature of the field line, are the components parallel and perpendicular to the field line.

9.2.1 Bernoulli's Equation – Steady Flow

For a fluid in steady flow, Bernoulli's equation along a streamline in incompressible flow is

$$\frac{1}{2} u^2 + \frac{p}{\rho} = \text{const} \tag{9.30}$$

Taking the scalar product of $\mathbf{u}$ with the steady equation of motion, Eq. (9.28), we obtain

$$\mathbf{u} \cdot \nabla \left[\frac{p}{\rho} + \frac{B^2}{2\,\mu_0\,\rho} + \frac{1}{2} u^2 \right] = \frac{1}{\mu_0\,\rho} \underbrace{\{ \mathbf{u} \cdot (\mathbf{B} \cdot \nabla) \mathbf{B} \}}_{\text{Maxwell stress}} \tag{9.31}$$

It is clear that Bernoulli's equation can only be applied if the Maxwell stress term is small. Integrating along a streamline, we find that under such conditions:

$$p + \frac{B^2}{2\mu_0} + \frac{1}{2}\,\rho\, u^2 = \text{const} \tag{9.32}$$

since the plasma is assumed incompressible. This equation is simply a generalisation of Bernoulli's equation to include the magnetic pressure.

9.2.2 Kelvin's Theorem – Circulation

Kelvin's theorem states that in the flow of an ideal fluid, the circulation around a loop fixed in the field given by

$$\Gamma = \oint \mathbf{u} \cdot \mathrm{d}\mathbf{l} = \oint \nabla \wedge \mathbf{u} \cdot \mathrm{d}\mathbf{s} = \oint \zeta \cdot \mathrm{d}\mathbf{s} \tag{9.33}$$

is constant, $\mathrm{d}\Gamma/\mathrm{d}t = 0$, where the vorticity $\zeta = \nabla \wedge \mathbf{u}$. Taking the curl of the equation of motion and noting that $\nabla \wedge \nabla\phi = 0$:

$$\rho\, \nabla \wedge \left[\frac{\partial \mathbf{u}}{\partial t} + \zeta \wedge \mathbf{u}\right] = \frac{1}{\mu_0} \nabla \wedge [(\mathbf{B} \cdot \nabla)\mathbf{B}] \tag{9.34}$$

Integrating the left-hand side over the loop when the magnetic field is zero, allows us to recover Kelvin's theorem for a fluid

$$\frac{d}{dt} \oint \zeta \cdot d\mathbf{s} = \frac{\mathrm{d}\Gamma}{\mathrm{d}t} = 0 \tag{9.35}$$

But in plasma the right-hand side gives a non-zero contribution, and Kelvin's theorem is no longer valid. Consequently, the plasma generates vorticity through the magnetic field. Therefore, in contrast to a fluid, a magnetised plasma can support shear, the field imposing a degree of 'stiffness' to the plasma. As a result, fluids can only support compressional longitudinal waves, namely sound waves, and no transverse waves (or shear waves).[4] Plasma, in contrast, permits transverse waves which have vorticity associated with them, in addition to possible longitudinal waves.

Solids support both longitudinal and transverse waves through their elasticity, which provides both longitudinal and shear elements. Familiar examples are provided by seismic disturbances, e.g. earthquakes.

9.2.3 Alfvén Waves

The simplest possible type of transverse wave in plasma are shear Alfvén waves, which occur in uniform incompressible infinite conductivity plasma, moving at a constant velocity in a uniform field, $\mathbf{B}_0$. Transforming to the rest frame of the plasma $\mathbf{u}_0 = 0$. The plasma is perturbed by small changes in velocity $\mathbf{u}'$ and field $\mathbf{B}'$. Retaining only terms of first order in the perturbation induced by the wave, the linearised equations are

$$\nabla \cdot \mathbf{u}' = 0 \qquad \text{and} \qquad \nabla \cdot \mathbf{B}' = 0 \tag{9.36}$$

---

4 Inhomogeneous fluids with a discontinuity allow transverse waves, e.g. gravity waves, as Kelvin's theorem is not valid for loops crossing the discontinuity. In this case, 'stiffness' is provided by gravity and surface tension at the surface.

from the condition of incompressibility and Maxwell's equations, respectively. The linearised Euler equation (9.28) is

$$\rho \frac{\partial \mathbf{u}'}{\partial t} = -\nabla p' + \frac{1}{\mu_0}(\nabla \wedge \mathbf{B}') \wedge \mathbf{B}_0 \tag{9.37}$$

and finally from Maxwell's equation and the infinite conductivity condition

$$\frac{\partial \mathbf{B}'}{\partial t} = -\nabla \wedge \mathbf{E}' = \nabla \wedge (\mathbf{u}' \wedge \mathbf{B}_0) = (\mathbf{B}_0 \cdot \nabla)\mathbf{u}' \tag{9.38}$$

Differentiating Eq. (9.37) with respect to time, and substituting from (9.38) gives

$$\rho \frac{\partial^2 \mathbf{u}'}{\partial t^2} = -\nabla \left(\frac{\partial p'}{\partial t} \right) + \frac{1}{\mu_0} \{\nabla \wedge [(\mathbf{B}_0 \cdot \nabla)\, \mathbf{u}']\} \wedge \mathbf{B}_0 \tag{9.39}$$

Noting that the operator curl ($\nabla\wedge$) commutes with $\mathbf{B}_0 \cdot \nabla$ since $\mathbf{B}$ is constant and that $\nabla \cdot \boldsymbol{\zeta}' = 0$, we obtain

$$\rho \frac{\partial^2 \boldsymbol{\zeta}'}{\partial t^2} = \frac{1}{\mu_0}(\mathbf{B}_0 \cdot \nabla)[(\mathbf{B}_0 \cdot \nabla)\boldsymbol{\zeta}'] \tag{9.40}$$

by taking the curl ($\nabla\wedge$) to eliminate ∇p.

Consider a plane wave

$$\boldsymbol{\zeta}' = \boldsymbol{\zeta}_0 \exp\left[i(\mathbf{k} \cdot \mathbf{r}) - \Omega\, t\right] \tag{9.41}$$

where $\boldsymbol{\zeta}$ is the vorticity, $\mathbf{k}$ the wave number, and Ω the angular frequency. Introducing the unit vectors $\hat{\mathbf{k}}$ and $\hat{\mathbf{b}}$ in the direction of propagation and the ambient field, respectively, the angle between $\hat{\mathbf{k}}$ and $\hat{\mathbf{b}}$ is the propagation angle with respect to the field lines, namely $\theta = \arccos\,(\hat{\mathbf{b}} \cdot \hat{\mathbf{k}})$. Substituting yields the wave equation:

$$\left(\frac{B^2}{\mu_0 \rho} k^2 \cos^2\theta - \Omega^2 \right) \boldsymbol{\zeta}' = 0 \tag{9.42}$$

whose phase velocity is

$$C = \frac{\Omega}{k} = V_A \cos\theta \tag{9.43}$$

where V_A is the Alfvén velocity,

$$V_A = \frac{B_0}{\sqrt{\mu_0 \rho}} \tag{9.44}$$

The phase velocity $C = V_A \ \cos\theta$ is characteristic of a series of waves with phase fronts inclined at an angle θ to the magnetic field moving with speed V_A along the field line (Figure 9.1). However, it

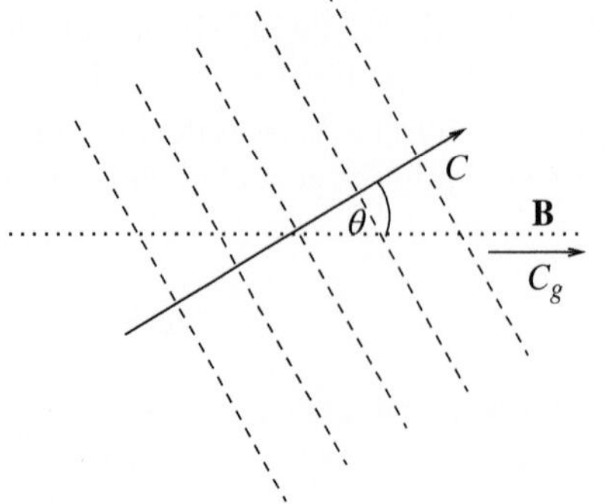

Figure 9.1 Sketch of Alfvén phase wave fronts propagating at an angle θ to the magnetic field **B**.

follows from Eq. (5.74) that the group velocity, C_g, which measures the velocity with which energy is transported by the wave train, is $V_A\hat{\mathbf{k}}$ directed along the field.

Since $\nabla \cdot \mathbf{u}' = \imath\, \mathbf{k} \cdot \mathbf{u}' = 0$, it follows that the plasma motion is perpendicular to the direction of propagation, as is characteristic of a transverse wave. These are shear waves where the stiffness required to generate the wave is provided by the stretching of the magnetic field lines. Since $\nabla p = \imath\, p\, \mathbf{k}$, the pressure gradient and the perturbation velocity are perpendicular to each other, and from Eq. (9.37) the former cannot drive the latter. The wave is therefore independent of the pressure, which takes a constant value. Taking the scalar product of Eq. (9.37) with **B**, it follows that $\mathbf{B} \cdot \mathbf{u}' = 0$ and that the velocity is also perpendicular to the magnetic field. Consequently, the velocity of the wave is normal to both the wavenumber of **k** and the ambient field **B** in the direction of $\mathbf{k} \wedge \mathbf{B}$. It is easily shown that the field perturbation $\mathbf{B}'$ is parallel to velocity perturbation $\mathbf{u}'$. The force is therefore also parallel to $\mathbf{k} \wedge \mathbf{B}$ and the forces are therefore consistent with the balance implied by Eq. (9.37).

Because the plasma and magnetic field lines are tied by the frozen field condition, a transverse displacement of the plasma leads to a transverse force induced by the Maxwell stress along the field line. The oscillation may be easily visualised in terms of a stretched string of unit cross section with tension $T = B^2/\mu_0$ and mass per unit length ρ, for which the wave speed is $\sqrt{T/\rho}$. The perturbation therefore propagates along the field lines with the Alfvén speed, the angular dependence simply reflects the inclination of the phase front to the field (Figure 9.1).

Since these waves are transverse $\nabla \cdot \mathbf{u}' = 0$, so their disturbance requires no density variation. As a consequence, there is no need to make the incompressible approximation, and we may expect that shear Alfvén waves are a general plasma phenomenon.

10

Waves in MHD Fluids

10.1 Introduction

In this section, we will examine the behaviour of waves in plasma within the ideal magnetohydrodynamics (MHD) approximation. This will allow the inclusion of effects due to compressibility neglected in Chapter 9. As a result, we find there are three distinct families of small amplitude waves. In addition, when the amplitude is large, magneto-hydrodynamic shock waves are formed due to the non-linear nature of the full equations, in direct analogy with classical fluid mechanics. However, due to the presence of a magnetic field, the shocks in this case are much more complex and may even form in the absence of collisions.

Our starting point will be the single fluid ideal MHD equations, assuming the plasma to be represented by a collisionless fluid polytropic equation of state with adiabatic index $\gamma = 5/3$

$$\frac{\partial \rho}{\partial t} + \nabla(\rho\, \mathbf{u}) = 0 \tag{10.1}$$

$$\frac{\mathrm{d}\mathbf{u}}{\mathrm{d}t} = \frac{\partial \mathbf{u}}{\partial t} - (\mathbf{u} \cdot \nabla)\, \mathbf{u} = \left\{ -\frac{1}{\rho} \nabla \left(p + \frac{1}{2}\frac{1}{\mu_0} B^2 \right) + \frac{1}{\mu_0\, \rho} (\mathbf{B} \cdot \nabla) \mathbf{B} \right\} = 0 \tag{10.2}$$

$$\frac{\partial}{\partial t} \left[\rho \left(\epsilon + \frac{1}{2} u^2 \right) + \frac{1}{2} \left(\epsilon_0 E^2 + \frac{1}{\mu_0} B^2 \right) \right] + \frac{\partial}{\partial \mathbf{r}} \left[\rho \mathbf{u} \left(h + \frac{1}{2} u^2 \right) + \frac{1}{\mu_0} \mathbf{E} \wedge \mathbf{B} \right] = 0 \tag{10.3}$$

together with Maxwell's equations

$$\begin{aligned} \nabla \wedge \mathbf{E} = -\frac{\partial \mathbf{B}}{\partial t} \quad &\text{and} \quad \nabla \wedge \mathbf{B} = \mu_0 \mathbf{j} \\ \nabla \cdot \mathbf{E} = 0 \quad &\text{and} \quad \nabla \cdot \mathbf{B} = 0 \end{aligned} \tag{8.11}$$

The equation of energy conservation Eq. (10.3) can be written for ideal MHD in the alternative form Eq. (8.33):

$$\frac{\mathrm{d}\epsilon}{\mathrm{d}t} = \frac{\partial \epsilon}{\partial t} + (\mathbf{u} \cdot \nabla)\epsilon = -\frac{1}{\rho} [\, p\, \nabla \cdot \mathbf{u} + \cancel{\sigma_{ij} \dot{e}_{ij} + \nabla \cdot \mathbf{q} - \mathbf{E} \cdot \mathbf{j}}\,] \tag{8.33}$$

From which making use of the internal energy and enthalpy of an ideal gas:

$$\epsilon = \frac{1}{(\gamma - 1)} \frac{p}{\rho} \qquad \text{and} \qquad h = \frac{\gamma}{(\gamma - 1)} \frac{p}{\rho}$$

we obtain the polytropic adiabatic equation of state for a fluid particle:

$$\frac{\mathrm{d}}{\mathrm{d}t} \left(\frac{p}{\rho^\gamma} \right) = 0 \qquad \text{or in more common form} \qquad \frac{p}{\rho^\gamma} = \text{const} \tag{10.4}$$

Foundations of Plasma Physics for Physicists and Mathematicians, First Edition. Geoffrey J. Pert.

10.2 Magneto-sonic Waves

Although an incompressible plasma can only support transverse waves (electromagnetic and Alfvén waves), plasmas are compressible. In analogy with compressible fluids, this gives rise to longitudinal (sonic) waves. Due to any applied magnetic field and the charged particle nature of plasma, we must expect that these new waves will contain both longitudinal and transverse elements. A full discussion of the generalisation to the theory of acoustic waves in compressible fluid mechanics introduced by the magnetic fields in an ideal magneto-hydrodynamic fluid is given by Friedrichs and Kranzer (1958). Consider a perturbation (denoted by primes) on an ambient state (denoted by the subscript 0) in the plasma rest frame[1]:

- Density: $\rho_0 + \rho'$ (ρ' is the perturbation, ρ_0 is the ambient condition)
- Pressure: $p_0 + p'$
- Magnetic field: $\mathbf{B}_0 + \mathbf{B}'$
- Electric field: $\mathbf{E}'$ ($\mathbf{E}_0 = 0$) initially
- Velocity: $\mathbf{u}'$ ($\mathbf{u}_0 = 0$ in the rest frame)
- Displacement: ξ', where $\mathbf{u}' = \mathrm{d}\xi'/\mathrm{d}t$ ($\xi_0' = 0$ initially)

Linearising the MHD equations, and assuming the flow is adiabatic

$$\frac{\partial \rho'}{\partial t} + \rho_0 \nabla \cdot \mathbf{u}' = 0 \tag{10.5}$$

$$\rho_0 \cdot \frac{\partial \mathbf{u}'}{\partial t} = -\nabla p' + \frac{1}{\mu_0}(\nabla \wedge \mathbf{B}') \wedge \mathbf{B}_0 \tag{10.6}$$

$$\frac{\partial p'}{\partial t} = \gamma \frac{p_0}{\rho_0} \frac{\partial \rho'}{\partial t} \tag{10.7}$$

$$\mathbf{E}' + \mathbf{u}' \wedge \mathbf{B}' = 0 \tag{10.8}$$

$$\frac{\partial \mathbf{B}'}{\partial t} = -\nabla \wedge \mathbf{E}' \tag{10.9}$$

$$\mathbf{u}' = \frac{d\xi'}{dt} \approx \frac{\partial \xi'}{\partial t} \tag{10.10}$$

Equation (10.5) becomes

$$\rho' + \rho_0 \nabla \cdot \xi' = 0 \tag{10.11}$$

Combining Eqs. (10.7) and (10.11) gives

$$p' = \gamma \frac{p_0}{\rho_0} \rho' = -\gamma\, p_0\, \nabla \cdot \xi' \tag{10.12}$$

Combining (10.8)–(10.10) gives

$$\frac{\partial \mathbf{B}'}{\partial t} = \left(\frac{\partial \xi'}{\partial t} \wedge \mathbf{B}_0 \right) \tag{10.13}$$

Integrating noting that the magnetic field perturbation and the displacement are zero at time $t = 0$:

$$\mathbf{B}' = \nabla \wedge (\xi' \wedge \mathbf{B}_0) \tag{10.14}$$

Substituting in Eq. (10.6):

$$\frac{\partial^2 \xi'}{\partial t^2} - \gamma \frac{p_0}{\rho_0} \nabla (\nabla \cdot \xi') - \frac{1}{\mu_0\, \rho_0}(\nabla \wedge \mathbf{B}') \wedge \mathbf{B}_0 = 0 \tag{10.15}$$

---

1 Strictly speaking the perturbation quantities should be taken at the Lagrangian fluid point $\mathbf{r} = (\mathbf{r}_0 + \xi)$. However, at the order of the perturbation, we may set $\mathbf{r} \approx \mathbf{r}_0$.

Using the standard vector relation and noting $\mathbf{B}_0$ is constant

$$\nabla \wedge (\xi' \wedge \mathbf{B}_0) = \xi' \nabla \cdot \mathbf{B}_0 - \mathbf{B}_0 \nabla \cdot \xi' + (\mathbf{B}_0 \cdot \nabla)\xi' - (\xi' \cdot \nabla)\mathbf{B}_0 \tag{10.16}$$

and

$$(\nabla \wedge \mathbf{B}') \wedge \mathbf{B}_0 = (\mathbf{B}_0 \cdot \nabla)\mathbf{B}' - \nabla(\mathbf{B}' \cdot \mathbf{B}_0) \tag{10.17}$$

we obtain

$$\begin{aligned}(\nabla \wedge \mathbf{B}') \wedge \mathbf{B}_0 \\ = \{\nabla[\mathbf{B}_0{}^2\ \nabla \cdot \xi' - (\mathbf{B}_0 \cdot \nabla)\ (\mathbf{B}_0 \cdot \xi')] + (\mathbf{B}_0 \cdot \nabla^2)\ \xi' - \mathbf{B}_0\ [(\mathbf{B}_0 \cdot \nabla)\ (\nabla \cdot \xi')]\}\end{aligned} \tag{10.18}$$

Therefore,

$$\begin{aligned}\rho_0 \frac{\partial^2 \xi'}{\partial t^2} - \gamma\ p_0\ \nabla(\nabla \cdot \xi') - \frac{1}{\mu_0}\nabla\ [\mathbf{B}_0{}^2\ \nabla \cdot \xi' - (\mathbf{B}_0 \cdot \nabla)(\mathbf{B}_0 \cdot \xi')] \\ - \frac{1}{\mu_0}\ (\mathbf{B}_0 \cdot \nabla)^2\ \xi' + \frac{1}{\mu_0}\ \mathbf{B}_0\ (\mathbf{B}_0 \cdot \nabla)(\nabla \cdot \xi') = 0\end{aligned} \tag{10.19}$$

We now introduce the Alfvén speed:

$$V_A{}^2 = \frac{B_0^2}{\mu_0 \rho_0} \tag{10.20}$$

and the sound speed:

$$V_s{}^2 = \frac{\gamma p_0}{\rho_0} \tag{10.21}$$

The ratio of the Alfvén speed to the sound speed:

$$\alpha = \frac{V_s{}^2}{V_A{}^2} = \left(\frac{\gamma}{2}\right)\beta \tag{10.22}$$

can be expressed in terms of the plasma $\beta = p_0/(B_0{}^2/2\mu_0)$

Substituting simplifies Eq. (10.19) to

$$\frac{1}{V_A{}^2}\frac{\partial^2 \xi'}{\partial t^2} - \nabla[(1+\alpha)\nabla \cdot \xi' - (\hat{\mathbf{b}} \cdot \nabla)(\hat{\mathbf{b}} \cdot \xi')] - (\hat{\mathbf{b}} \cdot \nabla)^2 \xi' + (\hat{\mathbf{b}} \cdot \nabla)(\nabla \cdot \xi')\hat{\mathbf{b}} = 0 \tag{10.23}$$

where $\hat{\mathbf{b}} = \mathbf{B}_0/|\mathbf{B}_0|$ is the unit vector in the direction of the ambient field $\mathbf{B}_0$. We note that the displacement is ξ', so the wave cannot be longitudinal unless the magnetic field $\mathbf{B}_0$ vanishes in which case, the wave reduces to a simple gas dynamic sound wave. In general, however, we expect that the dispersion relation for the wave must include an Alfvén wave, which is itself transverse.

We now introduce a plane wave: $\xi' = \xi'_0 \exp[i(\mathbf{k} \cdot \mathbf{r}) - \omega t]$ into Eq. (10.23)

$$\frac{V^2}{V_A{}^2}\xi' - (\hat{\mathbf{b}} \cdot \hat{\mathbf{k}})^2 \xi' - [(1+\alpha)(\hat{\mathbf{k}} \cdot \xi') - (\hat{\mathbf{b}} \cdot \hat{\mathbf{k}})(\hat{\mathbf{b}} \cdot \xi')]\,\hat{\mathbf{k}} + (\hat{\mathbf{b}} \cdot \hat{\mathbf{k}})\,(\hat{\mathbf{k}} \cdot \xi')\,\hat{\mathbf{b}} = 0 \tag{10.24}$$

where $V = \omega/k$ is the phase velocity and $\hat{\mathbf{k}} = \mathbf{k}/|\mathbf{k}|$ is the unit vector in the direction of propagation. The equation contains three vector components which we may express as a set of coupled equations in three Cartesian coordinates of the form:

$$\mathbf{A} \cdot \xi = \begin{pmatrix} V^2 - (V_A{}^2 + V_s{}^2 \sin^2\theta) & 0 & -V_s{}^2\ \sin\theta\ \cos\theta \\ 0 & V^2 - V_A{}^2 \cos^2\theta & 0 \\ -V_s{}^2\ \sin\theta\ \cos\theta & 0 & V^2 - V_s{}^2 \cos^2\theta \end{pmatrix} \begin{pmatrix} \xi'_x \\ \xi'_y \\ \xi'_z \end{pmatrix} = 0 \tag{10.25}$$

defined in the co-ordinates $(x,\ y,\ z)$ with z parallel to the magnetic field $\hat{\mathbf{b}}$, the direction of propagation $\hat{\mathbf{k}}$ in the plane $(x,\ z)$ and y perpendicular to both the magnetic field and the direction of propagation. θ is the angle between the magnetic field and the direction of propagation.

This set of equations has a non-trivial solution only if the determinant of the coefficients det **A** vanishes.

$$\det(\mathbf{A}) = (V^2 - {V_A}^2\cos^2\theta)\begin{vmatrix} C^2 - ({V_A}^2 + {V_s}^2\sin^2\theta) & -{V_s}^2\sin\theta\cos\theta \\ -{V_s}^2\sin\theta\cos\theta & V^2 - {V_s}^2\cos^2\theta \end{vmatrix} = 0 \tag{10.26}$$

Hence, we generate a cubic equation for the phase velocity V. Solving we obtain the dispersion relation $\omega(\mathbf{k})$ for the wave in terms of the phase velocity.[2]

One root is immediately apparent

$$V^2 = {V_A}^2\cos^2\theta \tag{10.28}$$

This dispersion relation is characteristic of shear Alfvén waves. The phase velocity $V = V_A\cos\theta$, where θ is the angle between the magnetic field $\hat{\mathbf{b}}$ and the propagation direction $\hat{\mathbf{k}}$. The displacement, $\boldsymbol{\xi}'$ is perpendicular to both $\hat{\mathbf{b}}$ and $\hat{\mathbf{k}}$, typical of a transverse wave. This result is consistent with that found earlier for shear Alfvén waves in incompressible plasma (Section 9.2.3).

The 2×2 determinant contains two coupled equations, which are consistent if

$$V^4 - ({V_A}^2 + {V_s}^2)V^2 + {V_A}^2{V_s}^2\cos^2\theta = 0 \tag{10.29}$$

giving a quadratic equation for V^2 with real roots. Therefore, there are two distinct waves, each of which can move forwards or backwards known as the *fast* and *slow* waves. It is easily shown that the consistency relation (Eq. (10.29)) can written as

$$(V^2 - {V_A}^2)(V^2 - {V_s}^2) = {V_A}^2{V_s}^2\sin^2\theta \tag{10.30}$$

The magnetic field, propagation vector, and the displacement are seen from Eq. (10.25) to be co-planar in both fast and slow waves. Each wave contains both longitudinal and transverse displacements.

$$V_\pm^2 = \frac{1}{2}\{({V_A}^2 + {V_s}^2) \pm \sqrt{\{({V_A}^2 + {V_s}^2) - 4\,{V_A}^2\,{V_s}^2\cos^2\theta\}}\} \tag{10.31}$$

The + sign corresponds to the fast wave, and the − to the slow. $\hat{\mathbf{k}}$, $\hat{\mathbf{b}}$ and ξ' are all co-planar (Figure 10.1).

---

2 A simpler method can be used to construct the dispersion relations by forming the components of the wave in three non-orthogonal directions: parallel to the wave vector $\hat{\mathbf{k}}$, parallel to the magnetic field $\hat{\mathbf{k}}$ in the plane of $\mathbf{k}$ and $\hat{\mathbf{b}}$, and perpendicular to both the propagation and field $\hat{\mathbf{k}}$ and $\hat{\mathbf{b}}$, where $\hat{\mathbf{k}}$ is the unit vector parallel to $\mathbf{k}$.

- $\hat{\mathbf{k}}\cdot\xi'$: gives the component parallel to $\hat{\mathbf{k}}$.
- $\hat{\mathbf{b}}\cdot\xi'$: gives the component parallel to $\hat{\mathbf{b}}$.
- $(\hat{\mathbf{b}}\wedge\hat{\mathbf{k}})\cdot\xi'$: gives the component perpendicular to both $\hat{\mathbf{b}}$ and $\hat{\mathbf{k}}$.

Performing these multiplications on Eq. (10.24), we obtain

$$\hat{\mathbf{k}}\cdot\xi' \Rightarrow \left[\frac{V^2}{V_A^2} - (1+\alpha)\right]\hat{\mathbf{k}}\cdot\xi' + (\hat{\mathbf{b}}\cdot\hat{\mathbf{k}})(\hat{\mathbf{b}}\cdot\xi') = 0 \tag{10.27a}$$

$$\hat{\mathbf{b}}\cdot\xi' \Rightarrow \frac{V^2}{V_A^2}\hat{\mathbf{b}}\cdot\xi' - \alpha(\mathbf{b}\cdot\hat{\mathbf{k}})(\hat{\mathbf{k}}\cdot\xi') = 0 \tag{10.27b}$$

$$(\hat{\mathbf{b}}\wedge\hat{\mathbf{k}})\cdot\xi' \Rightarrow \left[\frac{V^2}{V_A^2} - (\hat{\mathbf{b}}\cdot\hat{\mathbf{k}})^2\right]\hat{\mathbf{b}}\cdot\hat{\mathbf{k}}\wedge\xi' = 0 \tag{10.27c}$$

These three equations must be satisfied simultaneously by any allowed wave. It is easily shown that these reduce to the dispersion relations found already.

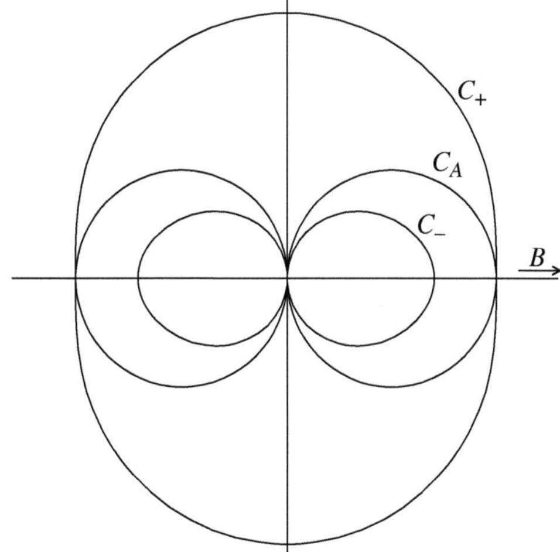

Figure 10.1 Polar plot of the fractional wave velocities of the fast, slow and Alfvén waves $V(\theta)/V_A$ for the case $\alpha = (V_s/V_A)^2 = 0.5$.

The two limiting cases are

- If the wave is propagated along the field $\hat{\mathbf{k}}$ is parallel to $\hat{\mathbf{b}}$ and $\theta = 0$, the phase velocities of the fast and slow waves are
$$V_+ = \frac{1}{2}\sqrt{\{V_A{}^2 + V_S{}^2 + |V_A{}^2 - V_s{}^2|\}} = \text{Max}\,(V_A,\ V_S) \tag{10.32}$$
$$V_- = \frac{1}{2}\sqrt{\{V_A{}^2 + V_S{}^2 - |V_A{}^2 - V_s{}^2|\}} = \text{Min}\,(V_A,\ V_S) \tag{10.33}$$
- If the wave is propagated normal to the field $\hat{\mathbf{k}}$ is perpendicular to $\hat{\mathbf{b}}$, and $\theta = \pi/2$, the phase velocities of the fast and slow waves are
$$V_+ = \sqrt{(V_A^2 + V_s^2)} \tag{10.34}$$
$$V_- = 0 \tag{10.35}$$
The fast wave is purely longitudinal, the magnetic pressure adding to the kinetic pressure, to give an enhanced 'sound wave'.
The slow wave does not propagate when the wave vector is perpendicular to the field $\hat{\mathbf{k}} \perp \hat{\mathbf{b}}$.

A polar plot of the wave velocities relative to the Alfvén speed for the three waves in the typical case $\alpha = (V_s/V_A)^2 = 1/2$, where the sound speed is a fraction 0.7 of the Alfvén speed. We note that the plot is also appropriate when the sound speed is greater than the Alfvén velocity by taking the velocities relative to the sound speed and $\alpha' = (V_A/V_s)^2 = 1/2$.

Since the vectors ξ', $\hat{\mathbf{k}}$, and $\hat{\mathbf{b}}$ are all co-planar, we construct the components parallel to and perpendicular to the magnetic field:

$$\xi'_{\parallel} = (\hat{\mathbf{b}} \cdot \xi')\,\hat{\mathbf{b}} \qquad \text{and} \qquad \xi'_{\perp} = \xi' - (\hat{\mathbf{b}} \cdot \xi')\,\hat{\mathbf{b}}$$
$$\hat{\mathbf{k}}_{\parallel} = (\hat{\mathbf{b}} \cdot \hat{\mathbf{k}})\,\hat{\mathbf{b}} \qquad \text{and} \qquad \hat{\mathbf{k}}_{\perp} = \hat{\mathbf{k}} - (\hat{\mathbf{b}} \cdot \hat{\mathbf{k}})\,\hat{\mathbf{b}} \tag{10.36}$$

Therefore,

$$\hat{\mathbf{k}}_{\parallel} \cdot \xi'_{\parallel} = (\hat{\mathbf{b}} \cdot \hat{\mathbf{k}})\,(\hat{\mathbf{b}} \cdot \xi') \qquad \text{and} \qquad \hat{\mathbf{k}}_{\perp} \cdot \xi'_{\perp} = (\hat{\mathbf{k}} \cdot \xi') - (\hat{\mathbf{b}} \cdot \hat{\mathbf{k}})\,(\hat{\mathbf{b}} \cdot \xi') \tag{10.37}$$

The difference between the fast and slow waves is most clearly seen by comparing the phase difference between the longitudinal (parallel to the direction of propagation) and transverse (perpendicular to the propagation) components of the displacement. The longitudinal and transverse components are given by $\hat{\mathbf{k}}_{\parallel}$ and $\xi'_{\parallel}$ for the longitudinal, and $\hat{\mathbf{k}}_{\perp}$ and $\xi'_{\perp}$ for transverse. From Eq. (10.27a), it follows that

$$\frac{\hat{\mathbf{k}}_{\perp}\cdot\hat{\xi}'_{\perp}}{\hat{\mathbf{k}}_{\parallel}\cdot\hat{\xi}'_{\parallel}} = \frac{\sin\theta}{\cos\theta}\frac{\xi'_x}{\xi'_z} = -\frac{[V^2/V_A^2 - \alpha]}{[V^2/{V_A}^2 - (1+\alpha)]} = \frac{(1-\alpha)\pm\sqrt{[(1+\alpha)^2 - 4\alpha\cos^2\theta]}}{(1+\alpha)\pm\sqrt{[(1+\alpha)^2 - 4\alpha\cos^2\theta]}} \tag{10.38}$$

This ratio is positive for a fast wave, and negative for a slow.

The major difference between these waves is therefore that in fast waves, the longitudinal and transverse displacements are in phase, and in slow waves, they are out of phase. Given an arbitrary initial condition, the displacement and its velocity will fix the relative phases of the initial fast and slow components amplitudes.

It is instructive to examine the pressure components driving the wave. The gas pressure

$$p'_G = -\gamma p_0 \nabla\cdot\xi = {V_s}^2\rho' \tag{10.39}$$

and the magnetic pressure

$$\begin{aligned} p'_M &= \frac{1}{\mu_0}\mathbf{B}_0\cdot\mathbf{B}' = \imath\frac{{\mathbf{B}_0}^2}{\mu_0}k\,[(\hat{\mathbf{b}}\cdot\hat{\mathbf{k}})(\hat{\mathbf{b}}\cdot\xi') - (\hat{\mathbf{k}}\cdot\xi')] \\ &= \imath\frac{{\mathbf{B}_0}^2}{\mu_0}k\,[(-V^2/{V_A}^2)+\alpha](\hat{\mathbf{k}}\cdot\xi') = (V^2 - {V_s}^2)\rho' \end{aligned} \tag{10.40}$$

using Eq. (10.27a). For the fast wave $V_+ > V_s$, whereas for the slow $V_- < V_s$ so that the kinetic and magnetic pressure modulations are in phase for the fast wave, but in anti-phase for the slow.

In plasma with small β, i.e. $p_M \gg p_S$ and $V_A \gg V_S$

i. *Fast wave*: It is purely compressional and moves with the Alfvén speed – compressional Alfvén wave (see Section 11.3.2). The wave propagation results from the longitudinal compression of the magnetic field.
ii. *Slow wave*: It is purely compressional with velocity $V_- \approx \sqrt{V_S}\,\cos\theta$. Its group velocity is the sound velocity V_S directed along the field (cf. Figure 9.1 – pure acoustic wave). The wave propagation is due to the compression of the plasma.

The *wave front* or *ray surface* generated by a magneto-sonic wave is complicated in an anisotropic medium. As shown in Section 5.5, in a medium with axial symmetry, as is the case here due to the magnetic field, the wave front is co-incidental with the propagation of the wave group. Since the magneto-sonic waves are dispersionless $V(\not{\omega}, \theta)$, the wavefront is given by Eq. (5.74). Figure 10.2 shows the wave fronts for fast, slow and Alfvén waves for different values of the parameter α generated by an isotropic point source at the origin.

- The behaviour of the fast wave front takes the form of an oblate spheroid, symmetric about the field line, reflecting the larger phase velocity perpendicular to the field.
- Since the Alfvén wave propagates along the field lines, the wave front cannot extend transversely. It must, therefore, take the form of two points, forward and backward, along the magnetic field line through the source.
- The slow wave similarly is propagated preferentially along the field. The wave front reflects this behaviour being confined to a narrow cone symmetric about the field line. The origin of the cusp is identified with a point of inflexion in the plot of $V(\theta)$ (Thompson, 1962, p. 86). It occurs at decreasing values of the angle θ as the parameter α increases, until $\alpha = 1$, when the cusp occurs at (1.0, −0.5) and $\theta = 0$.

The equations of ideal compressible MHD (8.7a), (8.13), (8.32), and (9.6) form a set of hyperbolic equations. Consequently, as in compressible hydrodynamics, the general solution in the absence

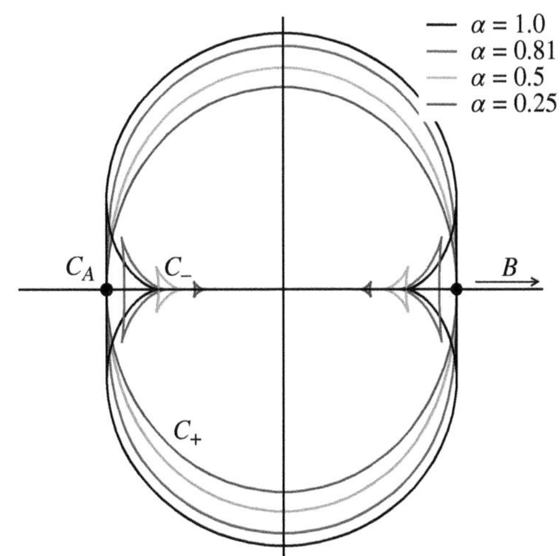

Figure 10.2 Polar plot of the wave fronts for the fast, slow and Alfvén waves for the cases $\alpha = 0.25,\ 0.5,\ 0.81,\ 1.0$.

of discontinuities may be found in terms of characteristics. However, in this case, there are three distinct families of waves, fast, Alfvén and slow, each of which will form a set of characteristics. The structure of the characteristics for each wave may be inferred from their wave front. In hydrodynamics, they are established by isotropic sound waves, which form a spherical wave front from a point source, the characteristics being the rays of the outgoing sound waves. In MHD, we may apply a similar construction, the wave fronts of the three families of characteristics are the wave fronts of the corresponding magneto-sonic waves, and the characteristics the corresponding rays. A full discussion of this behaviour is given by Grad (1959).

10.3 Discontinuities in Fluid Mechanics

10.3.1 Classical Fluids

In classical fluid mechanics, three different types of discontinuity may be identified:

1. *Contact surface*: A discontinuity involving a change of the material constituents across the interface. The density and the concomitant state variables are no longer continuous. The normal component of velocity is zero in the rest frame of the interface, the surface moving with the speed of the flow in the laboratory frame. Contact surfaces are stable in ideal flow, but are subject to broadening due to diffusion and thermal conduction when dissipation is included.
2. *Tangential discontinuity*: At a tangential discontinuity in ideal (dissipationless) flow, one layer of fluid slips over a second without any frictional restraint. The determining boundary conditions across the interface are the continuity of the pressure and the perpendicular component of the velocity; the parallel component of the velocity is discontinuous. Viscosity broadens tangential discontinuities in non-ideal flow. Tangential discontinuities are unstable to the Kelvin–Helmholtz instability, which normally leads to the rapid growth of turbulence.
3. *Shock waves*: The dynamics of compressible fluids has discontinuities, where a large amplitude compression steepens into a shock. Since the sound speed[3] in an adiabatic flow:

$$c^2 = \gamma p/\rho \quad \text{and} \quad p \propto \rho^{(1/\gamma)} \quad \text{therefore} \quad c^2 \propto \rho^{-(\gamma-1)/\gamma} \tag{10.41}$$

3 We use c and V_s interchangeably to denote the sound speed in gas dynamic and magneto-hydrodynamic flows, respectively.

Very simply as the density increases, the sound speed increases. When the flow velocity of the gas exceeds the sound speed, 'pressure' waves from the back build up to generate a sharply steepening front known as a *shock wave*. These 'pressure' waves are the characteristics of the flow and are associated with classical sound waves in the fluid. Across the shock wave only the component of velocity transverse along the shock front is continuous, all the remaining variables change discontinuously in the manner described by the well-known Rankine–Hugoniot relations.

10.3.2 Discontinuities in Magneto-hydrodynamic Fluids

In the more general case of discontinuities in MHD flows, the situation is more complex in that there are now three distinct families of waves, the *fast* and *slow* magneto-sonic waves, and *Alfvén* waves. Each will generate a family of characteristics, moving with speeds which increase non-linearly with the compression. By analogy with standard fluid mechanics, we may therefore anticipate a complex series of shock waves whose form depends on the relationship of the fluid velocity to the appropriate magneto-sonic sound speeds. Thus, we identify four conditions:

1-state	superfast	$u > V_+$
2-state	subfast	$V_+ > u > V_A$
3-state	superslow	$V_A > u > V_-$
4-state	subslow	$u < V_-$

where u is the appropriate upstream flow velocity.
Fast shocks connect super-Alfvén flows.
Slow shocks connect sub-Alfvén flows.
Intermediate shocks connect super-Alfvén flows to sub-Alfvén.

Shocks can be categorised by the transition from an i state to a j state ($i \mapsto j$) as defined above. It can be shown that the condition that the requirement that entropy increases across the shock requires that $i < j$ (Anderson, 1963; Liberman and Velikhovich, 1986). However, as we shall show further more stringent conditions (Section 10.8) limit the existence of planar shocks to the cases $1 \mapsto 2$ (*fast shock*) and $3 \mapsto 4$ (*slow shock*).

The solution to the shock equations is unique if one state (either upstream or downstream) and the type of the other are specified (Liberman and Velikhovich, 1986). There are no more than four possible solutions of the governing Rankine–Hugoniot equations allowed, one of which must be the initial state and the remaining three (when they exist) potential final states. Each solution is of a distinct type and represents a possible shock transitions provided it conforms to the entropy condition and to the evolutionary requirement (Section 10.8).

Some special, singular cases exist where one state lies at the boundary between two types:

- *Rotational discontinuity* ($2 = 3 \mapsto 2 = 3$) discussed later corresponds to flipping the magnetic field (Eq. (10.B.1)).
- *Switch-on shocks* ($1 \mapsto 2 = 3$) where the upstream flow is aligned with the shock normal and the downstream Alfvénic. Switch-on shocks are the limiting case of $1 \mapsto 2$ fast shocks and $1 \mapsto 3$ intermediate shocks.
- *Switch-off shocks* ($2 = 3 \mapsto 4$), where the upstream flow is Alfvénic and the downstream is aligned with the shock normal. Switch-off shocks are the limiting case of $2 \mapsto 3$ intermediate shocks and $2 \mapsto 4$ slow shocks.

10.4 The Rankine–Hugoniot Relations for MHD Flows

The discontinuous flow variables must satisfy a set of jump conditions in the rest frame of a planar stationary discontinuity. Define a set of co-ordinates: x perpendicular to the plane of the shock, and y and z along its surface, with velocity components (u, v, w), respectively. Assuming the local region of rapidly varying flow is infinitesimally thin, it can be represented by a discontinuity. Outside the discontinuity we assume the flow is dissipationless, i.e. ideal, and follow the equivalent problem in gas dynamics, which leads to the Rankine–Hugoniot equations. In the rest frame of the discontinuity, the flow through it must satisfy the conservation laws of MHD derived in Section 8.3.

Let the upstream variables be denoted by subscript 1 and the downstream by 2, and matching the flux of mass, momentum, energy, and magnetic flux through the discontinuity:

- *Conservation of mass*:

$$\rho_1 u_1 = \rho_2 u_2 \tag{A}$$

- *Conservation of momentum normal to the discontinuity*:

$$p_1 + \rho_1 {u_1}^2 + \frac{(B_{y1}^2 + B_{z1}^2 - B_{x1}^2)}{2\,\mu_0} = p_2 + \rho_2 {u_2}^2 + \frac{(B_{y2}^2 + B_{z2}^2 - B_{x2}^2)}{2\,\mu_0} \tag{B}$$

- *Conservation of momentum transverse to the discontinuity*:

$$\left.\begin{aligned} \rho_1 u_1 v_1 - \frac{B_{x1}B_{y1}}{\mu_0} &= \rho_2\, u_2 v_2 - \frac{B_{x2}B_{y2}}{\mu_0} \qquad y\text{-direction} \\ \rho_1 u_1 w_1 - \frac{B_{x1}B_{z1}}{\mu_0} &= \rho_2 u_2 w_2 - \frac{B_{x2}B_{z2}}{\mu_0} \qquad z\text{-direction} \end{aligned}\right\} \tag{C}$$

- *Conservation of energy*:

$$\rho_1 u_1 \left[\hbar_1 + \frac{1}{2}(u_1^2 + v_1^2 + w_1^2)\right] + \left.\frac{\mathbf{E}_1 \wedge \mathbf{B}_1}{\mu_0}\right|_x = \rho_2 u_2 \left[\hbar_2 + \frac{1}{2}(u_2^2 + v_2^2 + w_2^2)\right] + \left.\frac{\mathbf{E}_2 \wedge \mathbf{B}_2}{\mu_0}\right|_x \tag{D}$$

  where the enthalpy per unit mass, or specific enthalpy $\hbar = \epsilon + p/\rho$.
- *Magnetic field continuity*: Since the system is steady state, Maxwell's equations give

$$\nabla \wedge \mathbf{E} = 0 \qquad \text{and} \qquad \nabla \cdot \mathbf{B} = 0$$

  Integrating through the discontinuity, we obtain the result that the transverse electric and normal magnetic fields are both continuous across the discontinuity:

$$\mathbf{E}_{t_1} = \mathbf{E}_{t_2} \qquad \text{and} \qquad B_{x_1} = B_{x_2} \tag{E}$$

  Assuming as above that the flow outside the discontinuity, both upstream and downstream, is ideal MHD:

$$\mathbf{E} + \mathbf{v} \wedge \mathbf{B} = 0$$

  the transverse magnetic field conditions become

$$\left.\begin{aligned} u_1 B_{y1} - v_1 B_{x1} &= u_2 B_{y2} - v_2 B_{x2} \\ u_1 B_{z1} - w_1 B_{x1} &= u_2 B_{z2} - w_2 B_{x2} \end{aligned}\right\} \tag{F}$$

It is easily shown that the Rankine–Hugoniot conditions are consistent with a frame transformation due to an arbitrary uniform velocity parallel to the front $(0, v', w')$, i.e.

$$u \to u \quad v \to v + v' \quad w \to w + w'$$

The Rankine–Hugoniot transition $\underset{RH}{\mapsto}$ has some important if obvious properties:

- *Reflexive*: $A \underset{RH}{\mapsto} A$. Every state is connected to itself.
- *Symmetric*: if $A \underset{RH}{\mapsto} B \Longrightarrow B \underset{RH}{\mapsto} A$. If state A is connected to B, then state B is connected to A. Of course, entropy considerations do not allow such reversibility in practice.
- *Transitive*: $A \underset{RH}{\mapsto} B \wedge B \underset{RH}{\mapsto} C \Longrightarrow A \underset{RH}{\mapsto} C$. If state A is connected to B and B to C, then state A is connected to C.

10.5 Discontinuities in MHD Flows

1. *Contact surface*: If the mass flux $j\rho_1\, u_1 = 0$, but the density is discontinuous $\rho_1 \neq \rho_2$ across the surface, the velocity normal to the surface, then $u_1 = u_2 = 0$ is zero. The fluid moves parallel to the discontinuity. If the normal field $B_x \neq 0$ is non-zero, it follows from Eqs. (A)–(F) that the velocity, pressure, and magnetic field are continuous across the surface, but the density, temperature, and entropy may be different across the surface. A contact surface therefore corresponds to the interface between two different materials.
2. *Tangential discontinuity*: If the mass flux $j = 0$ and the normal field $B_x = 0$ are both zero, and the density is discontinuous $\rho_1 \neq \rho_2$ across the surface, $u_1 = u_2 = 0$ and the fluid moves parallel to the discontinuity. It follows from Eqs. (A)–(F) in this case that neither the tangential velocity $\mathbf{v}_t$ nor the tangential magnetic field $\mathbf{B}_t$[4] need be continuous. The pressure discontinuity and the tangential field discontinuity are related by Eq. (B),

   $$p_1 + {B_{t1}}^2/2\mu_0 = p_2 + {B_{t2}}^2/2\mu_0.$$

   In contrast to the tangential discontinuity in hydrodynamics, which is unstable due to the Kelvin–Helmholtz instability, the tangential magnetic field introduces stiffness into the flow and inhibits the instability (Landau and Lifshitz, 1984, p. 243).
3. *Rotational or Alfvén discontinuity*: If the density $\rho_1 = \rho_2$ is equal across the discontinuity, the normal velocity $u_1 = u_2$ is also continuous. From Eqs. (C) and (F), respectively, we obtain

   $$\mathbf{v}_{t2} - \mathbf{v}_{t1} = B_x\,(\mathbf{B}_{t2} - \mathbf{B}_{t1})/(\mu_0\,\rho\,u) \quad \text{and} \quad \mathbf{v}_{t2} - \mathbf{v}_{t1} = u\,(\mathbf{B}_{t2} - \mathbf{B}_{t1})/B_x$$

   Dividing these two equations:

   $$u = B_x/\sqrt{\mu_0\,\rho}$$

   and multiplying them together

   $$\mathbf{v}_{t2} - \mathbf{v}_{t1} = (\mathbf{B}_{t2} - \mathbf{B}_{t1})/\sqrt{\mu_0 \rho}$$

   Equation (D) may be written in terms of the internal energy

   $$\epsilon_2 + \frac{(p_2 + {\mathbf{B}_{t2}}^2/2\mu_0)}{\rho} + \frac{1}{2}\left(\mathbf{v}_2 - \frac{\mathbf{B}_{t2}}{\sqrt{\mu_0\,\rho}}\right)^2 = \epsilon_1 + \frac{(p_1 + {\mathbf{B}_{t1}}^2/2\mu_0)}{\rho} + \frac{1}{2}\left(\mathbf{v}_{t1} - \frac{\mathbf{B}_{t1}}{\sqrt{\mu_0\,\rho}}\right)^2$$

   making use of Eq. (B) to cancel terms as indicated.

4 We use the subscript t to indicate vector components lying in the plane of the discontinuity (y, z).

Thus, the internal energy is continuous, as is the density. Therefore, all the thermo-dynamic state variables are continuous. In particular, the pressure and therefore, from Eq. (B) the magnitude of the transverse magnetic field must be continuous.

$$p_1 = p_2 \qquad \text{and} \qquad B_{x1} = B_{x2}$$

The direction of the magnetic field may, however, be discontinuous across the surface, i.e. the transverse field is rotated through the surface. If we transform the frame by means of a velocity shift along the surface so that

$$\mathbf{v}_{t1} - \mathbf{B}_{t1}/\mu_0\,\rho = \mathbf{v}_{t2} - \mathbf{B}_{t2}/\mu_0\,\rho$$

the transverse velocity and the magnetic field changes become parallel. The velocity is then rotated through the surface by the same angle as the field. The change in the magnetic field is due to a thin current sheet flowing in the front.

In the rest frame of the front, the velocity of the fluid normal to the front, i.e. flowing through the surface is equal to the Alfvén velocity determined by the normal component of the magnetic field $B_x/\sqrt{\mu_0\,\rho}$. If the upstream plasma is at rest, the front moves with the Alfvén speed through it.

4. *Shock waves*: In a similar manner to classical hydrodynamics, large amplitude magneto-sonic waves can develop into shock waves. However, the presence of the magnetic field greatly increases the complexity of the relatively simple structures found in gas dynamics. In particular, the stability and uniqueness of the wave is no longer guaranteed. As we have seen, the wave is no longer only longitudinal, and as a consequence the tangential symmetry of gas dynamics is lost. Thus, we can get flows along the surface of the shock, which are not continuous through the surface.

10.6 MHD Shock Waves

The general solution of the shock discontinuity conditions from the set of Rankine–Hugoniot equations (A)–(F) from a known upstream flow, to calculate the downstream state is extremely complex involving eight unknowns, three velocity components, three magnetic fields components and two plasma state variables.

10.6.1 Simplifying Frame Transformations

Fortunately, the number of variables may be reduced by two transformations in the shock plane.

1. It follows from Eqs. (C) and (F) that provided the normal component of the magnetic field is non-zero $B_x \neq 0$, the tangential magnetic field differences $(\mathbf{B}_{t2} - \mathbf{B}_{t1})$ and $(\mathbf{B}_{t2}/\rho_2 - \mathbf{B}_{t1}/\rho_1)$ are both parallel to $(\mathbf{v}_{t2} - \mathbf{v}_{t1})$ and therefore to each other. Consequently, since $V_2 \neq V_1$, it follows that $\mathbf{B}_{t2} \parallel \mathbf{B}_1$. As a result, we may simplify the equations by a rotation of the transverse (y, z) plane about the normal to the shock to transform the co-ordinate system so that the magnetic field components lie in the (x, y) plane only.

   After the rotation $v_{z2} = v_{z1}$ so that the z component of the velocity is then constant through the transition and may be eliminated by a velocity shift in the z direction. The resultant velocity vectors then lie in the same plane as the field, i.e. the vectors $\mathbf{B}_1$ and $\mathbf{B}_2$ and $\mathbf{v}_1$ and $\mathbf{v}_2$ are all coplanar with the normal to the shock.

If the normal component of the magnetic field ($B_x = 0$) is zero, it follows from Eq. (C) that $\mathbf{v}_{t1} = \mathbf{v}_{t2}$. By a suitable velocity shift, the plasma may be made to move perpendicularly to the front only, i.e. $\mathbf{v}_t = 0$. A simple rotation about the axis then allows the magnetic field to lie in the y direction only, thereby recovering the above result.

2. Provided $B_x \neq 0$, the transverse electric field may be zeroed by a velocity change in the direction y in the shock plane $v \to v - uB_y/B_x$. Following the transformation to this *de Hoffmann–Teller frame*, the velocity and magnetic field are parallel to each other on both sides of the shock:

$$\mathbf{v}_1 \wedge \mathbf{B}_1 = u_1 B_{y_1} - v_1 B_x = u_2 B_{y_2} - v_2 B_x = \mathbf{v}_2 \wedge \mathbf{B}_2 = 0$$

There is an alternative transformation in which the transverse momentum flux, instead of the electric field, is zeroed, but this is infrequently used.

Transforming on to the (x, y) plane, the Rankine–Hugoniot equations simplify to

$$\rho_1\, u_1 = \rho_2\, u_2 \tag{A$'$}$$

$$p_1 + \rho_1\, {u_1}^2 + \frac{B_{y1}^2}{2\,\mu_0} = p_2 + \rho_2\, {u_2}^2 + \frac{B_{y2}^2}{2\,\mu_0} \tag{B$'$}$$

$$\rho_1\, u_1\, v_1 - \frac{B_x B_{y1}}{\mu_0} = \rho_2\, u_2\, v_2 - \frac{B_x B_{y2}}{\mu_0} \tag{C$'$}$$

$$\begin{aligned} &\rho_1\, u_1 \left[\hbar_1 + \tfrac{1}{2}\,(u_1^2 + v_1^2) \right] + \tfrac{u_1}{\mu_0} B_{y1}^2 - \tfrac{v_1}{\mu_0} B_x\, B_{y1} \\ &= \rho_2\, u_2 \left[\hbar_2 + \tfrac{1}{2}\,(u_2^2 + v_2^2) \right] + \tfrac{u_2}{\mu_0} B_{y2}^2 - \tfrac{v_2}{\mu_0} B_x\, B_{y2} \end{aligned} \tag{D$'$}$$

$$u_1\, B_{y_1} - v_1\, B_x = u_2\, B_{y_2} - v_2\, B_x = 0 \text{ (in the de Hoffmann–Teller frame).} \tag{F$'$}$$

10.7 Properties of MHD Shocks

MHD is described by a set of hyperbolic differential equations. As we have seen, these admit three distinct types of wave. In general, each set of waves supports a family of characteristics mapped by the wave fronts discussed in Section 10.2 and Section 5.5. Consequently, in any given situation, there exist three families of shock waves. Many of the characteristic features of shocks in hydrodynamics can be carried out with modification into MHD, which we examine.

Perhaps the most important similarity is that the entropy increases in a compressive shock provided the term $\partial^2 p/\partial V^2|_s > 0$ as is the case for nearly all materials (Lifshitz and Pitaevskii, 1981). Magneto-hydrodynamic shocks therefore exist in compression only.

10.7.1 Shock Hugoniot

Introducing the flux $j = \rho_1\, u_1 = \rho_2\, u_2$ and the specific volume $V = 1/\rho$, Eq. (B') takes the form, which may be directly compared with the equivalent relation for gases when $\mathbf{B} = 0$:

$$j^2 = \frac{[p_2 - p_1 + ({B_{y2}}^2 - {B_{y1}}^2)/2\mu_0]}{[V_1 - V_2]} \tag{B$''$}$$

From Eqs. (C') and (F') it follows that

$$j^2(V_2\, B_{y2} - V_1\, B_{y1}) = {B_x}^2\, (B_{y2} - B_{y1})/\mu_0 \tag{F''}$$

Equation (D') may be re-written as

$$h_2 - \hbar_1 + \tfrac{1}{2}j^2\ ({V_2}^2 - {V_1}^2) + \tfrac{1}{2}\left[\left(\upsilon_2 - \tfrac{B_x}{\mu_0 j}B_{y2}\right)^2 - \left(\upsilon_1 - \tfrac{B_x}{\mu_0 j}B_{y1}\right)^2\right]$$
$$+ (V_2{B_{y_2}}^2 - V_1{B_{y_1}}^2)/\mu_0 - {B_x}^2\,({B_{y2}}^2 - {B_{y1}}^2)/(2\,{\mu_0}^2\,j^2) = 0 \tag{D''}$$

Substituting from Eqs. (B") and (F") Eq. (D") becomes

$$h_2 - \hbar_1 \quad -\frac{1}{2}\,(V_1 + V_2)\,(p_2 - p_1) + (V_2 - V_1)\,(B_{y2} - B_{y1})^2/4\mu_0 = 0$$
$$\epsilon_2 - \epsilon_1 \quad -\frac{1}{2}\,(V_1 - V_2)\,(p_2 + p_1) + (V_2 - V_1)\,(B_{y2} - B_{y1})^2/4\mu_0 = 0 \tag{G''}$$

which are the equations of the shock adiabat modified to include the magnetic pressure, where $\hbar$ is the specific enthalpy of the plasma.

Equation (C') may be re-written as

$$\upsilon_2 - \upsilon_1 = B_x(B_{y2} - B_{y1})/\mu_0\, j \tag{C''}$$

Equations (B"), (C"), (F"), and (G") provide a complete set of equations for the solution of the behaviour of magneto-hydrodynamic shock waves. Equation (G") is the MHD generalisation of the usual Hugoniot curve from fluid mechanics, since it follows from Eq. (9.18) that the specific enthalpy of the plasma $\hbar = \gamma p/(\gamma - 1)\rho$.

In the case of a perpendicular shock, when the normal; field $B_x = 0$ the hydrodynamic form is closely reproduced as

$$j^2 = (p_2^* - p_1^*)/(V_1 - V_2) \quad \epsilon_2^* = \epsilon_1^* + \frac{1}{2}(p_2^* + p_1^*)(V_1 - V_2) \tag{10.42}$$

where $p^* = p + B^2/2\mu_0$ and $\epsilon^* = \epsilon + VB^2/2\mu_0$, where the magnetic field and energy are included with the kinetic pressure and energy.[5]

10.7.2 Shock Adiabat – General Solution for a Polytropic Gas

The polytropic equation of state relations for the sound speed, internal energy, and specific enthalpy, and the Alfvén speed defined with respect to the normal component of the magnetic field B_x are, respectively,

$$c^2 = \frac{\gamma\, p}{\rho} \qquad \epsilon = \frac{1}{(\gamma - 1)}\frac{p}{\rho} \qquad \hbar = \frac{\gamma}{(\gamma - 1)}\frac{p}{\rho} = \gamma\,\epsilon = \frac{1}{(\gamma - 1)}\,c^2 \qquad {V_A'}^2 = \frac{{B_x}^2}{\mu_0\,\rho} \tag{10.43}$$

We may obtain two independent expressions for the pressure ratio across the shock. First, by substituting the polytropic form for the enthalpy and internal energy and summing the Eq. (G"), and second, from Eq. (B")

5 Note that this form satisfies the thermodynamic relation $p^* = -\partial\epsilon^*/\partial V$.

$$\frac{p_2}{p_1} = \frac{[(\gamma+1)\,V_1 - (\gamma-1)\,V_2] - (\gamma-1)}{[(\gamma+1)\,V_2 - (\gamma-1)\,V_1]} - \frac{(V_1 - V_2)\,(B_{y_1} - B_{y_2})^2}{[(\gamma+1)\,V_2 - (\gamma-1)\,V_1]}$$
$$= 1 + j^2 \frac{(V_1 - V_2)}{p_1} - \frac{(B_{y_1}{}^2 - B_{y_2}{}^2)}{2\mu_0\, p_1} \tag{10.44}$$

Introducing a set of dimensionless variables $u = u_2/u_1 = V_2/V_1 = \rho_1/\rho_2$ is the reciprocal compression ratio, $b = B_{y2}/B_x$ the dimensionless downstream transverse field, $b_0 = B_{y1}/B_x$ the upstream dimensionless transverse field so that $\theta = \arctan b_0$ is the upstream field angle, and $M_s = u_1/c_1$ the upstream Mach number and $M'_A = u_1/C'_{A1}$ the upstream Alfvén Mach number. Substituting we obtain an expression, which must be obeyed by all solutions of the Rankine–Hugoniot equations

$$\frac{(\gamma+1)}{(\gamma-1)}\{u - u_{GD}\}(u-1) - \frac{1}{2}\frac{(u-1)(b-b_0)^2}{M'^2_A} + \frac{1}{2}\left\{\frac{(\gamma+1)}{(\gamma-1)}u - 1\right\}\frac{(b^2 - b_0{}^2)}{M'^2_A} = 0 \tag{10.45}$$

where $u_{GD} = [(\gamma-1)M_s{}^2 + 2]\big/(\gamma+1)M_s{}^2$ is the downstream velocity behind a gas dynamic shock with the same Mach number M_s.

The solutions must also satisfy a second condition on u and b, namely from Eq. (F'')

$$b\,(u - M'_{A1}{}^{-2}) = b_0\,(1 - M'_{A1}{}^{-2}) \tag{10.46}$$

since the normal field B_x is constant through the shock, the Alfvén Mach number scales as $M'_A = M'_{A1}\sqrt{u}$ and M'_{A1} is the Alfvén Mach number of the incoming flow. The hyperbola (10.46) has two branches: on the upper branch $u > 1/M'_{A1}{}^2$ the incoming flow is super-Alfvénic, and on the lower $u < 1/M'_{A1}{}^2$ sub-Alfvénic.

A solution to the Rankine–Hugoniot equation is obtained at the intersection of the curve (10.45) with the rectangular hyperbola (10.46), which expresses the condition of ideal MHD upstream and downstream of the shock. Figure 10.3 shows a typical plot of the two curves. It can be seen that there are only four possible intersections of the two curves, one of which corresponds to the upstream flow state. The remaining three intersections, should they all exist, represent corresponding solutions of the Rankine–Hugoniot relations, two are super-Alfvénic and two sub-Alfvénic. However, many of the possible transitions are not allowed. The points are numbered in the order of decreasing values of u: ①, ②, ③, ④ as shown. Since the root $u = 1$ must exist, being the initial condition, another root of the same type also exists. Clearly, from Figure 10.3 points ① and ② are super-Alfvénic ($M'_A > 1$) and ③ and ④ sub-Alfvénic ($M'_A < 1$).

The four roots are characterised by their sonic character with respect to fast and slow waves. The Mach number of the incoming flow with respect to fast or slow sonic waves is $M_\pm = u_1/C_\pm$. Since $C'_A = V_A\ \cos\theta$, the sonic waves can be described by Eq. (10.30), which can be re-written as

$$(V_\pm{}^2 - V_s^2)\,(V_\pm{}^2 - V'^2_A) = V_\pm{}^2\, V'^2_A\, b^2 \tag{10.47}$$

Dividing through by u_1, we obtain a relation for the magneto-sonic Mach numbers:

$$(M_\pm{}^{-2} - M_s{}^{-2})\,(M_\pm{}^{-2} - M'^{-2}_A) = M_\pm{}^{-2}\, M'^{-2}_A\, b^2 \tag{10.48}$$

The plane of (M_A, M_s) may be divided into supersonic and subsonic parts by the lines $M_\pm = 1$, respectively, such that if

$$M'^2_A > 1 + \frac{M_s^2\, b^2}{(M_s^2 - 1)} \tag{10.49a}$$

or more appropriately

$$(M'^2_A - 1)\,(M_s^2 - 1)\big/M_s^2\, b^2 > 1 \tag{10.49b}$$

the flow is supersonic with respect to the appropriate magneto-sonic wave.

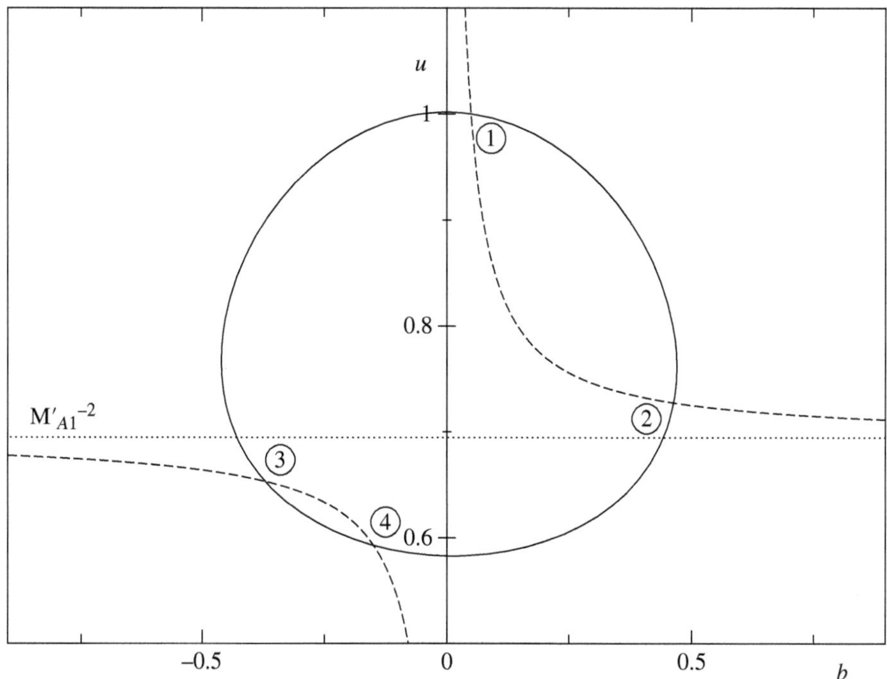

Figure 10.3 Plots of the relations for u vs. b from Eqs. (10.45) and (10.46) showing the allowed transition points and the asymptote ${M'_{A1}}^2$. The plots show the case when the initial flow $u = 1, \ldots$ is taken as the solution point ①.

To identify the condition of the four solutions, we note that each of points ①, ②, … represent a possible upstream condition of the Rankine–Hugoniot equations, the remaining three being possible downstream states. As we move through each case, the curves are adjusted to set the corresponding value of $u = 1$, the values of M_{s1}, M'_{A1} and b_1 being adjusted correspondingly. In particular for intersections ③ and ④, $M'_{A1} < 1$. Consider the gradients of the two curves at the initial point $u_1 = 1$ as it is moved successively around the four intersection points. The gradient of the line, u_E (10.45), and the hyperbola, u_H (10.46), at $u = 1$ are, respectively,

$$\dot{u}_E = -\frac{b_1\, {M_{s1}}^2}{{M'_{A1}}^2\, ({M_{s1}}^2 - 1)} \quad \text{and} \quad \dot{u}_H = -\frac{({M'_{A1}}^2 - 1)}{b_1\, {M'_{A1}}^2} \tag{10.50}$$

Hence,

$$\frac{({M'_A}^2 - 1)\, ({M_s}^2 - 1)}{{M_s}^2\, b^2} = \frac{\dot{u}_E}{\dot{u}_H} \tag{10.51}$$

Referring to Figure 10.3 we have

$$\begin{array}{llrl} ① & \dot{u}_E > \dot{u}_H & & u_1 > V_+ \\ ② & \dot{u}_E < \dot{u}_H & V'_A < & u_1 < V_+ \\ ③ & \dot{u}_E < \dot{u}_H & V'_A > & u_1 > V_+ \\ ④ & \dot{u}_E > \dot{u}_H & & u_1 < V_- \end{array} \tag{10.52}$$

which is the condition given earlier, Section 10.3.1. This result admits the following important conclusion that if the shock type and one state are known, the shock is unique.

If the 'ellipse' u_E touches the hyperbolic line u_H reducing the contact on either the super-Alfvénic or sub-Alfvénic branch to a single point, $\dot{u}_E = \dot{u}_H$, Eq. (10.48) reduces to (10.51) for $M_\pm = 1$, i.e. the tangent points are the corresponding sonic points.

A further important result due independently to Germain and Shercliff can be proved after a lengthy analysis (Anderson, 1963, §2.3) namely that the entropy of the state increases as the velocity decreases $S(i) > S(j)$ if >. We note that only shock transitions > involving an increase of entropy are allowed by the second law of thermodynamics.

10.8 Evolutionary Shocks

Thus far, we have not considered either the existence or stability of magneto-hydrodynamic shocks. In fluid hydrodynamics, shocks are only found in compression as a result of the thermo-dynamic requirement that entropy increase through the shock, which is due to the condition $\partial^2 V/\partial p^2|_S > 0$ obeyed by nearly all fluids (Bethe, 1942; Weyl, 1944; Landau and Lifshitz, 1959). However, although this conclusion allows the existence of steady shocks, it does not determine whether the shock will break up into a series of non-steady waves, as a consequence of the *evolutionary* nature (or 'stability' condition), which is most easily demonstrated by considering shocks in hydrodynamic fluids (Lax, 1957; Landau and Lifshitz, 1959).

Consider a planar discontinuity, which is perturbed uniformly across its surface,[6] for example by a weak sound wave incident on it from the downstream side. The disturbance gives rise to a series of outgoing characteristics (pressure/sound waves) propagating upstream and downstream away from the discontinuity, which are required to match the perturbation to the upstream and downstream flows. These waves have velocities $u_1 = v_1 \pm c_1$ and $u_2 = v_2 \pm c_2$ away from the discontinuity, where the subscript 1 refers to flow on the upstream side and 2 to that on the downstream and v and c are the flow velocity and sound speed in the rest frame respectively. Since the waves must be outgoing from the discontinuity we require $u_1 < 0$ for the downstream waves and $u_2 > 0$ for the upstream. The entropy perturbation is propagated with the flow velocity, v_2 downstream. The parameters ('boundary conditions') established by the perturbation in the discontinuity are typically two thermodynamic state parameters and one velocity, and must be matched by those associated with the outgoing waves. Each wave is characterised by a single parameter. Therefore, the number of outgoing waves must exactly match the number of variables associated with the perturbation within the discontinuity. Noting that waves are outgoing if $u_1 < 0$ upstream and $u_1 > 0$ downstream, the total number of outgoing waves (including the entropy wave) depending whether the flow is subsonic or supersonic in each case is

$v_1 > c_1$	$v_2 < c_2$	Number = 3
$v_1 > c_1$	$v_2 > c_2$	Number = 4
$v_1 > c_1$	$v_2 < c_2$	Number = 4
$v_1 < c_1$	$v_2 > c_2$	Number = 5

The 'boundary conditions' are determined by the three jump conditions for mass, momentum, and energy across the discontinuity (Rankine–Hugoniot conditions). Since only three equations, and therefore three parameters, are available, only the case where the upstream flow is supersonic $v_1 > c_1$ and the downstream subsonic $u_2 < c_2$ allow a unique solution. In the remaining cases, the number of variables exceeds the number of conditions, and an infinity of solutions are possible.

---

6 If there is a variation across the surface, the shock may suffer from instability. That is not the case considered here.

In consequence, we may expect that the discontinuity splits into a continuous form, exemplified by a rarefaction discontinuity $u_1 < c_1$ and $u_2 > c_2$, which is forbidden and not found in practice, where a non-steady expansion fan is formed. The other case, where the number of parameters is insufficient to match the conditions, also implies that no 'steady' flow is possible.

Shocks satisfying this condition in which the number of unknown parameters matches the number of boundary conditions are called *evolutionary*. Non-evolutionary discontinuities cannot occur in reality and would rapidly break up if initially formed. The evolutionary condition is not the same as instability which is normally associated with exponential growth terms.

10.8.1 Evolutionary MHD Shock Waves

The situation in MHD is complicated by the fact that there are three different categories of shocks corresponding to the fast, Alfvén and slow magneto-sonic waves. In addition the waves are non-planar, although as we have seen the field and velocity on either side of the discontinuity are co-planar with the normal to a planar shock. We give a brief introduction to this topic based on the account in Landau and Lifshitz (1984, §73). Fuller discussion will be found in Anderson (1963, Ch 3) and Liberman and Velikhovich (1986, § 3.3.5).

As before we consider the perturbation applied to a planar shock in the y, z plane. The velocities and fields associated with the shock are taken to be in the x, y plane. Therefore, perturbations associated with outgoing Alfvén waves lie in the direction z, and those associated with magneto-sonic waves are in the y, z plane. Thus, we have two independent groups of perturbations:

$$\delta B_z,\ \delta v_z$$

with variables transported by Alfvén waves subject to z components of the two Eqs. (C) and (F), and

$$\delta B_y,\ \delta v_x,\ \delta v_y,\ \delta \rho,\ \delta s$$

transported by the magneto-sonic waves with variables subject to four Eqs. (B), (D), and the y components of Eqs. (C) and (F) independent of the Alfvén set and finally, the entropy δs transported by the flow (one equation). Since $\nabla \cdot \mathbf{B} = 0$, it follows that $\partial B_x/\partial x = 0$ and δB_x is constant across the shock.

As before, the waves have upstream and downstream velocities $v_{x1} \pm V_{A1}$ and $v_{x2} \pm V_{A2}$ for the Alfvén waves, $v_{x1} \pm V_{+1}$ and $v_{x2} \pm V_{+2}$ for the fast waves, and $v_{x1} \pm V_{-1}$ and $v_{x2} \pm V_{-2}$ for the slow. An evolutionary shock, therefore, requires two outgoing Alfvén waves, i.e.

$$\begin{array}{lll} v_{x1} > V_{A1} & \text{and} & v_{x2} > V_{A2} \\ v_{x1} < V_{A1} & \text{and} & v_{x2} < V_{A2} \end{array} \tag{10.53}$$

and similarly, for the fast and slow waves. Remembering that $C_+ \geq C_A \geq C_-$ and assembling all these constraints, we obtain

$$\begin{array}{lll} \text{Fast shock waves} & v_{x1} > V_{+1} & V_{+2} > v_{x2} > V_{A2} \\ \text{Slow shock waves} & V_{A1} > v_{x1} > V_{-1} & V_{-2} > v_{x2} \end{array} \tag{10.54}$$

Figure 10.4, taken from Landau and Lifshitz (1984), shows the evolutionary limitations of MHD shocks for the case, where $V_{A1} > V_{-1}$. The vertically hatched areas represent the Alfvén wave condition (10.53) and the horizontally hatched the magneto-sonic waves condition (10.54). The overlap areas are the wave velocities allowed by an evolutionary condition.

The limitations imposed by the evolutionary condition are discussed by Landau and Lifshitz (1984, p. 251), where it is shown that

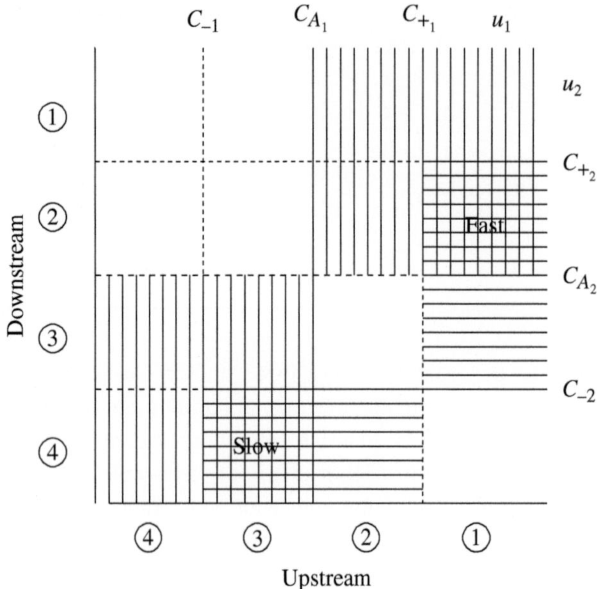

Figure 10.4 Sketch of the evolutionary limitations of MHD shocks, indicating the types of shock transition $1 \mapsto 2$ and $3 \mapsto 4$ allowed by evolution.

- Since the magnetic field increases through a fast shock but weakens through a slow, a fast shock cannot have a transition to slow shock or *vice versa*.
- Because the transverse field must reverse in an Alfvén discontinuity, it cannot transform to a shock.
- It follows from the inequalities (10.53), a fast shock cannot pass into a tangential discontinuity.

Consequently, there can be transitions between a tangential discontinuity and either a contact surface, an Alfvén discontinuity or a slow shock only.

10.8.2 Parallel Shock – Magnetic Field Normal to the Shock Plane

Consider the case where the plane of the shock is perpendicular to both the flow and the magnetic field, i.e.

$$v_1 = 0, \qquad B_{x1} = B_{x2} = B_x \qquad \text{and} \qquad B_{y1} = 0 \tag{10.55}$$

Making use of Eq. (C') or Eq. (F') to calculate the transverse magnetic field behind the shock in turn, we obtain

$$B_{y_2} = \mu_0\, \rho_2\, u_2\, \frac{v_2}{B_x} = \frac{v_2 B_x}{u_2} \tag{10.56}$$

Therefore,

- **either** $v_2 = B_{y2} = 0$, and it follows immediately that Eqs. (B'')–(G'') yield the conditions for a normal gas dynamic shock with the magnetic field unchanged.
- **or** $u_2{}^2 = B_x{}^2/\mu_0\, \rho_2 = V_{A2}{}^2 \Rightarrow B_{y2} \neq v_2 \neq 0$, known as a *switch on shock* in which a transverse field is spontaneously generated. The shock velocity

$$u_1{}^2 = y\; V_{A1}{}^2 \tag{10.57}$$

where $y = \rho_2/\rho_1 = u_1/u_2$ is the shock compression. Switch-on shocks clearly have $u_1 > V_{A1}$ and are therefore fast.

In the context of our analysis of MHD shocks stemming from Figure 10.3, the hyperbola (10.46) reduces to the limiting condition expressed by the pair of straight lines ($b = b_0 = 0$) and ($u = M_A'^{-2}$). The two intercepts ② and ③ are clearly degenerate, since the sign of $b_0 = 0$ is undefined. Switch-on and switch-off shocks, having ② = ③ as end-points, are therefore singular. The gas dynamic shock is the transition ①–④.

Across the parallel shock corresponding to the hydrodynamic solution, since the magnetic field is constant, we have

$$V_{A2} = \sqrt{y}\, V_{A1} = \sqrt{\frac{(\gamma+1)\, u_1^2}{(\gamma-1)\, u_1^2 + 2\, V_{s1}^2}}\, V_{A1} \tag{10.58}$$

$$V_{s2} = \frac{\sqrt{[2\,\gamma\, u_1^2 - (\gamma-1)\, V_{s1}^2][(\gamma-1)\, u_1^2 + 2\, V_{s1}^2]}}{(\gamma+1)\, u_1^2}\, V_{s1} \tag{10.59}$$

$$u_{f_2} = \mathrm{Max}\,(V_{A2}, V_{s2}) \qquad u_{s2} = \mathrm{Min}\,(V_{A2}, V_{s2}) \tag{10.60}$$

$$u_{f_1} = \mathrm{Max}\,(V_{A1}, V_{s1}) \qquad u_{s1} = \mathrm{Min}\,(V_{A1}, V_{s1}) \tag{10.61}$$

Figure 10.5 shows the state of the flow behind the shock wave for the case, where the sound speed is greater than the Alfvén speed ($V_{s1} > V_{A1}$). Behind the shock, the fast wave speed is equal to sound speed ($u_{f_2} - V_{s2}$) and the slow wave speed to the Alfvén speed ($u_{f_2} - V_{s2}$). The shock is clearly always fast and, referring to Figure 10.4, always evolutionary.

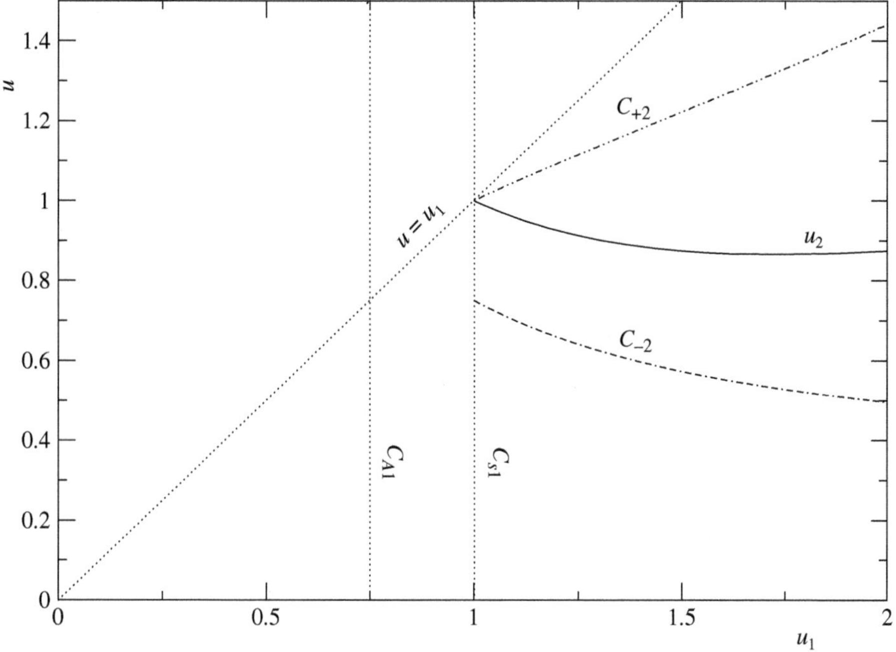

Figure 10.5 Plot of the flow (u_2), and fast (C_{+2}) and slow (C_{-2}) wave speeds behind a parallel MHD shock for the case, where the sound speed exceeds the Alfvén speed, $V_{s1} > V_{A1}$, with values ($V_{s1} = 1$, $V_{A1} = 3/4$, $\gamma = 5/3$).

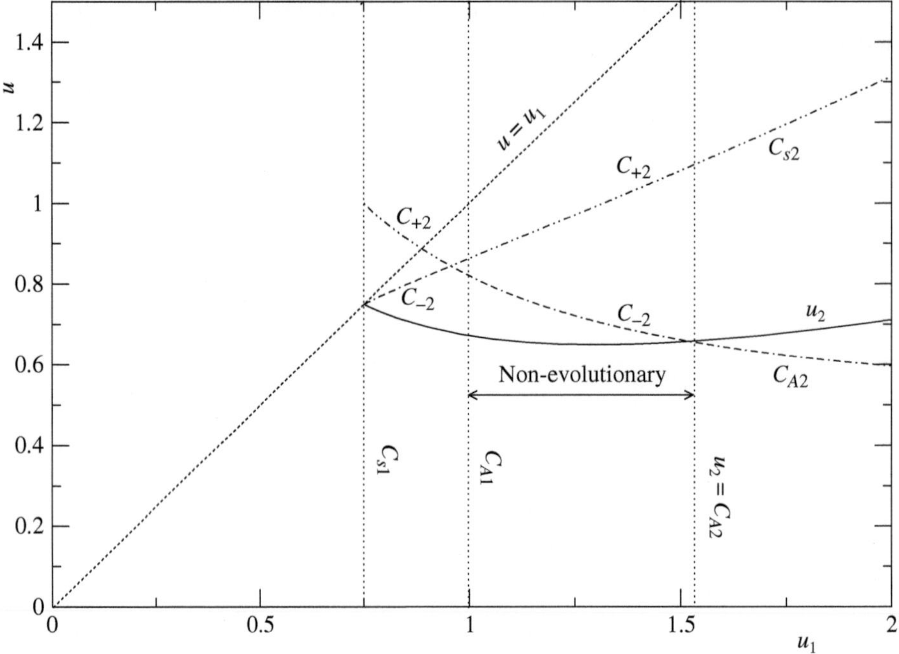

Figure 10.6 Plot of the flow (u_2), and fast (V_{+2}) and slow (V_{-2}) wave speeds behind a parallel MHD shock for the case, where the Alfvén speed exceeds the sound speed, $V_{A1} > V_{s1}$ with values ($V_{A1} = 1,\ V_{s1} = 3/4,\ \gamma = 5/3$).

However, when the Alfvén speed is greater than the sound speed ($V_{A1} > V_{s1}$), the situation is more complex as seen from Figure 10.6. Initially the shock is slow, but becomes fast once the shock speed exceeds the speed of the fast wave in the undisturbed plasma ($u_1 > V_{A1}$). Referring again to Figure 10.4, we see that shock is non-evolutionary over the range between $u_1 = V_{A1}$ and $u_2 = V_{A2}$, the latter point corresponding to $u_1 = \sqrt{\{[(\gamma+1)V_{A1}{}^2 - 2\,V_{s1}{}^2]/(\gamma-1)\}}$. For large values of the shock speed, the shock is evolutionary. This particular case illustrates an important difference between gas dynamic and MHD shocks. The former are always evolutionary, but the latter only over a restricted range, even though the changes in the values across the shock are identical. This is a consequence of the normal magnetic field, which induces Alfvén waves in MHD.

10.9 Switch-on and Switch-off Shocks

From Eq. (10.57), switch-on shocks satisfy the conditions $B_{y_1} = 0,\ u_1\,u_2 = V_{A1}{}^2$ and $V_2 = V_1\,[V_{A1}/u_1]^2$. Eliminating the pressure behind the shock p_2 between Eqs. (F") and (G"), we obtain the magnetic field behind the shock

$$B_{y_2}{}^2 = \frac{2\,\mu_0\,\rho_1}{V_{A1}{}^2}\{u_1{}^2 - V_{A1}{}^2\}\left\{\frac{(\gamma+1)}{2}V_{A1}{}^2 - V_{s1}{}^2 - \frac{(\gamma-1)}{2}u_1{}^2\right\} \tag{10.62}$$

Since $B_{y_2}{}^2 > 0$ it follows that the switch-on shock velocity

$$V_{A1} < u_1 < \sqrt{\{[(\gamma+1)V_{A1}{}^2 - 2V_{s1}{}^2]/(\gamma-1)\}} \tag{10.63}$$

Thus, it is in the restricted non-evolutionary region of 'hydrodynamic' shocks that switch-on shocks may be found. However, since the normal flow velocity and the Alfvén speed behind the shock are equal

$$u_2 = \frac{{V_{A1}}^2}{u_1} \qquad \text{and} \qquad V_{A2} = \sqrt{\frac{u_2}{u_1}}\, V_{A1} = u_2$$

switch-on shocks lie on the evolution/non-evolutionary border on the evolutionary diagram (Figure 10.4). The sound speed behind the shock is similarly calculated:

$${V_{s2}}^2 = \gamma \left\{ {V_{s1}}^2 - \frac{(\gamma-1)}{\gamma} \frac{{V_{s1}}^2 {V_{A1}}^2}{{u_1}^2} + \frac{(\gamma-1)}{2} \frac{({u_1}^2 - {V_{A1}}^2)^2}{{u_1}^2} \right\} \tag{10.64}$$

Switch-off shocks on the other hand propagate at the Alfvén speed into the medium. They are characterised by an initial transverse magnetic field, which is reduced to zero through the shock.

The origin of the switch-on and switch-off shocks can be understood as a result of the fact that parallel shocks propagate along the background magnetic field. MHD identifies two waves namely, magneto-sonic and Alfvén with this behaviour. Alfvén waves also contain transverse field components. As a result, if the wave steepen into a shock, then both sets of waves steepen together. The transition across it may contain transverse field components – *switch-on shock*.

As noted earlier, the Rankine–Hugoniot equations may allow a transition, which although it is itself forbidden by entropy considerations, the symmetric transition is allowed. The inverse transition to the forbidden 'switch-on' discontinuity at the slow/intermediate interface ($4 \mapsto 2 = 3$) is a *switch-off shock*. The physical mechanism by which the field is switched-off is therefore similar to that in switch-on shocks but 'inverted'. It is easily established that since the transverse magnetic field $B_{y_1} \neq 0$ and $B_{y_2} = 0$ that $u_1 = V_{A1}, u_1\, u_2 = V_{A2}$ and $u_2 = \sqrt{(\rho_1 \rho_2)}\, V_{A2}$ in the switch-off shock. Since the compression ratio $y = \rho_2/\rho_1 > 1$, it is clear that switch-off shocks are slow $u_1 < V_{A1}$.

Switch-on and switch-off shocks lie at the interface of evolutionary and non-evolutionary shocks. However, several observations of switch-on shocks have been made particularly in magnetospheric environments. Numerical studies establish that their evolution may be due to a weak level of resistivity in the plasma or to the fact that when the perturbation is applied, the field has a small tangential component.

10.10 Perpendicular Shock – Magnetic Field Lying in the Shock Plane

From Eqs. (A') and (F'), it follows that $v_2 = v_1$. By an appropriate frame transformation parallel to the shock front (Section 10.6.1), the velocity parallel to the shock both upstream and downstream may be set to zero:

$$u_1 \neq 0 \qquad \mathbf{v}_{t1} = \mathbf{v}_{t2} = 0 \qquad B_{x1} = B_{x2} = 0 \qquad \text{and} \qquad B_{y_1} \neq 0 \tag{10.65}$$

From Eqs. (A')–(F'), the following jump relations are obtained:

$$\frac{\rho_2}{\rho_2} = y \tag{10.66}$$

$$\frac{B_{y2}}{B_{y_1}} = y \tag{10.67}$$

$$\frac{u_2}{u_1} = y^{-1} \tag{10.68}$$

$$\frac{p_2}{p_1} = 1 + \frac{\gamma\, u_1{}^2\,(y-1)}{V_{s1}{}^2\, y} + \frac{\gamma}{2}\frac{V_{A1}{}^2}{V_{s1}{}^2}(1-y^2) \tag{10.69}$$

and the shock adiabat takes the quadratic form:

$$(2-\gamma)\,V_{A1}{}^2\,y^2 + [(\gamma\,V_{A1}{}^2 + 2\,V_{s1}{}^2) + (\gamma-1)\,u_1{}^2]\,y - (\gamma+1)\,u_1{}^2 = 0 \tag{10.70}$$

In principle, there are two solutions y_1 and y_2 whose product is

$$y_1\,y_2 = -\frac{(\gamma+1)u_1^2}{(2-\gamma)V_{A1}{}^2} \tag{10.71}$$

Since $1 < \gamma < 2$, it follows that one root is positive and the other negative. For strong shocks $u_1 \to \infty$, it follows that $y \to (\gamma+1)/(\gamma-1)$. If $y > (\gamma+1)/(\gamma-1)$, it is easily shown that $u_1{}^2 < 0$ and therefore that this value is again the maximum compression possible. In a weak shock $y \to 1$, the speed $u_1 = \sqrt{V_{A1}{}^2 + V_{s1}{}^2}$ corresponds to the fast magneto-sonic wave as expected since Alfvén and slow waves do not propagate perpendicularly. Since as $B_x \to 0$, $M'_A \to \infty$, it follows from Section 10.7.2 that perpendicular shocks are always fast.

10.11 Shock Structure and Stability

It is well known from studies of gas dynamic shocks that, on compression, downstream characteristics propagate faster behind a wave. As a result, the wave steepens eventually forming a shock if stabilised or breaking if not. Diffusion through the wave resulting from the strong gradient is usually sufficient to stabilise the shock and leads to the necessary increase in entropy. In gas, dynamic shocks dissipation from collisional processes such as viscosity and/or thermal conduction are normally sufficient to provide stabilisation.

Early studies of MHD shocks considered the role of collisional processes (Anderson, 1963; Liberman and Velikhovich, 1986). However, studies of the shock waves in the magnetosphere have shown that the diffusion lengths, typically mean free paths, are far too long to account for the necessary dissipation. As a result, it is now accepted that most MHD found in practice are collisionless. The necessary diffusion is provided by plasma instabilities and turbulence. This is a major topic not appropriate for a book of this nature. A modern review of these effects is given by Balogh and Treumann (2013) to which reference should be made.

Appendix 10.A Group Velocity of Magneto-sonic Waves

As an example of the group velocity in non-dispersive media (see Section 5.5), consider the velocity of a pulse of the magneto-sonic waves, where the system has axial symmetry about the magnetic field direction. The phase velocity depends on the angle θ between the field $\mathbf{B}_0$ and the propagation vector $\mathbf{k}$. Since $\omega = k\ V(\theta)$, where V is the phase velocity, and

$$\frac{\partial k}{\partial k_x} = \cos\theta \qquad \text{and} \qquad \frac{\partial k}{\partial k_y} = \sin\theta$$

$$\frac{\partial \theta}{\partial k_x} = -\frac{1}{k}\sin\theta \qquad \text{and} \qquad \frac{\partial \theta}{\partial k_y} = \frac{1}{k}\cos\theta$$

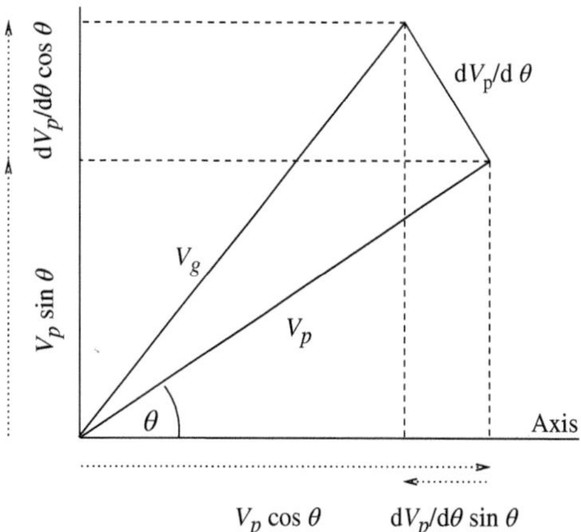

Figure 10.A.1 Geometrical interpretation of Eqs. (5.74) showing the relationship between the group (V_g) and phase (V) velocities.

it follows that the group velocity components parallel and perpendicular to the field are

$$\mathbf{V}_g|_{\parallel} = V\ \cos\theta - \frac{dV}{d\theta}\ \sin\theta \qquad \text{and} \qquad \mathbf{V}_g|_{\perp} = V\ \sin\theta + \frac{dV}{d\theta}\ \cos\theta \tag{10.A.1}$$

It is easily shown that there is a simple interpretation of this result in that on a polar plot the group velocity lies on the normal plane through the phase velocity (Figure 5.5) (Lighthill, 1960) as proved generally above. The group (ray) velocity is therefore always greater than the phase (wave) velocity. The peak amplitude of the pulse clearly relates to the situation where the different frequency components are in phase. As a result, waves of different frequency interfere coherently resulting in a large local amplitude and therefore energy. The propagation of a pulse is therefore accomplished along a ray defined by the group (ray) velocity. The wave (phase) velocity in contrast refers to the speed with which the phase of an infinite plane parallel wave changes as it is propagates (Figure 10.A.1).

In a similar fashion to the wave group, an ensemble of infinite monochromatic waves in space, each defined by a surface of constant phase, has a peak amplitude where the coherent spatial interference of many plane beams locally of slightly differing wave number takes place. The amplitudes of the members of the set of waves are determined by the appropriate spatial Fourier transform from a knowledge of the source distribution. The wave system is therefore represented by an infinite set of plane waves. It is evident that each wave front (normal plane) must be normal to the direction of motion of the phase (wave normal), given by the propagation vector $\mathbf{k}$ at the appropriate phase velocity. Coherent superposition of the waves therefore takes place at the envelope of the phase fronts of the parallel waves, i.e. along the normals to the phase trajectories. Let the normal to the trajectory be described by the functional form $f(x, y, \theta) = 0$, where x is parallel to the ambient field and y perpendicular to it, then the envelope is given by the simultaneous satisfaction of the derivative $\dot{f}(x, y, \theta) = 0$ with respect to the parameter θ defining the phase trajectory. Anderson (1963, p. 52) gives a full account of this phenomenon.

Consider a constant point source at the origin ($x = 0,\ y = 0$). The phase paths to which the envelope must be normal, are given by the straight lines ($Y = \tan\theta\ X$). Thus, the normal to the trajectory at ($X = V(\theta)\ t\ \cos\theta,\ Y = V(\theta)\ t\ \sin\theta$) at time t is

$$f(x,\ y,\ \theta) = y - V(\theta)\ t\ \sin\theta + \cot\theta\ [x - V(\theta)\ t\ \cos\theta] = 0 \tag{10.A.2}$$

where $t = 0$ is the time of emission. The derivative

$$\dot{f}(x,\ y,\ \theta) = -x\ \text{cosec}^2\theta + \cot\theta\ \text{cosec}\ \theta\ V(\theta)\ t - \text{cosec}\ \theta\ \frac{dV}{d\theta}\ t = 0 \tag{10.A.3}$$

Hence, solving for x and substituting in Eq. (10.A.2), we obtain the parametric equation of the envelope, namely the locus of the peak amplitude of waves emitted at time $t = 0$

$$x = V\ t\ \cos\theta - \frac{dV}{d\theta}\ t\ \sin\theta \qquad \text{and} \qquad y = V\ t\ \sin\theta + \frac{dV}{d\theta}\ t\ \cos\theta \tag{10.A.4}$$

which will be recognised as the motion of the pulse defined by the group velocity (Eq. (5.74)), consistent with the picture given in Section 5.6 defined by the optics of anisotropic crystals (Born and Wolf, 1965, ch.14).

Since the amplitudes of the waves are small, waves emitted from different sources add linearly. The solution for extended sources is therefore easily obtained.

Appendix 10.B Solution in de Hoffmann–Teller Frame

The exact Hugoniot form of the general solution is too cumbersome for many applications. There as in fluid mechanics it is useful to derive relations for the downstream flow in terms of the upstream values for polytropic gases. This is most easily carried out in the de Hoffmann–Teller frame where the energy equation (D') simplifies to

$$\frac{\gamma}{(\gamma-1)}\frac{p_1}{\rho_1} + \frac{1}{2}({u_1}^2 + {v_1}^2) = \frac{\gamma}{(\gamma-1)}\frac{p_2}{\rho_2} + \frac{1}{2}({u_2}^2 + {v_2}^2) \tag{D'''}$$

Equations (A')–(C') and (F') in the de Hoffmann–Teller frame taking into account the polytropic equation of state relations (10.43) yield expressions for the downstream state in terms of the upstream flow variables and the compression ratio y

$$\frac{\rho_2}{\rho_1} = y \tag{α}$$

$$\frac{B_{x2}}{B_{x1}} = 1 \tag{β}$$

$$\frac{B_{y2}}{B_{y1}} = y\ \frac{({u_1}^2 - {V_{A1}}^2 \cos^2\theta_1)}{({u_1}^2 - y\ {V_{A1}}^2 \cos^2\theta_1)} \tag{γ}$$

$$\frac{u_2}{u_1} = y^{-1} \tag{δ}$$

$$\frac{v_2}{v_1} = \frac{({u_1}^2 - {V_{A1}}^2 \cos^2\theta_1)}{({u_1}^2 - y\ {V_{A1}}^2 \cos^2\theta_1)} \tag{ϵ}$$

$$\frac{p_2}{p_1} = 1 + \frac{\gamma\ {u_1}^2\ (y-1)}{{V_{s1}}^2\ y}\left\{1 - \frac{y\ {V_{A1}}^2\,[(y+1)\,{u_1}^2 - 2\ y\ {V_{A1}}^2 \cos^2\theta_1]}{2\,({u_1}^2 - y\ {V_{A1}}^2\cos^2\theta_1)^2}\right\} \tag{ζ}$$

Finally, substituting in Eq. (D''') for p_2, we obtain an equation known as the *shock adiabat* relating the compression ratio y to the incoming flow velocity u_1:

$$\begin{aligned}&({u_1}^2 - y\ \cos^2\theta_1\ {V_{A1}}^2)^2\ \{[(\gamma+1)-(\gamma-1)y]\,{u_1}^2 - 2\ y\ {V_{s1}}^2\}\\ &-y\ {u_1}^2\ {V_{A1}}^2 \sin^2\theta_1\ \{[\gamma + (2-\gamma)\,y]\,{u_1}^2 - [(\gamma+1)-(\gamma-1)\,y]\ y\ {V_{A1}}^2\cos^2\theta_1\} = 0\end{aligned} \tag{η}$$

whose solutions y form the foundation of the Rankine–Hugoniot equations for polytropic MHD fluids. The remaining relations for density, pressure and field as function of the shock speed are directly obtained from the Eqs. (α)–(ζ).

The equivalence with the corresponding equations of hydrodynamics is easily demonstrated by setting $V_{A1} = 0$ when

$$y = [(\gamma+1)\,{M_{s1}}^2]/[(\gamma-1)\,{M_{s1}}^2 + 2]$$

where $M_{s1} = u_1/V_{s1}$ is the shock Mach number. The standard result for the pressure ratio is easily obtained from Eq. (ζ).

In the weak shock limit, $y \approx 1$ and the adiabat reduces to

$$(u_1{}^2 - V_{A1}{}^2 \cos^2\theta_1)\{(u_1{}^2 - V_{A1}{}^2 \cos^2\theta_1)(u_1{}^2 - V_{s1}{}^2) - u_1{}^2 V_{A1}{}^2 \sin^2\theta_1\} = 0$$

whose solutions are easily seen to be

$$u_1{}^2 = C_+{}^2 = \frac{V_{A1}{}^2 + V_{s1}{}^2 + \sqrt{(V_{A1}{}^2 + V_{s1}{}^2)^2 - 4\,V_{A1}{}^2 V_{s1}{}^2 \cos^2\theta_1}}{2}$$

$$u_1{}^2 = V_{A1}{}^2 \cos^2\theta_1$$

$$u_1{}^2 = C_-{}^2 = \frac{V_{A1}{}^2 + V_{s1}{}^2 - \sqrt{(V_{A1}{}^2 + V_{s1}{}^2)^2 - 4\,V_{A1}{}^2 V_{s1}{}^2 \cos^2\theta_1}}{2}$$

which are immediately recognised as the fast, Alfvén, and slow magneto-sonic waves, Eqs. (10.28) and (10.29).

Since the shock adiabat, Eq. (η), is a cubic in the compression ratio y, it is easily seen that for any particular initial incoming velocity u_1 it possesses three distinct solutions. These only represent physically allowed solutions if they are real. Since the coefficients of the cubic in (η) are real, we may have either three real or one real and two imaginary solutions. The three distinct solutions may be identified as the large amplitude forms of the magneto-sonic waves. They are categorised by the speed of the plasma flow normal to the front, or alternatively by the speed of propagation into plasma at rest as fast, intermediate and slow. Clearly allowed solutions only exist if the flow velocity is greater than the speed of the corresponding magneto-sonic wave.

$$\underbrace{u_1 > C_{+1}}_{\text{fast}} \qquad \underbrace{u_1 > V_{A1}\ \cos\theta}_{\text{intermediate}} \qquad \underbrace{u_1 > C_{-1}}_{\text{slow}} \tag{10.B.1}$$

The transverse magnetic field through the shock in the de Hoffmann–Teller frame is given by Eq. (γ) to satisfy

$$\underbrace{B_{y_2} > B_{y_1}}_{\text{fast}} \qquad \underbrace{B_{y_2} = -B_{y_1}}_{\text{intermediate}} \qquad \underbrace{B_{y_2} < B_{y_1}}_{\text{slow}} \tag{10.B.2}$$

the result for intermediate shocks being obtained by the use of l'Hôpital's rule and Eq. (η). Thus, in the de Hoffmann–Teller frame, slow and fast shocks refract the magnetic field lines towards and away from the shock normal, respectively. Consequently, the total magnetic field increases through a fast shock and decreases through a slow. In intermediate shocks, the field parallel to the shock plane is reversed, the magnitude of the total field is therefore constant.

In the strong shock limit, $u_1 \gg V_{A1},\ V_{s1}$, Eq. (η) has a dominant term $u_1{}^6$, which gives the single real solution $y = (\gamma+1)/(\gamma-1)$ as $u_1 \to \infty$, the same compression ratio as in a strong hydrodynamic shock. For a compression ratio slightly less than the maximum $y = (\gamma+1)/(\gamma-1) - \delta y$, the incoming flow speed is given by the term $u_1{}^4$

$$u_1 = \sqrt{\left\{\frac{2}{(\gamma-1)^3}\frac{[(\gamma-1)V_{s1}{}^2 + \sin^2\theta_1\,V_{A1}{}^2]}{\delta y}\right\}} \tag{10.B.3}$$

This solution is clearly a fast shock. It can be shown by numerical checks that the slow and intermediate solutions gradually merge and disappear as the shock strength is increased. Slow and intermediate shocks therefore have a limited strength, fast socks on the other hand have no limit to their strength. As with hydrodynamic shocks, the compression ratio and the transverse magnetic field ratio are both limited to the value $(\gamma + 1)/(\gamma - 1)$.

10.B.1 Parallel Shocks

Alternatively, setting $\theta = 0$ in Eq. (η) gives a cubic equation in y, whose solutions are a single root corresponding to the gas dynamic shock

$$y = \frac{(\gamma + 1)M_{s1}{}^2}{[2 + (\gamma - 1)M_{s1}{}^2]} \qquad \text{or} \qquad u_1{}^2 = \frac{2\, y\, V_{s1}{}^2}{[(\gamma + 1) - (\gamma - 1)\, y]} \tag{10.B.4}$$

and a double root due to the switch-on shock

$$y = M_{A1}{}^2 \qquad \text{or} \qquad u_1{}^2 = y\, V_{A1}{}^2 \tag{10.B.5}$$

11

Waves in Cold Magnetised Plasma

11.1 Introduction

Perhaps the most characteristic feature of plasma is the plethora of possible wave motions possible. This arises as a consequence of both the electro-magnetic forces between the charged particles and the short-range collisions. The particle motions may be transverse or longitudinal or a combination of the two. The waves are electro-magnetic with perturbations in the electric and magnetic field related by Maxwell's equations – a generalisation of the familiar electro-magnetic waves of optics modified by the presence of the plasma medium. In this chapter, we examine the waves in cold plasma occurring as a result of collective particle behaviour.

When the plasma is cold, there is zero random thermal motion of the particles. Consequently, all the particles of a particular species undergo a common motion in the electric and magnetic fields generated by their displacement. The behaviour of the plasma particles is therefore simply the motion of charged elements, each behaving as though they were single 'fluid particles' of appropriate mass and charge. The cold plasma wave model is therefore sometimes known as the 'fluid model'.

The introduction of thermal motions markedly complicates the modal structure of the wave configurations, which we study in Chapter 6. However, the study of cold plasma waves forms an important basis from which the general picture may be explored allowing for both the similarities and the differences introduced by the thermal motions.

11.2 Waves in Cold Plasma

To proceed, we must calculate the response of the plasma particles to the electric and magnetic fields in order to calculate the polarisation and hence the dielectric constant. Since the plasma waves are assumed to be weak, we neglect terms of second and higher order in the perturbation induced thereby.

The equation of motion of particle with mass m_q and charge $Z_q\, e$ is [1]

$$m_q \frac{\mathrm{d}\mathbf{v}_q}{\mathrm{d}t} = Z_q\, e\,(\mathbf{E} + \mathbf{v}_q \wedge \mathbf{B}) \approx Z_q\, e\,(\mathbf{E} + \mathbf{v}_q \wedge \mathbf{B}_0) \tag{11.1}$$

where $\mathbf{B}_0$ is the ambient magnetic field. To proceed, we introduce a monochromatic wave as before and take Cartesian co-ordinates with the ambient magnetic field parallel to the z axis, $\mathbf{B}_0 = B_0\, \hat{\mathbf{e}}_z$.

1 Note electrons are included with $Z_e = -1$.

Foundations of Plasma Physics for Physicists and Mathematicians, First Edition. Geoffrey J. Pert.

The propagation vector **k** is taken to lie in the x, z plane defining the x axis orthogonal to z and thus the y axis orthogonal to both x and z. The equation of motion is easily solved in the 'circular' basis set $\hat{\mathbf{e}}_{\pm}$, $\hat{\mathbf{e}}_z$, which is related to the normal Cartesian set $\hat{\mathbf{e}}_x$, $\hat{\mathbf{e}}_y$, $\hat{\mathbf{e}}_z$ by

$$\hat{\mathbf{e}}_{\pm} = \hat{\mathbf{e}}_x \pm \imath\, \hat{\mathbf{e}}_y \qquad \text{and} \qquad \hat{\mathbf{e}}_{\pm} \wedge \hat{\mathbf{e}}_z = \pm \imath\, \hat{\mathbf{e}}_{\pm} \tag{11.2}$$

$$v_{q\pm} = \frac{\imath}{B_0} \frac{\Omega_k}{(\omega \pm \Omega_k)} E_{\pm} \qquad v_{qz} = \frac{\imath}{B_0} \frac{\Omega_k}{\omega} E_z \tag{11.3}$$

where

$$v_{q\pm} = v_{qx} \pm \imath\, v_{qy} \qquad E_{\pm} = E_x \pm \imath\, E_y \tag{11.4}$$

and $\Omega_q = Z_q\, e\, B_0 / m_k$ is the cyclotron frequency for the particles q.

The total current density is therefore

$$\mathbf{j} = \sum_q Z_q\, e\, n_q\, \mathbf{v_q} = \imath\, \epsilon_0 \begin{pmatrix} \sum_q \frac{{\Pi_q}^2}{\omega(\omega + \Omega_q)} & 0 & 0 \\ 0 & \sum_q \frac{{\Pi_q}^2}{\omega(\omega - \Omega_q)} & 0 \\ 0 & 0 & \sum_q \frac{{\Pi_q}^2}{\omega^2} \end{pmatrix} \begin{pmatrix} E_+ \\ E_- \\ E_z \end{pmatrix} \tag{11.5}$$

where the qth particle plasma frequency

$${\Pi_q}^2 = \frac{{Z_q}^2 e^2 n_q}{\epsilon_0\, m_q} \tag{11.6}$$

The dielectric displacement is given by

$$\mathbf{D} = \epsilon_0\, \mathbb{K} \cdot \mathbf{E} = \epsilon_0 \mathbf{E} + \imath\, \frac{1}{\omega}\, \mathbf{j} = \epsilon_0 \begin{pmatrix} R & 0 & 0 \\ 0 & L & 0 \\ 0 & 0 & P \end{pmatrix} \begin{pmatrix} E_+ \\ E_- \\ E_z \end{pmatrix} \tag{11.7}$$

where

$$R = 1 - \sum_q \frac{{\Pi_q}^2}{\omega^2} \frac{\omega}{(\omega + \Omega_q)} \tag{11.8a}$$

$$L = 1 - \sum_q \frac{{\Pi_q}^2}{\omega^2} \frac{\omega}{(\omega - \Omega_q)} \tag{11.8b}$$

$$P = 1 - \sum_q \frac{{\Pi_q}^2}{\omega^2} \tag{11.8c}$$

A similarity transformation, $\mathbb{K} = \mathbb{U}^* \, \mathbb{K}_{\text{circ}} \, U$ using the unitary matrix

$$\mathbb{U} = \frac{1}{\sqrt{2}} \begin{pmatrix} 1 & \imath & 0 \\ 1 & -\imath & 0 \\ 0 & 0 & \sqrt{2} \end{pmatrix}$$

may be used to transform from the circular basis $\hat{\mathbf{e}}_{\pm}$ to the usual Cartesian set $\hat{\mathbf{e}}_{1,2}$ to give

$$\mathbb{K} = \begin{pmatrix} S & -\imath D & 0 \\ \imath\, D & S & 0 \\ 0 & 0 & P \end{pmatrix} \tag{11.9}$$

where

$$S = \frac{R+L}{2} \qquad D = \frac{R-L}{2} \tag{11.10}$$

To proceed, we introduce the vector

$$\mathbf{n} = \frac{\mathbf{k}c}{\omega} \tag{11.11}$$

parallel to the direction of propagation and magnitude equal to the refractive index. slightly re-defining the matrix $\mathbb{M}$, Eq. (13.33)

$$\mathbb{M} = \mathbf{nn} - n^2\mathbb{I} + \mathbb{K} \tag{11.12}$$

Without loss of generalisation we consider the wave propagation to be in the x, z plane. The eigenmode equation for waves in the plasma is then

$$\mathbb{M}\cdot\mathbf{E} = \begin{pmatrix} S - n^2\cos^2\theta & -\imath D & n^2\cos\theta\,\sin\theta \\ \imath D & S - n^2 & 0 \\ n^2\cos\theta\,\sin\theta & 0 & P - n^2\sin^2\theta \end{pmatrix} \cdot \begin{pmatrix} E_x \\ E_y \\ E_z \end{pmatrix} = 0 \tag{11.13}$$

where θ is the angle between the applied field and the direction of propagation.

The set of Eq. (11.13) form a homogeneous set of linear simultaneous equations, which only have a solution if

$$\det(\mathbb{M}) = \begin{vmatrix} S - n^2\cos^2\theta & -\imath D & n^2\cos\theta\,\sin\theta \\ \imath D & S - n^2 & 0 \\ n^2\cos\theta\,\sin\theta & 0 & P - n^2\sin^2\theta \end{vmatrix} = 0 \tag{11.14}$$

where θ is the angle between $\mathbf{k}$ and $\mathbf{B}_0$. This condition, called the *dispersion relation*, determines the values of the wavenumber possible at a given frequency, and therefore from Eq. (11.11), the *phase velocity* $V = \omega/k$.

Making use of the identity $S^2 - D^2 = R\,L$, the determinant can be expanded to give a quadratic in n^2

$$A\,n^4 - B\,n^2 + C = 0 \tag{11.15}$$

where

$$A = S\sin^2\theta + P\cos^2\theta \qquad B = R\,L\sin^2\theta + P\,S(1+\cos^2\theta) \qquad C = P\,R\,L \tag{11.16}$$

Solving for n^2, we obtain

$$n^2 = (B \pm F)/2\,A \tag{11.17}$$

where

$$F^2 = (R\,L - P\,S)^2\sin^4\theta + 4P^2\,D^2\cos^2\theta \geq 0 \tag{11.18}$$

so that n always has either purely real or purely imaginary roots.

- In the former case, the wave is oscillatory. There are two possible different wave motions of a specified oscillation frequency:
  i. with a negative sign in Eq. (11.17) *ordinary waves* or *O modes*
  ii. with a positive sign in Eq. (11.17) *extraordinary waves* or *X mode*.
- In the second case, there are two possible evanescent (decaying waves) with zero oscillation. The two roots of opposite sign represent motion in opposite directions.

The dispersion relation can be written in the alternative form

$$\tan^2\theta = \frac{P\,(n^2 - R)\,(n^2 - L)}{(S\,n^2 - R\,L)\,(n^2 - P)} \tag{11.19}$$

For propagation along the ambient magnetic field where $\theta = 0$

$$P = 0 \qquad \text{or} \qquad n^2 = R \qquad \text{or} \qquad n^2 = L \tag{11.20}$$

and for propagation perpendicular to the ambient field where $\theta = \pi/2$

$$n^2 = R\,L/S \qquad \text{or} \qquad n^2 = P \tag{11.21}$$

11.2.1 Cut-off and Resonance

There are two singular cases in which the reciprocal refractive index is either zero or infinity.

i. *Cut-off*: If $n = 0$, the phase velocity becomes infinite, which is clearly non-physical. In practice, this is usually a result of a changing physical inhomogeneous environment. It follows from Eq. (11.15) that cut-off occurs if $C = 0$ or from Eq. (11.16) if

$$P = 0 \qquad \text{or} \qquad R = 0 \qquad \text{or} \qquad L = 0 \tag{11.22}$$

If the wave is incident on a cut-off, the value of n passing from real $n^2 > 0$ to imaginary $n^2 < 0$, the wave is reflected at the layer $n = 0$ and becomes non-propagating, or evanescent, thereafter. We note that since $P,\ R,\ L$ are functions only of the particle density and the ambient magnetic field, the cut-off condition depends solely on the state of the plasma independent of the angle of propagation. Since $m_e \ll m_i$, the ion terms may be neglected in comparison with the electron. When $R = 0$

$$\omega^2 + \Omega_e\,\omega - \Pi_e{}^2 = 0 \tag{11.23}$$

whose positive solution yields the upper cut-off

$$\omega_R = |\Omega_e|/2 + \sqrt{(\Omega_e{}^2/4 + \Pi_e{}^2)} > |\Omega_e| \tag{11.24}$$

Similarly when $L = 0$

$$\omega^2 - \Omega_e\,\omega - \Pi_e{}^2 = 0 \tag{11.25}$$

whose positive solution yields the lower cut-off

$$\omega_L = -|\Omega_e|/2 + \sqrt{(\Omega_e{}^2/4 + \Pi_e{}^2)} \tag{11.26}$$

ii. *Resonance*: If $n \to \pm\infty$, the phase velocity reduces to zero. This occurs when the wave frequency matches that of one of the natural oscillations in the plasma to which it is coupled. The behaviour is then identical to that of a simple mechanical oscillator driven at its natural frequency. From Eq. (11.19), resonance occurs if

$$\tan^2\theta = -\frac{P}{S} \tag{11.27}$$

If propagation is parallel to the axis, $\theta = 0$, resonance occurs if $S = \frac{1}{2}\,(R + L) \to \infty$ when the frequency of the wave is co-incident with the cyclotron motion of one of the particles. If $R \to \pm\infty$, the resonant particles are electrons and if $L \to \pm\infty$ positively charged ions.
If propagation is perpendicular to axis, $\theta = \pi/2$, resonance occurs for $S \to 0$.
Resonances from parallel and perpendicular propagation are known as *principal resonances*

11.2.2 Polarisation

In a wave of right-handed circular polarisation defined with respect to the axis of the ambient magnetic field, the x components of the electric field of the wave and the velocity of the particles scale in time as E_x, $v_x \sim \cos(-\omega\, t) \sim \Re[\exp(-\imath\, \omega\, t)]$. Similarly, the y components as E_y, $v_y \sim \sin(-\omega\, t) \sim \Re\,[\imath\, \exp(-\imath\, \omega\, t)]$. In circular polarisation, the amplitudes in both directions are the same, only the phase is different. Thus for right-handed circular polarisation, $\imath\, E_y/E_x = 1$. Similarly for left-handed polarisation, $\imath\, E_y/E_x = -1$.

Taking the middle line of the matrix equation $\mathbf{M} \cdot \mathbf{E} = 0$ yields

$$\frac{\imath\, E_y}{E_x} = \frac{(n^2 - S)}{D} \tag{11.28}$$

Therefore, for propagation parallel to ambient magnetic field $\theta = 0$ we have $n^2 = R$ for right-handed circular polarisation and $n^2 = L$ for left-handed.

Using Eqs. (11.2), (11.3), and (11.28), we obtain

$$\frac{\imath\, v_y}{v_x} = \frac{(s_k\, \omega - \Omega_k)\,((n^2 - L) + (s_k\, \omega + \Omega_k)\,(n^2 - R)}{(s_k\, \omega - \Omega_k)\,((n^2 - L) - (s_k\, \omega + \Omega_k)\,(n^2 - R)} \tag{11.29}$$

where $Z_k = \mathrm{sgn}(z_k)$.

11.3 Cold Plasma Waves

The general discussion of the complete spectrum of waves in cold plasma is complex due to the large number of independent variables on which the dispersion matrix depends. It is not appropriate to discuss these in detail here, but refer the interested reader to the text by Stix (1962). We will however consider a representative set of solutions, which illustrate the behaviour shown in the generally. To further simplify matters, we will, in most cases, assume the only particles are a single species of ions and electrons, and furthermore that since $m_e \ll m_i$ the ion motion can be neglected in respect of the electrons.

11.3.1 Zero Applied Magnetic Field

In this simple case, $B_0 = \Omega_e = \Omega_i = 0$. Since there is no axis of symmetry, the waves are isotropic, and

$$R = L = P = 1 - \frac{\Pi_e{}^2}{\omega^2} - \frac{\Pi_i{}^2}{\omega^2} \approx 1 - \frac{\Pi_e{}^2}{\omega^2} \tag{11.30}$$

since $\Pi_e \gg \Pi_i$ and therefore

$$S = P \qquad \text{and} \qquad D = 0 \tag{11.31}$$

The governing equation for the electric fields reduces to

$$\mathbb{M} \cdot \mathbf{E} = \begin{pmatrix} P - n^2 \cos^2\theta & 0 & n^2\, \sin\theta\, \cos\theta \\ 0 & P - n^2 & 0 \\ n^2\, \sin\theta\, \cos\theta & 0 & P - n^2 \sin^2\theta \end{pmatrix} \cdot \begin{pmatrix} E_x \\ E_y \\ E_z \end{pmatrix} = 0 \tag{11.32}$$

and therefore, the dispersion relation reduces to

$$\det(\mathbb{M}) = \begin{vmatrix} P - n^2 \cos^2\theta & 0 & n^2\, \sin\theta\, \cos\theta \\ 0 & P - n^2 & 0 \\ n^2\, \sin\theta\, \cos\theta & 0 & P - n^2 \sin^2\theta \end{vmatrix} = P\,(P - n^2)^2 = 0 \tag{11.33}$$

The roots of this equation are

$$P = 0 \qquad \text{and} \qquad P = n^2 \quad \text{(double root)} \tag{11.34}$$

The solutions are readily identified by noting that since $\mathbf{B}_0 = 0$, the directions of the co-ordinate axes are arbitrary. Let the z direction be taken parallel to the wave vector $\mathbf{k}$, and $\theta = 0$. The allowed waves are therefore

i. The root $P = 0$ represents a wave with frequency $\omega = \Pi_e$, the plasma frequency independent of the wave length. Setting $\theta = 0$, it follows from Eq. (11.32) that $E_x = E_y = 0$ and $E_z \neq 0$, so that the wave is longitudinal, $\mathbf{E} \parallel \mathbf{k}$. The group velocity, $V_g = \partial\omega/\partial k$ is therefore zero, there is no energy flux and the wave is a non-propagating stationary oscillation. We recognise this as the cold plasma wave described earlier in Section 1.2.
ii. The double root $P = n^2$ represents two waves with phase velocity $V = c/\sqrt{(1 - \Pi_e^{\,2}/\omega^2)}$. The fields $E_x \neq 0$ and $E_y \neq 0$ and $E_z = 0$. The waves are therefore transverse $\mathbf{E} \perp \mathbf{k}$. Clearly, they represent the familiar electro-magnetic waves modified by the refractive index of the plasma n^{-1}. If $\omega > \Pi_e$, the refractive index is imaginary and the wave evanescent. The two solutions represent the two independent polarisations with fields E_x, E_y, at right angles to one another.

11.3.2 Low Frequency Velocity Waves

Consider waves passing through magnetised plasma when the frequency of the wave is much smaller than the characteristic frequencies in plasma, namely $\omega \ll \Pi_i$, $\Omega_i \ll \Pi_e$, Ω_e. Since $\Pi_e^{\,2}/\Omega_e \Omega_i = -\Pi_i^{\,2}/\Omega_i^{\,2}$, it follows from Eqs. (11.8a), (11.8b), (11.8c), and (11.10) that

$$S \approx 1 + \frac{\Pi_i^{\,2}}{\Omega_i^{\,2}} \qquad D \approx 0 \qquad P \approx -\frac{\Pi_e^{\,2}}{\omega^2} \tag{11.35}$$

and that in consequence the dispersion relation is

$$\begin{vmatrix} 1 + \Pi_i^{\,2}/\Omega_i^{\,2} - n^2 \cos^2\theta & 0 & -n^2 \cos\theta \sin\theta \\ 0 & 1 + \Pi_i^{\,2}/\Omega_i^{\,2} - n^2 & 0 \\ n^2 \cos\theta \sin\theta & 0 & -\Pi_e^{\,2}/\omega^2 - n^2 \sin^2\theta \end{vmatrix} = 0 \tag{11.36}$$

The element is M_{33} is much larger (in magnitude) than all the others subject to the ordering above. The expansion of the determinant approximately yields the product of the diagonal $M_{11}\, M_{22}\, M_{33}$, after neglecting the small terms. Since $M_{33} \neq 0$, we accordingly obtain two roots

$$n^2 \cos^2\theta = 1 + \Pi_i^{\,2}/\Omega_i^{\,2} \qquad n^2 = 1 + \Pi_i^{\,2}/\Omega_i^{\,2} \tag{11.37}$$

Introducing the Alfvén speed $V_A = \sqrt{(B_0^{\,2}/\mu_0\, \rho)}$, these roots may be written

$$\omega = \frac{k\, V_A \cos\theta}{\sqrt{1 - V_A^{\,2}/c^2}} \approx k\, V_A \cos\theta \equiv k_\parallel V_A \tag{11.38:a}$$

$$\omega = \frac{k\, V_A}{\sqrt{1 - V_A^{\,2}/c^2}} \approx k\, V_A \tag{11.38:b}$$

since $V_A \ll c$ in most plasmas.

These represent the familiar Alfvén waves normally discussed in the context of magneto-sonic waves. At low frequencies, the inertial term in the acceleration Eq. (11.1) is small compared to the remaining terms, and therefore, $\mathbf{E} + \mathbf{v} \wedge \mathbf{B}_0 = 0$ equivalent to the ideal MHD condition (Section 9.1). Taking the appropriate rows of the matrix equation $\mathbb{M} \cdot \mathbf{E} = 0$ we identify

- *Shear Alfvén wave* $\omega = k\, V_A\, \cos\theta$, Eq. (11.38:a) is the familiar shear Alfvén wave discussed later in Sections 9.2.3 and §10.2. These waves are incompressible with $E_y = E_z = 0$, and $E_x \neq 0$ and $v_x = v_z = 0$ and $v_y \neq 0$. They are transverse waves with the oscillation normal to the both magnetic field z and the direction of propagation x. The restoring force is due to the 'bending' of trapped magnetic field and the resulting torsion as in a vibrating string (Section 9.2.3).
- *Compressional Alfvén wave* $\omega = k\, V_A$, Eq. (11.38:b) is the fast magneto-sonic wave in cold plasma discussed in Section 10.2. These waves have $E_x = E_z = 0$ and $E_y \neq 0$, and therefore $v_y = v_z = 0$ and $v_x \neq 0$. They are therefore longitudinal waves with oscillation along the direction of propagation supported by the magnetic pressure differences due to the compression and expansion of the magnetic field.

11.3.3 Propagation of Waves Parallel to the Magnetic Field

Since $\theta = 0$ the determinental condition, Eq. (11.14)

$$\det(\mathbb{M}) = \begin{vmatrix} S - n^2 & -\imath D & 0 \\ \imath\, D & S - n^2 & 0 \\ 0 & 0 & P \end{vmatrix} = P \begin{vmatrix} \frac{1}{2}(R+L) - n^2 & -\frac{1}{2}\,\imath\,(R-L) \\ \frac{1}{2}\,\imath\,(R-L) & \frac{1}{2}(R+L) - n^2 \end{vmatrix} = 0 \tag{11.39}$$

whose roots are

$$P = 0 \quad \text{and} \quad n^2 = R \quad \text{and} \quad n^2 = L \tag{11.40}$$

We now consider each of these solutions in turn

i. $P = 0$: Making use of the Eq. (13.34) which determines the electric field, it is easily seen that $E_x,\ E_y = 0$ and $E_z \neq 0$. It thus follows that the velocity components $v_x,\ v_y = 0$ and $v_z \neq 0$. The wave is therefore a longitudinal oscillation along the magnetic field, with frequency

$$\omega^2 = \sum_q \Pi_q{}^2 \approx \Pi_e{}^2 \tag{11.41}$$

The wave is a plasma wave oscillating along the field independently of the magnetic field since the velocity and magnetic field are parallel. This mode propagates in the absence of the magnetic field (Section 11.3.1).

ii. $n^2 = R$: In Section 11.2.2, it is shown that this case represents a right-handed circularly polarised wave with frequency given by

$$n^2 = 1 - \sum_q \frac{\Pi_q{}^2}{\omega^2} \frac{\omega}{(\omega + \Omega_q)} \approx 1 - \frac{\Pi_e{}^2}{(\omega + \Omega_e)(\omega + \Omega_i)} \tag{11.42}$$

for a quasi-neutral two component plasma of electrons and ions when $n_e = Z\, n_i$ and

$$\frac{\Pi_q{}^2}{\Omega_q{}^2} = \frac{c^2}{B_0{}^2/\mu_0\,\rho} = \frac{c^2}{V_A{}^2} \tag{11.43}$$

A sketch of the general form of the dispersion relation for right hand circularly polarised waves propagating parallel to the ambient magnetic field is shown in Figure 11.1. The upper branch may exist in the absence of the magnetic field, but the lower branch only when a background magnetic field is present (Section 11.3.1).

We first identify any cut-offs or resonances. From Eq. (11.42) it follows that since a resonance occurs if $R = \infty$, namely when $\omega \approx |\Omega_e|$ (remembering that $\Omega_e < 0$). It is immediately seen

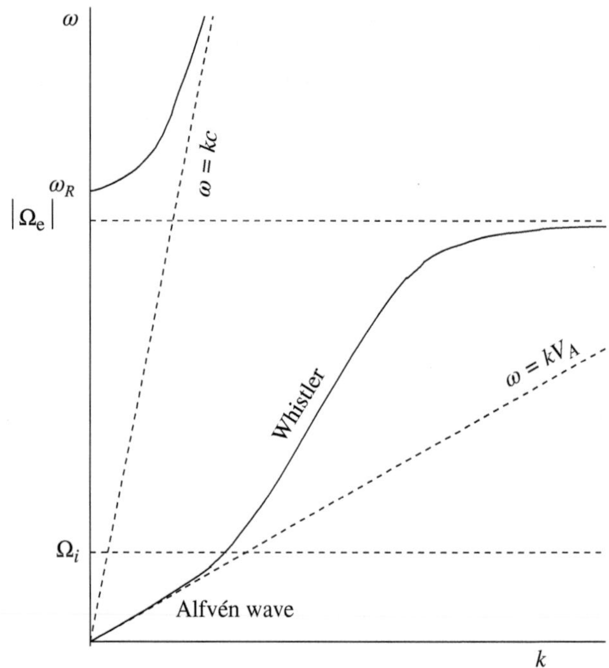

Figure 11.1 Sketch of the dispersion relation for right handed circularly polarised radiation propagating parallel to the magnetic field.

that this occurs when the wave electric field is in synchronism with the electron cyclotron rotation – *electron cyclotron resonance.* The electron rotates in the same sense at the same speed as the field and 'sees' a constant electric field. As the electron is accelerated, the necessary energy is extracted from the field, and gives rise to strong absorption of the wave. In practice the resonance is limited by non-linear effects neglected in this simple linear analysis.
The cut-off frequency when $n^2 = R = 0$ was found earlier

$$\omega_R \approx |\Omega_e|/2 + \sqrt{(\Omega_e{}^2/4 + \Pi_e{}^2)} > |\Omega_e| \tag{11.24:a}$$

For frequencies slightly below the resonance $\omega < |\Omega_e|$, the refractive index $n^2 > 0$ and the wave is propagates. Between the cyclotron frequency and the cut-off frequency $|\Omega_e| < \omega < \omega_R$, the refractive index $n^2 < 0$ and the wave is evanescent. Above the cutoff $n^2 > 0$ and the wave again propagates.
At low frequencies, the wave is a right-handed circularly polarised Alfvén wave discussed earlier. When propagation is along the magnetic field, the two components of the Alfvén wave degenerate into a single mode with electric field and velocity perpendicular to the magnetic field. As the frequency is increased above the ion cyclotron frequency Ω_i, the dispersion relation continues up to the electron cyclotron frequency Ω_e with increasing wave number – *electron cyclotron waves.* In the range $\Omega_i \ll \omega \ll |\Omega_e|$, the dispersion relation is approximately (Stix, 1962)

$$n^2 \approx 1 - \frac{\Pi_e{}^2}{\omega(\omega - |\Omega_e| \cos\theta)} \approx \frac{\Pi_e{}^2}{\omega\, |\Omega_e| \cos\theta} \tag{11.44}$$

when $n^2 > 1$ and $\omega \ll |\Omega_e|$. The group velocity, namely the speed of energy propagation and the angular spread of a beam, are obtained from Eq. (5.75), and may be shown to be $|V_g| \sim \sqrt{\omega}$

and the angular spread $|\psi| \approx \text{arccot}\,\sqrt{8} \approx 19°28'$. Thus, energy is propagated along the field lines within a tight beam with a speed which decreases with the frequency. Consequently, if a multi-frequency wave generated, for example by lightening in the north polar region, it is dispersed in frequency travelling along the earth's magnetic field lines and appears as wave of initially high frequency decreasing in tone. These waves are the well-known *whistler modes*.

At high frequency $\omega \gg \Pi_e$, the wave has $n^2 = 1$ and is simply an electromagnetic wave unmodified by the magnetic field. When $\omega > |\Omega_e|$, the wave is initially evanescent, but becomes propagating again when $n^2 > 0$ as the frequency increases. The frequency at which this takes place is the upper cut-off given by Eq. (11.24).

When the frequency is much greater that the electron cyclotron frequency $\omega \gg |\Omega_e|$, the dispersion relation becomes that of electro-magnetic waves in plasma found earlier, namely phase velocity $V = c/\sqrt{1 - {\Pi_e}^2/\omega^2}$.

iii. $n^2 = L$: In Section 11.2.2, it is shown that this case represents a left-handed circular polarised wave field with dispersion relation

$$n^2 = 1 - \sum_q \frac{{\Pi_q}^2}{\omega^2} \frac{\omega}{(\omega - \Omega_q)} \approx 1 - \frac{{\Pi_e}^2}{(\omega - \Omega_e)(\omega - \Omega_i)} \tag{11.45}$$

for a quasi-neutral two component plasma of electrons and ions. A sketch of the general form of the dispersion relation for right hand circularly polarised waves propagating parallel to the ambient magnetic field is shown in Figure 11.2. As for right handed polarised waves, the upper branch exists in the absence of the magnetic field, but the lower branch only if the background magnetic field is present (Section 11.3.1).

The general behaviour of waves (Figure 11.2) is similar to that for right-handed polarisation discussed above. At low frequencies, the wave is simply a left-hand circularly polarised Alfvén

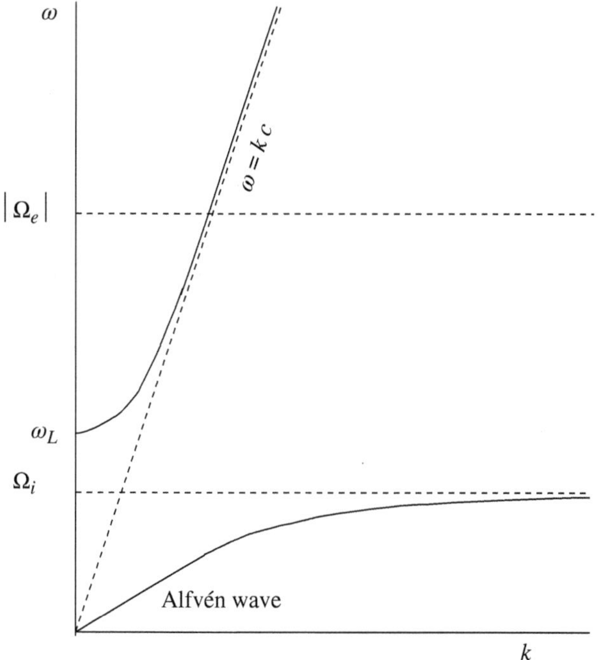

Figure 11.2 Sketch of the dispersion relation for left handed circularly polarised radiation propagating parallel to the magnetic field.

wave. Since $\Omega_i \ll |\Omega_e|$, always no whistler type mode is possible. As the frequency approaches the ion cyclotron frequency $\omega \to \Omega_i$, the wave approaches the resonance. At resonance ($\omega = \Omega_i$), the wave field is synchronous with the ion cyclotron rotation, which consequently strongly absorbs the wave energy as the ions increase their kinetic energy. At frequencies immediately above the resonance, the wave is cut-off and non-propagating. Propagation is resumed at the lower cut-off ω_L given by Eq. (11.26)

$$\omega_L = -|\Omega_e|/2 + \sqrt{({\Omega_e}^2/4 + {\Pi_e}^2)} > {\Pi_i}^2/\Omega_i \tag{11.26;a}$$

since ${\Pi_i}^2/\Omega_i = {\Pi_e}^2/|\Omega_e|$ and $\Pi_i \gg \Omega_i$ generally for plasma systems. At frequencies above the cut-off, the dispersion relation approaches the electro-magnetic wave value appropriate to plasma in the same way as right handed polarisation.

11.3.4 Propagation of Waves Perpendicular to the Magnetic Field

Since $\theta = \pi/2$ with the applied field $\mathbf{B}_0 \parallel \hat{\mathbf{e}}_z$ and the propagation vector $\mathbf{k} \parallel \hat{\mathbf{e}}_x$, the condition Eq. (11.14)

$$\det(\mathbb{M}) = \begin{vmatrix} S & -\imath D & 0 \\ \imath D & S - n^2 & 0 \\ 0 & 0 & P - n^2 \end{vmatrix} = (P - n^2) \begin{vmatrix} \frac{1}{2}(R+L) - n^2 & -\frac{1}{2}\,\imath\,(R-L) \\ \frac{1}{2}\,\imath\,(R-L) & \frac{1}{2}(R+L) - n^2 \end{vmatrix} = 0 \tag{11.46}$$

whose roots are

$$n^2 = P \qquad \text{and} \qquad n^2 = \frac{R\,L}{S} \tag{11.47}$$

A sketch of the resulting dispersion relation is shown in Figure 11.3. In this case, the coefficients in Eq. (11.15) are $A = S$, $B = (R\,L + P\,S)$, and $F = (P\,S - R\,L)$ which determines the nature of the wave. We consider each solution in turn

i $n^2 = P$: Making use of Eq. (11.14) which determines the electric field, it is easily seen that E_x, $E_y = 0$ and $E_z \neq 0$. It then follows that the velocity components $v_x = v_y = 0$ and $v_z \neq 0$. The wave is a transverse electron oscillation along the applied magnetic field. Consequently, there is no Lorentz force due to the applied magnetic field, and therefore, the wave propagates independently of the magnetic field. Substituting for n and P, we obtain the frequency

$$\omega^2 = \sum {\Pi_q}^2 + k^2\,c^2 \approx {\Pi_e}^2 + k^2\,c^2 \tag{11.48}$$

This is the dispersion relation for a polarised electro-magnetic wave propagating in a plasma with phase velocity $V_p = c\,\sqrt{(1 - {\Pi_e}^2/\omega^2)}$ with electric field parallel to the ambient magnetic field. This mode is the upper branch in Figure 11.3 is known as the *ordinary wave* or O-mode. It is easily seen from Section 11.3.1 that it occurs if the background field is absent.

ii $n^2 = R\,L\,/\,S$: Making use of Eq. (11.14), the electric field has non-zero eigenvector $\mathbf{E} = E_0\,(1, -\imath\,S\,/\,D,\,0)$. Correspondingly the velocity components are only non-zero in the plane perpendicular to the applied field.
The cut-offs are easily seen to occur when either $R = 0$ or $L = 0$. The cut-off frequencies are given by the expressions for ω_R and ω_L found earlier, namely Eqs. (11.24) and (11.26), respectively. The resonant frequencies are the roots of the expression

$$S = 1 - \frac{{\Pi_e}^2}{\omega^2 - {\Omega_e}^2} - \frac{{\Pi_i}^2}{\omega^2 - {\Omega_i}^2} = 0 \tag{11.49}$$

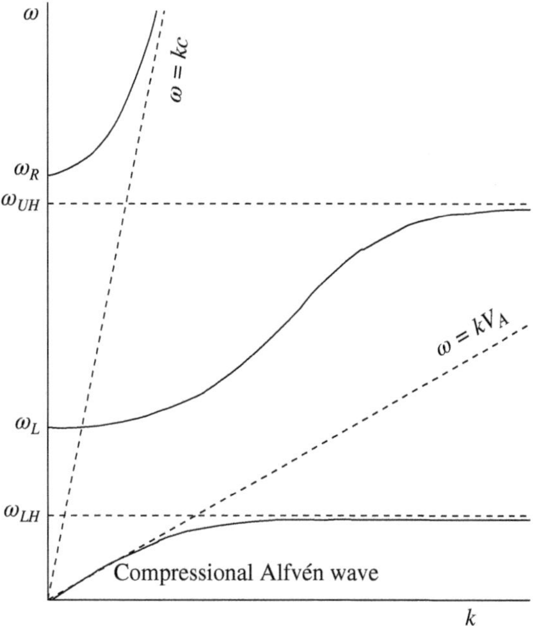

Figure 11.3 Sketch of the dispersion relation for radiation propagating perpendicular to the magnetic field.

If the third term in this expression is small compared to the other two, we obtain the root known as the *upper hybrid frequency*

$$\omega_{UH} = \sqrt{(\Pi_e{}^2 + \Omega_e{}^2)} \tag{11.50}$$

It is easily checked that since $m_e \ll m_i$, the value of the omitted term is small. The upper hybrid oscillation is an electron longitudinal mode aligned perpendicular to the applied magnetic field. It is closely related to the electron plasma wave due to the electro-static Coulomb restoring force modified by an addition restoring force due to the Lorentz term.[2]

To obtain the second root, we follow Cairns (1985) and note that the product of the roots is

$$\Omega_e{}^2\,\Omega_i{}^2 + \Pi_e{}^2\,\Omega_i{}^2 + \Pi_i{}^2\,\Omega_e{}^2 \approx \Omega_e{}^2\,\Omega_i{}^2 + \Pi_i{}^2\,\Omega_e{}^2 \tag{11.51}$$

to obtain the lower hybrid frequency

$$\omega_{LH} = \sqrt{\frac{\Omega_e{}^2\,\Omega_i{}^2 + \Pi_i{}^2\,\Omega_e{}^2}{\Pi_e{}^2 + \Omega_e{}^2}} \tag{11.52}$$

2 The nature of this wave is illustrated by the following model. The force due to the magnetic field is easily calculated. Let all wave quantities vary as exp $[\imath\,(k\,x - \omega\,t)]$ and let δ_x be the displacement of electrons in the direction of propagation x then

$$v_x = \imath\,\omega\,\delta_x \qquad f_y = -ev_xB_z = \imath\,\omega\,(-e\,B_z)\,\delta_x$$
$$v_y = -f_y/\imath\,\omega\,m_e = (-e\,B_z/m_e)\delta_x \qquad f_x = -v_y\,B_z = -m_e\,(-e\,B_z/m_e)^2\,\delta_x = -m_e\,\Omega_e{}^2\,\delta_x$$

Adding the electro-static restoring force $-m_e\,\Pi_e{}^2\,\delta_x$ to the Lorentz force, the frequency of the resulting oscillation

$$\omega_{UH} = \sqrt{\Pi_e{}^2 + \Omega_e{}^2} \tag{11.50:a}$$

The electron path is an ellipse in the plane normal to the magnetic field.

Making use of the condition $m_i \gg m_e$, this frequency may be expressed in a number of equivalent ways

$$\omega_{LH}{}^{-2} = (\Omega_e\,\Omega_i)^{-1} + (\Omega_i{}^2 + \Pi_i{}^2)^{-1} \approx (\Omega_e\,\Omega_i)^{-1} + \Pi_i{}^{-2} \tag{11.52:a}$$

The lower hybrid wave is a longitudinal oscillation of both electrons and ions. It is difficult to achieve in practice as the direction of propagation must be closely aligned perpendicular to the field lines (within about $\sqrt{(m_e/m_i)}$rdns), or the electrons can move sufficiently rapidly along the field lines to shield the oscillations and create an electrostatic ion cyclotron wave. Lower hybrid waves have been considered as a method to heat plasma in fusion devices.
It is not difficult to show that the characteristic frequencies satisfy the following inequalities

$$\omega_{LH} < \omega_L < \omega_{UH} < \omega_R \tag{11.53}$$

We may now interpret the dispersion diagram (Figure 11.3). At low frequencies ($\omega < \omega_{LH}$), the wave is a compressional Alfvén wave (torsional Alfvén waves do not propagate normal to the field). This wave vanishes if the field is absent.
At medium frequencies ($\omega_L < \omega < \omega_{UH}$), the wave becomes a plasma wave in the absence of a magnetic field.
At high frequencies ($\omega > \omega_L$), the wave is a polarised electro-magnetic wave with electric field normal to both the magnetic field and the direction of propagation (y direction).
These modes constitute the *extraordinary* or X-mode.

11.3.5 Resonance in Plasma Waves

In Chapter 5, we show the importance of damping in waves when the driving frequency matches the characteristic frequency of the wave. As we saw in Section 5.4.2, the energy absorbed by the resonance does not affect the energy absorbed. None the less the structure of the resonance is dependent on the ability of the wave to dissipate the energy absorbed from the pump. In plasma, the net energy flux is determined by the time averaged Poynting vector (see footnote 4 in Section 3.1.1).

$$\mathbf{S} = \frac{1}{2\mu_0}(\mathbf{E} \wedge \mathbf{B}^* + \mathbf{E}^* \wedge \mathbf{B}) \tag{11.54}$$

The energy delivered by the wave Q is equal to minus that gained by the wave per unit volume namely the divergence of the flux

$$Q = -\nabla \cdot \mathbf{S} = -\frac{\imath\,\omega}{4}(\epsilon - \epsilon^*)\,\mathbf{E} \cdot \mathbf{E}^* = \omega\,\Im(\epsilon)\,\overline{\mathbf{E}^2} \tag{11.55}$$

where $\overline{\mathbf{E}^2} = 1/2\,\mathbf{E} \cdot \mathbf{E}^*$ is the mean squared electric field averaged over a cycle.

To demonstrate this effect, consider the simple case of a wave travelling parallel to the background magnetic field $\hat{\mathbf{e}}_z$ polarised with the electric field in $\hat{\mathbf{e}}_y$ direction. The electric field satisfies Eq. (5.38). The refractive index is $b/(z + \imath\,\varepsilon)$ with $b > 0$ and $\varepsilon < 0$ if the resonance absorbs energy from the incoming wave. Making use of Eq. (11.55), it follows that the power absorbed by the plasma

$$Q = \frac{1}{2}\,\omega\,\epsilon_0\,\frac{b\,\varepsilon}{(z^2 + \varepsilon^2)}|E_y{}^2| = -\frac{k_0{}^2 b}{2\,\mu_0\,\omega}\,\frac{\varepsilon}{(z^2 + \varepsilon^2)}|E_y{}^2| \tag{11.56}$$

since $k_0{}^2 = \omega^2/c^2 = \epsilon_0\,\mu_0\,\omega^2$.

Hence it follows the plasma is absorbing if $\epsilon < 0$ in accord with Section 5.38, since $b > 0$. The width of the resonance is given by the parameter $|\epsilon|$, but the energy absorbed is independent of the damping mechanism. The profile of the resonance $|\epsilon|/(z^2 + \epsilon^2)$ is characteristic of resonance, familiar from basic studies of forced oscillations.

As noted earlier, the damping may arise from several effects: convection, collisionless damping, etc. characteristic of plasma with finite temperature. However, the simplest mechanism arises from collisions between the particles, which re-distribute energy from the oscillation into thermal modes. Adding a simple collisional term using the Drude model described earlier in Section 2.2.3 with the simple B.G.K. collision term Eq. (6.11). Adding the momentum loss to the momentum transfer equation for particle q, Eq. (11.1), we obtain

$$m_q \frac{d\mathbf{v}_q}{dt} =\approx Z_q\, e\,(\mathbf{E} + \mathbf{v}_q \wedge \mathbf{B}_0) - m_q\, \nu_q\, \mathbf{v}_q \tag{11.57}$$

where ν_q is the collision frequency for particle q. Assuming the collision frequency is independent of the field we may solve the dispersion equation following the same procedure as before by replacing ω by $\omega + \imath\, \nu_q$ to obtain

$$R = 1 - \sum_q \frac{\Pi_q^{\;2}}{(\omega + \imath\, \nu_q)(\omega + \Omega_q + \imath\, \nu_q)} \tag{11.58a}$$

$$L = 1 + \sum_q \frac{\Pi_q^{\;2}}{(\omega + \imath\, \nu_q)(\omega - \Omega_q + \imath\, \nu_q)} \tag{11.58b}$$

$$P = 1 - \sum_q \frac{\Pi_q^{\;2}}{(\omega + \imath\, \nu_q)^2} \tag{11.58c}$$

To illustrate the collisional behaviour, consider the electron plasma resonance in right-handed circularly polarised waves along the magnetic field. In this case, the frequency is much greater than the ion cyclotron frequency and the dispersion relation

$$n^2 = R \approx 1 - \frac{\Pi_e^{\;2}}{\omega - |\Omega_e| + \imath\, \nu_e} \tag{11.59}$$

Assuming that in neighbourhood of the resonance point $|\Omega_e| \approx \omega + |\Omega_e|'z$, the refractive index

$$n^2 \approx 1 + \frac{b}{z + \imath\, \epsilon} \tag{11.60}$$

where $b = \Pi_e^{\;2}/\omega\, |\Omega_e|'$ and $\epsilon = -\nu_e/|\Omega_e|'$. The width of the resonance is approximately $|\epsilon| = \nu_e/|\Omega_e|'$. The thickness of the resonant layer is therefore determined by damping mechanism.

12

Waves in Magnetised Warm Plasma

12.1 The Dielectric Properties of Unmagnetised Warm Dilute Plasma

Let us examine the response of warm plasma of temperature T to a weak imposed electromagnetic wave with wave vector $\mathbf{k}$ and frequency ω. If the frequency is sufficiently large ($\omega \gg \omega_i$), the ions are effectively stationary and we need consider only the response of the electrons. The ions provide a background charge which neutralises the electron charge associated with the unperturbed electron distribution. The net space charge is therefore generated by the perturbed electron distribution alone. At long wavelengths, this leads to a coherent, co-operative motion described by the Vlasov equation assuming collisions are weak.

The applied electric field may be described in a general manner as a wave of angular frequency ω and wave vector $\mathbf{k}$ where the relationship $\omega(\mathbf{k})$ is given by the appropriate dispersion relationship, which we wish to find:

$$\mathbf{E}(\mathbf{r}, t) = \Re\{\mathbf{E}_0 \exp[\imath(\mathbf{k}\cdot\mathbf{r} - \omega t)]\} \tag{12.1}$$

Any general waveform is simply represented by the summation of simple waves determined by its Fourier expansion, since the perturbation is weak. The response of the plasma to this field is given by the Vlasov equation:

$$\frac{\partial f}{\partial t} + \mathbf{v}\cdot\frac{\partial f}{\partial \mathbf{r}} + \frac{q}{m}\mathbf{E}\cdot\frac{\partial f}{\partial \mathbf{v}} = 0 \quad \left(\text{or } \left.\frac{\partial f}{\partial t}\right|_{\text{coll}}\right) \tag{12.2}$$

allowing for the possibility of a collision term.

Since we assume that the applied field is weak, its effect may be described as a first-order perturbation: $f = f_0 + f_1$, where f_0 is the ambient distribution (assumed Maxwellian) and f_1 the perturbation ($f_0 \ll f_1$). In general, this leads to separate equations for the electrons and ions; however, as noted earlier, since the ion mass is large, the effect of the field on the ions is much weaker than that on the electrons, and therefore can be neglected. Substituting and retaining terms up to first order

$$\frac{\partial f_0}{\partial t} = 0 \quad \text{and} \quad \frac{\partial f_1}{\partial t} + \mathbf{v}\cdot\frac{\partial f_1}{\partial \mathbf{r}} - \frac{e}{m}\mathbf{E}_0 \exp\{\imath(\mathbf{k}\cdot\mathbf{r}) - \omega t\}\cdot\frac{\partial f_0}{\partial \mathbf{v}} = 0 \tag{12.3}$$

Expressing the general form $f_1(\mathbf{v}, \mathbf{r}, t)$ as a Fourier transform $\tilde{f}_1(\mathbf{v}, \mathbf{k}, \omega)$

$$f_1(\mathbf{v}, \mathbf{r}, t) = \iiiint \mathrm{d}\omega\, \mathrm{d}\mathbf{k}\, \tilde{f}_1(\mathbf{v}, \mathbf{k}, \omega) \exp\{\imath(\mathbf{k}\cdot\mathbf{r} - \omega t)\} \tag{12.4}$$

Foundations of Plasma Physics for Physicists and Mathematicians, First Edition. Geoffrey J. Pert.

for which equation (12.3) yields

$$\imath\,(\mathbf{k}\cdot\mathbf{v}-\omega)\tilde{f}_1 = \frac{e}{m}\mathbf{E}_0\cdot\frac{\partial f_0}{\partial\mathbf{v}} \tag{12.5}$$

and thus the expression for the perturbation

$$\tilde{f}_1(\mathbf{v},\mathbf{k},\omega) = \frac{e\,\mathbf{E}_0\cdot\partial f_0/\partial\mathbf{v}}{\imath\,m\,(\mathbf{k}\cdot\mathbf{v}-\omega)} \tag{12.6}$$

From the perturbed distribution function $\tilde{f}(\mathbf{k},\omega)$, Eq. (12.6), we calculate the Fourier components of the charge and current densities

$$\left.\begin{array}{lll}\rho(\mathbf{r},t) & \rightarrow & \tilde{\rho}(\mathbf{k},\omega) = -e\int\tilde{f}_1(\mathbf{v},\mathbf{k},\omega)\,\mathrm{d}\mathbf{v}\\ \mathbf{j}(\mathbf{r},t) & \rightarrow & \tilde{\mathbf{j}}(\mathbf{k},\omega) = -e\int\tilde{f}_1(\mathbf{v},\mathbf{k},\omega)\,\mathbf{v}\,\mathrm{d}\mathbf{v}\end{array}\right\}\exp\left[\imath\,(\mathbf{k}\cdot\mathbf{r}-\omega\,t)\right] \tag{12.7}$$

by integration over the velocity. These are wave terms with the same frequency and wave number as the applied field.

12.1.1 Plasma Dispersion Relation

We seek to find a relationship between the frequency and the wave number of disturbances propagated as a result of the electron oscillations. This is usually expressed in the functional form $D(\omega,\mathbf{k}) = 0$, the starting point being Maxwell's equations:

$$\begin{array}{ll}\dfrac{1}{\mu_0}\nabla\wedge\mathbf{B} = \mathbf{j}+\epsilon_0\dfrac{\partial\mathbf{E}}{\partial t} & \nabla\wedge\mathbf{E} = -\dfrac{\partial\mathbf{B}}{\partial t}\\ \nabla\cdot\mathbf{B} = 0 & \nabla\cdot\mathbf{E} = \dfrac{\rho}{\epsilon_0}\end{array} \tag{12.8}$$

which may be reduced by successive substitutions to the familiar form of a wave equation

$$\nabla^2\mathbf{E}-\epsilon_0\mu_0\frac{\partial^2\mathbf{E}}{\partial t^2} = \mu_0\frac{\partial\mathbf{j}}{\partial t}+\frac{1}{\epsilon_0}\nabla\rho \tag{12.9}$$

For sinusoidally oscillating plane waves propagating consistently with Eq. (12.1), the electric field $\mathbf{E}$, current density $\mathbf{j}$, and charge density ρ

$$\left(k^2-\frac{\omega^2}{c^2}\right)\mathbf{E} = \frac{\imath}{\epsilon_0}\left(\frac{1}{c^2}\,\omega\mathbf{j}-\mathbf{k}\,\rho\right) \tag{12.10}$$

Transverse waves are characterised by the electric field $\mathbf{E}$ perpendicular to the wave vector $\mathbf{k}$, i.e. $\mathbf{k}\cdot\mathbf{E} = 0$ and $\mathbf{k}\wedge\mathbf{E}\neq 0$, or $\nabla\cdot\mathbf{E} = 0$ and $\nabla\wedge\mathbf{E}\neq 0$. Transverse waves are therefore electromagnetic waves familiar across the spectrum from X-rays to radio.

Similarly, longitudinal waves have $\mathbf{E}$ parallel to $\mathbf{k}$, i.e. $\mathbf{k}\cdot\mathbf{E}\neq 0$ and $\mathbf{k}\wedge\mathbf{E} = 0$, or $\nabla\wedge\mathbf{E} = 0$ and $\nabla\cdot\mathbf{E}\neq 0$. Longitudinal waves are therefore electrostatic with no magnetic field component. These waves are the warm plasma wave equivalent of cold plasma waves discussed in Section 1.2.

We can identify the dispersion relations for the two different classes of waves by taking the vector and scalar products of Eq. (12.10) with $\mathbf{k}$ to separate them

$$\left(k^2-\frac{\omega^2}{c^2}\right)\mathbf{k}\wedge\mathbf{E} = \frac{\imath}{\epsilon_0\,c^2}\,\omega\,\mathbf{k}\wedge\mathbf{j} = \frac{\omega^2}{c^2}\,\frac{\imath}{\omega\,\epsilon_0}\,\mathbf{k}\wedge\mathbf{j} \tag{12.11}$$

$$\left(k^2-\frac{\omega^2}{c^2}\right)\mathbf{k}\cdot\mathbf{E} = \frac{i}{\epsilon_0}\left(\frac{\omega}{c^2}\mathbf{k}\cdot\mathbf{j}-k^2\rho\right) \tag{12.12}$$

respectively. Both equations must be simultaneously obeyed for each class of wave.

12.1.1.1 Dispersion Relation for Transverse Waves

When $E_{\|} = 0, j_{\|} = -\imath\,\omega\,(\epsilon_{\|} - \epsilon_0)\,E_{\|} = 0$ and $\rho = -k\,j_{\|}/\omega = 0$. The scalar product of Eq. (12.10) with **k**, namely Eq. (12.12) is satisfied. The wave is therefore purely electromagnetic as deduced earlier. Turning to the vector product of Eq. (12.10) with **k**, namely Eq. (12.12), and making use of Eq. (13.3)

$$\epsilon_\perp \mathbf{E}_\perp = \epsilon_0 \mathbf{E}_\perp - \frac{1}{\imath\,\omega}\mathbf{j}_\perp \tag{12.13}$$

to eliminate **j** from Eq. (12.12) we obtain

$$\left[k^2 - \frac{\omega^2}{c^2}\frac{\epsilon_\perp}{\epsilon_0}\right]\mathbf{k}\wedge\mathbf{E} = 0 \tag{12.14}$$

Since the wave is transverse $\mathbf{k}\wedge\mathbf{E} \neq 0$ and the dispersion relation for transverse waves is therefore given by:

$$\frac{k^2\,c^2}{\omega^2} = \frac{\epsilon_\perp}{\epsilon_0} \tag{12.15}$$

12.1.1.2 Dispersion Relation for Longitudinal Waves

When $E_\perp = 0, j_\perp = 0$ and the vector product of Eq. (12.10) with **k** namely Eq. (12.12) is obeyed, the wave is purely electrostatic as already identified.

The equation of current continuity:

$$\frac{\partial\rho}{\partial t} + \nabla\cdot\mathbf{j} = 0 \quad\rightarrow\quad -\imath\,(\omega\rho - \mathbf{k}\cdot\mathbf{j}) = 0 \tag{12.16}$$

can be written as

$$\frac{\omega}{c^2}\mathbf{k}\cdot\mathbf{j} - k^2\rho = \frac{1}{\omega}\left[\frac{\omega^2}{c^2} - k^2\right]\mathbf{k}\cdot\mathbf{j} \tag{12.17}$$

Substituting in the scalar product of Eq. (12.10) with **k**, namely Eq. (12.12), we obtain

$$\left(k^2 - \frac{\omega^2}{c^2}\right)\mathbf{k}\cdot\mathbf{E} = -\frac{\imath}{\omega\,\epsilon_0}\left(k^2 - \frac{\omega^2}{c^2}\right)\mathbf{k}\cdot\mathbf{j} \tag{12.18}$$

Making use of Eq. (13.3)

$$\varepsilon_{\|}E_{\|} = \epsilon_0 E_{\|} - \frac{1}{\imath\,\omega}j_{\|} \tag{12.19}$$

and substituting for **j** we obtain

$$\left[k^2 - \frac{\omega^2}{c^2}\right]\epsilon_{\|}\,\mathbf{k}\cdot\mathbf{E} = 0 \tag{12.20}$$

Since we require $E_{\|} \neq 0$, and $\omega^2 = k^2\,c^2$ is an electromagnetic wave in free space and therefore not an appropriate solution, the condition

$$\epsilon_{\|}(k,\omega) = 0 \tag{12.21}$$

must therefore be the dispersion relationship for longitudinal waves.

12.1.2 Dielectric Constant of a Plasma

Since the dielectric constant $\epsilon(\mathbf{k},\omega)$ depends on the relationship between two vectors **E** and **j** (or **D**), it is a second-order tensor. In the absence of a magnetic field, we may identify two configurations, one with the electric field parallel to and one with the field perpendicular to the wave vector

(Section 13.1.1). The two resulting electron motions are uncoupled and give rise to two independent components of the dielectric constant which may be taken parallel $\epsilon_\parallel$ and perpendicular $\epsilon_\perp$ to the wave vector **k** and represent longitudinal and transverse waves, respectively.

Taking into account the oscillatory nature of the fields, the charge density and current amplitudes are from Eq. (13.3)

$$\rho_0 = -\imath\,(\epsilon - \epsilon_0)\,\mathbf{k}\cdot\mathbf{E}_0 \qquad \text{and} \qquad \mathbf{j}_0 = -\imath\,(\epsilon - \epsilon_0)\,\omega\,\mathbf{E}_0 \tag{12.22}$$

where ρ_0 and $\mathbf{j}_0$ are the amplitudes of the charge density and current, respectively.

Substituting from Eqs. (12.6) and (12.7).

$$\rho_0 = -\frac{e^2}{m}\int \frac{\mathbf{E}_0\cdot\partial f_0/\partial\mathbf{v}}{[\imath\,(\mathbf{k}\cdot\mathbf{v}-\omega)]}\,\mathrm{d}\mathbf{v} \tag{12.23}$$

$$\mathbf{j}_0 = -\frac{e^2}{m}\int \frac{\mathbf{E}_0\cdot\partial f_0/\partial\mathbf{v}}{[\imath\,(\mathbf{k}\cdot\mathbf{v}-\omega)]}\,\mathbf{v}\,\mathrm{d}\mathbf{v} \tag{12.24}$$

To evaluate the integrals, we note that the equilibrium distribution function f_0 is symmetric in **v**, and therefore, it is an even function of **v**. Consequently, the derivatives $\partial f_0/\partial v_\parallel$, $\partial f_0/\partial v_1$, and $\partial f_0/\partial v_2$ are all odd in $v_\parallel$, v_1, and v_2, respectively, where $(v_1, v_2) = \mathbf{v}_\perp$ are the orthogonal velocity components perpendicular to **k**. However, terms such as $v_\parallel\,\partial f_0/\partial v_\parallel$ are even.

Consider the charge density integral. Integrating the terms and noting that the odd terms vanish:

$$\int \frac{\mathbf{E}_0\cdot\partial f_0/\partial\mathbf{v}|_1}{[\imath\,(k\,v_\parallel-\omega)]}\,\mathrm{d}v_1 = \int \frac{\mathbf{E}_0\cdot\partial f_0/\partial\mathbf{v}|_2}{[\imath\,(k\,v_\parallel-\omega)]}\,\mathrm{d}v_2 = 0 \tag{12.25}$$

so that the only remaining term is over the parallel components

$$\rho_0 = -\frac{e^2}{m}E_{0\parallel}\int_L \frac{\partial f_0/\partial v_\parallel}{[\imath\,(k\,v_\parallel-\omega)]}\,\mathrm{d}v_\parallel \tag{12.26}$$

Proceeding in a similar fashion eliminating odd terms on integration, we obtain for the current density **j** the following expressions

$$j_{0\parallel} = -\frac{e^2}{m}E_{0\parallel}\int_L v_\parallel\,\frac{\partial f_0/\partial v_\parallel}{[\imath\,(k\cdot v_\parallel-\omega)]}\,\mathrm{d}\mathbf{v} \qquad \text{and} \qquad j_{01,2} = -\frac{e^2}{m}E_{01,2}\int_L v_{1,2}\,\frac{\partial f_0/\partial v_{1,2}}{[\imath\,(k\,v_\parallel-\omega)]}\,\mathrm{d}\mathbf{v} \tag{12.27}$$

The charge and current densities are not independent, but satisfy the equation of continuity

$$\frac{\partial\rho}{\partial t} + \nabla\cdot\mathbf{j} = \imath\,(\omega\rho_0 - \mathbf{k}\cdot\mathbf{j}_0) = -\frac{e^2}{m}E_{0_\parallel}\int \frac{\imath\,(\mathbf{k}\cdot\mathbf{v}-\omega)}{\imath\,(\mathbf{k}\cdot\mathbf{v}-\omega)}\frac{\partial f_0}{\partial v_\parallel}\,\mathrm{d}\mathbf{v} = 0 \tag{12.28}$$

From Eqs. (12.26) and (12.27), it is clear that, as noted earlier, there are two decoupled dielectric motions: ∥ longitudinal parallel to the field propagation direction, **k** and ⊥ transverse, perpendicular parallel to the field propagation direction, **k**. Since

$$\left.\begin{matrix}(\epsilon_\parallel - \epsilon_0)\\(\epsilon_\perp - \epsilon_0)\end{matrix}\right\} = \frac{1}{\imath\,\omega}\left\{\begin{matrix}j_\parallel\,/\,E_\parallel\\ j_{1/2}\,/\,E_{1/2}\end{matrix}\right.$$

The corresponding dielectric constants have components

$$\frac{\epsilon_\parallel}{\epsilon_0} = 1 + \frac{\omega_p^2}{\omega^2}\int \frac{v_\parallel\,\partial\hat{f}_0/\partial v_\parallel}{(1-k\,v_\parallel/\omega)}\mathrm{d}\mathbf{v} = 1 + \frac{\Pi^2}{\omega^2}\left[\left(\frac{\omega^2}{k^2}\right)\int \frac{\partial\hat{f}_0/\partial v_\parallel}{(\omega/k - v_\parallel)}\,\mathrm{d}\mathbf{v}\right] \tag{12.29}$$

$$\frac{\epsilon_\perp}{\epsilon_0} = 1 + \frac{\Pi^2}{\omega^2}\int \frac{\mathbf{v}_\perp\cdot\partial\hat{f}_0/\partial\mathbf{v}_\perp}{(1-k\,v_\parallel/\omega)}\,\mathrm{d}\mathbf{v} = 1 - \frac{\Pi^2}{\omega^2}\left[\frac{\omega}{k}\int \frac{\hat{f}_0}{(\omega/k - v_\parallel)}\,\mathrm{d}\mathbf{v}\right] \tag{12.30}$$

where $\Pi^2 = n_e e^2/\epsilon_0 m$ is the plasma frequency, $\hat{f}_0 = f_0/n_e$ the probability density of electrons, i.e. the probability of finding an electron per unit volume.

12.1.2.1 The Landau Contour Integration Around the Singularity

The denominators of these expressions are singular at the resonance condition when $v_\parallel$ (the component of $\mathbf{v}$ parallel to $\mathbf{k}$) satisfies:

$$kv_\parallel = \omega \tag{12.31}$$

Electrons satisfying this condition move with the phase velocity of the wave $v_\parallel = \omega/k$. Alternatively, the Doppler shift experienced by the particles moving with velocity $v_\parallel = \omega/k$ brings the field to rest relative to the particles, thereby inducing a resonance condition.

As we have seen, the Vlasov equation is reversible. However as we shall show, the introduction of causality necessarily introduces damping into the motion. Landau (1946) realised the resolution of this problem lay in the fact that the time development of the external field could not be simply defined by the infinite sinusoidal waveform (12.1). In a physical situation, the field must be switched in some way which gives rise to an imaginary component of the frequency term $\omega \to \omega + \imath\gamma$. The 'switch-on' term is chosen to satisfy causality, i.e. the field is initially zero. This can be achieved in one of three ways which maintain the necessary physical reality (see Thompson, 1962; Lifshitz and Pitaevskii, 1981, §29):

i. Slow switch on starting from time $t = -\infty$. The field is slowly increased by introducing a term $\exp(\gamma t)$ into Eq. (12.1) thereby making ω complex, and letting $\gamma \to 0$.
ii. Include weak collisions. Add a simple collision term, for example of the Bernstein–Greene–Kruskal (BGK) form, to Eq. (12.2) with relaxation time $\tau \to \infty$.
iii. An initial value problem. The field is switched at time $t = 0$. The field is described by a Laplace transform. This is Landau's original method and perhaps formally most satisfactory. It is treated in many standard texts (Stix, 1962; Montgomery and Tidman, 1964; Boyd and Sanderson, 2003), but since the mathematics is more complicated, we refer the reader to Landau's original paper (Landau, 1946), or to the works referenced above.

In each of these approaches, the denominator of (12.6) may be changed to $[\imath(\mathbf{k}\cdot\mathbf{v} - \omega) + \epsilon]$, where ϵ is a very small quantity. Provided $\epsilon > 0$, this term explicitly introduces causality, introducing a forward direction in time. Although as $\epsilon \to 0$, the system still becomes thermodynamically reversible.

$$\rho_0 = -\frac{e^2}{m}\lim_{\epsilon\to 0}\int \frac{\mathbf{E}_0 \cdot \partial f_0/\partial \mathbf{v}}{\{\imath[\mathbf{k}\cdot\mathbf{v} - (\omega + \imath\epsilon)]\}}\,\mathrm{d}\mathbf{v} \tag{12.32}$$

$$\mathbf{j}_0 = -\frac{e^2}{m}\lim_{\epsilon\to 0}\int \frac{\mathbf{E}_0 \cdot \partial f_0/\partial \mathbf{v}}{\{\imath[\mathbf{k}\cdot\mathbf{v} - (\omega + \imath\epsilon)]\}}\,\mathbf{v}\,\mathrm{d}\mathbf{v} \tag{12.33}$$

As we have seen in Section 13.1, these integrals are of a type characteristic of the introduction of causality, which involve a singularity at the phase matching point $v_\parallel = \omega/k$. We have already seen how to treat this singularity in Section 5.4.2 by introducing a branch-cut from the pole in the upper half-plane. This allows us to consider a real function $g(x)$ analytic in the lower half-plane as the integrand in

$$\lim_{\epsilon\to 0}\int_{-\infty}^{\infty} \frac{g(z)}{z - (x_0 + \imath\epsilon)}\,\mathrm{d}z \tag{12.34}$$

which has a pole at $x_0 + \imath\epsilon$. Taking the integral along the real axis avoiding the pole, the path of integration passes below the pole Figure 12.1a.

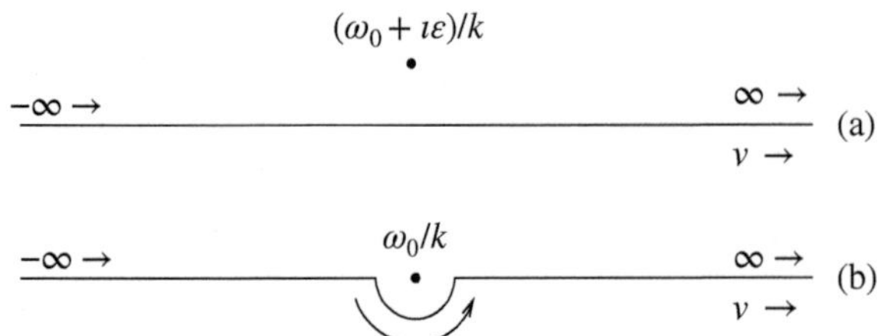

Figure 12.1 The path of integration for the Landau prescription for finite ϵ (a) and the limit $\epsilon \rightarrow 0$ (b).

As $\epsilon \rightarrow 0$, the pole approaches the axis until in the limit it lies on it, and we deform the path of the integral so that it always lies below the pole. Since f_0 is a Maxwellian, it is easily established that the integrand is analytic in the lower half-plane. Therefore, it follows from Cauchy's theorem that the value of the integral is unchanged (Section M.3.M.3.iv). Let the deformed path be along the real axis from $-\infty$ to a point $(x_0 - \delta)$, a circular path of radius δ around the pole and continuing from $(x_0 + \delta)$ along the axis to ∞ Figure 12.1b.

Let the circular path around the pole be $z = x_0 + \delta e^{\imath\,\theta}$ with θ running from $-\pi$ to 0, since the path is anti-clockwise (see Section M.3.M.3.x). Hence, in the limit $\epsilon \rightarrow 0$, the integral becomes as $\delta \rightarrow 0$

$$\lim_{\delta\to 0}\left\{\int_{-\infty}^{(x_0-\delta)} + \int_{-(x_0+\delta)}^{\infty}\right\}\frac{g(x)}{(x-x_0)}\,dx + \imath g(x_0)\int_{\pi}^{0} d\theta = \fint_{-\infty}^{\infty}\frac{g(x)}{(x-x_0)}\,dx + \imath\,\pi\,g(x_0) \tag{12.35}$$

since $g(x)$ is continuous at x_0. The first term is the Cauchy principal value integral, defined as

$$\fint_{-\infty}^{\infty}\frac{f(x)}{x}dx = \lim_{\epsilon\to 0}\int_{-\infty}^{\infty}\frac{xf(x)}{(x^2+\epsilon^2)}\,dx = \lim_{\delta\to 0}\left\{\int_{-\infty}^{-\delta}\frac{f(x)}{x}\,dx + \int_{\delta}^{\infty}\frac{f(x)}{x}\,dx\right\} \tag{12.36}$$

provided the function $f(x)$ is continuous at $x = 0$.[1]

Since the denominator of the integrand is $(\omega/k - v_{\|})$ leading to a change of sign of the term containing the delta function, we obtain the Landau prescription for these integrals:

$$\int_L \frac{1}{x} = \fint \frac{1}{x} + \imath\pi\delta(x) \tag{12.38}$$

where $\delta(x)$ is the Dirac delta function.

Integrating along the Landau contour, the space charge density for example becomes

$$\rho_0 = -e\int f\,\mathbf{dv} = -\frac{e^2}{m}\lim_{\epsilon\to 0}\int \frac{\mathbf{E}_0\cdot\partial f_0/\partial\mathbf{v}}{\{\imath\,[\mathbf{k}\cdot\mathbf{v} - (\omega + \imath\,\epsilon)]\}}\,\mathbf{dv} = -\frac{e^2}{m}\int_L\frac{\mathbf{E}_0\cdot\partial f_0/\partial\mathbf{v}}{\{\imath\,(\omega - \mathbf{k}\cdot\mathbf{v})\}}\,\mathbf{dv} \tag{12.39}$$

Defining the functions $\phi_{\|}(x)$ and $\phi_{\perp}(x)$, the permittivity can be expressed as

$$\phi_{\|}(x) = x^2\,G'(x) = x^2\int_L\frac{\partial\hat{f}_0/\partial v_{\|}}{(x - v_{\|})}\,\mathbf{dv} \qquad \text{and} \qquad \phi_{\perp}(x) = x\,G(x) = x\int_L\frac{\hat{f}_0}{(x - v_{\|})}\,\mathbf{dv} \tag{12.40}$$

---

1 This result is the linear form of the Sokhotski–Plemelj theorem (Section M.3.M.3.xi) which we will use later in the treatment of the Kramers–Kronig relations (Eqs. (13.11)). An alternative derivation avoiding the calculus of residues is as follows

$$\begin{aligned}\lim_{\epsilon\to 0}\int_{-\infty}^{\infty}\frac{g(x)}{x-(x_0\pm\imath\epsilon)}\,dx &= \lim_{\epsilon\to 0}\int_{-\infty}^{\infty}\frac{(x-x_0)\mp\imath\epsilon)}{(x-x_0)^2+\epsilon^2)}\,g(x)\,dx \\ &= \lim_{\epsilon\to 0}\int_{-\infty}^{\infty}\frac{(x-x_0)}{(x-x_0)^2+\epsilon^2}\,g(x)\,dx \mp \lim_{\epsilon\to 0}\int_{-\infty}^{\infty}\frac{\imath\,\epsilon}{(x-x_0)^2+\epsilon^2)}\,g(x)\,dx \\ &= \fint_{-\infty}^{\infty}\frac{g(x)}{(x-x_0)}\,dx \mp \imath\,\pi\,g(x_0)\end{aligned} \tag{12.37}$$

The second integral is performed by noting that for $\epsilon \approx 0$, $g(x) \approx g(x_0)$ and the integral reduces to the standard form $\int_{-\infty}^{\infty} dx/(1+x^2) = \pi$.

we obtain

$$\frac{\epsilon_{\parallel}}{\epsilon_0} = 1 + \frac{\Pi_e^{\ 2}}{\omega^2}\,\phi_{\parallel}\left(\frac{\omega}{k}\right) \qquad \text{and} \qquad \frac{\epsilon_{\perp}}{\epsilon_0} = 1 - \frac{\Pi_e^{\ 2}}{\omega^2}\,\phi_{\perp}\left(\frac{\omega}{k}\right) \tag{12.41}$$

where $\phi_{\parallel}$ and $\phi_{\perp}$ are components of the plasma dispersion function $G(x)$. Explicit expressions for the plasma dispersion function when the equilibrium distribution function is Maxwellian are given in Section 12.6.

These results may be easily generalised to consider a plasma made up of several species with different densities and temperatures

$$\frac{\epsilon_{\parallel}}{\epsilon_0} = 1 + \sum_s \frac{\Pi_s^{\ 2}}{\omega^2}\,\frac{\omega^2}{k^2}\,G_s'\left(\frac{\omega}{k}\right) \qquad \text{and} \qquad \frac{\epsilon_{\perp}}{\epsilon_0} = 1 - \sum_s \frac{\Pi_s^{\ 2}}{\omega^2}\,\frac{\omega}{k}\,G_s\left(\frac{\omega}{k}\right) \tag{12.41:a}$$

where $\Pi_s = q_s^2\, n_s/\epsilon_0\, m_s$ for particles of charges q_s, mass m_s and density n_s.

12.2 Transverse Waves

The dispersion relation for transverse waves can be written as a simple expression for the phase velocity

$$V_p = \frac{\omega}{k} = \frac{c}{\sqrt{\epsilon_{\perp}/\epsilon_0}} \tag{12.42}$$

The perpendicular component of the dielectric constant is given by Eq. (12.30). Assume initially that $V_p > c$, i.e. the phase velocity is therefore much larger than the typical electron velocity, namely the thermal velocity $V_p \gg (\sqrt{\kappa T/m})$, the velocity $v_{\parallel}$ may be neglected in comparison to $V_p = \omega/k$ in the denominator of Eq. (12.30) to give

$$\frac{\epsilon_{\perp}}{\epsilon_0} \approx 1 - \frac{\Pi_e^{\ 2}}{\omega^2} \tag{12.43}$$

The imaginary part is zero and the wave damping negligible. Thus, we obtain the phase velocity

$$V_p = \omega/k = c/\sqrt{(1 - \Pi_e^{\ 2}/\omega^2)} > c \tag{12.44}$$

consistent with our initial hypothesis.

This result holds generally for an arbitrary distribution because the phase velocity is greater than the velocity of light. Consequently, the distribution function at the resonance $f_0(V_p) = 0$. The term in $v_{\parallel}$ in Eq. (12.30) may be neglected and since $\int \tilde{f}_0(\mathbf{v})\, \mathrm{d}\mathbf{v} = 1$.

The phase velocity is positive only if $\omega > \Pi$, i.e. the frequency is greater than the plasma frequency. If this condition is not fulfilled, the electrons can respond faster than the applied electromagnetic wave and damp the wave. In this case, the wave is evanescent with k imaginary propagating only a limited distance into the plasma. This behaviour is found during the generation of plasma by lasers at solid surfaces, where the light penetrates into the plasma only as far as the critical density phase velocity is real, where the electron density is such that the local plasma frequency matches that of the wave

$$n_{e\text{crit}} = \epsilon_0\, m\, \omega^2/e^2 \tag{12.45}$$

Since the wave cannot travel beyond the point of reflection, the energy must be reflected back towards the vacuum. At normal incidence, the reflection occurs at the critical density, but at oblique incidence at a lower density. A wave incident on a plasma whose density increases to a value greater than the critical density is reflected.

Although the phase velocity exceeds the velocity of light $V_p > c$, the group velocity is

$$V_g = \mathrm{d}\omega/\mathrm{d}k = c\sqrt{(1 - \Pi_e{}^2/\omega^2)} < c \tag{12.46}$$

so that information is transferred slower than the speed of light, consistent with the theory of relativity.

Although the damping due to kinetic processes is negligibly small, electromagnetic waves in plasma are damped by collisions, a process known as inverse bremsstrahlung or collisional absorption. This is the dominant heating mechanism in many laser-plasma experiments.

12.3 Longitudinal Waves

The dispersion relation can be written as

$$\frac{\Pi_e{}^2}{\omega^2}\,\phi_\parallel = \frac{\Pi_e{}^2}{\omega^2}\,\frac{\omega^2}{k^2}\int_L \frac{\partial \tilde{f}_0/\partial v_\parallel}{\omega/k - v_\parallel}\,\mathbf{dv} = -1 \tag{12.47}$$

where $\phi_\parallel$ is complex due in part to the residue from the pole in the integrand and in part to the complex nature of the frequency

$$\omega = \Re(\omega) + \imath\,\Im(\omega) = \omega' - \imath\,\gamma \tag{12.48}$$

where γ is the damping rate. Provided $\omega' \gg \gamma$, the real part of the frequency is obtained directly from the principal part integral and the imaginary part by equating the complex part of the integral to the residue from the pole. Noting the integrals

$$\int \tilde{f}_0\,\mathbf{dv} = 1 \qquad \int v_\parallel \tilde{f}_0\,\mathbf{dv} = 0 \qquad \int m\,v_\parallel{}^2 \tilde{f}_0\,\mathbf{dv} = \kappa T \tag{12.49}$$

in the electron rest frame, it follows for long wavelengths $\omega/k \gg \kappa T$. Integrating by parts and approximating the divisor by the binomial expansion, we get

$$\begin{aligned}\Re\int_{-\infty}^{\infty} \frac{\partial \tilde{f}_0/\partial v_\parallel}{(\omega/k - v_\parallel)}\,\mathbf{dv} &\approx -\frac{k^2}{\omega^2}\int_{-\infty}^{\infty} \tilde{f}_0\left[1 + 2\left(\frac{k\,v_\parallel}{\omega}\right) + 3\left(\frac{k\,v_\parallel}{\omega}\right)^2 + \cdots\right]\mathbf{dv}\\ &= -\frac{k^2}{\omega^2}\left[1 + \frac{3\,k^2\,\kappa T}{m\,\omega^2} + \cdots\right] = -\frac{k^2}{\Pi_e{}^2}\end{aligned} \tag{12.50}$$

whence

$$\omega^2 \approx \Pi_e{}^2(1 + 3\,k^2\,\lambda_D{}^2) \tag{12.51}$$

This well-known result is known as the Bohm–Gross frequency (Bohm and Gross, 1949a). In a cold plasma $\omega^2 = \Pi_e{}^2$, the Bohm–Gross frequency reduces to that of a cold plasma wave, $3k^2\lambda_D^2$ being the correction for a warm plasma. Warm plasma changes the resonant frequency, introducing a wavelength dependence. Since the wave is now dispersive, the group velocity is non-zero and the wave propagates in contrast to the cold plasma case ($v_g \approx 3\,k\,\lambda_D\sqrt{\kappa T/m}$).

Turning now to the imaginary part of $\phi_\parallel$ and integrating by parts

$$\begin{aligned}&\Im \fint \mathbf{dv}\,\frac{\partial \tilde{f}_0/\partial v_\parallel}{(\omega/k - v_\parallel)} + \pi\int \mathbf{dv}_\perp\,\left.\frac{\partial \tilde{f}_0}{\partial v_\parallel}\right|_{v_\parallel=\omega/k}\\ &\approx \frac{2k^2\gamma}{\omega^3} + \pi\int \mathbf{dv}_\perp\,\left.\frac{\partial \tilde{f}_0}{\partial v_\parallel}\right|_{v_\parallel=\omega/k} = 0\end{aligned} \tag{12.52}$$

For a Maxwellian distribution,

$$\int \mathrm{d}\mathbf{v}_\perp \left.\frac{\partial \tilde{f}_0}{\partial v_\parallel}\right|_{v_\parallel=\omega/k} = -\sqrt{\frac{1}{2\,\pi}\frac{\omega}{k}}\sqrt{\frac{m^3}{(\kappa T)^3}}\exp\left[-\frac{m\,\omega^2}{2\,k^2\,\kappa T}\right] \tag{12.53}$$

Hence, substituting the Bohm–Gross frequency for ω and omitting small terms

$$\gamma = \frac{\pi}{2}\frac{\omega^3}{k^2}\int \mathrm{d}\mathbf{v}_\perp \left.\frac{\partial \tilde{f}_0}{\partial v_\parallel}\right|_{v_\parallel} \approx \sqrt{\frac{\pi}{8}}\frac{\Pi_e}{k^3\lambda_D{}^3}\exp\left[-\frac{1}{2\,k^2\lambda_D{}^2}-\frac{3}{2}\right] \tag{12.54}$$

This is weak for long wavelengths where $k\,\lambda_D \ll 1$. We get strong damping when $k\lambda_D > 1$ and the wavelength is short compared to the Debye length (Jackson, 1960). At these wavelengths, electrons are sufficiently mobile to neutralise the electric fields associated with the waves. Indeed no waves exist for very short wavelengths as they are damped before any oscillation is completed.

The Bohm–Gross frequency (12.51) is the modification to the plasma frequency (1.3) of cold plasma, which is introduced by the electron velocity distribution associated with the finite temperature. The distributed electron velocity thus introduces dispersion into the wave. As a consequence, the singularity at the plasma frequency of the cold plasma wave is removed. It follows from the Kramers–Kronig relations (13.12) that the varying dielectric constant implies damping, which we have identified as Landau damping. This behaviour is very general. We shall see that plasma waves in cold plasma (Section 11.2) with singularities are dispersed in thermal plasma and in consequence suffer damping losses.

12.4 Linear Landau Damping

The physical origin of Landau damping is obscure from the mathematical approach used to derive it. It is clear from the analysis that it is a consequence of the initial value condition used to treat the singularity at the resonance condition $v_\parallel = V_p = \omega/k$, and that it relies on the gradient of the distribution function at the resonance being negative $\partial\tilde{f}/\partial v_\parallel|_{\omega/k} < 0$. Indeed if this condition is not fulfilled it follows from Eq. (12.52) that the decay rate may be negative ($\gamma < 0$) and the wave grow rather than decay, i.e. the plasma is unstable. Like the Kramers–Kronig relation, to which it is closely allied, Landau's theory yields a damping rate, which is independent of the physical mechanism. If the field is weak, it is a consequence of the frequency response of the dispersion relation, that the wave must decay at the appropriate rate.

The absence of a physical mechanism in Landau's theory made the reality of Landau damping uncertain for many years. Bohm and Gross (1949a,b) identified the characteristic plasma wave solution and suggested that electron trapping (Section 12.5.1) accounted for the damping, but the rate did not agree with that calculated by Landau. Electron trapping was further analysed by Jackson (1960) using the model described in Section 12.5.1. For weak plasma waves, he obtained a damping rate qualitatively agreeing with Landau's formula. The situation was further clarified by van Kampen (1956). The issue was finally resolved by Dawson (1961) who derived the rate of resonant energy absorption which leads to the Landau damping rate (12.54).

12.4.1 Resonant Energy Absorption

At the resonant condition electrons have velocities which match the phase speed of the wave. Consequently, the electrons move with the wave seeing a nearly stationary longitudinal electric field which either accelerates them or decelerates them depending on their relationship with the phase of the wave. Electrons moving slightly slower than the wave in the accelerating phase are brought more closely into resonance, whereas electrons of the same speed at the decelerating phase are slowed. In fact the accelerated electrons experience the field for longer than the decelerated so that the net result is a gain in electron energy at the expense of the wave. Similarly, the electrons moving slightly faster than the wave experience a net loss of kinetic energy to the wave.[2] If the distribution function is decreasing at resonance $\partial\tilde{f}/\partial v_{\|}|_{\omega/k} < 0$, clearly more electrons gain energy than loose it, and the wave is damped. Furthermore, since the faster electrons gain energy and the slower loose it, the distribution is flattened to a plateau in the limit.

Following Dawson (1961), the linear phase of Landau damping is simply described by considering the perturbative motion of the near-resonant electrons in the wave field (see also Stix, 1962, Montgomery and Tidman, 1964, Kruer, 1988). The electric field and the velocity

$$E_{\|} = E_0 \cos(k\,z - \omega\,t) \qquad \text{and} \qquad m\frac{dv_{\|}}{dt} = e\,E_0 \cos(k\,z - \omega\,t) \tag{12.55}$$

To zero order electrons are nearly resonant and the velocity $v \equiv v_{\|} = v_0$ so that the position of the electron

$$z = z_0 + v_0\,t \tag{12.56}$$

The first-order perturbation velocity is obtained by substitution in Eq. (12.55) and integrating

$$m\frac{dv_1}{dt} = e\,E_0 \cos(k\,z_0 + k\,v_0\,t - \omega\,t) \quad \text{and} \quad v_1 = \frac{eE_0}{m}\frac{\sin(kz_0 + kv_0 t - \omega t) - \sin kz_0}{k\,v_0 - \omega} \tag{12.57}$$

As noted earlier, the damping rate must be evaluated as an initial value problem. We have therefore set the perturbation such that $v_1 = 0$ at time $t = 0$ in the preceding integration.

To calculate the next term in the perturbation, we correct for the first-order perturbation in Eq. (12.55). The perturbation to the position

$$z_1 = \int_0^t v_1\,dt = \frac{e\,E_0}{m}\left[\frac{-\cos(kz_0 + \alpha\,t) + \cos kz_0}{\alpha^2} - \frac{t\,\sin kz_0}{\alpha}\right] \tag{12.58}$$

where $\alpha = kv_0 - \omega$, and $v' = \alpha/k = v_0 - V_p$ is the velocity off-resonance. Since the perturbation is small

$$\begin{aligned} m\frac{dv_2}{dt} &= e\,E_0\{\cos\left[k\,(z_0 + z_1) + k\,v_0\,t - \omega\,t\right] - \cos\left[k\,z_0 + k\,v_0\,t - \omega\,t\right]\} \\ &\approx -\frac{ke^2{E_0}^2}{m}\sin\left(kz_0 + \alpha\,t\right)\left[\frac{-\cos(kz_0 + \alpha\,t) + \cos kz_0}{\alpha^2} - \frac{t\,\sin kz_0}{\alpha}\right] \end{aligned} \tag{12.59}$$

The change in the electron energy to the second order of approximation

$$\begin{aligned} \frac{d}{dt}\left(\frac{mv^2}{2}\right) &\approx v_1\frac{dv_1}{dt} + v_0\frac{dv_2}{dt} \\ &= \frac{e^2{E_0}^2}{m}\cos\left(kz_0 + \alpha t\right)\left[\frac{\sin(kz_0 + \alpha t) - \sin kz_0}{\alpha}\right] \\ &\quad - \frac{kv_0e^2{E_0}^2}{m}\sin\left(kz_0 + \alpha\,t\right)\left[\frac{-\cos(kz_0 + \alpha\,t) + \cos kz_0}{\alpha^2} - \frac{t\,\sin kz_0}{\alpha}\right] \end{aligned} \tag{12.60}$$

2 This mechanism is discussed in a graphical manner by Chen (1983) to which reference should be made.

is averaged over the starting position z_0 to obtain the mean energy gain for an electron whose initial velocity is v_0

$$\left\langle \frac{\mathrm{d}}{\mathrm{d}t}\left(\frac{mv^2}{2}\right)\right\rangle = \frac{e^2E_0^{\;2}}{2m}\left[-\frac{\omega\sin\alpha t}{\alpha^2} + t\cos\alpha t + \frac{\omega t\cos\alpha t}{\alpha}\right]$$
$$= \frac{e^2E_0^{\;2}}{2m}\frac{\mathrm{d}}{\mathrm{d}\alpha}\left[\left(1+\frac{\omega}{\alpha}\right)\sin\alpha t\right] \qquad (12.61)$$

by making use of the trigonometric identities for the sums and differences of angles, and noting that the averages $\langle\sin^2 kz_0\rangle = \langle\cos^2 kz_0\rangle = 1/2$ and $\langle\sin kz_0\, \cos\, kz_0\rangle = 0$. It is evident by inspection that the resulting expression is not singular at $\alpha = 0$. The integral over v_0 required to calculate the mean energy gain per electron is therefore straightforward. Since $v_0 = (\alpha+\omega)/k$, we obtain the energy gain per electron per unit time

$$\frac{\mathrm{d}W_e}{\mathrm{d}t} = \int\left\langle\frac{\mathrm{d}}{\mathrm{d}t}\left(\frac{mv_\parallel^{\;2}}{2}\right)\right\rangle\tilde{f}(\mathbf{v})\,\mathrm{d}\mathbf{v} = \frac{e^2E_0^{\;2}}{2m}\int\frac{\mathrm{d}}{\mathrm{d}\alpha}\left[\left(1+\frac{\omega}{\alpha}\right)\sin\alpha t\right]\bigg|_{\alpha=(kv-\omega)}\tilde{f}_\parallel(v)\,\mathrm{d}v \qquad (12.62)$$

where $\tilde{f}_\parallel(v_\parallel) = \int\tilde{f}(\mathbf{v})\,\mathrm{d}\mathbf{v}_\perp$ is the distribution function with respect to the parallel component of the velocity alone. Integrating by parts, and noting that $\lim_{t\to\infty}\sin\alpha t/\alpha = \pi\,\delta(\alpha)$, we obtain

$$\frac{\mathrm{d}W_e}{\mathrm{d}t} = -\frac{\epsilon_0\;E_0^{\;2}\,\Pi_e^{\;2}}{2\,k}\int v\frac{\mathrm{d}\tilde{f}_\parallel(v)}{\mathrm{d}v}\frac{\sin(k\,v-\omega)t}{(k\,v-\omega)}\,\mathrm{d}v_\parallel \qquad (12.63a)$$

$$= -\frac{\pi\,\epsilon_0\,\Pi_e^{\;2}\,E_0^{\;2}}{2\,k}\,v\frac{\mathrm{d}\tilde{f}_\parallel(v)}{\mathrm{d}v}\bigg|_{v=w/k} \qquad (12.63b)$$

The energy density in the wave is the sum of the mean values of the electrostatic energy and the electron kinetic energy. The electrostatic energy density is given by the familiar form $\epsilon_0\,E^2/2$. The kinetic energy associated with a particle is given by $\frac{1}{2}m\,[(v_0+v_1)^2) - v_0^{\;2}] = \cancel{m\,v_0\,v_1} + \frac{1}{2}m\,v_1^{\;2}$ – the term $v_0\,v_1$ vanishing on averaging over a period. Making use of Eqs. (12.55) and (12.57), the energy density in the wave is the sum of the mean values of the electrostatic energy and the electron kinetic energy. These terms are averaged over the period of the wave and the distribution to give the mean energy per unit volume in the wave

$$W = \frac{1}{2}\left[\frac{\epsilon_0E_0^{\;2}}{2} + \frac{n_emv^2}{2}\right] = \frac{1}{2}\left[\frac{\epsilon_0E_0^{\;2}}{2} + \frac{n_em\;e^2E_0^{\;2}}{2\;m^2\omega^2}\right] = \frac{\epsilon_0E_0^{\;2}}{4\,\pi}\left(1+\frac{\Pi_e^{\;2}}{\omega^2}\right) \approx \frac{\epsilon_0E_0^{\;2}}{2} \qquad (12.64)$$

Since the energy in the wave is proportional to the square of the amplitude, the energy decay rate is twice that of the wave amplitude decay rate and is given by

$$2\gamma = \frac{n_e\,\mathrm{d}W_e/\mathrm{d}t}{W} = \frac{\pi\,\Pi_e^{\;2}}{k}\frac{\omega}{k}\frac{\mathrm{d}\tilde{f}_\parallel(v)}{\mathrm{d}v}\bigg|_{v=\omega/k} \qquad (12.65)$$

Since $\omega \approx \Pi_e$, this result in agreement with our earlier one (12.54).

From this analysis, it is clear that the absorption results from changes in the velocity of individual electrons in the electrostatic field constituting the wave. It can be seen from Eq. (12.60) that the energy increase of any particular electron depends on its initial starting point on the wave and varies in time. In particular referring to Eq. (12.63a), we can see that the variation in the energy gain or loss of an individual electron at time t varies as the function $\mathrm{sinc}(k\,v'\,t) = \sin(k\,v'\,t)/k\,v't$ (Figure 12.2) familiar from diffraction theory. The argument of this function $kv't$ represents the number of wavelengths the particle has traversed since the onset at $t = 0$. As can be seen the

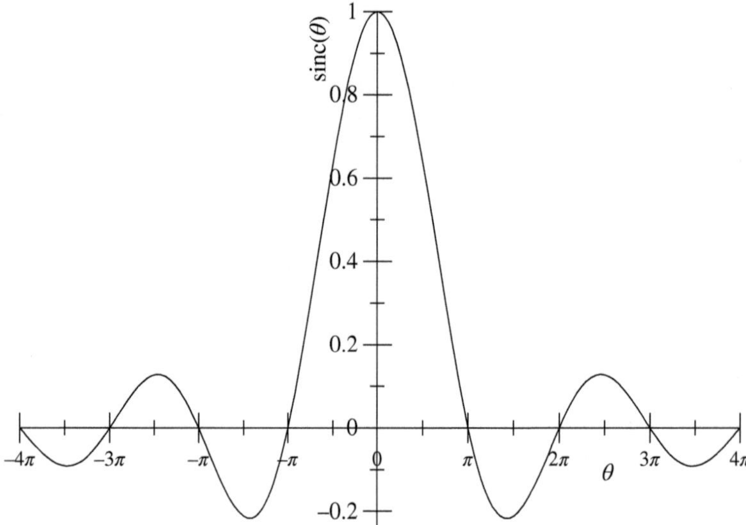

Figure 12.2 Plot of the function $\mathrm{sinc}(\theta) = \sin\theta/\theta$.

contribution varies in sign as this number increases with time. Consequently, as the time increases the width of the profile becomes smaller as the motions of the individual electrons become mixed. The integral of the energy gain over all electrons however is constant. The early phases of the wave motion are complicated if the wave profile is not sinusoidal. At late times, the motion becomes more homogeneous and the detail is lost. At this stage, the averaged energy transfer rate is constant and corresponds to the linear damping rate, 2γ.

12.5 Non-linear Landau Damping

12.5.1 Particle Trapping

The preceding analysis of Landau damping has assumed that the amplitude of the plasma wave is sufficiently small that the perturbation to the free streaming motion of an individual electron is weak and well represented by the term $|z_1| \ll 1/k$; the essential requirement for the validity of the theory being that the electron maintain its relationship with the phase of the electric field wave over times t long compared to the decay time γ^{-1}. Averaging over the initial starting point z_0, the mean squared value of the term kz_1 is obtained from Eq. (12.58)

$$\langle k^2 {z_1}^2 \rangle = \frac{k^2 e^2 {E_0}^2}{2m^2\alpha^4}[(1 - \cos\,\alpha t)^2 + (\alpha t - \sin\,\alpha t)^2] \tag{12.66}$$

It is easily shown that the maximum value of the term in square brackets, obtained for resonant particles $\alpha = 0$, is $(\alpha\, t)^4/4$ and that the condition of validity for linear Landau damping is

$$\omega_b\, t = \left(\frac{e\, E_0\, k}{m}\right)^{1/2} t \ll 8^{1/4} \tag{12.67}$$

This quantity ω_b represents bounce frequency of an electron trapped in the potential well of the wave. The electric field of the wave is sinusoidal so that $E = E_0 \sin(kx)$ in the rest frame of the wave where $x = z - V_p\, t$. At the bottom of the well, $E \approx E_0\, k\, x$. An electron, whose energy is close to the

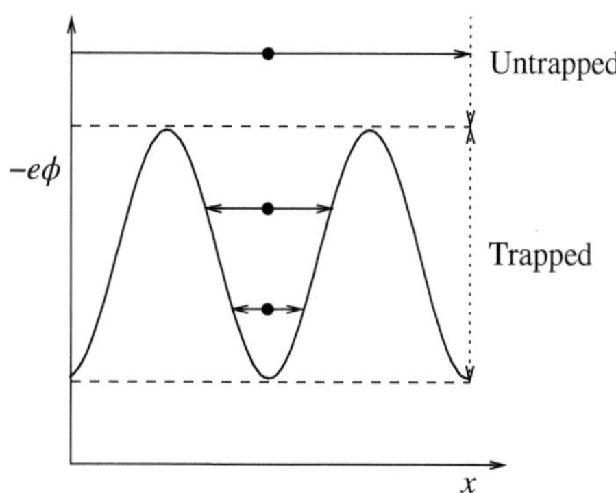

Figure 12.3 Sketch of the paths of particles in a potential well. Trapped or untrapped trajectories depend on the particle energy in the rest frame of the wave.

minimum of the well is trapped unable to escape from the well and is carried along with the wave (Figure 12.3). Within the potential well the electron bounces from wall to wall with a frequency given by

$$\ddot{x} = -\frac{e\,E_0}{m}\sin(kx) \approx -\frac{e\,k\,E_0}{m}x \tag{12.68}$$

namely an oscillation with frequency ω_b – the *bounce frequency*. Linear theory is clearly only valid if the wave damps before the electrons oscillate in the well $\gamma \gtrsim \omega_b$. Jackson (1960) showed that under these conditions the damping rate is approximately given by the Landau formula (12.54).[3]

When the wave field is sufficiently strong that $\gamma \ll \omega_b$, the electrons are trapped in the potential troughs before damping. In this case, the energy of an electron initially decays linearly at a rate γ for a time $\sim 2\pi/\omega_b$, releasing some of the trapped particles and then settles down to oscillate in the laboratory frame as the electron bounces between the walls of the well. In consequence the energy of the electrons, and therefore of the wave, shows a non-monotonic decay reflecting this motion, as the electrons and the wave exchange energy during bounces. Not all oscillations have the same period as the restoring force is sinusoidal rather than linear. Consequently, the phases of the oscillating particles vary and the coherent oscillation is lost due to phase mixing and the distribution function flattens to a plateau (see Davidson, 1972, §4.2).

We may modify the earlier picture (Section 6.3) of electron trapping to consider the trapping of particles in the rest frame of the plasma wave, consisting of a series of stationary wells (Figure 12.3). The distribution function of the background particles now includes a mean velocity modified to take into account the phase velocity of the wave V_p, so that the thermal velocity is changed from u to $u - V_p$ in the Maxwellian (Eq. 6.3). In practical situations, it is unlikely that the rate of growth of the field is sufficiently slow that the adiabatic condition is obeyed. As a result, the field may increase significantly whilst the electron is within the well. Although the particle has significant energy and is further accelerated on entry, it is unable to overcome the barrier and fails to exit from the well. As a result, electrons with energies near the resonance value become trapped. Also the electrons possess significant kinetic energy from the oscillating electric field. As a result, the trapped distribution can 'burrow' towards the core of distribution, as described in Section 12.5.2. Finally, when the number of trapped particles becomes large, they start to modify the shape of the well through their space charge.

3 Trapping must be clearly differentiated from plasma bunching discussed earlier Section 2.4.2.1. In the former case the trapped particles have velocity close to the phase velocity of the wave and move with the wave; in contrast in the latter the electrons 'surf ride' the wave with velocities dissimilar to that of the wave motion.

12.5.2 Plasma Wave Breaking

In the oscillating field of the plasma wave, the electric field $E = E_0 \sin(\omega t)$ and an electron oscillates with velocity $v = [eE_0/m\omega]\cos(\omega t)$. If E_0 is small, the oscillation velocity plays only the weak role exemplified by the linear theory, and only particles near the resonance are affected. However if the field is large, the oscillation velocity may be sufficiently large that particles from the 'cold' core of the distribution are brought into resonance, i.e. when $eE_0/m\omega \sim \omega/k$. A strong non-linear damping of the wave results from large numbers of electrons being accelerated. This phenomenon is known as *wavebreaking*. At the wave breaking field strength, large numbers of slow particles become strongly trapped, and the field energy is rapidly reduced as the particles are accelerated. At the wavebreaking condition $\omega = \sqrt{(e\,k\,E_0/m)} = \omega_b$, i.e. the bounce frequency is equal to that of the wave. The electrons are therefore trapped in approximately only one cycle.

We may compare this behaviour with that found in Section 1.2.1 for a cold plasma. Being non-dispersive, the wave in cold plasma is stationary in contrast to that in warm where the wave is progressive. However, if we transform into the stationary frame of the wave, breaking occurs as the fast moving particles in the wave frame overtake the more slowly moving ones, in complete analogy with the cold plasma model (Section 1.2.1). The result in both cases is that the ordered field structure of the wave is disturbed by a 'disordered' particle pattern.

In warm plasma, the threshold for wave breaking is reduced both by the additional electron velocity due to the thermal motion and by the pressure force acting on the particle as a result of the density variation introduced by the potential well. The effect of these terms can be taken into account by the *water bag model* introduced by Coffey (1971) (see also Kruer, 1988, §9.4).

The 'water bag' distribution replaces the equilibrium Maxwellian by a rectangular distribution with the same mean energy and pressure, i.e.

$$f(v) = \begin{cases} 1/v_{\max} & |v| < v_{\max} \\ 0 & \text{otherwise} \end{cases}$$

where $v_{\max} = \sqrt{3}\,\overline{v}_e = \sqrt{3\kappa T/m}$.

It easily shown that the pressure $p = n\,m\,\overline{v}_e^{\,2}$ of the one-dimensional Maxwellian (Gaussian) distribution is identical to that of the water-bag, $p = \frac{1}{3}\,n\,m\,v_{\max}^{\,2}$. Figure 12.4 compares the two distributions. The major difference is the absence of particles with $v \gtrsim \overline{v}_e$, which are crucial for linear Landau damping but play little role in wave-breaking. The model may be therefore expected to give an approximate estimate of the number of particles brought into resonance. The model assumes fixed ions and only the electrons are mobile. The number density (n) and mean velocity (u) of the particles are obtained by taking moments of the Vlasov equation

$$\frac{\partial n}{\partial t} + \frac{\partial}{\partial x}(n\,u) = 0$$

$$\frac{\partial u}{\partial t} + u\frac{\partial u}{\partial x} = -\frac{e}{m}E - \frac{1}{m\,n}\frac{\partial p}{\partial x}$$

where n_0 is the ambient electron density. Since the waves are high frequency, the pressure p is determined by the adiabatic equation of state $p = (m\,\overline{v}_e^{\,2}/n_0^{\,2})n^3$ where the adiabatic index $\gamma = 3$ is appropriate to one-dimensional motion, since there is only movement in the direction of propagation of the wave. Transforming into the rest frame of the wave moving with velocity V_p in the steady state

$$n\,u = n_0\,V_p \tag{12.69a}$$

$$u^2 - \frac{2\,e\,\phi}{m} + 3\,\overline{v}_e^{\,2}\frac{n^2}{n_0^{\,2}} = V_p^{\,2} + 3\,\overline{v}_e^{\,2} \tag{12.69b}$$

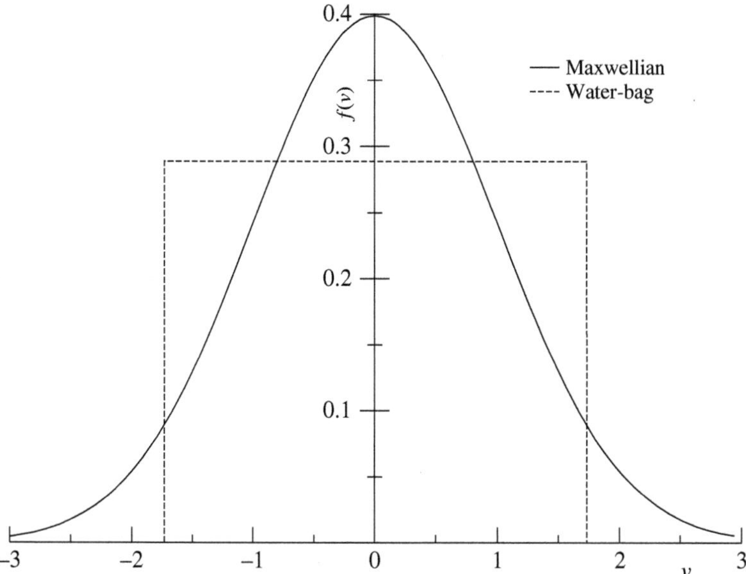

Figure 12.4 Comparison of the one-dimensional Maxwellian (full line) with the water-bag distribution (dashed line).

where $E = -\mathrm{d}\phi/\mathrm{d}x$ defines the potential ϕ and $p_0 = n_0\, m\, \overline{v}_e^{\,2}$ is the ambient pressure. The zero of the potential $\phi = 0$ being taken when $u = V_p$

Solving for the potential by eliminating n between Eqs. (12.69)

$$\frac{2\,e\,\phi}{m\,V_p^{\,2}} = \frac{u^2}{V_p^{\,2}} - 1 - \beta + \beta\frac{V_p^{\,2}}{u^2} \tag{12.70}$$

where $\beta = 3\,\overline{v}_e^{\,2}/V_p^{\,2} = v_{\mathrm{max}}^{\,2}/V_p^{\,2}$. If $\beta = 1$, then $v_{\mathrm{max}} = V_p$ and the fastest electrons in the water-bag distribution have zero velocity in the wave frame. Differentiating Eq. (12.70) with respect to u, we find that ϕ has an extremum when $u/V_p = \sqrt[4]{\beta}$,

$$\phi_{\mathrm{ex}} = -\frac{mV_p^{\,2}}{2\,e}(1 - \sqrt{\beta})^2 \tag{12.71}$$

where the electric field $E = 0$. Under this condition, electrons with the maximum water-bag velocity v_{max} have zero energy in the wave frame $\frac{1}{2}m(V_p - v_{\mathrm{max}})^2 + e\phi_{\mathrm{ex}} = 0$

Thus far, we have identified the conditions required by the electric field to support the dynamics of the electron motion, but have not taken account of its source in the space charge generated by the motion.

$$\frac{\partial E}{\partial x} = -\frac{1}{\epsilon_0}(n - n_0) \tag{12.72}$$

Multiplying Poisson's equation (12.72) by E in the form

$$E = -\frac{\mathrm{d}\phi}{\mathrm{d}x} = -\frac{1}{e\,n}\left(n_0\, m\, V_p\,\frac{\mathrm{d}u}{\mathrm{d}x} + \frac{\mathrm{d}p}{\mathrm{d}x}\right)$$

and integrating

$$\frac{1}{2}E^2 + \frac{1}{\epsilon_0}\left(n_0\, e\,\phi - n\, m\, u^2 - n_0\, m\,\overline{v}_e^{\,2}\frac{n^3}{n_0^{\,3}}\right) = -\frac{n_0\, m\, V_p^{\,2}}{2\epsilon_0}\left[(1 - \sqrt{\beta})^2 + \frac{8}{3}\sqrt[4]{\beta}\right] \tag{12.73}$$

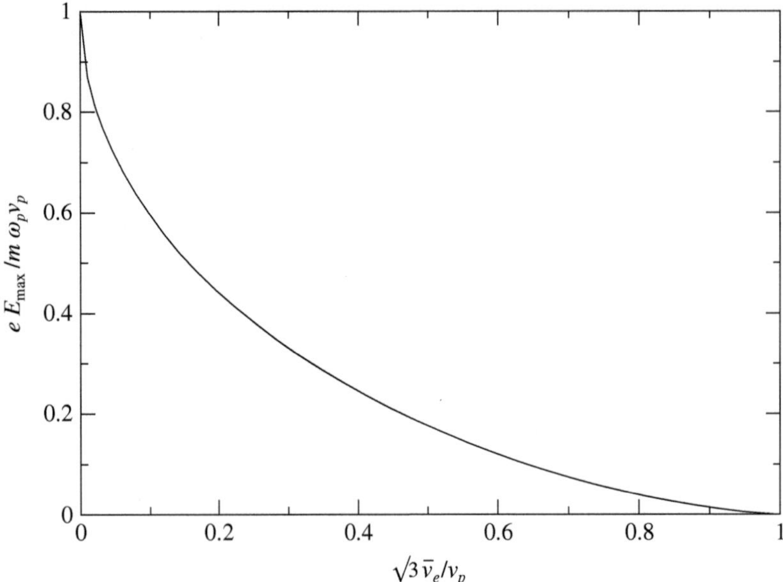

Figure 12.5 Wave breaking field E_{max} for trapped electrons using the water-bag model plotted as a fraction of the cold plasma value $m\,\Pi\,V_p/e$ as a function of the mean thermal speed plotted as a ratio of the maximum water-bag velocity $v_{max} = \sqrt{3}v_e$ to the phase speed V_p of the wave.

since $E = -d\phi/dx = 0$ when $\phi = \phi_{ex}$ from Eq. (12.71). At the maximum electric field is found when $dE/dx = 0$. Therefore, $n = n_0$ (Eq. (12.72)), $u = V_p$ (Eq. (12.69a)), and by definition the value $\phi = 0$. Hence

$$\frac{e^2 E_{max}^2}{m^2 \Pi^2 V_p{}^2} = 1 + 2\sqrt{\beta} - \frac{8}{3}\sqrt[4]{\beta} - \frac{\beta}{3} \tag{12.74}$$

The maximum field is the largest electric field in the potential well which can be maintained by the wave. If the field becomes larger than this value, the self-consistent steady state picture of the wave identified in this model can no longer be supported and the wave breaks. The consequence is the generation of a stream of fast electrons (in the laboratory frame) released from the collapsed well. This phenomenon is well known experimentally in studies of laser produced plasma at high irradiance.

The value of the maximum field in cold plasma $\overline{v}_e = 0$ is the same as that given by the cold plasma model. As might be expected the addition of the thermal velocity to the electrons markedly reduces the breaking field (Figure 12.5): for example a thermal velocity contribution as low as 0.2 of the phase velocity $\overline{v}_e = 0.2V_p$ reduces the breaking field to approximately 0.3 of the cold plasma value.

12.6 The Plasma Dispersion Function

In many cases, the equilibrium distribution is Maxwellian

$$\hat{f}_0 = \left(\frac{m}{(2\,kT)}\right)^{3/2} \exp\left[-\frac{m(v_{\parallel}{}^2 + v_1{}^2 + v_2{}^2)}{2\,kT}\right] \tag{12.75}$$

and the dispersion relation is conveniently expressed in terms of a general function which describes the plasma dispersion, the *plasma dispersion function* given by

$$\mathcal{Z}(\zeta) = \pi^{-1/2} \int_L \frac{e^{-\xi^2}}{(\xi - \zeta)} \, \mathrm{d}\xi \tag{12.76}$$

where $\Im(\zeta) > 0$. In practice,

$$\zeta = \frac{\omega}{k} \left(\frac{m}{2\mathcal{k}T} \right)^{1/2} = \frac{1}{\sqrt{2}} \frac{\omega}{\Pi} \frac{1}{k\,\lambda_D} \tag{12.77}$$

and T is the temperature

For real values of the argument z, the real part of the function

$$\Re[\mathcal{Z}(\zeta)] = I(\zeta) = \pi^{-1/2} \, ⨍ \, \frac{e^{-\xi^2}}{(\xi - \zeta)} \mathrm{d}\xi = -I(-\zeta) = \underbrace{-2e^{-\zeta^2} \int_0^{\zeta} e^{t^2} \, \mathrm{d}t}_{\text{Dawson's integral}} \tag{12.78}$$

and

$$\mathcal{Z}(\zeta) = -2e^{-\zeta^2} \int_0^{\zeta} e^{t^2} \, \mathrm{d}t + \imath \, \pi^{1/2} e^{-\zeta^2} \tag{12.79}$$

The real part of the plasma dispersion function is clearly the Hilbert transform of a Gaussian, which is in turn the imaginary part. The function is therefore consistent with the Kramers–Kronig relation (13.11) expressing its role in the complex permittivity of a Maxwellian plasma.

Thus, we obtain the derivative[4]

$$\mathcal{Z}'(\zeta) = -2\pi^{-1/2} \int_L \frac{\xi e^{-\xi^2}}{(\xi - \zeta)} \, \mathrm{d}\xi = -2(1 + \zeta \, \mathcal{Z}(\zeta)) \tag{12.82}$$

ζ real. In fact since $\mathcal{Z}(\zeta)$ is clearly analytic, we may continue it into the complex ζ plane. The function is analytic for $\Im(\zeta) > 0$ and may be analytically continued to $\Im(\zeta) \le 0$ by deforming the integration contour to pass below the pole at $\xi = \zeta$ to obtain the general alternative definition

$$\mathcal{Z}(\zeta) = 2\,\imath\, e^{-\zeta^2} \int_{-\imath\zeta}^{\infty} e^{-\xi^2} \, \mathrm{d}\xi = \imath \sqrt{\pi} e^{-\zeta^2} [1 - \operatorname{erf}(-\imath\,\zeta)] = \imath \sqrt{\pi} e^{-\zeta^2} \operatorname{erfc}(-\imath\,\zeta) \tag{12.83}$$

where erf ζ and erfc(ζ) are the error and complementary error functions, respectively.[5]

---

4 These results are obtained after an integration by parts

$$\mathcal{Z}'(\zeta) = \pi^{-1/2} \int_L \frac{1}{(\xi - \zeta)^2} e^{-\xi^2} \, \mathrm{d}\xi = -2\,\pi^{-1/2} \int_L \left[1 - \frac{\zeta}{\xi - \zeta} \right] e^{-\xi^2} \, \mathrm{d}\xi = -2[1 + \zeta \, \mathcal{Z}(\zeta)] \tag{12.80}$$

Solving the differential equation for ζ real subject to $\Re[\mathcal{Z}(0)] = 0$ and $\Im[\mathcal{Z}(0)] = \pi^{1/2}$ using $\exp(\xi^2)$ as integrating factor yields

$$e^{\zeta^2} \Re[\mathcal{Z}(\zeta)] = -2 \int_0^{\zeta} e^{t^2} \, \mathrm{d}t \qquad \text{and} \qquad e^{\zeta^2} \Im[\mathcal{Z}(\zeta)] = \pi^{1/2} \tag{12.81}$$

5 The function $w(\zeta) = e^{-\zeta^2} \operatorname{erfc}(-\imath\,\zeta)$ is known as the Fadeeva function. Hence, the plasma dispersion function can be written as

$$\mathcal{Z}(\zeta) = \imath \sqrt{\pi} \, w(\zeta)$$

For small ζ in the neighbourhood of $\zeta = 0$:

$$\mathcal{Z}(\zeta) = \imath\pi^{1/2}e^{-\zeta^2} - \sqrt{\pi}\,\zeta \sum_{n=0}^{\infty} \frac{(-\zeta^2)^n}{\Gamma(n+1/2)} \tag{12.84a}$$

$$\approx \imath\pi^{1/2}e^{-\zeta^2} - 2\zeta\left(1 - \frac{2\zeta^2}{3} + \frac{4\zeta^4}{15} - \cdots\right) \tag{12.84b}$$

and for large ζ the asymptotic form valid for complex ζ is:

$$\mathcal{Z}(\zeta) \approx \imath\pi^{1/2}\,\eta\, e^{-\zeta^2} - \frac{1}{\sqrt{\pi}} \sum_{n=0}^{\infty} \zeta^{-(2n+1)}\,\Gamma(n-1/2) \tag{12.85a}$$

$$= \imath\pi^{1/2}\,\eta\, e^{-\zeta^2} - \zeta^{-1}\left(1 + \frac{1}{2\zeta^2} + \frac{3}{4\zeta^4} + \cdots\right) \tag{12.85b}$$

where $\eta = 0,\ 1,\ 2$ if $\Im(\zeta) <, =, > 0$ respectively. $\Gamma(z)$ is the standard gamma function.

The plasma dispersion function has the following symmetry properties

$$\mathcal{Z}(-\zeta) = 2\,\imath\,\sqrt{\pi}\,e^{-\zeta^2} - \mathcal{Z}(\zeta) \tag{12.86a}$$

$$\mathcal{Z}(\zeta^*) = -[\mathcal{Z}(\zeta)]^* = Z^*(\zeta) + 2\,\imath\,\sqrt{\pi}\,e^{-\zeta^2} \tag{12.86b}$$

The complementary plasma dispersion function $\overline{\mathcal{Z}}(\zeta)$ is defined by Eq. (12.6) but with $\Im(\zeta) < 0$. Clearly,

$$\overline{\mathcal{Z}}(\zeta) = \mathcal{Z}(\zeta) - 2\,\imath\,\sqrt{\pi}\,e^{-\zeta^2} = -\mathcal{Z}(-\zeta) \tag{12.87}$$

A plot of the plasma dispersion function for real values of the argument z is plotted in Figure 12.6 using values for Dawson's integral taken from Abramowitz and Stegun (1965)

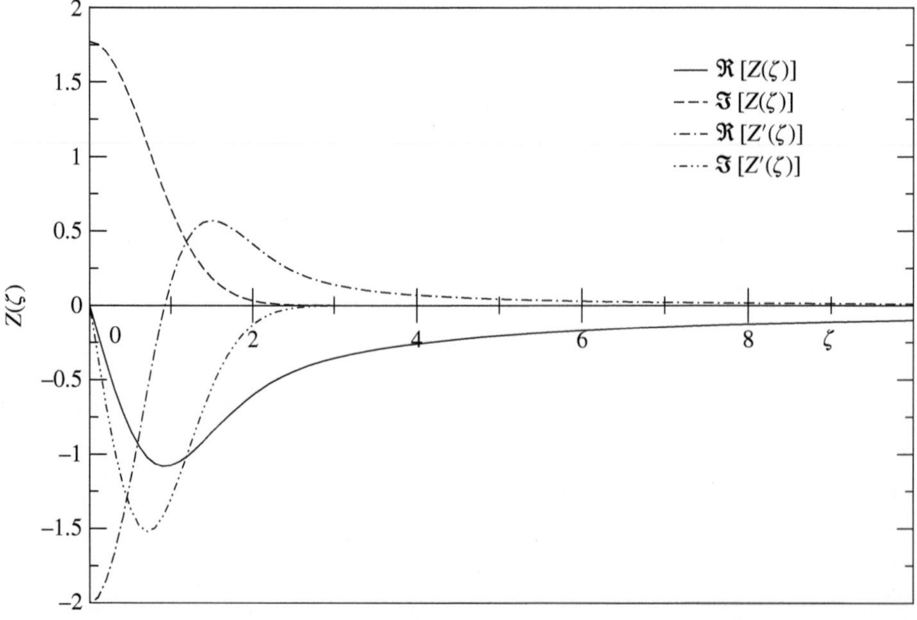

Figure 12.6 Plot of the real and imaginary components of the plasma dispersion function $\mathcal{Z}(\zeta)$ and its derivative $\mathcal{Z}'(\zeta)$ for real values of the argument ζ.

Using Eqs. (12.40) and (12.41), the dielectric constants for a Maxwellian plasma take the form

$$\frac{\epsilon_{\parallel}}{\epsilon_0} = 1 - \frac{{\Pi_e}^2}{\omega^2}\,\zeta^2\,\mathcal{Z}'(\zeta) \qquad \text{and} \qquad \frac{\epsilon_{\perp}}{\epsilon_0} = 1 + \frac{{\Pi_e}^2}{\omega^2}\,\zeta\,\mathcal{Z}(\zeta) \tag{12.88}$$

where $\zeta = \omega/k\sqrt{m/2\,\not{k}T}$.

The dispersion relation for transverse waves follows immediately since $\zeta \sim V_p/\overline{v}_e \gg 1$ and $|\zeta\,\mathcal{Z}(\zeta)| \approx 1$ giving the transverse wave phase velocity (12.44).

The dispersion relation for longitudinal waves has the simple form

$$\varepsilon_{\parallel} = 0 \qquad \frac{1}{2}\mathcal{Z}'(\zeta) - k^2\lambda_D^2 = 0 \qquad \therefore\ \zeta\mathcal{Z}(\zeta) = -(1 + k^2\lambda_D^2) \tag{12.89}$$

whose solution for $\zeta = \omega/\Pi\,k\,\lambda_D$ complex defines the frequency and damping of the wave. ζ is complex since $\omega = \omega_r + \imath\,\gamma$ is complex due to the damping γ.

Consider the case of waves whose wavelength is long compared to the Debye length, $\lambda \gg \lambda_D$. We anticipate that in this case the damping associated with the finite Debye length is weak. Since $k\,\lambda_D$ is small, ζ is large if $\omega \sim \Pi_e$ as we expect. Using the asymptotic expansion of $\mathcal{Z}'(\zeta)$ for large ζ

$$\mathcal{Z}'(\zeta) = \frac{1}{\zeta^2} + \frac{3}{2\,\zeta^4} + \cdots - 2\,\imath\,\pi^{-1/2}\,\zeta\,e^{-\zeta^2} \tag{12.90}$$

Considering only the first two terms for ζ real in the asymptotic expansion, we obtain the real part of the dispersion relation

$$\frac{1}{2\,\zeta^2} + \frac{3}{4\,\zeta^4} = k^2{\lambda_D}^2 \tag{12.91}$$

Taking the first term only since ζ is large, $\zeta^2 \approx [2k^2{\lambda_D}^2]^{-1}$ and $\omega \approx \Pi_e$. Using this as an approximation in the second term, we obtain

$$\zeta^2 \approx [2\,k^2\,\lambda_D^2]^{-1}\ \ (1 + 3k^2{\lambda_D}^2) \qquad \text{and} \qquad \omega^2 = {\Pi_e}^2(1 + 3k^2\,{\lambda_D}^2) + \cdots \tag{12.92}$$

in agreement with our earlier result (12.51).

The damping of the wave is given by the imaginary terms in the asymptotic expansion. Writing the damping term $\gamma = \Im\omega$ we obtain from the imaginary part of $\epsilon_{\parallel}$

$$\begin{aligned}\Im\left(\frac{1}{2\,\zeta^2}\right) &= \Im\left(\frac{{\Pi_e}^2\,k^2{\lambda_D}^2}{\omega^2}\right) \approx -2\left(\frac{\Pi_e\,k\,\lambda_D}{\omega}\right)^2\frac{\gamma}{\omega} \\ &= -2\,\pi^{1/2}\zeta\,e^{-\zeta^2} = -\sqrt{\frac{\pi}{2}}\,\frac{\omega}{\Pi_e\,k\,\lambda_D}\exp\left[-\frac{1}{2}\left(\frac{\omega}{\Pi_e\,k\,\lambda_D}\right)^2\right]\end{aligned} \tag{12.93}$$

Noting that $\Re(\omega) \approx \Pi$ and correcting for the Bohm–Gross frequency in the exponential term, we obtain the Landau damping rate at long wavelengths when $k\,\lambda_D \ll 1$

$$\gamma = \Im(\omega) = \left(\frac{\pi}{8}\right)^{\frac{1}{2}}\frac{\Pi_e}{(k^3\lambda_D^3)}\exp\left[-\frac{1}{2k^2\lambda_D^2} - \frac{3}{2}\right] \tag{12.94}$$

obtained earlier in agreement with Eq. (12.54).

12.7 Positive Ion Waves

Thus far, we have neglected the role of the positive ions necessary for charge neutrality. This has been a reasonable assumption since the mass of the ions is so much larger than they cannot respond to oscillations with frequency of the order of the electron plasma frequency. However, this does not exclude the possibility of slower waves associated with the ions at much lower frequency. As we now show this is possible, but only if the electrons are much hotter than the ions $T_e \gg T_i$.

Consider a quasi-neutral plasma with positive ions of charge $+Ze$, mass m_i, and temperature T_i at density $n_i = n_e/Z$. The total current and charge density is the sum of the ion and electron components

$$\rho = \rho_i + \rho_e \qquad \text{and} \qquad \mathbf{j} = \mathbf{j}_i + \mathbf{j}_e \tag{12.95}$$

Hence, it follows that the dielectric tensor ϵ of the two component medium

$$\mathbf{j} = \imath\,\omega\,(\epsilon - \epsilon_0\,\mathbf{I}) \cdot \mathbf{E} \tag{12.96}$$

where $\mathbf{I}$ is the identity tensor. Since the electron and ion currents add to give the total current, the net dielectric constant ϵ is given by

$$\epsilon - \epsilon_0\mathbf{I} = \epsilon_e - \epsilon_0\mathbf{I} + \epsilon_i - \epsilon_0\mathbf{I} \tag{12.97}$$

Since the individual dielectric constants are given by the earlier analysis for each particle in turn it follows from Eq. (12.88) that

$$\frac{\epsilon_\perp}{\epsilon_0} = 1 + \frac{\Pi_e^{\,2}}{\omega^2}\,\zeta_e\,\mathcal{Z}(\zeta_e) + \frac{\Pi_i^{\,2}}{\omega^2}\,\zeta_i\,\mathcal{Z}(\zeta_i) \quad \text{and} \quad \epsilon_\parallel = 1 - \frac{\Pi_e^{\,2}}{\omega^2}\,\zeta_e^{\,2}\,\mathcal{Z}'(\zeta_e) - \frac{\Pi_i^{\,2}}{\omega^2}\,\zeta_i^{\,2}\,\mathcal{Z}'(\zeta_i) \tag{12.98}$$

where $\Pi_e = \sqrt{n_e\,e^2/\epsilon_0\,m}$ and $\Pi_i = \sqrt{n_i\,(Ze)^2/\epsilon_0\,M}$ are the electron and ion plasma frequencies, respectively.

12.7.1 Transverse Waves

Due to the large mass imbalance $\Pi_e \gg \Pi_i$ and $\zeta_i \gg \zeta_e \gg 1$. For transverse waves, $\zeta\,|\mathcal{Z}(\zeta)| \approx 1$ and the ions terms make only a small change to the dielectric constant. This is not unexpected as due to their large mass the ions cannot respond the field whose frequency $\omega > \Pi_e$ if the wave is propagating.

12.7.2 Longitudinal Waves

Longitudinal waves are more complex. It is possible that the plasma may support two groups of waves. One is associated with the rapid electron motion and one much slower where the ions play a role. The waves must be consistent with the dispersion relation which may be simplified to

$$\mathcal{Z}'(\zeta_e) + \frac{T_e}{T_i}\,\mathcal{Z}'(\zeta_i) = 2\,k^2\,\lambda_D^{\,2} \tag{12.99}$$

and $\zeta_i = (M\,T_e/m\,T_i)^{1/2}\,\zeta_e$. When ζ is real, $\mathcal{Z}'(0.9241) \approx 0$ and two different modes are separated by the condition $\zeta_e \gtrapprox 1$, $\mathcal{Z}'(\zeta) < 0$, it is necessary that $T_e\,|\mathcal{Z}'(\zeta_i)/T_i\,| \gtrsim 2\,(1 + k^2\lambda_D^{\,2})$ to satisfy the dispersion condition.

12.7.2.1 Plasma Waves, $\zeta_e > 1$

It is clear that $\zeta_i \gg \zeta_e \gg 1$, except in the exceptional situation, $T_i \gg T_e$, which may be ignored. Hence, it can be seen from Figure 12.6 that $|\mathcal{Z}'(\zeta_e)| \gg |\mathcal{Z}'(\zeta_i)|$ and that the ion component may be neglected. The wave is therefore the high frequency ($\omega \sim \Pi_e$) electron plasma wave oscillating at the Bohm–Gross frequency, which was analysed earlier (Section 2.4.1).

12.7.2.2 Ion Waves $\zeta_e < 1$

No wave can exist unless $\zeta_i > 1$, but is not so large that $|\mathcal{Z}'(\zeta_i)|$ is small. In particular, if $T_e \sim T_i$ no solution to Eq. (12.99) is possible even if $\zeta_e << 1$ since the maximum of $|\mathcal{Z}'(\zeta)|$ is about 0.5. We must therefore consider the case $T_e \gg T_i$.

Assuming the phase velocity of any wave is small compared to the mean electron thermal speed $\overline{v}_e$, but large compared to that of the ions $\overline{v}_i$, namely $\overline{v}_e \gg V_p = \omega/k \gg \overline{v}_i$ it follows that $\zeta_e \ll 1$ and $\zeta_i \gg 1$. Hence, we may use the power series expansion of $\mathcal{Z}(\zeta)$ for the electrons and the asymptotic series for the ions the generate the dispersion relation

$$2(1 - \imath\sqrt{\pi}\,\zeta_e) - Z\frac{T_e}{T_i}\left[\frac{1}{\zeta_i^2} + \frac{3}{2}\frac{1}{\zeta_i^4} + 2\,\imath\sqrt{\pi}\,\zeta_i\,e^{-\zeta_i^2}\right] = -2\,k^2\,\lambda_{De}^{\;2} \tag{12.100}$$

whose solution gives the phase velocity and damping. Taking the real part of the equation only we obtain the real part of the frequency and hence the phase velocity

$$\left(\frac{\Re(\omega)}{k}\right)^2 = \frac{Z\,\cancel{k}\,T_e/M}{(1 + k^2\,\lambda_{De}^{\;2})} + \frac{3\,\cancel{k}T_i}{M} \approx \Pi_i^{\,2}\frac{\lambda_{De}^{\;2}}{(1 + k^2\lambda_{De}^{\;2})} \tag{12.101}$$

where the electron and ion Debye lengths, respectively, $\lambda_{De} \gg \lambda_{Di}$ when $T_e \gg T_i$.

The imaginary part which determines the Landau damping is given approximately by

$$\begin{aligned} \gamma &= \Im(\omega) \\ &\approx \left(\frac{\pi}{8}\right)^{1/2}\Re(\omega)\frac{1}{(1 + k^2\lambda_{De}^{\;2})}\left[\left(\frac{m}{M}\right)^{1/2} + \left(\frac{T_e}{T_i}\right)^{3/2}\exp\left(-\frac{T_e/2\,T_i}{1 + k^2\lambda_{De}^{\;2}} - \frac{3}{2}\right)\right] \end{aligned} \tag{12.102}$$

The first term in the square brackets gives the Landau damping due to the electrons and the second that due to the ions. Since $T_e \gg T_i$, the ion damping is weak and may be neglected.

In both these cases, it is easily established from Eq. (12.101) that the assumed condition $\overline{v}_e \gg V_p \gg \overline{v}_i$ is obeyed.

For long waves $k\lambda_{De} \ll 1$ and $T_e \gg T_i$, the phase velocity becomes $V_p = \sqrt{Z\,\cancel{k}T_e/M}$ and the damping is $\gamma = \omega\sqrt{\pi\,Z\,m/8\,M}$. These waves propagate as sound waves for which the speed is fixed by the electron pressure and the ion mass (Section 2.4.1) (Figure 12.7).

For shorter waves when $1/\lambda_{De} \ll k \ll 1/\lambda_{Di}$ then $\omega \approx \Pi_i$, the damping is still weak and the wave behaves as an ion plasma wave with an appropriate phase velocity given by Eq. (12.101). If $k\,\lambda_{Di}, \gtrsim 1$ ion Landau damping becomes sufficiently large to destroy the wave. In the allowed condition, Landau damping by the electrons is avoided because the phase velocity is near the centre of the electron distribution $V_p \ll \overline{v}_e$ and ion because it lies in the wings of the much narrower ion distribution $V_p \gg \overline{v}_i$.

The underlying physics behind these ion modes is readily understood. If the ion and electron temperatures are nearly equal $T_i \approx T_e$ a wave propagating as a sound wave with speed $\sqrt{\gamma p/\rho} \sim \sqrt{\cancel{k}T/M}$ as in a gas has speed comparable to the mean thermal speed of the ions and is therefore strongly damped. Consequently, the wave does not have the necessary coherence to transfer the excess pressure to the ions through the field – in a gas this is due to collisions which are negligible

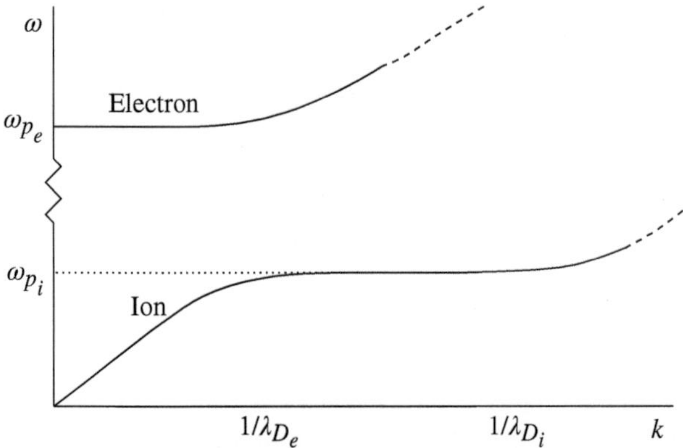

Figure 12.7 Sketch of the dispersion relation for an ion/electron plasma. The electron plasma wave exists for $k \lessapprox 1/\lambda_{De}$, and the ion waves with two branches the ion acoustic wave $k \lessapprox 1/\lambda_{De}$ and the ion plasma wave $1/\lambda_e \gtrapprox k \lessapprox 1/\lambda_{Di}$. At shorter wavelengths, both waves are damped by Landau damping (shown dashed).

here. When the ion temperature is reduced the electrons, being much more mobile, shield the ions maintaining approximate charge neutrality. The resulting electric fields acting on the ions provide the driving force although the ion pressure is negligible. The effective pressure on the massive ions is therefore that of the electrons as in an acoustic wave. The wave speed is now much larger than the ion thermal velocity and Landau damping avoided. Once the wavelength becomes larger than the electron Debye length, the electrons can no longer shield the ions from the fields, the electrons behaving as an inactive background charge. The ions then oscillate under their own space charge fields as a plasma wave with electron and ion roles reversed.

12.8 Microscopic Plasma Instability

Plasmas are inherently unstable and considerable effort must be made to control them. In Chapter 1, we examined the structure of waves and instabilities using the fluid model Section 2.4 and neglecting the kinetic particle motion. *Macroscopic instability* involves the development of perturbations spatially involving mass motion and changes in density and pressure as in Section 2.4.2. In addition to the instabilities associated with the fluid model, bulk mass motion MHD modes may also be unstable, as we show in Chapter 16. *Microscopic instability* is associated with the unstable growth of perturbations in the distribution function. As may be anticipated, it occurs at the resonance velocity in longitudinal waves ($v_\parallel \approx \omega/k$) and is specific to the thermal velocity. Thus, a perturbation grows from its initial weak linear phase into its non-linear form and ultimately becomes turbulent. Our interest will be only with the onset and conditions at which instability is established. The instability is therefore spatially local and reflects the velocity distribution of the particles. *Macroscopic instability* involves the development of perturbations spatially involving mass motion and changes in density and pressure. We shall investigate these in Chapter 16.

The source of microscopic instability can be found in the dispersion relation for longitudinal waves in the plasma, which for a single particle species takes the form of Eq. (12.47). From the imaginary part Eq. (12.52), we see the sign of the damping/growth of a wave at the resonance point, and therefore the onset of instability, depends on the sign of the parallel gradient of the

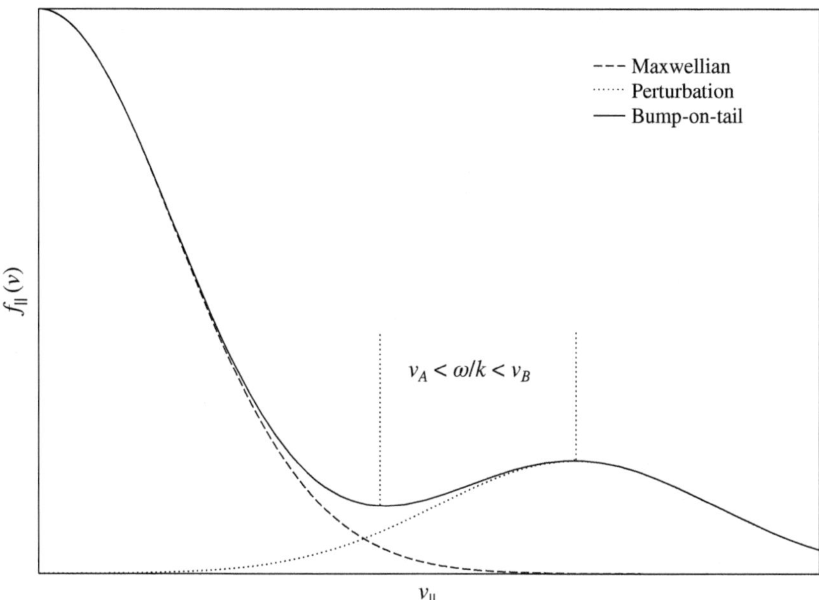

Figure 12.8 Schematic of a Maxwellian distribution perturbed by an additional fast component typical of a beam plasma.

distribution $\partial f_0/\partial v_{\parallel}|_{v_{\parallel}=\omega/k}$. In particular, if the unperturbed plasma has a 'bump on tail' distribution (Figure 12.8), there are velocities where the gradient of the distribution is positive and the plasma is likely to be unstable. An equilibrium Maxwell–Boltzmann distribution is always stable.

12.8.1 Nyquist Plot

Implicit in our derivation of the damping term γ in Eq. (12.52) is the condition that $|\gamma| \ll \omega$. As this may not be rigorously obeyed, the condition for instability that $\partial f_0/\partial v_{\parallel}\Big|_{v_{\parallel}=\omega/k} > 0$ is not strictly obeyed. To obviate this problem, we may apply the Nyquist analysis (Nyquist, 1932) from control theory.[6]

Consider longitudinal waves our starting point is the dispersion relation, Eq. (12.47)[7]

$$H(k,\omega) = 1 - \frac{\Pi_e^{\,2}}{k^2}\int_L \frac{\partial \hat{f}_0/\partial v_{\parallel}}{v_{\parallel} - \omega/k} \mathrm{d}v_{\parallel} = 0 \tag{12.104}$$

whose solutions yield the complex frequency ω for a real value of the wave number k. Remembering that instability is associated with a positive value of the imaginary part of the frequency, we can see

6 The closed loop transfer function of a *single input single output* (SISO) system can be written as

$$M(s) = \frac{G(s)}{1 + H(s)\,G(s)} \tag{12.103}$$

where $G(s)$ is the open-loop gain and $H(s)$ the feedback fraction. The system poles (instabilities) are at the zeroes of the denominator $D(s) = 1 + H(s)G(s) = 0$. The poles of $D(s)$ are the zeroes of $M(s)$ and are also the open-loop system poles. Nyquist constructs a plot of $D(s)$ (*Nyquist plot*) mapped around a loop in the unstable half-plane of s (Figure 12.9). Since instability occurs if a solution to the dispersion relation is found with $\Im(\omega) > 0$. It follows from Cauchy's principle of argument (Copson, 1935, §6.2, Flanigan, 1983, §.72) that the number of unstable solutions of (H) is equal to the number of unstable open loop poles P plus the number of times the path encircles the origin (N).

7 Henceforward, we will omit the || subscript and the circumflex superscript in dealing with longitudinal waves.

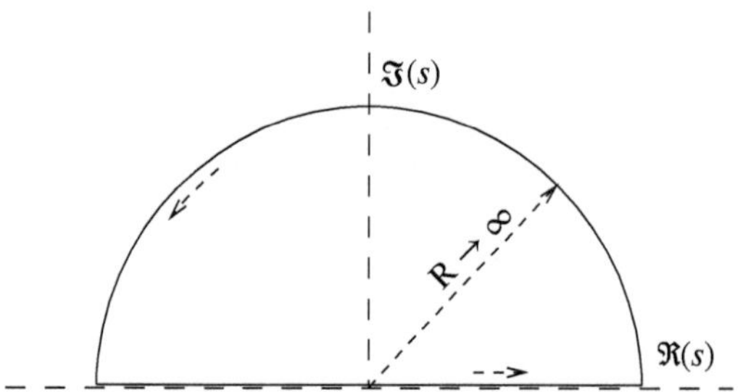

Figure 12.9 Path for s used to construct the Nyquist plot.

that unstable roots are associated with zeroes in the upper half of the imaginary plane of ω. We use the argument principle[8] (Copson, 1935, §6.2, Flanigan, 1983, §.72) with the Nyquist construction[9] to identify if there exist such zeroes of the function $H(k, \omega)$ in the upper complex plane of w and the plasma is therefore unstable. The dispersion Eq. (12.104) is analytic in the upper half-plane so that there are no poles within the applied contour. We note that we will not identify the unstable conditions, only if they exist. To do this, we construct a *Nyquist plot* in which we plot $\Im(H)$ against $\Re(H)$ around the Nyquist path, each point being for a different value of $s = \omega/k$, $\Im(s) > 0$. Since there are no poles $P = 0$ in the upper complex plane of ω, the number of zeroes Z, i.e. of unstable solutions $H = 0$, is simply the number of times the plot encircles the origin N. The calculation of the complex H around the Nyquist path is relatively straight forward in that the contribution from the return path at $R = \infty$ maps to the single point (1,0): the real contribution from $-\infty$ to $+\infty$ for values of $s = \omega/k + \imath 0$ is simply the principal part integral involving only real quantities, and the imaginary the contribution of the residue at $(v = \omega/k)$. Figure 12.9 shows typical plots for single and double peaks. The plot starts at $v = -\infty$ where $H = (1, 0)$ rotating in an anti-clockwise sense. The path crosses the $\Re(H)$ axis at points where the distribution function $f_0(v)$ has a turning point (see Figure 12.8). Finally, the path terminates back at the starting point $H = (1, 0)$ in the upper half-plane of H. Typical paths are shown in Figure 12.10:

- Figure 12.10a shows a typical single peaked distribution (e.g. a Maxwellian) whose Nyquist plot does not enclose the origin and we conclude is therefore stable.
- Figure 12.10b shows a stable double peaked distribution which also does not enclose the origin.
- Figure 12.10c shows a distribution which is marginally stable where the path nearly passes through the origin.
- Figure 12.10d shows an unstable distribution where the origin lies inside the plot.

12.8.1.1 Penrose's Criterion

The Nyquist method for investigating whether a plasma is unstable is unnecessarily complex involving the construction of a Nyquist plot. Penrose (1960) proposed a criterion based on the dispersion integral to be used as an alternative.

---

8 Cauchy's principle of argument considers $w(z)$ is a meromorphic function where a meromorphic function is one which is holomorphic (analytic) on all D except for a set of isolated points which are the poles of the function. Meromorphic functions can be expressed as the ratio of two holomorphic functions on D; any pole coincides with a zero of the denominator in the plane of z, is analytic except at a finite number of points (the poles of $w(z)$), and analytic at every point of the contour C. As $w(z)$ travels around the contour C in the anti-clockwise direction, the plot of the function $w(z)$ encircles the origin in the $\Re[w(z)]$, $\Im[w(z)]$ plane N times with N given by

$$N = Z - P \tag{12.105}$$

where Z is the number of zeroes and P the number of poles of $F(s)$ enclosed by the contour.

9 The Nyquist construction was developed by Harris (1959), Jackson (1960), and Penrose (1960).

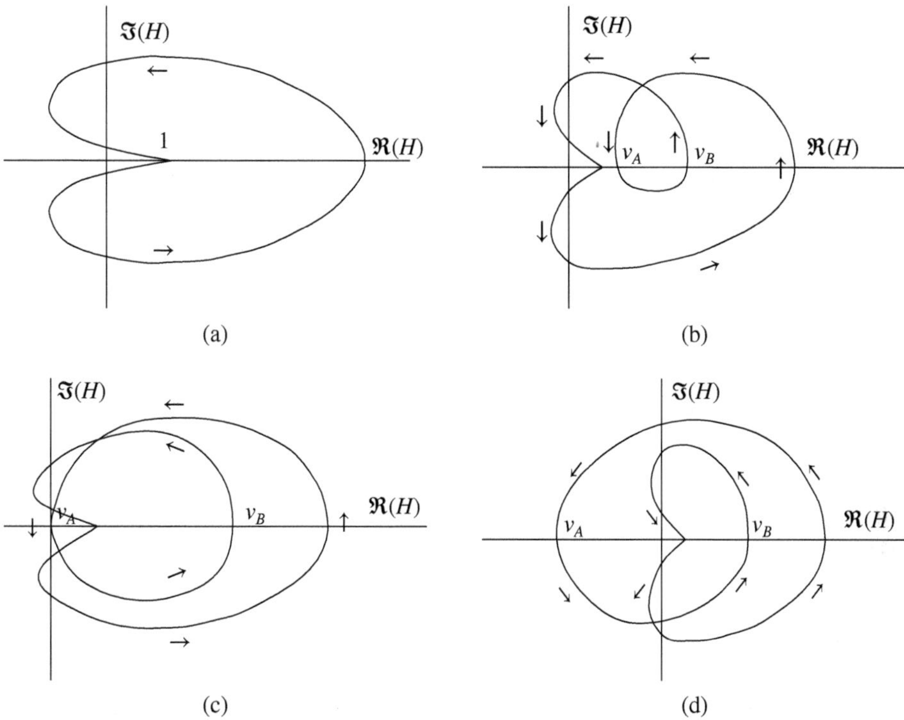

Figure 12.10 Schematic Nyquist plots showing stable and unstable single (Maxwellian) and two peaked distributions. The particle velocities V_A and v_B correspond to the turning points of the bump-on-tail distribution (Figure 12.8). The path in the complex H space is described as indicated. (a) Single peaked stable distribution, (b) double peaked stable distribution, (c) double peaked marginally stable distribution, and (d) double peaked unstable distribution.

The underlying idea is the simple one that if an unstable root of the dispersion equation exists the Nyquist contour must cross the $\Re(H)$ axis in its negative half-plane. The first minimum in the distribution at v_A (see Figure 12.8) marks the extreme of the plot in the $\Re(H)$ direction (see Figure 12.10). If this lies in the negative half-plane of $\Re(H)$, the distribution must (necessarily and sufficiently) be unstable. We therefore calculate H using $s = v_A$ to calculate its value at the minimum. Making use of the Landau prescription for the integral and noting that $\dot{f}_0(v_A) = 0$, Eq. (12.38) the dispersion integral may be written[10]

$$\begin{aligned} H &= 1 - \frac{\Pi_e^2}{k^2} \int_L \frac{\dot{f}_0(v)}{(v - v_A)} \mathrm{d}v \\ &= 1 - \frac{\Pi_e^2}{k^2} \left\{ \int_{-\infty}^{\infty} \frac{\mathrm{d}\{f_0(v) - f_0(v_A)\}}{(v - v_A)} + \imath\pi \cancel{[\dot{f}_0(v_A) - \dot{f}_0(v_A)]} \right\} \\ &= 1 - \frac{\Pi_e^2}{k^2} \int \frac{f_0(v) - f_0(v_A)}{(v - v_A)^2} \mathrm{d}v \end{aligned} \tag{12.106}$$

Since k can vary from 0 to ∞, this requires that the distribution

$$\int \frac{f_0(v) - f_0(v_A)}{(v - v_A)^2} \mathrm{d}v > 0 \tag{12.107}$$

10 The choice of the arbitrary constant in the integration by parts to be $f_0(v_A)$ removes the second-order pole at v_A and avoids the need for the integral to be a principal part.

Consequently, stability is determined by the following criterion (Penrose, 1960): *Exponentially growing modes exist if, and only if, there is a minimum of $f_0(v)$ at a value v_A such that $\int_{-\infty}^{\infty}(v-v_A)^2[f_0(v)-f_0(v_A)]\mathrm{d}v > 0$.*

If the plasma is made up of several different species and streams the dispersion integral becomes (Eq. 12.41:a)

$$H = 1 - \sum_s \frac{\Pi_s{}^2}{k^2} \int_L \frac{\dot{f}_s(v)}{(v-v_A)}\,\mathrm{d}v \tag{12.108}$$

which gives rise to the modified form of Penrose's criterion

$$\sum_s \Pi_s{}^2 \int \frac{f_{0s}(v)-f_{0s}(v_A)}{(v-v_A)^2}\mathrm{d}v > 0 \tag{12.107:a}$$

12.9 The Dielectric Properties of Warm Dilute Plasma in a Magnetic Field

We now turn our attention to the more complicated case of waves in warm plasma with a background magnetic field. The general pattern follows that for a cold plasma discussed earlier in Section 11.2. The families of waves are unchanged, but the addition of thermal motion leads to broadening of the resonance, analogous to Landau damping in plasma waves discussed above (Section 12.1.1). The analysis of the warm plasma case is however much more complex than the cold. The warm plasma thermal distribution of velocities means that we can no longer treat the particles as stationary, but must approach the problem through the solution of the corresponding Vlasov equation as in Section 12.1.1.

As before we consider the wave to a small perturbation to a uniform equilibrium plasma with a background steady magnetic field $\mathbf{B}_0$. The zero order steady solution of the Vlasov equation for particles of charge Ze and mass m is

$$\frac{\partial f_0}{\partial t} + \mathbf{v}\cdot\frac{\partial f_0}{\partial \mathbf{r}} + \frac{Ze}{m}(\mathbf{v}\wedge\mathbf{B}_0)\cdot\frac{\partial f_0}{\partial \mathbf{v}} = 0 \tag{12.109}$$

Therefore for an equilibrium plasma to exist, we require its distribution function f_0 satisfy the condition $(\mathbf{v}\wedge\mathbf{B}_0)\cdot\nabla_v f_0 = 0$. Most equilibrium distributions are even functions of the velocity components, and thereby satisfy this condition.

In a background magnetic field $\mathbf{B}_0$, the particles acquire a motion along and rotation about the field lines as described in Chapter 3. The particle position and velocity therefore vary in the unperturbed state in accord with this behaviour

$$\frac{\mathrm{d}\mathbf{r}'}{\mathrm{d}t} = \mathbf{v}' \qquad \text{and} \qquad \frac{\mathrm{d}\mathbf{v}'}{\mathrm{d}t} = \frac{Ze}{m}\mathbf{v}'\wedge\mathbf{B}_0 \tag{12.110}$$

since the zero order electric field $\mathbf{E}' = 0$. The first-order perturbation equation for distribution f_1 is therefore given by following the particles along their unperturbed paths

$$\frac{\partial f_1}{\partial t} + \mathbf{v}'\cdot\frac{\partial f_1}{\partial \mathbf{r}'} + \frac{Ze}{m}\mathbf{v}'\wedge\mathbf{B}_0\cdot\frac{\partial f_1}{\partial \mathbf{v}'} = -\frac{Ze}{m}(\mathbf{E}+\mathbf{v}\wedge\mathbf{B}_0)\cdot\frac{\partial f_0}{\partial \mathbf{v}'} \tag{12.111}$$

If we make the realistic assumption, used earlier, that the initial perturbation is zero, i.e. $f_1 \to 0$ as $t \to -\infty$, we may formally integrate Eq. (12.111) along an unperturbed particle trajectory to obtain

$$f_1(\mathbf{r}, \mathbf{v}, t) = \frac{Ze}{m} \int_{-\infty}^{t} [\mathbf{E}(\mathbf{r}', t') + (\mathbf{v}' \wedge \mathbf{B}'(\mathbf{r}', t')) \cdot \nabla_{\mathbf{v}} f_0(\mathbf{v}')] \, dt' \tag{12.112}$$

where $(\mathbf{r}'(t'), \mathbf{v}'(t'))$ is the trajectory which passes through $(\mathbf{r}(t), \mathbf{v}(t))$.[11]

In principle if we know the history of the applied magnetic field $\mathbf{B}_0(\mathbf{r}, t)$, we can integrate these relations to obtain the perturbed distribution in the presence of the applied wave. In practice, this is an impossible task unless the trajectory is simply defined. Such a case is when the magnetic field is constant in both time and space when the particle motions are simply helices moving along and rotating around the magnetic field (Section 3.2.1).

As before take a set of Cartesian axes with z parallel to the applied magnetic field, the wave vector in the plane of $x$and z and y completes the right handed set perpendicular to both x and z. The velocity components can be written in co-ordinate form $\mathbf{v} = (v_\perp \cos\theta, v_\perp \sin\theta, v_z)$. The trajectory is therefore

$$\mathbf{v}' = (v_\perp \cos[\Omega(t - t') + \theta], v_\perp \sin[\Omega(t - t') + \theta], v_\parallel) \tag{12.113}$$

where $\Omega = Z e B_0/m$ is the anti-clockwise cyclotron frequency, Eq. (3.3), and θ is the initial azimuthal phase angle. The velocities $v_\perp$ and $v_\parallel$ are both constants of motion since $\mathbf{B}_0$ is both uniform and constant. The particle velocities are easily integrated to give the positions

$$\mathbf{r}' - \mathbf{r} = \left(-\frac{v_\perp}{\Omega}\{\sin[\Omega(t - t') + \theta] - \sin\theta\}, \frac{v_\perp}{\Omega}\{\cos[\Omega(t - t') + \theta] - \cos\theta\}, v_\parallel(t' - t)\right) \tag{12.114}$$

Assuming the unperturbed distribution depends only on the constant velocities $v_\perp$and $v_\parallel$, i.e. the distribution is isotropic across the rotational plane (x, y) then $f_0(v) = f_0(v')$ along the trajectory. Hence, we can derive the required velocity gradient of the unperturbed distribution

$$\frac{\partial f_0}{\partial v'_x} = \frac{\partial f_0}{\partial v_\perp}\frac{\partial v_\perp}{\partial v'_x} = \frac{v'_x}{v_\perp}\frac{\partial f_0}{\partial v_\perp} = \cos[\Omega(t - t') + \theta]\frac{\partial f_0}{\partial v_\perp} \tag{12.115}$$

$$\frac{\partial f_0}{\partial v'_y} = \frac{\partial f_0}{\partial v_\perp}\frac{\partial v_\perp}{\partial v'_y} = \frac{v'_y}{v_\perp}\frac{\partial f_0}{\partial v_\perp} = \sin[\Omega(t - t') + \theta]\frac{\partial f_0}{\partial v_\perp} \tag{12.116}$$

$$\frac{\partial f_0}{\partial v'_z} = \frac{\partial f_0}{\partial v_z} \tag{12.117}$$

since $v_x'^2 + v_y'^2 = v_\perp{}^2$.

Consider that the perturbed quantities vary as a propagating wave with frequency ω and wave number $\mathbf{k}$, i.e. $\exp[\imath(\mathbf{k} \cdot \mathbf{r} - \omega t)]$ then the perturbed distribution function[12]

$$f_1(\mathbf{r}, \mathbf{v}, t) = \frac{Ze}{m} \int_{-\infty}^{t} \left[(E_x + v'_y B_z - v'_z B_y)\frac{\partial f_0}{\partial v'_x} + (E_y + v'_z B_x - v'_x B_z)\frac{\partial f_0}{\partial v'_y} + (E_z + v'_x B_y - v'_y B_x)\frac{\partial f_0}{\partial v'_z}\right] \exp[\imath\{\mathbf{k} \cdot (\mathbf{r}' - \mathbf{r}) - \omega(t' - t)\}] \, dt' \tag{12.118}$$

11 If the unperturbed distribution is isotropic $f_0 = f_0(v'^2)$, the perturbed distribution does not directly depend on the perturbed field $\mathbf{B}'$. This does not ensure that the waves are electrostatic since $\mathbf{E}' = -\nabla\phi - \dfrac{1}{c}\dfrac{\partial \mathbf{A}'}{\partial t}$ where $\mathbf{A}'$ is the magnetic vector potential. The waves are only electrostatic if the electric field can be written $\mathbf{E}' = -\nabla\phi$.

12 The wave propagation term $\exp[\imath(\mathbf{k} \cdot \mathbf{r} - \omega t)]$ is neglected henceforward.

We note that $v_y' \, [\partial f_0/\partial v_x'] = v_x' \, [\partial f_0/\partial v_y']$ and that $v_\perp$, $v_\parallel$. Using Eqs. (12.113) and (12.114) and Jacobi's expansion (Watson, 1944, §2.22)

$$\exp(\imath\,\alpha \sin x) = \sum_{n=-\infty}^{\infty} J_n(\alpha)\, \exp(\imath\, n\, x) \tag{12.119}$$

twice, we obtain

$$\begin{aligned} f_1 = \frac{Z\,e}{m} \int_{-\infty}^{t} &\Bigg[(E_x - v_\parallel B_y) \cos\chi \,\frac{\partial f_0}{\partial v_\perp} + (E_y + v_\parallel B_x) \sin\chi \,\frac{\partial f_0}{\partial v_\perp} \\ &+ (E_z + v_\perp B_y \cos\chi - v_\perp B_x \sin\chi) \frac{\partial f_0}{\partial v_\parallel} \Bigg] \\ &\times \sum_{n,m=-\infty}^{\infty} J_n\left(\frac{k_\perp v_\perp}{\Omega}\right) J_m\left(\frac{k_\perp v_\perp}{\Omega}\right) \exp\{\imath[(n\Omega + k_\parallel v_\parallel - \omega)(t' - t) + (m-n)\theta]\}\, \mathrm{d}t' \end{aligned} \tag{12.120}$$

where $\chi = \Omega(t - t') + \theta$ is the azimuthal angle at time t'.

The magnetic field $\mathbf{B} = \mathbf{k} \wedge \mathbf{E}/\omega$ is obtained from Maxwell's induction equation and is used to eliminate the magnetic field in Eq. (12.120). Therefore, since $k_y = 0$

$$B_x = -k_z\, E_y/\omega \qquad B_y = (k_z\, E_x - k_x\, E_z)/\omega \qquad B_z = k_x\, E_y/\omega \tag{12.121}$$

Substituting for $\mathbf{B}$ in Eq. (12.120)

$$\begin{aligned} f_1 = \frac{Z\,e}{m} \int_{-\infty}^{t} &\Bigg\{ \left[\frac{k_\parallel v_\perp}{\omega} \frac{\partial f_0}{\partial v_\parallel} + \left(1 - \frac{k_\parallel v_\parallel}{\omega}\right) \frac{\partial f_0}{\partial v_\perp} \right] [E_x \cos\chi + E_y \sin\chi] \\ &+ \left[\left(1 - \frac{k_\perp v_\perp}{\omega} \cos\chi\right) \frac{\partial f_0}{\partial v_\parallel} + \frac{k_\perp v_\parallel}{\omega} \frac{\partial f_0}{\partial v_\perp} \cos\chi \right] E_z \Bigg\} \\ &\times \sum_{n,m=-\infty}^{\infty} J_n\left(\frac{k_\perp v_\perp}{\Omega}\right) J_m\left(\frac{k_\perp v_\perp}{\Omega}\right) \exp\{\imath[(n\Omega + k_\parallel v_\parallel - \omega)(t' - t) + (m-n)\theta]\}\, \mathrm{d}t' \end{aligned} \tag{12.122}$$

To perform the integrals over the time t', we will need the two recurrence formulae for the Bessel function (Watson, 1944, §2.12, eqns 1 & 2), namely the sum and difference formulae, respectively

$$\frac{1}{2}\,[J_{n-1}(x) + J_{n+1}(x)] = \frac{n}{x}\, J_n(x) \qquad \text{and} \qquad \frac{1}{2}\,[J_{n-1}(x) - J_{n+1}(x)] = J_n'(x) \tag{12.123}$$

where $J_n'(x) = \mathrm{d}J_n(x)/\mathrm{d}x$ is the derivative of the Bessel function of order n.

Consider the term proportional to E_x. We may write

$$\cos\chi = \frac{1}{2}(e^{\imath\,\chi} + e^{-\imath\,\chi}) \tag{12.124}$$

so that setting $\tau = t' - t$ the integral becomes

$$\begin{aligned} &\frac{1}{2} \int_{-\infty}^{0} e^{\{\imath\, n\, \Omega\tau\}}\, e^{\{\imath\,(k_\parallel v_\parallel - \omega)\tau\}} [e^{\imath\,(\theta - \Omega\,\tau)} + e^{-\imath\,(\theta - \Omega\,\tau)}] \mathrm{d}\tau \\ &= \frac{1}{2}\,\imath \left[\frac{e^{\imath\,\theta}}{\omega - k_\parallel v_\parallel - (n-1)\,\Omega} + \frac{e^{-\imath\,\theta}}{\omega - k_\parallel v_\parallel - (n+1)\,\Omega} \right] \end{aligned} \tag{12.125}$$

Hence, the contribution to the perturbed distribution proportional to E_x is

$$\begin{aligned} f_{1x} = -\imath \frac{Z\,e}{2\,m\,\omega} &\sum_{n,m=-\infty}^{\infty} J_n\left(\frac{k_\perp v_\perp}{\Omega}\right) J_m\left(\frac{k_\perp v_\perp}{\Omega}\right) \\ &\times \left[\frac{e^{-\imath[(n-1) - m\theta]}}{\omega - k_\parallel v_\parallel - (n-1)\Omega} + \frac{e^{-\imath[(n+1) - m]\theta}}{\omega - k_\parallel v_\parallel - (n+1)\Omega} \right] U\, E_x \end{aligned} \tag{12.126}$$

Since the sum over n is taken over all values from $n = -\infty$ to $n = \infty$; we may adjust the numbering to obtain

$$\begin{aligned} f_{1x} &= -\imath \frac{Z e}{2\, m\, \omega} \sum_{n,m=-\infty}^{\infty} \frac{e^{-\imath(n-m)\theta}}{(\omega - k_\parallel\, v_\parallel - n\Omega)} (J_{n+1} + J_{n-1})\, U\, E_x \\ &= -\imath \frac{Z e}{m\, \omega} \sum_{n,m=-\infty}^{\infty} \frac{e^{-\imath(m-n)\theta}}{(\omega - k_\parallel\, v_\parallel - n\Omega)} \frac{n}{\Lambda} J_m(\Lambda)\, J_n(\Lambda)\, U\, E_x \end{aligned} \tag{12.127}$$

using the sum formula from Eq. (12.123). The argument $\Lambda = k_\perp\, v_\perp/\Omega$

The contribution from the term in E_y is calculated in a similar manner noting that

$$\sin \chi = \frac{1}{2\,\imath}(e^{\imath\, \chi} - e^{-\imath\, \chi}) \tag{12.128}$$

After performing the integrations, re-ordering the sums over n as in Eq. (12.127) and using the Bessel function difference formula (12.123), we obtain

$$f_{1y} = -\imath \frac{Z e}{m\, \omega} \sum_{n,m=-\infty}^{\infty} \frac{\imath e^{-\imath(m-n)\theta}}{(\omega - k_\parallel\, v_\parallel - n\Omega)} J_m(\Lambda)\, J_n'(\Lambda)\, U\, E_y \tag{12.129}$$

Finally, the term depending on E_z is easily evaluated in a similar manner

$$f_{1z} = -\imath \frac{Z e}{m\, \omega} \sum_{n,m=-\infty}^{\infty} \frac{e^{-\imath(m-n)\theta}}{(\omega - k_\parallel\, v_\parallel - n\Omega)} J_m(\Lambda)\, J_n(\Lambda)\, W\, E_z \tag{12.130}$$

Adding these terms together, we obtain an expression for the perturbed distribution function

$$f_1 = -\imath \frac{Z e}{m\, \omega} \sum_{n,m=-\infty}^{\infty} \frac{e^{-\imath(m-n)\theta}}{(\omega - k_\parallel\, v_\parallel - n\Omega)} J_m(\Lambda) \left[\frac{n}{a} J_n(\Lambda)\, U\, E_x + \imath\, J_n'(\Lambda)\, U\, E_y + J_n(\Lambda)\, W\, E_z\right] \tag{12.131}$$

where for particles s

$$U = (\omega - k_\parallel\, v_\parallel) \frac{\partial f_{0q}}{\partial v_\perp} + k_\parallel\, v_\perp \frac{\partial f_{0q}}{\partial v_\parallel} \tag{12.132a}$$

$$W = \frac{n\, \Omega_q\, v_\parallel}{v_\perp} \frac{\partial f_{0q}}{\partial v_\perp} + (\omega - n\, \Omega_q) \frac{\partial f_{0q}}{\partial v_\parallel} \tag{12.132b}$$

$$\Lambda_q = \frac{k_\perp\, v_\perp}{\Omega_q} \tag{12.132c}$$

We are now in a position to calculate the relative permittivity matrix provided we know the unperturbed distribution function $f_0(\mathbf{v})$. Most commonly this will be a Maxwellian, possibly with differing temperatures parallel $T_\parallel$ and perpendicular $T_\perp$ to the field. In either case, the distribution is uniform in the perpendicular plane, i.e. the angle θ is uniformly distributed.

The plasma dielectric constant $\mathbb{K}$ is given by the Maxwell equation $\mathbf{k} \wedge \mathbf{B} = -\frac{\omega}{c^2}\, \mathbf{E} - \imath\, \mu_0 \mathbf{j}$. Whence

$$K_{ij}\, E_j = E_i + \frac{\imath}{\omega\, \epsilon_0} j_i = E_i + \frac{\imath}{\omega\, \epsilon_0} \sum_q Z_q\, e \int v_i f_{1q}\, \mathbf{dv} \tag{12.133}$$

where the sum is taken over all the constituent particles q in the plasma.

Since

$$\int_0^{2\pi} e^{\imath\,(m-n)\,\theta}\mathrm{d}\theta = 2\,\pi\,\delta_{m,n}$$

$$\int_0^{2\pi} \cos\theta\, e^{\imath\,(m-n)\,\theta}\mathrm{d}\theta = \pi\,(\delta_{m,(n+1)} + \delta_{m(n-1)}) \tag{12.134}$$

$$\int_0^{2\pi} \sin\theta\, e^{\imath\,(m-n)\,\theta}\mathrm{d}\theta = \imath\,\pi\,(\delta_{m,(n+1)} - \delta_{m(n-1)})$$

and noting that $\mathbf{v} = (v_\perp\ \cos\theta,\ v_\perp\ \sin\theta,\ v_\parallel)$ we may average $v_i f_{1s}$ over the azimuthal angle θ for the perturbed distributions and add the contribution for each species from Eq. (12.131)

$$K_{ij} = \delta_{ij} + \sum_q \frac{Z_s^{\,2}\,e^2}{\omega^2\,\epsilon_0\,m_s}\sum_{n=-\infty}^{\infty}\int \frac{S_{ij}^n}{(\omega - k_\parallel v_\parallel - n\Omega_s)}\,\mathrm{d}\mathbf{v} \tag{12.135}$$

where the contribution to the dielectric constant from each species is given by

$$S_{ij}^n = \begin{pmatrix} v_\perp (n/\Lambda)^2\,J_n^{\,2}\,U & \imath\,v_\perp\,(n/\Lambda) J_n\,J_n'\,U & v_\perp (n/\Lambda) J_n^{\,2}\,W \\ -\imath\,v_\perp\,(n/\Lambda) J_n\,J_n'\,U & v_\perp J_n'^{\,2}\,U & -\imath\,v_\perp\,J_n\,J_n'\,W \\ v_\parallel (n/\Lambda)\,J_n^{\,2}\,U & \imath\,v_\parallel\,J_n\,J_n'\,U & v_\parallel\,J_n^{\,2}\,W \end{pmatrix} \tag{12.136}$$

The argument of all the Bessel functions is $\Lambda = k_\perp\, v_\perp/\Omega$. We observe that at this stage, when the unperturbed distribution is arbitrary, the permittivity does not satisfy the symmetry relations identified earlier, Eq. (13.23). In particular, $S_{xz} \neq S_{zx}$ and $S_{yz} \neq -S_{zy}$.

Examining Eq. (12.135), we can see that it represents a series of resonances of a particle with harmonics of the cyclotron frequency. These occur when the wave frequency Doppler shifted by the velocity $v_\parallel$ of particles, namely $\omega - k_\parallel\, v_\parallel$ matches that of the harmonic $n\,\Omega$. The occurrence of the harmonics is easily understood by considering the Larmor rotation of the particle, given by Eq. (12.114), causing the particle to change its position in the oscillating wave $\exp[\imath\,(\mathbf{k}\cdot\mathbf{r} - \omega t)]$. This introduces harmonics through Jacobi's expansion (12.119) in terms of Bessel functions. For electro-magnetic waves, the argument of the Bessel functions $k_\perp\, v_\perp/\Omega$ is small and the harmonic terms scale as $(k_\perp\, v_\perp/\Omega)^2/(2^n\, n!)$. Consequently, only the low harmonics are important in determining the damping of the wave resulting from the cyclotron motion – *cyclotron damping*. If propagation is perpendicular to the field $k_\parallel = 0$, and there is no Doppler shift. All the particles, independent of their velocity, are driven together as $k_\perp \to \infty$ as in cold plasma (Section 11.2.1, item ii). In practice, relativistic effects may lead to effective mass changes and introduce a velocity dependent term due to changes in the individual cyclotron frequency.

Thus far, we have evaluated the dielectric constant for a single particle of each species. We now need to integrate over the velocity distribution $f_0(\mathbf{v})$ to obtain the contribution from all the particles of species s. In the most common situation, the zero-order distribution function is Maxwellian possibly with differing temperatures for each species T_s, where the distribution is Maxwellian, but possibly with different temperatures $T_\perp$ perpendicular to the magnetic field and $T_\parallel$ parallel to it. The equilibrium distribution function for the general species is therefore

$$f_0(\mathbf{v})\,\mathrm{d}\mathbf{v} = \frac{m\,\mathfrak{n}}{2\,\pi\,\not{k}\,T_\perp}\sqrt{\frac{m}{2\,\pi\,\not{k}\,T_\parallel}}\exp\left[-\left(\frac{m v_\perp^{\,2}}{2\not{k}T_\perp} + \frac{m\,v_\parallel^{\,2}}{2\not{k}T_\parallel}\right)\right]\,2\pi v_\perp\,\mathrm{d}v_\perp \mathrm{d}v_\parallel \tag{12.137a}$$

$$\to \left(\frac{m\,\mathfrak{n}}{2\,\pi\,\not{k}T}\right)^{3/2}\exp\left[-\frac{m(v_\perp^{\,2} + v_\parallel^{\,2})}{2\,\not{k}T}\right]\,2\pi v_\perp\,\mathrm{d}v_\perp \mathrm{d}v_\parallel \tag{12.137b}$$

$$= \mathfrak{n}\,(\sqrt{\pi}\,\mathfrak{v})^{-3}\,\exp[-(v_\perp^{\,2} + v_\parallel^{\,2})/\mathfrak{v}^2]\,2\pi v_\perp\,\mathrm{d}v_\perp \mathrm{d}v_\parallel \tag{12.137c}$$

where $\mathfrak{n}$ is the particle density. In full thermodynamic equilibrium, the temperatures are equal $T_\perp = T_\parallel$ where $\mathfrak{v} = \sqrt{2\,kT/m}$, when an analytic solution is possible, and where the symmetry conditions discussed in Section 13.1.1 are valid . Differentiating the equilibrium distribution

$$\frac{\partial f_0}{\partial v_\perp} = -\frac{m\,\mathfrak{n}}{2\,\pi\,T}\frac{2}{\sqrt{\pi}\,\mathfrak{v}^3}\,v_\perp\,\exp\left[-\frac{{v_\perp}^2+{v_\parallel}^2}{\mathfrak{v}^2}\right] = A\,v_\perp \tag{12.138}$$

$$\frac{\partial f_0}{\partial v_\parallel} = -\frac{m\,\mathfrak{n}}{2\,\pi\,T}\frac{2}{\sqrt{\pi}\,\mathfrak{v}^3}\,v_\parallel\,\exp\left[-\frac{{v_\perp}^2+{v_\parallel}^2}{\mathfrak{v}^2}\right] = A\,v_\parallel \tag{12.139}$$

where

$$A = -\frac{m\,\mathfrak{n}}{2\,\pi\,T}\frac{2}{\sqrt{\pi}\,\mathfrak{v}^3}\,\exp\left[-\frac{{v_\perp}^2+{v_\parallel}^2}{\mathfrak{v}^2}\right]$$

and hence we obtain

$$U = \omega\,v_\perp\,A \qquad \text{and} \qquad W = \omega\,v_\parallel\,A \tag{12.140}$$

Substituting for U and W, we obtain

$$S_{ij}^n = \begin{pmatrix} (n/\Lambda)^2\,\omega\,{v_\perp}^2\,A\,{J_n}^2 & \imath\,(n/\Lambda)\omega\,{v_\perp}^2\,A\,J_n\,J_n' & (n/\Lambda)\omega\,v_\perp\,v_\parallel\,A\,{J_n}^2 \\ -\imath\,(n/\Lambda)\omega\,{v_\perp}^2\,A\,J_n\,J_n' & \omega\,{v_\perp}^2\,A\,{J_n'}^2 & -\imath\,\omega\,v_\perp\,v_\parallel\,A\,J_n\,J_n' \\ (n/\Lambda)\,\omega\,v_\perp\,v_\parallel\,A\,{J_n}^2 & \imath\,\omega\,v_\perp\,v_\parallel\,A\,J_n\,J_n' & \omega\,{v_\parallel}^2\,A\,{J_n}^2 \end{pmatrix} \tag{12.141}$$

We note that introducing a thermodynamic equilibrium distribution for the unperturbed plasma has recovered the permittivity matrix symmetry $S_{xy} = -S_{yx}$, $S_{xz} = S_{zx}$, and $Syz = -Szy$ required by Eq. (13.23).

To perform the integration over the equilibrium distribution, we note from Eq. (12.135) that the integrations over $v_\perp$ and $v_\parallel$ can be separated into sets of similar terms. We initially treat the integrals over $v_\perp$ which are based on the following standard forms. Using the integral (Watson, 1944, §13.31 eqn 1)

$$\int_0^\infty x\,J_n(\alpha\,x)\,J_n(\beta\,x)\,e^{-p\,x^2}\,\mathrm{d}x = \frac{1}{2\,p^2}e^{-(\alpha^2+\beta^2/4p^2}\,I_n(\alpha\,\beta/2\,p^2) \tag{12.142}$$

we obtain by differentiation and setting $\alpha = \beta$

$$\int_0^\infty t\,{J_n}^2(\alpha\,t)\,\exp(-p^2\,t^2)\,\mathrm{d}t = \frac{1}{2\,p^2}\,\exp\left(-\frac{\alpha^2}{2\,p^2}\right)\,I_n\left(\frac{\alpha^2}{2\,p^2}\right) \tag{12.143a}$$

$$\int_0^\infty t^2\,J_n(\alpha\,t)\,J_n'(\alpha\,t)\,\exp(-p^2\,t^2)\,\mathrm{d}t = \frac{\alpha}{4\,p^4}\,\exp\left(-\frac{\alpha^2}{2\,p^2}\right)\left[I_n'\left(\frac{\alpha^2}{2\,p^2}\right) - I_n\left(\frac{\alpha^2}{2\,p^2}\right)\right] \tag{12.143b}$$

$$\int_0^\infty t^3\,{J_n'}^2(\alpha\,t)\,\exp\left(-\frac{t^2}{2\,q^2}\right)\,\mathrm{d}t = q\exp(-q)\,[n^2 I_n(q) - 2\,q^2\,(I_n'(q) - I_n(q))] \tag{12.143c}$$

from which the following integrals follow

$$\frac{m}{2\pi\,kT}\int_0^\infty 2\,\pi\,v_\perp\,{J_n}^2\left(\frac{k_\perp\,v_\perp}{\Omega}\right)\exp\left(-\frac{m\,{v_\perp}^2}{2\,kT}\right)\,\mathrm{d}v_\perp$$
$$= \exp(-\lambda)I_n(\lambda) \tag{12.144}$$

$$\frac{m}{2\pi\,kT}\int_0^\infty 2\,\pi\,{v_\perp}^2\,J_n\left(\frac{k_\perp\,v_\perp}{\Omega}\right)\,J_n'\left(\frac{k_\perp\,v_\perp}{\Omega}\right)\exp\left(-\frac{m\,{v_\perp}^2}{2\,kT}\right)\,\mathrm{d}v_\perp$$
$$= \frac{k_\perp\,kT}{m\,\Omega}\exp(-\lambda)[I_n'(\lambda) - I_n(\lambda)] \tag{12.145}$$

$$\frac{m}{2\pi \mathcal{k}T}\int_0^{\infty} 2\pi\, v_\perp{}^3\, J'(n)^2\left(\frac{k_\perp v_\perp}{\Omega}\right)\exp\left(-\frac{m\, v_\perp^2}{2\mathcal{k}T}\right) dv_\perp$$
$$= \frac{1}{2}\frac{2\mathcal{k}T}{m}\exp(-\lambda)\left[\frac{n^2}{\lambda}I_n(\lambda) + 2\lambda I_n(\lambda) - 2\lambda I_n'(\lambda)\right] \quad (12.146)$$

where $\Lambda = k_\perp{}^2\,\mathcal{k}T/m\,\Omega^2$.

Turning now to the integral over $v_\parallel$, we note that the integrals contain a singularity $1/(\omega - k_\parallel v_\parallel - n\Omega)$ similar in form to that found for Landau damping, (Section 12.1.2) and there treated by an appropriate contour path. An identical method is appropriate here and reflects the treatment of resonances modified by the cyclotron motion. The requisite integrals are based on the plasma dispersion function $\mathcal{Z}(z)$ (Section 12.6). Since

$$\frac{1}{\sqrt{\pi}}\int_{-\infty}^{\infty}\frac{\exp(-v_\parallel{}^2/\mathfrak{v}^2)}{(\omega - k_\parallel v_\parallel - n\,\Omega)}\,dv_\parallel = -\frac{1}{k_\parallel}\mathcal{Z}(\zeta_n) \quad (12.147)$$

$$\frac{1}{\sqrt{\pi}}\int_{-\infty}^{\infty}\frac{v_\parallel\exp(-v_\parallel{}^2/\mathfrak{v}^2)}{(\omega - k_\parallel v_\parallel - n\,\Omega)}\,dv_\parallel = -\frac{\mathfrak{v}}{k_\parallel}[1+\zeta_n\,\mathcal{Z}(\zeta_n)] = \frac{\mathfrak{v}}{2\,k_\parallel}\mathcal{Z}'(\zeta_n) \quad (12.148)$$

where $\zeta_n = [(\omega - n\Omega)/k_\parallel]\sqrt{m/2\mathcal{k}T}$. We integrate Eq. (12.135) over the velocity distribution and sum over all particle species q to obtain

$$K_{ij} = \delta_{ij} + \sum_q \frac{\Pi_q{}^2}{k_\parallel\,\omega}\sqrt{\frac{m_q}{2\,\mathcal{k}T_q}}\exp\left(-\frac{k_\perp{}^2\,\mathcal{k}T_q}{m_q\Omega_q{}^2}\right)\sum_{n=-\infty}^{\infty} T_{ij}^{nq} \quad (12.149)$$

where

$$T_{ij}^{nq} = \begin{pmatrix} n^2\,\mathcal{Z}\,I_n/\Lambda_q & \imath\, n\,\mathcal{Z}\,(I_n' - I_n) & -n\,\mathcal{Z}'\,I_n/(2\Lambda_q)^{1/2} \\ -\imath\, n\,\mathcal{Z}\,(I_n' - I_n) & \mathcal{Z}\,(n^2 I_n/\Lambda_q + 2\Lambda_q I_n - 2\Lambda_q I_n') & \imath\,\Lambda_q^{1/2}\,\mathcal{Z}'\,(I_n' - I_n)/2^{1/2} \\ -n\,\mathcal{Z}'\,I_n'/(2\Lambda_q)^{1/2} & -\imath\,\Lambda_q^{1/2}\mathcal{Z}'\,(I_n' - I_n)/2^{1/2} & -\zeta_n\,\mathcal{Z}'\,I_n \end{pmatrix} \quad (12.150)$$

The argument of the Bessel functions I_n is Λ_q and that of the plasma dispersion function $\mathcal{Z}$ is ζ_{nq}.

- *Cold plasma*: When the temperature $T \to 0$, $\Lambda \to 0$ and $\zeta \to \infty$, but the product $\Lambda\,\zeta$ is finite. Since $I_n(\Lambda) \to (\Lambda/2)^{|n|}$ as $\Lambda \to 0$ and $\mathcal{Z}_n(\zeta) \to -1/\zeta$ it follows that only the terms $n = 0$ and $n = \pm 1$ contribute and the permittivity is

$$K_{ij} = \begin{pmatrix} 1 - \frac{1}{2}\sum_q \frac{\Pi_q{}^2}{\omega^2}\left(\frac{\omega}{\omega-\Omega_q} + \frac{\omega}{\omega+\Omega_q}\right) & -\frac{\imath}{2}\sum_q \frac{\Pi_q{}^2}{\omega^2}\left(\frac{\omega}{\omega-\Omega_q} + \frac{\omega}{\omega+\Omega_q}\right) & 0 \\ \frac{1}{2}\sum_q \frac{\Pi_q{}^2}{\omega^2}\left(\frac{\omega}{\omega-\Omega_q} + \frac{\omega}{\omega+\Omega_q}\right) & 1 - \frac{1}{2}\sum_q \frac{\Pi_q{}^2}{\omega^2}\left(\frac{\omega}{\omega-\Omega_q} + \frac{\omega}{\omega+\Omega_q}\right) & 0 \\ 0 & 0 & 1 - \sum_q \frac{\Pi_q{}^2}{\omega^2} \end{pmatrix} \quad (12.151)$$

identical to Eq. (11.9) found earlier.

- *Field free plasma*: When the ambient magnetic field $\mathbf{B}_0 = 0$, it follows that we may choose the axes as in Section 11.3.1 with the direction of propagation parallel to the z axis, so that

$k_\perp = 0$. Consequently $\Lambda = 0$. Using the small argument approximation for the modified Bessel function, it follows that the permittivity tensor

$$\epsilon_{xx} = \epsilon_{yy} = \epsilon_\perp = 1 + \sum_q \frac{\Pi_q^{\ 2}}{\omega^2} \zeta_q \, \mathcal{Z}(\zeta_q) \qquad \epsilon_{zz} = \epsilon_\parallel = 1 - \sum_q \frac{\Pi_q^{\ 2}}{\omega^2} \zeta_q^{\ 2} \, \mathcal{Z}'(\zeta_q) \qquad \epsilon_{ij} = 0 \quad \text{if} \quad i \neq j \tag{12.152}$$

where $\zeta_q = \omega/(k\sqrt{2\,\ķT/m_q})$ in agreement with Eq. (12.88).

12.9.1 Propagation Parallel to the Magnetic Field

When the wave travels along the ambient magnetic field, the wave vector $\mathbf{k} = k_\parallel \hat{\mathbf{z}}$, and $\Lambda = 0$. Since $I_n(\Lambda) \to (\Lambda/2)^{|n|}$, it follows that the only contributions to the dielectric constant matrix come from terms with $n = 0$or ± 1.

$$\begin{aligned} K_{11} = K_{22} &= 1 + \frac{1}{2}\sum_q \frac{\Pi_q^{\ 2}}{\omega\, k_\parallel\, \mathfrak{v}_q} \left[\mathcal{Z}\left(\frac{\omega - \Omega_q}{k_\parallel\, \mathfrak{v}_q}\right) + \mathcal{Z}\left(\frac{\omega + \Omega_q}{k_\parallel\, \mathfrak{v}_q}\right)\right] \\ K_{12} = -K_{21} &= -\sum_q \frac{\Pi_q^{\ 2}}{\omega\, k_\parallel\, \mathfrak{v}_q} \left[\mathcal{Z}\left(\frac{\omega - \Omega_q}{k_\parallel\, \mathfrak{v}_q}\right) + \mathcal{Z}\left(\frac{\omega + \Omega_q}{k_\parallel\, \mathfrak{v}_q}\right)\right] \end{aligned} \tag{12.153}$$

Hence, the dispersion relation is given by Eqs. (13.33)–(13.35)

$$\mathbb{M} \cdot \mathbf{E} = 0 \qquad \text{where} \qquad \mathbb{M} = \mathbb{K} - \frac{c^2}{\omega^2}(\mathbb{I}\, k^2 - \mathbf{kk}) \qquad \text{and} \qquad \therefore \det(\mathbb{M}) = 0 \tag{12.154}$$

and remembering that $k^2 = k_\perp^{\ 2} + k_\parallel^{\ 2}$

$$M_{11} = M_{22} = 1 - \frac{c^2 k_\parallel^{\ 2}}{\omega^2} + \frac{1}{2}\sum_q \frac{\Pi_q^{\ 2}}{\omega\, k_\parallel\, \mathfrak{v}_q} \left[\mathcal{Z}\left(\frac{\omega - \Omega_q}{k_\parallel\, \mathfrak{v}_q}\right) + \mathcal{Z}\left(\frac{\omega + \Omega_q}{k_\parallel\, \mathfrak{v}_q}\right)\right] \tag{12.155a}$$

$$M_{12} = -M_{21} = \imath\, \frac{1}{2}\sum_q \frac{\Pi_q^{\ 2}}{\omega\, k_\parallel\, \mathfrak{v}_q} \left[\mathcal{Z}\left(\frac{\omega - \Omega_q}{k_\parallel\, \mathfrak{v}_q}\right) + \mathcal{Z}\left(\frac{\omega + \Omega_q}{k_\parallel\, \mathfrak{v}_q}\right)\right] \tag{12.155b}$$

$$M_{33} = 1 - \sum_q \frac{\Pi_q^{\ 2}}{(k_\parallel\, \mathfrak{v}_q)^2}\, \mathcal{Z}'\left(\frac{\omega}{k_\parallel\, \mathfrak{v}_q}\right) \tag{12.155c}$$

$$M_{13} = M_{31} = M_{23} = M_{32} = 0 \tag{12.155d}$$

Comparing the matrix $\mathbb{M}$ for a thermal plasma (12.155) with that for a cold one Eq. (11.39), we can immediately see the similarities. As before there are three wave solutions given by

$$M_{33} \times \begin{vmatrix} M_{11} & M_{12} \\ -M_{12} & M_{11} \end{vmatrix} = 0 \tag{12.156}$$

i. *Longitudinal plasma wave*:

$$M_{33} = 1 + \sum_q \frac{2\Pi_q^{\ 2}}{(k_\parallel\, \mathfrak{v}_q)^2} \left[1 + \frac{\omega}{k_\parallel\, \mathfrak{v}_q}\, \mathcal{Z}\left(\frac{\omega}{k_\parallel\, \mathfrak{v}_q}\right)\right] = 0 \tag{12.157}$$

The eigenvector corresponding to this eigenvalue is $(0, 0, E_z)$, which is a longitudinal wave propagating with wave vector $k_\parallel$ along the magnetic field, the waves are therefore the warm plasma case of the cold plasma longitudinal waves propagating parallel to the field discussed in Section 11.3.3, item i. The longitudinal dispersion relation is independent of the magnetic field, the longitudinal waves parallel to the magnetic field allowed are similar to those when the field is absent discussed in Section 12.1. The waves are therefore either Landau damped electrostatic plasma waves Section 12.1.1 or under certain conditions ion plasma waves Section 12.7. These waves are strongly damped if $k_\parallel \lambda_{Dq} \gtrsim 1$ for any particle q, as before.

ii. *Circularly polarised waves*: The second determinant factorises to the two terms

$$M_{11} \pm \imath M_{12} = 1 - \frac{c^2 {k_\parallel}^2}{\omega^2} + \sum_q \frac{{\Pi_q}^2}{\omega k_\parallel \mathfrak{v}_q} \mathcal{Z}\left(\frac{\omega \pm \Omega_q}{k_\parallel \mathfrak{v}_q}\right) = 0 \tag{12.158}$$

The eigenvectors and eigenvalues of the two roots are

$$(E_x, \imath E_x, 0) \qquad \text{and} \qquad \frac{c^2 {k_\parallel}^2}{\omega^2} = 1 + \sum_q \frac{{\Pi_q}^2}{\omega k_\parallel \mathfrak{v}_q} \mathcal{Z}\left(\frac{\omega + \Omega_q}{k_\parallel \mathfrak{v}_q}\right) \tag{12.159}$$

corresponding to a right-handed circularly polarised wave – namely the *electron cyclotron wave* of Section 11.3.3, case ii.

$$(E_x, -\imath E_x, 0) \qquad \text{and} \qquad \frac{c^2 {k_\parallel}^2}{\omega^2} = 1 + \sum_q \frac{{\Pi_q}^2}{\omega k_\parallel \mathfrak{v}_q} \mathcal{Z}\left(\frac{\omega - \Omega_q}{k_\parallel \mathfrak{v}_q}\right) \tag{12.160}$$

corresponding to a left-handed circularly polarised wave – namely the *ion cyclotron wave* of Section 11.3.3, case iii.

Both these expressions reflect the resonances at the electron and ion cyclotron frequencies found for cold plasma in right- (Section 11.3.3, case ii) and left-handed (Section 11.3.3, case ii) circular polarisation, respectively.

At frequencies far from the resonance $|\omega - |\Omega_q|| \gg k_\parallel \mathfrak{v}_q$ where $Z(\zeta) \sim 1/\zeta$, it is easily seen that the dispersion relation reduces to the cold plasma forms, Eqs. (11.42) and (11.45), respectively.

However, near the resonance when $|\omega - |\Omega_q| \lessapprox k_\parallel \mathfrak{v}_q$, the characteristic behaviour associated with the dispersion function becomes important. Firstly, the resonance is broadened with a width $\sim k_\parallel \mathfrak{v}_q$, and is no longer a simple singularity. of zero width in frequency space. Importantly damping is introduced, similar to Landau damping in plasma waves. The mechanism in this case follows a similar pattern but is associated with the resonance between the particle gyration and the wave field. Electrons with angular frequency just below the wave frequency are accelerated into the wave, but those with higher rotational speed are slowed. Since in a Maxwellian distribution there are a more slow particles than fast, the net result is a transfer of energy from the wave to the particles – hence damping the wave (see Section 12.4.1).

12.9.2 Propagation Perpendicular to the Magnetic Field

When the wave is propagated perpendicular to the magnetic field, $k_\parallel = 0$ and consequently $\zeta_n \to \infty$. As a result, the plasma dispersion function and its derivative may be replaced by its asymptotic form (12.85). As a result, the elements of the matrix $\mathbb{M}$ become

$$M_{11} = 1 - \sum_q \frac{{\Pi_q}^2}{\omega} \frac{\exp(-\Lambda_q)}{\Lambda_q} \sum_{n=-\infty}^{\infty} \frac{n^2 I_n(\Lambda_q)}{\omega - n\,\Omega_q} \tag{12.161a}$$

$$M_{12} = -M_{21} = \imath \sum_q \frac{\Pi_q{}^2}{\omega} \exp(-\Lambda_q) \sum_{n=-\infty}^{\infty} \frac{n\,[I_n'(\Lambda_q) - I_n(\Lambda_q)]}{\omega - n\,\Omega_q} \tag{12.161b}$$

$$M_{22} = 1 - \frac{k_\perp{}^2 c^2}{\omega^2} - \sum_q \frac{\Pi_q{}^2}{\omega} \frac{\exp(-\Lambda_q)}{\Lambda_q} \sum_{n=-\infty}^{\infty} \frac{n^2 I_n(\Lambda_q) + 2\Lambda_q{}^2 I_n(\Lambda_q) - 2\Lambda_q^2 I_n'(\Lambda_q)}{\omega - n\,\Omega_q} \tag{12.161c}$$

$$M_{33} = 1 - \frac{k_\perp{}^2 c^2}{\omega^2} \sum_q \frac{\Pi_q{}^2}{\omega} \exp(-\Lambda_q) \sum_{n=-\infty}^{\infty} \frac{n^2 I_n(\Lambda_q)}{\omega - n\,\Omega_q} \tag{12.161d}$$

$$M_{13} = M_{31} = M_{23} = M_{32} = 0 \tag{12.162e}$$

The dispersion equation is the vanishing of the determinant det $(\mathbb{M})$ so that

$$M_{33} \times \begin{vmatrix} M_{11} & M_{12} \\ M_{21} & M_{22} \end{vmatrix} = 0 \tag{12.162}$$

i *O mode*:

The first root is clearly $M_{33} = 0$ with eigenvector $(0, 0, E_z)$ and is therefore a transverse wave with the electric field perpendicular to the direction of the wave. It is clearly the electro-magnetic *ordinary wave* found earlier for the cold plasma (case i, Section 11.3.3). The dispersion relation corresponding to these waves is

$$\frac{k_\perp{}^2 c^2}{\omega^2} = 1 - \sum_q \frac{\Pi_q{}^2}{\omega} \exp(-\Lambda_q) \sum_{n=-\infty}^{\infty} \frac{n^2 I_n(\Lambda_q)}{\omega - n\,\Omega_q} \tag{12.163}$$

Remembering that $I_n(\Lambda) \rightarrow (\Lambda/2)^{|n|}$ as $\Lambda \rightarrow 0$ it is easily seen that this dispersion relation reduces to that of cold plasma, Eq. (11.48).

There is a major difference with cold plasma evidenced by the terms in the sums involving resonances at the multiples of cyclotron frequency. Since $\Lambda = (k_\perp\,\rho(\mathfrak{v})/2)^2$, it is clear that the strength of the resonance depends on the Larmor radius based on the thermal speed $\rho(\mathfrak{v}) = m\,\mathfrak{v}/Z\,e\,B_0$. In fact the resonance arises as a finite Larmor radius effect, due to the phase variation over the cyclotron orbit. If the wavelength is large compared to the Larmor radius $k_\perp\,\rho(\mathfrak{v}) \ll 1$, the strength of the resonances fall off as $\Lambda^{|n|}$ so that higher resonances make little contribution. When the Larmor radius becomes comparable to the wavelength, higher order resonances make stronger contributions until eventually all resonances are active. In principle, since the dispersion function has zero argument, its effect is cancelled out, and these resonances do not suffer from collisionless damping (c.f. parallel propagation). The width of the resonance is correspondingly zero. Absorption of the wave by the resonance must then be due to collisions in the manner described in Section 5.4.2 with width equal to the collision frequency. In practice, this condition is only valid if the wave vector is exactly perpendicular to the field. Small angular deviation introduces the dispersion function with non-zero argument. This gives the resonance frequency width and corresponding absorption through the non-Hermitian part of the permittivity. Significant damping occurs if $\omega - n\,\Omega_q \lessapprox k_\parallel \mathfrak{v}$. Since the Larmor radius depends on the square root of the particle mass clearly these resonances are more important for ions.

These resonances have important practical application enabling heating in thermo-nuclear fusion plasma. Thus one has *electron cyclotron heating* and more commonly *ion cyclotron heating* quite widely applied.

ii *X mode*: The second solution is obtained by solving the 2×2 determinant $M_{11} M_{22} - M_{12} M_{21} = 0$. The waves have an eigenvector $(E_x, E_y, 0)$ and therefore is partly transverse and partly longitudinal. The roots of the dispersion equation satisfy

$$\left(1 - \sum_q \frac{\Pi_q{}^2}{\omega} \frac{\exp(-\Lambda_q)}{\Lambda_q} \sum_{n=-\infty}^{\infty} \frac{n^2 I_n(\Lambda_q)}{\omega - n\,\Omega_q}\right)$$
$$\times \left(1 - \frac{k_\perp{}^2 c^2}{\omega^2} - \sum_q \frac{\Pi_q{}^2}{\omega} \frac{\exp(-\Lambda_q)}{\Lambda_q} \sum_{n=-\infty}^{\infty} \frac{n^2 I_n(\Lambda_q) + 2\Lambda_q{}^2 [I_n(\Lambda_q) - I_n'(\Lambda_q)]}{\omega - n\,\Omega_q}\right)$$
$$= \left(\sum_q \frac{\Pi_q{}^2}{\omega} \exp(-\Lambda_q) \sum_{n=-\infty}^{\infty} \frac{n\,[I_n'(\Lambda_q) - I_n(\Lambda_q)]}{\omega - n\,\Omega_q}\right)^2 \tag{12.164}$$

In cold plasma, $\Lambda \to 0$ and the only term contributing to the sums comes from $n = 0$ and leads directly to the cold plasma dispersion relation. The dispersion relations therefore 'contain' the cold plasma X-modes, but modified by thermal effects. However in contrast to parallel propagation, the changes are not simply the introduction of resonances to the existing modes. In this case, the thermal motion introduces a new mode of vibration – the electrostatic *Bernstein modes*.[13] Since the waves are found when the propagation is perpendicular to the magnetic field, the argument of the dispersion function $\zeta_n \to \infty$ and the imaginary part of the function is infinitely small. Consequently, the waves are undamped.
Since these waves are electrostatic, it follows that they are slow moving so that $V_p/c = kc/\omega \ll 1$. Consequently, $M_{22} \to \infty$ and it follows that the dispersion relation is $M_{11} \approx 0$, i.e.

$$1 - \sum_q \frac{\Pi_q{}^2}{\omega} \frac{\exp(-\Lambda_q)}{\Lambda_q} \sum_{n=-\infty}^{\infty} \frac{n^2 I_n(\Lambda_q)}{\omega - n\,\Omega_q} = 0 \tag{12.165}$$

and the eigenvector is $(E_x, 0, 0)$ as is characteristic of a longitudinal wave. This can be shown to be the perpendicular component of the Harris dispersion relation for longitudinal waves (12.22). Since $\Lambda = k^2\mathfrak{V}^2/\Omega^2$, it is clear that as the temperature $T \to 0$, the thermal velocity becomes zero $\mathfrak{v} \to 0$ and therefore the wave number $k \to \infty$. Consequently, no wave of this type exists in cold plasma.
Due to the large mass difference between electrons and ions, the electron Bernstein waves have a much higher frequency than those of the ions. We may therefore consider electron Bernstein waves independently and for the moment neglect the contribution of the cold plasma X mode. Consider the basic equation for Bernstein waves of a single species, Eq. (12.165) is the simplified form

$$1 - \sum_{n=-\infty}^{\infty} \frac{\Pi_e{}^2}{\omega} \frac{\exp(-\Lambda)}{\Lambda} \frac{n^2 I_n(\Lambda)}{\omega - n\,\Omega} = 0 \tag{12.166}$$

The form of this relation is most simply found by examining its behaviour at small and large wave numbers, which correspond to large and small values of $\Lambda = (k\mathfrak{v}/\Omega)^2$. For small Λ, the modified Bessel function $I_n(\Lambda) \approx (\Lambda/2)^{|n|}$, so that the only direct contribution to the sum comes for the case $n = \pm 1$ when $\omega^2 = \Pi^2 + \Omega^2$ namely the upper hybrid frequency. For other values of $n \neq \pm 1$, the dispersion relation can only be satisfied when $\omega = n\,\Omega$, which are harmonics of the cyclotron frequency. For large Λ, the asymptotic expansion $I_n(\Lambda) \approx (2\,\pi\,\Lambda)^{1/2} \exp(\Lambda)$, so that $I_n(\Lambda)\, e^{-\Lambda} \to 0$. Hence for large values of Λ, solutions are only obtained when $\omega = n\,\Omega$ again at harmonics of the

13 The waves are named after their discoverer Bernstein (1958) in a general study of waves in magnetic fields using Landau's approach.

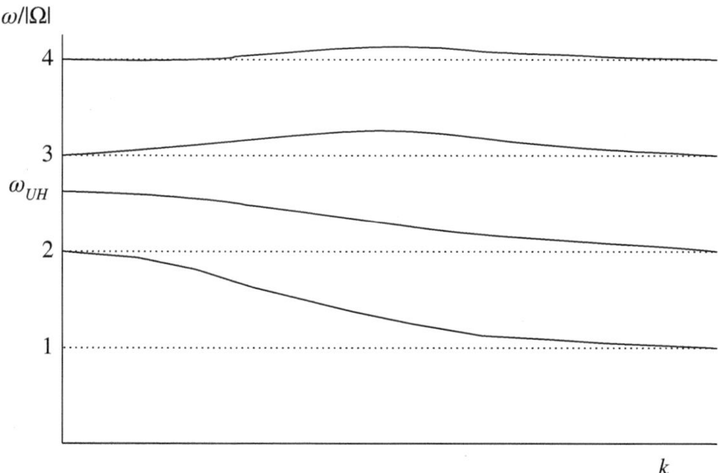

Figure 12.11 Sketch of the dispersion relation for a Bernstein mode in the case where $2\,\Omega < \omega_{UH} < 3\,\Omega$.

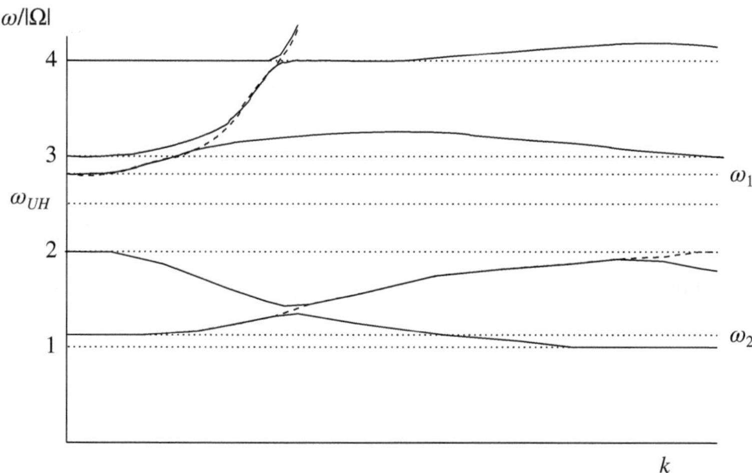

Figure 12.12 Sketch of the dispersion relation for waves modes propagating perpendicular to the magnetic field in the case where $2\,\Omega < \omega_{UH} < 3\,\Omega$. The dashed line shows the cold plasma extra-ordinary mode.

cyclotron frequency. Since the frequency must be regular, it may be concluded that the dispersion relation must be represented by a smooth curves in the region $(n-1)\,\Omega \leq \omega \leq n\,\Omega$. A plot of this function must be similar to Figure 12.11, which illustrates the case $2\,\Omega < \omega_{UH} < 3\,\Omega$.

In view of the large differences between the ion and electron masses, the frequencies of the Bernstein waves associated with the different species are well separated, and the electrons (at least) may be treated independently. However, for small values of the wave number, the phase velocity is no longer small and the role of the cold plasma extra-ordinary can longer be ignored. In particular, the electro-magnetic and the longitudinal waves cross over to give the pattern shown in Figure 12.12. Note the role of the cut-offs identified earlier, Eqs. (11.24) and (11.26) determining the frequency of the longest wavelength waves as found in cold plasma.

Appendix 12.A Landau's Solution of the Vlasov Equation

To obtain the perturbation form of the Vlasov equation (12.5), we Fourier transformed in both space and time and obtained a solution valid over all space and time. This introduced a singularity which was simply avoided using physical intuition, specifying the direction of time in an *ad hoc* fashion, thereby avoiding mathematical complexity. On the other hand, Landau directly avoided this problem by using a Laplace transformation (Section M.4) in time and starting the solution at time $t = 0$, but requiring considerably more complex mathematics, but providing a more general method.

Since we are considering longitudinal waves only, we integrate over the directions perpendicular to k, to obtain the perturbed longitudinal distribution function $F(v) \equiv \int f(\mathbf{v})\, \mathrm{d}\mathbf{v}_\perp$ where $v \equiv v_\parallel$.

As before we write the spatial variation as a Fourier transform (Eq. (12.4))

$$F_1(v, x, t) = \int \mathrm{d}k\, \tilde{F}_1(v, k, t) \exp(\imath\, k\, x) \tag{12.A.1}$$

and the electric field

$$E(x, t) = \int \mathrm{d}k\, \tilde{E}(k, t) \exp(\imath\, k\, x) \tag{12.A.2}$$

both switched on at time $t = 0$.

However, we are now treating the behaviour as an initial value problem, and the temporal variation is treated by a Laplace transform

$$\overline{F}_1(v, k, p) = \int_0^\infty \mathrm{d}t\, \tilde{F}_1(v, k, t)\, \exp(-pt) \tag{12.A.3}$$

$$\bar{E}(k, p) = \int_0^\infty \mathrm{d}t\, \tilde{E}(k, t)\, \exp(-pt) \tag{12.A.4}$$

the Laplace transform integral over t converges provided the growth of F_1 is no faster than exponential and defines F and E as an analytic functions of p in the right-hand half plane $p > 0$.

Noting that the Laplace derivative of $\mathrm{d}F/\mathrm{d}t$ is $p\,\bar{F} - \tilde{F}|_{t=0}$ as is easily shown by integration by parts, and applying the Fourier–Laplace transformation, Vlasov's equation (12.5) and Poisson's equation (12.8) are replaced by

$$(\imath\, k\, v + p)\overline{F}_1 = \frac{e}{m}\bar{E}\,\frac{\partial F_0}{\partial v} + \tilde{F}_1|_{t=0} \tag{12.A.5}$$

$$\imath\, k\, \epsilon_0\, \bar{E} = -e \int \overline{F}_1(v, p)\mathrm{d}v \tag{12.A.6}$$

Solving for $\overline{F}_1$, we obtain

$$\overline{F}_1 = \left\{ \frac{e}{m} E \frac{\partial F_0}{\partial v} + \tilde{F}_1|_{t=0} \right\} \Big/ (\imath k\, v + p) \tag{12.A.7}$$

Making use of Eqs. (12.7) and (13.3), we obtain the longitudinal permittivity and hence the dielectric function

$$D(k, p) = 1 - \frac{e^2}{\epsilon_0\, m\, k} \int_{-\infty}^{\infty} \frac{\partial F_0/\partial v}{(p + \imath\, k\, v)} \mathrm{d}v \tag{12.A.8}$$

The electric field

$$\tilde{E}(k, p) = \frac{\imath\, e}{\epsilon_0\, k\, D(k, p)} \int \frac{F_1(k, v, |_{t=0}}{p + \imath\, k\, v}\, \mathrm{d}v \tag{12.A.9}$$

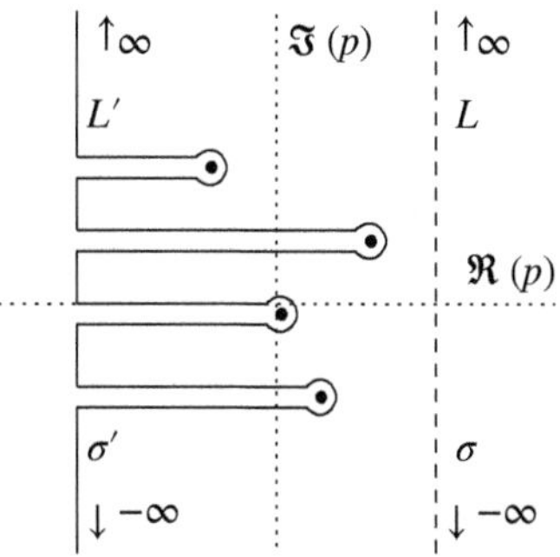

Figure 12.A.1 Analytic continuation of the Bromwich contour from σ to σ' avoiding poles by indenting the line integral using loops around the poles.

Finally,

$$\overline{F}_1 = \frac{1}{\imath k v + p} \left\{ \imath e \frac{\partial F_0/\partial v}{k m D} \int_{-\infty}^{\infty} \frac{F_1(v')|_{t=0}}{\imath k v' + p} \, \mathrm{d}v' + F_1(v)\Big|_{t=0} \right\} \tag{12.A.10}$$

We now need to invert the Laplace transform to obtain the solution back in laboratory variables. This involves a complex integration using the Bromwich line integral $\mathscr{L}$ (see Section M.4) which runs parallel to imaginary axis excluding any poles (Figure 12.A.1). In most cases, this can be extended to a contour enclosing the left-hand half-plane to yield a sum over the residues of the enclosed poles and any other singularities.

In this case, this is not a practical proposition. Instead Landau (1946) limited the solution to the asymptotic case as $t \to \infty$ so that most terms with $\Re(p) < 0$ give negligibly small contributions. This is accomplished by *analytic continuation* from the contour line $\mathscr{L}$: where $p = \sigma$ into the negative half-plane by encircling the poles in the usual way (Section M.3.M.3.viii). The integration line $\mathscr{L}$ may then be shifted to the indented line $\mathscr{L}'$ at $p = \sigma' < 0$ in the negative half-plane (Figure 12.A.1). The integral in the left-hand plane therefore reduces to integrals around the singularities lying between $\mathscr{L}'$ and $\mathscr{L}$. Again only the right-hand-most of those will give a non-negligible contribution. Typically, these are those where the dielectric function $D(k,p) = 0$. We assume, as may be proved *a posteriori*, that these correspond to a plasma decay so that $p < 0$.

Finally, we must consider the singularity in the integrals of the form

$$\int \frac{g(v)}{v - \imath p/k} \, \mathrm{d}v \tag{12.A.11}$$

where $g(v)$ is an *entire* function (i.e. holomorphic on the complex plane). There is therefore a pole at $v = \imath p/k$.

The Bromwich line integral has $p > 0$ and therefore before the line of integration is moved the pole lies above the line of the integral in v (Figure 12.A.2). However after shift to $\mathscr{L}'$ where $p =$

Figure 12.A.2 Integration paths around the pole at $\Im(v) = \Re(p)/k$ (filled dot). The pole lies: (a) above the line of integration; (b) on the line of integration; (c) below the line of integration. Note that the path of integration is deformed so as to remain always below the singularity.

$\sigma' < 0$, the pole is below the $\Re(v)$ axis and case (c) is applicable. In fact since the damping is weak $\sigma' \approx 0$ and the pole lies on the integration path (b). The integration of the dielectric function D is therefore along the Landau contour as we discussed earlier (Section 12.1.2.1). Consequently, the analysis may be reduced to that found earlier.

Appendix 12.B Electrostatic Waves

In calculating the family of waves in a plasma with a finite temperature, the electro-magnetic field has been calculated by the direct application of Maxwell's equations. Gauss's theorem is used implicitly via the conservation of charge, but does not explicitly appear. However if the time variation of the fields is very slow $\omega \ll c\,k$, then they may be treated as quasi-static. In this case, $\nabla \wedge \mathbf{E} = -\partial \mathbf{B}/\partial t \approx 0$ and the electric field is conservative described by a potential $\mathbf{E} = -\nabla\phi$. Furthermore, the perturbed magnetic field $\mathbf{B} \approx 0$ is zero. The electric field is given by Gauss' theorem

$$\nabla \cdot \mathbf{E} = -\nabla^2\phi = \rho/\epsilon_0 \tag{12.B.1}$$

Introducing a harmonic wave $\sim\exp[\imath(\mathbf{k}\cdot\mathbf{r} - \omega t)]$, the electric field in Faraday's law $\nabla \wedge \mathbf{E} = \imath\,\mathbf{k} \wedge \mathbf{E}$ can only be transverse $\mathbf{E} \perp \mathbf{k}$. The longitudinal component in the spatial term in the wave equation

$$\nabla \wedge (\nabla \wedge \mathbf{E}) = \nabla(\nabla \cdot \mathbf{E}) - \nabla^2\mathbf{E} = 0$$

obtained from Faraday's law is zero. The corresponding term for *longitudinal* or *electrostatic* waves arises directly from Gauss' law.

Electrostatic waves form an important sub-set of the general set of waves, where the perturbed magnetic field is zero, and the perturbation electric field is described by a scalar potential $\mathbf{E} = -\nabla\phi$. The dispersion relation is given by

$$\nabla \cdot (\nabla \wedge \mathbf{B}) = \mu_0\,\nabla \cdot \mathbf{D} = \frac{1}{c^2}\nabla \cdot (\mathbb{K} \cdot \mathbf{E}) = -\frac{1}{c^2}\nabla \cdot (\mathbb{K} \cdot \nabla\phi) = 0 \tag{12.B.2}$$

reflecting the fact that there is no free charge. Therefore, Eq. (13.32) simplifies to the alternative forms

$$\mathbf{k} \cdot \mathbb{K} \cdot \mathbf{E} = 0 \qquad \text{or} \qquad \mathbf{k} \cdot \mathbb{K} \cdot \mathbf{k} = 0 \tag{12.B.3}$$

and the dispersion relation follows from the solutions of

$$\det(\mathbb{K}) = 0 \tag{12.B.4}$$

To calculate the dispersion relation, we need to calculate the dielectric constant following the same approach as before but limiting the perturbed field to an electric field described by a scalar potential alone. The linearised equation for the perturbed distribution function is then given by Eq. (12.111) omitting the terms in the magnetic field

$$\frac{\partial f_1}{\partial t} + \mathbf{v}' \cdot \frac{\partial f_1}{\partial \mathbf{r}'} + \frac{Ze}{m}\mathbf{v}' \wedge \mathbf{B}_0 \cdot \frac{\partial f_1}{\partial \mathbf{v}'} = -\frac{Ze}{m}(\mathbf{E} + \mathbf{v} \wedge \mathbf{B}_0) \cdot \frac{\partial f_0}{\partial \mathbf{v}'} \tag{12.B.5}$$

which can be integrated as before to give Eq. (12.118) after omitting terms in $\mathbf{B}$ to give

$$f_1(\mathbf{r}, \mathbf{v}, t) = \frac{Ze}{m}\int_{-\infty}^{t}\left[E_x\frac{\partial f_0}{\partial v_x'} + E_y\frac{\partial f_0}{\partial v_y'} + E_z\frac{\partial f_0}{\partial v_z'}\right]\exp[\imath\{\mathbf{k}\cdot(\mathbf{r}' - \mathbf{r}) - \omega(t' - t)\}]\,\mathrm{d}t' \tag{12.B.6}$$

Following Eq. (12.120) using Jabobi's relation twice

$$f_1 = \frac{Z\,e}{m}\int_{-\infty}^{t}\left[E_x\cos\chi\frac{\partial f_0}{\partial v_\perp} + E_y\sin\chi\frac{\partial f_0}{\partial v_\perp} + E_z\frac{\partial f_0}{\partial v_\parallel}\right]$$
$$\times\sum_{n,m=-\infty}^{\infty} J_n\left(\frac{k_\perp v_\perp}{\Omega}\right) J_m\left(\frac{k_\perp v_\perp}{\Omega}\right)\exp\{\imath[(n\Omega + k_\parallel v_\parallel - \omega)(t'-t) + (m-n)\theta]\}\,\mathrm{d}t' \tag{12.B.7}$$

Performing the integrals as in Eq. (12.131), we obtain the perturbed distribution

$$f_1 = -\frac{Z\,e}{m}\sum_n\sum_m J_m(\Lambda)\frac{\exp[\imath(m-n)\,\theta]}{\omega - k_\parallel v_\parallel - n\,\Omega}$$
$$\times\left[\frac{n}{\Lambda}J_n(\Lambda)\frac{\partial f_0}{\partial v_\perp}k_x + J_n'(\Lambda)\frac{\partial f_0}{\partial v_\perp}k_y + J_n(\Lambda)\frac{\partial f_0}{\partial v_\perp}k_z\right]\phi \tag{12.B.8}$$

where $\Lambda = k_\perp v_\perp/\Omega$, and the electric field is written in terms of the potential $\mathbf{E} = -\nabla\phi - \imath\,\mathbf{k}\,\phi$.

We may now calculate the dielectric constant using Eq. (12.B.2) in the form

$$(\mathbf{k}\cdot\mathbb{K}\cdot\mathbf{k})\,\phi = k^2\phi - \rho/\epsilon_0 = 0 \tag{12.B.9}$$

Summing over all species of particles q, we obtain

$$\rho = \sum_q Z_q\,e\int f_{1q}\,\mathbf{dv} = \int_{-\infty}^{\infty}\mathrm{d}v_\parallel\int_0^{\infty}\mathrm{d}v_\perp\int_0^{2\pi}\mathrm{d}\theta\sum_q Z_q\,e f_{1q} \tag{12.B.10}$$

Averaging over the phase angle as before

$$f_1 = -\frac{Z\,e}{m}\sum_n\int_{-\infty}^{\infty}\mathrm{d}v_\parallel\frac{1}{(\omega - k_\parallel v_\parallel - n\Omega)}\int_0^{2\pi} 2\,\pi\,v_\perp\,\mathrm{d}v_\perp$$
$$\times\left[k_x\frac{n}{a}J_n^{\,2}(\Lambda)\frac{\partial f_0}{\partial v_\perp} + k_y J_n(\Lambda)J_n'(\Lambda)\frac{\partial f_0}{\partial v_\perp} + k_z J_n^{\,2}(\Lambda)\frac{\partial f_0}{\partial v_\parallel}\right]\phi \tag{12.B.11}$$

Taking axes as before so that the ambient field $\mathbf{B}_0$ is along z, and the wave vector $\mathbf{k}$ lies in the x, z plane, $k_y = 0$. The volume element in velocity space becomes $\mathbf{dv} = 2\,\pi\,v_\perp\,\mathrm{d}v_\perp\,\mathrm{d}v_\parallel$ and

$$\int f_1\,\mathbf{dv} = -\frac{Z\,e\,\phi}{m}\sum_n\int\frac{J_n^{\,2}(\Lambda)}{(\omega - k_\parallel v_\parallel - n\,\Omega)}\left[\frac{n\,\Omega}{v_\perp}\frac{\partial f_0}{\partial v_\perp} + k_\parallel\frac{\partial f_0}{\partial v_\parallel}\right]\mathbf{dv} \tag{12.B.12}$$

Making use of Eqs. (12.B.9) and (12.B.10), we obtain the Harris dispersion relation

$$k^2 + \sum_q \Pi_q^{\,2}\sum_n\int\frac{J_n^{\,2}(\Lambda_q)}{(\omega - k_\parallel v_\parallel - n\,\Omega_q)}\left[\frac{n\,\Omega_q}{v_\perp}\frac{\partial\tilde{f}_{0\,q}}{\partial v_\perp} + k_\parallel\frac{\partial\tilde{f}_{0\,q}}{\partial v_\parallel}\right]\mathbf{dv} = 0 \tag{12.B.13}$$

where $\tilde{f}_0 = f_0/\mathfrak{n}$ is the normalised distribution function and $\Pi_q^{\,2} = Z_q\,e/\epsilon_0\,m_q$.

The dielectric constant matrix is obtained from

$$k_\perp^{\,2}K_{xx} + k_\perp\,k_\parallel\,(K_{xz} + K_{zx}) + k_\parallel^{\,2}K_{zz} = 0 \tag{12.B.14}$$

whence

$$K_{xx} = 1 + \sum_q\frac{\Pi_q^{\,2}}{\omega^2}\sum_n\int\frac{v_\perp\,(n/\Lambda_q)^2\,J_n^{\,2}(\Lambda_q)}{(\omega - k_\parallel v_\parallel - n\,\Omega_q)}U\,\mathbf{dv} \tag{12.B.15}$$

$$K_{xz} = \sum_q\frac{\Pi_q^{\,2}}{\omega^2}\sum_n\int\frac{v_\perp\,(n/\Lambda_q)\,J_n^{\,2}(\Lambda_q)}{(\omega - k_\parallel v_\parallel - n\,\Omega_q)}W\,\mathbf{dv} \tag{12.B.16}$$

$$K_{zx} = \sum_q \frac{\Pi_q^{\ 2}}{\omega^2} \sum_n \int \frac{v_\parallel \,(n/\Lambda_q) J_n^{\ 2}(\Lambda_q)}{(\omega - k_\parallel\, v_\parallel - n\,\Omega_q)} U\, \mathbf{dv} \tag{12.B.17}$$

$$K_{zz} = 1 + \sum_q \frac{\Pi_q^{\ 2}}{\omega^2} \sum_n \int \frac{v_\parallel\, J_n^{\ 2}(\Lambda_q)}{(\omega - k_\parallel\, v_\parallel - n\,\Omega_q)} W\, \mathbf{dv} \tag{12.B.18}$$

where

$$U = (\omega - k_\parallel\, v_\parallel) \frac{\partial f_{0q}}{\partial v_\perp} + k_\parallel\, v_\perp \frac{\partial f_{0q}}{\partial v_\parallel} \tag{12.B.19a}$$

$$W = \frac{n\,\Omega_q\, v_\parallel}{v_\perp} \frac{\partial f_{0q}}{\partial v_\perp} + (\omega - n\,\Omega_q) \frac{\partial f_{0q}}{\partial v_\parallel} \tag{12.B.19b}$$

identical to the terms (12.133) found earlier. Substitution in Eq. (12.B.14) is easily seen to yield Eq. (12.B.9).

Following our earlier methodology for the general dispersion relation, we may now integrate over an unperturbed Maxwellian distribution. Since

$$\frac{\partial f_0}{\partial v_\perp} = -\frac{2v_\perp}{\pi^{3/2}\,\mathfrak{v}^5} \exp\left(-\frac{v^2}{\mathfrak{v}^2}\right) \qquad \text{and} \qquad \frac{\partial f_0}{\partial v_\parallel} = -\frac{2v_\parallel}{\pi^{3/2}\,\mathfrak{v}^5} \exp\left(-\frac{v^2}{\mathfrak{v}^2}\right) \tag{12.B.20}$$

and using Eq. (12.144a)

$$\begin{aligned} &\int \frac{J_n^{\ 2}(\Lambda_q)}{(\omega - k_\parallel\, v_\parallel - n\,\Omega_q)} \left[\frac{n\,\Omega_q}{v_\perp} \frac{\partial f_0}{\partial v_\perp} + k_\parallel \frac{\partial f_0}{\partial v_\parallel}\right] \mathbf{dv} \\ &= -\frac{2\mathfrak{n}}{\sqrt{\pi}\,\mathfrak{v}^3} \exp\left[-\frac{k_\perp^{\ 2}\,\mathfrak{v}^2}{2\Omega^2}\right] I_n\left(\frac{k_\perp^{\ 2}\,\mathfrak{v}^2}{2\Omega^2}\right) \int_{-\infty}^{\infty} \frac{n\Omega + k_\parallel\, v_\parallel}{\omega - k_\parallel\, v_\parallel - n\Omega} \exp\left(-\frac{v_\parallel^{\ 2}}{\mathfrak{v}^2}\right) \\ &= \frac{2\mathfrak{n}}{\pi^{3/2}\,\mathfrak{v}} \exp\left[-\frac{k_\perp^{\ 2}\,\mathfrak{v}^2}{2\Omega^2}\right] I_n\left(\frac{k_\perp^{\ 2}\,\mathfrak{v}^2}{2\Omega^2}\right) \left[1 + \frac{n\Omega}{k_\parallel\,\mathfrak{v}} + \left(\frac{\omega - n\Omega}{k_\parallel \mathfrak{v}}\right)\right] \mathcal{Z}\left(\frac{\omega - n\Omega}{k_\parallel \mathfrak{v}}\right) \end{aligned} \tag{12.B.21}$$

making use of the result, which follows from the generating function for modified Bessel functions, Abramowitz and Stegun (1965, eqn. 9.6.34), that $\sum_n I_n(x) = \exp(x)$. Summing over all particle species we obtain the final expression for the dispersion relation of the electrostatic modes

$$1 + \sum_q \frac{2\,\Pi_q^{\ 2}}{k^2 \mathfrak{v}_q^{\ 2}} \left[1 + \exp(-\Lambda_q) \sum_{n=-\infty}^{\infty} I_n(\Lambda_q) \frac{\omega}{k_\parallel \mathfrak{v}_q} \mathcal{Z}(\zeta_{nq})\right] = 0 \tag{12.B.22}$$

where $\zeta_n = (\omega - n\,\Omega)/(k_\parallel\, \mathfrak{v})$ and $\Lambda = k_\perp^{\ 2}\, \mathfrak{v}^2/2\,\Omega^2$.

Propagation parallel to the magnetic field where $k_\perp = 0$ and therefore $\Lambda \to 0$ only has non-zero terms in the sum over n for $n = 0\,(I(0) = 1,\ I_n(0) = 0 \text{ if } n \neq 0)$. The resulting equation is identical to Eq. (12.157) obtained earlier for the propagation of longitudinal plasma waves along the magnetic field. Since the waves are longitudinal, the field $\mathbf{E} \parallel \mathbf{k}$ and the eigenvector is $(0,\ 0,\ E_z)$.

Propagation perpendicular to the applied magnetic field where $k_\parallel = 0$ and therefore $\zeta \to \infty$, where $\mathcal{Z} \to -1/\zeta$. Making use of the reciprocal relation $I_{-n}(\Lambda) = I_n(\Lambda)$ and the sum relation $\sum_n \exp(-\Lambda)\, I_n(\Lambda) = 1$ it follows that the resulting equation is identical to Eq. (12.165) found earlier for the electrostatic longitudinal Bernstein modes, the field is perpendicular to the wave vector and the eigenvector consequently $(E_x,\ 0,\ 0)$. We have

$$\frac{2\,\Pi^2}{k^2\mathfrak{v}^2}\left[1+\exp(-\Lambda)\sum_{n=-\infty}^{\infty} I_n(\Lambda)\,\frac{\omega}{k_\parallel \mathfrak{v}}\mathcal{Z}(\zeta_n)\right] \tag{12.B.23}$$

$$= \frac{2\,\Pi^2}{k^2\mathfrak{v}^2}\sum_{n=-\infty}^{\infty}\left[\exp(-\Lambda)I_n(\Lambda)-\exp(-\Lambda)I_n(\Lambda)\,\frac{\omega}{\cancel{k_\parallel \mathfrak{v}}}\frac{\cancel{k_\parallel \mathfrak{v}}}{\omega-n\Omega}\right] \tag{12.B.24}$$

$$= -\frac{\Pi^2}{\Omega^2}\frac{\exp(-\Lambda)}{\Lambda}\sum_{n=-\infty}^{\infty}\frac{n\,\Omega}{\omega-n\Omega}I_n(\Lambda) \tag{12.B.25}$$

$$= -\frac{\Pi^2}{1}\frac{\exp(-\Lambda)}{\Lambda}\sum_{n=0}^{\infty}\frac{n^2}{\omega^2-n^2\Omega^2}\,I_n(\Lambda) \tag{12.B.26}$$

$$= -\frac{\Pi^2}{\omega^2}\frac{\exp(-\Lambda)}{\Lambda}\sum_{n=-\infty}^{\infty}\frac{n^2\,\omega}{\omega-n\,\Omega}\,I_n(\Lambda) \tag{12.B.27}$$

$$= -\frac{\Pi^2}{\omega}\frac{\exp(-\Lambda)}{\Lambda}\sum_{n=-\infty}^{\infty}\frac{n^2 I_n(\Lambda)}{\omega-n\Omega} \tag{12.B.28}$$

Consequently electrostatic plasma waves and Bernstein waves are merely different aspects of a more general type of electrostatic wave. This wave takes the form of a plasma wave when propagating parallel to the equilibrium magnetic field, or a Bernstein wave when propagating perpendicular to the magnetic field, and an intermediate form when propagating obliquely to the magnetic field.

13

Properties of Electro-magnetic Waves in Plasma

13.1 Plasma Permittivity and the Dielectric Constant

As was discussed earlier in Section 2.3, plasma is a linear medium in weak external fields. The electro-magnetic response of the plasma in weak fields is therefore linearly proportional to the applied electric field $\mathbf{E}_0$. This is similar to the familiar behaviour of non-conducting material in an electric field in electro-statics. In isotropic, linear dielectric materials, the induced polarisation charge or dipole moment per unit volume is proportional to the applied electric field, $\mathbf{E}$, i.e.

$$\mathbf{P} = \chi\, \mathbf{E} \tag{13.1}$$

where χ is dielectric susceptibility. Defining the dielectric displacement $\mathbf{D}$

$$\mathbf{D} = \epsilon_0 \mathbf{E} + \mathbf{P} = \varepsilon \mathbf{E} \tag{13.2}$$

and $\epsilon = \epsilon_0 + \chi$ the permittivity. If the medium is anisotropic, the polarisation is no longer parallel to the electric field and the susceptibility and the permittivity are tensors.[1] The polarisation $\mathbf{P}$ in plasma is due to the departure from exact charge neutrality, which is induced by the field and gives rise to the polarisation space charge of density $\rho = -\nabla \cdot \mathbf{P}$, as electron/ion pairs are separated, where ρ is the net space charge density.

At very high frequencies, the polarisation charge can no longer respond to the field and $\epsilon \to \epsilon_0$. Furthermore in plasma, there is no magnetisation due to the medium $\mathbf{M} = 0$ and the magnetic intensity $\mathbf{H}$ and the magnetic induction $\mathbf{B} = \mu_0 \mathbf{H} + \mathbf{M} = \mu_0 \mathbf{H}$ are equal. Applying these result to time dependent fields in plasma using Maxwell's equation, we treat the externally applied current explicitly, but the internal currents within the plasma implicitly as a component of the dielectric displacement, as discussed more fully in Section 2.3. In plasma, there are generally no external currents so that Maxwell's equations become

$$\mu_0{}^{-1}\, \nabla \wedge \mathbf{B} = \mathbf{j} + \epsilon_0 \frac{\partial \mathbf{E}}{\partial t} = \frac{\partial \mathbf{D}}{\partial t} = \epsilon \frac{\partial \mathbf{E}}{\partial t} \qquad \nabla \cdot \mathbf{D} = 0 \tag{13.3}$$

It follows from the equation of continuity for charge that the current due to variation of the polarisation charge is given

$$\frac{\partial \rho}{\partial t} + \nabla \cdot \mathbf{j} = 0 \to \mathbf{j} = \frac{\partial \mathbf{P}}{\partial t} \tag{13.4}$$

1 In plasma physics the relative permittivity $K = \epsilon/\epsilon_0$ or dielectric constant is often used.

Foundations of Plasma Physics for Physicists and Mathematicians, First Edition. Geoffrey J. Pert.

In general, the response of any linear medium expressed through the polarisation $\mathbf{P}(t)$ at time t is dependent on the electric field at earlier times $\tau \leq t$ which causes an effect known as the *causality condition*. This leads to a shift in the natural frequency of any oscillation *temporal dispersion*. Furthermore, since electro-dynamics of continuous media is a continuum theory, the definition of the polarisation and therefore the permittivity involves an averaging over the particles in a small region surrounding the point of measurement; they therefore also depend on the spatial path taken by the constituent particles *spatial dispersion* (Ginzburg, 1970; Lifshitz and Pitaevskii, 1981). In a steady state homogeneous medium, the permittivity takes the form.

$$\epsilon(\mathbf{r},\ \mathbf{r}',\ t,\ t') \rightarrow \epsilon[(\mathbf{r}-\mathbf{r}'),\ (t-t')]$$

In plasma, the scale lengths over which the averaging ($\sim \lambda_D$) is taken are small compared to the wavelengths of plasma waves; in particular, in cold plasma the Debye length $\lambda_D \rightarrow 0$. Therefore, we neglect spatial dispersion and the polarisation becomes

$$\mathbf{P}(t) = \int_{-\infty}^{\infty} \epsilon_0\ \chi(t,\tau)\ \mathbf{E}(\tau)\ \mathrm{d}\tau \tag{13.5}$$

When the system is *time-translation invariant*, the output is time-shifted by an equal amount to the input, but otherwise unchanged. Typical of this behaviour is one where $\chi(t,\tau) = \chi(t-\tau)$ is a function of the time difference alone, reflecting the relaxation of the medium to the driving field $\mathbf{E}(\tau)$. Thus

$$\chi(t) = \begin{cases} 0 & \text{if} \quad t < 0 \\ \chi(t) & \text{otherwise} \end{cases} \tag{13.6}$$

Taking the Fourier transforms of the field and the polarisation

$$\left.\begin{array}{c}\tilde{\mathbf{E}}(\omega) \\ \tilde{\mathbf{P}}(\omega)\end{array}\right\} = \frac{1}{\sqrt{2\,\pi}} \int \left\{\begin{array}{c}\mathbf{E}(t) \\ \mathbf{P}(t)\end{array}\right\} \exp\left(\imath\ \omega t\right)\ \mathrm{d}t \tag{13.7}$$

it follows from the convolution theorem that for such media

$$\tilde{\mathbf{P}}(\omega) = \epsilon_0\ \tilde{\chi}(\omega)\ \tilde{\mathbf{E}}(\omega) \tag{13.8}$$

where $\tilde{\chi}(\omega)$ is the complex spectral susceptibility, the real part representing an oscillation and the imaginary part damping whose Fourier transform

$$\tilde{\chi}(\omega) = \frac{1}{\sqrt{2\,\pi}} \int_0^{\infty} \mathrm{d}t\ \chi(t)\ \exp\left[\imath\ \omega\ t\right] \tag{13.9}$$

Since the susceptibility $\chi(t)$ is always finite, we conclude from Eq. (13.9) that the transform of the susceptibility is analytic in the upper half-plane $\Im(\omega) > 0$[2] and that $\tilde{\chi}(\omega) = \tilde{\chi}^*(-\omega)$. Since in most cases $\tilde{\chi} \rightarrow 0$ as $|\omega| \rightarrow \infty$, it follows that $|\chi(\omega)/\omega| \rightarrow 0$ faster than $1/|\omega|$ as $|\omega| \rightarrow \infty$ and the integral in (13.9) converges.

As a consequence of the analyticity in the upper half plane, we may use the Sokhotski–Plemelj theorem (Section M.3.(M.3.xi)) to relate the real and imaginary parts of the susceptibility. Using Cauchy's integral theorem and choosing a closed path in the upper half-plane along the real axis C_1, avoiding the pole on the real axis at ($\omega' = \omega$) by a small circular loop C_2 and returning by the

2 In the lower half-plane $\Im(\omega) < 0$, the integral (13.9) diverges and in general the function $\tilde{\chi}(\omega)$ may contain singularities.

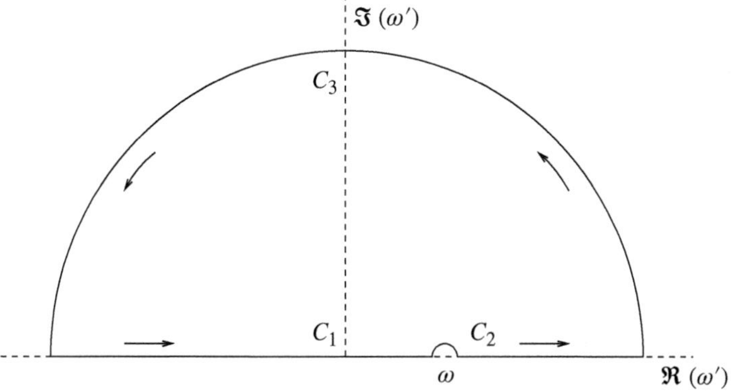

Figure 13.1 Contour used to evaluate the Kramers–Kronig integral.

semi-circle at ∞, C_3, where the length of the segment increases as $|\omega'|$, but applying the estimation lemma (Section M.3.(M.3.xv), p. 460), the integral vanishes as $\chi(\omega)$ vanishes at least as $1/|\omega'|$ (Figure 13.1) we evaluate

$$\begin{aligned}\oint_C \frac{\tilde{\chi}(\omega)}{\omega' - \omega}\, \mathrm{d}\omega' &= \int_{C_1} \frac{\tilde{\chi}(\omega)}{\omega' - \omega}\, \mathrm{d}\omega' + \int_{C_2} \frac{\tilde{\chi}(\omega)}{\omega' - \omega}\, \mathrm{d}\omega' + \cancel{\int_{C_3} \frac{\tilde{\chi}(\omega)}{\omega' - \omega}}\, \mathrm{d}\omega' \\ &= ⨍_{-\infty}^{\infty} \frac{\tilde{\chi}(\omega)}{\omega' - \omega}\, \mathrm{d}\omega' - \imath\, \pi \tilde{\chi}(\omega) = 0\end{aligned} \tag{13.10}$$

Equation (13.9) therefore becomes

$$\begin{aligned}\Re[\tilde{\chi}(\omega)] &= \frac{1}{\pi} ⨍_{-\infty}^{\infty} \mathrm{d}\omega' \, \frac{\Im[\tilde{\chi}(\omega')]}{\omega' - \omega} \\ \Im[\tilde{\chi}(\omega)] &= -\frac{1}{\pi} ⨍_{-\infty}^{\infty} \mathrm{d}\omega' \, \frac{\Re[\tilde{\chi}(\omega')]}{\omega' - \omega}\end{aligned} \tag{13.11}$$

where $⨍$ is the Cauchy principal value integral defined earlier in Eq. (12.36).

These equations are known as the Kramers–Kronig relations. They express a general result for media exhibiting a linear response function, which is analytic in the upper half plane. For electro-magnetic waves where $\epsilon = \epsilon_0(1 + \chi)$, they give a relationship between the real part of the dielectric constant (refractive index) and the imaginary part (absorption).

In general, $\chi(t)$ is real and consequently its Fourier transform is symmetric, i.e. $\tilde{\chi}(\omega) = \tilde{\chi}^*(-\omega)$ which allows us to write the integrals (13.11) over positive frequencies only

$$\begin{aligned}\Re[\tilde{\chi}(\omega)] &= \frac{2}{\pi} ⨍_{0}^{\infty} \mathrm{d}\omega' \, \frac{\omega' \, \Im[\tilde{\chi}(\omega')]}{\omega'^2 - \omega^2} \\ \Im[\tilde{\chi}(\omega)] &= -\frac{2}{\pi} ⨍_{0}^{\infty} \mathrm{d}\omega' \, \frac{\omega \, \Re[\tilde{\chi}(\omega')]}{\omega'^2 - \omega^2}\end{aligned} \tag{13.12}$$

The key element establishing the Kramers–Kronig relation is that of *causality*, i.e. the change in the polarisation must be preceded by the force causing it. We saw this important phenomenon earlier in establishing the damping of plasma waves (Section 12.1.1.2). The Kramers–Kronig relations express a remarkable general property of linear dielectric media, whose response function is analytic in the upper half plane, namely that the dispersion and absorption are related by a general expression, which *does not* depend on the detailed physical mechanism generating the response. The result is a direct result of causality, namely that the response must occur after the force producing it.

If the real part of the susceptibility $\Re[\tilde{\chi}(\omega)]$ is constant, it follows from either Eq. (13.11) or (13.12) that $\Im[\tilde{\chi}(\omega)] = 0$, i.e. there is no damping. However in general, media are dispersive, $\Re[\tilde{\chi}(\omega)]$ varies and $\Im[\tilde{\chi}(\omega)] \neq 0$ so that dispersion automatically implies damping, a result of some importance in regard to plasma waves.

Since the displacement is defined by $\mathbf{D} = \epsilon_0\mathbf{E} + \mathbf{P}$, it follows that the permittivity $\epsilon = \epsilon_0 + \chi$. The permittivity therefore similarly satisfies the Kramers–Kronig relations. In consequence, the real part of the permittivity (refractive index) is the negative of the Hilbert transform of the imaginary part (absorption), and the imaginary part the positive of the Hilbert transformation of the real.[3]

A more familiar quantity is the dielectric constant defined as $K = \epsilon/\epsilon_0$. Consistent with Eq. (13.5), the electric field of a monochromatic wave scales as $\mathbf{E_0} \exp(-\imath\,\omega\,t)$. From Eq. (13.3), it follows that the permittivity

$$\epsilon = \Re(\epsilon) + \imath\,\Im(\epsilon) = [\epsilon_0\,\mathbb{I} - \omega^{-1}\,\Im(\sigma)] + \imath\,\omega^{-1}\Re(\sigma) \tag{13.15}$$

where $\boldsymbol{\sigma}$ is the conductivity, a tensor quantity due to the resistivity of the medium, given by $\mathbf{j} = \boldsymbol{\sigma}\mathbf{E}$ (see Section 2.3).

13.1.1 The Properties of the Permittivity Matrix

When the medium is anisotropic, the polarisation is no longer collinear with the field. As a result, the susceptibility, permittivity, and conductivity are tensors and the governing linear relation for the fields become appropriate tensor relations. The susceptibility $\mathbb{K}$ is subject to a number of symmetry conditions. In general, the polarisation $\mathbf{P}$ and therefore the displacement $\mathbf{D}$ are linearly dependent on the summation of effects of the electric field at earlier times and neighbouring points in space. In a steady state, homogeneous medium

$$D_i(\mathbf{r}, t) = E_i(\mathbf{r}, t) + \int \mathrm{d}\mathbf{r}'\ \mathrm{d}t'\ \mathcal{K}_{ij}\ [(\mathbf{r} - \mathbf{r}'),\ (t - t')]\ E_j(\mathbf{r}',\ t') \tag{13.16}$$

Representing the field as a set of plane waves where $\mathbf{D}$ and $\mathbf{E}$ vary as $\exp[\imath\,(\mathbf{k}\cdot\mathbf{r} - \omega t)]$ by taking the Fourier transforms of each term, we obtain

$$D_i = \epsilon_{ij}(\mathbf{k}, \omega)\ E_j \tag{13.17}$$

Clearly, the permittivity tensor

$$\epsilon_{ij} = \delta_{ij} + \int_0^\infty \int \mathcal{K}_{ij}\,(\boldsymbol{\rho}, \tau)\ \exp\,[\imath\,(\mathbf{k}\cdot\boldsymbol{\rho} - \omega\,\tau)]\ \mathrm{d}\boldsymbol{\rho}\ \mathrm{d}\tau \tag{13.18}$$

---

3 The Kramers–Kronig relations apply to an isolated element such as an atomic absorption line where the imaginary part corresponding to the absorption profile, the frequency ω being measured from the line centre

$$\Im[\epsilon(\omega)] = -\sigma_0 \begin{cases} \dfrac{1}{\pi}\dfrac{\Delta}{(\Delta^2 + \omega^2)} & \text{Lorentz profile} \\ \dfrac{1}{\sqrt{\pi}}\dfrac{1}{\Delta}\exp(-\dfrac{\omega^2}{\Delta^2}) & \text{Doppler profile} \end{cases} \tag{13.13}$$

and the real part corresponding to the refractive index

$$\Re[\epsilon(\omega)] = \sigma_0 \begin{cases} \dfrac{1}{\pi}\dfrac{\omega}{(\Delta^2 + \omega^2)} & \text{Lorentz profile} \\ \dfrac{2}{\pi}\dfrac{1}{\Delta}F[\dfrac{\omega}{\Delta}] & \text{Doppler profile} \end{cases} \tag{13.14}$$

where $F(x) = e^{-x^2}\int_0^x e^{t^2}\,\mathrm{d}t$ is Dawson's function.

It follows immediately that

$$\epsilon_{ij}(\mathbf{k},\omega) = \epsilon_{ij}^*(-\mathbf{k},-\omega) \tag{13.19}$$

Spatial dispersion is manifest by the variation of the permittivity due to the wave vector $\mathbf{k}$: temporal dispersion through the term in frequency ω.

In an isotropic medium, the familiar result from optics namely that the permittivity and the refractive index are related should be noted

$$n^2 = K = \epsilon/\epsilon_0 \tag{13.20}$$

Isotropic plasma: Although as we have argued spatial dispersion is absent in cold plasma, the introduction of a non-zero Debye length due to a finite temperature gives rise to a dependence of the permittivity ϵ on the wave vector $\mathbf{k}$. As a consequence a characteristic direction, that of $\mathbf{k}$ is introduced into the system even if the medium is isotropic. The permittivity matrix becomes diagonal with two components $\epsilon_\parallel$ directed along the wave vector and $\epsilon_\perp$ uniformly in the plane normal to the wave vector:

$$\epsilon_{ij} = \begin{pmatrix} \epsilon_\perp & 0 & 0 \\ 0 & \epsilon_\perp & 0 \\ 0 & 0 & \epsilon_\parallel \end{pmatrix} \tag{13.21}$$

where the z axis is taken parallel $\mathbf{k}$ and x and y lie in the plane normal to $\mathbf{k}$. Examples of this effect are found in Sections 11.3.1 and 12.1.1.

Anisotropic plasma: The properties of the plasma medium are complicated by introduction of anisotropy in the plasma usually due to the presence of a background magnetic field $\mathbf{B}_0$. In consequence the spectrum of possible wave motions is extensive and complex even in cold plasma, where the absence of thermal pressure limits the interactive forces.

As a result of the plasma anisotropy, the permittivity matrix has well-defined symmetry, provided the plasma is locally in thermo-dynamic equilibrium. Specifying a set of orthogonal right-handed coordinates where $\hat{\mathbf{z}}$ is taken parallel to the magnetic field $\mathbf{B}_0$, the $\hat{\mathbf{x}}$, $\hat{\mathbf{z}}$ plane contains the wave vector $\mathbf{k}$ and $\hat{\mathbf{y}}$ is orthogonal to $\hat{\mathbf{x}}$ and $\hat{\mathbf{z}}$. It follows from Onsager's principle that $\epsilon_{ij}(\mathbf{B}_0) = \epsilon_{ji}(-\mathbf{B}_0)$ (Landau and Lifshitz, 1980). However, the directions of the axes are normally tied to those of the magnetic field, so that when the magnetic field is reversed the directions of the axes $\hat{\mathbf{z}}$ and $\hat{\mathbf{y}}$ are both reversed, but $\hat{\mathbf{x}}$ is unchanged. Therefore,

$$\epsilon_{xy}(\mathbf{B}_0) = -\epsilon_{yx}(-\mathbf{B}_0) \qquad \epsilon_{xz}(\mathbf{B}_0) = -\epsilon_{zx}(-\mathbf{B}_0) \qquad \epsilon_{yz}(\mathbf{B}_0) = \epsilon_{zy}(-\mathbf{B}_0) \tag{13.22}$$

Thus, the change $\mathbf{B_0} \to -\mathbf{B}_0$ gives rise to a change of sign in $\epsilon_{yx} = -\epsilon_{xy}$ and $\epsilon_{yz} = -\epsilon_{zy}$. Since the magnitude of $\epsilon(\mathbf{B}_0)$ is unchanged by the field reversal, these two terms must both be odd functions of $\mathbf{B}_0$; in contrast the remaining terms, which are all unchanged, must be even functions of $\mathbf{B}_0$. Hence, we obtain the symmetry relation

$$\epsilon_{xy} = -\epsilon_{yx} \qquad \epsilon_{xz} = \epsilon_{zx} \qquad \epsilon_{yz} = -\epsilon_{zy} \tag{13.23}$$

Identifying the real and imaginary parts of the permittivity tensor $\epsilon_{ij} = \epsilon'_{ij} + \imath\, \epsilon''_{ij}$. taking the above symmetry relations into account, the permittivity matrix may be written in terms of its Hermitian and anti-Hermitian components

$$\epsilon_{ij} = \underbrace{\begin{pmatrix} \epsilon'_{xx} & \imath\, \epsilon''_{xy} & \epsilon'_{xz} \\ -\imath\, \epsilon''_{xy} & \epsilon'_{yy} & \imath\, \epsilon''_{yz} \\ \epsilon'_{xz} & -\imath\, \epsilon''_{yz} & \epsilon'_{zz} \end{pmatrix}}_{\text{Hermitian}} + \underbrace{\begin{pmatrix} \imath\, \epsilon''_{xx} & \epsilon'_{xy} & \imath\, \epsilon''_{xz} \\ -\epsilon'_{xy} & \imath\, \epsilon''_{yy} & \epsilon'_{yz} \\ \imath\, \epsilon''_{xz} & -\epsilon'_{yz} & \imath\, \epsilon''_{zz} \end{pmatrix}}_{\text{Anti-Hermitian}} \tag{13.24}$$

The tensor components of the permittivity are themselves subject to the Kramers–Kronig relations between them.

The damping of the wave is given by the average work done by field **E** driving the currents given by $\partial \mathbf{D}/\partial t$ after averaging over a period of the field[4]

$$Q = \left\langle \mathbf{E} \cdot \frac{\partial \mathbf{D}}{\partial t} \right\rangle = \left\langle \epsilon_{ij}\, E_j\, \frac{\partial E_i}{\partial t} \right\rangle = \imath\, \frac{\omega}{4} \left[\epsilon^*_{ji}(\omega, \mathbf{k}) - \epsilon_{ij}(\omega, \mathbf{k}) \right] E^*_i\, E_j \tag{13.25}$$

It is obvious that only the anti-Hermitian component of the permittivity tensor contributes to the damping.

We note that the permittivity we obtained earlier from the Drude/Lorentz model, Eq. (2.69) is consistent with these relations.

13.2 Plane Waves in Homogeneous Plasma

The propagation of plane waves of small amplitude is easily developed for a single Fourier mode as a linear perturbation of Maxwell's equations which may be written

$$\nabla \wedge \mathbf{H} = \frac{\partial \mathbf{D}}{\partial t} \qquad \nabla \cdot \mathbf{B} = 0 \tag{13.26}$$

$$\nabla \wedge \mathbf{E} = -\frac{\partial \mathbf{B}}{\partial t} \qquad \nabla \cdot \mathbf{D} = 0 \tag{13.27}$$

since there are no externally applied currents or charges. In a linear medium, these equations are complimented by the properties of the medium itself

$$\mathbf{D} = \epsilon \cdot \mathbf{E} \qquad \mathbf{B} = \mu \cdot \mathbf{H} \tag{13.28}$$

where the dielectric permeability ϵ and the magnetic permittivity μ are in general tensors expressing the anisotropy of the medium.

Consider a single monochromatic plane wave travelling in the **k** direction with electric field

$$\mathbf{E}(\mathbf{r}, t) = \mathbf{E}_0\; \exp[\imath(\mathbf{k} \cdot \mathbf{r} - \omega t)] \tag{13.29}$$

and magnetic field

$$\mathbf{B}(\mathbf{r}, t) = -\frac{\mathbf{k} \wedge \mathbf{E}_0}{\omega} \exp[\imath(\mathbf{k} \cdot \mathbf{r} - \omega t)] \tag{13.30}$$

A backward wave travelling in −**k** direction is given by reversing the sign of **k**. Adding forward and backward waves of equal amplitude, we obtain a standing wave

$$\mathbf{E} = \mathbf{E}_0\; \cos(\mathbf{k} \cdot \mathbf{r})\; \cos(\omega t) \qquad \text{and} \qquad \mathbf{B} = \mathbf{B}_0\; \sin(\mathbf{k} \cdot \mathbf{r})\; \sin(\omega t) \tag{13.31}$$

where $\mathbf{B}_0 = -\imath \mathbf{k} \wedge \mathbf{E}_0/\omega$. We note the magnetic field is $\pi/2$ out of phase with the electric.

4 If two quantities A and B are sinusoidally oscillatory, we must use real values to calculate the average over the period so that

$$\Re(A)\; \Re(B) = \frac{1}{4}(A + A^*)\; (B + B^*)$$

and since A and B vary as $e^{\imath\, \omega\, t}$ terms varying as $e^{\imath\, 2\, \omega\, t}$ or $e^{-\imath\, 2\, \omega\, t}$ vanish on averaging

$$\langle A\; B \rangle = \frac{1}{4}(A\; B^* + A^*\; B)$$

Maxwell's equations simplify to a set of linear algebraic equations

$$\mathbf{k} \wedge \mathbf{B} = -\frac{\omega}{c^2}\,\mathbb{K}\cdot\mathbf{E} \qquad \mathbf{k}\wedge\mathbf{E} = \omega\,\mathbf{B} \tag{13.32}$$

where $\mathbb{K} = \epsilon/\epsilon_0 = (\mathbb{I} + \imath\,\sigma/\epsilon_0\,\omega)$ is the dielectric constant given by Eq. (2.69). Substituting for **B** and writing the tensor

$$\mathbb{M} = \mathbf{kk} - k^2\mathbb{I} + \frac{\omega^2}{c^2}\,\mathbb{K} \tag{13.33}$$

in dyadic form, we obtain the homogeneous simultaneous equations

$$\mathbb{M}\cdot\mathbf{E} = 0 \tag{13.34}$$

which only have a non-zero solution if

$$\mathcal{M}(\omega, \mathbf{k}) = \det(\mathbb{M}) = 0 \tag{13.35}$$

This equation known as the *dispersion relation* determines the frequency ω for a particular wave number **k**. Since the phase velocity

$$\mathbf{v}_p = \frac{\omega}{k}\,\hat{\mathbf{k}} \tag{13.36}$$

where $\hat{\mathbf{k}}$ is a unit vector in the direction of propagation of the wave, it follows that the dispersion relation determines the 'spreading' of different Fourier components of a wave packet.

13.2.1 Waves in Collisional Cold Plasma

As we have seen in Chapter 11, many of the waves in plasma are dominated by the motion of the electrons due to their mobility resulting from their small mass. As a result, the simple result for the permittivity which we derived in Section 2.3 may frequently be applied to these waves. In the general multi-species case, the total current is obtained from the sum of the contributions from the electrons and ions, as in Eq. (11.5). The inclusion of collisional damping is straightforward, but the damping collisional coefficients are often less well defined. The damping constant for each species is different, and may not be as simply estimated for ions as for electrons (see Ginzburg (1970, §6~) for a discussion of appropriate values of the collision frequency). If only electrons contribute to the wave, we may use the result for the permittivity derived earlier from the Drude model (2.69), which includes damping, in place of Eq. (11.9).

13.2.1.1 Isotropic Unmagnetised Plasma

When no magnetic field is present, the dielectric tensor becomes a simple scalar diagonal $\mathbb{K} = K\,\mathbb{I}$ where

$$K = \left\{1 - \sum_q \frac{\Pi_q{}^2}{\omega(\omega + \imath\,\nu_q)}\right\} = \left\{1 - \sum_q \frac{\Pi_q{}^2}{(\omega^2 + \nu_q{}^2)} + \imath \sum_q \frac{\Pi_q{}^2}{(\omega^2 + \nu_q{}^2)}\,\frac{\nu_q}{\omega}\right\} \tag{13.37}$$

the sums being taken over all charged species in the plasma.

Limiting ourselves to high frequency waves where the frequency of the wave is much greater than the ion plasma frequency $\omega \gg \Pi_i$, when the ions play no part, it is easy to see that Eq. (13.34) has three independent solutions.

i. *Longitudinal wave*: When the electric field is parallel to the wave vector ($\mathbf{E} \parallel \mathbf{k}$ then $\mathbb{M} = \omega^2/c^2\ \mathbf{K}$) and consequently the refractive index is similarly complex $n = k/k_0$

$$\Pi_e = \sqrt{\omega(\omega + \imath\nu_e)} \qquad \text{and} \qquad \omega \approx \Pi_e - \imath\, 0.5\ \nu_e \tag{13.38}$$

This is the familiar dispersion relation for cold plasma waves with an additional term for the damping by collisions (Section 11.3.1). In practice, this latter term is usually unimportant as the plasma is warm and Landau damping dominates.

ii. *Transverse waves*: There are two independent modes with the electric field perpendicular to each other $E_x \neq E_y$ and to the direction of propagation ($\mathbf{E} \perp \mathbf{k}$). Neglecting the ion term, a sinusoidal propagating wave $\mathbf{E} = \mathbf{E}_0 \exp[\imath\ (\mathbf{k} \cdot \mathbf{r} - \omega\ t)]$ has dispersion relation generating the wave number k

$$\frac{k^2\ c^2}{\omega^2} = K = 1 - \frac{\Pi_e^{\ 2}}{(\omega^2 + \imath\ \omega\ \nu_e)} = \left\{1 - \frac{\Pi_e^{\ 2}}{(\omega^2 + \nu_e^{\ 2})} + \imath\ \frac{\Pi_e^{\ 2}}{(\omega^2 + \nu_e^{\ 2})}\ \frac{\nu_e}{\omega}\right\} \tag{13.39}$$

In the absence of damping $\nu = 0$, this result clearly reduces to the phase relation for electro-magnetic transverse waves (Section 11.3.1).[5] Including the damping the wave number becomes complex $k = k' - \imath\ k''$ where $k_0 = \omega/c$ and

$$k'^2 - k''^2 = k_0^{\ 2}\ \frac{(\omega^2 - \Pi_e^{\ 2}) + \nu_e^{\ 2}}{(\omega^2 + \nu_e^{\ 2})} \qquad \text{and} \qquad 2\ k'\ k'' = k_0^{\ 2}\ \frac{\Pi_e^{\ 2}}{(\omega^2 + \nu_e^{\ 2})}\ \left(\frac{\nu_e}{\omega}\right) \tag{13.40}$$

Hence

$$\left(\frac{k'}{k_0}\right)^4 - \frac{(\omega^2 - \Pi_e^{\ 2}) + \nu_e^2}{(\omega^2 + \nu_e^{\ 2})}\left(\frac{k'}{k_0}\right)^2 - \frac{1}{4}\ \frac{\Pi_e^{\ 2}}{(\omega^2 + \nu_e^{\ 2})}\ \left(\frac{\nu}{\omega}\right) = 0 \tag{13.41a}$$

$$\left(\frac{k''}{k_0}\right)^4 + \frac{(\omega^2 - \Pi_e^{\ 2}) + \nu^2}{(\omega^2 + \nu_e^{\ 2})}\left(\frac{k''}{k_0}\right)^2 - \frac{1}{4}\ \frac{\Pi_e^{\ 2}}{(\omega^2 + \nu_e^{\ 2})}\ \left(\frac{\nu_e}{\omega}\right) = 0 \tag{13.41b}$$

Solving the quadratic equations and noting that only the positive sign has physical meaning

$$\left(\frac{k'}{k_0}\right)^2 = \frac{(\omega^2 - \Pi_e^{\ 2}) + \nu_e^{\ 2}}{(\omega^2 + \nu_e^{\ 2})}\left\{1 + \sqrt{1 + \frac{\Pi_e^{\ 2}\ (\omega^2 + \nu_e^{\ 2})}{[(\omega^2 - \Pi_e^{\ 2}) + \nu_e^2]^2}\ \left(\frac{\nu_e}{\omega}\right)}\right\} \tag{13.42a}$$

$$\left(\frac{k''}{k_0}\right)^2 = \frac{(\omega^2 - \Pi_e^{\ 2}) + \nu_e^{\ 2}}{(\omega^2 + \nu_e^{\ 2})}\left\{-1 + \sqrt{1 + \frac{\Pi_e^{\ 2}\ (\omega^2 + \nu_e^{\ 2})}{[(\omega^2 - \Pi_e^{\ 2}) + \nu_e^{\ 2}]^2}\ \left(\frac{\nu_e}{\omega}\right)}\right\} \tag{13.42b}$$

The collision frequency is generally much less than that of the electric field $\nu \ll \omega$ so that for most purposes

$$\frac{k'}{k_0} \approx \sqrt{1 - \frac{\Pi_e^{\ 2}}{\omega^2}} \qquad \text{and} \qquad \frac{k''}{k_0} \approx \frac{1}{2}\ \frac{\Pi_e^{\ 2}/\omega^2}{\sqrt{1 - \Pi_e^{\ 2}/\omega^2}}\ \frac{\nu_e}{\omega} \tag{13.43}$$

5 This is easily seen to be the familiar propagation relation for electro-magnetic waves $\omega/k = c/n = 1/\epsilon\ \mu$ for plasma.

The real part of the refractive index for electro-magnetic waves is consistent with the value for the phase velocity $V_p = c/n$ obtained earlier, Eq. (12.44) and Section 11.3.1, item ii. In the absence of a magnetic field and weak damping, the refractive index scales as $n \approx \sqrt{1 - \Pi_e^2/\omega^2}$ and therefore becomes zero $n \to 0$ as $\omega \to \Pi_e$. The wave is reflected at this point as the refractive index becomes imaginary and the wave evanescent, a situation discussed more fully in Section 5.4.1. At the reflection point the plasma polarisation current is equal and is in anti-phase with the displacement current associated with the applied electric field. This case is fully treated in Section 12.2.

Introducing the *irradiance* or energy flux of the wave[6]

$$I = n\, c\, \epsilon_0\, |E_0|^2/2 \tag{13.45}$$

where $n = \sqrt{1 - \Pi_e^2/\omega^2}$ is the refractive index in plasma. It is easily seen that in a plane parallel beam of radiation propagating in the z direction, the irradiance is attenuated as

$$\frac{dI}{dz} = -2\, k''\, I = -2\, \Im(k)\, I = \frac{\Pi_e^2\, \nu_e}{\omega^3}\, \frac{1}{2}\, c\, \epsilon_0\, |E_0|^2 \tag{13.46}$$

so that $\mu = -dI/dz\Big/I = 2\, k'' = \Pi_e^2\, \nu_e/n\, \omega^3$ is the narrow beam attenuation coefficient. dI/dz is the energy deposition rate per unit volume. Equation (13.46), is immediately identified with the general form, Eq. (13.25)

13.2.1.2 Anisotropic Magnetised Plasma

Remembering that the refractive index is proportional to the square root of the permittivity $n \sim \sqrt{\epsilon}$, several characteristic phenomena of the propagation of transverse electro-magnetic waves in plasma can be clearly observed in this result, namely.

We generalise the results of Section 2.3 to include both ions and electrons in the permittivity. In the presence of a magnetic field, the wave has a resonance at the Larmor frequency of the particle with charge q where $\omega = \Omega_q$ and in the absence of collisions $\epsilon \to \infty$. The resonance is between the rotation introduced by circularly polarised radiation and the cyclotron motion in the magnetic field (item ii, p.257). In this neighbourhood, the permittivity has a singularity of the form $\sim 1/[(\omega - \Omega_q) + \imath\, \nu_e]$. As a result, the damping introduced by collisions play the dominant role treated in Section 5.4.2 in the neighbourhood of the resonance.

When both collisions and magnetic field are active, the expression (2.69) for the permittivity is awkward to use. However, when the permittivity represents a travelling wave, either the collision frequency is large compared to the oscillation frequency and the situation is essentially the same as a d.c. field, or the collision frequency must be small compared to oscillation frequency $\nu_q \ll \omega$ for the wave to exist. As a result, the expression for the permittivity may be expanded in a series in

6 This result follows from the average of the Poynting flux over a period of the wave (see footnote 4)

$$I = |\langle \mathbf{E} \wedge \mathbf{H} \rangle| = \frac{1}{4}|(\mathbf{E} \wedge \mathbf{H}^* + \mathbf{E}^* \wedge \mathbf{H})| = \frac{1}{2}\frac{k}{\omega}\, \frac{\epsilon_0}{\epsilon\, \mu_0} E^2 \tag{13.44}$$

since $H = -k\, E/\mu_0\, \omega$ in a plane parallel beam and the phase velocity $V_p = \omega/k = 1/\epsilon\, \mu_0 = c/n$.

(ν_q/ω) into a term independent of the damping and one containing damping:

$$\epsilon_{ij} = \epsilon_0\, \delta_{ij} + \imath\, \frac{1}{\omega}\, \sigma_{ij}$$

$$= \epsilon_0 \begin{pmatrix} 1 - \sum_q \frac{\Pi_q{}^2\, \omega}{[\omega(\omega^2 - \Omega_q{}^2)]} & -\imath \sum_q \frac{\Pi_q{}^2\, \Omega_q}{[\omega(\omega^2 - \Omega_q{}^2)]} & 0 \\ \imath \sum_q \frac{\Pi_q{}^2\, \Omega_q}{[\omega\, (\omega^2 - \Omega_q{}^2)]} & 1 - \sum_q \frac{\Pi_q{}^2\, \omega}{[\omega(\omega^2 - \Omega_q{}^2)]} & 0 \\ 0 & 0 & \sum_q \frac{\Pi_q{}^2}{\omega^2} \end{pmatrix} \tag{2.69:a}$$

$$+ \imath\, \epsilon_0 \begin{pmatrix} \sum_q \frac{\Pi_q{}^2\, (\omega^2 + \Omega_q{}^2)\, \nu_q}{\omega\, (\omega^2 - \Omega_q{}^2)^2} & \imath \sum_q \frac{2\, \Pi_q{}^2\, \Omega_q\, \nu_q}{(\omega^2 - \Omega_q{}^2)^2} & 0 \\ -\imath \sum_q \frac{2\, \Pi_q{}^2\, \Omega_q\, \nu_q}{(\omega^2 - \Omega_q{}^2)^2} & \sum_q \frac{\Pi_e{}^2\, (\omega^2 + \Omega_q^2)\, \nu_q}{\omega\, (\omega^2 - \Omega_q{}^2)^2} & 0 \\ 0 & 0 & \sum_q \frac{\Pi_q{}^2\, \nu_q}{\omega^3} \end{pmatrix}$$

We note that the zero order array in the collision frequency ν_q is Hermitian and the first order anti-Hermitian. As we show in Section 13.1.1, the first order term is responsible for the damping of a wave. The zero order primarily determines the speed at which the wave propagates.

The dispersion relation generated from the permittivity matrix (2.69:a) considered the case of the magnetic field directed along the z axis. More general cases with the magnetic field directed along a different axis are easily obtained by a unitary transform for a rotation from one basis (x, y, x) to another (p, q, r) namely the matrix a_{jk}, which has the familiar orthogonal properties

$$a^{-1}{}_{jk} = a_{kj} \qquad a_{jk}\, a_{k\ell} = \delta_{j\ell} \qquad \det(a_{jk}) = 1$$

so that the permittivity matrix becomes $\epsilon'_{jk} = a_{jm}\, \epsilon_{mn}\, a_{nk}$ in the new basis. The matrix elements a_{jm} are the direction cosines between the axes j in the old basis and m in the new.

13.3 Plane Waves Incident Obliquely on a Refractive Index Gradient

Consider a stratified isotropic media in which the permittivity (and therefore the refractive index) varies as a function of the co-ordinate z (Ginzburg, 1970, §19). A plane electro-magnetic wave is incident obliquely at an angle α to the permittivity gradient in the plane of incidence (z, x): the direction y being normal to the plane of incidence. The plasma is assumed to be uniform normal to the plane of incidence so that all derivatives $\partial/\partial y \equiv 0$. Far from the cut-off, the wave is described by the WKB approximation Eq. (5.26). Near the point of reflection, we assume that the square of the refractive index is linearly dependent on the distance from the cut-off point (5.29).

Since the plasma is uniform in the x direction and the wave is monochromatic, the x derivative for all wave quantities is determined by the phase function $\partial/\partial x \equiv \imath\, k_\perp = \imath k_0 \sin \alpha_0$, and since the field oscillates with a frequency ω, $\partial/\partial t \equiv -\imath\, \omega$. The electric field is obtained from Maxwell's equations by eliminating $\mathbf{B}$

$$\nabla \wedge \mathbf{E} = \imath\, \omega\, \mathbf{B} \qquad \text{and} \qquad \nabla \wedge \mathbf{B} = -\imath\, \epsilon\, \mu_0\, \omega\, \mathbf{E} \tag{13.47}$$

to obtain

$$\nabla^2 \mathbf{E} + \left(\frac{n^2\,\omega^2}{c^2} \right) \mathbf{E} - \nabla(\nabla \cdot \mathbf{E}) = 0 \tag{13.48}$$

where the refractive index $n = \sqrt{\epsilon\,\mu_0}\,c$.

Alternatively solving for the magnetic field by eliminating **E**

$$\nabla^2 \mathbf{B} + \left(\frac{n^2\,\omega^2}{c^2} \right) \mathbf{B} + \frac{2}{n} \nabla n \wedge (\nabla \wedge \mathbf{B}) - \nabla(\nabla \cdot \mathbf{B}) = 0 \tag{13.49}$$

Expanding these results in Cartesian co-ordinates noting that the uni-directional refractive index gradient is in the z direction Eq. (13.48) becomes

$$\frac{\partial^2 E_z}{\partial z^2} + \frac{\partial^2 E_z}{\partial x^2} + \frac{\omega^2}{c^2}\, n^2(z)\, E_z - \frac{\partial}{\partial z} \nabla \cdot \mathbf{E} = 0 \tag{13.50}$$

$$\frac{\partial^2 E_x}{\partial z^2} + \frac{\partial^2 E_x}{\partial x^2} + \frac{\omega^2}{c^2}\, n^2(z)\, E_x - \frac{\partial}{\partial x} \nabla \cdot \mathbf{E} = 0 \tag{13.51}$$

$$\frac{\partial^2 E_y}{\partial z^2} + \frac{\partial^2 E_y}{\partial x^2} + \frac{\omega^2}{c^2}\, n^2(z)\, E_y = 0 \tag{13.52}$$

Since the field is uniform in the y direction normal to the plane of incidence, the term $\nabla \cdot \mathbf{E} = \partial E_z/\partial z + \partial E_x/\partial x$. The field components E_z and E_x are clearly coupled by the term $\nabla \cdot \mathbf{E}$, but the field E_y is independent.

The magnetic field is slightly more complicated, but satisfies the relations (13.49) which may be expanded to

$$\frac{\partial^2 B_z}{\partial z^2} + \frac{\partial^2 B_z}{\partial x^2} + \frac{n^2\,\omega^2}{c^2}\, B_z - \frac{1}{n^2} \frac{\partial n^2}{\partial z} \frac{\partial B_z}{\partial z} - \frac{\partial}{\partial z} \nabla \cdot \mathbf{B} = 0 \tag{13.53}$$

$$\frac{\partial^2 B_x}{\partial z^2} + \frac{\partial^2 B_x}{\partial x^2} + \frac{n^2\,\omega^2}{c^2}\, B_x - \frac{1}{n^2} \frac{\partial n^2}{\partial z} \frac{\partial B_x}{\partial z} - \frac{\partial}{\partial x} \nabla \cdot \mathbf{B} = 0 \tag{13.54}$$

$$\frac{\partial^2 B_y}{\partial z^2} + \frac{\partial^2 B_y}{\partial x^2} + \frac{n^2\,\omega^2}{c^2}\, B_y - \frac{1}{n^2} \frac{\partial n^2}{\partial z} \frac{\partial B_y}{\partial z} = 0 \tag{13.55}$$

Similarly the field components B_y and B_z are clearly coupled by the term $\nabla \cdot \mathbf{B}$, but the field B_x is independent.

The polarisation of the wave may therefore be established in two independent modes with fields (E_x, B_y, B_z) and (B_x, E_y, E_z).[7]

- *Perpendicular or s-polarisation*[8] (E_y, B_z, B_x): In this case, the electric field lies entirely in the normal plane perpendicular to the refractive index gradient. The electric field has a single component $E_x(z, x)$ normal to the plane of incidence. The magnetic field components $B_z(z, x)$ and $B_x(z, x)$ both lie in the plane of incidence.
  The electric field is therefore given by Eq. (13.52), which is form of Helmholtz's equation (5.2) to which the laws of geometrical optics apply, and to which more succinctly we may apply the WKB solution (5.25) for an oblique wave.

$$E_y = E_y(0) \left(\frac{\zeta(0)}{\zeta} \right)^{1/2} \exp\left[\pm\imath\, k_0 \int^z \zeta(z')\, \mathrm{d}z' + \imath\, k_\perp x \right] \tag{13.56}$$

7 The resolution of the incident plane wave into two independent polarisations is only possible in this case where there is single clearly defined direction to the refractive index gradient.

8 The name comes from the German word *senkrecht* meaning perpendicular since the electric field is perpendicular to the plane of incidence.

It is easily shown (Ginzburg, 1970) that the magnetic field does not satisfy this equation and contains a coupling term between the two components. Fortunately this is easily calculated by using Maxwell's equations directly to derive the magnetic field from the electric

$$c\,B_x = \imath\,k_0^{-1}\frac{\partial E_y}{\partial z} = \mp c B_x(0)\left(\frac{\zeta}{\zeta(0)}\right)^{1/2} \exp\left[\pm\imath\,k_0\int^z \zeta(z')\,\mathrm{d}z' + \imath k_\perp x\right] \tag{13.57}$$

$$c\,B_z = \xi E_y = c\,B_z(0)\left(\frac{\zeta(0)}{\zeta}\right)^{1/2} \exp\left[\pm\imath\,k_0\int^z \zeta(z')\,\mathrm{d}z' + \imath\,k_\perp x\right] \tag{13.58}$$

This solution to the wave equation using the WKB approximation is valid if the omitted terms in the governing equation are small

$$\frac{1}{(k_0{}^2\zeta(z)^2)}\left|\frac{3}{4}\left(\frac{1}{\zeta(z)}\frac{\mathrm{d}\zeta(z)}{\mathrm{d}z}\right)^2 - \frac{1}{2\,\zeta(z)}\frac{\mathrm{d}^2\zeta(z)}{\mathrm{d}z^2}\right| \ll 1 \tag{13.59}$$

- *Parallel or p polarisation* (B_x, E_z, E_y): In this case, the magnetic field lies entirely in the plane normal to the refractive index gradient. The magnetic field has the sole component B_y normal to the plane of incidence. The electric field components E_z and E_x are both in the plane of incidence. Equation (13.55), after consideration of the phase component along the x direction, simplifies to

$$\frac{\partial^2 B_y}{\partial z^2} - \frac{1}{n^2}\frac{\mathrm{d}n^2}{\mathrm{d}z}\frac{\partial B_y}{\partial z} + k_0{}^2\,\zeta^2\,B_y = 0 \tag{13.60}$$

To find the WKB approximation in this case we proceed as before by examining the trial solution $B_y = A\exp\left(\pm\imath\,k_0\int^z\zeta(z)\,\mathrm{d}z\right) = A\exp(\imath\,\phi)$ obtaining an equation for the phase

$$\imath\,\frac{\mathrm{d}^2\phi}{\mathrm{d}z^2} - \left(\frac{\mathrm{d}\phi}{\mathrm{d}z}\right)^2 - \imath\,\frac{1}{n^2}\frac{\mathrm{d}n^2}{\mathrm{d}z} + k_0{}^2\,\zeta^2\,\phi = 0 \tag{13.61}$$

Substituting the first approximations $\mathrm{d}\phi/\mathrm{d}z \approx \pm k_0\,\zeta(z)$ and $\mathrm{d}^2\phi/\mathrm{d}^2z \approx \pm k_0\mathrm{d}\zeta(z)/\mathrm{d}z$, we obtain a second approximation for the phase

$$\frac{\mathrm{d}\phi}{\mathrm{d}z} \approx \pm\left[k_0{}^2\zeta^2 \pm \imath\,k_0\left(\frac{\mathrm{d}\zeta}{\mathrm{d}z} - 2\,\zeta\,\frac{\mathrm{d}\ln(n)}{\mathrm{d}z}\right)\right]^{1/2} \approx \pm k_0\,\zeta(z) + \imath\left(\frac{\mathrm{d}\ln(\zeta(z)^{1/2})}{\mathrm{d}z} - \frac{\mathrm{d}\ln(n)}{\mathrm{d}z}\right) \tag{13.62}$$

after expanding the square root by the binomial theorem.

Hence integrating and substituting for ϕ in the equation $B_y = B_y(0)\exp(\pm\imath\,\phi)$ and using Maxwell's equations to give E_x and E_z, we obtain

$$B_y = B_y(0)\,\frac{n(z)}{n(0)}\sqrt{\frac{\zeta(0)}{\zeta(z)}}\,\exp\left(\pm\imath\,k_0\int^z\zeta(z')\mathrm{d}z\right) \tag{13.63}$$

$$E_x = \imath\,\frac{1}{n^2k_0}\frac{\partial(c\,B_y)}{\partial z} = E_x(0)\,\frac{n(0)}{n(z)}\sqrt{\frac{\zeta(z)}{\zeta(0)}}\,\exp\left(\pm k_0\,\imath\int^z\zeta(z')\mathrm{d}z\right) \tag{13.64}$$

$$E_z = -\frac{\xi}{n^2}\,c\,B_y = E_z(0)\frac{n(0)}{n(z)}\sqrt{\frac{\zeta(0)}{\zeta(z)}}\,\exp\left(\pm\imath\,k_0\int^z\zeta(z')\mathrm{d}z\right) \tag{13.65}$$

Substituting this solution for B_y back into Eq. (13.60), we find the omitted terms of magnitude, which must be small, have magnitude

$$\frac{1}{k_0^2\,\zeta(z)^2}\left|\frac{3}{4}\left(\frac{1}{\zeta(z)}\frac{\mathrm{d}\zeta(z)}{\mathrm{d}z}\right)^2 - \frac{1}{2\,\zeta(z)}\frac{\mathrm{d}^2\zeta(z)}{\mathrm{d}z^2} + \frac{1}{n}\frac{\mathrm{d}^2n}{\mathrm{d}z^2} - 2\left(\frac{1}{n}\frac{\mathrm{d}n}{\mathrm{d}z}\right)^2\right| \ll 1 \tag{13.66}$$

13.3.1 Oblique Incidence at a Cut-off Point – Resonance Absorption

It is shown in Chapter 5 in Eq. (5.26) the constancy of the transverse component of the wave number $\mathbf{k}_\perp$ allows a simple modification of the wave equation to take into account oblique incidence in a medium where the refractive index depends on one co-ordinate z alone. However, we note that the location of the turning point is changed from $k = 0$ to $k = k_\perp$, i.e. $k_\parallel = 0$. When we investigate electro-magnetic waves, we must consider two different polarisations

i. **E** *wave or s polarisation*: The electric field **E** perpendicular to the propagation plane y. The magnetic field **B** lies in the propagation plane $x,\ z$.
ii. **B** *wave or p polarisation*: The magnetic field **B** perpendicular to the propagation plane y. The electric field **E** lies in the propagation plane $x,\ z$.

As before we assume a linear permittivity gradient

$$\epsilon = n^2 = -a\,z + O(z^2) \tag{5.29'}$$

where $a > 0$

13.3.1.1 s Polarisation

Taking into account the travelling wave phase factor $\exp\,(\imath\, k_\perp x)$ perpendicular to the gradient due to oblique incidence and comparing Eq. (13.48) with Eq. (5.2) yields

$$\frac{\mathrm{d}^2 E_y}{\mathrm{d}z^2} + (k^2 - k_\perp{}^2)E_y = 0 \tag{13.67}$$

we note their equivalence, but with the wave number changed to $k_{\text{eff}} = \sqrt{k^2 - k_\perp{}^2}$. The turning point where $k_{\text{eff}} = 0$ therefore occurs when $k = k_\perp = k_0\ \sin\alpha_0$ where α_0 is the initial angle of incidence with respect to the refractive index gradient. Hence, the turning point $z_1 = \sin^2\alpha_0/a\ n_0{}^2$ in accordance with geometrical optics, α_0. Defining $\zeta = (k_0{}^2 a)^{1/3}(z - z_1)$, we obtain

$$\frac{\mathrm{d}^2 E_y}{\mathrm{d}\zeta^2} + \zeta\ E_y = 0 \tag{13.68}$$

which is Airy's equation as in Eq. (5.31). The solution thus follows that given in Section 5.4.1, but with the modified turning point and with a component of the wave vector perpendicular to the refractive index gradient.

We therefore conclude that the incoming wave is reflected at the appropriate turning point consistent with the laws of geometrical optics and with no loss of amplitude, and forms a standing wave along the gradient at the turning surface.

13.3.1.2 p Polarisation

Substituting for the refractive index from Eq. (5.29) in Eq. (13.49), we obtain

$$\frac{\mathrm{d}^2 B_y}{\mathrm{d}\zeta^2} + \frac{1}{\zeta}\frac{\mathrm{d}B_y}{\mathrm{d}\zeta} + \zeta\ B_y = 0 \tag{13.69}$$

The governing differential equation is modified by the presence of an additional term due to decreasing refractive index. Physically this is associated with the component of the electric field along the refractive index gradient E_z, as the evanescent electric 'tunnels' from the reflection point to the critical density. In the neighbourhood of the critical point, where the frequency of the electro-magnetic wave is equal to the local plasma frequency, the evanescent electric field, E_z is therefore in resonance with a plasma wave along the gradient. Consequently, there is an absorption of energy, typical of resonance. The energy is dissipated by the plasma wave either through collisions or by convection in warm plasma due to the finite plasma wave group velocity (Eq. (12.51)).

The solution of Eq. (13.69) comprises the arbitrary sum of two independent functions identified by Försterling by their series expansions, w hose asymptotes are Airy functions chosen to match the boundary condition $B_y \to 0$ as $z \to -\infty$. For small arguments, the two functions behave as

$$B_1(\zeta) = \zeta^2 + \cdots \qquad \text{and} \qquad B_2(\zeta) = B_1(\zeta)\ \ln(k_\perp\ \zeta) + 2/\zeta^2 + \cdots \tag{13.70}$$

The appropriate solution can therefore be written as a sum of B_1 and B_2

$$B_y = B_0\ \left[1 + \frac{1}{2}\ k_\perp{}^2\ z^2\ \ln(k_\perp\ z)\right] + \cdots \tag{13.71}$$

From Maxwell's equations, the electric field components

$$E_x = \frac{\imath\ c}{n^2\ \omega}\frac{\partial B_y}{\partial z} = -B_0\ \frac{\imath\ k_\perp{}^2}{a\ \omega}\ \ln(k_\perp\ x) \qquad \text{and} \qquad E_z = \frac{\imath\ c}{n^2\ \omega}\frac{\partial B_y}{\partial x} = -B_0\ \frac{k_\perp c}{a\ \omega}\ z^{-1} \tag{13.72}$$

Although the magnetic field is well behaved, the electric field is singular at the resonance point. To treat this condition, we may modify the permittivity to include an imaginary damping component $\epsilon = -a\ z + \imath\ \delta$ with $\delta \to +0$. The absorbed energy may be calculated following the approaches in either Section 5.4.2 or Section 11.3.5.

i. In the first case, the pole lies at $z = +\imath\ \delta/a$, a branch cut is taken from the pole to $z = +\imath\ \infty$, the function E_z is continuous on the real axis and its argument is 0 when $z > 0$ and $-\pi$ when $z < 0$. The averaged Poynting energy flux on the axis (see footnote 4 in Section 13.2) $\overline{S}_z = \Re(E_x\ B_y{}^*)/2\mu_0$ is zero for $z > 0$ and the flux approaching the resonance from $z < 0$ is the energy absorbed per unit area per unit time

$$W = \overline{S}_z = \frac{\pi\ k_\perp{}^2\ c^2\ B_0{}^2}{2\mu_0\ a\ \omega} \tag{13.73}$$

ii. In the second case, the absorbed power may be estimated from Eq. (11.55) since in the neighbourhood of the resonance, $E_z = \imath\ c/\epsilon\ \omega\ \partial B_y/\partial z$, the imaginary part of the permittivity being given by $\Im(\epsilon) = \delta/(a^2\ z^2 + \delta^2)$. As $z \to -0$, $|E_x| << |E_z|$, and using Eq. (11.55) the power delivered per unit volume is

$$Q = \frac{k_\perp{}^2\ c^2\ B_0{}^2}{2\ \mu_0\ \omega}\frac{\delta}{a^2\ z^2 + \delta^2} \tag{13.74}$$

Integrating over the resonance, we obtain the total power absorbed per unit area

$$W = \frac{\pi\ k_\perp{}^2\ c^2\ B_0{}^2}{2\ \mu_0\ a\ \omega} \tag{13.75}$$

However, we cannot calculate the reflectivity (or absorption coefficient) unless the coefficient B_0 is known. This is a difficult procedure and can only be accomplished numerically (Pert, 1978). It involves fitting the solution to the boundary conditions at $z = \pm\infty$ making use of the Försterling's solution (13.71) for sufficiently small values of z before fitting to the asymptotic WKB solutions by noting that the asymptotic approximations to the Fösterling functions, Eq. (13.70) are the Airy functions of Section 5.4.1. The absorption coefficient calculated in this way is shown in Figure 13.2 as a function of the angle parameter $q = (k_0\ a)^{2/3}\sin^2\theta$ where θ is the angle of incidence.

This phenomenon has a number of points of interest.

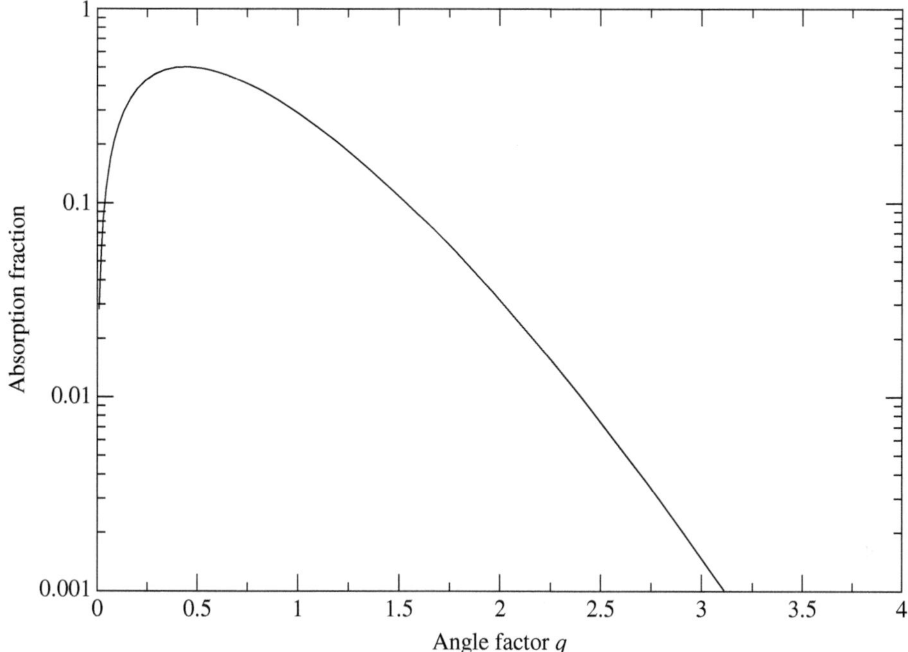

Figure 13.2 Plot showing the fractional absorption of linear resonant absorption plotted as a function of the angle parameter $q = (k_0\ a)^{2/3}\sin^2\theta$ from Pert (1978).

- The absorption process is important both in ionospheric physics and laser plasma interactions. In the latter case, the driven plasma wave is usually so strong that the absorption becomes non-linear and the pumped wave breaks.
- The present calculation has considered cold plasma. However, if the plasma has a small, but finite, temperature (warm plasma), the singularity at the resonance point is removed, although the E_z component of the electric field is greatly enhanced.
- Finally, the calculation is unusual in that the waves exhibit both cut-off and resonance.

13.4 Single Particle Model of Electrons in an Electro-magnetic Field

A useful starting point for understanding the behaviour of plasma in a strong electro-magnetic field is obtained from simple models based on the single particle model explored in Chapter 3.

13.4.1 Quiver Motion

Consider a uniform plane polarised electro-magnetic wave propagating in the $\mathbf{k}$ direction with vector potential $\mathbf{A} = \mathbf{A}_0\ \sin(\mathbf{k} \cdot \mathbf{r} - \omega\ t)$ oscillating a direction normal to $\mathbf{k}$ with frequency ω. The phase of the wave $\phi = \mathbf{k} \cdot \mathbf{r} - \omega t$. The motion of the particles due the electric field $\mathbf{E} = \mathbf{E}_0\ \cos(\mathbf{k} \cdot \mathbf{r} - \omega\ t)$ dominates that due to the magnetic field $\mathbf{B} = \mathbf{B}_0\ \cos(\mathbf{k} \cdot \mathbf{r} - \omega\ t)$, which lies in the direction normal to $\mathbf{k}$ and $\mathbf{E}$, and can be treated as a term of lower order. Thus, at the plane $z = 0$ the electron has an oscillation in velocity to first order

$$\mathbf{v}_1 = \mathbf{v_q}\ \sin(\mathbf{k}\cdot\mathbf{r} - \omega t) = -\frac{e\ \mathbf{E}_0}{m\ \omega}\ \sin(\mathbf{k}\cdot\mathbf{r} - \omega t) + \mathbf{v}_{d1} \tag{13.76}$$

where $\mathbf{v}_q = -e\ \mathbf{E}_0/m$ is the quiver velocity, i.e. the amplitude of the velocity induced in the electron by the field in the direction of the electric field. $\mathbf{v}_{d1}$ is the constant of integration determined by the initial phase of the wave when the electron is first influenced by the field. It represents a drift velocity needed to match the initial zero velocity to that imposed by the quiver, i.e. $\mathbf{v}_{d1} = e\mathbf{E}_0 \sin\phi/m\omega$ when ϕ is the phase at the starting time. In the usual case $\mathbf{v}_{d1} = 0$ when $\phi = 0$, the electron being freed at the peak electric field by multiphoton or tunnel ionisation (see Gibbon, 2005; Tallents, 2018). It is assumed that the field is sufficiently weak that the quiver motion is non-relativistic, i.e. the quiver velocity is much less than that of light $v_q \ll c$ (see Gibbon, 2005, for an analysis of relativistic behaviour).

If the field is very slow rising with a rise time T much larger than the period $\tau = 2\ \pi/\omega$, the drift velocity from alternate half-cycles reverses in sign and cancels out to zero over the full rise of the pulse. To demonstrate this, divide the pulse into successive cycles. Since the initial phase is not well-defined we are at liberty to choose the phase appropriately. During the nth cycle, the first half cycle from $n - \frac{1}{2}$ to n starting with an arbitrary phase at the start of the half-cycle $\phi_{(n-1/2)} = \phi$ the peak field increases from $E_{0(n-1/2)}$ to E_{0n} and the drift velocity $v_{d(n-1/2)}$ increases to v_{dn}. In the frame moving with $v_{d(n-1/2)}$, the electron is initially stationary and the change in the drift speed at the end of the half-cycle is approximately $-e\ (E_{0n} - E_{0(n-1/2)}) \sin\phi \Big/\ m\ \omega$ at the end of the first half-cycle. Now consider the second half-cycle with initial phase $\phi_n = \phi_{(n-1/2)} + \pi = \phi + \pi$, the additional drift is $e\ (E_{0(n+1/2)} - E_{0n}) \sin(\phi)/m\ \omega$. Adding the contribution from two successive half-cycles, the net increase in drift velocity

$$\delta v_d \approx e\,(E_{0(n+1/2)} - 2E_{0(n)} + E_{0(n-1/2)}) \sin(\phi)/m \approx 0 \tag{13.77}$$

accurate to a term of order $O(\tau/T)^4$. Summing over successive cycles, any drift is therefore very small.

Consider now the additional velocity induced by the magnetic field due to the quiver, which in a plane parallel beam is normal to the quiver and to the magnetic field,

$$\begin{aligned}\frac{d\mathbf{v}_2}{dt} = -\frac{e}{m}\ \mathbf{v}_1 \wedge \mathbf{B} &= -\frac{e^2}{m^2\ \omega}\ \mathbf{E}_0 \wedge \mathbf{B_0}\ \cos\,(\mathbf{k}\cdot\mathbf{r} - \omega\ t)\ \sin\,(\mathbf{k}\cdot\mathbf{r} - \omega\ t)\\ &= -\frac{e^2}{2m^2\ \omega}\ \mathbf{E}_0 \wedge \mathbf{B_0}\ \sin\,[2(\mathbf{k}\cdot\mathbf{r} - \omega\ t)]\end{aligned}$$

Hence integrating

$$\mathbf{v}_2 = \mathbf{v}_{d2} - \frac{e^2}{4\ m^2\ \omega^2}\mathbf{E}_0 \wedge \mathbf{B}_0\ \cos\,[2(\mathbf{k}\cdot\mathbf{r} - \omega t)] \tag{13.78}$$

where $\mathbf{v}_{d2}$ is the unknown constant of integration. The displacement of the electron

$$\mathbf{r}_q = \frac{e}{m\ \omega^2}\ \mathbf{E}_0\ \cos(\mathbf{k}\cdot\mathbf{r} - \omega\ t) - \frac{e^2}{8\ m^2\ \omega^3}\ \mathbf{E}_0 \wedge \mathbf{B}_0\ \sin[2(\mathbf{k}\cdot\mathbf{r} - \omega\ t)] + \mathbf{v}_{d2}\ t \tag{13.79}$$

The electron therefore exhibits a figure of eight motion in the plane containing the electric field and the normal to both the electric and magnetic fields. Together with an electron drift with velocity $\mathbf{v}_{d2}$.

The origin of the drift velocity $\mathbf{v}_{d2}$ is seen to be a consequence of the start-up phase, required to match the start-up velocity $v_2 = 0$ to that associated with the 'figure of eight' motion. Assuming that the electron is initially at rest $\mathbf{v} = \mathbf{v}_1 + \mathbf{v}_2 = 0$ when it first encounters the electric field with phase $\phi = 0$, then the drift velocity

$$\mathbf{v}_{d2} = \frac{e^2}{4\ m^2\ \omega^2} \mathbf{E_0} \wedge \mathbf{B_0} \tag{13.80}$$

which is the non-relativistic form of the result obtained elsewhere (Gibbon, 2005).

When the pulse is slowly rising, we may also use the construction above to obtain a reduced value for the drift.

13.4.2 Ponderomotive Force

If the electro-magnetic wave is non-uniform so that the fields possess gradients, the quiver motion of the electron causes the force on the electron to vary over the period of the oscillation. We shall assume that the time variation in the fields is due to external oscillation only, i.e. in operator form $\mathbf{v} \cdot \nabla \ll \partial/\partial t$. The ponderomotive force is the average of the force on the electron taken over the period of the wave. As a result, terms involving the product $\sin(\mathbf{k} \cdot \mathbf{r} - \omega t)\ \cos(\mathbf{k} \cdot \mathbf{r} - \omega t)$ average to zero.

Since the electric field

$$\mathbf{E} = \mathbf{E}_0 \cos(\mathbf{k} \cdot \mathbf{r} - \omega t) \tag{13.81}$$

it follows from Maxwell's equation $\partial \mathbf{B}/\partial t = -\nabla \wedge \mathbf{E}$ that the magnetic field

$$\mathbf{B} = \mathbf{B}_0 \sin(\mathbf{k} \cdot \mathbf{r} - \omega t) + \frac{1}{\omega} \mathbf{k} \wedge \mathbf{E}_0 \cos(\mathbf{k} \cdot \mathbf{r} - \omega t) \tag{13.82}$$

where $\mathbf{B}_0 = -1/\omega \nabla \wedge \mathbf{E}_0$.

Consider first the variation in the force parallel to the field direction due to the electric field gradient, the force on the electron

$$\begin{aligned} \mathbf{F} &= -e\ (\mathbf{E}_0 + [\mathbf{r}_q \cdot \nabla]\mathbf{E_0})\ \cos\ (\mathbf{k} \cdot \mathbf{r} - \omega\ t) \\ &= -e\ \mathbf{E}_0\ \cos\ (\mathbf{k} \cdot \mathbf{r} - \omega\ t) - \frac{e^2}{m\ \omega^2} (\mathbf{E}_0 \cdot \nabla)\mathbf{E_0}\ \cos^2\ (\mathbf{k} \cdot \mathbf{r} - \omega\ t) \end{aligned} \tag{13.83}$$

The contribution to the force on the electron from the magnetic field

$$\begin{aligned} \mathbf{F} = -e\ \mathbf{v}_1 \wedge \mathbf{B} &= -\frac{e^2}{m\ \omega^2} \mathbf{E_0} \wedge (\nabla \wedge \mathbf{E_0})\ \sin^2\ (\mathbf{k} \cdot \mathbf{r} - \omega t) \\ &+ \frac{e^2}{m\ \omega^2} \mathbf{E_0} \wedge (\mathbf{k} \wedge \mathbf{E_0})\ \sin\ (\mathbf{k} \cdot \mathbf{r} - \omega t)\ \cos(\mathbf{k} \cdot \mathbf{r} - \omega t) \end{aligned} \tag{13.84}$$

Adding these two components of the force together, averaging over a cycle of the field and using the vector identity $\frac{1}{2}\nabla|A|^2 = \mathbf{A} \wedge \nabla \wedge \mathbf{A} + (\mathbf{A} \cdot \nabla)\mathbf{A}$, we obtain the total force on an electron in the field of the wave

$$\mathbf{F} = -\frac{e^2}{4\ m\ \omega^2} \nabla|\mathbf{E}_0|^2 \tag{13.85}$$

Summing over the electrons contained in unit volume when the refractive index is n, we obtain the *ponderomotive force* per unit volume

$$\mathbf{F} = -\frac{\epsilon_0}{4} \frac{\Pi^2}{\omega^2}\ \nabla|\mathbf{E}_0|^2 = -\frac{1}{2c} \frac{\Pi^2}{\omega^2}\ \nabla I \tag{13.86}$$

where $I = c\ \epsilon_0\ |E_0{}^2|/2$ is the energy flux per unit area or *beam irradiance* (which is often loosely called intensity) of the electro-magnetic wave.

Clearly averaging over the period of the wave is a critical step in this analysis. This analysis is valid if two conditions are met,

i. Firstly, the variation in any quantity due to the motion must be slow compared to the oscillation.
ii. Secondly, the variation of the irradiance must be small over the period of the wave.

This derivation of the ponderomotive force is based on a simple physical picture appropriate to a wave incident normally on a plane surface. A similar physical picture was also developed by Hora (1971) based on the force on an electron undergoing the 'figure of eight' motion discussed earlier, Eq. (13.79). The general form of the ponderomotive force is given by consideration of the force on a dielectric in an electro-magnetic field as the gradient of the Maxwell stress (Appendix 13.A). Under the conditions of the present picture, it can be shown that Eq. (13.86) is obtained.

Although the ponderomotive force acts preferentially on the electrons,[9] the hydro-dynamic effect of the force influences the motion of the plasma as a whole. In a non-uniform plasma, charge separation fields induced by electron displacement give rise to an equivalent force on the ions. The plasma as a whole therefore experiences the stress given by Eq. (13.86).

An alternative expression for the ponderomotive force is obtained using the WKB approximation to describe the behaviour of the field when the density, and therefore the refractive index n, changes (see Eq. (13.A.10) Section 13.A). The ponderomotive force is written as

$$\mathbf{F} = \frac{\epsilon_0}{4} E_0'^{\,2} \frac{\Pi^2}{\omega^2 n^2} \nabla n \tag{13.87}$$

where E_0' is the incident electric field in vacuo. It follows that the ponderomotive force is directed from high to low intensity.

The normal reflection from a plane mirror generates a standing wave corresponding to forward and backward waves of equal amplitude. Consequently, there is no preferred direction and no ponderomotive force – the electric and magnetic fields being of equal magnitude and $\pi/2$ out of phase (Eq. (13.31)). The electromagnetic field stress $(\epsilon_0 E_0^2 + B_0^2/\mu_0)/2$ is therefore constant, independent of the propagation distance z. On the other hand, if the reflection occurs in plasma with a finite refractive index gradient, the reflected wave is again a standing wave, but with an Airy profile (see Eq. (5.37) and Figure 5.1). The plasma therefore experiences the ponderomotive force as the field stress reduces at lower plasma density.

13.4.3 The Impact Model for Collisional Absorption

Collisional, or alternatively inverse bremsstrahlung, absorption plays an important role in elucidating the interaction of lasers with plasma during the breakdown of solids and gases under high power irradiation. As a result, it has received considerable study using a variety of approaches both classical and quantum mechanical using the single particle picture. Within the range of interest in experimental studies, the resulting absorption coefficient is found to be independent of the model apart from a Coulomb logarithm. Similarly to the Spitzer term for the collision frequency introduced in Eq. (2.7), this is required to account for the logarithmic divergence in the final Coulomb integral. It is not appropriate to discuss in detail the various methods of calculating this term, but an asymptotic method yielding accurate results clearly illustrates the underlying physics. In the impact model, the quiver motion of electrons in the oscillating electric field is converted to random thermal motion by collisions.

9 There is a corresponding force on the ions, but it is a factor m/M times smaller.

We consider the electron motion due to the quiver motion $\mathbf{v}_1 = \mathbf{v}_q \sin \omega t$ superimposed on the random thermal velocity $\mathbf{v}_0$ so that the total velocity and energy of an electron are

$$\mathbf{v} = \mathbf{v}_0 + \mathbf{v}_1(t) \tag{13.88}$$

$$\frac{1}{2}mv^2 = \frac{1}{2}mv_0{}^2 + \frac{1}{2}mv_1(t)^2 + m\mathbf{v}_0 \cdot \mathbf{v}_1 \tag{13.89}$$

In an elastic collision between an electron and a stationary heavy particle, the electron velocity is unchanged in magnitude but is rotated through an angle χ. The central assumption of the impact model is that the collision duration is much shorter than the period of the wave, i.e. $\omega \ll b/v$ where a is the range of interaction force (as for example assumed in Section 4.5). The collision is therefore effectively instantaneous and the quiver velocity is unchanged by the collision. If this condition is not upheld, the electron executes several oscillations during the interaction and the total velocity is poorly defined at the conclusion of the collision. A detailed study of the electron/ion interaction in the presence of an electro-magnetic wave confirms this picture. The random thermal component of the velocity becomes

$$\mathbf{v}'_0 = \mathbf{v}' - \mathbf{v}_1 = \mathbb{O}\,\mathbf{v} - \mathbf{v}_1 \tag{13.90}$$

where $\mathbb{O}$ is the rotation operator for the scattering. Taking orthogonal co-ordinates defined by the unit vectors $\hat{\mathbf{i}}$ parallel to the incoming velocity $\mathbf{v}$, the plane $\hat{\mathbf{i}}$, $\hat{\mathbf{j}}$ containing the quiver velocity $\mathbf{v}_1$ and $\hat{\mathbf{k}}$ perpendicular to the $\hat{\mathbf{i}}$, $\hat{\mathbf{j}}$ plane, the scattered velocity is

$$\mathbb{O}\,\mathbf{v} = v\ \cos\chi\,\hat{\mathbf{i}} + v\ \sin\chi\ \cos\phi\,\hat{\mathbf{j}} + v\ \sin\chi\ \sin\phi\,\hat{\mathbf{k}} \tag{13.91}$$

where χ and ϕ are the angles of scatter.

Since the quiver velocity is unchanged, and the thermal velocity is changed to $\mathbf{v}'_0$, the gain in the thermal energy as a consequence is

$$\begin{aligned}\frac{1}{2}v_0'^2 - \frac{1}{2}v_0^2 &= \mathbf{v}_1 \cdot (\mathbf{v}_0 - \mathbf{v}'_0) = \mathbf{v}_1 \cdot (\mathbf{v} - \mathbf{v}') \\ &= |\mathbf{v}|\ |\mathbf{v}_1|\ [\cos\alpha\ \ (1-\cos\chi) - \sin\alpha\ \sin\chi\ \cos\phi]\end{aligned} \tag{13.92}$$

where α is the angle between the total velocity $\mathbf{v}$ and the quiver velocity $\mathbf{v}_1$. The thermal energy gain by electrons as a result of collisions with heavy particles per electron per unit time is given by the integral of the cross section σ over the solid angle of scattering[10]

$$q = 2\,\pi\,n\,m \int_0^{\pi} \mathbf{v}\cdot\mathbf{v}_1\ (1-\cos\chi)\ v\ \sigma(v,\chi)\ \sin\chi\ \mathrm{d}\chi = n\ m\ \mathbf{v}\cdot\mathbf{v}_1\ v\ \sigma_d(v) \tag{13.93}$$

where $\sigma_d(v)$ is the momentum transfer cross section, Eq. (2.6) and n the density of heavy particles. The heavy particles are assumed to be stationary and randomly distributed in space.

The mean energy absorption by isotropic electrons of thermal speed v_0 is completed by averaging over the angle θ between the quiver velocity $\mathbf{v}_1$ and the thermal speed $\mathbf{v}_0$. To proceed we need to know the variation of the momentum transfer cross section $\sigma_d(v)$ with the total velocity v. When the inter-particle force law is an inverse power law $F \sim r^s$, so that $\sigma_d = \sigma_0\ v^{-4/(s-1)}$, general results are obtained by performing the appropriate integral over the angle between the quiver and thermal velocities θ (Pert, 1972)

$$\overline{q} = \frac{1}{2}\ m\ n\ \sigma_0 \int_0^{\pi} (v_0\ v_1\ \cos\theta + v_1{}^2)\ (v_0{}^2 + v_1{}^2 + 2v_0\ v_1\ \cos\theta)^{[1-4/(s-1)]/2}\ \sin\theta\ \mathrm{d}\theta \tag{13.94}$$

10 This result is clearly independent of the manner by which the scattering angle θ is achieved. It is therefore applicable to multi-particle scattering (Section 2.1.3) provided the individual impacts are completed within the period of the field, and the Coulomb logarithm is cut-off at $b_{\max} \sim v/\omega$.

If the scattering particles are ions, the field is Coulombic $s = 2$ and

$$\overline{q} = \begin{cases} 0 & v_0 > v_1 \\ 2\, m\, n\, \sigma_0/|v_1| & v_0 < v_1 \end{cases} \quad s = 2 \qquad \text{Coulomb force} \tag{13.95}$$

Next we need to average over the oscillation of the quiver velocity, i.e. that of the electro-magnetic wave to obtain the final value for the energy gained by electrons per unit volume per unit time

$$\overline{q} = \begin{cases} 0 & v_0 \gg v_1 \\ \dfrac{1}{\pi}\dfrac{m\, n\, \sigma_0}{v_q} \ln \Delta(v_0)\, \ln\left\{\dfrac{(v_q + \sqrt{v_q{}^2 - v_0{}^2})}{(v_q + \sqrt{v_q{}^2 - v_0{}^2})}\right\} & v_0 < v_q \\ \dfrac{2}{\pi}\dfrac{m\, n\, \sigma_0}{v_q} \ln \Delta(\overline{v})\, \ln\left\{\dfrac{2 v_q}{v_0}\right\} & v_0 \ll v_q \end{cases} \tag{13.96}$$

For a Coulomb field $\sigma_0 = 4\,\pi\, Z^2\, e^4/(4\,\pi\,\epsilon_0)^2\, m^2\, \ln \Delta(v)$ where Z is the charge of the ion and $\ln \Delta(v)$ is the appropriate Coulomb logarithm; $\overline{v}$ is an average of the thermal and quiver speeds whose value we estimate later. Separating the weak velocity dependence in the Coulomb logarithm, we write $\sigma_0 = \sigma_0'\, \ln \Delta(v)$.

As it stands, the result for Coulomb interactions in weak fields, when $v_0 > v_1$, with constant Coulomb logarithm appears to be in error, since it is a familiar experimental result that absorption of weak electro-magnetic waves occurs in plasma. The failure is not a consequence of the model, but due to two factors.

i. The failure to include very slow moving electrons present for all reasonable distribution functions when $f(0) \neq 0$. As a result, there is non-zero population of electrons, $(4\,\pi/3)\, v_1{}^3\, f(0)$ per unit volume, which satisfy the condition $v_0 < v_1$. These give rise to an average absorption rate per electron

$$\overline{q} \approx \frac{4\,\pi}{3}\, m\, \frac{n}{n_e}\, \sigma_0'\, \ln \Delta(v_0)\, v_q{}^2\, f(0) \tag{13.97}$$

ii. The Coulomb logarithm is assumed to be constant. The weak dependence of the logarithmic cut-off on the particle velocity, can be taken into account. As we found in Section 2.1.2, it is necessary to introduce a cut-off to avoid the logarithmic singularity at large impact parameters, assumed to be at the Debye length. In the impact model, a more stringent cut-off is required following from the condition that the phase change of the field during the interaction must be small, i.e. the impact parameter $b < v/\omega$. The Coulomb logarithm therefore becomes $\ln \Lambda \Rightarrow \ln \sqrt{(1+\Delta^2)} \approx \ln \Delta$ where $\Delta = (b_{max}/b_{min}) = 4\,\pi\,\epsilon_0\, m\, v^3/Z\, e^2\, \omega$. Since the wave only propagates if $\omega > \omega_p$ and the Debye length $\lambda_d = v_0/\omega_p$, it follows that all the collisions lie inside the Debye sphere.[11] Therefore, provided the field is not too strong ($v_1 \lessapprox v_0$).

$$\ln \Delta(v) \approx \ln \Delta(v_0) + \frac{1}{\Delta(v_0)} \left.\frac{d\Delta}{dv}\right|_{v_0} (v - v_0) \approx \ln \Delta(v_0) + \frac{3}{v_0}(v - v_0) \tag{13.98}$$

Integrating over the angle θ as before we obtain

$$\overline{q} = \left\{ \frac{1}{3}\, n\, m\, \sigma_0'\, \frac{1}{\Delta(v_0)} \left.\frac{d\Delta}{dv}\right|_{v_0} \frac{v_q{}^2}{v_0} \qquad v_0 \gg v_q \right\} \qquad s = 2 \quad \text{Coulomb force} \tag{13.99}$$

11 Equation (13.93) has been derived from a classical model. However it can be shown (Pert, 1975) that the same result may also be obtained from the quantum mechanical Born approximation, but with a different Coulomb logarithm in which the minimum impact parameter is the electron wavelength, i.e. $\Delta = m\, v^2/\hbar\, \omega$.

The final step is an integral over the velocity distribution of the electrons. Once again the case of a Coulomb field the integration is logarithmically divergent in weak fields. However again this divergence may be simply circumvented by a simple cut-off. Noting that we require the impact parameters to satisfy the relation $b_{max} > b_{min}$ we cut-off the integral over velocity at the limiting velocity, and require thermal velocities $v_0 > v_{lim} = \sqrt[3]{Z\, e^2\, \omega/4\pi\, \epsilon_0\, m}$. For example if the electron velocity distribution is Maxwellian, the mean energy absorbed per electron per unit time

$$\overline{q} = 4\,\pi\left(\frac{m}{(2\,\pi\,\cancel{k}T)}\right)^{3/2}\int_{v_{lim}}^{\infty}\overline{q}(v_0)\,{v_0}^2\,e^{-m\,{v_0}^2/2\cancel{k}T}\,\mathrm{d}v_0 \tag{13.100}$$

$$\approx\begin{cases}\dfrac{4}{3\,\sqrt{\pi}}\,\dfrac{m\,n\,\sigma_0'}{v_T}\,\dfrac{{v_q}^2}{{v_T}^2}\left[\ln\Delta(v_T)-\dfrac{\eta\,\gamma}{2}\right] & v_T \gg v_q\\[2ex] \dfrac{2}{\pi}\,\dfrac{m\,n\,\sigma_0'}{v_q}\,\ln\Delta(\overline{v})\,\ln\left(\dfrac{2v_q}{v_T}\right) & v_T \ll v_q\end{cases} \tag{13.101}$$

where $v_T = \sqrt{2\,\cancel{k}Tm}$ the thermal speed, $\gamma = 0.577\,256\,65$ is the Euler–Mascheroni constant, and $\Delta(v) \propto v^\eta$ so that $\eta =$ 2or 3 for quantal or classical electrons, respectively. Equation (13.101) shows that when the field is weak, the thermal motion controls the absorption rate. However if the field is strong ($v_q \gg v_T$), the quiver motion totally dominates. When the Coulomb logarithm takes the approximate form $\Delta(v) \approx v/\omega\Big/Ze^2/mv^2 \gg 1$, an analytic closed form for a Maxwellian electron distribution encompassing Eq. (13.101) can be found, which is valid at all field strengths (see Pert, 1976a,b,1995), namely

$$\overline{q} = \frac{2}{3}\,n\,m\,\sigma_0'\,m\,{v_q}^2\{\ln\Delta(v_T)\,S_1(x) + \eta\,S_2(x)\} \tag{13.102}$$

where $x = m\,{v_q}^2/2\cancel{k}T$ and $\eta = v\,\mathrm{d}\Delta(v)/\mathrm{d}v|_{v_0}$. The term $S_1(x)$ clearly corresponds to the term in Eq. (13.98) and $S_2(x)$ to that in Eq. (13.99), and gives the correct values in weak fields ($v_T \gg v_q$). In strong fields, the asymptotic values of $S_1(x)$ and $S_2(x)$ agree with Eq. (13.101), provided the mean value of $\overline{v} = \sqrt{v_T\,v_q}$ in the Coulomb logarithm is taken to be the geometric mean of the thermal and quiver speeds.

As expected, in very large fields the collision rate is determined by the quiver velocity, the thermal component only appearing in the slowly varying logarithmic term. In this case, it follows from Eq. (13.96) that the absorption rate is given by Eq. (13.101) independent of the electron distribution provided it is isotropic – a result due to Silin (1965).

13.4.3.1 Electron–Electron Collisions

It follows from this simple non-relativistic model that there can be no heating in electron–electron collisions. Alternatively there is no dipole moment in electron–electron collisions to interact with the field. Provided the motion is non-relativistic, all electrons exhibit identical quiver motions in the field. Therefore, transforming to the frame of the quiver motion, electron–electron collisions reduce to their normal field-free thermal interaction, relaxing to an equilibrium Maxwellian distribution with a characteristic time given by Eq. (2.23) as discussed in Section 7.5.

13.4.4 Distribution Function of Electrons Subject to Inverse Bremsstrahlung Heating

As we saw in Appendix 8.C, the presence of an electric field heating the electrons can lead to a substantial modification of the electron distribution away from the simple Maxwellian. In that case it is a consequence of the heating of electrons by the electric field and the subsequent disbursement of the energy in collisions with molecules. Since this process does exchange energy between the electrons, electron–electron collisions being infrequent, the distribution departs from the Maxwellian. Similarly we may expect that in inverse bremsstrahlung (or collisional) absorption in scattering between electrons and ions, which is described by the Coulomb force law, electrons of differing speeds gain energy at rates, which do not conform to the Maxwellian. A Maxwellian distribution is only maintained if the electron–electron collision rate is sufficiently strong. This behaviour was first investigated by Langdon (1980) and subsequently independently by Balescu (1982a) and Jones and Lee (1982).

In an electron–ion collision, the energy gain by the electron is typically $\sim m\, {v_q}^2$. In contrast in an electron–electron collision energy $\sim m{v_T}^2$ is exchanged to maintain the Maxwellian. Since the electron density $n_e \approx Z\, n_i$, the electron–electron collision frequency scales as $\sim Z$, while the ion–electron collision scales as $\sim Z^2$, it follows that the Maxwellian is maintained if $Z\, {v_q}^2 \ll {v_T}^2$, i.e. if the field is weak or if the ion charge is small. However if the field is strong ($v_q \gg v_T$), the electron–ion collision rate is determined by the quiver velocity instead of the thermal and the maxwellianisation criterion is weakened to $v_q \gg Z\, v_T$ The modification of the distribution function away from Maxwellian and the resulting behaviour of the absorption coefficient is well demonstrated by considering a Lorentzian plasma. The consequent distribution is described by the Fokker–Planck equation retaining the time derivative and neglecting the Electron–electron collision term namely Eq. (8.95). Expanding the distribution function in Legendre polynomials as in Eq. (8.C.2), and truncating at the second order term provided the electric field is not too strong, i.e. $u_0 \ll v_T$. Assuming uniform ion density, so that spatial derivatives may be ignored, and neglecting the electron collision terms, we obtain expressions for the zero and first order terms

$$\frac{\partial f_0}{\partial t} - \frac{e\, E}{m}\frac{1}{3\, v^2}\left[\frac{\partial}{\partial v}(v^2 f_1)\right] = 0 + \cancel{C_e e_0} \tag{13.103:a}$$

$$\frac{\partial f_1}{\partial t} - \frac{e\, E}{m}\left[\frac{\partial f_0}{\partial v} + \cancel{\frac{2}{5\, v^3}\frac{\partial}{\partial v}(v^3 f_2)}\right] = -\frac{2\, \sigma_0}{v^3} f_1 + \cancel{C_e e_1} \tag{13.103:b}$$

where the Coulomb logarithm is assumed to be constant. The zero and first order electron–electron collision terms $C_e e_0$ and $C_e e_1$ are both set to zero, neglecting electron–electron relaxation to derive the most extreme case of distribution modification. Inspection of Eqs. (13.103) shows that the isotropic zero order term f_0 is comprised of a slowly varying term and terms of high order frequency terms and that f_1 essentially varies as $\cos \omega t$. Hence from Eq. (13.103:b), we obtain

$$f_1 = \frac{(eE_0/m)}{(2\, \sigma_0/v^3 - \imath\omega)}\frac{\partial f_0}{\partial v}\, e^{-\imath\, \omega t} = \left[u_0\, (\omega\tau_{ei})\, \frac{(1 + \imath\, \omega\, \tau_{ei})}{(1 + \omega^2\, {\tau_{ei}}^2)}\right]\frac{\partial f_0}{\partial v}\, e^{-\imath\, \omega\, t} \tag{13.104}$$

where $\nu_{ei}(v) = {\tau_{ei}}^{-1}(v) = 2\, \sigma_0/v^3$. Substituting $f_1(v)$ in Eq. (13.103:a) and noting footnote 4, we average over the wave period to obtain

$$\frac{\partial f_0}{\partial t} = \frac{\sigma_0\, {u_0}^2}{3}\frac{1}{v^2}\frac{\partial}{\partial v}\left(\frac{g(v)}{v}\frac{\partial f_0}{\partial v}\right) \tag{13.105}$$

where $g(v) = (1 + \omega^{-2}\, {\tau_{ei}}^{-2})^{-1}$.

The first order term given by Eq. (13.104) is the change in the distribution due to the quiver oscillation damped by the ion–electron collisions. The energy loss due to damping is converted into thermal motion and is simply the collisional heating, represented by the in phase resistive component; the reactive being the simple oscillation. This heating leads to the change in the zero order term expressed in Eq. (13.105). The current per unit area

$$\mathbf{J} = -e \int f(v)\, \mathbf{v}\, \mathbf{dv} = -e\left\{\cancel{\int f_0(v)\mathbf{v}\mathbf{dv}} + \int f_1(v)\, \cos\theta\, \mathbf{v}\, \mathbf{dv}\right\} \tag{13.106}$$

where θ is the angle between the velocity **v** and the electric field **E**. The work done per unit volume per unit time by the electric field averaged over the wave period to sustain the quiver motion is therefore consistent with Eq. (13.105)

$$\begin{aligned}\mathbf{E}\cdot\mathbf{J} &= -4\,\pi\,n_e\,m\,\sigma_0\,{u_0}^2\int_0^\infty \mathrm{d}v\int_0^\pi \mathrm{d}\theta\sin\theta\,\cos^2\theta\,g(v)\frac{\partial f_0(v)}{\partial v}\\ &= -\frac{4\,\pi}{3}\,m\,\sigma_0\,{u_0}^2\int g(v)\frac{\partial f_0(v)}{\partial v}\mathrm{d}v = \frac{4\,\pi}{3}\,m\,\sigma_0\,{u_0}^2\int \frac{1}{2}\,v^2\,\frac{\partial}{\partial v}\left(\frac{g(v)}{v}\frac{\partial f_0}{\partial v}\right)\,\mathrm{d}v\\ &= \frac{\partial}{\partial t}\left(\int \frac{1}{2}\,m\,v^2\,f_0(v)\,\mathbf{dv}\right)\end{aligned} \tag{13.107}$$

which is the rate of increase of the total kinetic energy.

In the particular case that the oscillation frequency is large compared to that of collisions $\omega \gg \nu_{ei}$ when $g(v) = 1$ it follows that the energy absorption rate per unit volume

$$\mathbf{E}\cdot\mathbf{J} = \frac{4\,\pi}{3}\,m\,\sigma_0\,{u_0}^2\,f_0(0) \tag{13.108}$$

which is identical to the value obtained earlier from the impact model, Eq. (13.97).

It is instructive to compare Eq. (13.104) with equivalent term from the Drude model, Eq. (2.53) where the collision frequency is assumed to be constant. In this simple model, the separation of conductivity into two regimes, namely resistive when $\omega\tau_{ei} < 1$ corresponding to a nearly constant electric field, and reactive when $\omega\tau_{ei} > 1$ and the electrons are oscillatory. In the general case, there may be a transition depending on the electron speed from reactive behaviour at slow speeds to resistive at fast. However the major interest is associated with collisional absorption by plasma irradiated by laser radiation, when $\omega\tau_{ei} \gg 1$ and $g(v) \approx 1$.

In the high frequency field case when $g(v) = 1$ over the entire distribution, it is easily shown by substitution that the distribution has a self-similar solution (known as a *Langdon distribution*)

$$f_0(v) = \frac{5^{2/5}}{4\,\pi\Gamma(3/5)}\,\frac{1}{U^3}\,\exp\left[-\frac{v^5}{5\,U^5}\right] \tag{13.109}$$

where

$$U(t)^5 = 5\,\frac{2\,\pi}{3}\,n\left(\frac{Z\,e^2}{m}\right)^2\,\ln\Lambda\,{u_0}^2\,t \tag{13.110}$$

The total energy per electron scales as

$$\epsilon = \frac{5^{2/5}}{\Gamma(3/5)}\,\frac{1}{2}\,m\,U^2 \tag{13.111}$$

The energy absorption rate per unit volume is therefore

$$\frac{\mathrm{d}\epsilon}{\mathrm{d}t} = \frac{4\,\pi}{3}\,m\,\sigma_0\,{u_0}^2\,\frac{5^{2/5}}{4\,\pi\Gamma(3/5)}\,\frac{1}{U^3} = \frac{4\,\pi}{3}\,m\,\sigma_0\,{u_0}^2\,\frac{5^{2/5}}{4\,\pi\Gamma(3/5)}\left(\frac{5^{2/5}}{2\Gamma(3/5)}\right)^{-3/2}\frac{1}{\epsilon^{3/2}} \tag{13.112}$$

Calculating the energy absorption by electrons in a plasma with a Langdon distribution we find its value is only a fraction 0.446 that of a Maxwellian with the same average particle energy. This is easily understood by comparing the profile of the distribution with that of a Maxwellian (Figure 13.3) where it is seen that the Maxwellian has a significantly greater fraction of slow particles, which are more readily scattered by the ions and consequently have a greater energy absorption. The situation in a real plasma is more complicated for as we have discussed the neglected electron–electron collisions drive the electron distribution towards an equilibrium

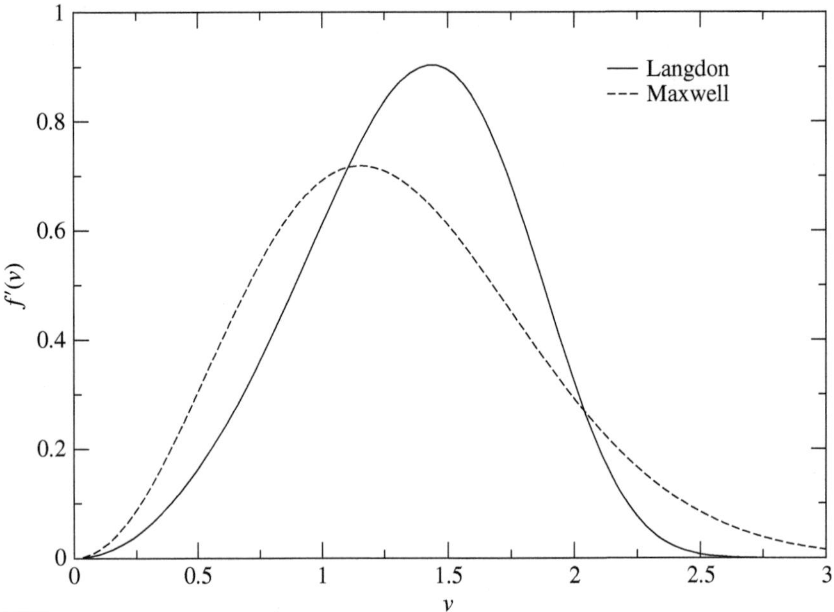

Figure 13.3 Comparison of the Langdon and Maxwell distributions of the same energy per particle, plotted as the term $f'(v) = 4\,\pi\,v^2\,f(v)$.

Maxwellian. The characteristic parameter $Z\,v_q^{\,2}/v_T^{\,2}$ determines the actual distribution and therefore the absorption coefficient. In practice the situation in an experiment is complicated by the dynamic nature of the plasma as a consequence of several factors.

- Increasing thermal energy, hence electron–electron collisional energy increases driving a Maxwellian.
- Delays in changes to the distribution due to the equilibrium relaxation time.
- Non-local effects due to gradients, for example in temperature, in the plasma.

As a result accurate values for any particular experimental configuration can only be achieved by detailed calculation, for example using Fokker–Planck models (see for example Town et al., 1994).

The changes in the distribution function induced by the absorption also give rise to modifications to transport coefficients implicit within the *a priori* calculations referenced above (Town et al., 1994). The modification to the thermal conductivity associated with a Langdon distribution has been calculated by Balescu (1982b) and Jones and Lee (1982).

The general solution for weak and strong fields neglecting electron–electron collisions has been investigated by Jones and Lee (1982). They find that in contrast to weak field case, the distribution in strong fields is not isotropic in the absence of electron–electron collisions. However as we have argued these are likely to play a stronger role in this case and isotropise the distribution. In this case also a self-similar distribution can be found whose form is Maxwellian. Consequently, the absorption rate may be found from our earlier result (13.102). For more details we refer the reader to the paper by Jones and Lee (1982).

Fortunately, an accurate value of the modification to the collisional absorption coefficient due to the change in the electron distribution is not usually required, and the approximate value given by a Maxwellian, Eq. (13.102) suffices.

13.5 Parametric Instabilities

13.5.1 Coupled Wave Interactions

In Chapters 11 and 12, we examined waves in plasma when the fields in the waves are small. As a result the waves could be developed in terms of a set of discrete modes whose frequency and wave number were related by a well-defined dispersion relation. However if this condition can no longer be upheld as the fields are larger, the distinct separation into non-interacting modes is lost as higher order terms in the dynamic equations introduce coupling between the different modes. In particular the condition which we introduced in Section 13.1 that the behaviour of electro-magnetic waves in linear media could be described by the polarisation, linearly proportional to the applied field, is no longer valid. When this occurs condition fails the acceleration, and therefore the displacement, of an electron in the medium is still proportional to the oscillating applied field, but additional terms, for example in plasma due to the ponderomotive force, induce a non-linear response where the susceptibility varies with the electric intensity

$$\chi = \chi_1 + \chi_2\, E + \chi_3\, E^2 + \cdots \tag{13.113}$$

It can be shown that the second order term χ_2 is only non-zero if the system has non-centrosymmetric symmetry, and therefore also normally in plasma (New, 2011). Consequently, the polarisation varies non-linearly in time. In the simplest case, a wave of frequency ω generates polarisation containing a term depending on $\cos^2(\omega t) = [1 + \cos(2\,\omega\, t)]/2$, namely the second harmonic of the fundamental wave. This polarisation will therefore generate an electro-magnetic wave at the second harmonic – *frequency doubling* – which is widely applied experimentally using a crystal lacking inversion symmetry as the non-linear medium.

More generally if two waves propagating with wave number $\mathbf{k}_2$ and $\mathbf{k}_3$, and frequency ω_2 and ω_3 are coupled through the non-linear polarisation resulting from the terms $\exp[\imath(\mathbf{k}_2 \cdot \mathbf{r} - \omega_2 t)]$ (②) and $\exp[\imath(\mathbf{k}_3 \cdot \mathbf{r} - \omega_3 t)]$ (③), we will generate a polarisation varying as $\exp[\imath(\mathbf{k}_1 \cdot \mathbf{r} - \omega_1 t)]$ (①) the product wave. Consequently, the two waves are mixed and generate a wave ① with wave number $\mathbf{k}_1$ and frequency ω_1 which obey the sum/difference phase matching relations

$$\mathbf{k}_1 = \mathbf{k}_2 + \mathbf{k}_3 \qquad \text{and} \qquad \omega_1 = \omega_2 + \omega_3 \tag{13.114}$$

This general effect is known as a *parametric process*; the creation of a new wave being due to the non-linear changes of a propagation parameter: in this case refractive index. We may alternatively regard Eq. (13.114) as the modulation of an intense pump beam ① to generate a signal ② and idler ③. If $\omega_1 > \omega_2$ the wave is *down-converted* and if $\omega_1 < \omega_2$ *up-converted*, the former being the principle underlying the development of *optical parametric amplification*. In plasma we shall be concerned with down conversion only, the strong *pump* beam (usually optical) converting into two *decay* modes (optical or plasma waves).[12]

Clearly a product waves ⓙ must satisfy the dispersion relation for the wave in the medium ($\omega_j = \omega(\mathbf{k}_j)$) to be able to propagate. If this requirement is not satisfied elements of the wave are damped over short distances by destructive interference from elements generated at different points. The general development of these phenomena using optical beams from lasers has led to the new science of *non-linear optics* discussed in detail in several texts such as Boyd (2008); New (2011), and Yariv (1997). An important consequence of this result is that the medium must be dispersive to allow the conditions (13.114) to be simultaneously satisfied.

---

12 Not all combinations of waves are allowed by the dispersion relation and the matching conditions. For example in plasma an optical wave cannot decay into two acoustic waves.

The wave matching conditions Eq. (13.114) multiplied by Planck's constant $\hbar$ has an important physical interpretation as the generation of a new wave ('photon') of appropriate energy and momentum from the union of the initiating ones or *vice versa*. From this relation, we can see that the creation of the waves ② and ③ involves the attenuation (or amplification) of the pump ①.

Provided the pump intensity is not too high the coupling between the modes can be expressed in terms of a quasi-linear relation of the form

$$\frac{\partial A_{\mathbf{k}_1}}{\partial t} = \iint d\mathbf{k}_2\, d\mathbf{k}_3\, \mathbb{K}(\mathbf{k}_1;\ \mathbf{k}_2,\ \mathbf{k}_3)\, A_{\mathbf{k}_2}\, A_{\mathbf{k}_3} \tag{13.115}$$

where $A_{(\mathbf{k})}$ is the wave amplitude and the kernel $\mathbb{K}(\mathbf{k}_1;\ \mathbf{k}_2,\ \mathbf{k}_3)$, the coupling coefficient, a tensor of rank 3 contains the matching conditions, Eq. (13.114). The kernel $\mathbb{K}$ contains the non-linear susceptibility; terms which must obey causality conditions similar to those for the linear term discussed in Section 13.1. As a result many of the terms, but not all (see Boyd, 2008, § 1.7.2) obey the Kramers–Kronig relation.

13.5.1.1 Manley–Rowe Relations

The calculation of the coupling coefficient is generally complex and not appropriate for this text. However some general relations have been found for cases of second order non-linearity which are widely applicable, as exemplified for optics (Boyd, 2008; New, 2011) and plasma (Davidson, 1972). A representative example of a dispersive, non-linear wave equation of second order ($\chi_2 = 0, \chi_3 \neq 0$) (Cairns, 1985) is given by the equation of motion (Cairns, 1985)

$$\frac{\partial^2 A}{\partial t^2} = \nabla^2 A - \nabla^2(\nabla^2 A) + C\, A^2 \tag{13.116}$$

where the term $\nabla^2(\nabla^2 A)$ allows dispersion[13] and A^2 provides the non-linear coupling with coefficient C. The linear modes are given by the dispersion relation

$$\omega^2 = k^2 + k^4 \tag{13.117}$$

so that the pump wave satisfies $k_1 < k_2 + k_3$ when $\omega_1 = \omega_2 + \omega_3$ and the wave number matching condition (13.114) may be satisfied for an arbitrary decay transition.

Assuming the wave amplitudes are not too strong we may consider the coupling term as a weak perturbation to the existing modes, which mixes them according tho Eq. (13.114) – it being assumed that wave mixing is slow compared to the period of the waves. Substituting in the wave Eq. (13.115), the coupling term A^2 contains terms such as $\Re\{A_2\, A_3\ \exp[\imath(\mathbf{k}_1 \cdot \mathbf{r} - \omega_1 t)]\}$ which are in phase with A_1 and therefore contribute to its growth.

Noting that the wave amplitude $A_1(t)$ is no longer constant due to the coupling and differentiating, we obtain

$$-2\,\omega_1\, \frac{dA_1}{dt} = \frac{1}{2}\, C\, A_2\, A_3 \tag{13.118a}$$

$$2\,\omega_2\, \frac{dA_2}{dt} = \frac{1}{2}\, C\, A_1\, A_2^* \tag{13.118b}$$

$$2\,\omega_3\, \frac{dA_3}{dt} = \frac{1}{2}\, C\, A_1\, A_3^* \tag{13.118c}$$

13 Note some dispersion is necessary to allow the matching condition to be obeyed.

having neglected terms of the form $\mathrm{d}^2 A_j/\mathrm{d}t^2$, as A_j is assumed to be slowly varying. This set of equations are readily seen to hold generally for any second order decay process where the coupling is accomplished through a term depending on the square of the field, i.e. $\chi_3 \neq 0$.

Introducing the symmetrised form of the equations in terms of the variables $A_j = \sqrt{\omega_j}\, a_j$

$$\frac{\mathrm{d}a_1}{\mathrm{d}t} = \imath\, \lambda\, a_2\, a_3 \tag{13.119a}$$

$$\frac{\mathrm{d}a_2}{\mathrm{d}t} = \imath\, \lambda\, a_1\, a_3^* \tag{13.119b}$$

$$\frac{\mathrm{d}a_3}{\mathrm{d}t} = \imath\, \lambda\, a_1\, a_2^* \tag{13.119c}$$

where $\lambda = \dfrac{1}{4}\, C\, \sqrt{\omega_1\, \omega_2\, \omega_3}$. Some important results immediately follow

$$\frac{\mathrm{d}}{\mathrm{d}t}\{\omega_1\, |a_1|^2 + \omega_2\, |a_2|^2 + \omega_3\, |a_3|^2\} = 0 \tag{13.120}$$

Since the energy density of a wave is quite generally proportional to the square of its amplitude, $|A_j|^2 \propto \omega_j\, |a_j|^2$ is the energy density of wave j to within a constant multiple. This equation therefore represents the conservation of energy amongst the waves,

$$\frac{\mathrm{d}}{\mathrm{d}t}\{|a_1|^2 + |a_2|^2\} = \frac{\mathrm{d}}{\mathrm{d}t}\{|a_1|^2 + |a_3|^2\} = \frac{\mathrm{d}}{\mathrm{d}t}\{|a_2|^2 - |a_3|^2\} = 0 \tag{13.121}$$

This set of equations, known as the *Manley–Rowe relations*, have an interpretation in terms of quantised waves 'particles'. Since the energy density of wave ⓙ is proportional to $|A_j|^2 \propto \omega_j\, |a_j|^2$ the density of particles of type j is $n_j \propto A_j^{\,2}/\hbar\, \omega \propto a_j^{\,2}$. Consequently, when one 'particle' of type ① is destroyed, one of types ② and ③ are simultaneously created.

If wave ① is the pump wave with initially the largest energy, the decay modes, ② and ③, grow, but are bounded by the conservation rule. In general, the long time solution is not steady, but oscillatory as energy shifts between the waves. Analytic solutions to the Eqs. (13.119) have been found in terms of Weierstrass elliptic functions (Bretherton, 1964; Martin and Segur, 2016).

These general results embodying the conservation of energy and quantised 'particles' apply very generally to non-linear wave coupling in many systems with third order non-linearity.

13.5.1.2 Parametric Instability

The Eqs. (13.119) describe the waves in the absence of losses through both damping and convection. These terms are easily added to the set

$$\frac{\mathrm{d}a_1}{\mathrm{d}t} + \nu_1\, a_1 + (\mathbf{v}_{g_1} \cdot \nabla)\, a_1 = \imath\, \lambda\, a_2\, a_3 \tag{13.122a}$$

$$\frac{\mathrm{d}a_2}{\mathrm{d}t} + \nu_2\, a_2 + (\mathbf{v}_{g_2} \cdot \nabla)\, a_2 = \imath\, \lambda\, a_1\, a_3^* \tag{13.122b}$$

$$\frac{\mathrm{d}a_3}{\mathrm{d}t} + \nu_3\, a_3 + (\mathbf{v}_{g_3} \cdot \nabla)\, a_3 = \imath\, \lambda\, a_1\, a_2^* \tag{13.122c}$$

where ν_j is the damping of wave j and $\mathbf{v}_{g_j}$ its group velocity, where the wave is made by a finite wave group.

Consider an infinite wave train, so that we may neglect the group velocity term. At early times when the intensity of the pump wave ① before the decay waves ② and ③ build up. The term $a_2\, a_3$

is small and a_1 is nearly constant. Try as a solution $a_1(t) = a_1(0)$ and $a_{2/3}(t) = a_{2/3}(0)\ \exp(\alpha t)$, then satisfy

$$(\alpha + \nu_2)a_2(0) = \imath\ \lambda\ a_1(0)a_3^*(0) \tag{13.123a}$$

$$(\alpha + \nu_2)a_3(0) = \imath\ \lambda\ a_1(0)a_2^*(0) \tag{13.123b}$$

which have a solution if

$$(\alpha + \nu_2)\ (\alpha + \nu_3) = \lambda^2\ |a_1(0)|^2 \tag{13.124}$$

Clearly, α is real and the decay modes grow exponentially, $\alpha > 0$, if

$$\lambda|a_1(0)| > \sqrt{\nu_2\ \nu_3} \tag{13.125}$$

This condition for the threshold of instability, namely growing pumped waves, is found when the drive provided by pump just exceeds the overall losses of the decay waves. The ultimate growth of the waves is ultimately limited by energy conservation condition (13.120).

If the frequencies of the waves are not perfectly matched so that

$$\Delta\omega = \omega_1 - \omega_2 - \omega_3 \tag{13.126}$$

is the small (much less than the wave frequencies themselves) mismatch, we may easily include a correction term to Eqs. (13.122)

$$\dot{a}_2 + \nu_2 a_2 = \imath\ \lambda\ a_1 a_3^*\ e^{\imath\ \Delta\omega\ t} \tag{13.127a}$$

$$\dot{a}_3 + \nu_3 a_3 = \imath\ \lambda\ a_1 a_2^*\ e^{\imath\ \Delta\omega\ t} \tag{13.127b}$$

Substituting $a_j = \tilde{a}_j\ e^{\imath\ \Delta\omega/2}$, we find a solution if

$$(\alpha + \nu_2 + \imath\ \Delta\omega/2)\ (\alpha + \nu_3 - \imath\ \Delta\omega/2) = \lambda^2\ |a_1(0)|^2 \tag{13.128}$$

which gives an increased instability threshold

$$\lambda|a_1(0)| > \sqrt{\{\nu_2\ \nu_3[1 + (\Delta\omega)^2/(\nu_1 + \nu_2)^2]\}} \tag{13.129}$$

As we have seen in Chapter 12 plasma supports a complex set of waves, even when the un-magnetised. As a result if the amplitudes are sufficiently large a wide of wave–wave couplings are possible. In general these lead to the generation of turbulence and a complex spectrum of waves. We shall not investigate this behaviour further.

13.5.2 Non-linear Laser-Plasma Absorption

An important set of parametric interactions are those in which a powerful transverse wave, a beam of electro-magnetic radiation interacts with longitudinal plasma waves, namely the plasma wave and the ion-acoustic wave. The dispersion relations for these waves are shown schematically in Figure 2.5. It is immediately apparent that there is a very large disparity between the phase velocities of these waves. Consequently that coupling involving an electro-magnetic driver can only satisfy the matching relations (13.114) if the transverse wave frequency is close to the plasma frequency, i.e. close to the critical density of the plasma.

There are four distinct mechanisms by which a transverse electro-magnetic wave can non-linearly couple with two different waves. These are illustrated in Figure 13.4 which shows graphically the possibilities allowed by the matching conditions and the plasma dispersion relations. The interactions fall into two different sub-groups.

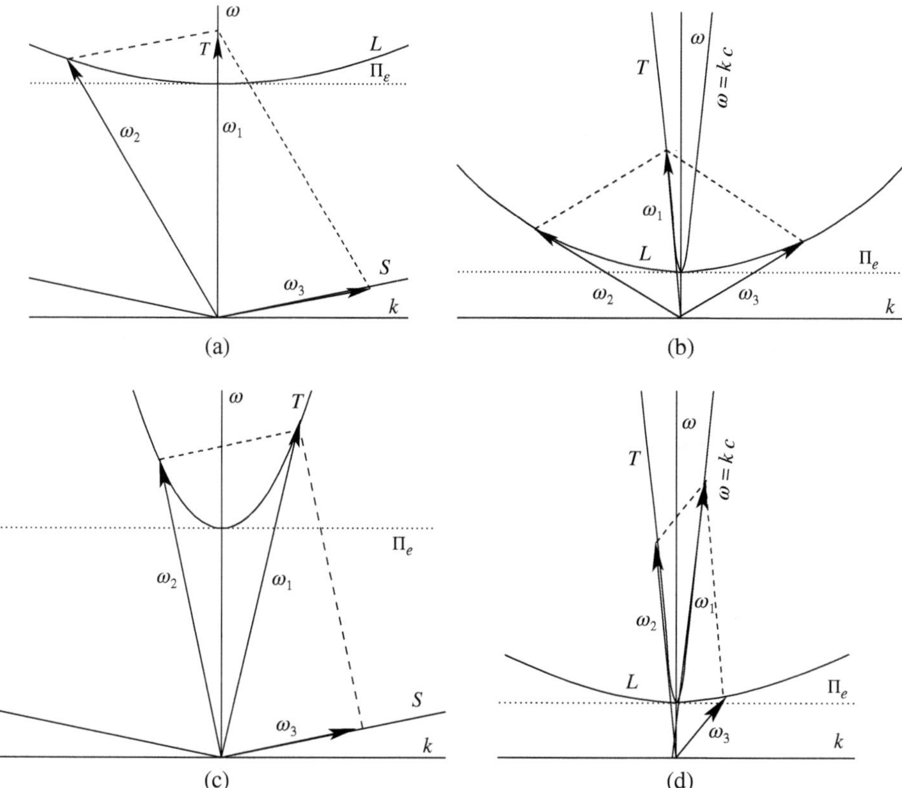

Figure 13.4 Sketches of the plasma dispersion relations showing the matching vector parallelograms for coupling the pump wave ω_1 to decay modes ω_2 and ω_3. (a) Plasma wave decay mode, (b) two plasmon decay mode, (c) stimulated Brillouin scattering, and (d) Stimulated Haman scattering.

13.5.2.1 Absorption Instabilities

1. *Parametric Decay Instability*: In this case, the transverse wave is coupled to two electro-static waves: a plasma wave and an ion sound wave. The low frequency of the sound wave $\omega_3 \ll \Pi_e$ demands that the transverse frequency and the plasma frequency are almost equal $\omega_1 \approx \Pi_e$ (Figure 13.4a). Hence, the instability occurs at the critical density where the pump frequency equals the plasma frequency. Clearly, the energy from the pump is distributed between the sound wave and the plasma wave: the larger fraction going to the latter. Since the wave number of the pump transverse electro-magnetic wave is very small, the wave numbers of the two decay waves must be nearly equal and opposite.
2. *Two plasmon decay instability*: In this case, the pump transverse wave couples to two longitudinal plasma waves. Due to its large phase velocity, the wave number of the pump wave is small. Consequently, the wave numbers of the decay waves must be nearly equal in magnitude but opposite in direction (Figure 13.4b). The frequencies of the two plasma waves are approximately equal to the local plasma frequency, and must therefore be half that of the pump. Consequently the instability occurs at quarter critical density of the pump and is consequently common known as the *quarter critical instability*.

13.5.2.2 Reflection Instabilities

a. *Stimulated Brillouin scattering*: In this case, two transverse waves is coupled to an ion sound wave. The incoming pump transverse wave ① splits into a backward (reflected) transverse wave ② and an ion sound wave ③. The low frequency of the sound wave $\omega_3 \ll \Pi_e$ demands that the transverse frequencies are almost equal $\omega_1 \approx \omega_2$ and exceed the plasma frequency to allow propagation (Figure 13.4c). Hence the instability occurs near the critical density where the pump frequency nearly equals the plasma frequency. The matching conditions require a forward going ion sound wave and a backward transverse.
b. *Stimulated Raman scattering*: In this case, the pump transverse wave ① couples to a second transverse wave ② and a longitudinal plasma wave ③. The pump wave is split into a reflected transverse wave and a plasma wave. The matching parallelogram is as shown in Figure 13.4d. This instability has the potential to be very damaging producing fast electrons through the collapse of the plasma wave. Stimulated Raman scattering plays an important role in many plasma accelerator, beat wave and wake field, schemes.

The driving non-linearity behind all these instabilities is the coupling induced by the ponderomotive force of the primary pump beam through the spatial density oscillation of one wave on the other. The derivation of the threshold for each of these instabilities is straightforward, but lengthy, involving the dynamics of electrons or ions subject to the ponderomotive forces induced by terms of the form $\langle|\mathbf{E}_1 \cdot \mathbf{E}_2|\rangle$ reflecting the coupling of the pump wave ① with one of the decay waves ② on the dynamics of the second decay wave ③. It is clear that the instability is driven when the resonance condition (13.114) is satisfied. As the mathematical details of the calculation of the instability thresholds are long, but not particularly difficult, we refer to the standard texts for their elucidation, Kruer (1988) and Lashmore-Davies (1975) (see also Lashmore-Davies, 1981, and references therein).

These instabilities are most important during the interaction of laser beams with plasma as for example required by inertial confinement fusion. Instabilities generating plasma waves are potentially damaging to the compression of the fuel pellet as a consequence of the fast electrons generated by the breaking plasma waves, which in the case of the two plasma decay mode may have phase velocities approaching that of light. Back scattering instabilities which give rise to a laser radiation losses are also damaging particularly in long scale plasmas used in 'hohlraum' for indirect direct drive. Of the two stimulated Brillouin scatter which in principle is not limited to the region close to critical density is the more severe. The interested reader should consult Kruer (1988) and Atzeni and Meher-ter-Vehn (2004) for more details.

Appendix 13.A Ponderomotive Force

Following Landau and Lifshitz (1984, §15) the force per unit volume in a fluid dielectric is given by a modified form of the Maxwell stress tensor, which takes into account the variation of the permittivity ϵ with density. Hora (1969) applied this result to plasma to obtain the general result for a plasma

$$\mathbf{F} = \nabla \cdot \mathbf{\Phi} \tag{13.A.1}$$

where the stress tensor

$$\mathbf{\Phi} = \left\{ -\frac{1}{2} \left[|\mathbf{E}|^2 \left(\epsilon - \rho \frac{\partial \epsilon}{\partial \rho} \right) + \frac{1}{\mu_0} |\mathbf{B}|^2 \right] I + \left(\epsilon\ \mathbf{EE} + \frac{1}{\mu_0} \mathbf{BB} \right) \right\} \tag{13.A.2}$$

where ρ is the mass density. The plasma permittivity is given by the familiar result (Eq. (13.20)) in terms of the refractive index $n^2 = \epsilon/\epsilon_0 = 1 - \Pi^2/\omega^2$ (Eq. (12.44)). Assuming the plasma is quasi-neutral, the electron density $n_e \propto \rho$ and the plasma frequency $\Pi^2 = n_e e^2/\epsilon_0 m$, we obtain

$$\rho \frac{\partial \epsilon}{\partial \rho} = -\frac{\Pi^2}{\omega^2} = 1 - \frac{\epsilon}{\epsilon_0} \tag{13.A.3}$$

so that the general expression for the stress tensor can be written

$$\boldsymbol{\Phi} = \left\{ -\frac{1}{2} \left[\epsilon_0 \, |\mathbf{E}|^2 + \frac{1}{\mu_0} |\mathbf{B}|^2 \right] I + \left(\epsilon \, \mathbf{EE} + \frac{1}{\mu_0} \mathbf{BB} \right) \right\} \tag{13.A.4}$$

which will be recognised as the Maxwell stress tensor

$$\mathsf{T} = \left\{ -\frac{1}{2} \left(\epsilon_0 |\mathbf{E}|^2 + \frac{1}{\mu_0} |\mathbf{B}|^2 \right) I + \epsilon_0 \mathbf{EE} + \frac{1}{\mu_0} \mathbf{BB} \right\} \tag{13.A.5}$$

for the vacuum stress plus a correction for the dielectric

$$\boldsymbol{\Phi} = \mathsf{T} - (\epsilon - \epsilon_0) \mathbf{EE} \tag{13.A.6}$$

In the absence of a dielectric $\epsilon = \epsilon_0$ and the stress is just the Maxwell stress $\boldsymbol{\Phi} = \mathsf{T}$ as expected.

To show the equivalence of this general result to our earlier result (Eq. (13.86)) for the ponderomotive stress we follow Hora (1971) and consider the behaviour of plane wave with amplitude E in a layer of varying refractive index $n(z)$ (Eq. (5.22))

$$E \approx E_0' \, n^{-1/2} \, \exp\left[\pm \imath \left(k_0' \int^z n \, \mathrm{d}z - \omega \, t \right) \right] \tag{13.A.7}$$

where E_0' and k_0' is the incident vacuum field and wave-number. The magnetic field is given by Maxwell's equation

$$H \approx H_0' \, n^{1/2} \, \exp\left[\pm \imath \left(k_0 \int^z n \, \mathrm{d}z - \omega \, t \right) \right] \tag{13.A.8}$$

where $H_0' = c \, E_0'$ is the vacuum magnetic field.

The ponderomotive force given by Eq. 13.A.4 is

$$\mathbf{F} = -\nabla \left[\frac{1}{2} \left(\epsilon_0 |\mathbf{E}|^2 + \frac{1}{\mu_0} |\mathbf{H}|^2 \right) \right] \tag{13.A.9}$$

Averaging over the period of the wave we obtain the time averaged ponderomotive force in agreement with Eq. (13.86)

$$\mathbf{F} = -\frac{\epsilon_0}{4} \nabla \left[\frac{(1+n^2)}{n} \, E_0'^2 \right] = \frac{\epsilon_0}{4} E_0'^2 \left[\frac{(1-n^2)}{n^2} \right] \nabla n = \frac{\epsilon_0}{4} \, E_0'^2 \, \frac{\Pi^2}{\omega^2 \, n^2} \nabla n = -\frac{\epsilon_0}{4} \, \frac{\Pi^2}{\omega^2} \nabla |\mathbf{E}'|^2 \tag{13.A.10}$$

The third expression is that obtained by Hora (1971) from the force associated with the drift due to the 'figure of eight' motion (Eq. (13.79)) in a plasma with a refractive index gradient.

14

Laser–Plasma Interaction

14.1 Introduction

The irradiation of gases or solids by intense laser radiation gives rise to local heating in which the target is vapourised and ionised. A dense plasma is formed within the gas or at the surface of a solid. The dynamics of the resulting plasma, although markedly different in the separate cases of gases and solids are very simply described by the methods of classical hydrodynamics (Pert, 2013, §13.I). The dissimilarities are due to the considerable differences in density in the plasma and their relationship with the critical density (Eq. (12.45)) at which the local plasma frequency equals the frequency of the incoming radiation, since electro-magnetic waves can only propagate when the electron density is less than the critical density (p. 277).

(a) *Gaseous targets*: When the target is a gas, the ionised medium normally has density less than the critical density. In consequence the gas initially breaks down at the point of maximum intensity, namely the focus in the rising pulse. As the intensity increases, the gas suffers a zone of energy deposition progressing towards the laser as the intensity successively exceeds the breakdown threshold. The hydrodynamic behaviour is therefore characteristic of a *detonation*, which is the model usually used to describe the hydrodynamics of *gas breakdown* supplemented by expressions for the breakdown threshold and collisional plasma absorption rate (Raizer, 1977, §20 and references therein). We shall not consider the behaviour during gas breakdown which is treated in detail by Raizer (1977).

(b) *Solid targets*: The density of a solid target is generally much larger than the critical density. As a consequence, the radiation cannot penetrate into the target and a zone of absorption is formed at the surface of the target. The resulting hot zone 'bores' its way into the target by thermal conduction vapourising and ionising the material. The hydrodynamics are essentially those of a classic *deflagration*. If the laser pulse is short the flow is not steady state as the time scale for thermal conduction and plasma evolution are comparable to that of the laser pulse.

At the basic hydrodynamic level, the theory of laser-matter interactions at low and moderate laser intensity can be treated as problems in fluid mechanics (see Pert (2013, Case study 13.1)). At this level laser-solid target interactions can be reduced to a series of self-similar models, each of which can be treated by an ordinary differential equation. In the first section of this chapter, we review and classify these models. Our aim is to identify the key physical phenomena which underlie the interaction of the laser beam with a solid target. We shall assume the mathematical aspects of the behaviour are treated in the series of papers referenced later. The models thus provide a simple qualitative background for laser-solid target interactions, but are not suitable for the detailed quantitative modelling essential for predictive purposes.

Foundations of Plasma Physics for Physicists and Mathematicians, First Edition. Geoffrey J. Pert.

Under high laser irradiance, these simple quasi-analytic models are no longer valid. Non-linear behaviour is induced by the high fields as we have briefly discussed in earlier chapters, amongst these are:

- The absorption of radiation from the electro-magnetic field into the plasma is no longer collisional and takes place through parametric other non-linear processes. These in turn lead to non-Maxwellian electron distributions.
- The large thermal gradients in the plasma induce non-Spitzer thermal conductivity and 'flux limitation'. The currents resulting from these electron flows give rise to magnetic fields.
- Mechanical forces induced by the strong fields such as the ponderomotive force modify the hydro-dynamics from that resulting from the particle pressure alone.
- Magnetic fields are generated by the currents associated with temperature and density gradients resulting from the focal structure of the laser beam, giving rise to magneto-hydrodynamic flows.

It will be appreciated that simple analytic models are inadequate to treat the plethora of interactions simultaneously taking place. The development of inertial confinement fusion (ICF) has required the construction of sophisticated computer models essential to represent the design and analyse experiments by including the complex simultaneous interaction of many processes (Atzeni and Meher-ter-Vehn (2004) give a useful survey of these applications).

14.2 The Classical Hydrodynamic Model of Laser-Solid Breakdown

High-intensity laser radiation breaks down cold unionised material initially by a multiphoton release of electrons leading to a low density of electrons. These are rapidly heated by local absorption or by inverse bremsstrahlung and in turn cause further ionisation. A rapid cascade thus occurs leading to a plasma with a high degree of ionisation. This breakdown takes different forms in gases and solids due to the markedly dissimilar background density with respect to the critical density.

When the material is a solid, the ambient density is typically about $10^2 - 10^3$ critical, so that $\rho_0 \gg \rho_{\text{crit}}$. The resulting plasma plume expands into vacuum, so that no downstream limitation on the flow occurs. The ambient temperature is very much less than that generated by the heating $T_1 \ll T_2$ and may be neglected. Absorption takes place in an expanding plasma plume as the material heats and expands. The flow therefore has the form of a deflagration. A shock, resulting from the high pressure at the front, precedes the heat front penetrating into the solid and initiates ionisation. A zone of heating and expansion from the high density follows the shock at the head of the deflagration structure, due to thermal conduction, before absorption takes place. The energy deposition is independent of the heat front and depends solely on the local thermodynamic state and the incident (external) energy flux.[1]

When an intense laser pulse is focused into gas or on to the surface of a solid, the material is initially ionised and rapidly forms a plasma within which strong absorption of the laser radiation occurs. The resulting high temperature gaseous medium subsequently undergoes rapid hydrodynamic motion with simultaneous heating from the absorbed laser pulse. A good approximation is obtained by treating the plasma as a polytropic gas with $\gamma = 5/3$ and constant ionisation Z. There are a range of possible flows depending on the wavelength and intensity of the laser and

1 This section is a summary of earlier work by the author Pert (1974, 1977, 1980, 1983, 1986a,b, 1987–1989, 1993) and references therein.

the nature of the material with which it interacts. These different flows take the form of steady and time dependent deflagrations depending on the conditions. The general structure of the hydrodynamic model of the laser–plasma interaction from solid targets may be developed through a series of self-similar models valid in different regimes. These simple models provide a qualitative understanding of the key factors and scalings which govern the development of plasma by laser irradiation, but are not sufficiently detailed for quantitative simulation.

14.2.1 Basic Parameters of Laser Breakdown

In order to have a broad understanding of this behaviour, we need to identify some key concepts controlling laser–plasma interaction, which are determined by the electron density and temperature

- In this section, we will treat the plasma as a single component fluid, the ion and electron temperatures being assumed equal.
- Energy is absorbed within the body of plasma from the laser beam, whose irradiance (energy flux) is w.
- Radiation can only propagate in plasma when the electron density is less than the critical density at which the laser frequency equals the characteristic oscillation frequency of the electrons in the plasma, Eq. (12.45).
- Provided the electron density is less than the critical density, the absorption coefficient is distributed through the plasma, due to electron/ion collisions, *inverse bremsstrahlung*, which in weak fields has value (13.101)

  $$\mu = b\,\rho^2\,T^{-3/2} = \mathfrak{b}\,\rho^2\,c^{-3} \tag{14.1}$$

  in terms of the mass density (ρ) and the temperature (T), which is represented by the isothermal sound speed $c \sim T^{1/2}$; the constant $\mathfrak{b}$ depends on the wavelength and plasma composition.
- In the neighbourhood of the critical density, absorption is local and due to a variety of non-linear mechanisms. Experimentally, this is found to result in strong absorption of typically 30% locally at the critical density of the incident radiation.
- Thermal conduction plays a crucial role transferring energy from the local absorption zone upstream to the ablation front, where the flow 'eats into' the solid material. The thermal conductivity (Eq. (2.44)) may be written in terms of the flow parameters as

  $$\kappa = a\,T^{5/2} = \mathfrak{a}\,c^5 \tag{14.2}$$

  As discussed in Section 2.2.2.1, the heat flux is limited to a value given by Eq. (2.47)

  $$q_{\text{lim}} = c\,\rho\,T^{3/2} = \mathfrak{c}\,\rho\,c^3 \tag{14.3}$$

  The laser heating pulse is specified by three parameters: the power W, the characteristic time of the heating pulse T, and a characteristic length, typically either the focal spot radius or the dimension of the target R. The irradiance or energy flux density is simply $w = W/r^\nu$, where ν is geometrical factor ($\nu = 0,\ 1,\ 2$ in planar, cylindrical, or spherical systems, respectively).

  It is useful at this stage to draw some elementary conclusions regarding the basic flow using dimensional analysis:

1. In the steady state, the energy flux of the outgoing plasma must balance the incoming heating flux $w \sim \rho c^3$

2. When the laser absorption is due to inverse bremsstrahlung distributed over the length R of the plasma, the optical depth $\mu R \sim 1$ must allow the beam to be fully absorbed (*self-regulating condition*)

$$\mu R \sim \mathfrak{b}[\rho^2/(w/\rho)]\, R \sim \mathfrak{b}\, \rho^3\, w^{-1}\, R \sim 1 \tag{14.4}$$

and the density is therefore

$$\rho \sim (w/\mathfrak{b}\, R)^{1/2} \tag{14.5}$$

3. Heat flow from the zone of absorption to the cold ablation front is due to thermal conduction whose scale length is easily established by an energy flow balance

$$\ell \sim T\Big/|\nabla T| \sim \kappa c^2\Big/\rho\, c^3 \sim \mathfrak{a}\, c^4\Big/\rho \sim \lambda_e \tag{14.6}$$

where λ_e is the electron mean free path. Since $w \sim \rho\, c^3$

$$\ell \sim \frac{\mathfrak{a}\,(w/\rho)^{4/3}}{\rho} \sim \mathfrak{a}\, \rho^{-7/3}\, w^{4/3} \tag{14.7}$$

4. In the simplest case, the absorption is local and due to non-linear processes at the critical density (*deflagration flow*). The appropriate numerical factor is easily found by numerical integration (Bobin, 1971; Pert, 1983). This upstream length is only weakly dependent on the flux limiting constant provided that its value $\mathfrak{c} > 3$. Smaller values, which give rise to discontinuity in the flow at the absorption surface (Pert, 1983), are not consistent with experiments.
5. Alternatively, if the heating is distributed by inverse bremsstrahlung the role of thermal conduction is still defined by the conduction length but with the density given appropriately by Eq. (14.5). We may eliminate the density ρ from Eq. (14.7) to get a value for the thermal conduction length in this case

$$\ell_{\text{dis}} \sim \mathfrak{a}\, \mathfrak{b}^{7/9}\, w^{5/9}\, R^{7/9} \tag{14.8}$$

From these results, we may form three dimensionless products whose values characterise the nature of the interaction as discussed in Section 14.2.2

$$\Phi = \rho/\mathfrak{d} = (w/\mathfrak{b}\, R)^{1/2}/\mathfrak{d}; \quad \Psi = \ell_{\text{loc}}/R = \mathfrak{a}\, \mathfrak{d}^{-7/3}\, w^{4/3}/R; \quad \Xi = \ell_{\text{dis}}/R = \mathfrak{a}\, \mathfrak{b}^{7/9}\, w^{5/9}/R^{2/9} \tag{14.9}$$

which represent in turn self-regulating/deflagration flow and thin/thick conduction layer in deflagration and self-regulating flows, respectively. These dimensionless products Φ, Ψ, Ξ therefore identify the different types of flow possible in this simple model. It will be obvious from our derivation that the dimensionless products are not independent. In fact it is easily shown that

$$\Phi^{7/3}\, \Psi^{-1}\, \Xi \sim 1 \tag{14.10}$$

14.2.2 The General Theory of the Interaction of Lasers with Solid Targets

Since there are two different absorption processes and two different thermal scale lengths, there are correspondingly four limiting forms of the deflagration structure depending on which parameters are dominant. Using dimensional analysis, we can identify which model presents the best approximation of the particular solution to the interaction of laser radiation with solid targets. Calculations of the flow in each of the limiting cases lead to soluble eigenvalue problems involving ordinary differential equations, and yield accurate numerical constants of proportionality. However, accurate calculations of the flow in the general case involving thermal conduction, and including flux limitation and absorption by both inverse bremsstrahlung and non-linear processes at the critical surface can be only be generated by numerical modelling. We may identify two important cases for which simple models can generate useful estimates of the plasma conditions in their appropriate regime and calculate the general physical characteristics expected in the interaction under different conditions. The values generated are broadly in agreement with those found in experiments within the regimes for which the models are applicable. However it should be noted that there are a number of plasma effects which are not included in the models and in many cases markedly change the nature of the interaction.

As before we identify a general set of characteristic constant parameters defining the interaction laser with plasma which may be summarised as follows:

- Material parameters
  a. Collisional absorption coefficient $\mathfrak{a}$
  b. Thermal conductivity coefficient $\mathfrak{b}$
  c. Flux limiting constant $\mathfrak{c}$
  d. Absorption (critical) density $\mathfrak{d}$
  e. Polytropic gas constant γ.
- Geometry. The experimental arrangement is taken into account in the three geometrical configurations by the parameter ν.
- Laser parameters
  a. Laser power W defined corresponding by the power density (flux), power per unit length and power per sterradian for each geometrical configuration, respectively.
  b. Scale length R – focal spot radius, cylinder or sphere radius.
  c. Temporal pulse length T.

Using dimensional analysis we may form from these variables, a set of dimensionless parameters:

$$\mathfrak{R} = \mathfrak{a}^3\ \mathfrak{b}^4\ \mathfrak{d}^5\ R$$
$$\mathfrak{T} = \mathfrak{a}^2\ \mathfrak{b}^3\ \mathfrak{c}^4\ T$$
$$\mathfrak{W} = \mathfrak{a}^{(3+3\nu)}\ \mathfrak{b}^{(3+4\nu)}\ \mathfrak{d}^{(2+5\nu)}\ W$$

to which we must add the flux limiting factor $\mathfrak{c}$, the polytropic constant γ, and the dimension factor ν. We may also include the alternative parameter, the dimensionless energy

$$\mathfrak{E} = \mathfrak{a}^{(5+3\nu)}\ \mathfrak{b}^{(6+4\nu)}\ \mathfrak{d}^{(6+5\nu)}\ W\ T = \mathfrak{W}\ \mathfrak{T}$$

It will be appreciated from our earlier discussion that we may identify a number of differing limiting cases, for example in steady state flow ($t \gg T$) we have the steady self-regulating (collisional absorption) and deflagration (local absorption) flows. Defining ρ_{coll} as the characteristic density at which collisional absorption would occur in a self-regulating flow, the distinctive flows are separated by the conditions $\rho_{coll} \underset{\ll}{\gg} \rho_{crit}$ and $R_0 \underset{\ll}{\gg}$, the first inequality identifying the absorption as either distributed collisional or local at the critical density and the second the dominant heat transport due to either absorbed energy or thermal conduction (thick or thin), respectively. Thus, we have after eliminating the appropriate variable the boundaries of the regime in which a particular physical effect dominates are given by[2]

$$\Psi^{3/4} = \mathfrak{W}\ \mathfrak{R}^{-(\nu+3/4)} \sim 1 \qquad \text{local/conduction}$$

2 Since $W = w\ R^{\nu}$ it is easy to see that these boundaries are identical to those identified in Eq. (14.9).

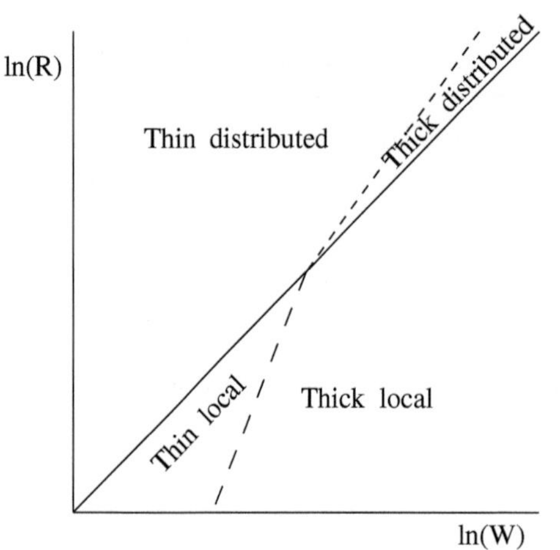

Figure 14.1 Sketch of the regime diagram for long pulses in the absence of flux limitation.

$$\Phi^2 = \mathfrak{W}\,\mathfrak{R}^{-(\nu+1)} \sim 1 \qquad \text{collisional/local}$$
$$\Xi^{9/5} = \mathfrak{W}\,\mathfrak{R}^{-(\nu+2/5)} \sim 1 \qquad \text{collisional/conduction}$$

The differing flow regimes may be simply illustrated by a diagram as plotted in Figure 14.1 for the long pulse steady flow condition (Pert, 1993). It will be appreciated that the regime boundaries cannot be exactly established as each model represents a limiting case and one merges into another. Representative numerical values for the boundaries can be determined by detailed modelling. The plot reveals some important general results

a. The most important regimes are thin distributed and thin local.
b. The thin local flow regime is enhanced by flux limitation decreasing the separation between the head of the heat front and the zone of deposition and any consequent transverse heat flow. If flux limitation is local and stronger than current experimental estimates, $\mathfrak{c} \leq 3$, a discontinuity is needed to account for the flow at the critical density (Pert, 1983).
c. The thick distributed flow regime is closely following that for local flow. This is due to weak absorption in the tail of the expansion fan, which drives the absorption density to approach its maximum value at $\rho \to \rho_{\text{crit}}$; and the heating therefore occurring near the critical density.
d. Flux limitation plays a role in thick distributed flow (Pert, 1993) as thermal conduction is necessary to support the downstream expansion fan; due to the decreasing density this rapidly becomes flux limited.

We now examine the two cases of practical importance in more detail.

14.2.3 Distributed Heating – Low Intensity, Self-regulating Flow

If the irradiation intensity is relatively weak, the temperature of the plasma in the plume is not sufficiently high that the downstream plasma is transparent. The entire incident energy flux w is deposited within the plasma body. The plume, being heated by inverse bremsstrahlung, delivers sufficient heat to maintain the expansion without the need for significant thermal conduction or absorption at the critical density. The absorption is determined by the optical depth, namely the product of the absorption coefficient and the length (μx). Since the plasma must be optically dense, the overall optical depth is of order unity. The spatial heat distribution within the plasma varies in time as the scale length of the plasma plume increases; and the flow is, in consequence, time varying. The density must decrease and the temperature increase with time along the plume. The flow self-regulates to maintain this relation. If the plasma is too hot or too tenuous, the optical depth is reduced, more radiation reaches the ablation surface, and increased ablation restores the *status quo*. Similarly *vice versa*, if the plasma is too cold or dense. The flow is therefore stabilised by the functional form of the absorption coefficient (Afanas'ev et al., 1966; Caruso et al., 1966).

14.2.3.1 Early Time Self-similar Solution

We consider the development of the plasma plume away from the solid surface initially under the influence of a beam of constant irradiance w. Since the problem involves only two characteristic parameters, namely b and w, it is expressible in a self-similar form with parameter

$$\xi = \frac{x}{b^{1/8}\, w^{1/4}\, t^{9/8}} \sim \frac{x}{c\, t} \tag{14.11}$$

representing the development of the plasma in space x and time t.[3] The overall flow may be expressed as a set of ordinary differential equations with independent variable ξ, subject to two-point boundary conditions at the head and tail of the expansion. These equations can be numerically integrated by standard methods and iterated on to the boundary conditions to obtain accurate scalings (Pert, 1986a,b). Simple scaling relations for the optical depth, flow speed, and energy follow from dimensional analysis

$$b\, \rho^2\, c^{-3}\, v\, t \sim 1 \qquad v \sim c \qquad w \sim \rho\, v\, c^2$$

The general scalings for the velocity, isothermal sound speed, density, and specific internal energy

$$v = b^{1/8}\, w^{1/4}\, t^{1/8}\mathfrak{V}(\xi) \quad c = b^{1/8}\, w^{1/4}\, t^{1/8}\mathfrak{C}(\xi) \quad \rho = b^{-3/8}\, w^{1/4}\, t^{-3/8}\mathfrak{D}(\xi) \quad I = b^{1/4}\, w^{1/2}\, t^{1/4}\mathfrak{I}(\xi) \tag{14.12}$$

In Figure 14.2, we show the general solution for the one-dimensional self-regulating expansion of a plasma into vacuum away from an infinite slab. The dynamic variables are given in terms of their dimensionless forms as functions of the dimensionless distance ξ. It is evident that the scaling constants are typically of order unity.

The self-similar description may be trivially generalised to treat cases where the irradiance varies as a power law with time. Thus if the total energy delivered per unit area $E = \varepsilon\, t^n$, the self-similar parameter is written as

$$\xi = \frac{x}{b^{1/8}\, E^{1/4}\, t^{7/8}} \tag{14.13}$$

The numerical constants in Eq. (14.12) take values appropriate to the value of n (Pert, 1986a). This extension to the basic theory is useful in dealing with the initial development of the plasma as the irradiance increases with time, for example assuming the irradiance increases linearly with time as $w = w'\, t$.

14.2.3.2 Late Time Steady-State Solution

In practical situations, the purely one-dimensional solution given above fails when the length of the plasma becomes comparable with the focal spot width R_0 (Caruso and Gratton, 1968; Puell, 1970). This occurs approximately at the sonic point where the flow velocity equals the (isothermal)

---

3 It is obvious that neither x nor t can be cast into dimensionless forms using b and w alone. The dimensionless forms of the dynamic variables, density, velocity, pressure, etc. must therefore depend on the single dimensionless variable ξ alone. The form of the dynamic variables retain the same shape, but their scales change with time; hence, they are similar to one another; for more details see Pert (2013).

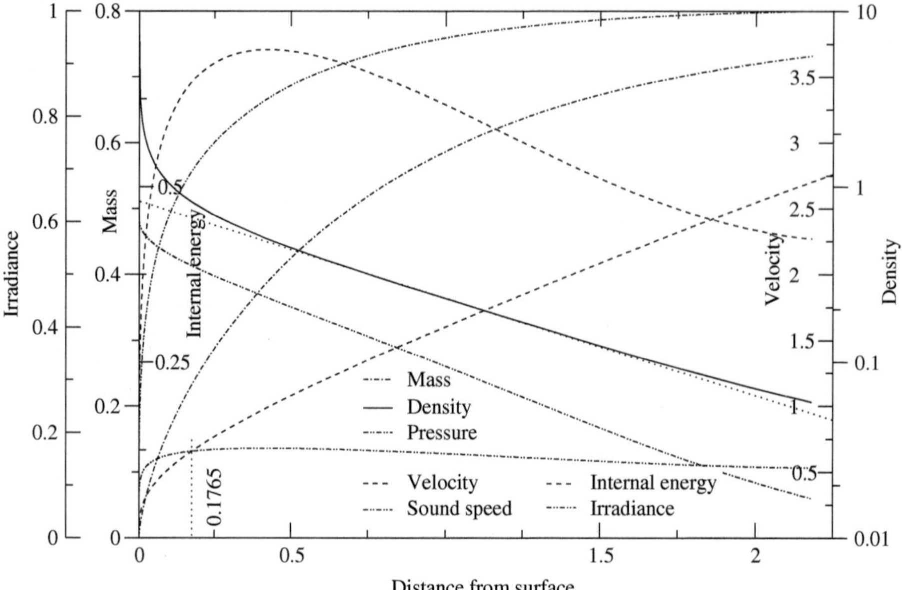

Figure 14.2 Solutions of the one-dimensional self-regulating expansion from a solid under constant irradiance. The local values of the flow parameters: mass integral $\mathfrak{M}$, density $\mathfrak{D}$, pressure $\mathfrak{P}$, velocity $\mathfrak{V}$, sound speed $\mathfrak{C}$, internal energy $\mathfrak{I}$, and irradiance $\mathfrak{W}$ are plotted in dimensionless form as functions of the dimensionless distance from the ablation front ξ. The isothermal sonic point where the velocity equals the isothermal sound speed is shown at $\xi = 0.1765$.

sound speed and the outward flow can change from a planar to a radial expansion; in so doing, the flow changes from subsonic to supersonic. The absorption in the plasma downstream of the sonic point rapidly decreases allowing a steady flow to form. The transition to radial flow at the sonic transition takes place at a distance of approximately equal to the focal spot radius from the target. Downstream the density and therefore the absorption coefficient decrease rapidly with distance. The optical depth is approximately limited to the value from the sonic point to the target. In a similar manner, the flow is stabilised by the self-regulating condition

$$b\,\rho^2\,c^{-3}\,R_0 \sim 1$$

and a steady state is in consequence established.[4] The situation is similar in cylindrical geometry ($\nu = 1$) where a linear beam is focused on to a fibre of radius R_0, or in spherical ($\nu = 2$) where a radially convergent beam impacts a sphere of radius R_0. An additional parameter namely the focal spot width R_0, is introduced into the scaling replacing the time and taking account of the steady flow:

$$\rho \sim b^{-1/3} W^{1/3} R_0^{\,-(\nu+1)/3} \qquad c \sim b^{1/3} W^{2/9} R_0^{\,-(2\nu-1)/9} \tag{14.14}$$

allowing for the geometry of the focus. The power W is modified accordingly to be power per unit area, power per unit length, or the total power. The scaling constants are of order unity, and values can be calculated from the hydrodynamic model (Pert, 1986b, 1989).

The characteristic time $t \sim r_0/c$ is a measure of the time taken to establish the steady state and its temporal stability.

4 A new dimensional parameter R_0 is introduced which allows a dimensionless form for the distance x/R_0. The self-similar form is lost and a steady state solution becomes possible.

14.2.4 Local Heating – High Intensity, Deflagration Flow

If the radiation is intense, the electron temperature is high and inverse bremsstrahlung plays little part in absorption. The energy deposition is therefore localised at the critical density. No energy can be deposited from the laser beam upstream of the critical density. Initially before a plasma plume has time to evolve, laser energy is deposited at the target surface. Heat diffuses through non-linear thermal conduction heating the plasma at solid density before the hydrodynamic motion has started. Ahead of the thermal front, a shock wave is generated, and behind a rarefaction is established in which the density falls to the critical density. Upstream of the critical density, a zone of strong thermal conduction with thickness $\ell = \kappa_{\text{crit}}\ T_{\text{crit}} / \frac{1}{2}\ \rho_{\text{crit}}\ c^{\ 3}_{\text{crit}} = 2\ \mathfrak{a}\ c^{\ 4}_{\text{crit}}$ heats the fluid behind the shock (Fauquignon and Floux, 1970; Bobin, 1971). The fluid is progressively heated within this region and expands to flow smoothly and steadily into the zone of local absorption at the critical density.

14.2.4.1 Early Time Thermal Front

The initial stages of a thermal front before the steady state sets in are due to direct heating at or near the solid surface before hydro-dynamic motion is established. The laser energy is absorbed at close to the solid density and the heat transported into the solid by thermal conduction. Assuming the material is rapidly ionised a steep thermal front is propagated into the solid with a speed which changes with time. On the other hand, the development of the hydro-dynamic front is determined by the first characteristic, whose speed varies with the temperature. If the characteristics propagate faster than the thermal front, the flow will develop into the characteristic deflagration as the critical density layer becomes well established. More commonly the thermal front initially moves ahead of the characteristics, until the hydrodynamic front overtake the thermal and eventually dominates the expansion (Babuel-Peyrissac et al., 1969; Caruso and Gratton, 1969; Anisimov, 1971; Salzmann, 1973).

The flow of heat in a non-linear medium such a plasma is described by the thermal conductivity, Eq. (14.2), which varies a simple power law in temperature. Expressed in terms of the specific internal energy ϵ the thermal diffusivity $\chi = \mathfrak{a}\,\epsilon^n$. Energy is absorbed at a rate depending on time such that the flux of energy per unit area at the surface increases the total specific internal energy in the medium at time t by $w = S_0\, t^m$ per unit area. Provided it is initially cold, the problem contains only two-dimensional parameters $\mathfrak{a}$ and S_0 and can be reduced to a self-similar form describing the propagation of internal energy ϵ into the medium. The self-similar variable takes the form

$$\xi = x / [\mathfrak{a}\ S_0^{\ n}\ t^{(nm+n+1)}]^{1/(n+2)} \tag{14.15}$$

and the internal energy parameter

$$\epsilon = (S_0^{\ 2}\ t^{(2m+1)} / \mathfrak{a})^{1/(n+2)}\ f(\xi) \tag{14.16}$$

Heat therefore diffuses with an unchanging spatial profile moving outwards with a speed decreasing in time as $t^{m(n-1)}$. When $n > 1$, this profile has a sharply rising front at a characteristic value of the similarity variable ξ_0 (Figure 14.3); Appendix 14.A summarises the analysis of this behaviour.

The speed at which the thermal front propagates into the plasma is given by

$$v_T = \left.\frac{dx}{dt}\right|_{\xi_0} = \frac{nm+n+1}{n+2}\ \xi_0\,[\mathfrak{a}\ S_0^{\ n}\ t^{(nm-1)}]^{1/(n+2)} \tag{14.17}$$

The mean specific energy

$$\bar{\epsilon} \approx [S_0 t^{m+1}/(m+1)] \Big/ \xi_0\,[\mathfrak{a} S_0^{\ n}\ t^{(nm+n+1)}]^{1/(n+2)} = (S_0^{\ 2}\ t^{(2m+1)}/\mathfrak{a})^{1/(n+2)} \Big/ [(m+1)\ \xi_0] \tag{14.18}$$

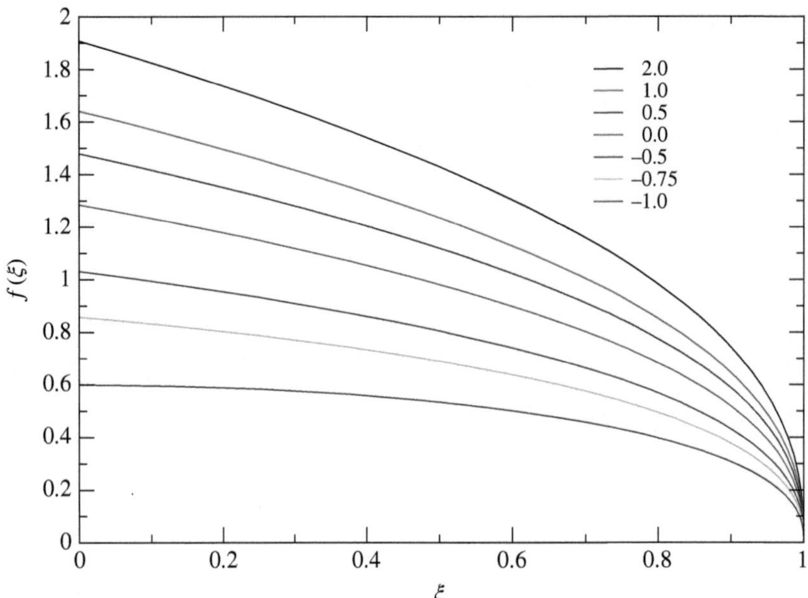

Figure 14.3 Solutions of the self-similar non-linear diffusion model in dimensionless units for different values of the time power m when the conductivity power $n = 5/2$ and the head of the thermal wave is set at $\xi_0 = 1$.

determines the sound speed $c \approx \sqrt{\gamma\,(\gamma-1)\,Z\,\overline{\epsilon}/M}$, where Z and M are the charge number and mass of the background ions, respectively and $\gamma = 5/3$ the polytropic constant. For plasma, the power index $n = 5/2$, hence the distance travelled by thermal wave

$$x_T = \xi_0\ \mathfrak{a}^{1/(n+2)}\ S_0^{n/(n+2)}\ t^{(nm+n+1)/(n+2)} = \xi_0\ \mathfrak{a}^{2/9}\ S_0^{5/9}\ t^{2(5m+1)/9} \tag{14.19}$$

and the distance travelled by leading characteristic

$$\begin{aligned} x_S &= \frac{2(n+2)}{(2n+2m+5)} \sqrt{\frac{\gamma\,(\gamma-1)\,Z}{M}}\ \mathfrak{a}^{-1/[2(n+2)]}\ S_0{}^{[n/2(n+2)]} t^{(2n+2m+5)/2(n+2)} \sqrt{\xi_0} \\ &= [9/2\,(m+5)]\ \sqrt{10/9}\ \mathfrak{a}^{-1/9}\ S_0{}^{2/9}\ \sqrt{Z\,\xi_0/M}\, t^{2(m+5)/9} \end{aligned} \tag{14.20}$$

The characteristic therefore meets the head of the diffusion front when the time

$$t' \approx [3\,\sqrt{10}/2\,(m+2)]\sqrt{Z/M\xi_0}[\mathfrak{a}\ S_0]^{-3/8(m-1)} \tag{14.21}$$

There are three regimes (see (Figure 14.4) depending on the value of m:

(a) $m < 1$: Thermal diffusion dominated flow is initially established. After time t' when the waves meet (the flow becomes a shock wave hydrodynamic motion) propagating into the solid. The thermal wave becomes subsumed into the hydrodynamic flow away from the solid, forming the deflagration discussed in Section 14.2.4.2.

(b) $m = 1$: The pulse rises linearly with time, it follows that $v_t \sim c \sim t^{1/3}$. The two waves are therefore 'parallel' and never meet. The leading wave is determined by the relative magnitudes of the coefficients of x_T and x_S.

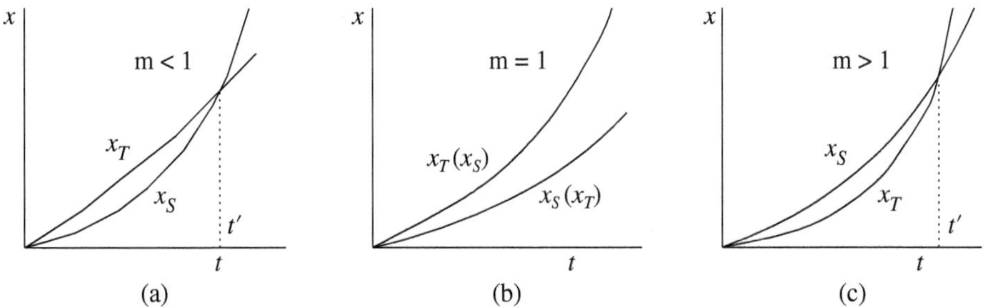

Figure 14.4 Sketch to illustrate the possible velocity arrangements of the diffusion and hydrodynamic waves during the initial phase of laser heating of a solid target.

(c) $m > 1$: The precursor thermal wave is not formed, and a shock wave (hydrodynamic motion) propagating into the solid is established from the start of the pulse. The thermal wave is totally subsumed into the hydrodynamic flow, forming a deflagration as the growth in the heating pulse decreases.

14.2.4.2 Late Time Steady-State Flow

When the flow is established and the pulse has become settled, a classical deflagration in which the upstream plasma is heated by thermal conduction from the absorption zone and drives a shock wave into the solid ahead of the heat front. It may be argued (Pert, 1974) that downstream the deflagration is a Chapman–Jouget flow and that the exit velocity is equal to the local sound speed. This model assumes radiation is absorbed only in the neighbourhood of the critical density, no heat is deposited further downstream, and the subsequent flow must be a rarefaction. Hence, following our earlier argument, the downstream flow must be sonic. However, the plasma is likely to be sufficiently hot that the rarefaction is maintained at uniform temperature by thermal conduction, i.e. it is an isothermal rarefaction, which is driven by a downstream heat flow $p_2 v_2$ and the downstream exit velocity is consequently the isothermal sound speed $v_2 = c_2$. The deflagration is described by the modified Rankine–Hugoniot equations with the Chapman–Jouget condition applied at the isothermal sound speed at its downstream end. Neglecting the heat loss to the rarefaction, the solution for the exit velocity and pressure are then

$$\rho_2 = \rho_{\text{crit}} \qquad v_2 = c_2 = \sqrt[3]{\frac{2\,(\gamma - 1)\,w}{(3\,\gamma - 1)\,\rho_{\text{crit}}}} \qquad p_2 = \frac{1}{\gamma}\left[\frac{2\,(\gamma-1)}{(3\,\gamma-1)}\right]^{2/3} \rho_{\text{crit}}^{1/3}\, w^{2/3}$$

w the heat deposited per unit time per unit area in the deflagration is a fraction of that incident due to reflection at the critical density.

Taking into account the heat conducted downstream necessary the isothermal expansion fan, namely $\rho_{\text{crit}}\, v_2\, \epsilon_2$

$$p_2 = \left[\frac{2(\gamma-1)}{(5\gamma-3)}\right]^{2/3} \rho_{\text{crit}}^{1/3}\, w^{2/3} \tag{14.22}$$

Provided the thickness of the conduction zone is small compared to any characteristic scale length, as is generally the case, the problem may be treated as one dimensional. In which case it is reasonably easy to calculate the flow structure (Bobin, 1971; Pert, 1983).

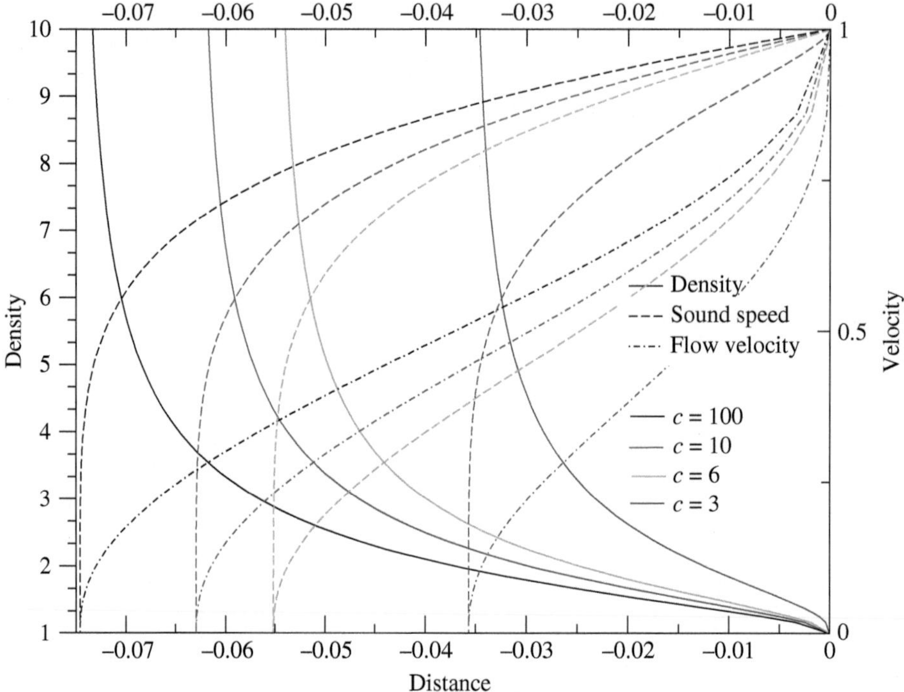

Figure 14.5 Solutions of the one dimensional steady deflagration model in dimensionless units. The distance is plotted as x/ℓ, the velocities as v/V_2 and the density as ρ/ρ_{crit}.

Figure 14.5 shows typical profiles for different values of the flux limiting parameter $\mathfrak{c}$ using the harmonic form Eq. (2.49). When $\mathfrak{c} > 3$, the thermal conduction flux upstream can accommodate the convective energy flow with an increased gradient as the flux limiting factor is reduced $\mathfrak{c} \to 3$ downstream. However, when $\mathfrak{c} < 3$ upstream conduction can no longer match the required convective flow and the temperature profile is discontinuous at the absorption point. Fortunately estimated values of the flux limiter normally exceed this value, and discontinuities do not generally occur due to this effect. For $\mathfrak{c} > 3$, we note that the thickness of the deflagration front is dependent on the degree flux limitation taking place.

The temperature profile shows the characteristic steep drop-off associated with non-linear thermal conduction as the temperature, and therefore the conductivity, become small at the head of the deflagration (Zel'dovich and Raizer, 1966; Pert, 1977).

14.2.5 Additional Simple Analytic Models

There are two further simple analytic models which have found quite wide application for treating simple laser expansions.

14.2.5.1 Short Pulse Heating

The behaviour when the incident laser pulse has a duration which is short compared to any thermal or hydrodynamic times, such as the meeting time t', Eq. (14.21). In this we may simply modify the analysis of Section 14.2.4.1 to consider the case where an energy Q_0 is deposited at the surface (Babuel-Peyrissac et al., 1969; Caruso and Gratton, 1969). As shown in Appendix 14.A, this corresponds to the singular case $m = -1$ and is easily treated by the transformations $S/B \to Q/b$ and $b = 1$ in the general formulae. Thus, the thermal wave is overtaken by the hydrodynamic wave when the time

$$t > [3\sqrt{10}/2]\sqrt{Z/M\xi_0}[a\ Q_0]^{3/8} \approx 3.8903(Z/M)^{1/2}\ (a\ Q_0)^{3/8\|} \tag{14.23}$$

After this time, the development is dominated by the hydrodynamic wave, which drives a strong shock wave upstream into the undisturbed gas and a rarefaction wave downstream (Anisimov, 1970). The total flow is described by the 'short impact model' of gas under the action of an impulsive load (Zel'dovich and Raizer, 1966, Ch.XII,§4).

14.2.5.2 Heating of Small Pellets – Homogeneous Self-similar Model

If the scale length for the heat front at the onset of developed hydrodynamic flow, namely $x_T(t')$ is larger than the initial dimension of the body x_0, for example a small pellet or foil, then the heat front penetrates through the body before it has time to disassemble. Although this condition is not usually met, the simple self-similar model in which the plasma expands isothermally with a uniform linear velocity gradient has been found to give good agreement with experiment in studies of the heating and expansion of small pellets.

The various properties make the model easy to apply to many experimental situations using relatively simple standard numerical integration methods (Appendix 14.B). Since the model is applicable to an arbitrary heating pulse, additional effects such as the development of ionisation are easily included. The solutions are asymptotic to the exact flow. It is necessary to apply a matching condition to generate an approximate model of the development from the initial configuration. This is usually accomplished by equating the mass of the real and model flows and setting the initial $1/e$ width to a fraction of the pellet radius. The value of this fraction is determined either by numerical simulation (Fader, 1968) or by entropy considerations (Pert, 1987): the differences between the resulting solutions being relatively unimportant (Pert, 1980).

14.3 Simulation of Laser-Solid Target Interaction

As has been pointed out, the simple models describe the key underlying hydrodynamic aspects of the laser-solid interaction process; however, most of the detailed physics is necessarily omitted to generate a simpler representative problem, which may be tackled analytically. As a result, we may identify key features which must be included in any simulation that can give accurate estimates of the plasma state. However to obtain an accurate representation of any experiment, we must include much of the physics described in Sections 7.6, 8.3, 8.4, 13.3, 13.4, and 13.5.

Figure 14.6 shows a schematic of the arrangement of a typical laser–plasma simulation code based on a hydrodynamic core. As may be expected from the simple models described above the key element is the hydrodynamic core simulation using a two temperature magneto-hydrodynamic description of the dynamics of the flow based on the equations derived in Chapter 8. These are integrated over a series of small time-steps on a computational mesh on which the fluid state and dynamic variables are specified at a finite set of points within the spatial computational domain. They may be used in a number of different finite difference schemes using a computational mesh to set up the basic simulation.

- *Eulerian scheme*: The mesh is stationary in the laboratory frame. Eulerian schemes are normally orthogonal and reasonably simple to set up in an arbitrary number of dimensions, but have problems if fine resolution is required, for example in shock-waves.

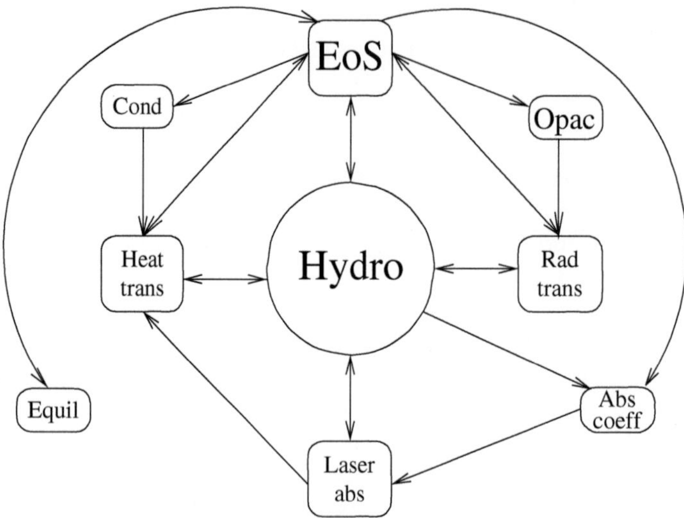

Figure 14.6 Sketch to illustrate the key interactions in the simulation of a laser–plasma interaction.

- *Lagrangian scheme*: The mesh is fixed in the fluid and consequently moves with it thereby mitigating the resolution problem. Lagrangian schemes are excellent in one dimension, but have problems in two or more with cell points overtaking. Periodic re-zoning is essential. The finite difference forms are more complex due to the non-orthogonality of the mesh.
- *Arbitrary Lagrangian–Eulerian schemes*: ALE (arbitrary Lagrangian–Eulerian) schemes attempt to combine the best features of the previous two methods using weak regular weak re-zoning on a non-orthogonal mesh. They are the best methods currently available, despite their obvious problems.

Hydro. The core of the code controls the simulation. Additional physics is 'hung on to it' as required in such a way that it interacts to influence the flow. This behaviour is clearly illustrated in Figure 14.6 which shows how most of the important additional terms interact often through a second key element

Equation of state (EoS). This key element generates pressures, and electron and ion temperatures more generally than using the ideal gas model as before. Since the state of system is specified by two thermodynamic variables through the equation of state (EoS), the density and specific internal energy for both electrons and ions generated by the hydro code are used to obtain the remaining terms. The EoS may be specified either by an analytic approximation or by interpolated tables. Ionisation/recombination may be included either in steady or a non-steady situation. These values of density and temperature are used to calculate

- *Thermal heat transport* through the thermal conductivity.
- *Radiative heat transport* through the opacity.
- *Ion-electron equilibration* via collisions
- *Laser absorption* including collisional and non-linear terms.

This list is not exclusive: for example as we have not included the effects associated with the magnetic field, which greatly complicate many of those outlined here through the non-parallel Hall terms.

Appendix 14.A Non-linear Diffusion

It follows from dimensional analysis that the general form of self-similar parameter can be written as

$$\xi = x \left\{ \left[\frac{(n+2)}{(2\ m+1)\ A} \right]^{(n+1)} \mathfrak{a} \left(\frac{S_0}{B} \right)^n t^{(n\ m+n+1)} \right\}^{1/(n+2)} \tag{14.A.1}$$

Separating the dependence on the similarity variable, we obtain that the internal energy can be expressed as

$$\epsilon = \left\{ \left[\frac{(n+2)}{(2m+1)\ \mathfrak{a}\ A} \right] \left(\frac{S_0}{B} \right)^2 t^{(2\ m+1)} \right\}^{1/(n+2)} f(\xi) \tag{14.A.2}$$

This form introduces two arbitrary scaling constants A and B , but reduces to the standard form, Eqs. (14.15) and (14.16) when $A = (n+2)/(2\ m+1)$ and $B = 1$.

Substituting for ϵ in the equation of thermal diffusion, $f(\xi)$ satisfies the differential equation

$$A \frac{\mathrm{d}}{\mathrm{d}\xi} \left(f^n \frac{\mathrm{d}f}{\mathrm{d}\xi} \right) = f - \frac{(nm+n+1)}{(2\ m+1)} \frac{\mathrm{d}f}{\mathrm{d}\xi} \tag{14.A.3}$$

subject to one of the two alternative forms of the boundary conditions

$$f^n \left. \frac{\mathrm{d}f}{\mathrm{d}\xi} \right|_{\xi=0} = -B \qquad \text{or} \qquad \int f\ \mathrm{d}\xi = \frac{2\ m+1}{(n+2)(m+1)} A\ B \tag{14.A.4}$$

It is easily shown that laboratory parameters x and ϵ are independent of the values of A and B,[5] which introduces a level of flexibility useful in calculation. In particular, it is convenient to integrate over the range $0 \leq \xi < 1$ (Pert, 1977), which may be converted to the primary set by the transformation $\xi \to B^{-n/(n+2)}\ \xi$ when B is given by Eq. (14.A.4) and $A = (n+2)/(2m+1)$. Values of B obtained when $\xi_0 = 1$ are given in Table 14.A.1

The characteristic behaviour of non-linear diffusion with a diffusivity parameter $\mathfrak{a} > 0$ is clearly seen from Figure 14.3 where the flow parameter rapidly falls to zero once its value becomes small, approaching the ξ abscissa at right angles. The limiting similarity parameter for the head of the wave ξ_0 is readily obtained from the equation for ξ, Eq. (14.A.1) once the value of B is known from either of the alternative forms of the boundary conditions, (14.A.4).

Table 14.A.1 Values of the similarity parameter at the head of the thermal wave ξ_0 when $n = 2.5$ for different values of the time power index m.

m	**−1.0**	**−0.75**	**−0.5**	**0.0**	**0.5**	**1.0**	**2.0**	**5.0**	**10.0**
B	0.0	0.160 95	0.377 45	0.923 34	1.582 2	2.330 6	4.046 1	10.500	24.340
ξ_0	—	2.758 8	1.718 2	1.045 3	0.775 0	0.625 0	0.460 0	0.270 8	0.169 8

5 Let us transform $A \to \alpha\ A$ and $B \to \beta\ B$. In consequence ξ and f transform to $\xi \to X\ \xi$ and $f \to Y\ f$. It follows from Eqs. (14.A.3) and (14.A.4) that $\alpha = X^2/Y^n$ and $\beta = Y^{(n+1)}/X$ or $\alpha\ \beta = X\ Y$. Hence $X = (\alpha^{(n+1)}\ \beta^n)^{1/(n+2)}$ and $Y = (\alpha\ \beta^2)^{1/(n+2)}$. The resultant transformations of x and ϵ are therefore obtained from Eqs. (14.A.1) and (14.A.2)

$$x \to X/\{\alpha^{(n+1)/(n+2)}\ \beta^{n/(n+2)}\}\ x = x \qquad \text{and} \qquad \epsilon \to Y/\{\alpha^{1/(n+2)}\ \beta^{2/(n+2)}\} = \epsilon$$

The dimensional parameters x and ϵ are therefore independent of the values of the arbitrary constants A and B: a result consistent with the tenets of dimensional analysis.

When the time index is negative $m < 0$, the power is initially infinite, which is clearly unphysical. However provided the index is larger than $m \geq -1$ the total energy delivered at the edge $Q_0 = S_0/(m+1)$ remains finite. The case $m = -1$, corresponding to a constant delivered energy, is important representing the instantaneous deposition of energy Q_0 per unit area on the surface. When $m = -1$, the power S_0 has the same dimensions as the total energy Q_0. Retaining the two arbitrary scaling constants A and B, we may eliminate the singularity by formally taking the limit $\lim_{m \to -1} S_0/B = Q_0/b$. The self-similar parameter ξ can therefore be expressed in terms of the total energy Q_0 instead of the delivered flux S_0

$$\xi = x \left\{ \left[\frac{(n+2)}{[2\,(-A)]} \right]^{(n+1)} \mathfrak{a} \left(\frac{Q_0}{b} \right)^n t \right\} \tag{14.A.5}$$

and the internal energy

$$\epsilon = \left\{ \left[\frac{(n+2)}{\mathfrak{a}\,(-A)} \right] \left(\frac{Q_0}{b} \right)^2 t^{-1} \right\}^{1/(n+2)} f(\xi) \tag{14.A.6}$$

Substituting for $S_0 = \lim_{m \to -1} Q_0\,(m+1)$ as before and introducing the scaling constants A and $B = \lim_{m \to -1} b\,(m+1)$ we obtain the modified form of Eq. (14.A.3)

$$A \frac{\mathrm{d}}{\mathrm{d}\xi} \left(f^n \frac{\mathrm{d}f}{\mathrm{d}\xi} \right) = f + \frac{\mathrm{d}f}{\mathrm{d}\xi} = \frac{\mathrm{d}(\xi f)}{\mathrm{d}\xi} \tag{14.A.3:a}$$

subject to $f(\xi_0) = 0$ and the modified integrated form of Eq. (14.A.4)

$$\int f\,\mathrm{d}\xi = \frac{1}{(n+2)}\,b\,(-A) \tag{14.A.4:a}$$

As before the constants A and b are arbitrary and chosen for convenience. The normal laboratory forms are obtained with $b = 1$ and $A = -(n+2)$. Equation (14.A.3:a) has an analytic solution

$$A\,f^n \frac{\mathrm{d}f}{\mathrm{d}\xi} = \xi\,f + \cancelto{0}{\text{const.}} \tag{14.A.7}$$

Integrating again

$$\frac{(-A)}{n} f^n = \frac{1}{2}(\xi_0^2 - \xi^2) \tag{14.A.8}$$

the boundary condition

$$\begin{aligned} \int f\,\mathrm{d}\xi &= \int_0^{\xi_0} \left[\frac{n}{2\,(-A)} ({\xi_0}^2 - \xi^2) \right]^{1/n} \mathrm{d}\xi = \frac{1}{2} \left(\frac{n}{2\,(-A)} \right)^{1/n} {\xi_0}^{(1+2/n)} B(1/2, (1+1/n)) \\ &= \frac{1}{n+2}\,b\,(-A) \end{aligned} \tag{14.A.9}$$

where $B(a,b) = \Gamma(a)\,\Gamma(b)/\Gamma(a+b)$ is the beta function. If $n = 2.5$ and $(-A) = (n+2)$

$$\int f\,\mathrm{d}\xi = 0.489\,787\,128\,7\,{\xi_0}^{9/5} = b \tag{14.A.10}$$

Hence, if $b = 1$ then $\xi_0 = 1.486\,682\,156$

The average energy per particle at time t

$$\overline{f(\xi)} = \int_0^{\xi_0} f\,\mathrm{d}\xi \Big/ \xi_0 = \frac{2m+1}{(n+2)\,(m+1)}\,A\,B \Big/ \xi_0$$

$$= \frac{2m+1}{(n+2)\,(m+1)}\,A\,B\left\{\left[\frac{(n+2)}{(2\,m+1)\,A}\right]^{(n+1)}\,\mathfrak{a}\,\left(\frac{S_0}{B}\right)^n\,t^{(n\,m+n+1)}\right\}^{1/(n+2)}\Big/x_0$$

$$= \left(\frac{(2m+1)}{(n+2)}\,\mathfrak{a}\,A\right)^{1/(n+2)}\,B\left(\frac{S_0}{B}\right)^{n/(n+2)}\,t^{(nm+n+1)/(n+2)}\Big/[(m+1)\,x_0] \tag{14.A.11}$$

$$\overline{\epsilon} = h(t)\,\overline{f(\xi)}$$

$$= \left[\frac{n+2}{(2m+1)\,\mathfrak{a}\,A}\left(\frac{S_0}{B}\right)^2\,t^{2m+1}\right]^{1/(n+2)}\,\left[\frac{(2m+1)}{(n+2)\,(m+1)}\,A\,B\right]\Big/\xi_0$$

$$= \left[\frac{n+2}{(2m+1)\,\mathfrak{a}\,A}\left(\frac{S_0}{B}\right)^2\,t^{2m+1}\right]^{1/(n+2)}$$

$$\times\left(\frac{(2m+1)}{(n+2)}\,\mathfrak{a}\,A\right)^{1/(n+2)}\,B\left(\frac{S_0}{B}\right)^{n/(n+2)}\,t^{(nm+n+1)/(n+2)}\Big/[(m+1)\,x_0]$$

$$= \frac{S_0}{B}\,t^{m+1}\Big/[(m+1)\,x_0] \tag{14.A.12}$$

Appendix 14.B Self-similar Flows with Uniform Velocity Gradient

The model is based on a self-similarity parameter such that we are able to separate the dynamic variables into the product of a function depending only on time and a second on the similarity variable

$$\xi_i = x_i \Big/ X(t) \tag{14.B.1}$$

where the set x_i are the position vectors of the Lagrangian fluid cells ζ_i. Clearly, the set of variables ξ_i are fixed for a particular fluid element, i.e. the set ξ_i are a Lagrangian variable and its value identifies a particular fluid element.

General properties of the model follow:

1. *The fluid velocity v_i is linearly dependent on the spatial co-ordinate x_i:*
   The velocity of a fluid particle ξ_i is

$$v_i = \left.\frac{\mathrm{d}x_i}{\mathrm{d}t}\right|_{\xi_i} = \xi_i\frac{\mathrm{d}X_i}{\mathrm{d}t} = \frac{x_i}{X_i}\,V_i \tag{14.B.2}$$

   where $V_i = \mathrm{d}X_i/\mathrm{d}t$.
2. *The flow is symmetric in the co-ordinates ξ_i:*
   The flow must have symmetry in the co-ordinates x_i, since $v_i(-x_i) = -v_1(x_i)$ etc.
3. *The model is homogeneous*:
   The mass of a cell dM fixed in the fluid of volume $\mathrm{d}\tau_x$ and the elementary volume $\mathrm{d}\tau_\xi$ of the Lagrangian co-ordinate element ξ_i are constant in time. The Jacobian of the transformation between the configuration and Lagrangian spaces

$$\frac{\mathrm{d}\tau_x}{\mathrm{d}\tau_\xi} = J = \frac{\partial(x_1, x_2, \dots)}{\partial(\xi_1, \xi_2, \dots)} = \prod_{i=1}^{\nu} X_i \tag{14.B.3}$$

Since $\mathrm{d}M = \rho \, \mathrm{d}\tau_x$ = const it follows that $\rho \, J$ is an invariant of motion. Furthermore, since J is a function of time alone, the density ρ is a separable function of time and the Lagrangian space co-ordinates (similarity variables)

$$\rho = \rho_0(t) \, f({\xi_i}^2)$$

where $f({\xi_1}^2, {\xi_2}^2, \dots)$ is represented as $f({\xi_i}^2)$ allowing for the symmetry of the function $f({\xi_i}^2)$ noted above. The temporal variation is determined by the Jacobian $\rho_0(t) = J(0)/J(t) \cdot \rho_0(0)$. The rate of dilation $\mathrm{d}[\ln(\rho)]/\mathrm{d}t = \dot{J}(t)/J(t)$ is a function of time alone and therefore has the same spatial value everywhere. The flow is therefore homogeneous as the ratio of the volume elements in configuration space for fluid elements of identical size depends only on factor varying in time.

4. *The thermodynamic variables are all separable*:
   Therefore, density, pressure, temperature take the form $\rho(x_i, t) = \rho_0(t) f \xi_i)$ Euler's equation determines the density and also the pressure distribution. The equation becomes

$$\frac{\partial p}{\partial \xi_i} = -\xi_i \, f({\xi_i}^2) \, \rho_0(t) \, X_i(t) \, \frac{\mathrm{d}V_i}{\mathrm{d}t} \tag{14.B.4}$$

   and the pressure is necessarily a separable function.

$$p = p_0(t) \, \phi({\xi_i}^2)$$

   where $\phi({\xi_1}^2, \, {\xi_2}^2, \dots)$ is represented by $\phi({\xi_i}^2)$. It follows from the EoS that all other thermodynamic state variables are separable as well. Therefore,

$$\frac{\partial[\phi({\xi_i}^2)]}{\partial({\xi_i}^2)} = -\frac{1}{2} \, \lambda_i \, f({\xi_i}^2) \qquad \text{and} \qquad p_0(t) = {\lambda_i}^{-1} \, \rho_0(t) \, X_i \, \frac{\mathrm{d}V_i}{\mathrm{d}t} \tag{14.B.5}$$

   for all i. λ_i are separation constants, which have an arbitrary value and usually determine the relationship of the characteristic scale width X_i to the fluid boundary.

5. *The profiles of the thermodynamic variables are determined by the heating pulse*:
   The equation of energy conservation in Lagrangian co-ordinates can be written

$$\frac{\mathrm{d}\epsilon}{\mathrm{d}t} - \frac{p}{\rho^2} \frac{\mathrm{d}\rho}{\mathrm{d}t} = Q \tag{14.B.6}$$

   where Q is the heat release rate per unit mass and ϵ the specific internal energy. Clearly Q must be separable

$$Q = Q_0(t) \, q({\xi_i}^2) \tag{14.B.7}$$

   For a polytropic gas where $\epsilon = \frac{p}{[(\gamma-1) \, \rho}]$

$$q({\xi_i}^2) = \mu \, \phi({\xi_i}^2)/f({\xi_i}^2)$$

$$Q_0(t) = \mu^{-1} \left\{ \frac{1}{(\gamma-1)} \, \rho_0(t) \, \dot{p}_0(t) - \frac{\gamma}{(\gamma-1)} \, p_0(t) \, \dot{\rho}(t)/[\rho(t)]^2 \right\} \tag{14.B.8}$$

   where μ is a separation constant. This equation together with Eq. (14.B.5) define the pressure and density distributions from $q({\xi_i}^2)$.

The condition that the heat release function be separable into the product of a function of time $Q_0(t)$ and one of the variable $q({\xi_i}^2)$ is therefore clearly a necessary condition for the self-similar condition (14.B.1) to be upheld, and can also be shown to be sufficient. Flows of this type are therefore self-similar even though the number of dimensional variables may be larger than is normally the case.[6]

The mass M and the total energy $E(t)$ in the gas at any time, which is the sum of the kinetic E_k and thermal E_t energies, are easily found

$$M = \int \rho \mathrm{d}\tau = [J\rho_0(t)] \int f(\xi_i^2)\, \mathrm{d}\tau_\xi$$

$$E_k = \int \rho \frac{1}{2} \sum_{i=1}^{\nu} v_i^2\, \mathrm{d}\tau = \frac{1}{2}[J\rho_0(t)] \int f(\xi_i^2) \sum_{i=1}^{\nu} \xi_i^2\, V_i^2\, \mathrm{d}\tau_\xi \tag{14.B.9}$$

$$E_t = \int \rho\, c_v\, T\, \mathrm{d}\tau = \frac{1}{(\gamma - 1)} \int p\, \mathrm{d}t = \frac{1}{\nu}[J\, p_0(t)] \int f(\xi_i^2) \sum_{i=1}^{\nu} \xi_i^2 X_i \frac{\mathrm{d}V_i}{\mathrm{d}\tau}\, \mathrm{d}\tau_\xi$$

since the pressure is zero at the gas boundary.

Ellipsoidal flows are a useful sub-class of flows, which are particularly easy to evaluate. Provided the initially specified variables have no preferred direction, the flow is spherical in ξ_i space but ellipsoidal in configuration space x_i. The initial shape of the body may be ellipsoidal, rather than spherical, so that X_i and V_i take different values, but the heat distribution function and the state variables have no preferred direction in Lagrangian space. Thus, these variables are functions of the Lagrangian 'radius'

$$\zeta^2 = \sum_{i=1}^{\nu} \xi_i^2 = \sum_{i=1}^{\nu} \left(\frac{x_i}{X_i}\right)^2$$

The density distribution function f depends on ζ^2 only

$$\int \xi_i^2\, f(\xi_i^2)\, \mathrm{d}\tau_\xi = \frac{1}{\nu} \int \zeta^2\, f(\zeta^2)\, \mathrm{d}\tau_\xi$$

Defining

$$\Psi = \frac{1}{2} \frac{\int \zeta^2\, f(\zeta^2)\, \mathrm{d}\tau_\xi}{\int f(\zeta^2)\, \mathrm{d}\tau_\xi}$$

we obtain

$$\sum_{i=1}^{3} \left\{ V_i^2 + \frac{2}{\nu\,(\gamma - 1)}\, X_i \frac{\mathrm{d}V_i}{\mathrm{d}t} \right\} = \nu\, \frac{E(t)}{\Psi\, M} \tag{14.B.10}$$

where $E(t)$ is the total energy of the gas.

This equation taken together with Eq. (14.B.5) in the form

$$X_1 \frac{\mathrm{d}V_1}{\mathrm{d}t} = X_2 \frac{\mathrm{d}V_2}{\mathrm{d}t} = \cdots = \lambda\, \frac{p_0(t)}{\rho_0(t)} \tag{14.B.11}$$

uniquely defines the flow, in a convenient form for numerical integration. The fluid may have a finite initial energy $E(0)$, and the temporal form of the heating pulse $Q_0(t)$ is subject to the condition that $\lim_{(t\to 0)} Q_0(t) \to A\, t^n$, $n > -1$.

---

6 The dimensional parameters include the initial dimensions of the pellet and the heating pulse duration. The simpler problem with an infinitely small mass and heating pulse increasing as a power law in time is easily shown by dimensional analysis to yield the same self-similar form.

The condition that the spatial heat release function $q({\xi_i}^2)$ is constant is met in two simple cases. In ellipsoidal flow, defining the ellipsoidal variable $\zeta^2 = \sum {\xi_i}^2$, there are two simple conditions consistent with the separability of the heating pulse.

- *Adiabatic expansion*: The pellet is instantaneously heated isentropically and expands adiabatically $Q(x, t) = 0$ so that

$$f(\zeta^2) = f_0\,(1 - \zeta^2/{\zeta_0}^2)^{1/(\gamma-1)} \tag{14.B.12}$$

- *Isothermal expansion*: If thermal conduction is strong the expanding pellet maintains a uniform temperature and the density profile takes the form of a Gaussian

$$f(\zeta^2) = f_0 \exp(-\zeta^2/{\zeta_0}^2) \tag{14.B.13}$$

and ζ_0 is the half $1/e$ width of the profile.

Case Study 14.1

The Fluid Dynamics of Inertial Confinement Fusion

14.1.1 Basic Principles

In this section, we briefly discuss the underlying physics of ICF. This is essentially a problem of applied hydrodynamics and involves many topics covered earlier. Plasma physics also plays a role, but is generally subordinate to the hydrodynamics at the introductory level outlined here.

ICF relies on the ability of a pellet of fusion material (usually a 50:50 mixture of deuterium and tritium, DT) to undergo fusion before disassembling by expansion. This requires the fluid to be at high temperature $\gtrsim$ 10 keV $\approx 10^7\,°K$. Making use of Eq. (14.B.10) we may estimate the disassembly time of a sphere of uniformly heated plasma of sound speed c and radius R from the time taken for the density to decrease by half as

$$\tau \approx R/3\,c \tag{14.14}$$

since the fluid is a plasma with equal numbers of ions and electrons under these conditions $(E/M = 3c^2)$.

The rate at which fusion reactions occur in the plasma is determined by the average of the rate product $\langle \sigma v \rangle$, where σ is the cross section for fusion and v the random thermal velocity, over the velocity distributions of the deuterium and tritium ions. It is found that in the temperature range 20–50 keV, the ratio of $\langle \sigma v \rangle$ and c is approximately constant and consequently the fractional burn-up of the fusion constituents is

$$f \approx \rho R/(6 + \rho R) \tag{14.15}$$

taking into account the depletion during the burn, where the density ρ is measured in g/cm^3 and the radius R in cm. In practice, depletion limits the fractional burn to about 35%. To achieve these values requires a density/radius product $\rho R > 0.3$ g/cm^2. The burn-time can be increased by restricting the expansion of the fuel by enclosing it in a heavy shell, known as a *tamp*. However, this requires considerable input energy and is consequently inefficient, although other design constraints may introduce a measure of tamping.

The thermo-nuclear gain is defined as

$$g = \frac{\text{Fusion energy yield}}{\text{Initial thermal energy}} \tag{14.16}$$

A simple uniformly heated sphere has a gain of only about 50, which is insufficient to overcome the losses inherent in the pumping power source and generator. Typically values $\sim 10^4$ are required for an effective power plant.

The parameter ρR is a measure of the collision probability for particles within the plasma body. If a region with $\rho R > 0.3$ is generated at the centre of the pellet, escaping α-particles and photons suffer collisions and transfer energy to heat a colder surround of DT mixture to fusion temperatures and thereby induce a burn, which propagates through the pellet, *hot spot ignition*. In this case, only the central region need be heated to fusion temperatures, the outer parts remaining relatively cold. The gain may thus be greatly increased.

The energy released by burning 1 g of DT fuel is 3×10^5 MJ equivalent to about 75 Mtonnes of the high explosive trinitrotoluene (TNT). The largest amount of energy which can be expected to be handled routinely and safely is about 100 MJ (25 kg of TNT). This corresponds to a pellet mass of 0.3 mg of fuel. Since we have argued that we require a ρR product of not less than 0.3 g/cm^2 and $M \approx \rho R^3 = (\rho R)^3/\rho^2$, we see that the density of the pellet must be $\approx 10^2$ g/cm^3 or a compression over liquid DT of about 5×10^2. Under these conditions, the pressure in the hot spot is about 10^{12} atm and even in the cooler outer regions about 10^9 atm. These values are clearly far in excess of those achievable mechanically.

14.1.1.1 Hydrodynamic Compression

The material in this section covers the underlying principles of ICF which were treated in the author's earlier work (Pert, 2013). The following is consequently an almost direct quotation from that text.

Table 14.1 Compression ratio for a collapsing shock.

γ	δ	ρ_4/ρ_1	ρ_R/ρ_1	γ	δ	ρ_4/ρ_1	ρ_R/ρ_1
5/3	0.688 38	9.55	30.92	1.6	0.694 19	11.06	41.11
1.5	0.704 43	14.39	69.87	7/5	0.717 18	20.07	143.06
1.3	0.733 78	31.27	402.37	9/7	0.736 65	33.76	439.65

We have seen it is possible to achieve high compression during the collapse of spheres and shells. The compression achievable in a collapsing shock is only about 30, Table 14.1. On the other hand if the collapse is achieved adiabatically very large compression may be obtained. This is a consequence of the fact that the entropy generated in a shock leads to a temperature rather than a density increase, by which the pressure is raised. The generation of shocks in the collapsing pellet therefore leads to a reduction in compression. A purely adiabatic collapse leads to a cold core, which is ideal for the outer cold region into which the burn may propagate. However it is essential to design a central hot spot, for example by generating a shock propagating into this region by a suitable design of the drive pulse. Although the original proposal

(Continued)

Case Study 14.1 (Continued)

for inertial fusion was based on spheres, it was quickly realised that multi-layered shells were a more effective approach, and have been used subsequently. It can be shown that collapsing spheres are much more sensitive to deviations from the ideal pressure pulse than shells (a consequence of the fact that significant compression can occur during the coasting compression phase); and, that provided it is smooth, the profile of the applied pulse is not too critical. In addition the energy required to compress a sphere is significantly greater than that needed for a shell. Current designs for inertial fusion targets are all based on multiply layered shells.

A typical fusion pellet consists of a thin inner layer of frozen DT on the inner shell. The outer layers of the shells are in two parts. An inner layer of relatively heavy material to act as a shield preventing X-rays or fast electrons penetrating the fuel and raising its temperature to prevent high compression being achieved. An outer layer of lighter material is used to generate the pressure. The outer layers also act as a tamp, helping to increase the disassembly time of DT fuel.

The pellet and drive pulse must be carefully engineered to overcome a number of problems, essentially related to uniformity and stability. The shells and the drive beams must be uniform to a few percent to allow a uniform collapse and high compression. Hydrodynamic instabilities, principally the Rayleigh–Taylor instability (Section 16.4) at interfaces undergoing acceleration, destroy the uniformity of the collapse, and therefore present a serious problem and must be limited or avoided.

In Section 14.2.4 we showed that large pressures can be generated by the ablation of material away from a surface. This is essentially the same as the rocket effect where the momentum transfer from the escaping burnt fuel generates the pressure. In principle any source of heat rapidly deposited at the surface generates such a pressure, which may be used to compress the fuel, provided the rate is sufficiently large. However the pressure generated by these methods is not sufficiently large to balance that needed the pressure generated in the burning fuel. The required pressure multiplication is the result of two factors, namely the convergence of the collapsing shell and the rapid release of momentum accumulated over a long drive period. Several methods have been proposed to provide the drive source

a. *Direct drive laser heating*: In this case the laser directly irradiates the surface of the outer layer of the shell. From Eq. (14.22) we see that higher pressure is obtained with higher critical density, i.e. shorter wavelength lasers. As a result recent design of direct drive fusion is based on laser pulses of 0.35 μm wavelength. Due to the fact that the drive is applied at the critical density which is significantly less than solid, the interface is susceptible to the Rayleigh–Taylor instability. Direct drive also suffers from a number of problems associated with plasma instabilities and care must be taken with the magnitude of the pulse intensity. These effects are also mitigated by the use of short wavelength lasers.
b. *Indirect drive laser heating*: As we noted above higher pressures are achieved with shorter wavelength radiation. This raises the possibility of using soft X-rays for the drive, where the absorption is due to photo-ionisation and the critical density much greater than solid.

The Rayleigh–Taylor instability is thereby avoided. By placing the pellet in a hot enclosure, namely a *hohlraum*, filled with thermal black body radiation, the uniformity of illumination may (in principle) be improved (Lindl, 1993). A major disadvantage of this approach is the inefficiency resulting from the energy required to generate the X-rays.

In recent years research has tended to be concentrated around indirect drive, but significant interest is maintained in direct drive. A book describing the status of ICF is by Atzeni and Meher-ter-Vehn (2004) (see also Hooper, 1995).

15

Magnetically Confined Plasma

15.1 Introduction

The most important problem in current plasma physics research concerns the ability to maintain a plasma at a sufficiently high density and temperature that fusion reactions can take place over a long enough time that electrical power can be generated on a commercial basis. At present, there are two distinct methods envisaged by which such a scenario may be obtained. If the plasma is maintained in a stable, quasi-static equilibrium by confining magnetic fields, magnetic confinement fusion (MCF) may occur. Alternatively, if the plasma is highly compressed, the necessary lifetime of the plasma may be sufficiently short that the burn is completed before the plasma disassembles: inertial confinement fusion (ICF).

In this chapter, we examine the conditions necessary to allow magnetic confinement to occur using the ideal magnetohydrodynamics model of plasma (Chapter 9). We have discussed inertial confinement in Chapter 14. Magnetic confinement encompasses two aspects both of which must be satisfied. Firstly the plasma must be held in equilibrium by the magnetic field. Secondly the plasma must be stable against the growth of small perturbations (instabilities). The establishment of an equilibrium plasma does not ensure its stability. To avoid the problems associated with cooling at the plasma ends, linear devices are impracticable. As a result, experimental devices are toroidal, with two main approaches: tokamaks and stellarators. The former have a relatively simple magnetic configuration and are considered to be the most likely practicable device. The latter are characterised by complex magnetic geometry, but have the advantage that an internal plasma current is not needed and the plasma may be steady state. We shall not discuss stellarators in this book and refer to accounts elsewhere. As we have seen in Section 9.1.3.1, the virial theorem requires that steady equilibrium magnetic confinement can only be achieved in systems where an external magnetic field is provided.

15.2 Equilibrium Plasma Configurations

Equilibrium in a plasma constrained by the magnetic field requires the following conditions to be satisfied by the magnetic induction and the current density

$$\nabla \wedge \mathbf{B} = \mu_0 \mathbf{j} \qquad \nabla \cdot \mathbf{B} = 0 \qquad \nabla \cdot \mathbf{j} = 0 \qquad \mathbf{j} \wedge \mathbf{B} = \nabla p \tag{15.1}$$

Foundations of Plasma Physics for Physicists and Mathematicians, First Edition. Geoffrey J. Pert.

It follows directly from these conditions that

$$\nabla \cdot \mathbf{j} = 0 \qquad \mathbf{B} \cdot \nabla p = 0 \qquad \mathbf{j} \cdot \nabla p = 0 \qquad (\mathbf{B} \cdot \nabla)\mathbf{B}/\mu_0 = \nabla(p + B^2/2\,\mu_0) \tag{15.2}$$

It can be seen that equilibrium can only be obtained if the gradient of the plasma pressure is normal to both the current and the magnetic field. We may therefore define an *isobaric surface* on which the pressure is constant, and both the current and the magnetic field lie in the surface. The resulting *magnetic surface* encloses a confined plasma in equilibrium. More generally, we may imagine the plasma to be made of a series of nested isobaric or magnetic surfaces lying inside one another (like an onion) as the pressure increases through the plasma.

Since both the current and magnetic induction are everywhere single valued, their field lines cannot intersect, but both must lie in the magnetic surface. Clearly, lines of $\mathbf{j}$ and $\mathbf{B}$ cannot be parallel. If the surface is simply connected (e.g. spherical), Ampère's theorem implies that there must be a net current through any cut through the surface, which is impossible. More generally if such a surface lies in a bounded region of space, when neither $\mathbf{B}$ nor $\mathbf{j}$ vanish on the surface, the isobaric surfaces are tori. The surfaces are therefore multiply connected, the simplest forms being an infinite linear cylinder and a simple toroid. Clearly, inside the magnetic surface there must lie a set of isobaric surfaces, each of toroidal form mirroring the external surface. As we move deeper into the contained plasma, the pressure at each surface increases until the *magnetic axis* is reached at which the pressure maximises. Clearly, the Lorentz force $\mathbf{j} \wedge \mathbf{B}$ must be directed inwards. A more complete introduction to the properties of magnetic surfaces is given by Rose and Clark (1961, §7.2).

15.3 Linear Devices

It is easy to see that when the plasma is contained within a uniform cylinder of radius a an equilibrium configuration is obtained when the magnetic field and current both lie on the surface of the cylinder. Assuming cylindrical symmetry pressure balance requires

$$\frac{\mathrm{d}}{\mathrm{d}r}\left[p + \frac{(B_\theta{}^2 + B_z{}^2)}{2\mu_0}\right] = -\frac{B_\theta{}^2}{\mu_0\, r} \tag{15.3}$$

and the current is

$$\mu_0 \mathbf{j} = \left[0, -\frac{\mathrm{d}B_z(r)}{\mathrm{d}r}, -\frac{1}{r}\frac{\mathrm{d}B_\theta(r)}{\mathrm{d}r}\right] \tag{15.4}$$

in a cylindrical co-ordinate system $(r,\ \theta,\ z)$ with r the radial co-ordinate normal to the axis and z along the axis of symmetry. An alternative form of the pressure balance is obtained from Eq. (15.3) by multiplying the equation by r^2 and integrating over the radius a

$$\overline{p} + \frac{1}{2\,\mu_0}(\overline{B_z{}^2} - B_z{}^2(a)) = \frac{1}{2\,\mu_0} B_\theta{}^2(a) \tag{15.5}$$

where the averages of the kinetic and magnetic pressures are

$$\overline{p} = \frac{2\,\pi}{\pi\, a^2}\int_0^a p(r)\, r\, \mathrm{d}r \qquad \text{and} \qquad \overline{B_z{}^2} = \frac{2\,\pi}{\pi\, a^2}\int_0^a B_z{}^2\, r\, \mathrm{d}r \tag{15.6}$$

This useful result is also applicable to toroidal plasmas when the curvature of the field lines is small, i.e. the major radius is much larger than the minor $R_0 \gg a$. Alternatively, we may regard it as the zeroth approximation of the equilibrium condition in a toroid in an expansion in terms of $a/R_0 \ll 1$ (Appendix 15.B).

In the absence of any initial field, the current flows in a narrow layer at the outer edge of the plasma, a consequence of the high plasma conductivity. Two basic configurations are allowed.

a. *Theta pinch*: When the current flows azimuthally around the cylinder (θ direction), the self-generated magnetic field is parallel to the axis of the cylinder (z direction). The Lorentz force is directed inwards, opposing the thermal pressure of the plasma. In practical devices, a rapid discharge from a high voltage capacitor bank is taken through a single turn coil surrounding a gas filled enclosure. Breakdown occurs in the gas near the outer wall of the surrounding vacuum vessel and the resulting plasma rapidly moves inwards to the axis driving a shock wave into the gas. The collapsing plasma at the axis forms a high temperature, dense plasma. In the absence of an initial internal magnetic field, the current at the plasma surface is equal to and opposite that in the coil and the internal field remains zero; the plasma therefore behaves as a perfect diamagnetic fluid. The theta-pinch is a relatively simple device in principle capable of generating high temperature, high density plasma suitable for thermo-nuclear application. Unfortunately, it suffers from a number of problems, which severely limit any such application. The use of magnetic mirrors to mitigate the loss of plasma through the open ends has not proved successful. Rotation induced in the plasma during the collapse, for example by the asymmetry in the current drive to the coils, leads to instability in the plasma column. Both these effects and the transient nature of the current drive limit the lifetime of the plasma below that needed for practical fusion (Glasstone and Lovberg (1960) give a good survey of early experimental work on these devices). Although little used now, the device has found application in fields, such as spectroscopy, where the transient nature of the plasma is not inconvenient.
b. *Z pinch*: When the current and field are interchanged so that the current is parallel to the axis (z direction) and the resulting magnetic field is azimuthal (θ direction), the Lorentz force is again directed inwards. An extremely simple device in which the current is obtained from a pair of electrodes connected to rapid high current discharge. In this case also the current flows in a narrow layer at the surface of the plasma provided there is no internal field. The lifetime of the plasma is clearly limited by the duration of the discharge. The operation of this device is very similar in that the gas breaks down at the surface of the containing vessel, and the resulting plasma is driven into the axis. At the collapse, a hot, dense plasma is formed (see problem). In principle, after several bounces the plasma settles down to a steady state confined by the magnetic field, at which time a simple relation connects the current and temperature (problem). Unfortunately, the plasma column is unstable and the lifetime of the plasma is consequently short. Although the device has no direct value for fusion applications, very rapid discharge Z pinches find uses in a number of areas, where the transient nature of the plasma is unimportant. Very fast, high current Z-pinches have proved to be very efficient X-ray generators. A thorough review of the status of Z-pinches is given by Haines (2011).
c. *Screw pinch*: A generalisation of the theta and Z-pinches is the case where the magnetic field has both B_θ and B_z non-zero is known as a screw pinch. The gradient of the field lines

$$r\frac{\mathrm{d}\theta}{\mathrm{d}z} = \Theta = \frac{B_\theta(r)}{B_z(r)} \tag{15.7}$$

defines the rotational transform[1] over a length $2\pi R_0$, namely the angle a field line at a radius r rotates through about the z axis over a length $2\,\pi\,R_0$

$$\iota = \int_0^{2\pi R_0} \frac{\mathrm{d}\theta}{\mathrm{d}z}\mathrm{d}z = \frac{2\,\pi\,R_0\,B_\theta(r)}{r\,B_z(r)} \tag{15.8}$$

1 Confusingly the rotational transform is conventionally denoted by the Greek letter ι which should not be confused with the mathematical symbol for imaginary quantities $\imath$. The appropriate term is normally easily distinguished by context.

which is used to define the safety factor[2]

$$q(r) = 2\pi/\iota = r\,B_z(r)/R_0\,B_\theta(r) \tag{15.9}$$

Screw pinches introduce greater flexibility to the magnetic field configuration and may be designed with improved stability over the simple B_θ and B_z devices. However they find little application in a linear form due to the unconfined nature of the plasma at the ends.

15.4 Toroidal Devices

The principal objection to the use of linear systems for an equilibrium plasma system is the open ends of the cylinder in a finite device. The most obvious solution to this problem is to bend the cylinder into a torus, thereby automatically closing the ends.

A torus is formed by the rotation of a circle of radius a about a circle of radius R_0 forming a tube resembling a tyre (Figure 15.1).[3] The radius R_0 is known as the major radius and its axis the major axis; the radius a is the minor radius (see Figure 15.4) and its centre describes the minor axis along a circle of the major radius. The ratio of the major radius to the minor radius is known as the aspect ratio, so that the inverse aspect ratio $\varepsilon = a/R_0$ is normally a small parameter. The global volume distributed about the major axis defines the toroidal surface. Cross sections normal to the minor axis define the poloidal surfaces, which are assumed to be identical around the torus.

The designation of a point within the torus can be described by a set of global co-ordinates (R, ϕ, Z) or equivalently by those associated with the poloidal surface (r, θ, ϕ).[4] The two sets are entirely equivalent right-handed orthogonal sets of unit vectors. ϕ is known as the *toroidal angle* and is the rotational angle in the plane of the torus. θ is known as the *poloidal angle* and is the polar angle in the poloidal plane with respect to the toroidal plane. The position of a point within the toroid is therefore given by the co-ordinates (R, ϕ, θ) where R is the distance from the centre $R = R_0 + r\cos\theta$, where the distance from the minor radius to the point is r.

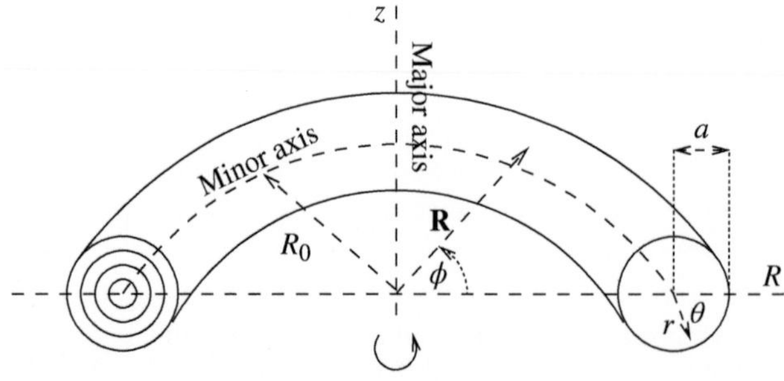

Figure 15.1 Torus showing the principal geometrical parameters. The unit vectors $(\hat{\mathbf{R}}, \hat{\phi}, \hat{\mathbf{z}})$ and $(\hat{\mathbf{r}}, \hat{\theta}, \hat{\phi})$ form two linked right-handed orthogonal sets.

---

2 The safety factor is conventionally denoted by the symbol q as is the general symbol for charge q. We have used different fonts to distinguish them.

3 As we shall see, in practical devices the tube which forms the toroid, may not be circular. However circular systems are easier to treat and the essential physics is unchanged in more general geometries.

4 There are several alternate orthogonal co-ordinate systems which may be used to represent the points in a torus. The one shown in Figure 15.1 $[(\hat{\mathbf{R}}, \hat{\phi}, \hat{z}); (\hat{\mathbf{r}}, \hat{\theta}, \hat{\phi})]$ is probably the most convenient and is widely used (Sauter and Medvedev, 2003).

At its simplest such a device may be imagined behaving as a linear system discussed earlier. Unfortunately, the curvature of the 'cylinder' axis introduces an obvious complication, which arises because the magnetic fields on the inside and outside of the toroid are no longer equal. Consequently, a magnetic field gradient exists across the plasma volume, which tends to push the plasma to the outside wall.

15.4.1 Pressure Balance

Simple toroidal devices suffer from the disadvantage that the plasma has a tendency to move outwards due to the gradients introduced by a combination of the gas dynamic pressure and the magnetic fields. Within the MHD approximation, these are easily understood by simple models.

Consider the magnetic surface to be a toroid of circular cross section whose major radius is much greater than its minor $R_0 \gg a$, then we may use simple qualitative arguments to identify these drifts.

a. *Kinetic pressure* : Remembering that the magnetic surface is also an isobaric surface, the area on the inside of the tube is less than that on the outside giving rise to a net outwards force. Consider an element of the surface defined by a toroidal segment $\delta\phi$ and a poloidal $\delta\theta$, the area of the element is $a\, r\, \delta\phi\, \delta\theta$ where $r = R_0 + a\ \cos\theta$ is the distance from the major axis to the element. The total outwards force on the toroid exerted by the plasma inside the isobaric surface is obtained by summing the contributions from elements around a minor radius

$$\int_0^{2\pi} p\, a\, (R_0 + a\ \cos\theta)\ \cos\theta\ d\theta\ \delta\phi = \pi\, a^2\, p\, \delta\phi \tag{15.10}$$

b. *Magnetic pressure* : The toroidal geometry imposed on the magnetic field induces gradients, which lead to differences in the magnetic pressure between the inner and outer surfaces. These may be identified for the poloidal and toroidal fields in a qualitative manner as follows:
   (i) *Toroidal field* : The toroidal magnetic field is normally generated by a current perpendicular to the plane of the toroid, usually generated by external coils wrapped around the torus in the poloidal plane. Taking into account the toroidal symmetry, Ampère's theorem gives the magnetic field at a point (R, ϕ, θ) on the magnetic surface as

$$B_\phi \approx \mu_0\, I_z/2\, \pi\, (R_0 + a\ \cos\theta) \approx \mu_0\, I_z/2\, \pi\, R_0(1 - a/R_0\ \cos\theta) \tag{15.11}$$

   where I_z is the current through the plane of the torus enclosed by the radius R. The magnetic pressure on the surface of the toroid is therefore

$$\frac{1}{2\mu_0} B_\phi{}^2 \approx \frac{1}{2\mu_0} B_{\phi 0}{}^2 \left(1 - \frac{a}{R_0}\ \cos\theta\right)^2 \tag{15.12}$$

   There is clearly a magnetic pressure difference $\sim (a/R_0)\, B_{\phi 0}{}^2/2\mu_0$ between the inner and outer surfaces of the toroid. The net force acting on the magnetic surface is easily obtained by the application of Eq. (15.10)

$$\begin{aligned} F_\phi &= \int_0^{2\pi} \frac{1}{2\mu_0} B_{\phi 0}{}^2 \left(1 - \frac{a}{R_0}\ \cos\theta\right)^2 a\, (R_0 + a\ \cos\theta)\ \cos\theta\ d\theta\ \delta\phi \\ &= -\pi\, \frac{1}{2\mu_0} B_{\phi 0}{}^2\, a^2\, \delta\phi \end{aligned} \tag{15.13}$$

   where $B_{\phi 0} = \mu_0\, I_z/2\, \pi R_0$.

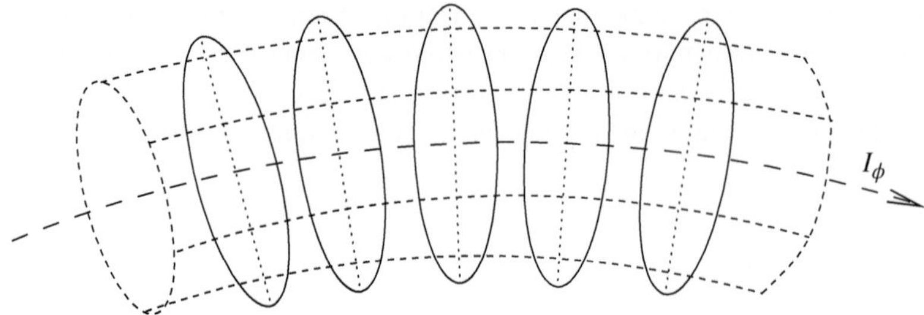

Figure 15.2 Poloidal flux loops (full line) distributed around the minor axis (dashed), bent by the toroidal curvature of the magnetic surface (dotted). The axial current flows along the minor axis of the toroid.

(ii) *Poloidal field* : The poloidal field is generated by toroidal currents I_ϕ, which are usually obtained either by induction from an externally applied field generated by wrapping coils around the torus or from the plasma currents. For simplicity, this current may be imagined to flow along the minor axis of the toroidal surface (Figure 15.2). In the absence of toroidal curvature, this would give rise to equally spaced circular magnetic field loops of induction $B_\theta = I_\phi/2\mu_0 r$ lying in the plane normal to and encircling the minor axis. Introducing the curvature causes the loops to bunch on the inside and spread out on the outside (Figure 15.2). The symmetry of the system requires that the magnetic field flux lines still form a series of loops normal to the minor axis, their plane parallel to the radius of curvature R_0. The cross section of each tube of flux varies as $R\,\delta\phi\,\delta a$, where R is the distance from the centre of curvature (i.e. the minor axis) and δa the width of the flux tube in the poloidal plane. Remembering that the intensity of the field is proportional to the density of the lines of flux, it follows that due to convergence along the radius of curvature, the field on the surface of the toroid at a radius R from the major axis of the toroid varies as

$$B_\theta \approx B_{\theta 0}\,R_0/R = B_{\theta 0}\,R_0/(R_0 + a\,\cos\theta) \approx B_{\theta 0}(1 - a/R_0\,\cos\theta) \tag{15.14}$$

where $B_{\theta 0} = \mu_0\,I_\phi/2\,\pi\,a$.
The similarity with the toroidal field imbalance relation (15.11) allows us to immediately deduce that the net outwards force on the element $\delta\phi$ is

$$F_\theta = -\pi\,\frac{1}{2\mu_0}{B_{\theta 0}}^2\,a^2\,\delta\phi \tag{15.15}$$

The similarity of these two expressions for the magnetic field components suggests that we may consider the total field varying as $B \approx B_0\,(1 - \varepsilon\cos\theta)$ across the torus, with gradient $\dot{B}_0 \approx B_0/R_0$ along the minor axis.

$$\begin{aligned} F &= -\int_0^{2\pi} \frac{1}{2\mu_0}(B_0 - a\,\dot{B}_0\,\cos\theta)^2\,a\,(R + a\,\cos\theta)\,\cos\theta\,\mathrm{d}\theta\,\delta\phi \\ &\approx -\frac{\pi}{2\mu_0}B_0\,(B_0 - 2\dot{B}_0\,R_0)\,a^2\,\delta\phi \end{aligned} \tag{15.16}$$

15.4.1.1 Pressure Imbalance Mitigation

The movement of the plasma body associated with the magnetic field gradients may be mitigated in two ways:

a. A cylinder of high conductivity material around the outside of the torus will contain the induction flux within it. Therefore, the poloidal flux B_θ will be compressed on the outward side of the torus and the concomitant magnetic pressure increase. However, the toroidal B_ϕ flux lines are not restrained by the toroid and may slip from the outer to the inner side and therefore are not compressed.

The circular poloidal field lines B_θ are parallel to the major axis Z on the outer surface of the toroid, and anti-parallel on the inner. Consequently, the addition of a field parallel to the major axis B_z of appropriate sign adds to the poloidal field on the outer side and reduces that on the inner changing the magnetic field gradient and consequently the magnetic pressure in a controlled manner.

b. A magnetic field applied parallel to the major axis interacting with the toroidal current will provide a force to balance that generated by the pressure gradients. A simple derivation of the required magnetic field is obtained by considering the force resulting from the gradient of the magnetic energy in the plasma system in Appendix 15.A.

In equilibrium, the balancing force generated by the Lorentz force is provided by the toroidal current and the vertical magnetic field namely $2\,\pi\,R_0\,I_\phi\,B_Z$. The vertical magnetic field required to ensure equilibrium obtained in Appendix 15.A is consequently

$$B_z = \frac{a}{2\,R_0} B_\theta(a) \left[\ln\left(\frac{8\,R_0}{a}\right) + \beta_\theta + \frac{\ell_i}{2} - \frac{3}{2} \right] \tag{15.17}$$

where $\beta_\theta = 2\mu_0\,\overline{p}/B_\theta(a)^2$ is the poloidal β. and the internal inductance parameter $\ell_i = \overline{B_\theta{}^2}/B_\theta{}^2(a)$.

15.4.2 Guiding Centre Drift

It is easy to see that in a toroidal system with both toroidal and poloidal magnetic fields, the magnetic field exhibits all the possible forms of variation identified in Section 3.5: field line gradients, curvature, and twists. Referring to Eq. (3.42), it is clear that the drift velocities imposed on the electrons and ions are in opposite directions leading to charge separation. This results in an electric field normal to the central plane across the toroid (z direction). Consequently, an $\mathbf{E} \wedge \mathbf{B}$ drift is established which drives the particles outwards from the toroidal chamber independent of the sign of their charge.

To overcome this problem, the magnetic field lines are designed to spiral around the toroid with the dominant toroidal flux to which is added a weaker poloidal field. The particles following the field lines therefore spend some time in the outer region and less in the inner. The drifts in the outer part are compensated by those in the inner so that the particle nearly returns to its original position after a rotation around the toroid cancelling out outward motion.

We define the geometrical characteristics of the toroidal magnetic field by similar terms to those we used for a screw pinch. The field line pitch is given by Eq. (15.7). The tilt of the magnetic field with respect to minor axis of the toroid is

$$\frac{R_0}{r}\frac{d\phi}{d\theta} = \frac{B_\phi}{B_\theta}$$

Since the rotation may be different for different starting points, the safety factor is defined as the average number of toroidal circuits required to complete one poloidal rotation of the field line

$$q(r) = \frac{1}{2\,\pi} \oint \frac{B_\phi}{R_0\,B_\theta}\, r\, d\theta \tag{15.18}$$

the integral being taken over a single toroidal circuit around the flux surface. In general the safety factor is not an integer or a rational fraction, so that the field lines do not close on themselves.

An average over a large number of toroidal rotations is therefore necessary to specify the rotational transform. A flux surface on which the field lines close is said to be *ergodic*. We note that the rotational transform and the safety factor normally vary across the poloidal cross section.

The safety factor can also be interpreted in terms of the flux between two neighbouring toroidal flux surfaces of separation Δ, width $r\,\delta\theta$ in the poloidal plane, and a complete toroidal rotation, the poloidal and toroidal fluxes are respectively

$$\Delta\Psi = 2\,\pi\,R_0\,B_\theta\,\Delta \qquad \text{and} \qquad \Delta\Phi = \oint B_\phi\,r\,\mathrm{d}\theta\,\Delta \tag{15.19}$$

Taking the ratio we obtain

$$q = \frac{\mathrm{d}\Phi}{\mathrm{d}\Psi} \tag{15.20}$$

15.5 The General Problem: The Grad–Shafranov Equation

In general, the magnetic surfaces are not circular, but due to the plasma diamagnetism typically have D shape; the minor axis is defined by the field line of zero poloidal field $B_\theta = 0$. Most toroidal confinement systems are symmetric about the major axis. In this case, the equilibrium condition of the plasma in a torus can be described by imposing toroidal symmetry ($\partial/\partial\phi = 0$) and using (toroidal) cylindrical co-ordinates $(R,\ \phi,\ Z)$ with respect to the toroidal plane (Figure 15.1). Introducing the toroidal component of the magnetic potential $\mathbf{B} = \nabla\wedge\mathbf{A}$, we define the *flux function* $\psi = 2\,\pi\int_0^R \mathrm{d}\Psi = R\,A_\phi$, so that the poloidal field components[5]

$$B_R = -\frac{1}{R}\frac{\partial\psi}{\partial Z} \qquad \text{and} \qquad B_Z = \frac{1}{R}\frac{\partial\psi}{\partial R} \tag{15.21}$$

Clearly, $(\mathbf{B}\cdot\nabla)\psi = 0$. Consequently, the flux function is constant along lines of induction.

The general configuration of a toroidally symmetric, equilibrium configuration is obtained by applying the symmetry condition to the basic equilibrium relations (15.1) and (15.2).

Since $\mathbf{B}\cdot\nabla p = 0$, the Jacobian

$$\frac{\partial(p,\psi)}{\partial(R,Z)} = \frac{\partial p}{\partial R}\frac{\partial\psi}{\partial Z} - \frac{\partial p}{\partial Z}\frac{\partial\psi}{\partial R} = -R\left[B_R\frac{\partial p}{\partial R} + B_Z\frac{\partial p}{\partial Z}\right] = 0 \tag{15.22}$$

and consequently $p = p(\psi)$. When their Jacobian is zero, two functions are dependent (see for example Lowry and Hayden, 1957, p. 239). Consequently, the pressure is a function of the flux function only, $p = p(\psi)$. Thus, the surfaces of constant ψ are also isobaric surfaces $p = const$ and therefore the magnetic surface, on which the lies of magnetic induction and current lie, is as well. As a result the parameter ψ is a parameter marking individual magnetic surfaces.

Similarly since $\partial p/\partial\phi = 0$ and therefore $(\nabla\wedge\mathbf{B})\wedge\mathbf{B}|_\phi = 0$, the Jacobian

$$\frac{\partial((R\,B_\phi),\psi)}{\partial(R,Z)} = \left[\frac{\partial(R\,B_\phi)}{\partial R}\frac{\partial\psi}{\partial Z} - \frac{\partial(R\,B_\phi)}{\partial Z}\frac{\partial\,\psi}{\partial R}\right] = -R\left[\frac{\partial(R\,B_\phi)}{\partial R}B_R + \frac{\partial(R\,B_\phi)}{\partial Z}B_Z\right] = 0 \tag{15.23}$$

Hence, $RB_\phi = f(\psi)$. It follows that the surfaces of constant toroidal field RB_ϕ are also surfaces of constant ψ. The current densities in the poloidal plane are given by Ampère's theorem

$$j_R = -\mu_0\,\frac{1}{R}\frac{\partial f}{\partial Z} \qquad \text{and} \qquad j_Z = \mu_0\,\frac{1}{R}\frac{\partial f}{\partial R} \tag{15.24}$$

5 Note the similarity of the flux function to the stream function in incompressible fluid mechanics, stemming from the facts that $\nabla\cdot\mathbf{B} = 0$, and $\partial B_\phi/\partial\phi = 0$.

and from the integral form of the theorem it follows that $f(\psi) = \mu_0 I_\theta(\psi)/2\pi$ where $I_\theta(\psi)$ is the total poloidal current through a disc of radius R normal to the major axis through the magnetic surface ψ due to both the plasma and the field coils.

Making use of the pressure balance equation from Eq. (15.1) together with the functional relations derived above (see problem 8), we obtain the Grad–Shafranov equation (Grad and Rubin, 1958; Shafranov, 1958, 1966)

$$\Delta^*\psi = -f(\psi)\dot{f}(\psi) - \mu_0 R^2 \dot{p}(\psi) \tag{15.25}$$

where the superscript dot indicates derivatives with respect to ψ and the operator $\Delta^*\psi$ is

$$\Delta^*\psi = R\frac{\partial}{\partial R}\left(\frac{1}{R}\frac{\partial\psi}{\partial R}\right) + \frac{\partial^2\psi}{\partial Z^2} \tag{15.26}$$

The magnetic induction is given by the field in the poloidal plane from Eq. (15.21) and the toroidal component from $f(\psi)$ namely

$$\mathbf{B} = \frac{1}{R}[\nabla\psi \wedge \hat{\mathbf{e}}_\phi + f(\psi)\,\hat{\mathbf{e}}_\phi] \tag{15.27}$$

where $\hat{\mathbf{e}}_\phi$ is the unit vector in the ϕ direction. It follows from Eq. (15.19) that the toroidal field is not determined by the Grad–Shafranov equation or *vice versa*. Its value determines the safety factor of the configuration, but does not influence the magnetic or pressure profiles.

Taking the toroidal component of the curl of this term we obtain (problem 8)

$$\Delta^*\psi = -\mu_0 R j_\phi \tag{15.28}$$

so that the current density also is not affected by the toroidal field.

15.6 Boundary Conditions

The boundary conditions applied to the problem by the containment vessel and applied fields may take one of a number of forms; the most important being:

a. *Perfectly conducting wall*: The simplest boundary case is that where the plasma extends to a stationary perfectly conducting wall with pre-defined surface $S_w(r, \theta, z)$. The electromagnetic fields at the wall require the tangential electric field and the normal magnetic field components to be zero

$$\mathbf{n} \wedge \mathbf{E} = 0 \qquad \text{and} \qquad \mathbf{n} \cdot \mathbf{B} = 0 \tag{15.29}$$

where $\mathbf{n}$ is the outward normal at the surface.
Finally from the infinite conductivity form of Ohm's law $\mathbf{E} + \mathbf{v} \wedge \mathbf{B} = 0$, it follows by taking the vector product with $\mathbf{n}$ that

$$\mathbf{n} \cdot \mathbf{v}|_{S_w} = 0 \tag{15.30}$$

and the tangential velocity component at the wall is undefined.

b. *Insulating vacuum region*:
This case is more difficult to formulate involving a vacuum region between the plasma and a conducting wall. There are therefore a set of electromagnetic field equations in the vacuum which must be solved between the plasma bounding surface S_p and the wall S_w. In this region, Maxwell's equations have the simple form

$$\nabla \wedge \tilde{\mathbf{B}} = 0 \qquad \text{and} \qquad \nabla \cdot \tilde{\mathbf{B}} = 0 \tag{15.31}$$

where the tilde indicates quantities measured in the vacuum. Consequently, the magnetic field can be expressed as a scalar potential $\tilde{\mathbf{B}} = \nabla V$ which satisfies Laplace's equation $\nabla^2 V = 0$ with boundary condition $\mathbf{n} \cdot \nabla V|_{S_w} = 0$.

Unlike the conducting wall, the plasma boundary S_p is free to move. Hence, the velocity component normal to the boundary is arbitrary $\mathbf{n} \cdot \mathbf{v}|_{S_p} \neq 0$. As a result, we must identify a set of jump conditions to treat quantities across the interface, similar to those across MHD shocks . To calculate the jump relations, we may use the method applied in Section 10.4 where the transition is treated in the rest frame of the flow and subsequently transformed into the laboratory frame. We assume that the boundary is moving with a velocity normal to the interface

$$v_n \mathbf{n} = (\mathbf{n} \cdot \mathbf{v})\mathbf{v}$$

and transform into the frame moving with the interface so that the flow across the interface appears stationary, $\mathbf{n}$ being the unit normal to the interface. Consider the behaviour in the stationary frame, the jump conditions are obtained by integrating Maxwell's equations in either a small box surrounding the wall or around a closed loop surrounding the wall (Figure 15.3) as the depth of the box or the loop tends to zero $\lim d \to 0$

$$[[\mathbf{n} \cdot \mathbf{B}']] = 0 \qquad [[\mathbf{n} \wedge \mathbf{B}']] = \mu_0 \mathbf{J}_s \qquad \text{and} \qquad [[\mathbf{n} \wedge \mathbf{E}']] = 0 \tag{15.32}$$

where $[[\cdots]]$ is the difference in the values across the interface, and the prime indicates values in the frame moving with the wall. In a perfect conductor, the thickness of the skin current at the interface becomes infinitely small and the current per unit width $\mathbf{J}_s$ lies in the surface. We shall argue that the magnetic field lines must lie in the interface and that consequently $(\mathbf{B} \cdot \nabla)$ is a surface derivative. Integrating the equation of momentum conservation (9.5) in the pill-box, we obtain

$$\int_V \rho \frac{\mathrm{d}\mathbf{v}'}{\mathrm{d}t} \, \mathrm{d}V = -\int_S [p' + B'^2/2\mu_0] \, \mathrm{d}s + \int_S (\mathbf{B}' \cdot \mathbf{n})\mathbf{B}'/\mu_0 \, \mathrm{d}s \tag{15.33}$$

and letting the depth $d \to 0$. Since $[[B'_n]] = 0$ it follows that, unless there is a surface mass, the magnetic field lines must lie within the interface. Therefore, the normal component of the magnetic field must be zero across the interface

$$\mathbf{n} \cdot \mathbf{B}' = 0 \tag{15.34}$$

to avoid 'refraction' of the field lines which would give rise to infinite acceleration. This is consistent with the standard result from electro-magnetic theory for a perfect conductor (Maxwell, 1904, Arts 654-655).

Finally, we obtain the equation of pressure balance across the interface

$$[[p' + B'^2/2\mu_0 = 0]] \tag{15.35}$$

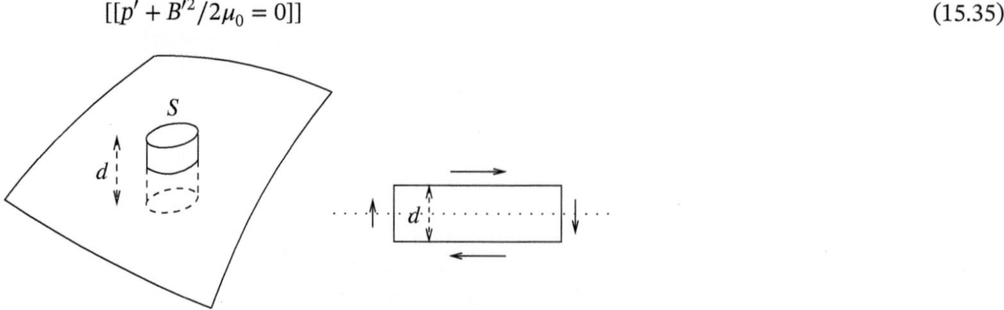

Figure 15.3 Pill-box volume and closed loop across interface for treating *divergence* and *curl* term, respectively.

which completes the analysis of the flow in the rest frame of the interface.[6]
Transforming back into the laboratory frame using the appropriate Galilean transformation

$$p' = p \qquad \mathbf{B}' = \mathbf{B} \qquad \mathbf{n} \wedge \mathbf{E}' = \mathbf{n} \wedge (\mathbf{E} + \mathbf{v} \wedge \mathbf{B}) = \mathbf{n} \wedge \mathbf{E} - (\mathbf{n} \cdot \mathbf{v})\mathbf{B}$$

the jump conditions may be written in the laboratory frame

$$\begin{aligned} [[\mathbf{n} \cdot \mathbf{B}]]_{S_p} &= 0 \qquad \text{and} \quad [[\mathbf{n} \wedge \mathbf{B}]]_{S_p} = \mathbf{J}_S \\ &\text{and} \\ [[\mathbf{n} \wedge \mathbf{E} - (\mathbf{n} \cdot \mathbf{v})\mathbf{B}]]_{S_p} &= 0 \qquad \text{and} \quad [[p + B^2/2\mu_0]]_{S_p} = 0 \end{aligned} \tag{15.36}$$

In the ideal MHD approximation, the plasma is a perfect conductor so that $\mathbf{E} + \mathbf{v} \wedge \mathbf{B} = 0$, and therefore within the plasma $\mathbf{n} \cdot \mathbf{B}|_{S_p} = 0$ and $[\mathbf{n} \wedge \mathbf{E} - (\mathbf{n} \cdot \mathbf{v})\mathbf{B}]|_{S_p} = 0$. Hence the set (15.36) reduce to

$$\mathbf{n} \cdot \tilde{\mathbf{B}}|_{S_p} = 0 \quad \text{and} \quad [\mathbf{n} \wedge \tilde{\mathbf{E}} - (\mathbf{n} \cdot \mathbf{v})\tilde{\mathbf{B}}]|_{S_p} = 0 \quad \text{and} \quad [[p + B^2/2\mu_0]]_{S_p} = 0 \tag{15.37}$$

Finally if there are no surface currents $[[\mathbf{n} \wedge \mathbf{B}]]_{S_p} = 0$, and the pressure falls smoothly to zero at the plasma edge $p|_{S_p} = 0$. The jump conditions finally reduce to

$$\mathbf{n} \cdot \tilde{\mathbf{B}}|_{S_p} = 0 \quad \text{and} \quad \mathbf{n} \wedge \tilde{\mathbf{E}}|_{S_p} = 0 \quad \text{and} \quad [[B^2/2\mu_0]]_{S_p} = 0 \tag{15.38}$$

Suppose we have a solution for the field in the plasma. The solution in the vacuum region involves the solution of Laplace's equation for the magnetic scalar potential V. As is well known the solution of Laplace's equation is determined by either value or the normal gradient of the variable on the surface. In this case, the normal gradient of V is determined on both the outer (conducting) and inner (plasma) surfaces and the field uniquely determined. The tangential magnetic field and therefore the total field B^2 in the vacuum are known, but these values may not be consistent with the final jump condition. The solution to this apparently *over determined problem* is to introduce as a new variable, the surface of the plasma S_p and to adjust the interface to the achieve the matching conditions – *free boundary problem* which are determined within the overall solution. In practice, this is difficult to achieve and more usual practice is to preset the plasma boundary and then adjust the outer vacuum surface to match the interface – *fixed boundary problem.*

15.7 Equilibrium Plasma Configurations

The plasma is maintained in equilibrium by the magnetic pressures associated by the poloidal and toroidal fields balanced against the kinetic pressure. In a system with toroidal symmetry, such configurations are obtained from solutions to the Grad–Shafranov equation. The detailed equilibrium configuration is determined subject to input from the externally applied fields and the geometrical boundary. In many cases, this general problem is simplified through a knowledge of pre-determined values of the pressure $p(\psi)$ and toroidal field functions $f(\psi)$, which are obtained either by experiment or otherwise.

6 These results can also be obtained from the general treatment of interfaces in magneto-hydrodynamic flow as detailed in Section 10.4 for the particular case when the downstream density is zero ($\rho_2 = 0$). The mass flux is consequently zero in Eqs. (A), (B), and (C). Assuming the transverse magnetic field across the interface is discontinuous $B_{y_1} \neq B_{y_2}$, Eq. (C) requires the normal magnetic field to be zero $B_{x1} = B_{x2} = 0$. The remaining results follow directly.

The most general solutions of the Grad-Shafranov equation are obtained by numerical integration of the Grad–Shafranov equation supplemented by Maxwell's equation and conductivity model. However more specific solutions, which are in equilibrium, may be obtained by analytic methods if the geometrical profile of the plasma is known. Freidberg (2014) gives an extensive review of these various approaches, which we outline below.

15.7.1 Perturbation Methods

The large aspect ratio of many tokamaks allows an expansion in terms the dimension ratio (inverse aspect ratio) $\varepsilon = a/R_0 \ll 1$ when the plasma $\beta \sim 1$. The full application of the perturbation methods is described by Freidberg (2014).

Here we will present in Appendix 15.B a relatively simple analytic solution to the Grad–Shafranov equation generated using the zero and first order approximations obtained by a perturbation expansion using the Solov'ev (1968) values for the pressure and flux function gradients in the form

$$p(\psi) = p_0 \left(1 + k_1 \frac{\psi}{\psi_0}\right) \qquad \text{and} \qquad {I_\theta}^2(\psi) = {I_0}^2(1 + k_2\, \psi/\psi_0) \tag{15.39}$$

The pressure and the poloidal magnetic flux function $2\,\pi\, f(\psi) = \mu_0\, I_\theta(\psi)$ are therefore linear functions of the parameter ψ.[7]

The detailed analysis of this perturbation solution up to the first order is given in Appendix 15.B. Some important characteristics found in more general solutions are obtained

a. A 'vertical' magnetic field is essential to eliminate the outwards drift of the plasma in the absence of alternative limiters. The value of the field

$$B_Z = B_\theta(a)\, \frac{a}{2\,R} \left\{ \ln\left(\frac{8\,R}{r}\right) + \beta_\theta + \frac{1}{2}\,\ell_i - \frac{3}{2} \right\} \tag{15.40}$$

depends on the plasma poloidal β_θ and on the internal inductance parameter ℓ_i both of which can be calculated if the current distribution within the plasma is known. Unfortunately, the lack of information on the resistivity distribution across the toroid makes this an almost insuperable problem. Long duration tokamaks often use feedback control to maintain the equilibrium taking the term $\beta_\theta + \ell_i/2$ into account.

b. *Shafranov shift*: The magnetic surfaces are found to be displaced outwards from the minor axis by

$$\Delta = \frac{1}{2\,R_0} \left(\beta_\theta(r) + \frac{\ell_i(r)}{2}\right) (a^2 - r^2) \tag{15.41}$$

where r is the radial distance from the minor axis, namely the radius from the minor axis of the approximately circular surface ψ, namely $r_0 \approx 2\,a\,\sqrt{(\psi/\psi_0)}$ in zero order. where a is the minor radius. The shift Δ is linearly dependent on the value of the poloidal β_θ in the plasma. This factor in turn relates to the aspect ratio $\varepsilon^{-1} = R_0/a$. When $\beta_\theta \sim \varepsilon$ and $k_2 < 0$, the current and magnetic field are nearly collinear so that the plasma is nearly force free. The flux surfaces are consequently nearly circular and the shift small. As the plasma beta increases until $\beta_\theta \sim 1$, the increase in plasma pressure and the diamagnetic current, which nullifies the poloidal current, drives the plasma to the wall. Finally, when $k_2 \gg 1$, and $\beta_\theta \sim A$ the poloidal current is reversed

7 This linear assumption is unphysical as the current varies in the plasma due to a non-uniform electrical conductivity and discontinuously at the plasma edge. The assumption is made for simplicity to demonstrate the perturbation method. We note the same approximation is made as the particular integral used in the analytic expansion of the Grad–Shafranov equation Section 15.7.2.

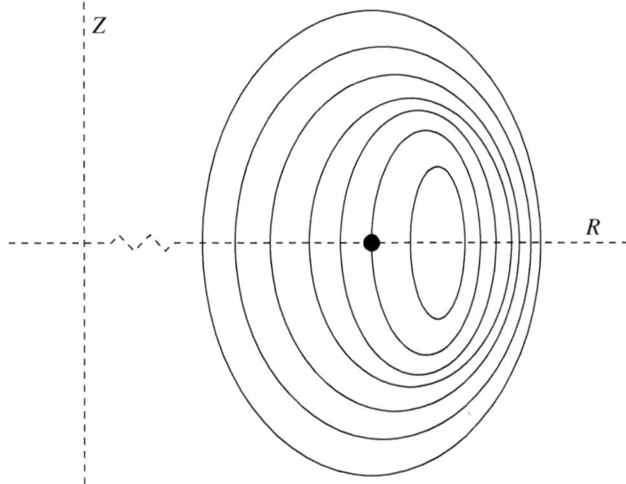

Figure 15.4 Sketch of the magnetic surfaces illustrating the shift away from the minor axis, represented by a circular dot.

and a magnetic well is formed in the toroidal field comparable to that in a theta pinch forming a well in which the plasma is contained. The value of beta is limited by the aspect ratio to the *equilibrium plasma beta* $\sim R_0/a$. This limit is also the stability limit for tokamaks and thus gives an upper bound of the value of β_θ usable.

15.7.2 Analytical Solutions of the Grad–Shafranov Equation

Exact analytical solutions to the Grad–Shafranov equation may be obtained by the familiar method using the general solution to the homogeneous equation $\Delta^* \psi = 0$ and a particular integral (Solov'ev, 1968, 1975). Although not as general in application as numerical methods, these solutions are useful for developing the scaling of devices, and for providing a ready analytic forms for use in calculating stability. We illustrate the method by following Zheng et al. (1996) and Cerfon and Freidberg (2010). A thorough discussion of the method is given by Freidberg (2014, p. 177).

A simple exact form of the particular integral due to Solov'ev (1968) is generated by setting

$$-\mu_0 \dot{p} = A_1 \qquad \text{and} \qquad f\dot{f} = A_2 \tag{15.42}$$

The constants A_1 and A_2 normally define the toroidal plasma current I_A and the average value of the poloidal beta β_θ. The particular integral satisfies

$$\Delta^* \psi = -[A_1 R^2 - A_2] \tag{15.43}$$

whose solution is

$$\psi = \psi_0 + \frac{1}{8} A_1 R^4 - \frac{1}{2} A_2 Z^2 \tag{15.44}$$

The boundary conditions which determine the shape of the magnetic surface are included by means of a polynomial solution to the homogeneous equation which includes an appropriate number of undetermined constants. Assuming up/down symmetry only even powers of Z are included (Zheng et al., 1996). We may write the homogeneous solution as a polynomial

$$\psi = \sum_{n=0,2,\ldots} \alpha_n(R) Z^n = \sum_{n=0,2,\ldots} \sum_{k=0}^{k=n/2} G(n,k,R) Z^{n-2k} \tag{15.45}$$

where

$$G(n,k,R) = g_1\, G_1(n,k,R) + g_2\, G_2(n,k,R) \tag{15.46}$$

which have exact integrals (see Appendix 15.C)

$$\begin{aligned}
&G_1(n,0,R) = 1\\
&G_1(n,k>0,R) = (-1)^k \frac{n!}{(n-2k)!)}\,\frac{1}{2^{2k}k!(k-1)!}\,R^{2k}\left(2\ln(R) + \frac{1}{k} - 2\sum_{j=1}^{k}\frac{1}{j}\right)\\
&G_2(n,k,R) = (-1)^k \frac{n!}{(n-2k)!}\,\frac{1}{2^{2k}k!(k+1)!}\,R^{2k+2}
\end{aligned} \tag{15.47}$$

The sum is truncated after a suitable number of terms depending on the number of geometric parameters specified for the magnetic surface. The terms $G(n,k,R)$ are a set of terms of terms of each order. The set used by Solov'ev (1968); Zheng et al. (1996); Cerfon and Freidberg (2010) are

$$\begin{aligned}
&\psi_{01} = 1 \qquad \psi_{21} = R^2 \qquad \psi_{22} = R^2\,\ln(R) - Z^2 \qquad \psi_{41} = R^4 - 4\,R^2\,Z^2\\
&\psi_{42} = 2\,Z^4 - 9\,R^2\,Z^2 + 3\,R^4\,\ln(R) - 12R^2\,Z^2\,\ln(R)\\
&\psi_{61} = R^6 - 12R^4\,Z^2 + 8\,R^2\,Z^4\\
&\psi_{62} = 8\,Z^6 - 140\,R^2\,Z^4 + 75\,R^4\,Z^2 - 15\,R^6\,\ln R + 180\,R^4\,Z^2\ln(R) - 120R^2\,Z^4\,\ln(R)
\end{aligned} \tag{15.48}$$

so that the general solution,[8] which determines the magnetic surface $\psi = 0$ is

$$\psi = \sum_{n=0}^{N} c_{n1}\psi_{n1} + c_{n2}\,\psi_{n2} + \frac{1}{8}A_1\,R^4 - \frac{1}{2}A_2\,Z^2 = 0 \tag{15.49}$$

up to the required order N.

Given the geometrical profile of the magnetic surface $R(Z)$, the constants c_{n1} and c_{n2} are calculated by the solution of the simultaneous equations using current values of A_1 and A_2. The undetermined constants A_1 and A_2 are found by iteration from experimental parameters, for example the toroidal current

$$I_\phi = \int j_\phi\,\mathrm{d}R\,\mathrm{d}Z = -\frac{1}{R}\int (A_1\,R^2 - A_2)\,\mathrm{d}R\,\mathrm{d}Z \tag{15.50}$$

and the averaged poloidal plasma beta

$$\beta_\phi = \frac{8\pi}{\mu_0}\frac{1}{{I_\phi}^2}\int \mathrm{p}\,\mathrm{d}R\,\mathrm{d}Z = \frac{8\pi\,A_1}{{\mu_0}^2}\,\frac{1}{{I_\phi}^2}\int \psi\,\mathrm{d}R\,\mathrm{d}Z \tag{15.51}$$

to which the solution is driven by iteration.

15.7.3 Numerical Solutions of the Grad–Shafranov Equation

The Grad–Shafranov equation is in the form of a non-linear elliptic equation ($\Delta^*(\psi) = S(R,\psi)$), where $S(R,\psi)$ is a non-linear functional incorporating the flux $f(\psi)$ and pressure $p(\psi)$ functions. Finite difference methods of solution are available involving the inversion of a large sparse matrix. The equation is solved subject to appropriate boundary conditions, which may be formulated in one of two ways

a. The outer magnetic surface of the plasma is specified, either by experimental measurements or, for example, by a conducting shell on which ψ is specified, known as the *closed boundary form*.

8 This relatively simple solution is based on a polynomial basis set. Alternative sets of functions which generate solutions of the homogeneous equation may be used instead.

The numerically generated solution of the equations then allows the internal magnetic surfaces and the currents and fields to be calculated.

b. The plasma is separated from the physical boundary by a vacuum region, subject to experimentally determined currents and magnetic fields, the *open boundary form.*

Following Cerfon and Freidberg (2010), it is convenient to normalise the Grad–Shafranov equation into dimensionless terms

$$\overline{R} = \frac{R}{R_0} \qquad \overline{Z} = \frac{Z}{R_0} \qquad \overline{\psi} = \frac{\psi}{\psi_0} \qquad C = -\frac{{R_0}^2}{\psi_0^2}A_1 \qquad A = -\frac{{R_0}^2}{{\psi_0}^2}A_2 \tag{15.52}$$

so that the Grad–Shafranov equation becomes

$$\overline{R}\frac{\partial}{\partial \overline{R}}\left(\frac{1}{\overline{R}}\frac{\partial \overline{\psi}}{\partial \overline{R}}\right) + \frac{\partial \overline{\psi}}{\partial \overline{Z}^2)} = [(A+C)-A]\,\overline{R}^2 + A = (1-A)\,\overline{R}^2 + A \tag{15.53}$$

setting $\overline{\psi}^2 \to (A+C)\,\overline{\psi}^2$. The scaling parameters R_0 and ψ_0 are obtained from the external experimental parameters of the device studied, for example R_0 may be the major radius of the plasma.

15.8 Classical Magnetic Cross Field Diffusion

Diffusion is critical in fusion plasma confined by a closed magnetic field. Of particular importance is the cross field (perpendicular) term $D_\perp = D_0/(1+\omega^2\,\tau^2)$ when the collisionality is small $\Omega\,\tau \gg 1$. This term represents particle loss across the field lines. Under conditions of low collisionality the classical diffusion rate, Eq. (2.65), scales as

$$D_\perp = \frac{kT\,\tau}{m}\frac{1}{(1+q\,B\,\tau/m)^2} \approx \frac{m\,kT}{q^2\,\tau}\frac{1}{B^2} \tag{15.54}$$

In a strong field charged particles rotate on cyclotron orbits about the magnetic field line (see Chapter 3). The motion along the field line is uninhibited, exemplified by the term $D_\parallel = D_0 \times 1$, i.e. the field free diffusivity. However, the cross field terms $D_\perp$ are reduced by the factor $(1+\Omega^2\,\tau^2)$ being tied to the field line. The origin of this term is easily understood from the random walk model (Section 2.2.1). In a strong field, the charged particles spiral around the field line. Their path is a circular in the plane normal to the field with *Larmor radius* $\rho = v_\perp/\Omega$, where $v_\perp$ is the velocity perpendicular to the field line. In a collision with a second particle the 'guiding centre' of the motion, which moves along the field line, is displaced to a new line and the radius changed. As a result, the particles execute a random walk through the field; the net result being a drift in the direction opposite to the density gradient. The characteristic 'jump' length is the Larmor radius rather than the mean free path. The cross field diffusion rate is therefore

$$D_\perp \sim \rho^2/\tau \sim m\,kT/q^2\,\tau\,B^2 \tag{15.54:a}$$

as above.

Particle transport only occurs in *unlike* collisions, i.e. electron/ion or ion/electron collisions. To demonstrate this consider the change in the guiding centre position due to a collision. The Larmor radius depends only on the component of velocity perpendicular to the field line $\mathbf{v}_\perp$. At collision the particle position is unchanged but the guiding centre moves. Therefore, considering the position of the guiding centre of the particle before ($\mathbf{r}$) and after ($\mathbf{r}'$) collision

$$\mathbf{r}' + \rho' = \mathbf{r} + \rho \tag{15.55}$$

where $\mathbf{r}$ is the position of the particle guiding centre and

$$\rho = \mathbf{B} \wedge \mathbf{p}/qB^2 \tag{15.56}$$

is the Larmor radius vector of the particle where $\mathbf{p} = m\mathbf{v}$ is the particle momentum. Since the particle's position is unchanged at the collision, the shift of the guiding centre

$$\Delta\mathbf{r} = \mathbf{r}' - \mathbf{r} = -(\rho' - \rho) \tag{15.57}$$

$$= -\frac{\mathbf{B}}{qB^2} \wedge (\mathbf{p}' - \mathbf{p}) = -\frac{\mathbf{B}}{qB^2} \wedge \Delta\mathbf{p} \tag{15.58}$$

In a collision between *like particles* it follows that since momentum is conserved, $\Delta\mathbf{v}_1 + \Delta\mathbf{v}_2 = 0$, and therefore the mean guiding centre $(\mathbf{r}_1 + \mathbf{r}_2)/2$ is unchanged. Consequently, there is zero net cross field diffusion. For unlike particles in a fusion plasma comprising hydrogen ions ① and electrons ② where $q_1 = -q_2 = e$ and $Z_1 = 1$, the ions and electrons have equal temperatures. It also follows from Eq. (15.54) and momentum conservation that $\Delta\mathbf{p}_1 = -\Delta\mathbf{p}_2$. Hence, from Eq. (15.54) $\Delta\mathbf{v}_1 = \Delta\mathbf{v}_2$ and the cross field diffusion rates are equal $D_{\perp 1} \approx D_{\perp 2}$. Their diffusion is ambipolar without the establishment of an electric field (*intrinsically ambipolar*).[9]

In contrast calculating the thermal diffusivity collisional terms between like particles plays a role. The presence of a thermal gradient ensures that the argument above for like particles is no longer valid due the difference in speed and energy of particles above and below the plane normal to the gradient, leading to different Larmor radii. Similarly the ion and electron terms are not. However, applying the random walk model and noting that $\tau_{ee} \sim \tau_{ei}$ we conclude the electron thermal diffusivity is approximately equal to the electron diffusivity $\chi_e \sim D_e \sim D_i$. The ion thermal diffusivity is determined by the same random walk arguments as before except that allowing the ion–ion terms to be present introduces the ion–ion collision time. From Eq. (2.25), $\tau_{ii}/\tau_{ie} \sim \sqrt{m/M} \ll 1$ dominates. Hence, a larger value of the ion thermal diffusivity results from ion–ion collisions and ion cross field thermal conduction dominates electronic $\chi_i/\chi_e \sim \sqrt{M/m} \gg 1$. This is a consequence of the fact that the large ion Larmor radius more than compensates for the longer collision time.

15.9 Trapped Particles and Banana Orbits

The path traced out by the guiding centre of charged particle orbits in a toroid with circular cross section, and twisted magnetic field lines can be pictured very simply in the poloidal plane of a rotation due to circular poloidal field and the drift resulting from the gradients of the toroidal field (Eq. 15.D.1). An analysis of the guiding centre motions is given in Appendix 15.D.

In Section 3.3, we showed that the guiding centre is tied to the magnetic field lines moving with the component of the velocity parallel to the field modified by any parallel electric field component and variation along the field, Eq. (3.41). Guiding centre motion is due to various drifts. The perpendicular component of the particle velocity simply changes its gyromotion.

As the charged particles move backwards and forwards along the field lines between any trapping points, they will be subjected to drifts due to both the curvature and the field gradient from Eqs. (3.63) and (3.64) namely

$$\mathbf{U} = \frac{(v_\parallel{}^2 + v_\perp{}^2/2)}{R\Omega}\,\hat{\mathbf{k}} \tag{15.59}$$

9 Alternatively, the Larmor radii factors scale as $\rho_1{}^2/\rho_2{}^2 = m_1/m_2$, and from Eq. (2.A.11), the collision frequencies as $\nu_{12}/\nu_{21} \approx m_2/m_1$. Hence, the ratio $D_{\perp 1}/D_{\perp 2}$ is independent of the mass of the particles, the smaller Larmor radius of the lighter particles being compensated by their larger collision frequency.

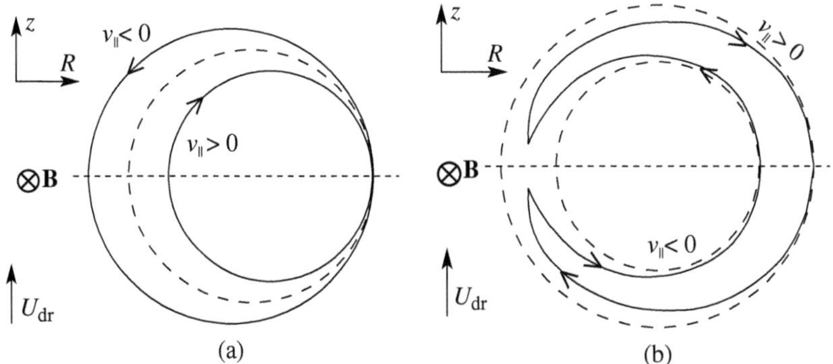

Figure 15.5 Sketch of a projection in the poloidal plane of the magnetic surfaces illustrating the paths of particles. The dashed lines represent the poloidal flux lines. The toroidal ϕ direction and therefore the toroidal current and flux are all directed into the paper and the corresponding poloidal flux lines are clockwise. (a) Passing particles and (b) trapped particles.

perpendicular to the mid-plane of the torus in the (z) direction. It follows that the behaviour on either side of the mid-plane with respect to field lines is different. Due to the curvature, the trajectory moves inward when $z < 0$ and outward when $z > 0$ (Figure 15.5).

At the simplest level, we may consider the motion due to transit along the toroidal field lines with a poloidal rotation superimposed and an added drift **U** in the $-z$ direction perpendicular to the toroidal plane. Projecting the motion into the poloidal plane

$$\frac{dR}{dt} = \omega z \qquad \text{and} \qquad \frac{dz}{dt} = -\omega(R - R_0) + U \tag{15.60}$$

representing a circular rotation anti-clockwise about the point on the mid-point plane a distance $R - R_0$ from the minor axis with angular frequency ω determined by the transit time of one helical rotation, $\omega = \iota\, v_{\parallel}/R_0$. Solving we find

$$[R - (R_0 - U/\omega)]^2 + z^2 = (R_0 - U/\omega)^2 \tag{15.61}$$

i.e. a circular motion with the centre displaced by the drift towards the centre by U/ω.

As a result of the poloidal field, the magnetic field lines spiral around the torus forming toroidal helices. Provided $\iota\varepsilon \ll 1$, (where $\varepsilon - r/R_0$) the total field is approximately the toroidal field $B \approx B_\phi$. Assuming a circular poloidal cross section, the field line minor radius r is approximately constant and the field varies along the helix such that $B_0/(1 + r/R_0) < B < B_0(/1 - r/R_0)$. The particle guiding centres therefore follow the field lines, provided they have a sufficiently large parallel velocity to avoid reflection by the increasing field. Since the total velocity $v_0{}^2 = v_{\parallel 0}^2 + v_{\perp 0}^2$ is constant, it follows taking the reference point at the field minimum we deduce from Eq. (3.96) that the particle is reflected at a poloidal angle θ_{ref} when

$$\begin{aligned} v_{\perp 0}^2 &= v_0{}^2/[1 + 2\,r/R_0 \sin^2(\theta_{\text{ref}}/2)] \\ v_{\parallel 0}^2 &= 2\,r/R_0 \sin^2(\theta_{\text{ref}}/2)\, v_0{}^2/[1 + 2\,r/R_0 \sin^2(\theta_{\text{ref}}/2)] \end{aligned} \tag{15.62}$$

where the velocities v_0, $v_{\parallel}$, and $v_{\perp 0}$ are taken at the field minimum $\theta = 0$. Consequently, particles with $v_{\parallel 0}/v_{\perp 0} \leq \sqrt{2\,r/R_0}$ will be trapped. The remainder *passing particles* follow the field lines, but with their paths distorted by guiding centre drifts due to the gradients in the field.

- *Passing particles*: Typical paths of passing particles projected on to the poloidal plane are shown in Figure 15.5a. Clearly, there is a difference in the behaviour of particles travelling parallel to or anti-parallel to the flux lines due to the drift. In the lower half-plane $z > 0$, the upward guiding centre drift takes the particles to inner flux surfaces. In contrast in the upper half-plane $z < 0$, the drift takes the particles in the opposite direction to outer flux surfaces. Over a full rotation along the helix, the poloidal drifts cancel out returning the particle to its initial flux surface. The time taken by a passing particle for a full rotation is

$$T = 2\,\pi\,R_0\,q/v_\parallel \tag{15.63}$$

where q is the safety factor.
and the displacement of the path over half a period

$$\Delta r \approx \frac{1}{R_0\,\Omega}\left\{v_\parallel^2 + \frac{1}{2}\,{v_\perp}^2\right\}\frac{\pi\,R_0\,q}{v_\parallel} \tag{15.64}$$

- *Trapped particles*: A similar pattern of behaviour is found for trapped particles except that the particles are reflected at their turning point and return along a displaced path in the same half-plane in the poloidal plane projection (Figure 15.5b). After completing their orbit in the other half-plane, the particles return to their original flux surface. The characteristic *banana orbit* in the poloidal projection is easily understood in terms of the outward and inward drift in the upper and lower half-planers, respectively.
For the *trapped particles*, it follows that the velocity ratio of trapped particles at the field minimum $\theta = 0$ is limited to

$$\frac{v_{\parallel 0}}{v_{\perp 0}} \leq \sqrt{2\varepsilon} \tag{15.65}$$

where $\varepsilon = r/R_0$.

Trapped particles oscillate along the toroidal field line from $\theta = 0$ via $\theta = \theta_{\text{ref}}$ though $\theta = 0$ to $\theta = -\theta_{\text{ref}}$ returning to $\theta = 0$ with approximate velocity $\pm v_\parallel$. The distance travelled along the field line in each quarter cycle is approximately $R_0\,\theta_{\text{ref}}\,q$. The period is therefore approximately

$$T = 4\,R_0\,\pi\,q/v_\parallel = 2\sqrt{2}\,R_0\,\pi\,q/\sqrt{\varepsilon}v_\perp \tag{15.66}$$

Barely trapped particles have reflection points $\theta_{\text{ref}} \approx \pi$ and velocity $v_\parallel \approx \sqrt{2\,\varepsilon}v_\perp$ and period $4\,R_0\,\theta_{\text{ref}}\,q/\sqrt{2\varepsilon}v_\perp$. Due to the gradient drift, particle is displaced from the initial flux line on its outward and inward paths over a quarter period given by Eq. (15.59) for barely trapped particles

$$\Delta r \approx \frac{1}{R_0\,\Omega}\left\{2\,\varepsilon\,v_\perp^2 + \frac{1}{2}\,{v_\perp}^2\right\}\frac{\pi\,R_0\,q}{\sqrt{2\,\varepsilon}\,v_\perp} \approx \frac{1}{\sqrt{\varepsilon}}\,q\,\frac{v_T}{\Omega} = \frac{1}{\sqrt{\varepsilon}}\,q\,\rho \tag{15.67}$$

These results are in good agreement with the more analytic approach based on guiding centre motions in Appendix 15.D, see Eqs. 15.D.18 and (15.D.15:b).

15.9.1 Collisionless Banana Regime ($\nu_* \ll 1$)

The collision time must be adjusted to take into account the small angle needed to transfer a trapped orbit to passing one, namely a fractional change in the parallel velocity $\sim \sqrt{\varepsilon}$. Making use of Eq. (2.17), the effective collision time is reduced to $\tau_{\text{eff}} \approx \tau\,\varepsilon$. A useful parameter is the collisionality ν_*, the average number of times a particle is scattered into a passing particle before completing a banana orbit.

$$\nu_* = \frac{T}{\tau_{\text{eff}}} = \frac{1}{\varepsilon\,\tau}\,\frac{R\,q}{\sqrt{\varepsilon}\,v_\perp} = \frac{R\,q}{\varepsilon^{3/2}\,\tau\,v_\perp} \tag{15.68}$$

Trapped particles must complete a full orbit, i.e. therefore exist only if $\nu_* < 1$.

From Eq. (2.25), it follows that $\nu_{*i}/\nu_{*e} \sim \sqrt{m_e/m_i} \ll 1$. However since $v_\parallel \ll v_\perp$, it follows that $v_\perp \approx v_T$, and we obtain from Eq. (2.A.11) that

$$\nu_{*e} = \frac{n\,R_0\,q}{\varepsilon^{3/2}T_e^{\,2}} \qquad \text{and} \qquad \nu_{*i} = \frac{n\,R_0\,q}{\varepsilon^{3/2}T_i^{\,2}} \tag{15.69}$$

The collisionality for ions and electrons are therefore independent of the mass provided the ion and electron temperatures are equal. We also note that since $\nu_* \propto n/T^{3/2}$ the collisionality decreases for hot plasma, and therefore trapped particles are important in hot plasma environments such a tokamaks. When $v_* \ll 1$ trapped particles exist in banana orbits for many cycles – *banana regime*.

15.9.1.1 Diffusion in the Banana Regime

Let us consider the effect of collisions on the particles as they move around the torus along the field lines. As we discussed in Section 15.8, collisions flip the guiding centre from one field line to another: the average shift in that case being approximately equal to the radius of the orbit about the field line, namely the Larmor radius. Similarly collisions will transfer a particle from a trapped orbit to a passing one with a shift corresponding to the trapped orbit width $\Delta r \approx \pi\,\rho\,q/\sqrt{2\,\varepsilon}$, Eq. (15.67). To estimate the corresponding cross field diffusion rate, we may use the random walk model Section 2.2.1 with the mean free path replaced by trapped particle displacement distance Δr.

Assuming that the ions and electrons are each in a Maxwellian distribution, we can estimate the fractional number of trapped particles as $\Delta n/n \approx v_\parallel/v_T = \sqrt{2\varepsilon}$ since $v_\parallel \ll v_T$.

Applying the random walk model – in N time steps the particles diffuse $\sqrt{N}$ space steps. Taking into account the small fraction of trapped particles, which contribute to the diffusion, we obtain in unit time

$$N \approx \frac{1}{\tau_{\text{eff}}}\sqrt{2\,\varepsilon} \tag{15.70}$$

The particle diffusion coefficient is therefore

$$D_B \approx N\frac{\Delta r^2}{\tau_{\text{eff}}} \sim \frac{\sqrt{2\varepsilon}}{\tau_{\text{eff}}}\Delta r^2 \approx \frac{q}{\varepsilon^{3/2}}\frac{(v_T/\Omega)^2}{\tau} \approx \frac{q^2}{\varepsilon^{3/2}}\underbrace{\frac{\rho^2}{\tau}}_{\substack{\text{Classical}\\ \text{diffusivity}}} \sim q\frac{2}{\varepsilon^{3/2}}D_{\perp\,\text{clas}} \tag{15.71}$$

and we note that the neo-classical diffusion rate is $q^2/\varepsilon^{3/2}$ greater than the classical value.

Since the classical diffusivity (Eq. (15.54)) is the same for both ions and electrons (p. 396),we conclude that neo-classical diffusion is also intrinsically ambipolar.

15.9.1.2 Bootstrap Current ($\nu_* \ll 1$)

In a uniform plasma, the banana orbits rotate in opposite directions on the outward and return paths, which are separated by the orbit width Δr. Therefore, the current carried by an outward orbit is cancelled by the return orbit on a flux line $\Delta r = \pi/\sqrt{2\varepsilon}\;q\;\rho$ away. Consequently in an outward density gradient $\mathrm{d}n/\mathrm{d}r$, there is a net current along the field line in a manner similar to the diamagnetic current, Section 3.9 (see also problem 3). Taking account of the fact that a fraction $\sqrt{\varepsilon}$ of particles are trapped and move along the field line with velocity $\sqrt{2\varepsilon}\,v_\perp$, this current along the field line is

$$j' \sim q\sqrt{\varepsilon}\,\sqrt{2\varepsilon}\,\Delta r\,\frac{\mathrm{d}n}{\mathrm{d}r} \sim q\,\frac{\sqrt{\varepsilon}}{B}\,\underset{\sim}{k}T\,\frac{\mathrm{d}n}{\mathrm{d}r} \tag{15.72}$$

The untrapped passing particles similarly carry a diamagnetic current due to the density gradient, but in contrast this current is perpendicular to the field line. Due to collisions, there is a transfer of momentum from the trapped particles to the passing which adjust their velocities. The bootstrap current, j_B arises from the difference in velocity between the passing ions and electrons.

The momentum loss by the passing electrons to the passing ions is given by resistive term $\nu_{ei} m_e j_B/e$. This is balanced by the momentum transfer to the passing electrons from the trapped electrons. Due to the slow velocity of the trapped electrons along the field lines, their distribution is limited to a region $\sim \sqrt{\epsilon}$ of velocity space, and the effective collision frequency determined by the diffusion time of this region $\nu_{\text{eff}}^{-1} \sim (\sqrt{\varepsilon})^2\, \nu_{ee}^{-1}$. The momentum exchange from trapped to passing electrons is $(\nu_{ee}/\varepsilon)\, m_e\, j'/e$. Hence

$$j_B \sim \nu_{ee}/(\varepsilon,\ \nu_{ei}) j' \sim \frac{\nu_{ee}}{\nu_{ei}} \frac{q}{\sqrt{\varepsilon}} \frac{kT}{B} \frac{\mathrm{d}n}{\mathrm{d}r} \tag{15.73}$$

Since the additional momentum required to drive the bootstrap current is supplied by the trapped electrons moving along the field lines, the resulting current is toroidal.

15.9.2 Resistive Plasma Diffusion – Collisional Pfirsch–Schlüter Regime

Turning now to the case where the collisionality is large $\upsilon_* \gg 1$, we consider first the simple cylindrical plasma. The system is described by modified MHD equations including the effect of collisions, thus the equilibrium equations include

$$\mathbf{E} + \mathbf{v} \wedge \mathbf{B} = \boldsymbol{\eta} \cdot \mathbf{j} \qquad \text{and} \qquad \mathbf{j} \wedge \mathbf{B} = \nabla p \tag{15.74}$$

where $\boldsymbol{\eta}$ is the resistivity tensor. Taking the vector product with $\mathbf{B}$

$$\mathbf{v} = \frac{1}{B^2} (\mathbf{E} \wedge \mathbf{B} - \boldsymbol{\eta}\, \nabla p) \tag{15.75}$$

The first term corresponds to the frozen field condition in ideal MHD discussed in Section 9.1.1. The second term represents a drift due collisions driven by the pressure gradient, corresponding to a diffusion coefficient across the field lines

$$D = \frac{\eta_\perp\, n\, kT}{B^2} = \frac{\eta_\perp\, \beta}{2\mu_0} \approx \frac{\rho^2}{\tau} = D_{\perp\text{clas}} \tag{15.76}$$

as in Section 15.8. $\beta = \frac{p}{B^2/2\,\mu_0}$ is the plasma β parameter, Eq. (9.26)

$$\beta_P = \frac{2\int_0^a p\, r\, \mathrm{d}r}{a^2\, B_\theta^2/2\mu_0} \tag{15.77}$$

In a cylinder the radial velocity,

$$\upsilon_r = \frac{1}{B^2} \left(E_\theta B_z - E_z\, B_\theta - \eta_\perp \frac{\mathrm{d}p}{\mathrm{d}r} \right) \tag{15.78}$$

In a steady state, $\nabla \wedge \mathbf{E} = 0$ and therefore $E_\theta = 0$ and E_z is constant. If the radial velocity is zero $\upsilon_r = 0$ and

$$\frac{\mathrm{d}p}{\mathrm{d}r} = -\frac{E_z\, B_\theta}{\eta_\perp} \tag{15.79}$$

In a tokamak, the toroidal field is much larger than the poloidal so that $E_z \approx E_\parallel = \eta_\parallel j_\parallel$, so that

$$\frac{\mathrm{d}p}{\mathrm{d}r} = -\frac{\eta_\parallel}{\eta_\perp} j_z\, B_\theta \tag{15.80}$$

Integrating by parts and using Ampère's equation

$$2\int_0^a p\, r\, \mathrm{d}r = -\int_0^a \frac{\mathrm{d}p}{\mathrm{d}r}\, r^2\, \mathrm{d}r = \frac{\eta_\parallel}{\eta_\perp}\int \frac{1}{\mu_0}\frac{1}{r}\frac{\mathrm{d}}{\mathrm{d}r}(r\,B_\theta)\,B_\theta\, \mathrm{d}r = \frac{\eta_\parallel}{\eta_\perp}\frac{B_\theta(a)^2}{2\mu_0}a^2 \tag{15.81}$$

Hence, $\beta_P = \eta_\parallel/\eta_\perp$. In a hydrogen plasma, the ratio $\eta_\parallel/\eta_\perp \approx 0.51$.

15.9.2.1 Pfirsch–Schlüter Current ($\nu_* \gg 1$)

We now consider the case of a twisted toroidal discharge when the collisionality is large, in particular when $T/\tau = R\,q/\tau\, v_T > 1$, the particles fail to complete a banana orbit before suffering a collision. As in the banana case, a toroidal current is established to maintain current balance with the drift generated by the pressure gradient.

If the plasma is in equilibrium, the pressure gradient is balanced by the Lorentz force associated with the internal current $\mathbf{j}_\perp = \dfrac{\mathbf{B}}{B^2}\wedge\nabla p$ perpendicular to the magnetic field $\mathbf{B}$. However, the condition that the current be conserved $\nabla\cdot\mathbf{j} = 0$ requires a current parallel to the field $\mathbf{j}_\parallel$

$$\nabla\cdot\mathbf{j}_\parallel = -\nabla\cdot\mathbf{j}_\perp = -\nabla\cdot\left(\frac{\mathbf{B}}{B^2}\wedge\nabla p\right) = -\nabla p\cdot\nabla\wedge\left(\frac{\mathbf{B}}{B^2}\right) \tag{15.82}$$

$$= -\nabla p\cdot\left[\nabla\left(\frac{1}{B^2}\right)\wedge\mathbf{B} + \frac{\mu_0\mathbf{j}}{B^2}\right] = 2\,\nabla p\cdot\frac{\nabla B}{B^3}\wedge\mathbf{B} \tag{15.83}$$

We approximate $B = B_0(1 - r/R\,\cos\theta)$. Let $\mathrm{d}s$ be an element of length along the field line, then $\mathrm{d}/\mathrm{d}s = \mathrm{d}\theta/\mathrm{d}s\cdot\mathrm{d}/\mathrm{d}\theta = (1/q)\,\mathrm{d}/\mathrm{d}\theta$. Taking care with the sign

$$\frac{\mathrm{d}j_\parallel}{\mathrm{d}s} = -2\nabla p\cdot\frac{\nabla B}{B^3}\wedge\mathbf{B} \tag{15.84}$$

Finally integrating, we obtain the parallel (Pfirsch–Schlüter) current

$$j_{PS} = -\frac{2}{qB}\frac{\mathrm{d}p}{\mathrm{d}r}\cos\theta = -2\frac{1}{B_\theta}\frac{r}{R}\frac{\mathrm{d}p}{\mathrm{d}r}\cos\theta \tag{15.85}$$

15.9.2.2 Diffusion in the Pfirsch–Sclüter Regime

When collisions are strong, the particles do not complete a full banana orbit and their paths are therefore those of passing particles. The particles are deviated from the flux surface by

$$\Delta r \sim q\, v_T/\Omega = q\,\rho \tag{15.86}$$

and the diffusion coefficient

$$D_{PS} \sim \Delta r^2/\tau \sim q^2\,(v_T/\Omega)^2/\tau = q^2\rho^2/\tau = q^2\,D_{\perp\mathrm{clas}} \tag{15.87}$$

is q^2 greater than the classical value. As before the dependence on the classical diffusivity guarantees that Pfirsch–Schlüter diffusion is intrinsically ambipolar.

A more accurate calculation (Section 15.E.1) shows that there is an additional fraction to the diffusive flux of the passing particles $2\,q^2\,\eta_\parallel/\eta_\perp \approx q^2$ to the classical diffusion rate, since in a hydrogen plasma $\eta_\parallel/\eta_\perp \approx 1/2$; the simple result above is in good agreement with more accurate calculation.

15.9.3 Plateau Regime

The bootstrap regime is applicable when the collision frequency $\nu_c (= 1/\tau) < \varepsilon^{3/2}\, v_T/qR$ and the Pfirsch–Schlüter when $\nu > v_T/qR$. At the boundaries of the banana and Pfirsh–Schlüter regimes, the diffusion coefficients are respectively

$$D_\perp \sim \left\{\begin{matrix} \varepsilon^{-3/2} & q^2\,\rho^2\,\nu_c\,(= \varepsilon^{3/2}\,v_T/qR) \\ & q^2\,\rho^2\{\nu_c\,(= v_T/qR)\end{matrix}\right\} = q\,\rho^2\,v_T/R \tag{15.88}$$

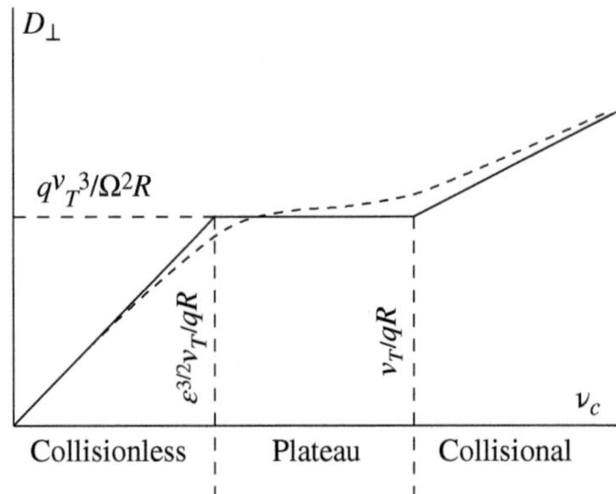

Figure 15.6 Sketch of the variation of cross field diffusion coefficient in toroidal plasmas with collision frequency to illustrate the change from collisionless (banana) to collisional (Pfisch–Schlüter) regimes through an intermediate plateau phase. The dotted line indicates the behaviour of more accurate kinetic calculations.

This result suggests that between the collisionless banana and the collisional Pfirsch–Schlüter regimes there is a region where the diffusion is approximately constant (Figure 15.6 – the *plateau regime*. This conclusion is confirmed by more accurate kinetic calculations.

$$D_P \sim q\, v_T{}^3/\Omega^2\, R \tag{15.89}$$

However, unlike this simple picture where the diffusivity in the plateau region is constant, the more accurate calculations show a slow variation with the collision frequency as illustrated in Figure 15.6 by the dotted line.

15.9.4 Diffusion in Tokamak Plasmas

In both the collisionless banana and the collisional Pfirsch–Schlüter regimes, the jump is a constant factor of Larmor radius. It follows from the random walk model that the conditions imposed for like (p. 396) and unlike particles (p. 396) by the conservation of momentum must hold generally in a tokamak plasma. Therefore, the electronic and ionic diffusion is a constant factor multiplied by the classical rate. Hence, the cross field diffusion rate is the same for both ions and electrons $D_{\perp i} = D_{\perp e}$ and we directly obtain for both ions and electrons

$$\left.\begin{array}{lcl} D_{\perp e} & \approx & q/\varepsilon^{3/2} \\ D_{\perp PS} & \approx & q^2 \end{array}\right\} \frac{\rho_e^2}{\tau_{ei}} \tag{15.90}$$

where $\rho_e = v_{Te}/\sqrt{m_e}\, e\, B$ and $\tau = \tau_{ei}$. As before, the plasma is intrinsically ambipolar.

The thermal diffusivity is determined by a random walk in exactly the same way as the diffusion coefficient except that as noted earlier the electron and ion components are not equal. Taking the different relaxation times into account

$$\begin{array}{lcl} \left.\begin{array}{lcl} \chi_{\perp Be} & \approx & q/\varepsilon^{3/2} \\ \chi_{\perp PSe} & \approx & q^2 \end{array}\right\} \frac{\rho_e^2}{\tau_{ei}} \\ \left.\begin{array}{lcl} \chi_{\perp Be} & \approx & q/\varepsilon^{3/2} \\ \chi_{\perp PSe} & \approx & q^2 \end{array}\right\} \frac{\rho_i^2}{\tau_{ii}} \end{array} \tag{15.91}$$

Using Eq. (2.25), the cross field ion thermal diffusivity is therefore approximately $\sqrt{2\,M/m} \approx 60$ times that of the electrons.

Appendix 15.A Equilibrium Maintaining 'Vertical' Field

A clear picture of the magnetic field in the vertical direction is obtained by considering the force resulting from the gradient in the magnetic energy of the plasma system. This is particularly simple in the case of a large aspect ratio torus, where the major radius R_0 is much larger than the minor a, i.e. $\varepsilon^{-1} = R_0/a \gg 1$. The inductance of a thin toroidal ring, radius R_0, of circular cross section of radius a is the sum of the internal and external self-inductances. As the curvature is small, the conductor may be treated as a long wire and therefore $B_\theta = \mu_0 I_\phi / 2\,\pi\,a$. The internal self-inductance of the conductor itself L_{in} is calculated from the magnetic energy stored in the inductance $\frac{1}{2} L_{\text{in}}\, I_\phi{}^2$ which is equated to the magnetic energy in the field $\overline{B_\theta{}^2}/2\,\mu_0\,V$ where $V = 2\,\pi^2\,a^2\,R_0$ is the volume of the field

$$L_{\text{in}} = \frac{2\,\pi^2\,a^2}{\mu_0}\overline{B_\theta{}^2}\,R_0 \Big/ \frac{4\,\pi^2\,a^2}{\mu_0^2}\,B_\theta^2 = \frac{\mu_0}{2}\,R_0\,\frac{\overline{B_\theta{}^2}}{B_\theta{}^2(a)} = \frac{\mu_0}{2}\,R_0\,\ell_i \tag{15.A.1}$$

The internal inductance parameter ℓ_i is defined by the mean field squared as

$$\ell_i = \frac{\overline{B_\theta{}^2}}{B_\theta{}^2(a)} = \frac{2\,\pi\,\int_0^a B_\theta{}^2(r)\,r\,\mathrm{d}r}{\pi\,a^2 B_\theta{}^2(a)} \tag{15.A.2}$$

and is clearly determined by the current distribution across the torus, for example if the current is uniformly distributed $\ell_i = 1/2$.

The external self-inductance of a thin circular ring of radius R_0 and thickness a given by (Smythe, 1968, §8.10)

$$L_{ex} = \mu_0\,R_0\left[\ln\left(\frac{8\,R_0}{a}\right) - 2\right] \tag{15.A.3}$$

The total self-inductance of a toroid treated as thin circular ring of major radius R_0 and minor radius a is therefore

$$L_\phi = \mu_0\,R_0\left[\ln\left(\frac{8\,R_0}{a}\right) - 2 + \frac{\ell_i}{2}\right] \tag{15.A.4}$$

and gives rise to stored energy

$$U_1 = \frac{1}{2}\,L_\phi\,I_\phi{}^2 \tag{15.A.5}$$

Calculating the radial variation of the magnetic energy due to the poloidal fields, we must take into account the expansion of the field against the field at the outer edge, so that the effective magnetic energy per unit volume

$$U_2 = \frac{1}{2\,\mu_0}(\overline{B_\phi{}^2} - B_\phi{}^2(a)) = \left(\frac{1}{2\,\mu_0}B_\theta{}^2(a) - \overline{p}\right) \tag{15.A.6}$$

where the final equality makes use of Eq. (15.6) since the aspect ratio is large.

The radial force due to the energy term U_1 is

$$F_{R_1} = \frac{\partial U_1}{\partial R_0} = \mu_0\left[\ln\left(\frac{8\,R_0}{a}\right) - 1 + \frac{\ell_i}{2}\right] I_\phi{}^2 \tag{15.A.7}$$

and due to the term U_2

$$F_{R_2} = \frac{\partial U_2}{\partial R_0} = -\frac{\partial}{\partial R}\left[\frac{1}{2\,\mu_0}(\overline{B_\phi{}^2} - B_\phi{}^2(a))V\right] = \frac{2\,\pi^2\,a^2}{2\,\mu_0}(\overline{B_\phi{}^2} - B_\phi{}^2(a)) \tag{15.A.8}$$

and finally due to the pressure

$$F_{R_p} = \frac{\partial}{\partial R}(pV) = 2\,\pi^2\,a^2\,\overline{p} \tag{15.A.9}$$

where $V = 2\,\pi^2 a^2 R_0$ is the plasma volume.

Summing the force terms and making use of Eq. (15.2) we obtain

$$F_R = \mu_0 {I_\phi}^2 \left[\ln\left(\frac{8\,R_0}{a}\right) + \frac{\ell_i}{2} - \frac{3}{2} + \frac{2\,\mu_0\,\overline{p}}{{B_\theta}^2(a)}\right] \tag{15.A.10}$$

In equilibrium, this force is balanced by the Lorentz force generated by the toroidal current and the vertical magnetic field: $2\,\pi\,R_0\,I_\phi\,B_Z$. Hence, the vertical magnetic field required to ensure equilibrium is

$$B_Z = \frac{a}{2\,R_0} B_\theta(a) \left[\ln\left(\frac{8\,R_0}{a}\right) + \beta_\theta + \frac{\ell_i}{2} - \frac{3}{2}\right] \tag{15.A.11}$$

in agreement with the later derivation Eq. (15.B.19) using the perturbation approach in Appendix 15.B: $\beta_\theta = 2\mu_0\,\overline{p}/B_\theta(a)^2$ is the poloidal β.

Appendix 15.B Perturbation Solution of the Grad–Shafranov Equation

A relatively simple solution to the Grad–Shafranov equation is obtained by expanding the differential operator as a series in the inverse aspect ratio, which is assumed to be small $a/R_0 \ll 1$. The solution assumes that the outer magnetic surface, where the pressure p and toroidal current density j_ϕ are zero, is circular, i.e. a circle of radius a in the poloidal plane.

A simple solution using the perturbation approach is obtained by using the simple Solov'ev (1968) form for the pressure $p(\psi)$ and poloidal current $I_\phi(\psi)$ flowing inside the magnetic surface ψ which we write as

$$p(\psi) = p_0\left(1 + k_1\,\frac{\psi}{\psi_0}\right) \qquad \text{and} \qquad {I_\phi}^2(\psi) = {I_0}^2(1 + k_2\,\psi/\psi_0) \tag{15.39}$$

The pressure and the poloidal magnetic field term $RB_\theta \sim I_p hi(\psi)$ are therefore linear functions of the parameter ψ.[10]

We transform to co-ordinates in the poloidal plane (r, θ, z) where $(R = R_0 + r\cos\theta)$ (Figure 15.1) to obtain the modified Grad–Shafranov equation

$$\begin{aligned}\mathcal{L}(\psi) &= \left[\frac{\partial^2}{\partial r^2} + \frac{1}{r}\frac{\partial}{\partial r} + \frac{1}{r^2}\frac{\partial^2}{\partial \theta^2} - \frac{1}{(R_0 + r\,\cos\theta)}\left(\cos\theta\,\frac{\partial}{\partial r} - \frac{\sin\theta}{r}\frac{\partial}{\partial\,\theta}\right)\right]\psi \\ &= -\mu_0(R_0 + r\,\cos\theta)^2\,\frac{\mathrm{d}p}{\mathrm{d}\,\psi} + \frac{{\mu_0}^2}{8\pi^2}\frac{\mathrm{d}I^2}{\mathrm{d}\psi}\end{aligned} \tag{15.B.1}$$

The constant ψ_0 is as yet undefined so we set

$$\psi_0 = \mu_0\,{R_0}^2\,\frac{k_1\,p_0}{\psi_0} + \frac{{\mu_0}^2}{8\,\pi^2}\,\frac{k_2\,{I_0}^2}{\psi_0} \tag{15.B.2}$$

10 This linear assumption is unphysical as the current varies in the plasma due to a non-uniform electrical conductivity. The assumption is made for simplicity to demonstrate the perturbation method. We note the same approximation is made as the particular integral used in the analytic expansion of the Grad–Shafranov equation Section 15.7.2.

Substituting the approximate forms (15.39) for the inner ($r < a$) region containing plasma and the outer ($r > a$) in vacuo

$$\mathcal{L}(\psi) = \begin{cases} -\mathcal{K}\left[\dfrac{(R_0 + r\cos\theta)^2}{{R_0}^2} - 1\right] - \dfrac{\psi_0}{a^2} & \text{inner} \\ 0 & \text{outer} \end{cases} \tag{15.B.3}$$

where $\mathcal{K} = \mu_0\, k_1\, {R_0}^2\, p_0/\psi_0$.

We assume the torus has a large aspect ratio so that $a \ll R_0$ and we may expand the differential equation in terms of the ratio $a/R_0 \ll 1$. To lowest order

$$\mathcal{L}(\psi^{(0)}) = \frac{\partial^2\psi^{(0)}}{\partial r^2} + \frac{1}{r}\frac{\partial\psi^{(0)}}{\partial r} = \begin{cases} -\dfrac{\psi_0}{a^2} & \text{inner} \\ 0 & \text{outer} \end{cases} \tag{15.B.4}$$

whose solution subject to the boundary conditions that ψ is continuous at the boundary $r = a$ and that the pressure and toroidal current are both zero $p = j_\phi = 0$ at $r = a$ giving

$$\psi^{(0)} = \begin{cases} -\psi_0/4\,a^2\,r^2 & \text{inner} \\ -\left[\frac{1}{4} + \frac{1}{2}\ln(r/a)\right]\psi_0 & \text{outer} \end{cases} \tag{15.B.5}$$

Turning now to the terms in first order in a/R_0, we obtain

$$\mathcal{L}(\psi^{(1)}) = \left(\frac{\partial^2}{\partial r^2} + \frac{1}{r}\frac{\partial}{\partial r} + \frac{1}{r^2}\frac{\partial}{\partial\theta^2}\right)\psi^{(1)} = \begin{cases} \dfrac{1}{R_0}\cos\theta\,\dfrac{\partial\psi^{(0)}}{\partial r} - 2\,\mathcal{K}\,\dfrac{r}{R_0} & \text{inner} \\ \dfrac{1}{R_0}\cos\theta\,\dfrac{\partial\psi^{(0)}}{\partial r} & \text{outer} \end{cases} \tag{15.B.6}$$

These equations may be solved by assuming that equations are separable

$$\psi^{(1)}(r,\theta) = \psi'^{(1)}(r)\cos\theta \tag{15.B.7}$$

Integrating the first order equations subject to the condition of continuity at the boundary $r = a$, we obtain

$$\psi(r,\theta) = \begin{cases} -\dfrac{r^2}{4\,a^2}\psi_0 + \dfrac{a}{8\,R_0}\left(\dfrac{\psi_0}{2} + 2\,a^2\,\mathcal{K}\right)\left(\dfrac{r}{a} - \dfrac{r^3}{a^3}\right)\cos\theta & \text{inner} \\ -\left[\dfrac{1}{4} + \dfrac{1}{2}\ln\left(\dfrac{r}{a}\right)\right]\psi_0 & \\ +\frac{r}{8\,R_0}\left[\left(\frac{\psi_0}{2} - 2\,a^2\,\mathcal{K}\right)\left(1 - \frac{a^2}{r^2}\right) + 2\,\psi_0\ln\left(\frac{a}{r}\right)\right]\cos\theta & \text{outer} \end{cases} \tag{15.B.8}$$

The poloidal field is calculated from Eq. (15.21)

$$B_r = -\frac{1}{(R + r\cos\theta)}\frac{1}{r}\frac{\partial\psi}{\partial\theta} \quad \text{and} \quad B_\theta = \frac{1}{(R + r\cos\theta)}\frac{\partial\psi}{\partial r} \tag{15.B.9}$$

Hence, the θ component of the poloidal field is

$$B_\theta = \begin{cases} B_\theta(a)\left[\dfrac{r}{a} + \dfrac{a}{8R_0}\left(1 + \dfrac{4\,a^2\,\mathcal{K}}{\psi_0}\right)\left(1 - \dfrac{3r^2}{a^2}\right)\cos\theta\right] & \text{inner} \\ B_0(a)\left[\dfrac{a}{r} - \left\{\dfrac{a}{8R_0}\left(1 - \dfrac{4a^2\mathcal{K}}{\psi_0}\right)\left(1 + \dfrac{a^2}{r^2}\right) + \dfrac{a}{2R_0}\left(\ln\left(\frac{e}{r}\right) + 1\right)\right\}\cos\theta\right] & \text{outer} \end{cases} \tag{15.B.10}$$

where

$$B_\theta(a) = -\psi_0/(2\,R_0\,a) \tag{15.B.11}$$

is the poloidal field at the plasma surface in the zero order approximation.

To zero order the pressure variation across the poloidal disc is given by Eq. (15.B.5). Substituting $k_1 = 4$, we obtain

$$p(r) = p_0\,(1 - r^2/a^2) \tag{15.B.12}$$

Therefore, the mean pressure $\overline{p} = 2\,\pi\,\int_0^a p(r)\,r\,\mathrm{d}r/\pi\,a^2 = \frac{1}{2}\,p_0$. We define the poloidal plasma beta factor at the plasma minor radius by

$$\beta_\theta = \frac{2\,\mu_0\,\overline{p}}{B_\theta{}^2(a)} = \frac{a^2\,\mathcal{K}}{\psi_0} \tag{15.B.13}$$

Consequently, the general plasma β factor across the poloidal surface is given by

$$\beta_\theta(r) = \frac{2\,\mu_0}{B_\theta{}^2(a)}(\overline{p} - p(r)) \tag{15.B.14}$$

Making use of Eq. (15.28) and the zero order Grad–Shafranov equation (15.B.4) together with the zero order solution (15.B.5) it follows that to zero order in this approximation the current density j_ϕ is uniform and the magnetic field $B_\theta(r) \propto r$. Therefore, $\beta_\theta(r)$ is constant in this case.

In a similar fashion, we define the internal inductance parameter, which is simply the normalised internal inductance of a long straight conductor modified to take into account the current distribution. Using Eq. (15.A.2) we have for the internal inductance

$$\ell_i = \frac{2\,\pi\,\int_0^a B_\theta{}^2(r)\,r\,\mathrm{d}r}{\pi\,a^2 B_\theta{}^2(a)}$$

which in this particular case where the current density is uniform has the value $\ell_i = \frac{1}{2}$. If the current is peaked near the centre of the discharge $\ell_i > \frac{1}{2}$, otherwise if near the edge $\ell_i < \frac{1}{2}$. We shall show that the replacement $\frac{1}{2} \Rightarrow \ell_i$ allows us to generalise these results for non-uniform current distributions.

Returning to Eq. (15.B.10) for the first order poloidal magnetic fields, we can rewrite the outer field in the form

$$\begin{aligned} B_\theta(r,\theta) = {} & B_\theta(a)\left\{\frac{a}{r} - \frac{a}{2\,R}\left[\ln\left(\frac{8\,R}{r}\right) - \left(\beta_\theta + \frac{1}{2}\ell_i - \frac{1}{2}\right)\left(\frac{a}{r}\right)^2\right]\,\cos\theta\right\} \\ & \underbrace{+\,B_\theta(a)\,\frac{a}{2\,R}\left\{\ln\left(\frac{8\,R}{r}\right) + \beta_\theta + \frac{1}{2}\,\ell_i - \frac{3}{2}\right\}\cos\theta}_{\text{Constant vertical magnetic field: } B_Z} \end{aligned} \tag{15.B.15}$$

generalised to allow for non-uniform current density. The last term is independent of the radius r is a constant field. The factor $\cos\theta$ ensures that this field is in the $\hat{\mathbf{z}}$ direction, i.e. at an angle θ to the azimuthal direction $\hat{\mathbf{e}}_\theta$. It is therefore the vertical field component B_Z, which is required to ensure equilibrium.

Turning now to the inner flux surfaces ψ = const, we obtain from Eq. (15.B.8)

$$r^2 + \frac{1}{R_0}\left(\frac{1}{4} + \frac{a^2\,\mathcal{K}}{\psi_0}\right)(a^2 - r^2)\,r\,\cos\theta = 4\,a^2\,\frac{\psi}{\psi_0} \tag{15.B.16}$$

The equation of a circle radius ρ whose centre is displaced by $\boldsymbol{\Delta}$ is given by

$$r^2 + 2\,r\,\Delta\,\cos\theta + \Delta^2 = \rho^2 \tag{15.B.17}$$

from which it can be seen that flux surfaces are approximately circles displaced by a distance

$$\Delta = \frac{1}{2\,R_0}\left(\beta_\theta(r) + \frac{\ell_i(r)}{2}\right)(a^2 - r^2) \approx \frac{1}{2\,R_0}\left(\beta_\theta(r) + \frac{\ell_i(r)}{2}\right)(a^2 - r_0^2) \tag{15.B.18}$$

from the minor axis outwards away from the major axis, where $r_0 = \sqrt{4\,a^2\psi/\psi_0}$ is the zero order radius of the magnetic surface ψ. The local internal inductance parameter at radius r is defined as

$$\ell_i(r) = \frac{2\,\pi\,\int_0^r B_\theta{}^2(r)\,r\,\mathrm{d}r}{\pi\,r^2 B_\theta{}^2(r)} \tag{15.A.2:a}$$

This shift of the centre of the magnetic surface is known as the *Shafranov shift*.

The constant vertical magnetic field

$$B_Z = B_\theta(a)\,\frac{a}{2\,R}\left\{\ln\left(\frac{8\,R}{r}\right) + \beta_\theta + \frac{1}{2}\,\ell_i - \frac{3}{2}\right\} \tag{15.B.19}$$

is required to support the plasma by the Lorentz force against the expansion due to gradients in the kinetic and magnetic pressures discussed in Section 15.4, items (a) and (b) respectively. We note that this calculation is in agreement with that given earlier Section 15.4, item 15.A

The agreement of these results enables some useful conclusions:

a. The earlier derivation is more general, independent of the assumption made regarding the terms in Eq. (15.39) which infer a uniform distribution of pressure and current across the cross section.
b. This more general calculation leads naturally to the inclusion of the internal inductance parameter, rather than its uniform current value $\frac{1}{2}$ in the calculation of the 'vertical' magnetic field, and consequently for consistency in the external, Eq. (15.B.15), where the term $(\beta_\theta + \ell_i/2)$ consequently appears consistently.

Appendix 15.C Analytic Solutions of the Homogeneous Grad–Shafranov Equation

Following Zheng et al. (1996), we express the solutions of the homogeneous Grad–Shafranov equation symmetric Z in the form

$$\psi = \sum_{n=0,2,\ldots} \alpha_n(R)\,Z^n = \sum_{n=0,2,\ldots}\;\sum_{k=0}^{k=n/2} G(n,k,R)Z^{n-2k} \tag{15.C.1}$$

Differentiating with respect to Z the Grad–Shafranov equation becomes

$$R\frac{\partial}{\partial R}\left(\frac{1}{R}\frac{\partial\psi}{\partial R}\right) = -\frac{\partial^2\psi}{\partial Z^2} = -\sum_n\sum_k (n-2k)(n-2k-1)G(n,k,R)Z^{(n-2k-2)} \tag{15.C.2}$$

Hence equating terms $Z^{(n-2k)}$, we obtain the differential equations for $G(n,k,R)$

$$R\frac{\partial}{\partial R}\left[\frac{1}{R}\frac{\partial G(n,0,R)}{\partial R}\right] = 0 \tag{15.C.3a}$$

$$R\frac{\partial}{\partial R}\left[\frac{1}{R}\frac{\partial G(n,k,R)}{\partial R}\right] = -(n-2k+1)(n-2k+2)G(n,(k-1),R) \quad \left\{\frac{n}{2} \geq k > 0\right\} \tag{15.C.3b}$$

It is obvious that Eq. (15.C.3a) allows two independent solutions

$$G_1(n,0,R) = 1 \qquad \text{and} \qquad G_2(n,0,R) = R^2 \tag{15.C.4}$$

and therefore that the general solution for $G(n,k,R)$ contains two independent families of terms, each of which is individually a solutions of the homogeneous Grad–Shafranov equation:

$$G(n,k,R) = g_1 G_1(n,k,R) + g_2 G_2(n,k,R) \tag{15.C.5}$$

The integrals for $G(n, k, R)$ are most easily established by induction to be

$$G_1(n, 0, R) = 1 \qquad \text{and} \qquad G_2(n, 0, R) = R^2 \tag{15.C.6a}$$

$$G_1(n, k > 0, R) = (-1)^k \frac{n!}{(n-2k)!)} \frac{1}{2^{2k}k!(k-1)!} R^{2k} \times \left(2\ln(R) + \frac{1}{k} - 2\sum_{j=1}^{k} \frac{1}{j}\right) \tag{15.C.6b}$$

$$G_2(n, k, R) = (-1)^k \frac{n!}{(n-2k)!} \frac{1}{2^{2k}k!(k+1)!} R^{2k+2} \tag{15.C.6c}$$

For general values of $k \geq 1$, the corresponding derivatives are

$$\frac{1}{R}\frac{\partial G_1}{\partial R} = (-1)^k \frac{n!}{(n-2k)!} \frac{1}{2^{(2k-1)}[(k-1)!]^2} R^{2(k-1)} \left(\left[2\ln R - 2\sum_{j=1}^{k-1}\frac{1}{j}\right]\right)$$

$$\begin{aligned} R\frac{\partial}{\partial R}\left(\frac{1}{R}\frac{\partial G_1}{\partial R}\right) &= (-1)^k \frac{n!}{(n-2k)!} \frac{1}{2^{(2(k-1))}(k-1)!\,(k-2)!} \\ &\quad \times R^{2(k-1)} \left\{ 2\left(\left[\ln R - \sum_{j=1}^{k-1}\frac{1}{j}\right]\right) + \frac{1}{(k-1)} \right\} \\ &= -[n-(2k-1)][n-2(k-1)]G_1(n,(k-1),R) \end{aligned} \tag{15.C.3}$$

and

$$\frac{1}{R}\frac{\partial G_2}{\partial R} = (-1)^k \frac{n!}{(n-2k)!} \frac{1}{2^{(2k-1)}[k!]^2} R^{2k}$$

$$\begin{aligned} R\frac{\partial}{\partial R}\left(\frac{1}{R}\frac{\partial G_2}{\partial R}\right) &= (-1)^k \frac{n!}{(n-2k)!} \frac{1}{2^{2(k-1)}[(k-1)!k!} R^{2k} \\ &= -[n-(2k-1)][n-2(k-1)]G_2(n,(k-1),R) \end{aligned} \tag{15.C.4}$$

Therefore, since Eqs. (15.C.6b) and (15.C.6c) satisfy Eq. (15.C.3) for $k = 1$ by Eq. (15.C.6a), they do so for all values of k.

It is clear that the sums $\sum_{k=1}^{n/2} G_1(n, k, R)$ and $\sum_{k=1}^{n/2} G_2(n, k, R)$ are each of terms of constant total order[11] each yield an independent exact integral of the Grad–Shafranov equation. Any solution is therefore obtained by the general sum over terms of increasing order

$$\psi = \sum_{n,2,\ldots} \left[g_{1n} \sum_{k=1}^{n/2} G_1(n,k,R) + g_{2n} \sum_{k=1}^{n/2} G_2(n,k,R) \right] Z^{(n-2k)} \tag{15.C.5}$$

with appropriate values of the coefficients g_{1n} and g_{2n} determined by the geometrical configuration.

Appendix 15.D Guiding Centre Motion in a Twisted Circular Toroidal Plasma

Let us consider the path traced out by the guiding centre of charged particle orbits in a toroid with circular cross section,[12] and twisted magnetic field lines with rotational transform ι (Eq. (15.8)).

11 The total order being the sum of the powers of the terms in R and Z, namely n and $(n + 2)$, respectively for the terms G_1 and G_2.

12 A circular cross section allows a straightforward quantitative analytic treatment. The results may be expected to be qualitatively accurate for more complex forms.

The toroid flux line is assumed to have a large aspect ratio $\varepsilon^{-1} = R_0/r \gg 1$. As discussed earlier the field in a tokamak is made up with two components, the primary toroidal magnetic field B_ϕ is due to current along the major axis. The poloidal field B_θ is generated by toroidal currents flowing in the plasma, and induces the rotation of the field lines

$$B_\phi = B_0 / \left(1 + \frac{r}{R_0}\cos\theta\right) \approx B_0\left(1 - \frac{r}{R_0}\cos\theta\right) \qquad \text{and} \qquad B_\theta \approx \frac{\iota\, r}{2\,\pi\, R_0} B_0 \quad (15.D.1)$$

The rotational transform $\iota(r)$ is assumed to be a function of the poloidal radius r only. In a tokamak the toroidal field is much stronger than the poloidal so that $\Theta = B_\theta / B_\phi = \iota\,\epsilon/2\,\pi \ll 1$. Provided $\Theta^2 \ll \epsilon$ the variation in the total field due to the poloidal field can be neglected. We also assume that the plasma β is small so that

$$\beta = n(\not{k}T_i + \not{k}T_e)/(\mu_0 {B_0}^2/2) \ll 1$$

so that the magnetic field is only weakly perturbed by the plasma ($Z = 1$).

In a helical field, the guiding centres of charged particles move along the magnetic surface formed by field lines at a fixed minor radius r with varying angle θ; therefore, projected on to the poloidal plane, the guiding centre trajectories appear as circles of radius r. In a toroidal plasma, the curvature induces drift motions across the lines of force in the z direction.. Since in the approximation $B_\theta \ll B_\phi$ the radius of the field lines varies only slowly along the major axis, we treat the problem as symmetric in ϕ and consider the projection of the guiding centre motion in the poloidal plane alone. Furthermore, we assume the electric potential $\Phi(r)$ is a function of r alone and is sufficiently small that the corresponding drift is much less than the particle speed. The guiding centre equations of motion are then

$$\frac{\mathrm{d}r}{\mathrm{d}t} = -\frac{(\frac{1}{2}{v_\perp}^2 + {v_\parallel}^2)\sin\theta}{\Omega\, R_0} \qquad (15.D.2a)$$

$$r\frac{\mathrm{d}\theta}{\mathrm{d}t} = -\frac{(\frac{1}{2}{v_\perp}^2 + {v_\parallel}^2)\cos\theta}{\Omega\, R_0} + \frac{q}{m\,\Omega}\frac{\mathrm{d}\Phi}{\mathrm{d}r} + \frac{\iota\, v_\parallel\, r}{2\,\pi\, R_0} \qquad (15.D.2b)$$

where the term containing ι in Eq. (15.D.2b) takes into account the rotational transform about the minor axis of the toroid. It is easily seen that if the drift terms are zero, the rotational transform gives rise to the circular motion in the poloidal plane with constant radius r and angular velocity $\omega = \mathrm{d}\theta/\mathrm{d}t = \iota v_\parallel / R_0$ as in Eq. (15.60).

Eliminating the time

$$\left(\frac{1}{2}{v_\perp}^2 + {v_\parallel}^2\right)\sin\theta\, r\frac{\mathrm{d}\theta}{\mathrm{d}r} + \left(\frac{1}{2}{v_\perp}^2 + {v_\parallel}^2\right)\cos\theta = \left(\frac{1}{2}{v_\perp}^2 + {v_\parallel}^2\right)\frac{\mathrm{d}}{\mathrm{d}r}(r\cos\theta)$$
$$= \frac{q}{m} R_0 \frac{\mathrm{d}\Phi}{\mathrm{d}r} + \frac{\iota}{2\,\pi}\,\Omega\, v_\parallel\, r \qquad (15.D.3)$$

Both the energy $E = \frac{1}{2}({v_\perp}^2 + {v_\parallel}^2) + q\,\Phi(r)\,/\,m$ and the first adiabatic invariant (magnetic moment) $\tilde{\mu} = {v_\perp}^2\,/\,2\,B_\phi(r)$ are constants of motion, and approximating the toroidal field as $B_\phi(r) \approx B_0(1 - r/R_0\cos\theta)$, differentiating the energy invariant it follows that

$$\frac{q}{m}\frac{\mathrm{d}\Phi}{\mathrm{d}r} = -v_\parallel\frac{\mathrm{d}v_\parallel}{\mathrm{d}r} + \frac{\tilde{\mu}B_0}{R_0}\frac{\mathrm{d}}{\mathrm{d}r}(r\cos\theta) \qquad (15.D.4)$$

Substituting we obtain

$$\frac{v_\parallel}{\Omega\, R_0}\frac{\mathrm{d}}{\mathrm{d}r}(r\cos\theta) = \frac{\iota r}{2\,\pi\, R_0} - \frac{1}{\Omega}\frac{\mathrm{d}v_\parallel}{\mathrm{d}r} \qquad (15.D.5)$$

Assuming the normal condition $\mathrm{d}v_{\|}/dr \ll v_{\|}/r$, we may integrate this equation holding $v_{\|}$ constant on the left-hand side to obtain a constant of motion

$$J = v_{\|}\left(1 + \frac{r}{R_0}\cos\theta\right) - \frac{\Omega}{2\pi R_0}\int_{r_0}^{r} \iota\, r\,\mathrm{d}r \qquad (15.D.6)$$

which is analogous to the usual longitudinal invariant.

Since the deviation from the undrifted circular orbit is small, an expansion of the invariant (15.D.6) between the reference point (r_0, θ_0) and an arbitrary point on the orbit (r, θ) may be made to terms of second order in $(r - r_0)$. Neglecting the terms and making use of the energy invariant, Berk and Galeev (1967) obtain[13]

$$\frac{1}{2}\{\Omega\,\Theta^2 + [-v_{\|}\,\dot{\Theta} + \dot{v}_E]\}\,(r - r_0)^2 + \Delta v_{\|}\,\Theta\,(r - r_0) - v_g\,(r\,\cos\theta - r_0\,\cos\theta_0) = 0 \qquad (15.D.10)$$

The derivative terms in square brackets $[\cdots]$, associated with the gradients in the rotational transform $\dot{\Theta}$ and electric field drift $\dot{v}_E$, are normally small and usually omitted. We will follow this practice.

The magnetic field ratio $\Theta = B_\theta/B_\phi = \varepsilon\,\iota/2\,\pi = \varepsilon/q$ and the electric field drift speed in the poloidal plane around the minor axis $v_E = \dfrac{q}{m\Omega}\dfrac{\mathrm{d}\Phi}{dr} = \dfrac{1}{B_\phi}\dfrac{\mathrm{d}\Phi}{dr}$. The toroidal electric field drift velocity along the minor axis is therefore v_E/Θ. We have set

$$v_{\|0} = \pm\{2\,[E - \tilde{\mu}\,B_0 - (q/m)\Phi(r_0)]\}^{1/2} \qquad \Delta v_{\|} = v_{\|0} - v_{E0}/\Theta \qquad v_g = \frac{\tilde{\mu}B_0 + v_{\|0}^{\;2}}{\Omega R_0} \qquad (15.D.11)$$

Letting the reference point be on the mid-plane $(r_0, 0)$; it is clear that the motion is symmetric about the mid-plane in the poloidal projection, i.e. $r(-\theta) = r(\theta)$.

Solving the quadratic equation

$$r - r_0 = \frac{1}{\Omega\,\Theta}\{-\Delta v_{\|} \pm \sqrt{\Delta v_{\|}^2 + 2\,\Omega\,v_g\,r_0\,(\cos\theta - \cos\theta_0)}\} \qquad (15.D.12)$$

if

$$4\,\Omega\,v_g\,r_0\,\sin^2(\theta/2) \geq \Delta v_{\|}^{\;2} \qquad (15.D.13)$$

---

13 To establish the equation of path, we integrate Eq. (15.D.10) using the trapezoidal rule to generate an approximation to the integral in second order in $(r - r_0)$

$$J(r,\theta) - J(r_0,\theta_0) \approx \frac{1}{2}\left\{\left.\frac{\mathrm{d}J}{\mathrm{d}r}\right|_r + \left.\frac{\mathrm{d}J}{\mathrm{d}r}\right|_{r_0}\right\}(r - r_0) = 0 \qquad (15.D.7)$$

But it follows from the derivation leading to Eq. (15.D.6) that the derivative of the invariant J is given by Eq. (15.D.3)

$$\frac{\mathrm{d}J}{\mathrm{d}r} = \frac{\iota}{2\pi}\frac{1}{2}(v_{\|} + v_{\|0})\frac{r}{R_0} + \frac{q}{m\,\Omega}\frac{\mathrm{d}\Phi}{\mathrm{d}r} - v_g\frac{\mathrm{d}}{\mathrm{d}r}(r\,\cos\theta)$$

$$\approx \Theta\left\{\frac{1}{2}(v_{\|} + v_{\|0}) + v_E\right\} - v_g\frac{(r\cos\theta - r_0\cos\theta_0)}{(r - r_0)} \qquad (15.D.8)$$

and we set $v_E = [v_E|_{r_0} + \dot{v}_E|_{r_0}(r - r_0)/2]$ etc. to maintain the second order expansion and obtain the square bracketed terms in Eq. (15.D.10), noting that since $v_\perp \gg v_{\|}$, v_g does not depend on B_ϕ and therefore $\dot{v}_g \approx 0$. But since $\varepsilon = r/R_0 \ll 1$

$$J(r,\theta) - J(r_0,\theta_0) = v_{\|} - v_{\|0} - \frac{1}{2}\frac{\iota}{2\pi R_0}\Omega(r^2 - r_0^{\;2}) = 0 \qquad (15.D.9)$$

Eliminating $v_{\|}$ Eq. (15.D.10) follows.

If the velocity $\Delta v_{\parallel} > 0$ is positive, the sign of the surd in Eq. (15.D.12) is positive and *vice versa*. When the polar angle $\theta = \theta_{\text{ref}}$, where $\theta_{\text{ref}} = 2\arcsin\left\{\sqrt{\Delta v_{\parallel}{}^2 / 4\,\Omega\, v_g\, r_0}\right\}$, the particle is reflected and its path reverses to form a banana orbit (in the poloidal plane). Since for trapped particles $v_{\parallel}/v_{\perp} \lessapprox \sqrt{2\varepsilon}$ and $\varepsilon \ll 1$ (equation 15.65) it follows that $v_g \approx \frac{1}{2} v_{\perp}^2 / \Omega\, R_0$. Hence Eq. (15.62) follows directly from Eq. (15.13).

Away from the turning point, $\Delta v_{\parallel}$ is not too small and the displacement takes the simpler form

$$r - r_0 \approx \text{sgn}(\Delta v_{\parallel})\, (\tilde{\mu} B_0 + v_{\parallel 0}^{\,2})/(\Delta v_{\parallel}\, \Omega\, \Theta\, R_0) r_0\, (\cos\theta - \cos\theta_0) \tag{15.D.14}$$

which may be applied to passing particles and to trapped particles not near the turning point. If the particle is trapped, its motion is along the positive branch from $\theta = 0$ to $\theta = \theta_{\text{ref}}$, returning along the negative branch to $\theta = 0$, and then by the similar, but converse, path in the other half-plane as in Figure 15.5. The maximum deviation occurs on the return branch at $\theta = 0$.

Let us contrast the deviation suffered by particles whose velocity is only slightly different from the trapping limit when $\theta_{\text{ref}} \approx \pi$ and $\Delta v_{\parallel}{}^2 = 4\,\Omega\, r_0\, v_g \approx 4\,\tilde{\mu}\, B_0\, r_0 / R_0$ (since $v_{\parallel}/v_{\perp} \approx \sqrt{2\epsilon} \ll 1$).

- **Passing particle** which is just untrapped, the maximum deviation at $\theta = \pi$ is

$$r - r_0 = \Delta v_{\parallel} \,/\, \Omega\, \Theta \approx 2(\tilde{\mu}\, B_0\, r_0 / R_0)^{1/2} / \Omega\, \Theta \tag{15.D.15:a}$$

- **Trapped particle** which is just trapped , the maximum deviation at $\theta = 0$ is

$$r - r_0 = 2\,\Delta v_{\parallel} \,/\, \Omega\, \Theta \approx 4(\tilde{\mu}\, B_0\, r_0 / R_0)^{1/2} / \Omega\, \Theta \tag{15.D.15:b}$$

The equivalence of these two values should not surprise us simply reflecting the fact that the trapped and untrapped paths are almost identical over the first part of their paths from $\theta = 0$ to $\theta = \pi$, and that the trapped particle travelling from $\theta = \pi$ back to $\theta = 0$ along the other branch adds the same term a second time.

We extend the guiding centre calculation to obtain an expression for the period of the oscillation of the trapped particles. Writing Eq. (15.D.2b) in the form

$$r\frac{\mathrm{d}\theta}{\mathrm{d}t} = \frac{\iota\, r}{2\,\pi\, R_0}\, \Delta v_{\parallel} \tag{15.D.16}$$

neglecting the gradient and rotational drift terms. Since the drift terms are neglected, the particle motion is along the field line in the toroidal space; but follows the radial lines $r = r_0$ in the poloidal plane from $\theta = 0$ to $\theta = \theta_{\text{ref}}$ is reflected and retraces its path from θ_{ref} to 0 with the parallel component of velocity $v_{\parallel} \to -v_{\parallel}$. Unlike the velocity parallel to the field lines $v_{\parallel}$, the electric field drift v_E/Θ does not reverse on reflection, and therefore nor does the total velocity $\Delta v_{\parallel}$. Kadomtsev and Pogu'tsev (1967) circumvented the problem caused thereby setting the potential to zero $\Phi(r) = 0$ and $\Delta v_{\parallel} = v_{\parallel}$, so that

$$\begin{aligned} v_{\parallel} &= \sqrt{v_{\parallel 0}^{\,2} + v_{\perp 0}^{\,2} - v_{\perp}^2} = \sqrt{v_{\parallel 0}^{\,2} - \tilde{\mu}\, B_0\, \varepsilon (1 - \cos\theta)} \\ &= \sqrt{\tilde{\mu}\, B_0\, \varepsilon}\, \sqrt{2\varkappa^2 - (1 - \cos\theta)} = \sqrt{2\,\tilde{\mu}\, B_0\, \varepsilon}\, \sqrt{\varkappa^2 - \sin^2(\theta/2)} \end{aligned} \tag{15.D.17}$$

where $2\,\varkappa^2 = v_{\parallel}{}^2 / \varepsilon\, \tilde{\mu}\, B_0$. The value of the poloidal angle at reflection, when the argument of the surd becomes zero, is conveniently expressed as $\theta_{\text{ref}} = 2\arcsin \varkappa$.

Hence, the period from Eq. (15.D.16)

$$\tau = 4\frac{r}{\Theta\sqrt{2\,\tilde{\mu}\, B_0\, \varepsilon}} \int_0^{\theta_{\text{ref}}} \frac{\mathrm{d}\theta}{\sqrt{\varkappa^2 - \sin^2(\theta/2)}} = \frac{4\sqrt{2}\, r}{\Theta\sqrt{2\,\tilde{\mu}\, B_0\, \varepsilon}} K(\varkappa) \approx \frac{4\sqrt{2}\, r}{\Theta\, v\sqrt{\varepsilon}} K(\varkappa) \tag{15.D.18}$$

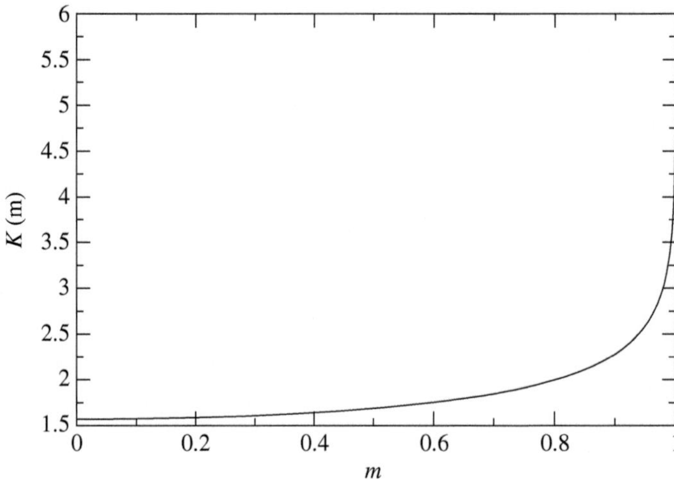

Figure 15.D.1 Plot of the complete elliptic integral of the first kind.

since $v_{\|} \sim \sqrt{2\,\varepsilon}\, v_{\perp} \ll v_{\perp} = \sqrt{2\,\tilde{\mu}\,B_0}$ and $K(\varkappa)$ is a complete elliptic integral of the first kind.[14] Noting that if $\varkappa \ll 1$ is small $K(\varkappa) \approx \pi/2$ (Figure 15.D.1) and the period is identical to that given by the simple model, Eq. (15.66).

Galeev and Sagdeev (1968) modify this result to take into account electric field drift to give the period

$$\tau = \frac{4\sqrt{2}\,r}{\Theta\, v\sqrt{\varepsilon(v^2 + v_E{}^2/\Theta^2)}} K(\varkappa) \tag{15.D.19}$$

changing the value of $\varkappa$ accordingly

$$2\,\varkappa^2 = [\Delta v_{\|}(r_0, 0)]^2/\varepsilon\,[v^2 + v_E{}^2/\Theta^2] \tag{15.D.20}$$

In a similar fashion, we may calculate the second adiabatic invariant for the banana orbit

$$J_{\|} = \oint v_{\|} \mathrm{d}\ell_{\|} = 4\frac{r}{\Theta}\sqrt{2\,\tilde{\mu}\,B_0}\int_0^{\theta_{\text{ref}}} \sqrt{\varkappa^2 - \sin^2(\theta/2)}\,\mathrm{d}\theta = 8\frac{r\sqrt{2\,\tilde{\mu}\,B_0\,\varkappa}}{\Theta} E(\varkappa) \tag{15.D.21}$$

Appendix 15.E The Pfirsch–Schlüter Regime

The poloidal current density is the sum of components parallel and perpendicular to the magnetic field

$$j_P = \frac{B_P}{B} j_{\|} - \frac{B_\phi}{B} j_{\perp} = \frac{\mathrm{d}f}{\mathrm{d}\psi} B_P \tag{15.E.1}$$

making use of Eqs. (15.24) and (15.27).

14 Elliptic integrals are usually expressed in Legendre's form (Abramowitz and Stegun, 1965, Ch.17) where the complete integral of the first kind is

$$K(m) = \int_0^{\pi/2} (1 - m\sin^2\theta)^{-1/2}\,\mathrm{d}\theta = \int_0^{\arcsin(\sqrt{m})} (m - \sin^2\phi)^{-1/2}\,\mathrm{d}\phi$$

when $\sin\phi = \sqrt{m}\sin\theta$.

From the equilibrium condition using Eq. (15.21)

$$j_{\perp} = -\frac{1}{B}|\nabla p| = -\frac{1}{B}\frac{\mathrm{d}p}{\mathrm{d}\psi}|\nabla\psi| = -\frac{RB_P}{B}\frac{\mathrm{d}p}{\mathrm{d}\psi} \tag{15.E.2}$$

Since from Eq. (15.27) $R\,B_{\phi} = f(\psi)$, we have

$$j_{\parallel} = \dot{f}\,B - f\,\dot{p}/B \tag{15.E.3}$$

where the dot denotes differentiation with respect to ψ.

Taking Ohm's law along the field line

$$\eta_{\parallel} j_{\parallel} = \frac{B_P}{B}E_p + \frac{B_{\phi}}{B}E_{\phi} \tag{15.E.4}$$

where E_P is the electric field component parallel to the poloidal magnetic field.

Since the electric field is conservative $\oint \mathbf{E}_P \cdot \mathbf{d}\boldsymbol{\ell} = 0$ and substituting for $j_{\parallel}$ and E_P, we obtain

$$\dot{f} = f\dot{p}\frac{\langle 1/B_P\rangle}{\langle B^2/B_P\rangle} + \frac{\langle E_{\phi}\,B_{\phi}/B_P\rangle}{\eta_{\parallel}\langle B^2/B_P\rangle} \tag{15.E.5}$$

where the mean $\langle x\rangle = \oint x\,\mathrm{d}\ell / \oint \mathrm{d}\ell$.

Finally, substituting for $\dot{f}$ we obtain an expression for the current parallel to the field line

$$j_{\parallel} = \underbrace{-f\dot{p}\left(\frac{1}{B} - \frac{\langle 1/B_P\rangle}{\langle B^2/B_P\rangle}\right)}_{\text{Pfirsch–Schlüter current}} + \underbrace{\frac{\langle E_{\phi}\,B_{\phi}/B_P\rangle}{\eta_{\parallel}\langle B^2/B_P\rangle}B}_{\text{Parallel electric field driven current}} \tag{15.E.6}$$

Using Eq. (15.D.1) for the magnetic fields with $B_P \approx \varepsilon B_0$, the Pfirsch–Schlüter current in a circular large aspect ratio torus is easily calculated

$$j_{PS} \approx -f\,\dot{p}\,\frac{2\,\varepsilon}{B}\cos\theta \tag{15.E.7}$$

Since

$$f = RB \qquad \text{and} \qquad \dot{p} = \frac{1}{RB}\frac{\mathrm{d}p}{\mathrm{d}r} \tag{15.E.8}$$

we obtain the familiar expression

$$j_{PS} = -2\frac{1}{B_{\theta}}\frac{r}{R}\frac{\mathrm{d}p}{\mathrm{d}r}\cos\theta = -\frac{2}{qB}\frac{\mathrm{d}p}{\mathrm{d}r}\cos\theta \tag{15.E.9}$$

15.E.1 Diffusion in the Pfirsch–Schlüter Regime

The velocity normal to the flux surface is given by the toroidal form of Eq. (15.75) whence substituting from E_P from Eq. (15.E.4) and Eqs. (15.E.6) and (15.E.7)

$$v_{\perp} = \frac{E_P B_{\phi} - e_{\phi} B_P}{B^2} - \eta_{\perp}\frac{\nabla_{\perp}p}{B^2} \tag{15.E.10}$$

$$= \frac{B_{\phi}}{B\,B_P}\eta_{\parallel}j_{\parallel} - \eta_{\perp}\frac{\nabla_{\perp}p}{B^2} - \frac{E_{\phi}}{B_P} \tag{15.E.11}$$

$$= \frac{B_{\phi}}{B\,B_P}\eta_{\parallel}j_{PS} - \eta_{\perp}\frac{\nabla_{\perp}p}{B^2} + \frac{1}{B_P}\left(\frac{\langle E_{\phi}\,B_{\phi}/B_P\rangle}{\langle B^2/B_P\rangle}B_{\phi} - E_{\phi}\right) \tag{15.E.12}$$

To obtain a value for the diffusion flux, we must integrate over the flux surface

$$\Gamma = 2\,\pi\,n\oint v_{\perp}R\,\mathrm{d}\ell = 2\,\pi\,n\,\langle v_{\perp}\,R\rangle\oint \mathrm{d}\ell \tag{15.E.13}$$

where the average of the Pfirsch–Schlüter term

$$\langle v_\perp R\rangle_{PS} = -\eta_\| f\,\dot{p}\left\langle \frac{R\,B_\phi}{B_P}\left(\frac{1}{B^2} - \frac{\langle 1/B_P\rangle}{\langle B^2/B_P\rangle}\right)\right\rangle \approx -2\left(\frac{r}{R}\right)^2 \eta_\| \frac{\mathrm{d}p/\mathrm{d}r}{B_\theta^2} \tag{15.E.14}$$

in the circular large aspect ratio approximation. Adding the cross field gradient and electric field terms, we finally obtain (Wesson, 2004)

$$\frac{\langle v_\perp R\rangle}{R_0} = \frac{1}{B^2}(\eta_\perp + 2q^2\eta_\|)\frac{\mathrm{d}p}{\mathrm{d}r} - \frac{E_\phi\,B_\theta}{B^2} \tag{15.E.15}$$

The diffusion coefficient

$$D = m\,kT\,T/q^2\,\tau B^2\,(1 + 2\,q^2\,\eta_\|/\eta_\perp) \approx \rho^2/\tau(1 + 2\,q^2\,\eta_\|/\eta_\perp){\sim}\ q^2\,D_{\perp\mathrm{clas}} \tag{15.E.16}$$

since $\eta_\| \approx 0.51\eta_\perp$. The diffusion rate is therefore a factor $(1 + 2\,q^2\,\eta_\|/\eta_\perp)$ greater than the classical value associated with the perpendicular resistivity $\eta_\perp$ alone due to the curvature inherent in the toroidal geometry of the plasma.

16

Instability of an Equilibrium Confined Plasma

16.1 Introduction

In Chapter 15, we examined the conditions under which an equilibrium state of plasma with the kinetic pressure is supported by a magnetic fields. The major difficulty in achieving controlled fusion confined within such a 'magnetic bottle' is maintaining the plasma against the inherent instabilities for a sufficiently long times to enable the fusion reactions to take place. Instabilities lead to the destruction of the structured plasma described in Chapter 15. Essentially there occur two forms of instability:

- **Macroscopic instability** is unstable behaviour arising within the magneto-hydrodynamic description of the plasma. The classic forms namely the sausage and kink instabilities occur in linear cylindrical plasmas such as Z- or θ-pinches. The introduction of tokamak toroidal geometry further complicates the situation leading to additional instability modes such as ballooning modes due to particle interchange, and tearing modes due to the variation in the current distribution across the plasma.
- **Microscopic instability** is associated with the kinetic behaviour of the particles. They typically develop from departures of the distribution functions of the particles away from Maxwellians which grow exponentially in time. In due course they lead to the development of turbulence within the plasma.

16.2 Ideal MHD Instability

In Chapter 15, we developed the conditions for an equilibrium plasma within the ideal magneto-hydrodynamic approximation. However, this analysis does not guarantee the long time maintenance of this steady configuration; but the development of a magnetically confined fusion reactor demands that the plasma lifetime be sufficiently long to ensure that the a significant fraction of the fuel is burnt. To answer this question requires an analysis of the response of the equilibrium plasma to perturbations. In this section, we will examine the temporal response of the plasma to small departures from equilibrium within the ideal magneto-hydrodynamic approximation.

16.2.1 Linearised Stability Equations

We consider the temporal response of a small displacement ξ of a plasma fluid element from the equilibrium state. The system is stable provided the temporal development of the displacement is

Foundations of Plasma Physics for Physicists and Mathematicians, First Edition. Geoffrey J. Pert.

everywhere bounded, but unstable otherwise. This procedure involves the linearisation of the ideal MHD equations; proceeding in a similar manner to Section 10.2 where we linearised the ideal MHD equations to an uniform plasma to identify the three characteristic families of MHD plasma waves. Here we apply the same methods to study the development of perturbations in static equilibrium plasma when the plasma resistivity can be neglected. In this case, the ambient plasma must satisfy the equilibrium conditions:

$$\nabla \wedge \mathbf{B}_0 = \mu_0 \mathbf{J}_0 \qquad \nabla \cdot \mathbf{B}_0 = 0 \qquad \mathbf{j}_0 \wedge \mathbf{B}_0 = \nabla p_0 \qquad \mathbf{u}_0 = 0 \tag{16.1}$$

As before we introduce the perturbed (dashed) quantities:

$$\rho = \rho_0 + \rho' \qquad p = p_0 + p' \qquad \mathbf{B} = \mathbf{B}_0 + \mathbf{B}' \qquad \mathbf{E} = 0 + \mathbf{E}' \qquad \mathbf{v} = 0 + \mathbf{v}' \qquad \xi' = 0 + \int \mathbf{v}' \, dt \tag{16.2}$$

The equation of continuity, adiabatic equation of state determined by the pressure perturbation and the magnetic perturbation induced by Faraday's law yield

$$\rho' = -\nabla \cdot (\rho_0\, \xi') \qquad p' = -\xi' \cdot \nabla p_0 - \gamma\, p_0\, \nabla \cdot \xi' \qquad \mathbf{B}' = \nabla \wedge (\xi' \wedge \mathbf{B}_0) \tag{16.3}$$

Defining the force operator for the perturbation

$$\begin{aligned} \mathbf{F}(\xi') = \rho_0 \frac{\partial^2 \xi'}{\partial t^2} &= \mathbf{j}_0 \wedge \mathbf{B}' + \mathbf{j}' \wedge \mathbf{B}_0 - \nabla p' \\ &= \frac{1}{\mu_0}(\nabla \wedge \mathbf{B}_0) \wedge \mathbf{B}' + \frac{1}{\mu_0}(\nabla \wedge \mathbf{B}') \wedge \mathbf{B}_0 - \nabla p' \end{aligned} \tag{16.4}$$

which is seen to be identical to Eq. (10.16) when the plasma is uniform. From Eq. (16.3), it is immediately apparent that the force operator $\mathbf{F}(\xi')$ is a linear function of the displacement alone (and not of the rate of change of the displacement). More generally, we observe that all the perturbation variables ρ', p', $\mathbf{B}'$ as well as $F(\xi')$ depend on the perturbation alone ξ' and not on the perturbation velocity $\dot{\xi}'$.

The general problem of calculating the onset (or otherwise) of instability can be attacked by two different approaches

(a) **Direct integration** of Eq. (16.4) for the displacement as an initial value problem subject to the appropriate boundary and initial conditions. Such a direct approach requiring detailing computer modelling presents practical difficulties as it may require a long development of the solution if the instability is slow growing.
Alternatively the more common method of investigating the temporal development of the perturbations is to use a normal mode formulation (Section 16.2.2), thereby allowing the temporal development of the perturbation to be expressed as an eigenvalue problem.

(b) **Potential energy extremum** : Alternatively we may attack the onset of instability by investigating the potential energy of the system (see Section 6.4). As we have seen the threshold for instability is found when potential energy changes from a maximum to a minimum (or *vice versa*). This method is often simpler.

We consider the energy balance of the plasma during the perturbation in which the development of the kinetic energy is associated with a decrease in the potential energy of the system as a whole. Following Bernstein et al. (1958) (see also Rose and Clark, 1961, §12.6), we develop the theory by a simple heuristic method, which nonetheless is mathematically sound. More detailed (and lengthy) formal analyses are given by Kadomtsev and Pogu'tsev (1967) and Freidberg (2014).

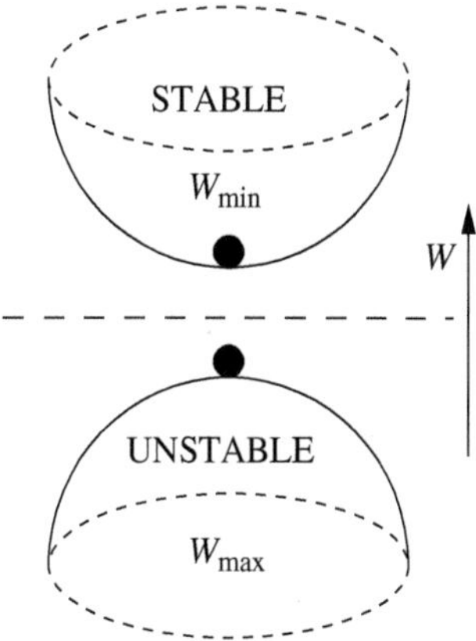

Figure 16.1 Mechanical analogy of stable (upper) and unstable (lower) configurations of a ball and bowl under gravity.

In particular, the condition for stability will be based on the change of potential energy associated with the displaced perturbation. It is a familiar result from simple mechanics namely: if the potential energy of a particle, measured as zero at the equilibrium position, varies as a function of its position, $W(z)$, about an equilibrium position. The potential has an extremum $\mathrm{d}W/\mathrm{d}z = 0$ at the equilibrium: and the equilibrium is stable if the potential is a minimum, i.e. a small displacement generates a force which returns the particle to its unperturbed position $\mathrm{d}^2W/\mathrm{d}z^2 < 0$; and is unstable if it is a maximum $\mathrm{d}^2W/\mathrm{d}^2z > 0$. This situation is simply exemplified by the familiar case of a ball inside or outside a bowl (see Figure 16.1).

It follows from Eq. (8.32) that the total energy of the plasma in the system can be expressed as the sum of the kinetic and the potential energies

$$\int \mathrm{d}\mathbf{r} \left(\underbrace{\frac{1}{2}\,\rho\, v^2}_{\text{Kinetic}} + \underbrace{\frac{1}{2}\,\epsilon_0\, E^2 + \frac{1}{2\,\mu_0}\, B^2 + \frac{1}{(\gamma-1)} p}_{\text{Potential}} \right) \tag{16.5}$$

which we note is constant in the absence of an external energy source, as the total energy flux out of the system is zero.

First let us heuristically calculate a value for the change in the potential energy about the equilibrium due to the perturbation ξ' and the force $\mathbf{F}(\xi')$. As the disturbance in the plasma is only a small perturbation in the plasma, the force per unit volume on a plasma element $\mathbf{F}(\xi')$ is linear in the displacement. The potential energy must be a single valued function of the particle displacements and independent of the path by which these changes are achieved. We may therefore consider the development of the perturbed system $\xi'(\mathbf{r})$ to be established through a series of steps in which the displacement is everywhere a fraction α of the final value and the force function correspondingly $\mathbf{F}(\alpha\xi') = \alpha\ \mathbf{F}(\xi')$. Therefore, the work done per unit volume increasing the displacements by a fraction $\delta\alpha$ is $\rho_0\ \xi' \cdot\ \mathbf{F}(\xi')\ \alpha\ \delta\alpha$, which must be at the expense of a corresponding reduction in the potential energy. Since the potential energy is measured with respect to equilibrium position, it follows that it must be a quadratic form of the type

$$W(\xi', \xi') = \frac{1}{2} \sum c_{ij}\ \xi'_i\ \xi'_j \tag{16.6}$$

where the sum is taken over all components of all fluid elements, and from Taylor's theorem the coefficients c_{ij} are the corresponding second derivatives.

Summing over the steps $\delta\alpha$, the total potential energy of the system due to the perturbation is the quadratic form, which does not contain $\dot{\xi}'$

$$W(\xi', \xi') = -\frac{1}{2}\int \xi' \cdot \mathbf{F}(\xi')\mathrm{d}\mathbf{r} \tag{16.7}$$

A more satisfactory approach following Bernstein et al. (1958) to the calculation of the potential energy near an extremum is to note that, as above, from Taylor's theorem we may write the total potential

$$W_{\mathrm{tot}}(\mathbf{r}) = W_0 + \cancel{\left.\frac{\mathrm{d}W}{\mathrm{d}x_i}\right|_0 \xi'_i} + \frac{1}{2}\left.\frac{\mathrm{d}^2 W}{\mathrm{d}x_i\,\mathrm{d}x_j}\right|_0 \xi'_i\,\xi'_j + \cdots \tag{16.8}$$

where $W_0 = 0$ is the total potential energy in the equilibrium state. The perturbation potential is therefore a symmetric bilinear function of the displacement $W(\xi, \xi)$.

The ambient velocity of the plasma is zero, consequently the total kinetic energy is due to the velocity of the perturbation alone

$$K(\dot{\xi}' \cdot \dot{\xi}') = \frac{1}{2}\int \rho_0\,\dot{\xi}' \cdot \dot{\xi}'\,\mathrm{d}\mathbf{r} \tag{16.9}$$

Since energy is conserved the rate of change of potential energy must be balanced against the rate of change of kinetic energy so that

$$\begin{aligned}\dot{K} &= \rho_0 \int \dot{\xi}'\,\ddot{\xi}'\,\mathrm{d}\mathbf{r} = -\dot{W}(\xi', \xi')\\ &= \int \dot{\xi}' \cdot \mathbf{F}(\xi')\,\mathrm{d}\mathbf{r} = -W(\dot{\xi}', \xi') - W(\xi', \dot{\xi}')\end{aligned} \tag{16.10}$$

The functions ξ' and $\dot{\xi}'$ are independent satisfying the same boundary conditions (see Rose and Clark (1961, §12.6) for a brief discussion of this point). Equation (16.10) may therefore be considered as an identity between any two displacements ξ' and $\dot{\xi}' \equiv \eta'$ which are subject to the same boundary conditions. Thus, we may replace $\dot{\xi}'$ by any function satisfying the same boundary conditions, for example ξ' in which case we obtain directly from Eq. (16.10) that

$$W(\xi', \xi') = -\frac{1}{2}\int \xi' \cdot \mathbf{F}(\xi')\,\mathrm{d}\mathbf{r} \tag{16.7'}$$

Interchanging ξ' and $\eta' = \dot{\xi}'$ in Eq. (16.10), we obtain

$$\int \eta' \cdot \mathbf{F}(\xi')\,\mathrm{d}\mathbf{r} = \int \xi' \cdot \mathbf{F}(\eta')\,\mathrm{d}\mathbf{r} \tag{16.11}$$

for an arbitrary pair of independent functions ξ' and η'. The force function $\mathbf{F}(\xi')$ is therefore *self-adjoint*.

Since Eq. (16.4) is linear in the perturbation ξ, we may seek solutions in which the temporal development can be described by an oscillatory/exponential term $\sim \exp(\imath\omega t)$. In particular, we may seek to find the solution from a series of the eigenfunctions of the force Eq. (16.4)

$$\mathbf{F}(\xi_n) = -\rho_0\,\omega_n{}^2\,\xi_n \tag{16.12}$$

where ξ_n is the nth eigenfunction with eigenvalue $\omega_n{}^2$.

It follows from the self-adjoint property of the operator $\mathbf{F}$ that two such eigenfunctions have the property

$$\int \xi_m \cdot \mathbf{F}(\xi_n)\,\mathrm{d}\mathbf{r} - \int \xi_n \cdot \mathbf{F}(\xi_m)\,\mathrm{d}\mathbf{r} = [\omega_m{}^2 - \omega_n{}^2]\int \rho_0\,\xi_m\,\xi_n\,\mathrm{d}\mathbf{r} = 0 \tag{16.13}$$

from which we draw two important results.

1. Let $\xi_m = \xi_n^*$ so that ${\omega_m}^2 = \omega_n^{2*}$. The eigenvalue ω^2 is real. The solution corresponding to the eigenfunction is either purely oscillatory $\omega^2 > 0$ or purely exponential growing $\omega^2 < 0$ solution.
2. Let ${\omega_m}^2 \neq {\omega_n}^2$ then $\int \rho_0\ \xi_m\ \xi_n\ \mathbf{dr} = 0$ and the eigenfunctions are orthogonal provided there is no degeneracy, as is indeed the case. The general solution may therefore be obtained by an expansion in the ortho-normal eigenfunctions if the set is assumed to be complete (see Freidberg, 2014, p. 345).

16.2.2 Normal Mode Analysis – The Stability of a Cylindrical Plasma Column

As an example of the normal mode analysis, we consider the stability of a uniform plasma cylinder (Figure 16.2) with longitudinal (z direction) magnetic fields $\mathbf{B}_i$ and $\mathbf{B}_e$ inside and outside the sharp plasma boundary. Assuming there is a current I flowing in the longitudinal direction at the plasma boundary, there is an azimuthal field $B_\theta = \mu_0\ I/2\ \pi\ r$ outside the column. The perturbation displacement is assumed to expressed as a set of Fourier modes

$$\xi(r)\ \exp(\imath(k\ z + m\ \theta)) \tag{16.14}$$

To further simplify the problem, the plasma is assumed to be incompressible $\nabla \cdot \xi = 0$. Remembering that

$$\nabla \wedge (\mathbf{A} \wedge \mathbf{B}) = (\mathbf{B} \cdot \nabla)\mathbf{A} - (\mathbf{A} \cdot \nabla)\mathbf{B} + \mathbf{A}(\nabla \cdot \mathbf{B}) - \mathbf{B}(\nabla \cdot \mathbf{A})$$

it follows from Eq. (16.3) that

$$\mathbf{B}' = -\imath\ k\ B_i\ \xi \tag{16.15}$$

and that the pressure perturbation $p' = 0$.

Taking the divergence of Eq. (16.4) and making use of the vector relations

$$\nabla \cdot (\mathbf{A} \wedge \mathbf{B}) = \mathbf{B} \cdot (\nabla \wedge \mathbf{A}) - \mathbf{A} \cdot (\nabla \wedge \mathbf{B})$$

$$\nabla \wedge (\nabla \wedge \mathbf{A}) = \nabla(\nabla \cdot \mathbf{A}) - \nabla \cdot \nabla\mathbf{A}$$

we obtain

$$\mathbf{B}_i \cdot \nabla^2 \xi = 0 \qquad \text{or} \qquad \nabla^2 \xi_z = 0 \tag{16.16}$$

Substituting the displacement, Eq. (16.14), we obtain

$$\left[\frac{\mathrm{d}^2}{\mathrm{d}r^2} + \frac{1}{r}\frac{\mathrm{d}}{\mathrm{d}r} - \left(k^2 + \frac{m^2}{r^2}\right)\right] \xi_z(r) = 0 \tag{16.17}$$

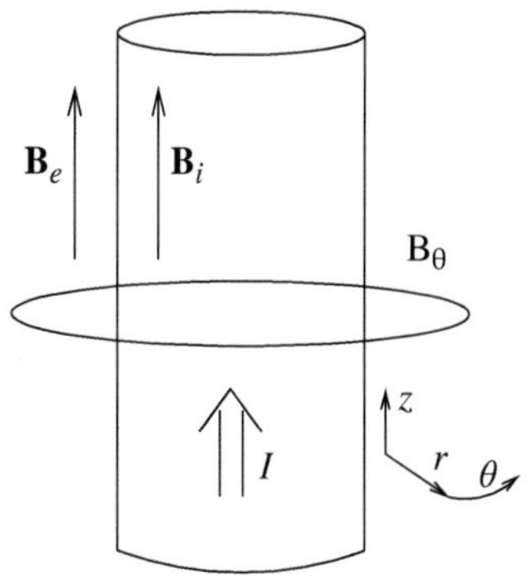

Figure 16.2 Change in the normal $\delta\mathbf{n}$ to the interface surface as the interface elements are moved from surface S to surface $S + \delta S$ by the perturbation ξ.

which will be recognised as the modified form of Bessel's equation, which has solutions in terms of modified Bessel functions of order m (Abramowitz and Stegun, 1965, § 9.6). Inside the cylinder, we require the solution $I_m(k\ r)$ which is well behaved as $r \to 0$, but singular as $r \to \infty$; outside the cylinder, we require the solution $K_m(k\ r)$ which tends to zero as $r \to \infty$ and is singular as $r \to 0$ (see Figure 16.5). Inside the cylinder

$$\xi_z(r) = \xi_z(a)\ I_m(k\ r)/I_m(k\ a) \tag{16.18}$$

where a is the radius of the cylinder.

Substituting for the perturbed field $\mathbf{B}'$ from Eq. (16.15) in the rth component of Eq. (16.4) we obtain

$$\xi_r = \frac{\imath\ k\ B_i^2/\mu_0}{[\omega^2\ \rho_0 - k^2\ B_i^2/\mu_0]}\ \frac{\mathrm{d}}{\mathrm{d}r}\ \xi_z(r) \tag{16.19}$$

Outside the plasma cylinder in vacuo, the current is zero, and the magnetic field may be expressed as a potential $\mathbf{B} = \nabla V$ given by $\nabla^2 V = 0$ as in Section 15.6 (b). Since the potential must take the same form as the perturbation: namely $V(r,\theta,z) \equiv V(r)\ \exp(\imath(k\ z + m\ \theta))$ (Eq. (16.14)), it follows, as for Eq. (16.16), that $V(r)$ is a solution of the modified Bessel's equation, Eq. (16.17) with a solution which is well behaved for large arguments, namely $K_m(r)$

$$V(r) = V(a)\ \frac{K_m(kr)}{K_m(ka)} \tag{16.20}$$

where $V(a)$ is a constant to be determined.

The solution is completed by matching the conditions (Eq. (15.38)) across the plasma/vacuum boundary as in Section 15.6(b).

The continuity of the magnetic induction normal to the boundary includes terms from the perturbation on both the plasma and vacuum sides of the interface plus a term due to the change of the surface, which results in a net change in each term $\mathbf{B}' \cdot \mathbf{n} + \mathbf{B} \cdot \delta\mathbf{n} + (\xi \cdot \nabla)B_n$ namely:

$$\underbrace{B_r'(a) + B_e\ \delta n_z + \xi_r\ \partial B_\theta/\partial r|_a}_{\text{Outside}} = \underbrace{B_r'(a) + B_i\ \delta n_z}_{\text{Inside}} \tag{16.21}$$

The change in the normal to the surface as the Lagrangian elements are moved is easily calculated.

If the infinitesimal vector $\Delta\mathbf{r}$ lies in the surface $S(\mathbf{r}) = 0$, then $\Delta\mathbf{r} \cdot \mathbf{n} = 0$. Referring to Figure 16.3 for small displacements, the change in the normal vector $\delta\mathbf{n}$ at the displaced surface $S + \delta S \equiv [S(\mathbf{r} + \xi(\mathbf{r})) = 0]$ is given by

$$[\Delta\mathbf{r} + (\Delta\mathbf{r} \cdot \nabla)\ \xi] \cdot [\mathbf{n} + \delta\mathbf{n}] = \Delta\mathbf{r} \cdot \mathbf{n} = 0 \tag{16.22}$$

Therefore, noting that $\Delta\mathbf{r} \cdot \mathbf{n} = 0$ for all such $\Delta\mathbf{r}$ lying in the surface and letting $\boldsymbol{\delta}\mathbf{r} \to 0$ we obtain

$$\Delta\mathbf{r} \cdot \boldsymbol{\delta}\mathbf{n} + (\Delta\mathbf{r} \cdot \nabla)\xi \cdot \mathbf{n} = 0 \tag{16.23}$$

which must be valid for all allowed values of $\Delta\mathbf{r}$.[1]

---

1 Taking a basis set $\mathbf{e}_1$ and $\mathbf{e}_2$ in the tangential plane of S and letting $i = 1$ and $i = 2$ successively it follows that

$$\cancel{\Delta r_i}\ \delta n_i = -\cancel{\Delta r_i}\ \frac{\partial \xi_j}{\partial x_i}\ n_j \equiv -(\nabla^T\ \xi) \cdot \mathbf{n} \qquad (i = 1,\ 2) \tag{16.24}$$

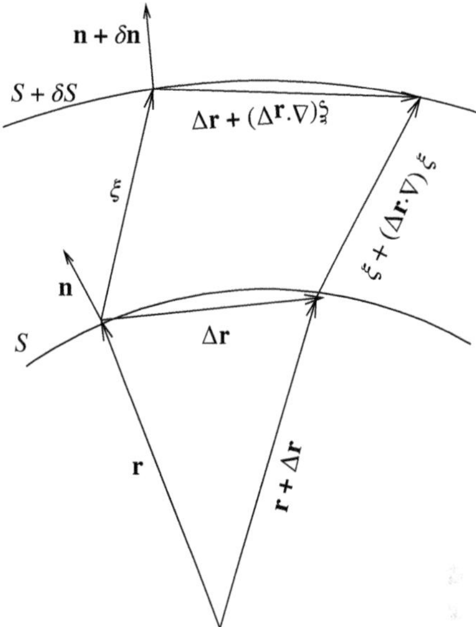

Figure 16.3 Change in the normal $\delta\mathbf{n}$ to the interface surface as the interface elements are moved from surface S to surface $S + \delta S$ by the perturbation ξ.

Since the interface in the un-perturbed plasma is cylindrical, the normal vector is $\mathbf{n} = (1, 0, 0)$ and

$$\delta\mathbf{n} = -(\nabla^T \xi) \cdot \mathbf{n} = -\nabla^T \xi_r \tag{16.25}$$

where ∇^T is the gradient operator on the tangential plane. [2]

Making use of Eq. (16.15), the normal components of the magnetic field at the outer and inner surfaces of the interface, respectively, in Eq. (16.21) take the values

$$\begin{aligned} \left.\frac{\mathrm{d}V}{\mathrm{d}r}\right|_a - B_e \frac{\partial \xi_r}{\partial z} - B_\theta(a)\frac{1}{a}\frac{\partial \xi_r}{\partial \theta} &= V(a)\, k\, \frac{\dot{K}_m(a)}{K_m(a)} - \imath \left(k\, B_e + \frac{m}{a} B_\theta(a)\right) \xi_r(a) \\ &= B_r'(a) - B_i(\imath\, k\, \xi_r(a)) = 0 \end{aligned} \tag{16.26}$$

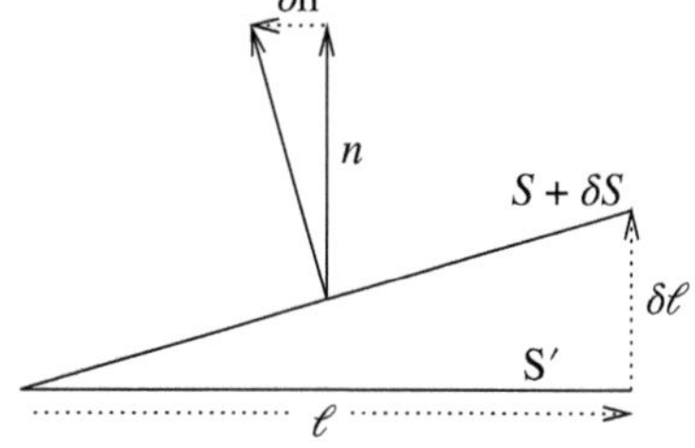

Figure 16.4 The plane of the normals showing the change $\delta\mathbf{n}$ of the normal as the interface elements are moved from surface S' to surface $S + \delta S$ by the perturbation.

2 A simple construction illustrates this result. The interface at S is shifted to meet the perturbed surface $S + \delta S$; the latter being the original surface which has been displaced and inclined. The surface S is displaced by ξ to S' and meets $S + \delta S$ (Figure 16.4). The tangential planes of S' and $S + \delta S$ are both normal to the plane formed by the normals n and $n + \delta n$ (normals plane). The intersection of the normals plane with the tangent plane of S' along the line $\boldsymbol{\ell}$ is the direction of largest perturbation ξ and represents the greatest separation between the tangential planes, namely $\boldsymbol{\delta\ell}$ at ℓ; $\boldsymbol{\delta\ell}$ is clearly co-planar with the normals. The angle between the tangential planes S' and $S + \delta S$ is clearly by $\delta\ell/\ell = \mathrm{d}\xi_n/\mathrm{d}\ell$ (see Figure 16.3) and hence for small displacements $\boldsymbol{\delta n} = -n\,\delta\ell/\ell\hat{\boldsymbol{\ell}} = -\nabla_T \xi \cdot n$ where ∇_T is the gradient operator in the tangential plane. The negative sign in Eq. (16.25) is a consequence of the difference in direction of $\boldsymbol{\delta n}$ and $\boldsymbol{\ell}$ (Figure 16.4).

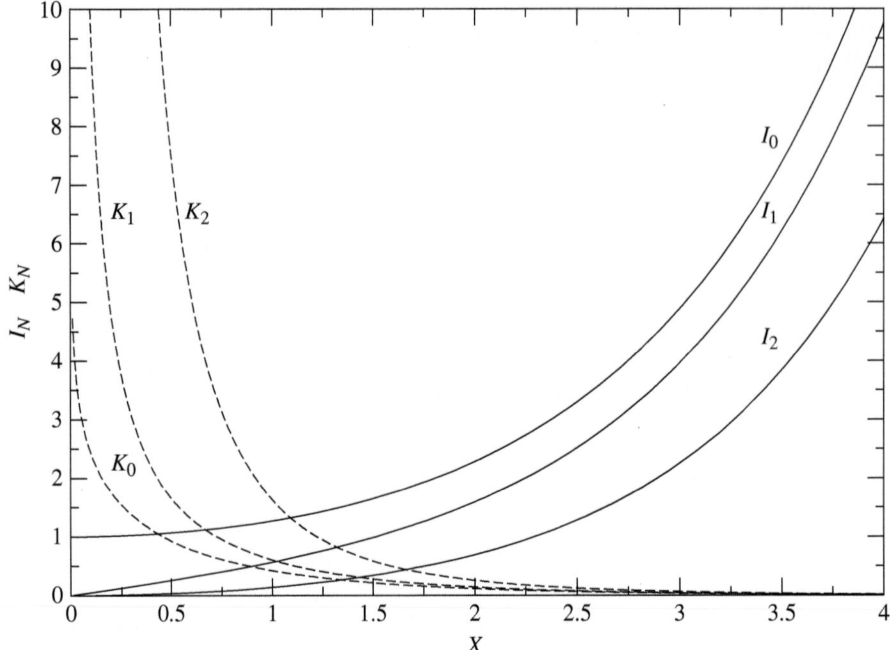

Figure 16.5 Plots of the modified Bessel functions $I_N(X)$ (full lines) and $K_N(X)$ (dashed lines) of order N and argument X.

Suppressing the exponential dependencies, the magnetic pressure balance is

$$\underbrace{\frac{1}{\mu_0}(B_e\, B'_z + B_\theta\, B'_\theta) = \frac{1}{\mu_0}\frac{1}{a}\,(\imath\, k\, a\, B_e + \imath\, m\, B_\theta)V(a)}_{\text{Vacuum}} = \underbrace{\frac{1}{\mu_0} B_i\, B'_z = \frac{\imath\, B_i^{\,2}}{\mu_0}\xi_z(a)}_{\text{Plasma}} \tag{16.27}$$

to which must be added the term due to the change in the azimuthal field in the vacuum associated with the perturbation, namely $(\partial B_\theta/\partial r|_a)\,\xi_r(a) = -(B_\theta(a)/a)\,\xi_r(a)$; yielding an addition pressure perturbation on the vacuum side $-(B_\theta^{\,2}(a)/\mu_0\, a)\,\xi_r(a)$. The total pressure balance is therefore

$$\frac{\imath\, B_i^{\,2}}{\mu_0}\xi_z(a) = \frac{1}{\mu_0}\,\frac{k\, V(a)}{a}\,(\imath\, k\, a\, B_e + \imath\, m\, B_\theta) - \frac{1}{\mu_0\, a}B_\theta^{\,2}(a)\,\xi_r(a) \tag{16.28}$$

From Eqs. (16.26) and (16.28) we obtain by eliminating $V(a)$

$$\imath\,\frac{\xi_z(a)}{\xi_r(a)} = \frac{1}{\mu_0}\,\frac{(k^2\, B_e + m\, B_\theta)^2}{k\, a}\,\frac{K_m(a)}{\dot{K}_m(a)} - \frac{1}{\mu_0\, a}B_\theta^{\,2} \tag{16.29}$$

Finally solving Eq. (16.19) for ω^2, we obtain the eigenvalue equation $\omega = \omega(\mathbf{k})$

$$\begin{aligned}\frac{\omega^2}{k^2} &= \frac{B_i^{\,2}}{\mu_0\,\rho_0} - \imath\,\frac{k\, B_i^{\,2}}{\mu_0\,\rho_0}\,\frac{1}{k\,\xi_r(a)}\,\frac{\mathrm{d}\,\xi_z}{\mathrm{d}\,r}\Big|_a = \frac{B_i^{\,2}}{\mu_0\,\rho_0} - \imath\,\frac{B_i^{\,2}}{\mu_0\,\rho_0}\,\frac{\xi_z(a)\,\dot{I}_m(k\,a)}{\xi_r(a)\,I_m(k\,a)}\\ &= \frac{1}{\mu_0\,\rho_0}\left[B_i^{\,2} - \frac{(B_e + m\, B_\theta)^2}{(k\,a)^2}\,\frac{K_m(k\,a)}{\dot{K}_m(k\,a)}\,\frac{\dot{I}_m(ka)}{I_m(ka)} - \frac{B_\theta^{\,2}\,\dot{I}_m(ka)}{k\,a\,I_m(ka)}\right]\end{aligned} \tag{16.30}$$

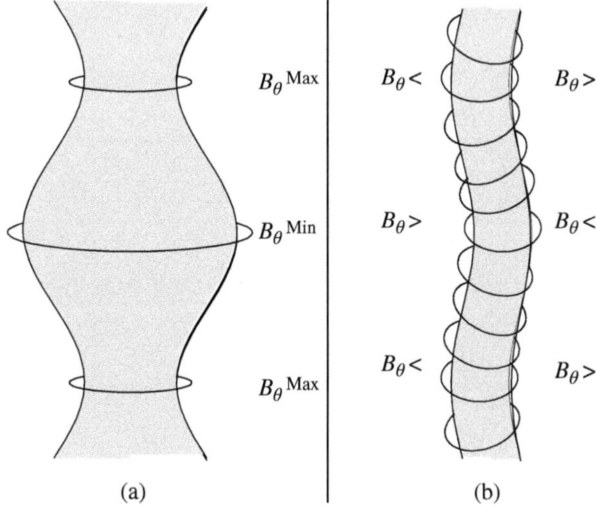

Figure 16.6 Sketches of the characteristic profiles of MHD instabilities in a cylindrical plasma. The role of the azimuthal magnetic field driving the $m = 0$ and $m = 1$ instabilities is demonstrated. (a) $m = 0$ Sausage instability. (b) $m = 1$ Kink instability.

As expected ω^2 is real and the plasma is either stable if $\omega^2 > 0$ or unstable with a purely growing mode if $\omega^2 < 0$. Furthermore, since $\dot{K}(x) < 0$ the first two terms in this expression are both positive. Instability is therefore a consequence of the third term due to the azimuthal magnetic field generated by the internal current along the cylinder.

The profiles of the plasma column subjected to the lower order perturbations are shown diagrammatically in Figure 16.6.

16.2.3 $m = 0$ Sausage Instability

The general structure of the $m = 0$ instability is shown in Figure 16.6a. The instability is azimuthally symmetric and the perturbation solely due to the $e^{\imath\, k\, z}$ longitudinal perturbation alone. It can be seen to result from the $B_\theta\, r^{-1}$ dependence of the azimuthal field where constricted regions suffer greater inwards magnetic pressure forces further enhancing the constriction.

When $m = 0$ Eq. (16.30) takes the simpler form

$$\frac{\omega^2}{k^2} = \frac{1}{\mu_0\, \rho_0} \left[B_i{}^2 - \frac{B_e^2}{(k\, a)^2} \frac{K_m(k\, a)}{\dot{K}_m(k\, a)} \frac{\dot{I}_m(ka)}{I_m(ka)} - \frac{B_\theta{}^2}{k\, a} \frac{\dot{I}_m(ka)}{I_m(ka)} \right] \tag{16.31}$$

Instability growth is clearly inhibited by the internal and external fields due to stretching of the longitudinal field lines and bunching which occurs due to the trapped field at the constriction. When the external field $B_e = 0$ is zero, the column is stable if

$$B_i{}^2 > \frac{B_\theta{}^2}{k\, a} \frac{\dot{I}_m(ka)}{I_m(ka)} > \frac{1}{2} B_\theta(a)^2 \tag{16.32}$$

since $\dot{I}_0(x)/x I_0(x) < 1/2$. Making use of the equilibrium relation

$$p + B_i{}^2/2\mu_0 = B_\theta{}^2/2\mu_0 + B_e{}^2/2\mu_0 \tag{16.33}$$

this result may be conveniently written in terms of the plasma β

$$\beta_\theta = \frac{p}{B_\theta(a)^2/2\,\mu_0} = 1 - \frac{\dot{I}_0(ka)}{k\,a\,I_0(ka)} \begin{cases} 1/2 & k\,a \ll 1 \\ 1 - 1/ka & k\,a \gg 1 \end{cases} \tag{16.34}$$

For most applications, we may assume that the wavelength is long compared to the plasma radius $k\,a \ll 1$ and

$$I_m(x) \approx \frac{(x/2)^m}{m!} \qquad K_m(x) \approx 2^{m-1}\,(m-1)!\,x^{-m} \qquad x \ll 1 \tag{16.35}$$

Substituting in Eq. (16.31) and differentiating to find the minimum value of ω^2 and using the pressure balance we find

$$\omega^2_{\text{min}} = \frac{mB_\theta(a)^2}{\mu_0\,\rho\,a^2}\left[\frac{(m-1)(1 + B_\theta(a)^2\big/B_e^{\;2} - \beta_t) - 1}{2 + B_\theta(a)^2\big/B_e^{\;2} - \beta_t}\right] \tag{16.36}$$

where $\beta_t = 2\,\mu_0\,p/B_e^{\;2}$ is the *toroidal beta*. It follows in low β plasma $|B_\theta/B_e| \ll 1$ only the $m = 1$ and $m = 2$ modes may be unstable.

16.2.4 $m = 1$ Kink Instability

The instability takes the form of a spiral perturbation of the cylindrical column (Figure 16.6b) resulting from the term $e^{\imath(kz+m\theta)}$. The instability is driven by the curvature of the cylinder which causes bunching of the azimuthal field lines on the concave side and hence an increased magnetic pressure acting inwards (see item b.ii in Section 15.4.1). As with the $m = 0$ instability, the stretching of the longitudinal field lines as the plasma is perturbed gives rise to a tension which tends to restore the plasma to its original cylindrical equilibrium profile. The stability of the plasma column against $m = 1$ perturbations is easily investigated by setting $m = 1$ in Eq. (16.30) and eliminating the internal magnetic field using the pressure balance, Eq. (16.33) to obtain

$$\omega^2 = \frac{k^2\,B_e^{\;2}}{\mu_0\,\rho}\left[2\left(1 + \frac{B_\theta(a)}{k\,a\,B_e}\right) + \frac{B_\theta^2(a)}{B_e^{\;2}} - \beta_t\right] \tag{16.37}$$

The important case when the azimuthal field is small, $|B_\theta(a)/B_e| \ll 1$ and the toroidal β small, $\beta_t \ll 1$, it follows that the plasma is stable ($\omega^2 > 0$) provided $|B_\theta/B_e| < |ka|$. However the wavelength is limited by the length of the plasma column L, so that the plasma is stable if $|B_\theta/B_e| < 2\,\pi\,a/L$. Applying this result to a toroid where the plasma length $L = 2\pi\,R_0$, the stability ratio of the toroidal to the poloidal field in a tokamak is limited by the safety factor

$$q(a) = a\,B_z/R_0\,B_\theta > 1 \tag{16.38}$$

a result independently due to Kruskal and Shafranov, which is applicable more generally to diffuse pinch as well as the sharp edged form discussed here. (A readable introduction to the kink modes by S Cowley is to be found in http://home.physics.ucla.edu/calendar/conferences/cmpd/talks/cowley.pdf)

Tokamaks are designed to have a limit on the toroidal current, and therefore the poloidal field to satisfy the Kruskal–Shafranov criterion. Furthermore as we have seen whenever the external field is concave towards the plasma small ripples grow and lead to *ballooning instabilities*. The growth of such instability is restricted by limiting β_t. Conversely, convex plasma smooths out such ripple instabilities.

16.3 Potential Energy

It is clear from this example that the calculation of stability via the solution of the eigenvalue equation is complex and involves more parameters than necessary for a strict evaluation of the stability of a configuration. A much simpler approach is through the identification of the nature of the extrema of the potential energy function as discussed earlier Section 16.2.1.[3] To calculate the total change in the potential energy of the plasma due to the perturbation, we return to Eq. (16.7), where $\mathbf{F}(r)$ is given by the linearised force Eq. (16.4)

$$W(\xi', \xi') = -\frac{1}{2}\int \xi' \cdot \mathbf{F}(\xi')\, \mathrm{d}\mathbf{r} \tag{16.39}$$

$$= -\frac{1}{2}\int \xi' \cdot \left\{ \frac{1}{\mu_0}(\nabla \wedge \mathbf{B}_0) \wedge \mathbf{B}' + \frac{1}{\mu_0}(\nabla \wedge \mathbf{B}') \wedge \mathbf{B}_0 + \nabla(\xi' \cdot \nabla p + \gamma\, p\, \nabla \cdot \xi') \right\} \mathrm{d}\mathbf{r} \tag{16.40}$$

To perform this integration, we look to develop the result as an integration by parts in terms of a volume integral and a surface term. To achieve this, we identify the elements of the potential which can be expressed as a divergence and thus integrated by Gauss' theorem. Thus taking the pressure terms

$$\begin{aligned} &\xi' \cdot \nabla(\gamma\, p\, \nabla \cdot \xi' + \xi' \cdot \nabla p) \\ &\quad = \nabla \cdot [\xi'(\gamma\, p\, \nabla \cdot \xi' + \xi' \cdot \nabla p)] - \{\gamma\, p\, (\nabla \cdot \xi')^2 + (\nabla \cdot \xi')\, \xi' \cdot \nabla p\} \end{aligned} \tag{16.41}$$

and the magnetic field[4]

$$-\frac{1}{2\,\mu_0}\xi' \cdot (\nabla \wedge \mathbf{B}') \wedge \mathbf{B}_0 = \frac{1}{2\,\mu_0}\{\nabla \cdot [\mathbf{B}' \wedge (\xi' \wedge \mathbf{B}_0)] + |\mathbf{B}'|^2\} \tag{16.43}$$

Integrating over the plasma volume V with surface S using Gauss' theorem, we obtain

$$\begin{aligned} W(\xi', \xi') &= \frac{1}{2}\int_V \left\{ \frac{1}{\mu_0}[B'^2 - (\nabla \wedge \mathbf{B}_0) \cdot (\mathbf{B}' \wedge \xi')] + \gamma\, p_0\, (\nabla \cdot \xi')^2 + (\nabla \cdot \xi')\xi' \cdot \nabla p \right\} \mathrm{d}V \\ &\quad + \frac{1}{2}\int_S \left\{ \frac{1}{\mu_0}\mathbf{B}_0 \cdot \mathbf{B}' - \gamma p_0 \nabla \cdot \xi - \xi' \cdot \nabla p_0 \right\} \xi' \cdot \mathrm{d}S \end{aligned} \tag{16.44}$$

The surface integral may be handled in two different ways depending on the nature of the plasma boundary.

1. For perturbations limited either by a conducting wall or entirely within the plasma, the fixed boundary condition may be imposed $\xi' \cdot \mathrm{d}S = 0$. In this case, the surface contribution is zero.
2. If the plasma bounds a vacuum, the surface integral terms from the plasma cancel some of those from the boundary when the boundary conditions (Section 15.6) are imposed, and the vacuum

3 The potential energy and the energy principle were first introduced in a seminal paper by Bernstein et al. (1958). The method has since been greatly refined, see for example Freidberg (2014) but the original paper remains worthy of study.

4 We use Eq. (16.3) for the perturbed magnetic field and then the properties of the scalar triple product, the expansion of the vector triple product and finally the vector identity $\nabla \cdot (\mathbf{A} \wedge \mathbf{B}) = \mathbf{B} \cdot \nabla \wedge \mathbf{A} - \mathbf{A} \cdot \nabla \wedge \mathbf{B}$ as follows:

$$\begin{aligned} -\xi' \cdot (\nabla \wedge \mathbf{B}') \wedge \mathbf{B}_0 &= (\nabla \wedge \mathbf{B}') \cdot (\xi' \wedge \mathbf{B}_0) \\ &= \{\nabla \cdot [\mathbf{B}' \wedge (\xi' \wedge \mathbf{B}_0)] + \mathbf{B}' \cdot \nabla \wedge (\xi' \wedge \mathbf{B}_0)\} \\ &= \{\nabla \cdot [\mathbf{B}' \wedge (\xi' \wedge \mathbf{B}_0)] + |\mathbf{B}'|^2\} \end{aligned} \tag{16.42}$$

volume term is introduced. The final result can be written as[5]

$$W = W_p + W_s + W_V \tag{16.45}$$

where

$$W_p = \frac{1}{2}\int_{V_p}\left[\frac{1}{\mu_0}\mathbf{B}'^2 - p'\,(\nabla\cdot\xi') - \xi'\cdot(\mathbf{J}\wedge\mathbf{B}')\right]\,\mathrm{d}V \tag{16.46:p}$$

$$W_s = \frac{1}{2}\int_S [[p'_{me} - p'_m - p']]\;\xi'\cdot\mathrm{d}S = \frac{1}{2}\int_S \frac{\partial}{\partial n}\left(\frac{B_e'^2}{2\mu_0} - \frac{B'^2}{2\mu_0} - p'\right)\;\xi'\cdot\mathrm{d}S \tag{16.46:s}$$

$$W_V = \frac{1}{2}\int_{V_e}\frac{B_e'^2}{\mu_0}\,\mathrm{d}V \tag{16.46:V}$$

The energy principle is completed by the stability condition:

$$W(\xi,\xi) \geq 0 \tag{16.47}$$

for all possible ξ.

As the derivation of these expressions is straightforward, but involving much algebraic manipulation, we will not derive them here (see Freidberg, 2014), these three terms can be easily understood physically: The plasma volume term W_p has contributions from firstly the magnetic field perturbation, secondly the work done in expansion, and thirdly the work done by the magnetic force generated by the equilibrium current. The boundary surface term is the work done by the difference of the magnetic + kinetic pressure across the surface. Finally, W_V is the potential energy of the vacuum field perturbation $\mathbf{B}'_e$.

From Eq. (16.44), we see that the stabilising term associated with ξ' is zero if the perturbation is parallel to the current since $\xi'\cdot(\mathbf{J}\wedge\mathbf{B}') = (\xi'\wedge\mathbf{J})\cdot\mathbf{B}' = 0$.

16.4 Interchange Instabilities

It is not appropriate to give an extensive discussion of MHD instabilities which can be identified using the energy principle, for which the reader is referred to Wesson (2004) and Freidberg (2014). However a few examples of the extensive group of interchange instabilities provide exemplars of the approach.

The archetype of the instability comes from fluid mechanics before the development of plasma physics, namely the *Rayleigh–Taylor instability* (Rayleigh, 1883; Taylor, 1950). A heavier fluid, density ρ_2, is in equilibrium above a lighter, density ρ_1, in a gravitational field g. If a volume ΔV of the heavy fluid, is interchanged with an equal volume of the lighter with a net displacement z, then the change in the potential energy $W(z) = (\rho_2 - \rho_1)\,g\,z\,\Delta V$ is only positive if $\rho_1 > \rho_2$, and is stable only if the light fluid is on top of the heavy, consistent with experience. The instability is unchanged if the gravitational field is replaced by an inertial field due to acceleration of the interface.

The Rayleigh–Taylor instability occurs in an equivalent form in plasma physics where the heavy/light fluid interface is replaced by the surface between the plasma and a massless fluid supported by the magnetic pressure. In an accelerating field this is unstable due to the interchange of plasma and 'vacuum fluid'. The mechanics of the instability are more complex (Rosenbluth and Longmire, 1957) than in the fluid case as the electric field $\mathbf{E}$ is replaced by the inertial field $\mathbf{g}$

5 These important results, known as the *energy principle* were originally derived by Bernstein et al. (1958) in a seminal paper, and may be found in many standard text books: Rose and Clark (1961) and Freidberg (2014).

(see also Chandrasekhar and Trehan, 1960; Kruskal and Schwarzschild, 1954). This gives rise to a $\mathbf{g} \wedge \mathbf{B}$ drift leading to charge separation between the ions and electrons, and which in turn gives rise to a $\mathbf{E} \wedge \mathbf{B}$ drift, and finally to a motion of plasma perpendicular to the interface, resulting in an interchange of plasma and vacuum. Although of the same basic type, these instabilities take a number of forms known as *ripple* or *flute* instabilities. Rayleigh–Taylor instability in the acceleration phase was found to be severe in early Z pinch experiments (Curzon et al., 1960). As a result of these discouraging results, work on Z-pinches lapsed. However in recent years work on fast rising pulsed high density Z-pinches has successfully resumed. Haines (2011) presents an extensive review of this work.

As noted in Section 3.5.2, particles are accelerated by the centrifugal force, which gives rise to curvature drift. The effective value of the gravitational acceleration is given by Eq. (3.65)

$$\mathbf{g} \rightarrow \frac{m}{q\,B^2}\left(v_{\parallel}{}^2 + \frac{1}{2}v_{\perp}{}^2\right)\frac{\mathbf{R}}{R^2} \tag{16.48}$$

which is parallel to the vector radius of curvature measured *from* the centre of rotation. As a result, the plasma is unstable if the curvature is directed from the plasma to the vacuum. Since the instability is caused by the $\mathbf{E} \wedge \mathbf{B}$ resulting from the charge separation field established by the gravitational drift $\mathbf{g} \wedge \mathbf{B}$, the overall effect is to generate a parallel drift of both ions and electrons along the magnetic field lines. This takes the form of a *ripple* across the plane of the magnetic field or as *flutes* along the flux tubes in a cylindrical plasma.

Thus far, the instabilities have been at the plasma/vacuum interface. However, internal instability within the plasma can also occur in similar conditions. In a toroidal system such as a tokamak, the configuration is generally favourable to avoid instability on the inside and unfavourable on the outside. In general, plasma at curved flux surfaces suffer instability if the curvature is convex outwards (favourable) or unstable (unfavourable) if concave. These are generically known as *ballooning* instabilities.

Referring to the set of Eqs. (16.46) the instabilities described thus far have all been *magnetically driven*, i.e. associated with the magnetic field. However, the kinetic pressure terms can also give rise to *pressure driven* instability. As an illustration of pressure driven instability, we give simple example of the standard interchange instability; the avoidance of which leads to the *minimum B*, 'magnetic field trap', particularly applied to 'magnetic mirror' devices (Section 3.10) (Taylor, 1963). Consider the interchange along the magnetic field of two adjacent flux tubes whose magnetic fluxes $\Phi = BS$ are equal. If the tube ① of pressure p_1 and volume V_1 interchanges with ② at p_2 and V_2, the total magnetic energy is unchanged but the internal energy in tube ① is changed adiabatically by $p_2 V_2{}' - p_1 V_1 = p_1[(p_1/p_2)^{\gamma} - 1]$, with a similar change for the tube ②. Thus writing $\delta p = p_2 - p_1$ and $\delta V = V_2 - V_1$, the total energy change due to the interchange is approximately

$$\delta p\ \delta V + \gamma\ p_1\ (\delta V)^2/V_1 \tag{16.49}$$

which is obviously positive if $\delta p\ \delta V > 0$, and the system is stable.

In a low β plasma as in a magnetic mirror where the field is provided by external coils. $\nabla \wedge \mathbf{B} \approx 0$ implies that if the field lines are concave towards the plasma, as in a mirror configuration. The field strength increases towards the plasma. This in the unfavourable case of Figure 16.7; the plasma will be unstable as proved to be the case. The instability takes the form of ripples or flutes along the field line. In a tokamak, the curvature is favourable on the inside of the torus, but unfavourable on the outside. Therefore, by rotating the plasma around the toroidal surface, the unstable element may be cancelled out.

In general in a cylindrical or toroidal discharge, pressure decreases outwards from the centre, so that to ensure stability we would like the volume occupied by the flux tube to decrease away from

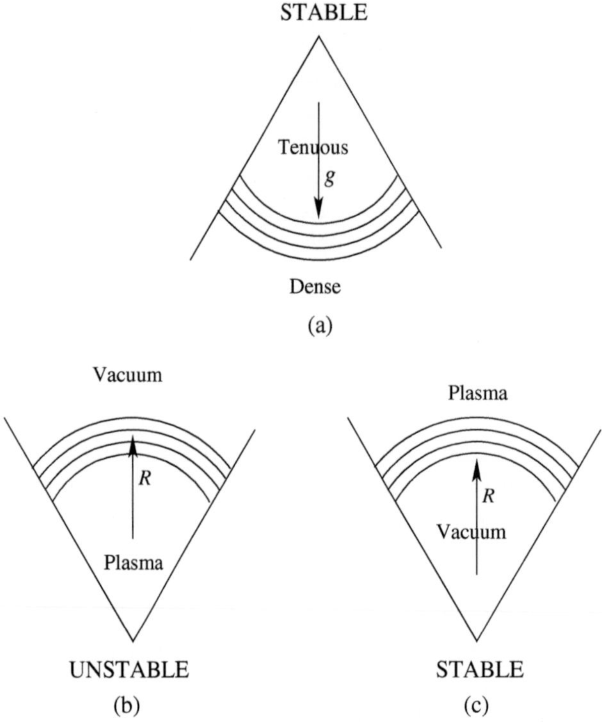

Figure 16.7 Stability of plasma in a curved magnetic field. b The field lines are convex with respect to the plasma, which is unstable. Panel (c) shows the concave, stable case. For reference (c) shows the arrangement of fluids in the hydrodynamic Rayleigh–Taylor instability.

the centre to satisfy the above condition. In an ideal plasma, the flux in the tube $\Phi = B\ S$ is constant where S is the cross section of the tube. Hence, $V = \int S\ \mathrm{d}l$ integrating along the flux tube, and we require that ideally $\int \mathrm{d}l/B$ should decrease away from the centre. This implies that the average of the field along the field lines should increase away from the plasma centre, or alternatively that the plasma occupy a region of average minimum B.

Generally, plasma at curved flux surfaces suffers instability if the curvature is convex outwards (unfavourable) rather than concave, which is stable (favourable) (Figure 16.7). These are generically known as *ballooning* instabilities. These are important in tokamaks providing a limiting maximum value of β at which the device is stable.

We will not pursue this topic further, which remains one of active study, but refer the reader to the extensive discussion which may be found in many texts for example Wesson (2004) or Freidberg (2014).

It was found in early studies of toroidal discharges that discharges were generally unstable and tended to become disorganised. However, it was found in experiments such as ZETA that under appropriate conditions an organised plasma environment with reversal of the internal toroidal magnetic field was established (Bodin and Newton, 1980). This relation was successfully attributed to an asymptotic relaxation process by Taylor (1974). These systems could be long lived and were initially thought to have potential as controlled fusion machines. However as the plasmas tend to be more susceptible to non-linear effects and turbulence, this has proved to be optimistic. However, it also makes it a useful system for studying non-ideal (resistive) magnetohydrodynamics. Reverse

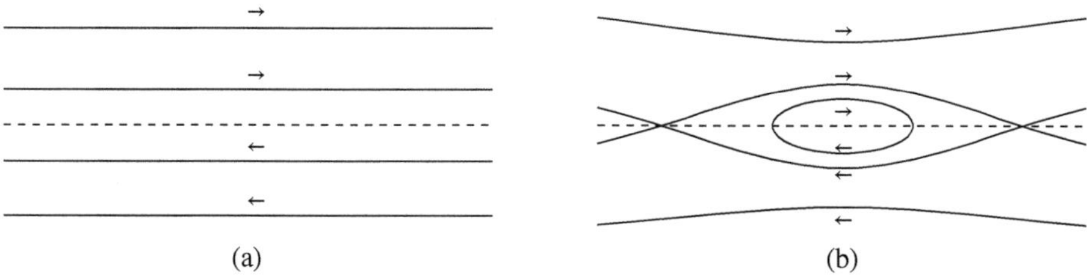

Figure 16.8 Break up of counter streaming plasma flows (a) into magnetic filaments (b) resulting from the tearing mode instability.

field pinches (RFPs) are also used in studying astrophysical plasmas, which share many common features. (A valuable review of RFPs is given by Bodin (1990).)

The instabilities discussed thus far are all generated in ideal collisionless plasma. However another set of behaviour occur as a result of resistivity in the plasma, the *resistive modes*. Of particular importance are the *tearing modes* where oppositely directed plasma streams breakup into filaments. Parallel but oppositely directed current streams attract each other. A lower potential energy is obtained by the current breaking up into filaments. This is due to the breaking and rejoining of the field lines by the resistivity (Figure 16.8).

Supplementary Material

M.1 Breakdown and Discharges in d.c. Electric Fields

M.1.1 Gas Breakdown and Paschen's Law

Paschen (1889) published a paper which detailed the results of experiments on the breakdown voltage of gases between two plane parallel electrodes of different materials set a distance d apart in different gases at various pressures p. This study represents the first quantitative measurements of the breakdown of gases, i.e. low temperature plasma physics. He found that his results could be expressed by an analytic equation in terms of the proper variable $p\,d$

$$V_B = \frac{B\,p\,d}{\ln(A\,p\,d) - \ln\left[\ln(1 + 1/\gamma)\right]} \tag{M.1.1}$$

where V_B is the breakdown voltage, p the gas pressure, d the gap separation, α the first Townsend coefficient, the number of ion pairs generated by an electron per unit path length, and γ Townsend's second coefficient or the secondary emission coefficient is the number of electrons released at the cathode per ion; (alternatively as β defined by Townsend (1915) the number of ion pairs formed per unit ion path). Values of α and γ are found experimentally, γ being generally poorly known. α and γ are found to be nearly constant over a restricted range.

Townsend (1915) made a major advance on Paschen (1889) work with the introduction of Townsend's sparking criterion which accurately predicted the Paschen breakdown law. He found that the breakdown voltage could be expressed in terms of proper variables (E/p, $\alpha\,d$, and α/p) where α is the first Townsend coefficient. Townsend (1915) also introduced the secondary ionisation coefficient now known to be due to secondary ionisation by ionic collision. As is now known the ionisation cross section is a function of particle relative velocity not energy (as assumed in classical hard sphere models). It is now understood that introducing a quantum picture the ionisation can be visualised as photo-ionisation by a virtual photon resulting from the rate of change (velocity) of the electric field due to the passing charged particle (McDaniel, 1964). As a result by the early 1920s, it became clear that the secondary electron emission γ at the cathode is due to ions as a result of secondary processes. Loeb (1936,1955, and references therein) gives a comprehensive critique of secondary emission processes.[1]

At high pressure when the electron mean free path is short, the breakdown is described by the avalanche process and the Townsend mechanism §M.1.4 is appropriate. In air for gaps of $\sim$ mm,

1 This led to a famous controversy between Townsend (1915) at Oxford (secondary ion collisions) and Thomson and Thomson (1933) at Cambridge (ion-cathode effects) supporting the different pictures, both of which have very similar functional forms. Sadly, for such a distinguished scientist, Townsend (1947) was still arguing the case for ion collisional ionisation when the matter was well resolved in the community – he died in 1957.

Foundations of Plasma Physics for Physicists and Mathematicians, First Edition. Geoffrey J. Pert.

Paschen found that the breakdown voltage is approximately a linear function of $p\,d$ namely $V_B \approx 30\,(p\,d)$ atm.cm. + 1.35(kV). At pressures below the minimum electron collisions become more infrequent and a higher field is required to generate the breakdown.

M.1.2 Similarity and Proper Variables

Self-similar relations between the governing variables in a problem are extremely useful enabling the solution under one set of conditions to be related to those of similar one. Normally these are found by identifying dimensionally similar groups. In discharges, such relationships also play an important role, but in contrast to the general approach not all the dimensional quantities may be identified and the set of dimensionless products is not therefore complete. As a result, the similar quantities are identified empirically, and consequently only a limited sub-set of all the possible combinations is found.

Two discharges between two electrodes of the same material in the same gas are said to be similar if they require the same potential difference at the same current, even if the geometry of the gaps is different.

We approach the problem of identifying the 'dimensionless products' by considering the response of the base variables to a proportional increase. For example geometrical similarity requires that the linear dimensions of discharge ① is n times that of ② ($d_1 = nd_2$), and that at the same time the gas pressure in tube ① is $1/n$ times that in tube ② . Under the conditions for similarity the following relations are obeyed:

Mean free path $\lambda_1 = n\lambda_2$ Gas density $\delta_1 = \delta_2/n$ Pressure $p_1 = p_2/n$ Electric fields at geometrical similar points $E_1 = E_2/n$ Townsend first coefficient $\alpha_1 = \alpha_2/n$ Charge density $\rho_1 = \rho_2/n^2$ Current density $j_1 = j_2/n^2$ Total current $i_1 = j_1 {d_1}^2/j_2 {d_2}^2 = i_2$ Time interval $\mathrm{d}t_1 = n\ \mathrm{d}t_2$ Changes in charge density $\left.\frac{\mathrm{d}N}{\mathrm{d}t}\right|_1 = \frac{1}{n^3}\left.\frac{\mathrm{d}N}{\mathrm{d}t}\right|_2$

Since $\lambda_1\ E_1 = \lambda_2\ E_2$, drift and random velocities at geometrically similar points are equal.

Within a discharge, the limited set of processes which follow these conditions exhibit similar behaviour, e.g. two body ionising collision or diffusion, whereas others such as stepwise ionisation or recombination do not (see von Engel (1965, appendix 1) or Cobine (1958, §8.3)). High pressure recombination is allowed, but low pressure and direct ion recombination are forbidden. Surface and wall processes must satisfy the condition $\frac{\mathrm{d}\sigma_1}{\mathrm{d}t} = \frac{1}{n^2}\frac{\mathrm{d}\sigma_2}{\mathrm{d}t}$ which is normally obeyed.

Only similar processes are allowed in similar discharges. Fortunately most simple discharges, such as glows, obey these rules, although many of the more complex systems found in industrial application do not.

Proper variables, which apply to similar discharges, are found by eliminating the term n in the above set. For example $E/p, p\ R, p\ \lambda$ where for historical reasons the ambient gas pressure p is used instead of the gas density δ. These variables depend on the electrode configuration but not their dimensions. These variables are consequently frequently used to identify general scaling relations, for example the breakdown field in Paschen's law $V_B(p\ d)$.

M.1.3 Townsend's First Coefficient

Townsend (1915) model for this quantity is based on the hard sphere collision (Drude) model Section 2.2.3 of electron conductivity. In each collision the electron is assumed, over averaging, to be brought to rest. Thus in the free path λ between collisions, the electron gains energy $eE\lambda$ from the electric field E. If this energy is greater than the ionisation energy I, the collision gives rise to the formation of an ion pair. The minimum free path for ionisation is therefore $\lambda = I/eE$. Since the probability of a free path $> \lambda$ is $\exp(-\lambda/\bar{\lambda})$ where $\bar{\lambda}$ is the mean free path, a fraction $\exp(-\lambda/\bar{\lambda})$ of the $1/\bar{\lambda}$ collisions per unit path cause ionisation. Thus, for simple hard sphere collisions $z = \bar{\lambda}^{-1} \exp -(\lambda/\bar{\lambda})$. The scaling of the mean free path with pressure (density) $\bar{\lambda} = p_1/p\, \overline{\lambda_1}$ where the subscript 1 indicates values at standard temperature and pressure (STP). Hence, the number of ionisation per unit path length can be estimated.

For many simple gases, it can be shown that a good fit for Townsend's first coefficient is obtained by this simple classical argument provided the values of A and B are measured empirically (von Engel, 1965, p. 180)

$$\alpha(E/p) \approx Ae^{-B\,p/E} \tag{M.1.2}$$

where A and B are tabulated values for different gases obtained by experiment.

M.1.4 Townsend's Breakdown Criterion

We derive Townsend's breakdown criterion in a gas between plane parallel electrodes a distance d apart, where p is the gas pressure, E the electric field, and V the potential between the electrodes. We define Townsend's first coefficient α as the ionisation generated by an electron per unit distance of travel and Townsend's second coefficient $\gamma < 1$ as the average number of electrons released from the cathode by a single ion. We calculate the condition for a sparking breakdown when the number of electrons released by secondary processes at the cathode equals the number freed in the previous generation.

Let electron flux towards the anode be j a distance x from the cathode. Then over distance δx, the electron flux is increased by $\delta j = \alpha\, j\, \delta x$. Hence, it follows that a single electron leaving the cathode gives rise to a flux

$$j(d) = \exp\left[\int_0^d \alpha(x)\, \mathrm{d}x\right] - 1 \tag{M.1.3}$$

Since ions and electrons are created in pairs, the total ion flux generated in the gas must equal the electron flux at the anode. Hence, the flux of secondary electrons released by the ions at the cathode to start the next generation is

$$j^1(0) = \gamma j^0(d) = \gamma\left(\exp\left[\int_0^d \alpha(x)\, \mathrm{d}x\right] - 1\right) \tag{M.1.4}$$

If $j^1(0) > 1$ clearly the flux carried by succeeding generations increases and an avalanche is established. The breakdown condition is therefore

$$\gamma\left(\exp\left[\int_0^d \alpha(x)\, \mathrm{d}x\right] - 1\right) > 1 \qquad \text{or simply} \qquad \alpha d \geq \ln(1 + 1/\gamma) \tag{M.1.5}$$

the latter result being between a pair of plane parallel electrodes.

The Townsend coefficients are generally established by experiment. Although the first is relatively simple to evaluate, the second involves multiple complicated processes at the cathode.

M.1.5 Paschen Curve and Paschen Minimum

Solving for the field from Eq. (M.1.2) and adding Townsend's sparking criterion (M.1.5)

$$E/p = B/(\ln A - \ln \alpha) \qquad \text{and} \qquad \alpha/p \cdot pd - \ln(1 + 1/\gamma) \tag{M.1.6}$$

we obtain

$$V_B = E/p \cdot pd = Bpd/(\ln A - \ln\alpha) = Bpd/(\ln A - \ln(\ln(1+1/\gamma)/pd) \tag{M.1.7}$$

which was the form of the breakdown voltage vs. (pd) found experimentally by Paschen.

$$V_B = \frac{Bpd}{\ln(Apd) - \ln(1+1/\gamma))} = \frac{B\,p\,d}{C+\ln(p\,d)} \tag{M.1.8}$$

where $C = \ln\{A/\ln(1+1/\gamma)\}$,

$$\frac{\mathrm{d}V_B}{\mathrm{d}(pd)} = \frac{B}{C+\ln(pd)} - \frac{B}{[C+\ln(pd)]^2} \tag{M.1.9}$$

Using this result, it is easily shown that the breakdown curve of V_B vs. $p\,d$ has a minimum when

$$C+\ln(p\,d) = 1 \qquad \text{or} \qquad p\,d = e^1 \ln(1+1/\gamma)/A \tag{M.1.10}$$

M.1.6 Radial Profile of Glow Discharge

A glow discharge involving an electrical discharge through a gas normally is a complex structure involving a structured cathode region, a quiescent positive column, and a short region adjacent to the anode. We examine the radial profile quiescent part of the discharge *positive column* which has found wide application for example in processing and lighting.

Consider a cylindrical tube of radius R. Gas is ionised in the plasma body at a rate $z\,n_e$ per unit volume per unit time, where n_e is the electron density which is much less than the gas density, which is uniform across the tube. The resultant electrons and ions drift by ambipolar diffusion to the tube walls where they recombine. It is assumed that recombination in the bulk is small and takes place after ambipolar drift to the walls of the tube. The loss of ions from the change in the ambipolar diffusion flux Γ is balanced by ion gain by electron collisional ionisation. Since the positive column is quiescent, we solve the resultant differential equation in the steady state and hence calculate the electron density profile across the tube. We assume the ionisation rate and the ambipolar diffusion coefficient are constant.

Clearly, the electron density is symmetric about the centre line and zero due to recombination at the wall. Since the problem is cylindrically symmetric and making use of problem 1, the rate of change of the diffusive flux across the radius must equal the ionisation rate

$$\nabla\cdot\mathbf{\Gamma} = -\frac{1}{r}\frac{\partial}{\partial r}\left(rn_e\right) = z\,n_e \tag{M.1.11}$$

whose solution is

$$n_e(r) = n_e(0)J_0(x) \tag{M.1.12}$$

where $x = \sqrt{z/D_a}r$ and $J_0(x)$ is the Bessel function of zero order of the first kind. Since the first zero of the Bessel function is at $x = 2.405$, the boundary condition $n_e(R) =$ yields an eigenvalue condition

$$\sqrt{z/D_a}\,R = \Lambda/R = 2.405 \tag{M.1.13}$$

which defines the diffusion length and determines the ionisation required to maintain a steady discharge.

This balance equation linking ionisation and diffusive loss was first derived by Schottky in 1925, and the consequent density distribution Eq. (M.1.12) is known as the Schottky equation.

M.1.7 Collisional Ionisation Rate for Low Temperature Electrons

We derive an expression for the ionisation rate z for warm electrons with temperature T. Since the temperature is less than ionisation energy I only electrons with energy E just above the threshold cause ionisation $E > I$ with a cross section which can be approximated as $\sigma_i \approx a\,(E - I)$ where a is constant specific to the gas in the positive column.[2] Assume the electron distribution is Maxwellian with the number of electrons per unit volume in the energy interval $F(E)$ dE

$$F(E) = 2\,\pi \left(\frac{1}{\pi\,\text{k}T}\right)^{3/2} n_e\,\sqrt{E}\,e^{-E/\text{k}\,T} \tag{M.1.14}$$

The number of ionising collisions per unit volume is therefore

$$\begin{aligned} z\,n_e &= \int_I^\infty N\,\sigma_i\,v\,F(E)\,\mathrm{d}E = 2\,\pi n_e \int_I^\infty N\,a\,(E-I)\,\sqrt{\frac{2\,E}{m}}\left(\frac{1}{\pi\,\text{k}T}\right)^{3/2}\sqrt{E}e^{-E/\text{k}\,T}\,\mathrm{d}E \\ &= 2\sqrt{\frac{2\,\pi}{m}}\left(\frac{1}{\text{k}T}\right)^{1/2} n_e\,N\,a\,e^{-I/\text{k}T}\int_0^\infty E'\,(E'+I)\,e^{-E'/\text{k}\,T}\,\mathrm{d}E' \\ &= 2\sqrt{\frac{2}{\pi\,m}}\,(\text{k}T)^{3/2}\,n_e\,N\,a\,\left(1 + I/2\text{k}T\right)\,e^{-I/\text{k}T} \approx \overline{v}\,n_e\,N\,a\,e^{-I/\text{k}T} \end{aligned} \tag{M.1.15}$$

where N is density of neutral molecules and $\overline{v} = \sqrt{8\text{k}T/\pi m}$ is the mean thermal speed.

Making use of the Schottky condition (M.1.13) and the ambipolar diffusion coefficient, Eq. (1.30), the ionisation rate required to support the ambipolar diffusion loss determines the electron temperature through

$$\left(\frac{\text{k}T}{I}\right)^{1/2} \exp\left(\frac{I}{\text{k}T}\right) = \frac{a}{\mu_+ p}\left(\frac{8I}{\pi m}\right)^{1/2}\frac{N}{p}(p\Lambda)^2 = \text{const}\,(pR)^2 \tag{M.1.16}$$

which yields a universal relationship giving good agreement with experiment between the proper variables T, the electron temperature, and pR the pressure radius product. The constant which depends on the gas is extensively tabulated for noble gases..

This final result forms the basis of the von Engel–Steenbeck theory of the positive column (von Engel, 1965; Cobine, 1958). The further development of the model calculates the current and electric field required to support this temperature by Ohmic heating. This is achieved by balancing the energy gain by an electron as it drifts in the electric field against the energy loss by the electron in collisions with molecules (Raizer, 1997). This energy transfer however is difficult to calculate as collisions may be either frequent in elastic collisions with a small energy transfer per collision, or inelastic with a large energy transfer, but infrequent. Both of these types of collisions distort the distribution away from a Maxwellian. For example in the case when the collisions are all elastic, the distribution tends towards the Druyvestyn form (Appendix 8.C) where the electric field and the mean energy are related by an expression

$$\overline{\epsilon} \sim \sqrt{\frac{M}{m}}\,e\,E\,\lambda \tag{M.1.17}$$

where $e\,E\,\lambda$ is the energy gained in one mean free path.

On the other hand, if inelastic collisions dominate a modification of Eq. (M.1.16) is appropriate, but with a modified energy level to account for the bound states.

2 It should be noted that although for small values of E near threshold the cross section varies proportionally to $E - I$, but drops at values of E well above I.

At low electron densities, electron–electron collisions are few so that the necessary thermalising collisions are lacking and a Maxwellian distribution is not correct. A full kinetic theory model is needed to obtain good agreement with experiment in this case (Raizer, 1997; Smirnov, 2015). Smirnov (2015) in particular details the calculation of the distribution in gas discharges in helium and argon.

M.1.8 Radio Frequency and Microwave Discharges

Radio frequency and microwave fields have become increasingly popular as sources for discharge plasma. In general, their behaviour closely follows that in d.c. fields and we will only briefly touch on them here, but refer the reader to one of several texts Brown (1966), Raizer (1997), Lieberman and Lichtenberg (2005), and Smirnov (2015) which cover this material in detail. Electrons are heated in the electro-magnetic field through the resistive term in Eq. (2.53) to temperatures at which collisional ionisation takes place as discussed for d.c. fields (Section M.1.7). As a result, an ionising avalanche takes place limited by the diffusion of ion/electron pairs out of the field volume. Initially when the electron density is very small free electron diffusion will be the limiting process, but once breakdown is established ambi-polar diffusion will be dominant requiring a lower field limit. A quiescent plasma column following the same principles as the d.c. positive column generates a stable, uniform low temperature plasma, suitable for many applications.

M.2 Key Facts Governing Nuclear Fusion

M.2.1 Fusion Rate

It is important in estimating the fusion power to be able to get an estimate of the fusion rate in the fuel, for example an equal proportion deuterium–tritium nuclei, which is obtained by integration of the fusion cross section taken over the velocity distributions of the deuterons and tritons. Since these are approximately equal mass, energy transfer between them is rapid and we may without significant error assume that each has a Maxwell/Boltzmann distribution with the same temperature T.

Since the particle energy required to overcome the Coulomb potential barrier between the nuclei is large compared to the particles thermal energy, only particles in the high energy tail of the distribution can undergo fusion, and only the low energy part of the cross section is important. Since the fusion interaction must be invariant under simple translations, we conclude that the cross section for fusion in a collision between two nuclei ① and ② is the product of three terms, all of which are determined by the motion of the two particles within their centre of mass frame:

1. A geometric term proportional to the de Broglie wavelength squared $\propto \lambda^2 = (\hbar^2/\overline{m}v)^2 \propto 1/\varepsilon$ where ε is the reduced mass energy and $\overline{m}$ the reduced mass (3).
2. The barrier transparency $\exp(-\sqrt{\varepsilon_G/\varepsilon})$, where $\varepsilon_G = 2\overline{m}c^2(\pi\alpha Z_1 Z_2)^2$, α is the fine structure constant $e^2/4\;\pi\;\epsilon_0\;\hbar\;c$, was calculated by Gamow (1928) by reduced the motion to the centre of mass frame and considering the penetration of the single particle of mass $\overline{m}$ through the barrier due to the joint Coulombic fields.
3. The reaction constant $S(\varepsilon)$ – a slowly varying factor characteristic of the reaction known as the *astrophysical S constant*, which takes account of the specific nuclei.

The reaction rate per unit volume is obtained from the product of the cross section and the relative velocities by integrating over the number of particles per unit volume. Hence, $\langle\sigma v\rangle$ is a function of u alone independent of V, reflecting the Galilean invariance of the fusion process

$$
\begin{aligned}
r &= \iint f_1(\mathbf{v}_1)\, f_2(\mathbf{v}_2)|\mathbf{v}_1 - \mathbf{v}_2|\;\sigma(\mathbf{v}_1 - \mathbf{v}_2)\mathrm{d}\mathbf{v}_1\mathrm{d}\mathbf{v}_2 \\
&= n_1 n_2\left(\frac{m_1}{2\;\pi k T}\right)^{3/2}\left(\frac{m_2}{2\;\pi k T}\right)^{3/2}\iint \mathrm{d}\mathbf{v}_1\mathrm{d}\mathbf{v}_2\;|\mathbf{v}_1 - \mathbf{v}_2|\;\sigma(\mathbf{v}_1 - \mathbf{v}_2) \\
&\quad\times \exp\left\{-\frac{\frac{1}{2}\;m_1\;{v_1}^2 + \frac{1}{2}\;m_2\;{v_2}^2}{kT}\right\}
\end{aligned}
\tag{M.2.1}
$$

To perform this integral, we transform to the centre of mass frame as in Appendix 2.A. We write the differential velocity $\mathbf{u} = \mathbf{v}_1 - \mathbf{v}_2$ and the centre of mass velocity $\mathbf{V} = (m_1\mathbf{v}_1 + m_2\mathbf{v}_2)/(m_1 + m_2)$ to reduce the integral to

$$
r = n_1 n_2\;4\;\pi\left(\frac{m_1\;m_2}{2\;\pi k T(m_1 + m_2)}\right)^{3/2}\int \mathrm{d}u u^3\sigma(u)\exp\left\{-\frac{1}{2}\frac{m_1 m_2}{(m_1 + m_2)}\frac{u^2}{kT}\right\}
\tag{M.2.2}
$$

The fusion cross section is a function of the total energy of the two colliding particles which is specified in two different, but equivalent, ways

1. In terms of the relative velocity in the centre of mass frame $\mathbf{u}$ when the energy $E = \frac{1}{2}\overline{m}u^2$.
2. In terms of the energy of the projectile ① striking the target ② at rest $E' = \frac{1}{2}m_1 u^2$.

Clearly, $E' = \frac{\overline{m}}{m_1} E$, writing $Y = \frac{\overline{m}}{m_1}\, \text{ƙ}T$ transforming the energy variable in Eq. M.2.1 from E to E'

$$r = \left(\frac{8}{\pi\, m_1}\right)^{1/2} n_1 n_2 \frac{1}{Y^{3/2}} \int dE'\, E'\, \sigma(E') \exp\left\{-\frac{E'}{Y}\right\} \tag{M.2.3}$$

The values of the fusion cross section may be obtained in several ways. Firstly a simple approximate analytic form derived from the Gamow (1928) formula may be useful if the results are only required for approximate estimates. The alternative is to use some of the published tabular data based on both experiment and theory.

a. *Gamow–Thompson approximation*:

Putting these terms together, the cross section can be estimated from the Gamow tunnelling approximation, which allows a simple analytic estimate of the rate. From Gamow (1928), write the cross section as

$$\sigma(E) = AE^{-1} \exp(-BE^{-1/2}) \tag{M.2.4}$$

A and B are found by the best fit to experimental data, when E is the relative energy of the two particles. We transform to the energy of the projectile particle E'

$$\sigma(E') = A'E'^{-1} \exp(-B'E'^{-1/2}) \tag{M.2.5}$$

where the constants $A' = \frac{\overline{m}}{m_1} A$ and $B' = \left(\frac{\overline{m}}{m_1}\right)^{1/2} B$ (see Table M.2.1). Substituting for the Maxwell–Boltzmann distribution

$$\begin{aligned}\langle \sigma\, v\rangle &= \left(\frac{8}{\pi\, m_1}\right)^{1/2} \frac{1}{Y^{3/2}} \int dE'\, E'\, \sigma(E') \exp\left\{-\frac{E'}{Y}\right\} \\ &= \left(\frac{8}{\pi\, m_1}\right)^{1/2} \frac{1}{Y^{3/2}} \int dE'\, E'\, [A'E'^{-1} \exp(-B'E'^{-1/2})] \exp\left\{-\frac{E'}{Y}\right\} \\ &= \left(\frac{8}{\pi\, m_1}\right)^{1/2} A'Y^{-1/2} \int dX \exp[-(X + B'(YX)^{-1/2})]\end{aligned} \tag{M.2.6}$$

where $X = E'/Y$. Performing the integral by the method of steepest descent

$$\int dx \exp[-(x + a\, x^{1/2})] \approx 2\left(\frac{1}{3}\pi\right)^{1/2} \left(\frac{1}{2}\, a\right)^{1/3} \exp\left\{-3\left(\frac{1}{2}a\right)^{2/3}\right\} \tag{M.2.7}$$

where $a = B'/Y^{1/2}$

$$\begin{aligned}\langle \sigma\, v\rangle &= \left(\frac{8}{\pi\, m_1}\right)^{1/2} A'\, Y^{-1/2} 2\left(\frac{1}{3}\pi\right)^{1/2} \left(\frac{1}{2}\, a\right)^{1/3} \exp\left\{-3\left(\frac{1}{2}a\right)^{2/3}\right\} \\ &= \left[4\left(\frac{2}{27}\right)^{1/6}\right] \left(\frac{\text{ƙ}T}{\overline{m}}\right)^{1/2} \frac{A}{\text{ƙ}T} \left(\frac{B^2}{(\text{ƙ}T)}\right)^{1/6} \exp\left\{-3\left(\frac{1}{4}\frac{B^2}{\text{ƙ}T}\right)^{1/3}\right\}\end{aligned} \tag{M.2.8}$$

This result is accurate to at least a factor of 2. It serves both to demonstrate the methodology and to provide order of magnitude results.

Table M.2.1 Gamow constants A' and B' from Thompson (1957).

Fuel	DT	DD
A' (barns keV)	7660	198
B' ($keV^{1/2}$)	46.9	41.6

Inspection of this result shows that the rate coefficient has the form

$$\langle \sigma v \rangle / T \sim X^n \exp(-X) \tag{M.2.9}$$

where $X = 3(B^2/4kT)^{1/3}$, or $kT = (B^2/4](3/X]^3$ and $n = 2$. The function (M.2.9) has a maximum value as T varies when $X = n$ or $kT = (B^2/4](3/n]^3$.

Therefore, the graph of the rate as a function of temperature has a maximum when $X = 7/2$, i.e. when $kT = 343/216B^2$. The existence of this turning point in the rate function is a general result is applicable to arbitrary expressions for the cross section, it is due to the structure of the cross section for fusion discussed on page 445where the particle wavelength decreases with increasing energy and the transparency increases. The effect on the overall rate for a thermal distribution being further enhanced by velocity of particles with sufficient energy increasing with temperature. More generally, functions $\langle \sigma v \rangle / T^n$ all have such a maximum; being zero at $T = 0$ as the nuclei cannot cross the barrier to approach each other, and zero as $T \to \infty$ as the particles cannot interact to achieve fusion.

b. *Tabular data*: In view of the importance of these rates for determining the performance of potential fusion reactors several data sets of the fusion cross section for various possible low energy fusion reactions have been prepared from experimental and theoretical analyses. A useful accessible reference is the Huba (2019) plasma formulary from where Tables (M.2.2) and (M.2.3) are taken. Fusion cross-sections in barns are given by

$$\sigma(\varepsilon) = \frac{A_5 + [((A_4 - A_3\varepsilon)^2 + 1)^{-1}]A_2}{\varepsilon[\exp(A_1(\varepsilon^{-1/2}) - 1]} \tag{M.2.10}$$

where ε is the incident particle (projectile) energy (in keV) towards the target at rest. The coefficient A_j are the Duane coefficients for each pair of nuclei (see Table M.2.2.).

Table M.2.2 Values of the fusion cross section (in barn $= 10^{-24} cm^2$) for particles of the target ion bombarded by the projectile.

	A1	A2	A3	A4	A5
$T(d,n)He^4(2)$	45.95	50200	$1.368e^{-2}$	1.076	409
$D(d,p)He^4(1)$	46.097	372	$4.36e^{-4}$	1, 22	0
$D(d,n)He^3$	47.88	482	$3.08e^{-4}$	1.177	0
$D(He^3,p)He^4(3)$	89.27	25900	$3.98e^{-3}$	1.297	647
$T(T,2n)He^4(4)$	38.39	448	$1.02e^{-3}$	2.00	0
$He^3(T,p,n)He^4(5)$	123.1	11250	0		0

Table M.2.3 Integrated fusion rates in (cm^3 s^{-1}) from Huba (2019).

Temperature (keV)	D–D (1)	D–T (2)	D–He^3 (3)	T–T (4)	T–He^3 (5)
1.0	$1.5e^{-22}$	$5.5e^{-21}$	$1.0e^{-26}$	$3.3e^{-22}$	$1.0e^{-28}$
2.0	$5.4e^{-21}$	$2.6e^{-19}$	$1.4e{-}^{23}$	$7.1e^{-21}$	$1.0e^{-25}$
5.0	$1.8e^{-19}$	$1.3e^{-17}$	$6.7e^{-21}$	$1.4e^{-19}$	$2.1e^{-22}$
10.0	$1.2e^{-18}$	$1.1e^{-16}$	$2.3e^{-19}$	$7.2e^{-19}$	$1.2e^{-20}$
20.0	$5.2e^{-18}$	$4.2e^{-16}$	$3.8e^{-18}$	$2.5e^{-18}$	$2.6e^{-19}$
50.0	$2.1e^{-17}$	$8.7e^{-16}$	$5.4e^{-17}$	$8.7e^{-18}$	$5.3e^{-18}$
100.0	$4.5e^{-17}$	$8.5e^{-16}$	$1.6e^{-16}$	$1.9e^{-17}$	$2.7e^{-17}$
200.0	$8.8e^{-17}$	$6.3e^{-16}$	$2.4e^{-16}$	$4.2e^{-17}$	$9.2e^{-17}$
500.0	$1.8e^{-16}$	$3.7e^{-16}$	$2.3e^{-16}$	$8.4e^{-17}$	$2.9e^{-16}$
1000.0	$2.2e^{-16}$	$2.7e^{-16}$	$1.8e^{-16}$	$8.0e^{-17}$	$5.2e^{-16}$

The reaction identifiers in Table M.2.2 are listed as standard with first: upper case projectile ion, second: lower case target ion, third: lower case light product ion (usually either proton or neutron), and fourth: Upper case heavy product ion (usually alpha particle).
Reaction rates $\langle \sigma v \rangle$ (in cm^3 s^{-1}) as a function of temperature averaged over a Maxwellian distribution are given in Table M.2.3:
For low energies ($T \lessapprox 25$ keV) the data may be represented by

$$\langle \sigma\ v \rangle_{DD} = 2.33e^{-14} T^{-2/3} \exp(-18.76 T^{-1/3})\ \text{cm}^3\ \text{s}^{-1} \tag{M.2.11}$$

$$\langle \sigma\ v \rangle_{DT} = 3.68e^{-12} T^{-2/3} \exp(-19.94 T^{-1/3})\ \text{cm}^3\ \text{s}^{-1} \tag{M.2.12}$$

where T is measured in keV. Note these values are essentially Gamow's expression, Eq. (M.2.8) with slightly modified values of the constants A and B valid in the asymptotic limit of the steepest descent approximation.

M.2.2 Lawson's Criterion

Lawson (1957) examined the energy balance of a fusion plasma to identify relatively simple criteria under which a plasma at high temperature of suitable nuclear fuels could be induced to burn at a sufficiently rapid rate to yield a net energy gain taking account the power losses and the initial heat content of the plasma. He found that there are two conditions which must be satisfied if the objectives were to be met. These are expressed in the following form:

I. Firstly identify when there would be net output power

$$\text{Net power out} = \text{Efficiency} \times (\text{Fusion} - \text{Radiation loss} - \text{Conduction loss}) \tag{M.2.13}$$

where

(a) **Net power** is the excess power required output above that needed to drive the reactions.
(b) **Efficiency** reflects the ability to convert the output back from the product particles to that required.

(c) **Fusion** is the power generated by the fusion reactions

$$\langle \sigma v \rangle n_1 n_2 \Delta\epsilon$$

where $\Delta\epsilon$ is the energy released in a single fusion reaction between nuclei ① and ②.[3]

(d) **Radiation loss** is the power lost by emission of radiation as light, X-rays, etc. Typically in these light nuclei plasma this will be bremsstahlung with a representative rate

$$P_B \approx 1.4 \times 10^{-34} N^2 T^{1/2} \qquad \text{W/cm}^3$$

n is the plasma number density in cm^{-3} and T the temperature in degrees.

(e) **Conduction** is the heat loss as particles leave the plasma body. Normally neglected at this level of approximation.

The necessary condition derived from this analysis is that the plasma temperature must be sufficiently high to offset the bremsstahlung loss, which represents the minimum allowable temperature.

Temperatures for fusion are generally measured in eV, the thermal energy $\mathcal{k}T$ specified in eV (1 eV ≡ 11 600 K).

The deuterium/tritium reaction has the lowest onset temperature and is therefore the basis for current experimentation but is offset by the fact the triton decays with half-life of 12.32 years. The deuterium/tritium reaction can be written

$${}^2_1D + {}^3_1T \longrightarrow {}^4_2He\ (3.5\ \text{Mev}) + {}^1_0n (14.1\ \text{Mev})$$

Alternatively for the deuterium/deuterium reaction, the reactants are relatively easily obtained being stable, but the onset temperature is substantially higher:

$${}^2_1D + {}^2_1D \rightarrow \begin{cases} {}^3_1T\ (1.01\ \text{Mev}) & + \ {}^1_1p\ (3.02\ \text{Mev}) \qquad 50\% \\ {}^3_2He\ (0.82\ \text{Mev}) & + \ {}^1_0n\ (2.45\ \text{Mev}) \qquad 50\% \end{cases}$$

The temperature when the steady state fusion rate ($\langle \sigma v \rangle\ n_1\ n_2\ \Delta\epsilon$) equals the rate of radiation ($1.4 \times 10^{-34}\ n^2\ T^{1/2}$) defines the minimum temperature at which useful fusion may occur. For DT reactions this takes the value of 2.6 keV. The DD reactions require considerably higher temperature (12.9 keV).

II. **Confinement time** is defined as the rate at which the system loses energy to the environment typically by radiation or conduction

$$\tau_E = \frac{W}{P_{\text{loss}}} \tag{M.2.14}$$

where W is the energy per unit volume and P_{loss} the power loss. In a DT mixture the plasma is 50% deuterons and 50% tritons ($n_D = n_T = \frac{1}{2}n$); since both ions have unit charge, the electron density equals the total ion charge n and the energy density is the total thermal energy $W = 3n\mathcal{k}T$.

In a steady-state configuration, the energy lost by the plasma equals that delivered by the charged particles in the fusion reactions – the neutrons being uncharged normally escape without slowing: the Lawson condition being that, the fusion energy delivered to the plasma exceed the loss

$$\frac{1}{4}\ \langle \sigma v \rangle \Delta E_C \geq P_{\text{loss}} = \frac{3n\mathcal{k}T}{\tau_E} \tag{M.2.15}$$

3 If the nuclei are identical ① = ②, then the density product must be modified to avoid double counting $n_1\ n_2 \rightarrow \frac{1}{2}{n_1}^2$.

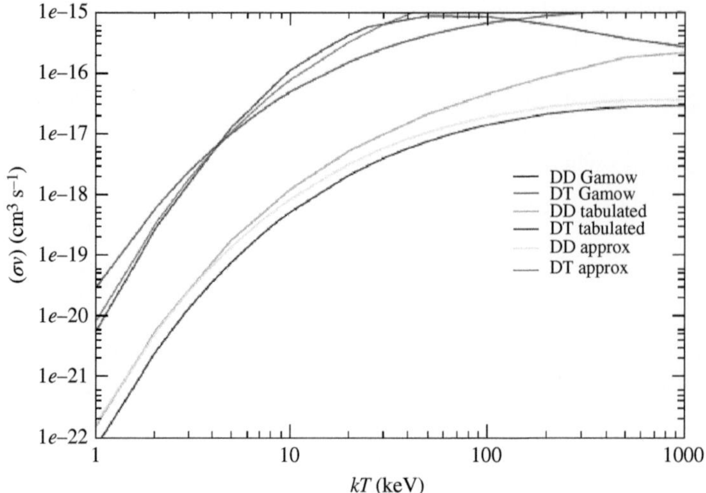

Figure M.2.1 Plots of the Maxwell–Boltzmann integrated fusion rate $\langle \sigma v \rangle$ culled from various sources.

For the DT reaction, the charged particles are α particles and $\Delta E_{\mathrm{Ch}} = 3.5$ MeV. Substituting we obtain the Lawson number subject to the condition

$$L = n\ \tau_E \geq \frac{12 k T}{\langle \sigma v \rangle\ E_{\mathrm{Ch}}} \tag{M.2.16}$$

the term $T/\langle \sigma v \rangle$ has a minimum at a temperature about 25 keV, which provides the lower limit for the Lawson number (Figure M.2.1). For a DT mixture, we obtain the classic form

$$n\tau_E \geq 1.5 \times 10^{14}\ \mathrm{s\ cm^{-3}} \tag{M.2.17}$$

Lawson (1957) considered cyclic operation of the device so that τ_E was simply the plasma cooling times. Nowadays τ_E is usually taken to be the lifetime of the plasma in normal operation.

M.2.3 Triple Product

In recent years, it has become apparent that a more useful figure of merit is the triple product $NT\tau_e$. This relies on the fact that the maximum attainable pressure is a constant even though the individual variables: density, temperature may vary significantly. Hence, multiplying Eq. (M.2.17) by the temperature, we introduce the pressure

$$nT\tau_E \geq \frac{12 k}{E_{Ch}} \frac{T^2}{\langle \sigma v \rangle} \tag{M.2.18}$$

As we have shown, the term $\langle \sigma v \rangle / T^2$ vs. T has a minimum (Figure M.2.2). For the DT reaction this occurs at $T = 14$ keV near where $\sigma v >\approx 1.1 \times 19^{-24} T\mathrm{keV}^2\ \mathrm{m^3\ s^{-1}}$. Substituting the minimum value of the triple product at 14 keV (Lawson condition DT)

$$nT\tau_E \geq 3.10^{21}\ \mathrm{keV\ s\ m^{-3}} \tag{M.2.19}$$

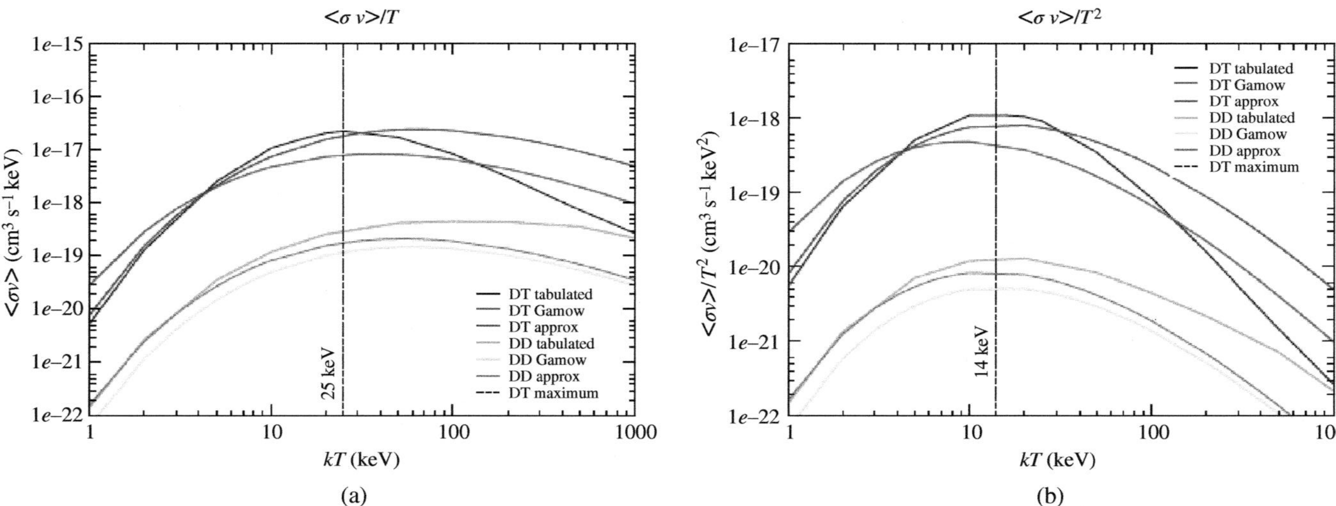

Figure M.2.2 Plots of the ratio of fusion rates for DT and DD to temperature. (a) Rates 1 $\langle \sigma v \rangle / T$. (b) Rates 1 $\langle \sigma v \rangle / T^2$.

M.3 A Short Introduction to Functions of a Complex Variable

In this brief summary, we establish many of the key elements of the theory of complex functions. As these are extensively used in plasma physics, it may be helpful to summarise their key properties, which are formally derived in many standard text-books, e.g. Copson (1935), Flanigan (1983), and Morse and Feshbach (1953, Ch.4).

It will be useful to initially define some key terms used in the theory of complex functions

- *Analytic function*: A complex single valued function of a complex variable locally given by a convergent power series and which can be differentiated to any order.
- *Holomorphic function*: A complex single valued function that is complex differentiable everywhere on its domain. Holomorphic functions satisfy the Cauchy–Riemann relations. Holomorphic functions are analytic and *vice versa*. A function holomorphic everywhere in a domain $\mathscr{D}$ is said to *regular* within the domain $\mathscr{D}$.
- *Meromorphic function*: A function which is holomorphic on a subset of the complex plane except at a set of isolated points which are poles of the function.
- *Singularity*: Any point at which the function $f(z)$ is no longer not defined or 'not well behaved' is called a singularity.
- *Isolated singularity*: A singularity that has no other singularities close to it. Such points may be excluded from the region of analyticity by drawing a small circle around them. Isolated singularities take one of three forms
  (1) *Removable singularity*: Removable singularity at z_0 is a point at which the function $f(z)$ is not formally defined but a which it can be modified to be continuous, e.g. at $z = 0$

$$f(x) = \begin{cases} \dfrac{\sin(z)}{z} & z \neq 0 \\ 1 & z = 0 \end{cases}$$

  is regular (analytic) at $z = 0$.
  (2) *Pole*: Let z_0 be a point at which the holomorphic function $f(z_0)$ has a **Zero**. The reciprocal function $1/f(z_0)$ is singular and has a **Pole** at z_0. If $f(z) \sim (z - z_0)^n$ near z_0, the pole is of the nth degree. Let the function be $1/f(z) \approx g(z)/(z - z_0)^n$ then its residue at z_0 is $g(z_0)$. Near the pole the function behaves relatively regularly in contrast to:
  (3) *Essential singularity*: Examples are $\exp(1/z)$ at $z = 0$ and branch points.
- *Branch point*: A point where a multivalued function is discontinuous going around a small loop, e.g. $\log z$ at $z = 0$ (logarithmic branch point), or $\sqrt{z}$ at $z = 0$ (transcendental branch point). Formally, the function moves from one Riemann surface to another during the rotation.
- *Branch cut*: Branch cut is usually a line joining two neighbouring branch points, e.g. $\sqrt{(1 - z^2)}$ between $z = -1$ and $z = 1$ across which the function is discontinuous in the complex plane. A branch cut is a curve in the complex plane such that it is possible to define a single analytic branch of a multi-valued function on the plane minus that curve. In the case of a single branch point, there is usually a second branch point at infinity, e.g. $\sqrt{z}$ where the point at infinity is the appropriate 'other' branch point to define the branch cut from 0 to connect to, e.g. along the whole negative real axis.
- *Complex differentiation*: We define complex differential of a function $f(z)$ at some point z_0 (if it exists) by the limiting process

$$\dot{f}(z) \equiv \left.\frac{df}{dz}\right|_{z_0} = \lim_{z \to z_0} \frac{f(z) - f(z_0)}{z - z_0} \tag{M.3.1}$$

- *Argand diagram*: The complex variable z is expressed by two real numbers $\Re(z)$ and $\Im(z)$, which may be considered as the Cartesian co-ordinates (x, y) of z on a planar plot.
- *Jordan curve*: A Jordan curve, sometimes called a plane simple closed curve, is a non-self-intersecting continuous loop in the plane. The Jordan curve theorem asserts that every Jordan curve divides the plane into an 'interior' region bounded by the curve and an 'exterior' region containing all of the nearby and far away exterior points respectively, so that every continuous path connecting a point of one region to a point of the other intersects with that loop somewhere. We shall consider a series of contours all of which are *Jordan curves*.
- *Rectifiable arc*: The length of a Jordan arc passing through the set of points $z(t) = x(t) + \imath\, y(t)$, where the parameter $t_0 < t_1, \ldots, t_n = T$. The polygonal arc can be approximated by drawing straight line elements between the points $z(t)$ with total length

$$\Sigma = \sum_{r=1}^{n} \left(z_r - z_{r-1}\right)$$

If Σ tends to a limit $L = \lim_{n=1}^{\infty} \Sigma$ as $n \to \infty$ and the greatest number of the set $\max(t_r - t_{r-1}) \to 0$, the arc is *rectifiable with length L*.
- *Riemann integration*: Consider a complex function $f(z)$, not necessarily holomorphic, but with a definite value at each point on the rectifiable arc $\mathscr{L}$. Following the same procedure as above, we divide the arc into elements t_r and form the sum

$$\Sigma = \sum_{r=1}^{n} f(\zeta_r)\, \left(z_r - z_{r-1}\right) \tag{M.3.2}$$

where ζ_r is a point on $\mathscr{L}$ lying between z_{r-1} and z_r.
As before if Σ tends to a limit $J = \lim_{n=1}^{\infty} \Sigma$ as $n \to \infty$ and the greatest number of the set $\max(t_r - t_{r-1}) \to 0$, *$f(z)$ is integrable along $\mathscr{L}$ with integral $\int_{\mathscr{L}} f(z) \mathrm{d}z = J$.*

M.3.1 Cauchy–Riemann Relations

Since $z = x + \imath\, y$, it follows that the real and imaginary parts can be written in terms of complex conjugate of $z^* = x - \imath\, y$, where $x = \frac{1}{2}(z + z^*)$ and $y = -\frac{1}{2}\imath\,(z - z^*)$. The complex variables z and z^* are therefore independent complex variables reflecting the independence of x and y. A general complex function has the form $f(z, z^*)$. However a holomorphic function satisfying Eq. (M.3.2) cannot be a function of z^*. Consider the complex holomorphic function $f(z) = u(x, y) + \imath\, v(x, y)$ in the complex plane $z = x + \imath\, y$, where u, v, x, and y are all real, then it is required that the complex derivative be independent of the direction of $\mathrm{d}z$. Then

$$\frac{\partial f}{\partial z^*} = \left(\frac{\partial u}{\partial x} + \imath\, \frac{\partial v}{\partial x}\right)\frac{\partial x}{\partial z^*} + \left(\frac{\partial u}{\partial y} + \imath\, \frac{\partial v}{\partial y}\right)\frac{\partial y}{\partial z^*} = \frac{1}{2}\left(\frac{\partial u}{\partial x} + \imath\, \frac{\partial v}{\partial x}\right) + \frac{1}{2}\,\imath\left(\frac{\partial u}{\partial y} + \imath\, \frac{\partial v}{\partial y}\right) = 0 \tag{M.3.3}$$

Equating real and imaginary parts

$$\boxed{\frac{\partial u}{\partial x} = \frac{\partial v}{\partial y} \quad \text{and} \quad \frac{\partial u}{\partial y} = -\frac{\partial v}{\partial x}} \tag{M.3.4}$$

we obtain the Cauchy–Riemann relations. These are readily shown to be both a necessary and sufficient condition for analyticity in a domain $\mathscr{D}$ provided u and v are everywhere single valued and continuous and differentiable in $\mathscr{D}$.

More generally, it follows that since $f(z)$ only, then $\dot{f}(z)$ also does not depend on z^* and successive derivatives of $f^{(n)}(z)$ exist.

M.3.2 Harmonic Functions

Since the real $u(x, y)$ and imaginary $v(x, y)$ parts of a holomorphic function satisfy the Cauchy–Riemann relations, it follows that

$$\frac{\partial^2 u}{\partial x^2} + \frac{\partial^2 u}{\partial y^2} = 0 \qquad \text{and} \qquad \frac{\partial^2 v}{\partial x^2} + \frac{\partial^2 v}{\partial y^2} = 0 \tag{M.3.5}$$

and $u(x, y)$ and $v(x, y)$ are conjugate functions satisfying Laplace's equation in two dimensions.

It follows from the Cauchy–Riemann relations that the contours $u(x, y) = \text{const}$ and $v(x, y) = \text{const}$ intersect at right angles. As a result, the lines $u(x, y) = \text{const}$ and $v(x, y) = \text{const}$ may be used to generate an orthogonal co-ordinate set, which form the basis of *conformal mapping*, transforming from the co-ordinate set (x, y) to (u, v).

M.3.3 Area Rule

Consider the line integral of a general function $f(z, z^*)$ around a closed contour $\mathscr{C}$ enclosing the domain $\mathscr{D}$. We transform the line integral into an integral over the area enclosed by $\mathscr{C}$ by noting that, for example, the difference in u on opposite sides of the contour at $y_2(x) - y_1(x)$ at x is $u_2 - u_1 = \int_1^2 \partial u/\partial y dy$. The integral around the loop is taken anti-clockwise as shown in Figure M.3.1, so that the area element δx must be taken with a minus sign as shown. Therefore, making use of the derivative from Eq. (M.3.3)

$$\begin{aligned}\int_{\mathscr{C}} f(z)\, \mathrm{d}z &= \oint_{\mathscr{C}} (u\, \mathrm{d}x - v\, \mathrm{d}y) + \imath\ (v\, \mathrm{d}x + u\, \mathrm{d}y) \\ &= \oiint_{\mathscr{C}} \left\{ \left(\frac{\partial u}{\partial y} + \frac{\partial v}{\partial x} \right) - \imath \left(\frac{\partial u}{\partial x} - \frac{\partial v}{\partial y} \right) \right\} \mathrm{d}x\, \mathrm{d}y = \oiint_{\mathscr{D}} \frac{\partial f}{\partial z^*}\, \mathrm{d}s\end{aligned} \tag{M.3.6}$$

where the integral is taken over the area of the contour $\mathscr{D}$.

M.3.4 Cauchy Integral Theorem

When $f(z)$ is a holomorphic function, analytic within the domain $\mathscr{D}$ enclosed by and on the contour $\mathscr{C}$ the integrand $\partial f/\partial z^* = 0$ and we obtain Cauchy's integral theorem directly from Eq. (M.3.6)

$$\boxed{\int_{\mathscr{C}} f(z)\, \mathrm{d}z = 0} \tag{M.3.7}$$

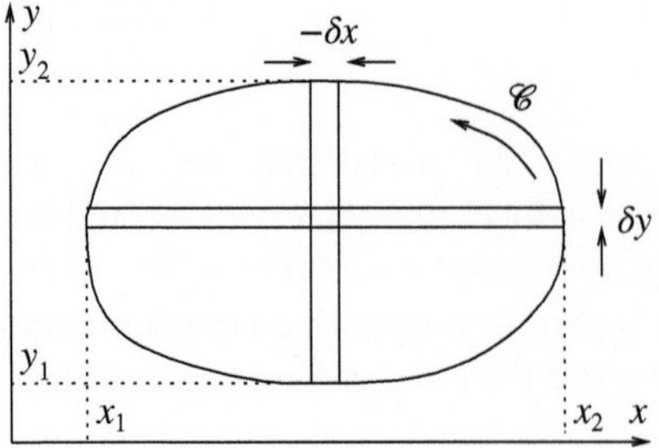

Figure M.3.1 The line integral is taken anti-clockwise around the loop $\mathscr{C}$ in the complex plane $z = x + \imath y$.

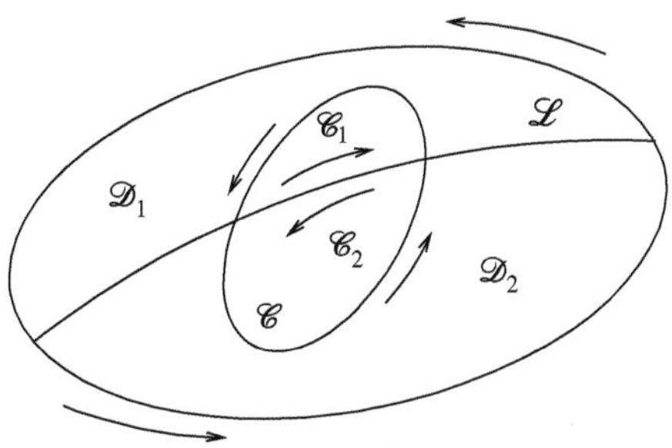

Figure M.3.2 The line integral $\int f(z)\mathrm{d}z$ is taken anti-clockwise around the loop $\mathscr{C}$ passing through the junction $\mathscr{L}$ of the domains $\mathscr{D}_1$ and $\mathscr{D}_2$. Taking the integral as the sum along the paths $\mathscr{C}_1$ and $\mathscr{C}_2$ the contribution along the arc $\mathscr{L}$ cancels out.

Suppose two rectifiable arcs $\mathscr{L}_1$ and $\mathscr{L}_2$ run from z_1 to z_2, then considering the closed contour $\mathscr{C} = \mathscr{L}_1 - \mathscr{L}_2$ it follows that provided the function $f(z)$ is holomorphic within the domain containing $\mathscr{L}_1$ and $\mathscr{L}_2$

$$\int_{\mathscr{L}_1} f(z)\ \mathrm{d}z = \int_{\mathscr{L}_1} f(z)\ \mathrm{d}z \tag{M.3.8}$$

i.e. the integral $\int_{z_1}^{z_2} f(z)\mathrm{d}z$ is dependent only on its limits z_1 and z_2.

Since the complex derivative $\mathrm{d}f/\mathrm{d}z$ is holomorphic if $F(z)$ is regular, it follows from equation (M.3.1) and (M.3.2) that

$$\int_{\mathscr{L}} \frac{\mathrm{d}f}{\mathrm{d}z}\mathrm{d}z = \sum_{r=0}^{\infty} \left\{ \frac{f(z_r) - f(z_{r-1})}{z_r - z_{r-1}} \right\} \left(z_r - z_{r-1}\right) = f(z_2) - f(z_1) \tag{M.3.9}$$

independent of the path $\mathscr{L}$, and therefore that differentiation and integration are inverse operations as in real algebras.

Furthermore, when $f(z)$ and $g(z)$ are two holomorphic functions, then since $\mathrm{d}[f(z)\ g(z)]/\mathrm{d}z = f(z)\ \mathrm{d}g(z)/\mathrm{d}z + g(z)\ \mathrm{d}f/\mathrm{d}z$, integration by parts is allowed with complex functions.

M.3.5 Morera's Theorem

It is the converse of the Cauchy integral theorem, namely that if every integral $\int_{\mathscr{C}} f(z)\ \mathrm{d}z = 0$ within the domain $\mathscr{D}$ then $f(z)$ is a regular function throughout $\mathscr{D}$. The proof follows by noting that since $\oint f(z)\ \mathrm{d}z = \oint \partial f/\partial z^{*} \mathrm{d}s = 0$ for every contour in $\mathscr{D}$ it follows that provided $f(z)$ is continuous that $\partial f/\partial z^{*} = 0$ and that the function $f(z)$ is therefore holomorphic.

M.3.6 Analytic Continuation

Suppose we have two neighbouring domains $\mathscr{D}_1$ and $\mathscr{D}_2$ separated by the line $\mathscr{L}$ on which $f_1(z) = f_2(z)$ where $f_1(z)$ and $f_2(z)$ are holomorphic functions in $\mathscr{D}_1$ and $\mathscr{D}_2$, respectively, then the function

$$f(z) = \begin{cases} f_1(z) & \text{in } \mathscr{D}_1 \\ f_2(z) & \text{in } \mathscr{D}_2 \end{cases} \tag{M.3.10}$$

is holomorphic throughout $\mathscr{D} = \mathscr{D}_1 \cup \mathscr{D}_2$.

Consider the integral around a contour $\mathscr{C}$. The proof is obvious if $\mathscr{C}$ does not intersect $\mathscr{L}$. Otherwise we form the line integral along $\mathscr{C}$ (Figure M.3.2) by constructing loops in lying $\mathscr{C}_1$ and $\mathscr{C}_2$ entirely within each domain such that the integral along the intersection cancels since $f_1(z) = f_2(z)$ along it. Since $f_1(z)$ and $f_2(z)$ are both holomorphic within $\mathscr{D}_1$ and $\mathscr{D}_2$ respectively it follows that $\int f(z)\ \mathrm{d}z = 0$ around any arbitrary loop in $\mathscr{D}$, and therefore applying Morera's theorem, the function (M.3.10) yields the desired analytical continuation.

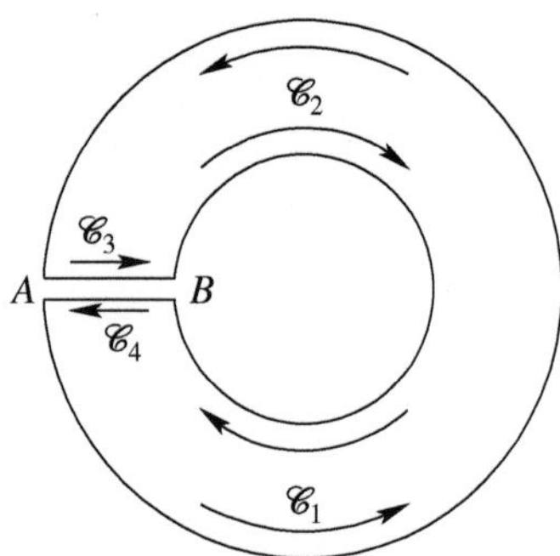

Figure M.3.3 The two contours $\mathscr{C}_1$ and $\mathscr{C}_2$ are joined by the adjacent links $\mathscr{C}_3$ and $\mathscr{C}_4$ whose net contribution cancels out. The line integral along the outer contour when the integrand $f(z)$ is holomorphic is equal to that taken along $\mathscr{C}_2$ taken in the same sense.

M.3.7 Extension or Contraction of a Contour

Consider the contours $\mathscr{C}_1$ and $\mathscr{C}_2$ such that the function $f(z)$ is holomorphic in the domain $\mathscr{D}$ separating them (Figure M.3.3). Suppose the line integral around $\mathscr{C}_2$ is known. Join the contours by a line AB and integrate along the full contour from $A - \mathscr{C}_1$, $\mathscr{C}_3$, $\mathscr{C}_2$, $\mathscr{C}_4$ back to A. Since $f(z)$ is single valued the contributions from AB cancel and we obtain

$$\int_{\mathscr{C}_1} f(z)\, \mathrm{d}z = \int_{\mathscr{C}_2} f(z)\, \mathrm{d}z \tag{M.3.11}$$

when the correct anti-clockwise sense around both contours is taken.

M.3.8 Inclusion of Isolated Singularities

The form of the integrand inside $\mathscr{C}_2$ need not be holomorphic. The preceding construction may be used to calculate the line integral around a number of points where the function $f(z)$ is no longer holomorphic, for example at singularities, by surrounding them by small isolated circular loops around each of which the loop integrals may be calculated. Thus,

$$\int_{\mathscr{C}} f(z)\, \mathrm{d}z = \sum_s \int_{\mathscr{C}_s} f(z)\, \mathrm{d}z \tag{M.3.12}$$

where s are the isolated singularities of $f(z)$ contained within $\mathscr{C}$.

M.3.9 Cauchy Formula

M.3.9.1 Interior Domain

Consider the integral $\oint_{\mathscr{C}} f(\zeta)/(\zeta - z)\, \mathrm{d}\zeta$ where $f(\zeta)$ is a holomorphic function inside and on the contour $\mathscr{C}$. The integrand has an isolated singularity (specifically a pole) at $\zeta = z$ lying inside $\mathscr{C}$. We exclude the singularity by a small circular contour $\mathscr{C}'$ centred at z of infinitely small radius r. Since the integrand is holomorphic in the domain $\mathscr{C} - \mathscr{C}'$

$$\begin{aligned}\oint_{\mathscr{C}} \frac{f(\zeta)}{(\zeta - z)}\, \mathrm{d}\zeta &= \oint_{\mathscr{C}'} \frac{f(\zeta)}{(\zeta - z)}\, \mathrm{d}\zeta = \lim_{r \to 0} \oint f(z)\, \frac{\imath\, r \exp(\imath\, \theta)}{r \exp(\imath\, \theta)}\, \mathrm{d}\theta \\ &= 2\, \pi\, \imath\, f(z)\end{aligned} \tag{M.3.13}$$

$f(z)$ is known as the (residue) of the function $f(\zeta)$ at the pole z.

If the pole at z lies outside the contour, $f(\zeta)/(\zeta - z)$ is holomorphic throughout the contour $\mathscr{C}$, so that $\oint_{\mathscr{C}} f(\zeta)/(\zeta - z)\, \mathrm{d}z = 0$. Hence

$$\boxed{\oint_{\mathscr{C}} \frac{f(\zeta)}{(\zeta - z)}\, \mathrm{d}\zeta = \begin{cases} 2\, \pi\, \imath\, f(z) & \text{if } z \text{ inside } \mathscr{C} \\ 0 & \text{otherwise} \end{cases}} \tag{M.3.14}$$

M.3.9.2 Exterior Domain

The function $f(\zeta)$ is regular in the domain exterior to the clockwise contour $\mathscr{C}$. Applying Cauchy's integral between $\mathscr{C}$ and an external counterclockwise contour $\mathscr{C}_\infty$ at infinity where $f(\zeta)$ is everywhere regular

$$\boxed{\oint_{\mathscr{C}} \frac{f(\zeta)}{(\zeta-z)} \mathrm{d}\zeta = \begin{cases} 2\pi\imath\;(f(z)-f(\infty)) & \text{if z outside } \mathscr{C} \\ -2\pi\imath\, f(\infty) & \text{if zinside } \mathscr{C} \end{cases}} \tag{M.3.15}$$

since

$$\oint_{\mathscr{C}_\infty} \frac{f(\zeta)}{(\zeta-z)} \mathrm{d}\zeta = \lim_{R\to\infty} \frac{1}{2\,\pi\,\imath} \int_0^{2\pi} \frac{f\left(R\, e^{\imath\,\theta}\right)}{1-\left(z/R\, e^{\imath\,\theta}\right)} = f(\infty)$$

This result may be generalised to poles of higher order (n) by an integration by parts after noting that $f(z)$ is single valued by definition

$$\begin{aligned} \frac{f(\zeta)}{(\zeta-z)^n}\,\mathrm{d}\zeta &= -\frac{1}{(n-1)} \left\{ \left[\frac{f(\zeta)}{(\zeta-z)^{(n-1)}} \right] - \oint_{\mathscr{C}} \frac{\dot{f}(\zeta)}{(\zeta-z)^{(n-1)}}\,\mathrm{d}\zeta \right\} = \cdots \\ &= \frac{1}{(n-1!)} \oint_{\mathscr{C}} \frac{f^{(n-1)}(\zeta)}{(\zeta-z)}\,\mathrm{d}\zeta = \frac{2\pi\,\imath}{(n-1)!}\, f^{(n-1)}(z) \end{aligned} \tag{M.3.16}$$

remembering that the derivatives $f^{(n)}$ of all orders of an analytic function exist.

M.3.10 Treatment of Improper Integrals

Consider the integral

$$\int_{\mathscr{L}} \frac{f(\zeta)}{(\zeta-z)}\,\mathrm{d}\zeta \tag{M.3.17}$$

when the point z lies on the arc of integration $\mathscr{L}$. We may avoid the singularity at z by *indenting* the path of integration by a small loop ℓ of radius ϵ avoiding z. The integral is said to exist as a *principal value* if integral

$$⨍_{\mathscr{L}} \frac{f(\zeta)}{(\zeta-z)}\,\mathrm{d}\zeta = \lim_{\epsilon\to 0} \left\{ \int^{z-\epsilon} \frac{f(\zeta)}{(\zeta-z)}\,\mathrm{d}\zeta + \int_{z+\epsilon} \frac{f(\zeta)}{(\zeta-z)}\,\mathrm{d}\zeta \right\} \tag{M.3.18}$$

exists. To which must be added the contribution from the semi-circular loop

$$\lim_{\epsilon\to 0} \int_{\ell} \frac{f(\zeta)}{\zeta-z}\,\mathrm{d}z = f(z) \int_0^{\mp\pi} \frac{\imath\epsilon\, e^{\imath\,\theta}}{\epsilon\, e^{\imath\,\theta}} \mathrm{d}r = \mp\imath\,\pi\, f(z) \tag{M.3.19}$$

so that in total

$$\int_{\mathscr{L}} \frac{f(\zeta)}{(\zeta-z)}\,\mathrm{d}\zeta = ⨍_{\mathscr{L}} \frac{f(\zeta)}{(\zeta-z)}\,\mathrm{d}\zeta \;\mp \pi\,\imath f(z) \tag{M.3.20}$$

The sign depending whether the path around the indentation is taken clockwise or anti-clockwise around the singularity, i.e. on the left- or right-hand side relative to the direction of integration (see Figure M.3.4). This problem is discussed further in Section M.3.11.

Figure M.3.4 Indented path around the pole on the integration contour $\mathscr{C}$. The left and right hand paths are shown displaced.

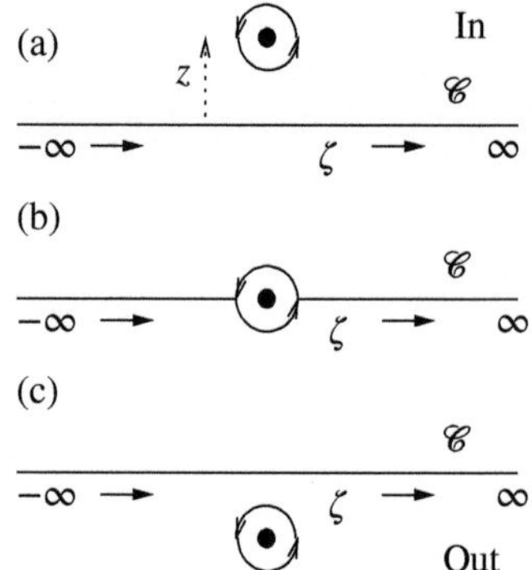

Figure M.3.5 Relationship of the pole at $\zeta = z$ (filled dot) with the contour $\mathscr{C}$. The pole lies: (a) internal to the contour line $\Phi_i(z)$; (b) on the contour line; (c) external to the contour line $\Phi_e(z)$. In these diagrams the interior of the loop is up and the exterior down with respect to the contour line $\mathscr{C}$ so that all contours run anti-clockwise.

M.3.11 Sokhotski–Plemelj Theorem

We consider the function defined by the integral

$$\Phi(z) = \frac{1}{2\pi \iota}\oint_{\mathscr{C}} \frac{\phi(\zeta)}{(\zeta - z)}\mathrm{d}\zeta = \frac{1}{2\pi \iota}\oint_{\mathscr{C}} \frac{\phi(\zeta)-\phi(z)}{(\zeta - z)}\mathrm{d}\zeta + \frac{1}{2\pi \iota}\oint_{\mathscr{C}} \frac{\phi(z)}{(\zeta - z)}\mathrm{d}\zeta \tag{M.3.21}$$

which exists, at least as a principal value.[4] The closed contour $\mathscr{C}$ divides the domain $\mathscr{D}$ into two spaces: interior and exterior to $\mathscr{C}$. For a regular function $\phi(z)$ in $\mathscr{D}$ the Cauchy integral Eq. (M.3.0) consequently defines two functions Φ_e outside and Φ_i inside $\mathscr{C}$

$$\begin{aligned}\Phi_e(z) &= \frac{1}{2\pi \iota}\oint_{\mathscr{C}} \frac{\phi(\zeta)-\phi(z)}{(\zeta - z)}\,\mathrm{d}\zeta \\ \Phi_i(z) &= \frac{1}{2\pi \iota}\oint_{\mathscr{C}} \frac{\phi(\zeta)-\phi(z)}{(\zeta - z)}\,\mathrm{d}\zeta + \phi(z) = \Phi_e(z) + \phi(z)\end{aligned} \tag{M.3.22}$$

This case is illustrated in Figure M.3.5 where path (a) corresponds to the pole internal to the contour Φ_i and path (c) to the external pole Φ_e. We note the jump across the contour due to the pole.

Turning now to path (b) where the pole z lies on the contour, we indent the contour into external or internal spaces so that the pole lies off the appropriate contour. Make use of the results of the Section M.3.10, equation (M.3.20), in the neighbourhood of the pole, we allow z to approach ζ on the contour from opposite sides and introduce the principal value integral

$$\begin{aligned}\Phi_e(z) &= \frac{1}{2\pi \iota}\fint_{\mathscr{C}} \frac{\phi(\zeta)}{(\zeta - z)}\,\mathrm{d}\zeta - \frac{1}{2}\,\phi(z) && \text{exterior} \\ \Phi_i(z) &= \frac{1}{2\pi \iota}\fint_{\mathscr{C}} \frac{\phi(\zeta)}{(\zeta - z)}\,\mathrm{d}\zeta + \frac{1}{2}\,\phi(z) && \text{interior}\end{aligned} \tag{M.3.23}$$

Thus close to the contour, allowing $z \to \zeta_0$, we have the first and second Plemelj formulae

$$\begin{aligned}\Phi_i(\zeta_0) - \Phi_e(\zeta_0) &= \phi(\zeta_0) \\ \Phi_i(\zeta_0) + \Phi_e(\zeta_0) &= \frac{1}{\pi \iota}\fint_{\mathscr{C}} \frac{\phi(\zeta)}{(\zeta - \zeta_0)}\,\mathrm{d}\zeta\end{aligned} \tag{M.3.24}$$

Respectively.[5]

These results will be readily recognised in terms of the difference and sum of the two integrals identified in Eq. (M.3.20).

---

4 If $\phi(\zeta)$ is well-behaved, $(\phi(\zeta) - \phi(z))/(\zeta - z)$ is an entire function and its integral along a closed contour is zero: $\int_{\mathscr{C}} (\phi(\zeta) - \phi(z))/(\zeta - z)\,\mathrm{d}z = 0$.

5 Taking $\mathscr{C}$ to be anti-clockwise, the internal and external sides are $\Im(z) < 0$ and $\Im(z) > 0$, respectively.

M.3.12 Improper Integral Along a Real Line

An important and simple version of the Sokhotski–Plemelj theorem for integrals along the real line was derived in footnote 1 on p. 274

$$\lim_{\epsilon\to 0}\int_a^b \frac{f(x)}{x \pm \imath\,\epsilon}\,\mathrm{d}x = \lim_{\epsilon\to 0}\int_a^b \frac{x^2}{(x^2+\epsilon^2)}\frac{f(x)}{x}\mathrm{d}x \mp \imath\,\pi \lim_{\epsilon\to 0}\int_a^b \frac{\epsilon}{\pi\,(x^2+\epsilon^2)}\,f(x)\,\mathrm{d}x$$
$$= ⨍_a^b \frac{f(x)}{x}\,\mathrm{d}x \mp \imath\,\pi\,f(0) \qquad \text{(M.3.25)}$$

A useful result is that if we define

$$\Phi(z) = \begin{cases} \displaystyle\int_{-\infty}^{\infty} \frac{\phi(\zeta)}{\zeta - z}\,\mathrm{d}\zeta & \Im(z) > 0 \\ \displaystyle\int_{-\infty}^{\infty} \frac{\phi(\zeta)}{\zeta - z}\,\mathrm{d}\zeta - 2\,\pi\,\imath\,\phi(z) & \Im(z) < 0 \end{cases} \qquad \text{(M.3.26)}$$

$\Phi(z)$ is an entire function formed by analytic continuation across the line $\Im(z) = 0$.

Although integrals of the form (M.3.21) occur frequently in physics, it is clear that the theory of complex functions cannot offer a definitive answer to their value. To do so requires further information from the physical context, as exemplified by the following typical cases which invoke representative solutions.

- *Kramers–Kronig relations*: Equation (13.9) on p. 319 is cast into the Sokhotski–Plemelj form with conditions of analyticity defining which contour is appropriate.
- *Landau prescription*: Physical argument is used to identify the nature of the singularity on the real axis by means of a small, non-zero imaginary part, which is reduced to zero on the path, but identifies the contour used; see Eq. (12.38) on p. 274.

M.3.13 Taylor and Laurent Series

Taylor's series for functions of real variables, which is familiar from elementary calculus, is derived using the mean value theorem. Functions of a complex variable also obey the Taylor's expansion most simply derived from Cauchy's theorem as follows. Consider the expansion about the point z_0, namely $f(z_0 + h)$

$$f(z_0+h) = \frac{1}{2\,\pi\,\imath}\oint \frac{f(z)}{z - z_0 - h}\,\mathrm{d}z = \frac{1}{2\,\pi\,\imath}\sum_{n\geq 0}\oint f(z)\left[\frac{h}{z-z_0}\right]^n\,\mathrm{d}z$$
$$= \sum_{n\geq 0}\frac{1}{n!}\,f^{(n)}(z_0)\,h^n \qquad \text{(M.3.27)}$$

The series is convergent within its radius of convergence where the function $f(z)$ is holomorphic. The outer radius of convergence is therefore defined by the 'distance' to the nearest singularity.

Consider now the case where the function $f(z)$ has an isolated singularity at z_0, e.g. a pole, and is regular inside radius R and outside radius $R' < R$, the function may be expanded in a *Laurent* series in terms of powers of the distance $(z - z_0)$

$$f(z) = \sum a_n(z-z_0)^n + \sum b_n(z-z_0)^{-n} \qquad \text{(M.3.28)}$$

M.3.14 The Argument Principle

Cauchy's principle of argument states that if $w(z)$ is a meromorphic function in the plane of z, analytic except at a finite number of points (the poles of $w(z)$), and is analytic at every point of the contour C, then as $w(z)$ travels around the contour in the anti-clockwise direction, the plot of the function $w(z)$ encircles the origin in the $\Re[w(z)]$, $\Im[w(z)]$ plane N times with N given by

$$\boxed{N = Z - P} \tag{M.3.29}$$

where Z is the number of zeroes and P the number of poles of $F(s)$ enclosed by the contour. The proof is in two parts:

(i) Let $w(z)$ be a meromorphic function inside and on a closed contour C which has no zeros or poles of $w(z)$ on it, then the number of zeros Z and poles P are related to the integral taken counter-clockwise around the simple contour

$$\oint_C \frac{\dot{w}(z)}{w(z)}\,\mathrm{d}z = 2\,\pi\,\imath\,(Z - P) \tag{M.3.30}$$

The zeros and poles are counted with their multiplicity and order respectively.
The proof is as follows.
Let z_Z be a zero, then $w(z) = (z - z_Z)^k\, f(z)$ where k is the multiplicity of the zero, and $f(z_Z) \neq 0$, then

$$\frac{\dot{w}(z)}{w(z)} = \frac{k}{(z - z_Z)} + \frac{\dot{f}(z)}{f(z)} \tag{M.3.31}$$

Since $f(z)$ is analytic at z_Z the residue is k.
Similarly at a pole z_P we have $w(z) = (z - z_P)^{-m}\, g(z)$ where m is the order of the pole and $g(z_P) \neq 0$. Proceeding as before

$$\frac{\dot{w}(z)}{w(z)} = \frac{k}{(z - z_Z)} + \frac{\dot{g}(z)}{g(z)} \tag{M.3.32}$$

the residue is $-m$. Thus each zero creates a pole of residue k and each pole a residue $-m$. The total number of zeros is therefore $Z = \sum k_Z$ and the total number of poles $P = \sum m_P$. Applying Cauchy's residue theorem to the integral around the contour C we obtain Eq. (M.3.0).

(ii) Map the contour C on to the space of $w \equiv W$ to generate the contour D in W, then

$$\frac{1}{2\,\pi\,\imath}\oint_C \frac{\dot{w}(z)}{w(z)}\mathrm{d}z = \frac{1}{2\,\pi\,\imath}\oint_D \frac{\mathrm{d}w}{w} = N \tag{M.3.33}$$

M.3.15 Estimation Lemma

If $f(z)$ is a complex continuous function on an arc $\mathscr{L}$ and is its absolute value $|f(z)| < M$ is bounded on $\mathscr{L}$, i.e. $M \equiv \max_{z \in \mathscr{L}} |f(z)|$ then

$$\int |F(z)|_{\mathscr{L}}\ \mathrm{d}z| = \left|\int_\alpha^\beta f(\gamma(t))\dot{\gamma}(t)\mathrm{d}t\right| \leq \left|\int_\alpha^\beta f(\gamma(t))\right|\ |\dot{\gamma}(t)\mathrm{d}t| \leq M\int_\alpha^\beta\ |\dot{\gamma}(t)|\ \mathrm{d}t = M\ell(\mathscr{L}) \tag{M.3.34}$$

This result is very simple to understand in terms of a set of smaller contour segments $\mathrm{d}t$ each with its own maximum max $|f(z)|$ with an overall maximum. Summed over the entire path, then the integral of $f(z)$ must be less than or equal to the product of the path length ℓ and the maximum M.

M.3.16 Jordan's Lemma

Consider the continuous complex function $f(z)$ defined on the semi-circular contour in the upper half-plane $\mathscr{C}_R = \{R\, e^{\imath\,\theta}\ |\theta \in [0,\pi]\}$. The function $f(z) = e^{\imath\, a\, z}\, g(z)$ when $z \in \mathscr{C}_R$ with $a > 0$.

The complex line integral

$$I_R = \int_{\mathscr{C}_R} f(z)\, \mathrm{d}z = \int_0^{\pi} g(R\, e^{\imath\,\theta}) e^{\imath\, R(\cos\theta + \imath\,\theta)}\ \imath\, Re^{\imath\,\theta}\ \mathrm{d}\theta \tag{M.3.38}$$

Noting the inequality $\left|\int_{\mathscr{C}_R} f(z)\mathrm{d}z\right| \leq \int_{\mathscr{C}_R} |f(z)|\ \mathrm{d}z$, we obtain

$$\begin{aligned} I_R = R\int_0^{\pi} \left|g(R\, e^{\imath\,\theta})\, e^{a\ R(\imath\cos\theta - \sin\theta)})\ \imath\, e^{\imath\,\theta}\right|\ \mathrm{d}\theta = R\int_0^{\pi} \left|g(R)e^{\imath\,\theta}\right|\ e^{-a\ R\sin\theta}\ \mathrm{d}\theta \\ \leq R\ M_R \int_0^{\pi} e^{-a\ R\ \sin\theta}\ \mathrm{d}\theta = 2\ R\ M_R \int_0^{\pi/2} e^{-a\ R\ \sin\theta}\ \mathrm{d}\theta \end{aligned} \tag{M.3.36}$$

where $M_R = \max_{\theta\in[0,\pi]} \left|g\left(R\, e^{\imath\,\theta}\right)\right|$ is the maximum value of $|g(z)|$ in $\mathscr{C}_R$.

Finally, we note that it can easily be shown that the graph of $\sin\theta$ is concave in the interval $\theta[0,\pi/2]$ and that the graph of $\sin\theta$ therefore lies above the line between its end-points and therefore we get *Jordan's inequality* namely $2/\pi \leq \sin\theta/\theta \leq 1$. Hence

$$\boxed{I_R \leq 2\ R\ M_R \int_0^{\pi/2} e^{-2\ aR\ \theta/\pi} \mathrm{d}\theta = \frac{\pi}{a}\left(1 - e^{-a\ R}\right) M_R \leq \frac{\pi}{a} M_R} \tag{M.3.37}$$

If $a \to 0$ then $(1 - e^{(a\ R)}) \to a\ R$ and we obtain the estimation lemma (Section M.3.M.3.15).

If $a > 0$ then provided $\lim_{R\to\infty} f(R\, e^{\imath\ R}) = 0$ uniformly with respect to θ when $0 \leq \theta \leq \pi$

$$\lim_{R\to\infty} \int_{\mathscr{C}_R} e^{\imath\ a\ z} f(z)\ \mathrm{d}z = 0 \tag{M.3.38}$$

M.3.17 Conformal Mapping

Let $w(z) = u(x,y) + \imath v(x,y)$ be an analytic function which has a non-zero derivative $\mathrm{d}w\big/\mathrm{d}z \neq 0$. A point z in the (x,y) plane maps into a corresponding one $w(z)$ in the u, v plane. The mapping is conformal, i.e. preserves both angles and their sense between the prototype and its map. Consider two elements $\mathrm{d}z_1$ and $\mathrm{d}z_2$ whose maps are $\mathrm{d}w_{1/2} = (\mathrm{d}w\big/\mathrm{d}z)\ \mathrm{d}z_{1/2}$. Since the angle between the elements is given by the argument of their ratio $\arg(\mathrm{d}z_1/\mathrm{d}z_2) = \arg(\mathrm{d}w_1/\mathrm{d}w_2)$, the result follows. Conformal maps are used extensively in physics and engineering, but infrequently in plasma physics.

M.4 Laplace Transform

The Laplace transform of a function $f(t)$ which is zero for $t < 0$ is

$$F(s) = \int_0^{\infty} f(t)\, e^{-st}\, \mathrm{d}t = \int_0^{\infty} (f(t)\, e^{-\sigma t})\, e^{\imath\omega t}\, \mathrm{d}t \tag{M.4.1}$$

where $s = \sigma + \imath\,\omega$.

Fourier transforming

$$\hat{f}(\omega) = \int_{-\infty}^{\infty} f(t)\, e^{-\imath\,\omega t}\, \mathrm{d}t \tag{M.4.2}$$

with inverse transform

$$f(t) = \int_{-\infty}^{\infty} \hat{f}(\omega)\, e^{\imath\,\omega t}\, \mathrm{d}t \tag{M.4.3}$$

In the Laplace transform, let

$$\phi(t) = \begin{cases} f(t)\, e^{-\sigma t} & t \geq 0 \\ 0 & t < 0 \end{cases} \tag{M.4.4}$$

The Fourier transform

$$\hat{\phi}(\omega) = \int_0^{\infty} \phi(t)\, e^{-\imath\,\omega\, t}\, \mathrm{d}t = \int_0^{\infty} f(t)\, e^{\sigma\, t}\, e^{-\imath\,\omega\, t}\, \mathrm{d}t = F(\sigma + \imath\,\omega) \tag{M.4.5}$$

Take the inverse Fourier transform of $\hat{\phi}(\omega)$

$$\phi(t) = f(t)\, e^{-\sigma\, t} = \frac{1}{2\,\pi} \int_{-\infty}^{\infty} F(\sigma + \imath\,\omega)\, e^{\imath\,\omega\, t}\, \mathrm{d}\omega \tag{M.4.6}$$

to get

$$f(t) = \frac{1}{2\pi} \int_{-\infty}^{\infty} F(\sigma + \imath\,\omega)\, e^{\sigma t}\, e^{\imath\,\omega\, t}\, \mathrm{d}\omega \tag{M.4.7}$$

Hence substituting we obtain the inverse of the Laplace transform called the *Bromwich integral*

$$f(t) = \frac{1}{2\,\pi\,\imath} \begin{cases} \int_{\sigma-\imath\,\infty}^{\sigma+\imath\,\infty} F(s)\, e^{st}\, \mathrm{d}s & t \geq 0 \\ 0 & t < 0 \end{cases} \tag{M.4.8}$$

M.4.1 Bromwich Contour

The inverse transform is normally evaluated as an integral around the Bromwich contour which consists of the closure of the line $\mathscr{L} = \Re(\omega) = \sigma$ from $\Im(\omega) = -\infty \rightarrow \infty$ (Figure M.4.1) by the semi-circular arcs of an infinitely large circle $\mathscr{C} = \mathscr{C}_1 \cup \mathscr{C}_2$. It normally follows from Jordan's lemma (Section M.3.M.3.xvi)that the contribution from the arcs at large radius is zero, and therefore the only contribution to the contour integral is from Bromwich's integral along $\mathscr{L}$.

For $t < 0$, the integral is only convergent on the right-hand contour $\mathscr{C}_2 \cup \mathscr{L}$. Hence by Morera's theorem, $F(s)$ is holomorphic in $\mathscr{C}_2$ since by definition the Cauchy integral is zero.

For $t > 0$, the integral is only convergent on the contour $\mathscr{C}_1 \cup \mathscr{L}$, i.e. the left-hand arc, which encloses the singularities of $F(s)$. If the only singularities are poles, the inverse transform of $F(s)$ is therefore the sum of the residues $\sum_i F(s_i)\, e^{s_i\, t}$. Since the integral of $F(s)$ must enclose all poles σ must be chosen greater than the largest real value of any pole or other singularity.

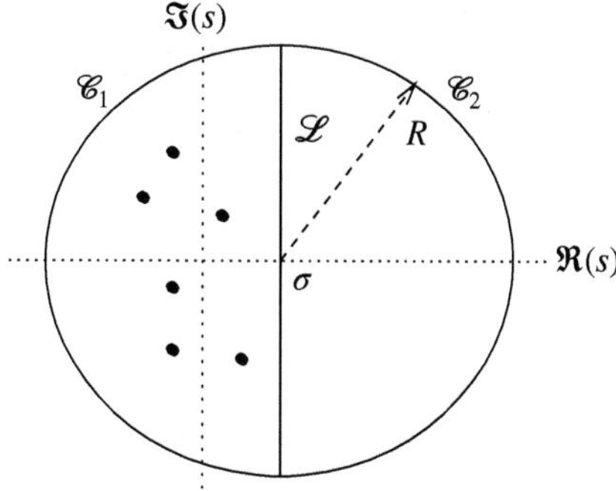

Figure M.4.1 Bromwich contour. All the singularities of the transform function $F(s)$ are contained in the left-hand infinite radius semi-circular contour $\mathscr{L} \cup \mathscr{C}_1$.

Problems

Problem 1: Show that the isotropic instantaneous release of particles from a point source and the subsequent diffusive expansion is described by the differential equation

$$\frac{\partial n}{\partial t} = D \frac{1}{r^2} \frac{\partial}{\partial r} \left[r^2 \frac{\partial n}{\partial r} \right] \tag{P.1}$$

where n is the density. Prove that the solution is given by

$$n(r) = \frac{4\,\pi}{(4\,D\,t)^{3/2}} \exp\left[-\frac{r^2}{4\,D\,t} \right] \tag{P.2}$$

Solution: Consider a closed volume V with surface S. It follows from the conservation of particles that an increase in the number of particles in the volume $\int_V n\ \mathrm{d}V$ must be balanced by the incoming flux through the surface $-\int_S \boldsymbol{\Gamma} \cdot \mathrm{ds}$. Therefore, applying Fick's law for the diffusion flux

$$\int_V \frac{\partial n}{\partial t}\ \mathrm{d}V = -\int_S \boldsymbol{\Gamma} \cdot \mathrm{ds} = \int_S D\ \nabla n \cdot \mathrm{ds} \tag{P.3}$$

Applying Gauss' theorem assuming the diffusivity is constant

$$\frac{\partial n}{\partial t} = D\ \nabla^2 n \tag{P.4}$$

For an isotropic expansion, we differentiate the solution to obtain

$$\frac{\partial n}{\partial t} = -n\ \frac{3}{2}\ \frac{1}{t} + n\ \frac{r^2}{4\,D\,t^2} \tag{P.5}$$

$$r^2\ \frac{\partial n}{\partial r} = -n\ \frac{2\,r^3}{4\,D\,t} \tag{P.6}$$

$$\frac{1}{r^2}\ \frac{\partial}{\partial r}\left[r^2\ \frac{\partial n}{\partial r} \right] = -n\ \frac{6}{4\,D\,t} + n\ \frac{4\,r^2}{16\,D^2\,t^2} \tag{P.7}$$

which solves the problem.

Problem 2: Assuming that the field gradient is small, show that the gradient drift velocity of particles of magnetic moment μ, cyclotron frequency Ω, mass m, charge q, and velocity v in a perpendicular gradient of the magnetic field B is approximately

$$U = \frac{m\ v^2\ \nabla_\perp \mathbf{B}}{2\ q\ B^2} = \frac{\mu}{m\ \Omega} \hat{\mathbf{b}} \wedge \nabla \mathbf{B} \tag{3.59:a}$$

in agreement with Eq. (3.59)

Solution: Use the following simple approximation in which the motion is separated into two parts (Figure P.1), one in a lower magnetic field B_1 and one in higher B_2 with cyclotron radii ρ_1 and ρ_2 and

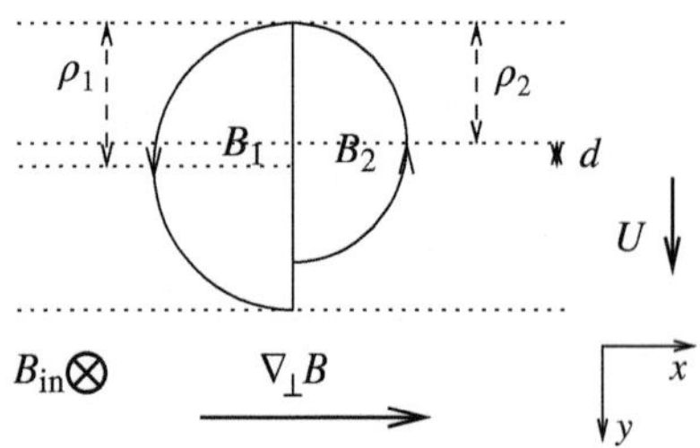

Figure P.1 Approximate model of gradient drift in which the motion is divided into high and low field rotations, positive charge shown.

corresponding cyclotron rotation times τ_1 and τ_2, respectively. The net drift of the guiding centre in the time $\frac{1}{2}(\tau_1 + \tau_2)$ is $d = (\rho_2 - \rho_1)$. Using the standard values for the cyclotron radius and the cyclotron frequency

$$U = \frac{(\rho_2 - \rho_1)}{\frac{1}{2}(\tau_1 + \tau_2)} = \frac{v_\perp}{\pi} \frac{(B_1 - B_2)}{(B_1 + B_2)}$$

Taking the average of the distance from the diameter in a hemisphere

$$\bar{y} = \int_{-1}^{1} y \, \mathrm{d}x \Big/ \int_{-1}^{1} \mathrm{d}x = \frac{\text{Area}}{\text{Diameter}} = \frac{\pi}{4} \times \text{Radius}$$

to form the gradient between segments ① and ② of the path

$$\nabla_\perp B \approx \frac{(B_2 - B_1)}{\frac{\pi}{4}(\rho_2 + \rho_1)}$$

and approximating

$$B_2 + B_1 \approx 2B \qquad \text{and} \qquad B_2 B_1 \approx B^2$$

we may write

$$B_2 - B_1 \approx \frac{\pi}{4} (\rho_2 + \rho_1) \nabla_\perp B = \frac{\pi}{4} \frac{m v_\perp}{q} \frac{(B_2 + B_1)}{(B_2 \, B_1)} \nabla_\perp B \approx \frac{\pi}{2} \frac{m \, v_\perp}{q \, B} \nabla_\perp B$$

and the drift velocity

$$U = \frac{\frac{1}{2} m \, v_\perp^2}{q \, B^2} \nabla_\perp B = \frac{\mu}{m \, \Omega} \hat{\mathbf{b}} \wedge \nabla B \tag{3.59:a}$$

Problem 3: Derive the expression (3.91) for the diamagnetic drift in a uniform magnetic field with pressure gradient using the guiding centre model.

Solution: Initially consider particles with a density gradient in the x direction normal to the magnetic field direction z. The diamagnetic current is expected to be in the direction of the mutual perpendicular y (Figure P.2).

The particles are assumed have of charge q and speed $v_\perp$ so that their Larmor radius in the field B is $\rho = m v_\perp / qB$ and lies in the plane x, z normal to B. Motion along the field is therefore unimportant and we may consider the plane normal the field alone. Consider a small element at the origin $x = 0$. Only particles lying on the circle $r = \rho$ will pass through the element. Defining the polar angle θ with respect to the x axis, clearly orbits starting at elements with $0 \leq \theta < \pi$ cancel out those from $\pi \leq \theta < 2\pi$ if the medium is uniform. However if the density varies as $n(x)$, this cancelling no longer takes place. Taking into account the anti-clockwise motion of positive charges about the magnetic field, the number of orbits from a small cell $\delta\theta \; \mathrm{d}\ell$ lying on the guiding centre

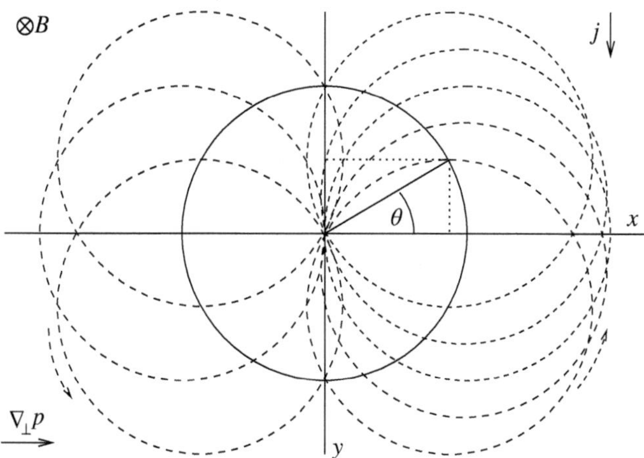

Figure P.2 Generation of the diamagnetic current from the cyclotron orbits in a density gradient. The density/pressure gradient is in the x direction, the magnetic field in the z (into the plane) and the diamagnetic current along y through an element at $x = 0, z = 0$. The full line is the locus of guiding centres whose cyclotron orbits (dashed) pass through the element at the origin.

circle at angle θ moving along the y direction at $x = 0$ is $-n(x)\ v_\perp\ \cos\theta$. Assuming the particles are uniformly distributed around their orbit, the current at $x = 0$ is $q\ v_\perp\ n(x)\ \delta\theta/2\ \pi$. Assuming the change in density is small over the Larmor radius, we may write

$$n(x) \approx n(0) + \frac{dn}{dx}|_x\, x \qquad \text{where} \qquad x = \rho\ \cos\theta$$

and obtain the current density

$$j_y = \int_0^{2\pi} q\ v_\perp\ n(x)\ \cos\theta\ d\theta/2\ \pi = q\ \rho\ v_\perp\ \frac{dn}{dx} \int_0^{2\pi} \cos^2\theta\ d\theta/2\ \pi = \frac{1}{2}\ q\ \rho\ v_\perp\ \frac{dn}{dx} \tag{P.8}$$

To include the thermal velocity variation, we replace the density $n(x)$ at x by the distribution for the perpendicular velocity component $f(n, T, v_\perp)$ where T is the temperature, and velocity element $dv_\perp$. Neglecting the zero order term as before and integrating over the angle θ, integration over the perpendicular velocity yields

$$\begin{aligned} j_y &= \frac{1}{2}\frac{m}{B} \int v_\perp^2 \frac{d}{dx} f(n, T, v_\perp)\ dv \perp = \frac{1}{2}\frac{m}{B}\ \frac{d}{dx} \int v_\perp^2 f(n, T, v_\perp)\ dv_\perp \\ &= \frac{1}{2}\frac{m}{B} \frac{d(n\langle v_\perp^2\rangle)}{dx} = \frac{1}{B}\frac{dp_\perp(x)}{dx} \end{aligned} \tag{P.9}$$

where $\frac{1}{2}\ m\langle v_\perp^2\rangle = \not{k}T$. Since the pressure $p_\perp = n\ \not{k}T = \frac{1}{2} \int v_\perp^2\ f(v_\perp)\ dv_\perp$ giving the current in vector form in accordance with Eq. (3.91)

$$\mathbf{j}_M = --\ \frac{1}{B^2}\ \nabla\ p_\perp \wedge \mathbf{B}. \tag{P.10}$$

Problem 4: Show that there is zero net current in a uniform plasma due to a magnetic field gradient perpendicular to the field by showing that the cyclotron orbital motion leads to a current equal and opposite to that due to the guiding centre drift.

Solution: We may approach this problem in a similar manner to that used to calculate the diamagnetic current (problem P.1) by considering the orbits of a positive charge q passing through

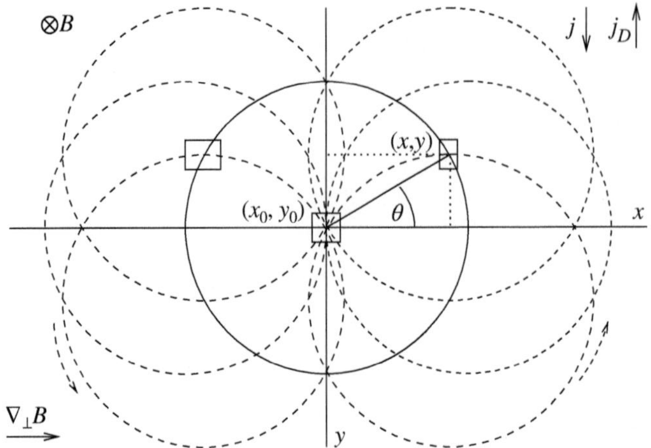

Figure P.3 Generation of the diamagnetic current from the cyclotron orbits in a magnetic field gradient. The field gradient is in the x direction, the magnetic field in the z (into the plane), and the current along y through an element at (x_0, y_0). The full line is the locus of guiding centres whose cyclotron orbits (dashed) pass through the element at (x_0, y_0).

an element at the origin (x_0, y_0) and working in a frame moving with the drift velocity v_D. Let the field gradient be in the x direction and the field itself in the z (Figure P.3). Due to the variation of the magnetic field the Larmor radius varies from point to point on the locus of the guiding centre. We assume that the value of the magnetic field appropriate to a particular orbit is that at its guiding centre (x, y). Consider the orbits with guiding centre in a small cell at $[(x = x_0 + \rho\ \cos\theta), (y = y_0 + \rho\ \sin\theta)]$. The resulting orbits pass through a cell $\mathrm{d}x_0$, $\mathrm{d}y_0$ at the origin. The ratio of the volume of the cell at the centre to that at the guiding centre is determined by the corresponding Jacobian

$$\frac{\mathrm{d}x_0\ \mathrm{d}y_0}{\mathrm{d}x\ \mathrm{d}y} = \frac{\partial(x_0, y_0)}{\partial(x, y)} = \begin{vmatrix} \partial x_0/\partial x & \partial y_0/\partial x \\ \partial y_0/\partial x & \partial y_0/\partial y \end{vmatrix} = \begin{vmatrix} 1 - \dot{\rho}\ \cos\theta & 0 \\ -\dot{\rho}\ \sin\theta & 1 \end{vmatrix} = 1 - \dot{\rho}\ \cos\theta \tag{P.11}$$

where $\dot{\rho} \equiv \mathrm{d}\rho/\mathrm{d}x = -mv_\perp \dot{B}/qB^2$. The element at the origin is therefore larger than that at the guiding centre where the field is increased $(0 \le \theta < \pi)$ and the density of orbits correspondingly smaller. The current in the y direction due to orbits with guiding centres in the element $\mathrm{d}\theta$ at angle θ is therefore

$$j_y = n\ q\ v_\perp \int_0^{2\pi} (1 + \dot{\rho}\ \cos\theta)\cos\theta\ \mathrm{d}\theta/2\ \pi = -n\ \not{q}\ m\ v_\perp^2 \dot{B}/\not{q}\ 2\ B^2 \tag{P.12}$$

The current due to the guiding centre drift is given by Eq. (3.59)

$$j_D = n\ m\ v_\perp^2\ \dot{B}_\perp/2\ B^2 \tag{P.13}$$

The current due to the drift of the guiding centres therefore exactly balances that due to the orbital motion. We conclude that there is no net current due to a magnetic field gradient.

Problem 5: Show that Eqs. (10.26) and (10.28) are equivalent.

Solution: The set of Eq. (10.26) may be written explicitly as

$$[V^2 - (V_A{}^2 + V_s{}^2 \sin^2\theta)]\xi_x' - [V_s{}^2\ \sin\theta\ \cos\theta]\xi_z' \tag{P.14x}$$

$$[V^2 - V_A{}^2 \cos^2\theta]\xi_y' = 0 \tag{P.14y}$$

$$[-V_s{}^2\ \sin\theta\ \cos\theta]\xi_x' + [V^2 - V_s{}^2 \cos^2\theta]\xi_z' = 0 \tag{P.14z}$$

and Eq. (10.28) as

$$[V^2 - (V_A{}^2 + V_s{}^2)]\ (\xi'_x\ \sin\theta + \xi'_z\ \cos\theta) + V_A{}^2\ \xi'_z\ \cos\theta = 0 \tag{P.15a}$$

$$[V^2 - V_A{}^2\ \cos^2\theta]\xi'_y = 0 \tag{P.15b}$$

$$V^2\xi'_z - V_s{}^2 \cos\theta\ (\xi'_x\ \sin\theta + \xi'_z\ \cos\theta) \tag{P.15c}$$

Clearly, Eqs. (P.14:y) and (P.15b) are identical, as are Eqs. (P.14:z) and (P.15c).

$$\begin{aligned}
&[V^2 - (V_A{}^2 + V_s{}^2)]\ (\xi'_x\ \sin\theta + \xi'_z\ \cos\theta) + \\
&\qquad V_A{}^2\ \xi'_z\ \cos\theta - \cos\theta\ [V^2\ \xi'_z - V_s{}^2\ \cos\theta\ (\xi'_x\ \sin\theta + \xi'_z\ \cos\theta)] \\
&= V^2\ \sin\theta\xi'_z - V_A{}^2\ \sin\theta\ \xi'_x - V_s{}^2(\xi'_x\ \sin\theta + \xi'_z\ \cos\theta)\ \sin^2\theta \\
&= \sin\theta\ \{V^2\xi'_z - V_A{}^2\ \xi_x - V_s{}^2\sin^2\theta\ \xi'_x + V_s{}^2\ \sin\theta\ \cos\theta\ \xi'_z\} = 0
\end{aligned} \tag{P.16}$$

Problem 6: Appleton–Hartree dispersion relation.
The general dispersion relation Eq. (11.14) yields a complex set of relationships even when the plasma is cold. Simpler forms are therefore useful when the plasma conditions may be limited. A useful case occurs when the wave frequency is sufficiently large that the ion motion may be neglected, for example the propagation of radio waves in the F-layer of the ionosphere investigated by Appleton and Hartree although the solution was given earlier by Lassen. Find the simplified form of the dispersion relation Eq. (11.14) when the ion motion is neglected.

Solution: When the ion motion is neglected, the terms A, B, and C take the simplified forms

$$R = 1 - \frac{\Pi_e{}^2}{\omega^2}\frac{\omega}{(\omega + \Omega_e)} \qquad L = 1 - \frac{\Pi_e{}^2}{\omega^2}\frac{\omega}{(\omega - \Omega_e)} \qquad P = 1 - \frac{\Pi_e{}^2}{\omega^2} \tag{P.17}$$

Hence the terms

$$A = S\sin^2\theta + P\cos^2\theta \qquad B = RL\sin^2\theta + PS(1 + \cos^2\theta) \qquad C = PRL \tag{P.18}$$

as in Eq. (11.16). As before the dispersion relation is reduced to the quadratic form in n^2

$$An^4 - Bn^2 + C = 0 \tag{P.19}$$

whose solution may be written in several equivalent forms

$$n^2 = \frac{-B \pm \sqrt{B^2 - 4\ A\ C}}{2\ A} = \frac{2\ C}{-B \pm \sqrt{B^2 - 4\ A\ C}} = 1 - \frac{2(A - B + C)}{2A - B \pm \sqrt{B^2 - 4\ A\ C}} \tag{P.20}$$

Substituting for A, B, and C, we obtain the Appleton–Hartree dispersion relation for electron waves

$$n^2 = 1 - \frac{2\ \alpha\ \omega^2\ (1 - \alpha)}{2\ \omega^2(1 - \alpha) - \Omega_e{}^2\ \sin^2\theta \pm \Omega_e\ \Gamma} \tag{P.21}$$

where $\Gamma = \sqrt{(\Omega_e{}^2\sin^4\theta + 4\omega^2(1 - \alpha)^2\cos^2\theta)}$ and $\alpha = \Pi_e{}^2/\omega^2$.

Problem 7: Bennett pinch relation.
If the axial electron current density j_z is uniform across a simple linear, static cylindrical plasma of radius r_0, show that the radial electron density is

$$n_e = \frac{n_0}{[1 + (r/r_0)^2]^2} \tag{P.22}$$

and that the temperature $kT = \mu_0\ e^2\ r_0{}^2\ n_0\ v_z/16$ where $v_z = -j_z/n_e\ e$ is the electron drift velocity.

Solution: It follows from Eq. (9.5) that since the plasma is static, the radial force balance reduces to

$$\nabla\left(p+\frac{1}{2\,\mu_0}B^2\right)=0 \qquad \text{or} \qquad p(r)+\frac{1}{2\mu_0}B(r)^2=\text{const}=p_0=\frac{1}{2\,\mu_0}{B_e}^2 \tag{P.23}$$

where $p_0 = 2\,n_0 \mathit{k}T$ is the gas pressure on axis, and from Ampère's theorem $B_e = \frac{1}{2}\,\mu_0\,r_0$ is the magnetic field at the plasma edge. Therefore, since the total pressure is constant

$$2\,n_0\,\mathit{k}T=\frac{1}{4}\,{\mu_0}^2\,\frac{{r_0}^2}{2\mu_0}=\frac{1}{16}\,\mu_0\,{r_0}^2 \tag{P.24}$$

Problem 8: Derive the Grad–Shafranov equation for plasma equilibrium in a system with cylindrical symmetry

$$\Delta^*\psi=-[f(\psi)\dot{f}(\psi)+\mu_0\,R^2\,\dot{p}(\psi)] \tag{15.25}$$

and the current density

$$\Delta^*\psi=-\mu_0\,R\,j_\phi \tag{15.28}$$

Solution: Calculate $\nabla\wedge\mathbf{B}$

$$\nabla\wedge B|_R=\frac{1}{R}\frac{\partial B_z}{\partial\phi}-\frac{\partial B_\phi}{\partial Z}=-\frac{1}{R}\,\dot{f}(\psi)\,\frac{\partial\psi}{\partial Z} \tag{P.25a}$$

$$\nabla\wedge B|_\phi=\frac{\partial B_R}{\partial Z}-\frac{\partial B_z}{\partial R}=-\frac{1}{R}\frac{\partial^2\psi}{\partial Z^2}-\frac{\partial\left(\frac{1}{R}\frac{\partial\psi}{\partial R}\right)}{\partial R}=-\frac{1}{R}\,\Delta^*\psi \tag{P.25b}$$

$$\nabla\wedge B|_Z=\frac{1}{R}\left(\frac{\partial(R\,B_\phi)}{\partial R}-\frac{\partial B_R}{\partial\phi}\right)=\frac{1}{R}\,\dot{f}(\psi)\,\frac{\partial\psi}{\partial R} \tag{P.25c}$$

and hence since $\mathbf{j}\wedge\mathbf{B}=\frac{1}{\mu_0}(\nabla\wedge\mathbf{B})\wedge\mathbf{B}$

$$(\nabla\wedge\mathbf{B})\wedge\mathbf{B}|_R=-\frac{1}{R}\Delta^*\psi\,B_Z-{\mu_0}^2\,\frac{1}{R^2}\,\dot{f}(\psi)\,f(\psi)\,\frac{\partial\psi}{\partial R}=-\frac{1}{R}(\dot{f}(\psi)\,f(\psi)+\Delta^*\psi)\,B_Z \tag{P.26a}$$

$$(\nabla\wedge\mathbf{B})\wedge\mathbf{B}|_Z=-{\mu_0}^2\,\frac{1}{R^2}\,\dot{f}(\psi)\,f(\psi)\,\frac{\partial\psi}{\partial Z}+\frac{1}{R}\,\Delta^*\psi\,B_R=\frac{1}{R}\,(\dot{f}(\psi)\,f(\psi)+\Delta^*\psi)\,B_R \tag{P.26b}$$

and the term ∇p

$$\nabla p|_R=\dot{p}(\psi)\,\frac{\partial\psi}{\partial R}=\dot{p}(\psi)\,R\,B_Z \tag{P.27a}$$

$$\nabla p|_Z=\dot{p}(\psi)\,\frac{\partial\psi}{\partial Z}=-\dot{p}(\psi)\,R\,B_R \tag{P.27b}$$

Equation (15.25) follows from $\mathbf{j}\wedge\mathbf{B}-\nabla p=0$ and Eqs. (P.26) and (P.27), and Eq. (15.28) from $j_\phi=\frac{1}{\mu_0}(\nabla\wedge\mathbf{B})|_\phi$ and Eq. (P.25b).

Problem 9: Calculate the Zheng et al. (1996) expansion coefficients.

Solution: We set $G_(n,k,R)=G(n,k,R);Z^{2k}$.
Calculating the terms $G_1'(n,k,R)$ up to terms of the sixth total order

$$G_1'(0,0,R)=1 \qquad G_1'(2,0,R)=Z^2 \qquad G_1'(4,0,R)=Z^4 \qquad G_1'(6,0,R)=Z^6$$

$$G_1'(2,1,R)=-R^2[\ln R-1/2] \qquad G_1'(4,1,R)=-\frac{24}{2\times 4}R^2[2\ln R-1]\,Z^2$$

$$= -6R^2\, Z^2\, \ln R + 3R^2\, Z^2$$

$$G'(6,1,R) = -\frac{720}{24\times 4\times 1} R^2 (2\ln R - 1) Z^4 = -15R^2 Z^4 \ln R + \frac{15}{2} R^2 Z^4$$

$$G_1'(4,2,R) = \frac{24}{16\times 2\times 1}\, R^4 \left[2\ln R - \frac{5}{2}\right] = \frac{3}{2}\, R^4\, \ln R - \frac{15}{8}\, R^4$$

$$G_1'(6,2,R) = \frac{720}{4\times 16\times 2\times 1} R^4 \left[2\ln R - \frac{5}{2}\right] Z^2 = \frac{45}{2} R^4 Z^2 \ln R - \frac{225}{8} R^4 Z^2$$

$$G_1'(6,3,R) = -\frac{720}{1\times 64\times 6\times 2} R^6 \left[2\ln R - \frac{10}{3}\right] = -\frac{15}{8} R^6\, \ln R + \frac{25}{8}\, R^6 \qquad \text{(P.28)}$$

and the terms $G_2'(n,k,R)$

$$G_2'(0,0,R) = R^2 \qquad G_2'(2,0,R) = R^2\, Z^2 \qquad G_2(4,0,R) = R^2\, Z^4$$

$$G_2'(2,1,R) = -\frac{1}{4} R^4 \qquad G_2'(4,1,R) = -\frac{24}{2\times 4\times 2} R^4\, Z^2 = -\frac{3}{2}\, R^2\, Z^2 \qquad \text{(P.29)}$$

$$G_2'(4,2,R) = \frac{24}{16\times 2\times 6}\, R^4 = \frac{1}{8}\, R^4$$

Collecting terms together and summing

$$1 = G_1'(0,0,R) \qquad R^2 = G_2'(0,0,R) \quad Z^2 - R^2 \ln(R)$$

$$= \underline{[G_1'(2,0,R) + G_1'(2,1,R)]} - \frac{1}{2}\, \underline{G_2'(0,0,R)}$$

$$R^4 - 4\, R^2\, Z^2 = -4\, \underline{[G_2'(2,0,R) + G_2'(2,1,R)]}$$

$$2\, Z^4 - 9\, R^2\, Z^2 + 3\, R^4\, \ln(R) - 12R^2\, Z^2\, \ln(R)$$

$$= 2\, \underline{[G_1'(4,0,R) + G_1'(4,1,R) + G_1'(4,2,R)]} - 15\, \underline{[G_2'(2,0,R) + G_2'(2,1,R)]}$$

$$R^6 - 12\, R^4\, Z^2 + 8\, R^2\, Z^4 = 8\, \underline{[G_2'(4,0,R) + G_2'(4,1,R) + G_2'(4,2,R)]}$$

$$8\, Z^6 - 140\, R^2\, Z^4 + 75\, R^4\, Z^2 - 15\, R^6\, \ln(R) + 180\, R^4\, Z^2\, \ln(R) - 120\, R^2\, Z^4\, \ln(R)$$

$$= 8\, \underline{[G_1'(6,0,R) + G_1'(6,1,R) + G_1'(6,2,R) + G_1'(6,3,R)]}$$

$$-\, 200\, \underline{[G_2'(4,0,R) + G_2'(4,1,R) + G_2'(4,3,R)]} \qquad \text{(P.30)}$$

Problem 10: Calculate the zero order Rosenbluth potentials for an isotropic distribution.

Solution: We use the Rosenbluth potentials $\mathcal{G}(\mathbf{v})$ and $\mathcal{H}(\mathbf{v})$ defined in Eq. (7.14) to form the Fokker–Planck equation for an isotropic distribution through Eq. (7.27). We consider only collisions between particles m with speed v and distribution $f(v)$ with particles m', speed v', and distribution $f'(v')$. The required partial derivatives are transformed to ordinary derivatives with respect to the speed v by making use of the results

$$\frac{\partial}{\partial v_i} = \nabla_v = \frac{v_i}{v}\frac{\mathrm{d}}{\mathrm{d}v}$$

and

$$\frac{\partial^2}{\partial v_i\, \partial v_j} = \left\{\frac{\delta_{ij}}{v} - \frac{v_i\, v_j}{v^3}\right\}\frac{\mathrm{d}}{\mathrm{d}v} + \frac{v_i\, v_j}{v^2}\frac{\mathrm{d}^2}{\mathrm{d}v^2}$$

and remembering that repeated indices imply summation, we obtain successively

$$\frac{\partial^2}{\partial v_i\, \partial v_i} = \nabla_v{}^2 = \frac{\mathrm{d}^2}{\mathrm{d}v^2} + 2\, \frac{1}{v}\frac{\mathrm{d}}{\mathrm{d}v} \qquad \text{(P.31)}$$

$$\frac{\partial^3}{\partial v_i\,\partial v_j\,\partial v_j} = \nabla_v(\nabla_v{}^2) = \frac{v_i}{v}\,\frac{\mathrm{d}^3}{\mathrm{d}v^3} + 2\,\frac{v_i}{v^2}\,\frac{\mathrm{d}^2}{\mathrm{d}v^2} - 2\,\frac{v_i}{v^3}\,\frac{\mathrm{d}}{\mathrm{d}v} \tag{P.32}$$

$$\frac{\partial^4}{\partial v_i\,\partial v_i\,\partial v_j\,\partial v_j} = \nabla_v{}^2(\nabla_v{}^2) = \frac{\mathrm{d}^4}{\mathrm{d}v^4} + 4\,\frac{1}{v}\,\frac{\mathrm{d}^3}{\mathrm{d}v^3} \tag{P.33}$$

Substituting in the terms in Eq. (7.27), we obtain

$$\frac{\partial^2}{\partial v_i\,\partial v_j}\left[f(v)\,\frac{\partial^2\mathcal{G}}{\partial v_i\,\partial v_j}\right] = \frac{1}{v^2}\frac{\mathrm{d}}{\mathrm{d}v^2}\left[v^2\,f(v)\,\frac{\mathrm{d}^2\mathcal{G}}{\mathrm{d}v^2}\right] - 2\,\frac{1}{v^2}\,\frac{\mathrm{d}}{\mathrm{d}v}\left[f(v)\,\frac{\mathrm{d}g}{\mathrm{d}v}\right] \tag{P.34}$$

$$\frac{\partial}{\partial v_i}\left[f(\mathbf{v})\frac{\partial\mathcal{H}}{\partial v_i}\right] = \nabla_v\cdot[f\ \nabla_v h] = \frac{1}{v^2}\,\frac{\mathrm{d}}{\mathrm{d}v}\left[v^2\,f\,\frac{\mathrm{d}h}{\mathrm{d}v}\right] \tag{P.35}$$

Turning now to the terms of the collision integral

$$\begin{aligned} I &= -\frac{\partial}{\partial v_i}\left[f(\mathbf{v})\frac{\partial\mathcal{H}}{\partial v_i}\right] + \frac{1}{2}\,\frac{\partial^2}{\partial v_i\,\partial v_j}\left[f(v)\,\frac{\partial^2\mathcal{G}}{\partial v_i\,\partial v_j}\right] \\ &= \frac{1}{v^2}\frac{\mathrm{d}}{\mathrm{d}v}\left\{-v^2\,f\,\frac{\mathrm{d}\mathcal{H}}{\mathrm{d}v} + \frac{1}{2}\frac{\mathrm{d}}{\mathrm{d}v}\left[v^2 f\frac{\mathrm{d}^2\mathcal{G}}{\mathrm{d}v^2}\right] - f\frac{\mathrm{d}\mathcal{G}}{\mathrm{d}v}\right\} \\ &= \frac{1}{v^2}\frac{\mathrm{d}}{\mathrm{d}v}\left\{-\left(\frac{\mathrm{d}\mathcal{G}}{\mathrm{d}v} + v^2\,\frac{\mathrm{d}\mathcal{H}}{\mathrm{d}v}\right)f + \left[v\,\frac{\mathrm{d}^2\mathcal{G}}{\mathrm{d}v^2} + \frac{1}{2}\,v^2\,\frac{\mathrm{d}^3\mathcal{G}}{\mathrm{d}v^3}\right]f + \frac{1}{2}\,v^2\,\frac{\mathrm{d}^2\mathcal{G}}{\mathrm{d}v^2}\,\frac{\mathrm{d}f}{\mathrm{d}v}\right\} \end{aligned} \tag{P.36}$$

To perform the integrals, we take an axial co-ordinate parallel to **v** so that the terms $\mathcal{G}(v)$ and $\mathcal{H}(v)$ corresponding to the equilibrium distribution $f(v)$ are themselves isotropic and take the values

$$\begin{aligned} \mathcal{G}(v) &= \iiint v'^2\,\mathrm{d}v'\ \sin\theta\,\mathrm{d}\theta\,\mathrm{d}\phi\,f'(v')(v^2 - 2v\,v'\ \cos\theta + v'^2)^{1/2} \\ &= \frac{2\pi}{3}\int \mathrm{d}v'\,\frac{v'}{v}\,f'(v')\left\{(v+v')^3 - \left[\begin{matrix}(v-v')^3 & \text{if}\ \ v > v' \\ (v'-v)^3 & \text{otherwise}\end{matrix}\right]\right\} \\ &= \frac{4\,\pi}{3}\left\{\int_v^\infty v'\,[v^2 + 3v'^2]\,f'(v')\,\mathrm{d}v' + \int_0^v \frac{v'^2}{v}\,[v'^2 + 3v^2]\,f'(v')\,\mathrm{d}v'\right\} \end{aligned} \tag{P.37}$$

and

$$\begin{aligned} \mathcal{H}(v) &= \frac{(m+m')}{m'}\iiint v'^2\,\mathrm{d}v'\ \sin\theta\,\mathrm{d}\theta\,\mathrm{d}\phi\,f'(v')(v^2 - 2v\,v'\ \cos\theta + v'^2)^{-1/2} \\ &= 2\,\pi\,\frac{(m+m')}{m'}\int \mathrm{d}v'\,\frac{v'}{v}\,f'(v')\left\{(v+v') - \left[\begin{matrix}(v-v') & \text{if}\ \ v > v' \\ (v'-v) & \text{otherwise}\end{matrix}\right]\right\} \\ &= 4\,\pi\,\frac{(m+m')}{m'}\left\{\int_v^\infty v'\,f'(v')\mathrm{d}v' + \int_0^v \frac{v'^2}{v} f'(v')\mathrm{d}v'\right\} \end{aligned} \tag{P.38}$$

consistent with Eqs. (8.111:b) and (8.112:b)
The derivatives of the functions $\mathcal{G}(v)$ and $\mathcal{H}(v)$ are calculated directly

$$\frac{\mathrm{d}\mathcal{G}}{\mathrm{d}v} = \frac{4\,\pi}{3}\left\{\cancel{-4v^3 f'(v)} + 2\,v\int_v^\infty v'\,f'(v')\mathrm{d}v' + \cancel{4v^3 f'(v)} + \int_0^v v^2 f'(v')\left(3 - \frac{v^2}{v^2}\right)\mathrm{d}v'\right\}$$

$$\frac{\mathrm{d}^2\mathcal{G}}{\mathrm{d}v^2} = \frac{4\,\pi}{3}\left\{\cancel{-2v^2 f'(v)} + 2\int_v^\infty v'\,f'(v')\mathrm{d}v' + \cancel{2v^2 f'(v)} + \int_0^v \frac{v^4}{v^3}\,f'(v')\mathrm{d}v'\right\}$$

$$\frac{\mathrm{d}^3\mathcal{G}}{\mathrm{d}v^3} = \frac{4\,\pi}{3}\left\{\cancel{-2vf'(v)} + \cancel{2vf'(v)} - 6\int_0^v \frac{v*^4}{v^4}\,f'(v')\,\mathrm{d}v'\right\}$$

$$\frac{\mathrm{d}^4\mathcal{G}}{\mathrm{d}v^4} = \frac{4\,\pi}{3}\left\{-6f'(v) + 24\int_0^v \frac{v*^4}{v^5} f'(v')\mathrm{d}v'\right\}$$

$$\frac{\mathrm{d}\mathcal{H}}{\mathrm{d}v} = 4\,\pi\,\frac{(m+m')}{m'}\left\{\cancel{-v^2 f'(v) + v^2 f'(v)} - \int_0^v \mathrm{d}v' v' f'(v')\right\} \tag{P.39}$$

Finally substituting for $\mathcal{G}$ and $\mathcal{H}$ we obtain

$$I = \frac{4\,\pi}{v^2}\frac{\mathrm{d}}{\mathrm{d}v}\left\{\frac{m}{m'}\left[\int_0^v v'^2\, f'(v')\,\mathrm{d}v'\right] f(v) + \frac{1}{3}v^2\left[\int_0^v \frac{v'^4}{v^3}\, f'(v')\,\mathrm{d}v' + \int_v^\infty v'\, f'(v')\,\mathrm{d}v'\right]\frac{\mathrm{d}f}{\mathrm{d}v}\right\} \tag{P.40}$$

consistent with Eq. (7.49) when the particles are identical $m' = m$ etc.

Problem 11: Calculate the first order Rosenbluth potentials for electron–electron scattering.

Solution: Consider a perturbed distribution function in Cartesian tensor form consisting of an isotropic unperturbed function disturbed by a small directed term

$$f(\mathbf{v}) = f_0(v) + \mathbf{v}\cdot f_1(v) \tag{P.41}$$

The Rosenbluth potentials include two terms associated with the vector distribution functions $f_1'(v)$ namely $\mathcal{G}_1(v)$ and $\mathcal{H}_1(v)$ parallel to f_1'. Retaining only terms of first order, the Fokker–Planck collision integral becomes

$$\begin{aligned}
I &= \left\{-\frac{\partial}{\partial v_i}\left[f(\mathbf{v})\frac{\partial \mathcal{H}}{\partial v_i}\right] + \frac{1}{2}\frac{\partial^2}{\partial v_i \partial v_j}\left[f(\mathbf{v})\frac{\partial^2 \mathcal{G}}{\partial v_i \partial v_j}\right]\right\}\\
&\approx \left\{-\frac{\partial}{\partial v_i}\left[f_0(\mathbf{v})\frac{\partial \mathcal{H}_0}{\partial v_i}\right] + \frac{1}{2}\frac{\partial^2}{\partial v_i \partial v_j}\left[f_0(\mathbf{v})\frac{\partial^2 \mathcal{G}_0}{\partial v_i \partial v_j}\right]\right\}\\
&+ \left\{-\frac{\partial}{\partial v_i}\left[\mathbf{v}\cdot f_1(v)\frac{\partial \mathcal{H}_0}{\partial v_i}\right] + \frac{1}{2}\frac{\partial^2}{\partial v_i \partial v_j}\left[\mathbf{v}\cdot f_1(v)\frac{\partial^2 \mathcal{G}_0}{\partial v_i \partial v_j}\right]\right\}\\
&+ \left\{-\frac{\partial}{\partial v_i}\left[f_0(v)\frac{\partial(\mathbf{v}\cdot\mathcal{H}_1)}{\partial v_i}\right] + \frac{1}{2}\frac{\partial^2}{\partial v_i \partial v_j}\left[f_0(v)\frac{\partial^2(\mathbf{v}\cdot\mathcal{G}_1)}{\partial v_i \partial v_j}\right]\right\}
\end{aligned} \tag{P.42}$$

The Rosenbluth potentials associated with the isotropic component $\mathcal{G}_0(v)$ and $\mathcal{H}_0(v)$ have already been calculated in Eqs. (P.37) and (P.38). The functions corresponding to the perturbation are calculated similarly taking into account the vector nature of the perturbation. Taking local axes along $\mathbf{v}$, we write $\mathbf{v}' = (v', \theta', \phi')$ and $f_1' = (f_1', \theta, \phi)$ so that

$$\mathbf{v}'\cdot f_1' = v\, f_1'[\cos\theta\ \cos\theta' + \sin\theta\ \sin\phi'(\cos\phi\ \cos\phi' + \sin\phi\ \sin\phi')]$$

Noting that the integrals over the azimuthal angle ϕ' of both the $\cos\phi'$ and $\sin\phi'$ terms yield zero, we may obtain the magnitude of the Rosenbluth potential $\mathcal{G}_1$ from the integrals

$$\begin{aligned}
\mathcal{G}_1(v) &= \iiint v'^2\mathrm{d}v'\ \sin\theta'\ \mathrm{d}\theta'\ \mathrm{d}\phi'\mathbf{v}'\cdot f_1'(v')(v^2 - 2v\, v'\ \cos\theta' + v'^2)^{1/2}\\
&= \iiint v'^2\mathrm{d}v'\ \sin\theta'\ \mathrm{d}\theta'\ \mathrm{d}\phi' v'\ f_1'(v')[\cos\theta\ \cos\theta'\\
&\qquad + \sin\theta\sin\phi'(\cos\phi\ \cos\phi' + \sin\phi\ \sin\phi')]\times(v^2 - 2v\, v'\ \cos\theta' + v'^2)^{1/2}\\
&= \frac{\pi\ \cos\theta}{v}\iiint v'\mathrm{d}v'\ \sin\theta'\ \mathrm{d}\theta'\ v'\ f_1'(v')\\
&\qquad \times[(v^2+v'^2)(v^2 - 2v\, v'\ \cos\theta' + v'^2)^{1/2} - (v^2 - 2v\, v'\ \cos\theta' + v'^2)^{3/2}]
\end{aligned}$$

$$= \frac{4\,\pi}{15}\cos\theta\left\{\int_0^v \mathrm{d}v'\ \frac{v'^4}{v^2}\ f_1'(v')\ (v'^2 - 5\ v^2) + \int_v^\infty \mathrm{d}v'\ v'\ v\ f_1'(v')\ (v^2 - 5v'^2)\right\} \tag{P.43}$$

and the magnitude of $\mathcal{H}_1(v)$

$$\begin{aligned}
\mathcal{H}_1(v) &= \frac{(m+m'}{m'}\iiint v'^2\mathrm{d}v'\ \ \sin\theta'\ \mathrm{d}\theta'\ \mathrm{d}\phi'\mathbf{v}'\cdot f_1'(v')(v^2 - 2v\ v'\ \ \cos\theta' + v'^2)^{1/2}\\
&= \frac{(m+m')}{m'}\iiint v'^2\mathrm{d}v'\ \ \sin\theta'\ \mathrm{d}\theta'\ \mathrm{d}\phi' v'\ f_1'(v')[\cos\theta\ \cos\theta'\\
&\qquad + \sin\theta\sin\phi'(\cos\phi\ \cos\phi' + \sin\phi\ \sin\phi')]\\
&\qquad \times(v^2 - 2v\ v'\ \ \cos\theta' + v'^2)^{-1/2}\\
&= \pi\ \frac{(m+m')}{m'}\ \frac{\cos\theta}{v}\iiint v'^2\ \mathrm{d}v'\ \ \sin\theta'\ \mathrm{d}\theta'\ f_1'(v')\\
&\qquad \times[(v^2+v'^2)(v^2 - 2v\ v'\ \ \cos\theta' + v'^2)^{-1/2} - (v^2 - 2v\ v'\ \ \cos\theta' + v'^2)^{1/2}]\\
&= \frac{4\,\pi}{3}\ \frac{(m+m')}{m'}\ \cos\theta\left\{\int_0^v \mathrm{d}v'\ \frac{v'^4}{v^2}\ f_1'(v') + \int_v^\infty \mathrm{d}v'\ v'\ v\ f_1'(v')\right\}
\end{aligned} \tag{P.44}$$

Therefore, the perturbation Rosenbluth potentials may be written in vector form as $\mathbf{v}\cdot\mathcal{G}_1(v)$ and $\mathbf{v}\cdot\mathcal{H}_1(v)$, where

$$\begin{aligned}
\mathcal{G}_1(\mathbf{v}) &= \frac{4\,\pi}{15}\left\{\int_0^v \mathrm{d}v'\ \frac{v'^4}{v^3}\ f_1'(v')\ (v'^2 - 5\ v^2) + \int_v^\infty \mathrm{d}v'\ v'\ f_1'(v')\ (v^2 - 5v'^2)\right\}\\
\mathcal{H}_1(\mathbf{v}) &= \frac{4\,\pi}{3}\ \frac{(m+m')}{m'}\left\{\int_0^v \mathrm{d}v'\ \frac{v'^4}{v^3}\ f_1'(v') + \int_v^\infty \mathrm{d}v'\ v'\ f_1'(v')\right\}
\end{aligned} \tag{P.45}$$

consistent with Eqs. (8.111b) and (8.112b).

The derivatives of the general form $\mathbf{v}\cdot\mathbf{A}(v)$ derived from a general vector function $\mathbf{A}(v)$ are easily obtained

$$\frac{\partial}{\partial v_i}(v_j\ A_j) = A_i + \frac{v_i\ v_j}{v}\ \frac{\mathrm{d}A_j}{\mathrm{d}v} \tag{P.46}$$

$$\frac{\partial^2}{\partial v_i\ \partial v_j}(v_k\ A_k) = \frac{v_j}{v}\ \frac{\mathrm{d}A_i}{\mathrm{d}v} + \frac{v_i}{v}\ \frac{\mathrm{d}A_j}{\mathrm{d}v} + \frac{v_i\ v_j\ v_k}{v^2}\ \frac{\mathrm{d}^2A_k}{\mathrm{d}v^2} \tag{P.47}$$

$$\frac{\partial^2}{\partial v_i\ \partial v_i}(v_jA_j) = v_i\ \frac{\mathrm{d}^2A_i}{\mathrm{d}v^2} + 2\ \frac{v_i}{v}\ \frac{\mathrm{d}A_i}{\mathrm{d}v} \tag{P.48}$$

$$\frac{\partial^3}{\partial v_i\ \partial v_i\ \partial v_j}(v_kA_k) = \frac{v_j\ v_k}{v}\ \frac{\mathrm{d}^3A_k}{\mathrm{d}v^3} + \left(\delta_{jk} + 2\ \frac{v_j\ v_k}{v^2}\right)\ \frac{\mathrm{d}^2A_k}{\mathrm{d}v^2} + 2\frac{1}{v}\left(\delta_{jk} - \frac{v_j\ v_k}{v^2}\right)\ \frac{\mathrm{d}A_k}{\mathrm{d}v} \tag{P.49}$$

$$\frac{\partial^4}{\partial v_i\ \partial v_i\ \partial v_j\ \partial v_j}(v_kA_k) = v_k\ \frac{\mathrm{d}^4A_k}{\mathrm{d}v^4} + 2\frac{v_k}{v}\frac{\mathrm{d}^3A_k}{\mathrm{d}v^3} + \frac{v_j}{v}\frac{\mathrm{d}^3A_j}{\mathrm{d}v^3} + 4\ \frac{v_k}{v^2}\frac{\mathrm{d}^2A_k}{\mathrm{d}v^2} \tag{P.50}$$

Problem 12: Calculate the mean energy gain in a collision between an electron in an oscillating electric field and a particle using the impact model when the law of force has an inverse power law form.

Solution: If the inter-particle force law is an inverse power law $F \sim r^s$, when $\sigma_d = \gamma\ v^{-4/(s-1)}$

$$\overline{q} = \frac{1}{2}\ m\ n\ \gamma \int_0^{\pi} (v_0\ v_1\ \cos\theta + {v_1}^2)\ ({v_0}^2 + {v_1}^2 + 2v_0\ v_1\ \cos\theta)^{[1-4/(s-1)]/2}\ \sin\theta\ d\theta \tag{P.51}$$

$$= \left(\begin{array}{lll} \left\{ \begin{array}{ll} m\ n\ \sigma_d\ {v_1}^2\ v_0(4/3 + 4{v_1}^2/15\ {v_0}^2) & v_0 > v_1 \\ m\ n\ \sigma_d\ {v_1}^3(1 + 2{v_0}^2/3\ {v_1}^2 - {v_0}^4/15\ {v_1}^4) & v_0 < v_1 \end{array} \right\} & s = \infty & \text{Hard sphere} \\ \left\{ m\ n\ (\sigma_d\ v){v_1}^2 = m\ n\ {v_1}^2\ \nu \right\} & s = 5 & \text{Maxwell mol.} \\ \left\{ \begin{array}{ll} 0 & v_0 > v_1 \\ 2\ m\ n\ \gamma/|v_1| & v_0 < v_1 \end{array} \right\} & s = 2 & \text{Coulomb force} \end{array} \right. \tag{P.52}$$

Next we need to average over the oscillation of the quiver velocity, i.e. that of the electro-magnetic wave to obtain the final value for the energy gained by electrons per unit volume per unit time

$$\overline{q} = \left(\begin{array}{lll} \left\{ \begin{array}{ll} \frac{2}{3}\ m\ n\ \sigma_d\ {v_q}^2\ v_0 = \frac{2}{3}\ m\ {v_q}^2 \nu_{\text{eff}} & v_0 \gg v_q \\ \frac{4}{3}\ \pi\ m\ n\ \sigma_d\ {v_q}^3 & v_0 \ll v_q \end{array} \right\} & s = \infty & \text{Hard sphere} \\ \left\{ \frac{1}{2}\ m\ n\ (\sigma_d\ v)\ {v_q}^2 = \frac{1}{2}\ m\ {v_q}^2\ \nu \right\} & s = 5 & \text{Maxwell mol.} \\ \left\{ \begin{array}{ll} 0 & v_0 \gg v_1 \\ \frac{1}{\pi}\ \frac{m\ n\ A'}{v_q}\ \ln\left\{ \frac{(v_q + \sqrt{{v_q}^2 - {v_0}^2})}{(v_q - \sqrt{{v_q}^2 - {v_0}^2})} \right\} & v_0 < v_q \\ \frac{2}{\pi}\ \frac{m\ n\ A'}{v_q}\ \ln\left\{ \frac{2v_q}{v_0} \right\} & v_0 \ll v_q \end{array} \right\} & s = 2 & \text{Coulomb force} \end{array} \right. \tag{P.53}$$

For a Coulomb field $A = 4\ \pi\ Z^2\ e^4/(4\ \pi\ \epsilon_0)^2\ m^2$ and $A' = A\ \ln\Delta(v_q)$, where Z is the charge of the ion.

Problem 13: Derive the dimensional matrix for the variables used by the hydrodynamic model of laser–plasma interaction $\mathfrak{a}, \mathfrak{b}, \mathfrak{d}, w$, and R.
Hence obtain the dimensionless parameters

$$q = a\ T^{5/2}\ (\mathrm{d}T/\mathrm{d}x) \qquad [a] = [M][T^{-3}] \Big/ [M]^{7/2}[L]^6[T]^{-7} = [M]^{-5/2}[L]^{-6}[T]^4$$

$$b = \mu\ \rho^{-2}\ T^{3/2} \qquad [b] = [L]^{-1} \cdot [M]^{-2}[L]^6 \cdot [M]^{3/2}[L]^3[T]^{-3} = [M]^{-1/2}[L]^8[T]^{-3}$$

Solution: Energy flux $w = W/r^{\nu}$

	a	b	d	w	R
[M]	−5/2	−3/2	1	1	0
[L]	−6	8	−3	0	1
[T]	4	−3	0	−3	0
	k_1	k_2	k_3	k_4	k_5

(P.54)

Take $k_5 = 1$ and $k_2 = 0$

$$\begin{aligned} -5k_1/2 + k_3 + k_4 &= 0 \\ -6k_1 - 3k_3 + 1 &= 0 \\ 4k_1 - 3k_4 &= 0 \end{aligned}$$

$$k_4 = 4/3k_1 \qquad k_3 = -4/3k_1 + 5/2k_1 = 7/6k_1 - 6k_1 - 7/2k_1 = -19/2k_1 = -1$$

$$k_1 = 2/19 \qquad k_3 = 7/37 \qquad k_4 = 8/37 \qquad a^{2/19} \quad d^{7/37} \quad w^{8/37} \quad R$$

Problem 14: Consider a plasma consisting of several streams (of different species) with each stream having a velocity distribution

$$f_s(v) = \frac{n_s\,\Delta_s}{\pi}\left\{\frac{1}{(v-u_s)^2 + {\Delta_s}^2}\right\} \tag{P.55}$$

where n_s is the number density, u_s the velocity and Δ_s is the 'thermal' width of the distribution of stream s.
Show that $f_0(v) = \sum_s q^2 f_s/m_s$ has a minimum at $v = 0$ if

$$\sum_s \frac{{\Pi_s}^2\, u_s\, \Delta_s}{({u_s}^2 + {\Delta_s}^2)^2} = 0 \tag{P.56}$$

and

$$\sum_s \frac{{\Pi_s}^2\, \Delta_s(3{u_s}^2 - {\Delta_s}^2)}{({u_s}^2 + {\Delta_s}^2)^3} > 0 \tag{P.57}$$

Solution: Differentiating each term in turn

$$\dot{f}_s(v) = -\frac{n_s\,\Delta_s}{\pi}\frac{(v-u_s)}{[(v-u_s)^2 + \Delta_s^2]^2} \quad : \qquad \dot{f}_s(0) = \frac{n_s\, u_s\, \Delta_s}{\pi\,({u_s}^2 + {\Delta_s}^2)} \tag{P.58}$$

$$\ddot{f}_s(v) = -\frac{n_s\,\Delta_s}{\pi}\left\{\frac{1}{[(v-u_s)^2 + \Delta_s^2]^2} - \frac{4(v-u_s)^2}{[(v-u_s)^2 + \Delta_s^2]^3}\right\} : \quad \ddot{f}_0(0) = \frac{n_s\,\Delta}{\pi}\left\{\frac{3{u_s}^2 - {\Delta_s}^2)}{{u_s}^2 + {\Delta_s}^2)^3}\right\} \tag{P.59}$$

Multiply the distribution of each stream by ${q_s}^2/\epsilon_0\, m_s$. If the sum over the derivatives at $v = 0$ is zero $v = 0$ is an extremum. This requires

$$\sum_s \frac{{q_s}^2\, n_s \Delta_s}{\epsilon_0\, m_s}\frac{u_s}{({u_s}^2 + {\Delta_s}^2)^2} = \sum_s \frac{{\Pi_s}^2\, u_s\, \Delta_s}{({u_s}^2 + {\Delta_s}^2)^2} = 0 \tag{P.60}$$

and if the second derivative is positive, it is a minimum if the second derivative is positive

$$\sum_s \frac{{\Pi_s}^2 \Delta_s\{3{u_s}^2 - {\Delta_s}^2)\}}{({u_s}^2 + {\Delta_s}^2)^3} > 0 \tag{P.61}$$

To apply the Penrose criterion, we require the integral which may be performed either by partial fractions or by a contour integration in the upper half plane of complex v. Integrating the dispersion integral by parts

$$\mathcal{I}(s) = \int_{-\infty}^{\infty} \frac{\dot{f}_0(v)}{(v-s)}\mathrm{d}v = \int_{-\infty}^{\infty} \frac{(f_0(v) - f_0(s))}{(v-s)^2}\mathrm{d}v \tag{P.62}$$

where $s = 0$ to avoid the principal part form of the integral. The required integral is

$$\mathcal{Z}(0) = \int_L \sum_s \frac{n_s\,\Delta_s}{\pi} \left\{ \frac{1}{(v - u_s)^2 + \Delta_s^2} \right\} \frac{1}{v^2}\, dv \tag{P.63}$$

- *Partial fractions*:
  Write arguments as partial fractions

$$\frac{1}{[(v-u)^2 + \Delta^2)]v^2} = \frac{Av + B}{(v-u)^2 + \Delta^2)} + \frac{Cv + D}{v^2} \tag{P.64}$$

Therefore

$$\begin{aligned}&(Av + B)v^2 + (Cv + D)[v - u)^2 + \Delta^2]\\ &= (A + C)v^3 + (B - 2uC + D)v^2 + (u^2C - 2uD + C\Delta^2)v + D(u^2 + \Delta^2) = 1\end{aligned} \tag{P.65}$$

and hence

$$D = \frac{1}{u^2 + \Delta^2} \qquad C = \frac{2uD}{u^2 + \Delta^2} = -A \qquad B = 2uC - D = \left(\frac{4u^2}{u^2 + \Delta^2} - 1\right) D = \frac{3u^2 - \Delta^2}{u^2 + \Delta^2} D \tag{P.66}$$

Noting that $\sum_s \Pi_s^{\,2}\,\Delta_s C_s = \sum_s 2\,\Pi_s^{\,2}\,u_s\,\Delta_s/(u_s^{\,2} + \Delta_s^{\,2})^2 = 0$, and that $f_s(0) = 1/(u_s^{\,2} + \Delta_s^{\,2}) = D$ The required dispersion integral becomes

$$\mathcal{Z}(0) = \sum_s \Pi_s^{\,2}\,\frac{\Delta_s}{\pi} \int_{-\infty}^{\infty} \frac{A_s\,v + B_s}{[(v - u_s)^2 + \Delta_s^{\,2}]} = \sum_s \Pi_s^{\,2}(A_s\,u_s + B_s) \tag{P.67}$$

Substituting for A and B

$$\mathcal{Z}(0) = \sum_s \Pi_s^{\,2} \left\{ \frac{3u_s^{\,2} - \Delta_s^2}{u_s^{\,2} + \Delta^2} D_s - \frac{2u^2}{(u^2 + \Delta^2)} D_s \right\} dv = \sum_s \Pi_s^{\,2} \frac{u_s^{\,2} - \Delta_s^{\,2}}{(u_s^{\,2} + \Delta_s^{\,2})^2} \tag{P.68}$$

- *Contour integration*:
  Alternatively as a contour integral along the Landau path and completed by a semi-circular loop at infinity. Factorising $(v - u)^2 + \Delta^2 = [(v - u) - \imath\,\Delta][(v - u) + \imath\,\Delta]$ there is one pole in the upper half plane at $v = u + \imath\,\Delta$ for each stream with residue $1/2\,\imath\,\Delta(u + \imath\,\Delta)^2$. Noting that the integral around the semi-circular path at $R \to \infty$ is zero, the integral along the Landau path, required by Penrose condition, is

$$\begin{aligned}\mathscr{Z}(0) &= \sum_s \Pi_s^{\,2} \int \frac{f_{0s}(v) - f_{0s}(0)}{v^2} dv = \sum_s \frac{\cancel{2\pi\imath\Delta_s}\,\Pi_s^{\,2}}{\cancel{2\pi\imath\Delta_s}} \frac{\left(u_s^{\,2} - \cancelto{0}{2\imath u_s\Delta_s} - \Delta_s^{\,2}\right)}{\left(u_s^{\,2} + \Delta_s^{\,2}\right)^2}\\ &= \sum_s \Pi_s^{\,2} \frac{(u_s^{\,2} - \Delta_s^{\,2})}{(u_s^{\,2} + \Delta_s^{\,2})^2}\end{aligned} \tag{P.69}$$

Applying Penrose's criterion, we conclude that if the following conditions are met

$$\sum_s \frac{\Pi_s^{\,2}\,u_s\,\Delta_s}{(u_s^{\,2} + \Delta_s^{\,2})^2} = 0 \qquad \sum_s \frac{\Pi_s^{\,2}\,\Delta_s(3u_s^{\,2} - \Delta_s^{\,2})}{(u_s^{\,2} + \Delta_s^{\,2})^3} > 0 \qquad \sum_s \Pi_s^{\,2} \frac{(u_s^{\,2} - \Delta_s^{\,2})}{(u_s^{\,2} + \Delta_s^{\,2})^2} > 0 \tag{P.70}$$

the plasma is unstable.

Bibliography

Abramowitz, M. and Stegun, I.A. (1965). *Handbook of Mathematical Functions with Formulas, Graphs and Mathematical Tables*. Washington, DC: National Bureau of Standards.

Afanas'ev, I.V., Krol, V.M., Krokhin, O.N., and Nemchinov, I.V. (1966). Gasdynamic processes in heating of a substance by laser radiation. *Appl Math Mech* 30: 1218–1225.

Alfvén, H. (1950). *Cosmical Electrodynamics, International Series of Monographs on Physics*. Oxford: Oxford University Press.

Allis, W.P. (1956). Motion of ions and electrons. In: S. Flügge. *Handbuch der Physik*, vol. 21, 383–444. Berlin: Springer-Verlag.

Allis, W.P., Buchsbaum, S.J., and Bers, A. (1963). *Waves in Anisotropic Plasmas*. Cambridge, MA: MIT Press.

Anderson, J.E. (1963). *Magnetohydrodynamic Shock Waves*. Cambridge, MA: MIT Press.

Anisimov, S.I. (1970). Effect of very short pulses on absorbing substances. *Sov Phys JETP* 31: 181–182.

Anisimov, S.I. (1971). Self-similar therma wave in a two temperature plasma heated by a laser pulse. *Sov Phys JETP Lett* 12: 287–289.

Atzeni, S. and Meher-ter-Vehn, J.M. (2004). *The Physics of Inertial Fusion*. Oxford: Oxford University press.

Babuel-Peyrissac, J.P., Fauquignon, C., and Floux, F. (1969). Effect of powerful laser pulse on low Z solid material. *Phys Lett A* 30: 290–291.

Balescu, R. (1960). Irreversible processes in ionized gases. *Phys Fluids* 3: 52–63.

Balescu, R. (1982a). Heating of electrons and ions by inverse bremsstrahlung absorption: a self-similar state of the plasma. *J Plasma Phys* 27: 553–570.

Balescu, R. (1982b). Thermal conductivity in a laser-created plasma heated by inverse bremsstrahlung absorption. *J Plasma Phys* 28: 65–92.

Balogh, A. and Treumann, R.A. (2013). *Physics of Collisionless Shocks: Space Plasma Shock Waves*. ISSI Scientific Report 12. Berlin: Springer-Verlag.

Bell, A.J., Evans, R.G., and Nicholas, D.J. (1981). Electron energy transport in steep temperature gradients in laser-produced plasmas. *Phys Rev Lett* 46: 243–246.

Berk, H.L. and Galeev, A.A. (1967). Velocity space instabilities in a toroidal geometry. *Phys Fluids* 10: 441–450.

Bernstein, I.B. (1958). Waves in a plasma in a magnetic field. *Phys Rev* 109: 10–21.

Bernstein, I.B., Frieman, E.A., Kruskal, N.D., and Kulsrud, R.M. (1958). An energy principle for hydromagnetic stability problems. *Proc R Soc A* 244: 17–40.

Bethe, H.A. (1942). On the Theory of Shock Waves for An Arbitrary Equation of State. *Report 545*, Office of Scientific Research and Development. Included in Johnson & Chéret (1998, p. 421–498).

Foundations of Plasma Physics for Physicists and Mathematicians, First Edition. Geoffrey J. Pert.

Bobin, J.L. (1971). Flame propagation and overdense heating in a laser created plasma. *Phys Fluids* 14: 2341–2354.

Bodin, H.A.B. (1990). Reversed field pinch. *Nucl Fusion* 30: 1717–1737

Bodin, H.A.B. and Newton, A.A. (1980). Reversed-field-pinch research. *Nucl Fusion* 20: 1255–1324.

Bogoliubov, N.N. and Mitropolsky, Y.A. (1961). *Asymptotic Methods in the Theory of Nonlinear Oscillations*. New York: Gordon and Breach.

Bohm, D. and Gross, E.P. (1949a). Theory of plasma oscillations. A. Origin of medium-like behaviour. *Phys Rev* 75: 1851–1864.

Bohm, D. and Gross, E.P. (1949b). Theory of plasma oscillations. B. Excitation and damping of oscillations. *Phys Rev* 75: 1864–1876.

Boltzmann, L. (1896). *Vorlesungen über Gastheorie*, Barth, Leipzig. Tarnslated as "Lectures on Gas Theory" by S.G. Brush. Berkeley, CA: University of California Press, 1964.

Born, M. and Wolf, E. (1965). *Principles of Optics*, 3e. Oxford: Pergamon.

Boyd, R.W. (2008). *Nonlinear Optics*, 3e. Burlington, MA: Academic Press.

Boyd, T.J.M. and Sanderson, J.J. (2003). *The Physics of Plasmas*. Cambridge: Cambridge University Press.

Braginskii, S.I. (1965). Transport processes in plasma. In: *Reviews of Plasma Physics*, vol. 1 (ed. M.A. Leontovich), 205–311. New York: Consultants Bureau.

Brandstatter, J.J. (1963). *An Introduction to Waves, Rays and Radiation in Plasma Media*. New York: McGraw-Hill.

Bretherton, F.P. (1964). Resonant interactions between waves. The case of discrete oscillations. *J Fluid Mech* 20: 457–479.

Briggs, R.J. (1964). *Electron-Stream Interaction with Plasmas, MIT Research Monograph 29*. Cambridge, MA: MIT Press.

Brillouin, L. (1960). *Wave Propagation and Group Velocity*. New York: Academic Press.

Brown, S.C. (1966). *Introduction to Electrical Discharges in Gases*. New York: Wiley.

Brown, S.C. (1967). *Basic Data of Plasma Physics*, 2e. Cambridge, MA: MIT Press.

Cairns, R.A. (1985). *Plasma Physics*. Glasgow: Blackie.

Callen, H.B. (1948). The application of Onsager's reciprocal relations to thermoelectricity, thermomagnetic and galvanomagnetic effects. *Phys Rev* 73: 1349–1358.

Caruso, A. and Gratton, R. (1968). Some properties of the plasmas produced by irradiating light solids by laser pulses. *Plasma Phys* 10: 867–877.

Caruso, A. and Gratton, R. (1969). Interaction of short laser pulses with solid materials. *Plasma Phys* 11: 839–847.

Caruso, A., Bertotti, B., and Giupponi, P. (1966). Ionization and heating of solid material by means of a laser pulse. *Nuovo Cimento B* 45: 176–189.

Cerfon, A.J. and Freidberg, J.P. (2010). "One size fits all" analytic solutions of the Grad–Shavronov equation. *Phys Plasmas* 17: 032502-1–032502-9.

Chandrasekhar, S. (1943). Stochastic problems in physics and astronomy. *Rev Mod Phys* 15: 1–89.

Chandrasekhar, S. (1958). Adiabatic invariants in the motion of charged particles. In: *The Plasma in a Magnetic Field* (ed. R.K.M. Landsdoff), 3–22. Stanford, CA: Stanford University Press.

Chandrasekhar, S. (1960). *Principles of Stellar Dynamics*. New York: Dover Publications.

Chandrasekhar, S. and Trehan, S.K. (1960). *Plasma Physics*. Chicago: University of Chicago Press.

Chapman, S. and Cowling, T.G. (1952). *The Mathematical Theory of Non-Uniform Gases*, 2e. Cambridge: Cambridge University Press.

Chen, F.F. (1983). *Introduction to Plasma Physics and Controlled Fusion*, 2e. New York: Plenum Press.

Clark, J.S., Fisher, H.N., and Mason, R.J. (1973). Laser-driven implosion of spherical DT targets to thermonuclear burn conditions. *Phys Rev Lett* 30: 89–92.

Clemmow, P.C. and Dougherty, J.D. (1969). *Electrodynamics of Particles and Plasmas*. Cambridge: Cambridge University Press.

Cobine, J.D. (1958). *Gaseous Conductors*. New York: Dover Publications.

Coffey, T.P. (1971). Breaking of large amplitude plasma oscillations. *Phys Fluids* 14: 1402–1406.

Cohen, R.H. (1976). Runaway electrons in an impure plasma. *Phys Fluids* 19: 239–244.

Cohen, R.S., Spitzer, L. Jr., and Routly, P.M.R. (1950). The electrical conductivity of an ionized gas. *Phys Rev* 80: 230–238.

Connor, J.W. and Hastie, R.J. (1975). Relativistic limitations on runaway electrons. *Nucl Fusion* 15: 415–424.

Copson, E.T. (1935). *An Introduction to the Theory of Functions of a Complex Variable*. Oxford: Oxford University Press.

Courant, R. and Hilbert, D. (1953). *Methods of Mathematical Physics*, vol. 1. New York: Wiley Interscience.

Curzon, F.L., Folkierski, A., Latham, R., and Nation, J.A. (1960). Experiments on the growth rate of surface instabilities in a linear pinch discharge. *Proc R Soc A* 257: 386–401.

David, N. and Hooker, S.M. (2003). Molecular-dynamic calculation of the relaxation of the electron energy distribution function in a plasma. *Phys Rev E* 68: 056401-1–056401-8.

Davidson, R.C. (1972). *Methods in Nonlinear Plasma Theory*. New York: Academic Press.

Dawson, J.M. (1959). Nonlinear electron oscillations in a cold plasma. *Phys Rev* 113: 383–387.

Dawson, J.M. (1961). On Landau damping. *Phys Fluids* 4: 869–874.

de Groot, S.R. and Mazur, P. (1984). *Non-Equilibrium Thermodynamics*. New York: Dover Publications.

Debye, P. and Hückel, E. (1923). The theory of electrolytes. I. Lowering of freezing point and related phenomena. *Phys Z* 24: 185–206.

Dolinsky, A. (1965). Numerical integration of kinetic equations. *Phys Fluids* 8: 436–443.

Drude, P. (1900a). Zur Elektronentheorie der Metalle. *Ann Phys* 306: 566–613.

Drude, P. (1900b). Zur Elektronentheorie der Metalle; II Teil Galvanomagnetishe und thermomagnetische Effecte. *Ann Phys* 308: 369–402.

Druyvestyn, M.J. (1930). De invloed der energieverliezen bij elastische botsingen in de theorie der electronendiffusie. *Physics* 10: 61–70.

Druyvestyn, M.J. and Penning, F.M. (1940). The mechanism of electrical discharges in gases of low pressure. *Rev Mod Phys* 12: 87–174.

Dupree, T.H. (1961). Dynamics of ionized gases. *Phys Fluids* 4: 696–702.

Eckhardt, R. (1989). Stan Ulam, John von Neumann and the Monte Carlo method. In: *From Cardinals to Chaos* (ed. N.G. Grant), 131–136. Cambridge: Cambridge University press.

Eddington, A.S. (1959). *The Internal Constitution of Stars*. New York: Dover Publications.

Epperlein, E.M. (1984). The accuracy of Braginkii's transport coefficients for a Lorentz plasma. *J Phys D Appl Phys* 17: 1823–1827.

Epperlein, E.M. and Haines, M.G. (1986). Plasma transport coefficients in a magnetic field by direct numerical solution of the Fokker–Planck equation. *Phys Fluids* 29: 1029–1041.

Fader, W.J. (1968). Hydrodynamic model of a spherical plasma produced by Q-spoiled laser irradiation of a solid particle. *Phys Fluids* 11: 2200–2208.

Fauquignon, C. and Floux, F. (1970). Hydrodynamic behaviour of solid deuterium under laser heating. *Phys Fluids* 13: 386–391.

Fermi, E. (1949). On the origin of cosmic radiation. *Phys Rev* 75: 1169–1174.

Fitzpatrick, R. (2015). *Plasma Physics - An Introduction*. Boca Raton, FL: CRC Press. Based on Lecture notes from University of Texas issued as Fitzpatrick (2009).

Flanigan, F.J. (1983). *Complex Variables Harmonic and Analytic Functions*. New York: Dover Publications.

Forrest, M.J. (2011). *Lasers Across the Cherry Orchards*. Forrest. Popular account of the important Tokamak validation experiments conducted by a team from UKAEA Culham during the 'cold war'.

Freidberg, J.P. (2014). *Ideal MHD*. Cambridge: Cambridge University Press.

Friedrichs, K.O. and Kranzer, H. (1958). Notes on Magneto-Hydrodynamics. VIII. Non-Linear Wave Motion. *New York University Report NYO 6486-VIII*. Institute of Mathematical Sciences.

Galeev, A.A. and Sagdeev, R.Z. (1968). Transport phenomena in a collisionless plasma in a toroidal magnetic system. *Sov Phys JETP* 26: 233–240.

Gamow, G. (1928). Quantum theory of the atomic nucleus. *Z Phys* 51: 204–213. Translation by the author available as http://web.ihep.su/dbserv/compas/src/gamow28/eng.pdf.

Gibbon, P. (2005). *Short Pulse Laser Interactions with Matter*. London: Imperial College Press.

Ginzburg, V.L. (1970). *The Propagation of Electromagnetic Waves in Plasmas*, 2e. Oxford: Pergamon.

Glasstone, S. and Lovberg, R.H. (1960). *Controlled Thermonuclear Reactions*. Princeton, NJ: Van Nostrand.

Grad, H. (1959). Propagation of magnetohydrodynamic waves without radial attenuation. In: *The Magnetohydynamics of Conductiong Fluids* (ed. D. Bershader), 37–60. Stanford, CA: Stanford University Press.

Grad, H. and Rubin, H. (1958). Hydromagnetic equilibria and force free fields. *Proceedings of the 2nd UN Conference on the Peaceful Uses of Atomic Energy*, Volume 31. Geneva: IAEA, pp. 190–196.

Haines, M.G. (2011). A review of the dense Z-pinch. *Plasma Phys Control Fusion* 53 (093001): 1–168.

Hammersley, J.M. and Handscomb, D.C. (1964). *Monte Carlo Methods*. London: Methuen.

Harris, E.G. (1959). Unstable plasma oscillations in a magnetic field. *Phys Rev Lett* 2: 34–36.

Hastie, R.J. (1981). Particle orbit theory. In: *Plasma Physics and Nuclear Fusion Research* (ed. R.D. Gill), 71–90. London: Academic Press.

Hendry, J. and Lawson, J.D. (1993). Fusion Research in the UK 1945–1960. *Technical Report AHO1*. UKAEA Technology.

Hertz, G. (1925). Über die Diffusion langsamer Elektronen im elektrischen Felde. *Z Phys* 32: 298–306.

Hinton, F.L. (1983). Collisional trnasport in plasma. In: *Handbook of Plasma Physics*, vol. 1 (ed. M.N. Rosenbluth and R.Z. Sagdeev), 147–197. Amsterdam: North Holland. physics.ucsd.edu/students/courses/fall2009/physics218a/Collisional (accessed 17 September 2020).

Hirschfelder, J.O., Curtiss, F., and Bird, R.B. (1964). *Molecular Theory of Gases and Liquids*. New York: Wiley.

Hooper, M.B. (ed.) (1995). *Laser-Plasma Interactions 5: Proceedings of the 45th Scottish Universities Summer School in Physics*, St Andrews, August 1994. Edinburgh: Scottish Universities Summer School Press.

Hora, H. (1969). Nonlinear confining and deconfining forces associated with the interaction of laser radiation with plasma. *Phys Fluids* 12: 182–191.

Hora, H. (1971). Nonlinear effect of expansion of laser produced plasmas. In: *Laser Interaction and Related Plasma Phenomena*, vol. 1 (ed. H.J. Schwartz and H. Hora), 383–426. New York: Plenum Press.

Huba, J.D. (2019). NRL Plasma Formulary. Technical report NRL/PU/6770-19-652. Plasma Physics Division, Naval Research Laboratory, Wahington, DC 20375, USA.

Hubbard, J. (1961a). The friction and diffusion coefficients of the Fokker–Planck equation in a plasma. *Proc R Soc London* 260: 114–126.

Hubbard, J. (1961b). The friction and diffusion coefficients of the Fokker–Planck equation in a plasma II. *Proc R Soc London* 261: 371–387.

Jackson, J.D. (1960). Longitudinal plasma oscillations. *J Nucl Energy, Part C, Plasma Phys* 1: 171–189.

Jeans, J.H. (1954). *The Dynamical Theory of Gases*, 4e. New York: Dover Publications.

Jeans, J.H. (1961). *Astronomy and Cosmogony*, 2e. New York: Dover Publications.

Johnston, T.W. (1960). Cartesian tensor scalar product and spherical harmonic expansions in Boltzmann's equation. *Phys Rev A* 110: 1103–1111.

Jones, R.D. and Lee, K. (1982). Kinetic theory, transport and hydrodynamics of a high-Z plasma in the presence of an intense laser field. *Phys Fluids* 25: 2307–2323.

Jorna, S. and Wood, L. (1987a). Fokker–Planck calculations on heat flow in plasmas. *J Plasma Phys* 38: 317–333.

Jorna, S. and Wood, L. (1987b). Fokker–Planck calculations on relaxation of anisotropic velocity distributions in plasma. *Phys Rev A* 36: 397–399.

Kadomtsev, B.B. and Pogu'tsev, O.P. (1967). Plasma instability due to particle trapping in a toroidal geometry. *Sov Phys JETP* 24: 1172–1179.

Kaufman, A.N. (1960). Plasma viscosity in a magnetic field. *Phys Fluids* 3: 610–616.

Knoepfel, H. and Spong, D.A. (1979). Runaway electrons in toroidal discharges. *Nucl Fusion* 19: 785–829.

Kruer, W.L. (1988). *The Physics of Laser Plasma Interactions*. Addison-Wesly.

Kruskal, M. (1962). Asymptotic theory of Hamiltonian and other systems with all solutions nearly periodic. *J Math Phys* 3: 806–828.

Kruskal, M. and Schwarzschild, M. (1954). Some instabilities of a completely ionized plasma. *Proc R Soc A* 223: 348–360.

Kulsrud, R.M., Sun, Y.-C., Winsor, N.K., and Fallon, H.A. (1975). Runaway electrons in a plasma. *Phys Rev Lett* 31: 690–693.

Kurchatov, I.V. (1956). The possibility of producing thermonuclear reactions in a gaseous discharge. Lecture given at UKAEA Harwell. https://www.iter.org/doc/www/content/com/Lists/MagStories/Attachments/64/kurchatov_1956.pdf (accessed 17 September 2020).

Landau, L.D. (1936). The transport equation in the case of Coulomb interaction. *Phys Z Sowjet* 10: 154–164. Translation in collected papers of L.D. Landau (Oxford: Pergamon) 2: 163–170, 1965.

Landau, L.D. (1946). On the vibrations of the electronic plasma. *J Phys* 10: 25–34.

Landau, L.D. and Lifshitz, E.M. (1959). *Quantum Mechanics, Course of Theoretical Physics*, 2e, vol. 3. Oxford: Pergamon.

Landau, L.D. and Lifshitz, E.M. (1976). *Mechanics, Course of Theoretical Physics*, 3e, vol. 1. Oxford: Butterworth-Heinenmann.

Landau, L.D. and Lifshitz, E.M. (1980). *Statistical Physics: Part 1, Course of Theoretical Physics*, 3e, vol. 5. Oxford: Butterworth-Heinenmann.

Landau, L.D. and Lifshitz, E.M. (1984). *Electrodynamics of Continuous Media, Course of Theoretical Physics*, 2e, vol. 8. Oxford: Butterworth-Heinenmann.

Landshoff, R. (1949). Transport phenomena in a completely ionized gas in presence of a magnetic field. *Phys Rev* 76: 904–908.

Landshoff, R. (1951). Convergence of the Chapman–Enskog method for a completely ionized gas. *Phys Rev* 82: 442–442.

Langdon, A.B. (1980). Nonlinear inverse bremsstrahlung and heated electron distributions. *Phys Rev Lett* 44: 575–579.

Langmuir, I. (1928). Oscillations in ionized gases. *Proc Natl Acad Sci U S A* 14: 627–637.

Lashmore-Davies, C.N. (1975). The coupled mode approach to non-linear wave interactions and parametric instabilities. *Plasma Phys* 15: 281–303.

Lashmore-Davies, C.N. (1981). Nonlinear laser plasma interaction theory. In: *Plasma Physics and Nuclear Fusion Research* (ed. R.D. Gill), 319–356. London: Academic Press.

Lawson, J.D. (1955). Some Criteria for a Useful Thermonuclear Reactor. Technical report AERE GP/R 1807. Harwell Berks, UK: Atomic Energy Research Establishment.

Lawson, J.D. (1957). Some criteria for a power producing reactor. *Proc Phys Soc B* 70: 6–10.

Lax, P. (1957). Hyperbolic systems of conservation laws II. *Commun Pure Appl Math* 10: 537–566.

Lenard, A. (1960). On Bogoliubov's kinetic equation for a spatially homgeneous equation. *Ann Phys (NY)* 3: 390–400.

Lieberman, M.A. and Lichtenberg, A.J. (2005). *Principles of Plasma Discharges and Materials Processing*, 2e. Hoboken, NJ: Wiley-Interscience.

Liberman, M.A. and Velikhovich, A.L. (1986). *Physics of Shock Waves in Gases and Plasmas.* New York: Springer.

Lifshitz, E.M. and Pitaevskii, L.P. (1981). *Physical Kinetics, Course of Theoretical Physics*, 2e, vol. 10. Oxford: Butterworth-Heinenmann.

Lighthill, M.J. (1960). Studies of magneto-hydrodynamic waves and other anisotropic wave motion. *Philos Trans R Soc* 252: 397–423.

Lighthill, M.J. (1978). *Waves in Fluids.* Cambridge: Cambridge University Press.

Lindl, J. (1993). The Edward Teller Medal Lecture: The Evolution Toward Indirect Drive and Two Decades of Progress Toward ICF Ignition and Burn. *Technical Report UCRL -JC 115197*. Lawrence Livermore National Laboratory.

Loeb, L.B. (1936). The problem of the mechanism of static spark discharge. *Rev Mod Phys* 8: 267–293.

Loeb, L.B. (1955). *Basic Processes of Gaseous Electronics.* Berkeley, CA: University of California Press.

Lorentz, H.A. (2004). *Theory of Electrons*, 2e. New York: Dover Publications.

Lowry, H.V. and Hayden, H.A. (1957). *Advanced Mathematics for Technical Students: Part Two*, 2e. London: Longmans.

MacDonald, W.M., Rosenbluth, M.N., and Chuck, W. (1957). Relaxation of a system of particles with coulomb interactions. *Phys Rev* 107: 350–353.

MacKeown, P.K. (1997). *Stochastic Simulation in Physics.* Singapore: Springer.

Margenau, H. (1946). Conduction and dispersion of ionized gases at high frequencies. *Phys Rev* 69: 506–513.

Marshall, W. (1960). The Kinetic Theory of An Ionized Gas. *Technical Report AERE T/R 2247, T/R 2352 and T/R 2419.* United Kingdom Atomic Energy Authority.

Martin, R.A. and Segur, H. (2016). Toward a general solution of the three-wave partial differential equations. *Stud Appl Math* 173: 70–92.

Maxwell, J.C. (1865). A dynamical theory of the electromagnetic field. *Philos Trans* 155: 459–512.

Maxwell, J.C. (1867). On the dynamical theory of gases. *Philos Trans* 157: 49–88.

Maxwell, J.C. (1904). *A Treatise of Electricity and Magnetism*, 3e. Oxford: Oxford University Press.

McDaniel, E.W. (1964). *Collision Phenomena in Ionized Gases.* New York: Wiley.

McGowan, A.D. and Sanderson, J.J. (1992). On the relaxation of non-thermal plasma. *J Plasma Phys* 47: 373–387.

Metropolis, N. (1989). The beginning of the Monte Carlo method. In: *From Cardinals to Chaos* (ed. N.G. Grant), 126–130. Cambridge: Cambridge University Press.

Montgomery, D.C. and Tidman, D.A. (1964). *Plasma Kinetic Theory, Advanced Physics Monograph Series.* New York: McGraw-Hill.

Morozov, A.I. and Solev'ev, L.S. (1966). Motion of charged particles in electromagnetic fileds. In: *Reviews of Plasma Physics*, vol. 2 (ed. M.A. Leontovich), 201–297. New York: Consultants Bureau.

Morse, P.M. and Feshbach, H. (1953). *Methods of Theoretical Physics.* New York: McGraw-Hill.

Morse, P.M., Allis, W.P., and Lamar, E.S. (1935). Velocity distribution for elastically colliding electrons. *Phys Rev* 42: 412–419.

Mott-Smith, H.M. (1971). History of plasmas. *Nature* 233: 219.

New, G. (2011). *Introduction to Non-Linear Optics*. Cambridge: Cambridge University Press.

Northrop, T.G. (1963). *The Adiabatic Motion of Charged Particles*. New York: Wiley Interscience.

Nuckolls, J., Wood, L., Thiessen, A., and Zimmerman, G. (1972). Laser compression of matter to super-high densities: thermonuclear (CTR) applications. Nature 239: 139–142.

Nyquist, H. (1932). Regeneration theory. *Bell System Tech J* 11: 126–147.

Page, L. and Adams, N.I. Jr (1945). Action and reaction between moving charges. *Am J Phys* 13: 141–147.

Parail, V.V. and Pogutse, O.P. (1986). Runaway electrons in a Tokammak. In: *Reviews of Plasma Physics*, vol. 11 (ed. M.A. Leontovich), 1–63. New York: Consultants Bureau.

Paschen, F. (1889). Ueber die zum funkenübergang in luft, wasserstoff und kohlensäure bei verschiedenen drucken erforderliche potentialdifferenz. *Ann Phys* 273: 69–96.

Paul, W. (1990). Electromagnetic traps for charged and neutral particles. *Rev Mod Phys* 62: 531–540. Lecture delivered 8 December 1989 on presentation of 1989 Nobel Prize in Physics.

Penrose, O. (1960). Electrostatic instabilities of a uniform non-maxwellian plasma. *Phys Fluids* 3: 258–265.

Persico, E. (1926). On the kinetic theory of a highly ionised gas. *Mon Not R Astron Soc* 86: 93–98.

Pert, G.J. (1972). Inverse bremsstrahlung absorption in large radiation fields during binary collisions-classical theory. *J Phys A Gen Phys* 5: 506–515.

Pert, G.J. (1974). Thermal conduction effects in lassr solid target interaction. *Plasma Phys* 16: 1019–1033.

Pert, G.J. (1975). Inverse bremsstrahlung absorption in large radiation fields during binary collisions - the classical theory as an asymptotic limit of the Born approximation. *J Phys B At Mol Opt Phys* 8: 3069–3077.

Pert, G.J. (1976a). Inverse bremsstrahlung absorption in large radiation fields during binary collisions-classical theory. II(b). Summed rate coefficients for coulomb collisions. *J Phys A Math Gen* 9: 1797–1800.

Pert, G.J. (1976b). Inverse bremsstrahpung absorption in large radiation fields during binary collisions-classical theory. II. Integrated rate coefficients for coulomb collisions. *J Phys A Math Gen* 9: 463–471.

Pert, G.J. (1977). A class of similar solutions of the non-linear diffusion equation. *J Phys A Math Gen* 10: 583–593.

Pert, G.J. (1978). The analytic theory of linear resonant absorption. *Plasma Phys* 20: 175–188.

Pert, G.J. (1980). Self-similar flows with uniform velocity gradient and their use in modelling the free expaqnsion of polytropic gases. *J Fluid Mech* 100: 257–277.

Pert, G.J. (1983). Deflagrations and discontinuities in laser-produced plasmas. *J Plasma Phys* 29: 415–438.

Pert, G.J. (1986a). Models of laser-plasma ablation. *J Plasma Phys* 35: 43–74.

Pert, G.J. (1986b). Models of laser-plasma ablation. Part 2. Steady-state theory: self-regulating flow. *J Plasma Phys* 36: 415–446.

Pert, G.J. (1987). The use of flows with uniform velocity gradient in modelling free expansion of a polytropic gas. *Laser Part Beams* 5: 643–658.

Pert, G.J. (1988). Models of laser-plasma ablation. Part 3. Steady-state theory: deflagration flow. *J Plasma Phys* 39: 241–276.

Pert, G.J. (1989). Two-dimensional hydrodynamic models of laser-produced plasmas. *J Plasma Phys* 41: 263–280.

Pert, G.J. (1993). Models of laser plasma ablation. Part 4. Steady-state theory: collisional absorption flow. *J Plasma Phys* 49: 295–316.

Pert, G.J. (1995). Inverse bremsstrahlung in strong radiation fields at low temperatures. *Phys Rev E* 51: 4778–4789.

Pert, G.J. (1999). Electron distributions generated by tunnelling ionization during gas breakdown by high intensity laser radiation. *J Phys B At Mol Opt Phys* 32: 27–52.

Pert, G.J. (2001). The calculation of the electron distribution function following tunnel ionisation using a Fokker–Planck method. *J Phys B At Mol Opt Phys* 34: 881–908.

Pert, G.J. (2013). *Introductory Fluid Mechanics for Physicists and Mathematicians*. Chicester: Wiley.

Phelps, A.V. (1990). The diffusion of charged particles in collisional plasmas: free and ambipolar diffusion at low and moderate pressures. *J. Res Natl Inst Stand Technol* 95: 407–431.

Puell, H. (1970). Heating of laser produced plasmas generated at plane solid targets I. theory. *Z Naturforsch A* 25: 1807–1815.

Putvinski, S. (2011). Runaway electrons in tokamaks and their mitigation in iter. IAEA TCM on Energetic Particles, Austin, September 2011. w3fusion.ph.utexas.edu/ifs/iaeaep/talks/s11-i11-putvinski-sergei-ep-talk.pdf (accessed 17 September 2020).

Raizer, Y.P. (1977). *Laser-Induced Discharge Phenomena*. New York: Consultants Bureau.

Raizer, Y.P. (1997). *Gas Discharge Physics*. Berlin: Springer-Verlag.

Rayleigh, J.W.S.L. (1883). Investigation of the character of the equilibrium of an incompressible heavy fluid of variable density. *Proc London Math Soc* 14: 170–177. See also Rayleigh (1896).

Rayleigh, J.W.S.L. (1896). *Theory of Sound*. Cambridge: Cambridge University Press. Second edition reprinted in Rayleigh (1945).

Rose, D.J. and Clark, Jr, M. (1961). *Plasmas and Controlled Fusion*. New York: MIT Press and Wiley.

Rosenbluth, M.N. and Longmire, C.L. (1957). Stability of plasmas confined by magnetic fields. *Ann Phys* 1: 120–140.

Rosenbluth, M.N. and Putvinski, S.V. (1997). Theory for avalanche of runaway electrons in tokamaks. *Nucl Fusion* 37 1355–1362.

Rosenbluth, M.N., MacDonald, W.M., and Judd, D.L. (1957). Fokker–Planck equation for an inverse square force. *Phys Rev* 107: 1–6.

Sakharov, A. (1990). *Memoirs*. London: Hutchinson.

Salzmann, H. (1973). Production of hot plasma of solid-state density by ultrashort laser pulses. *J Appl Phys* 44: 113–124.

Sauter, O. and Medvedev, S.Y. (2003). Tokamak coordinate conventions: COCOS. *Comput Phys Commun* 184: 293–302.

Shafranov, V.D. (1958). On magnetohydrodynamical configurations. *Sov Phys JETP* 6: 645–654.

Shafranov, V.D. (1966). Plasma equilibrium in a magnetic field. In: *Reviews of Plasma Physics*, vol. 2 (ed. M.A. Leontovich), 103–151. New York: Consultants Bureau.

Shafranov, V.D. (2001). On the history of research into controlled thermonuclear fusion. *Phys Usp* 44: 835–843.

Shkarofsky, I.P. (1961). Values of the transport coefficents in a plasma for any degree of ionization based on a Maxwellian ditribution. *Can J Phys* 39: 1619–1703.

Shkarofsky, I.P., Bernstein, I.B., and Robinson, B.R. (1963). Condensed presentation of transport coefficients in a fully ionized plasma. *Phys Fluids* 6: 40–47.

Shkarofsky, I.P., Johnston, T.W., and Bachynski, M.P. (1966). *The Particle Kinetics of Plasmas*. Reading, MA: Addison-Wesley.

Silin, V.P. (1965). Non-linear high frequency plasma conductivity. *Sov Phys JETP* 20: 1510–1516.

Smirnov, B.M. (2015). *Theory of Gas Discharge Plasma*. Heidelberg: Springer Cham.

Smythe, W.R. (1968). *Static and Dynamic Electricity*, 3e. New York: McGraw-Hill.

Sokolov, Y.A. (1979). "Multiplication" of accelerated electrons in a tokamak. *Sov Phys JETP Lett* 29: 218–221.

Solov'ev, L.S. (1968). The theory of hydromagnetic stability in toroidal plasma configurations. *Sov Phys JETP* 26: 400–407.

Solov'ev, L.S. (1975). Hydromagnetic stability of closed plasma configurations. In: *Reviews of Plasma Physics*, vol. 3 (ed. M.A. Leontovich), 239–331. New York: Consultants Bureau.

Sommerfeld, A. (1964). *Optics, Lectures on Theoretical Physics*, vol. 4. New York: Academic Press.

Spitzer, L. Jr. (1940). The stability of isolated galaxies. *Mon Not R Astron Soc* 100: 396–413.

Spitzer, L. Jr. (1956). *Physics of Fully Ionized Gases*. New York: Interscience.

Spitzer, L. and Härm, R. Jr (1953). Transport phenomena in a completely ionized gas. *Phys Rev* 89: 977–981.

Stix, T.H. (1962). *The Theory of Plasma Waves, Advanced Physics Monograph Series*. New York: McGraw-Hill.

Sturrock, P.A. (1958). Kinematics of growing waves. *Phys Rev* 112: 1488–1503.

Tallents, G.J. (2018). *An introduction to the atomic and radiation physics of plasmas*. Cambridge: Cambridge University Press.

Taylor, G.I. (1950). The instability of liquid surfaces when accelerated in a direction perpendicular to their planes I. *Proc R Soc A* 201: 192–196.

Taylor, J.B. (1963). Some stable plasma equilibria in combined mirror-cusp fields. *Phys Fluids* 6: 1529–1536.

Taylor, J.B. (1974). Relaxation of toroidal plasma and generation of reverse magnetic fields, *Phys Rev Lett* 33, 1139–1141.

Ter Haar, D. (2006). Adiabatic invariance and adiabatic invariants. *Contemp Phys* 7: 447–458.

Thompson, W.B. (1957). An introduction to plasma physics. *Proc Phys Soc B* 70: 1–5.

Thompson, W.B. (1962). *An Introduction to Plasma Physics*. Oxford: Pergamon.

Thomson, J.J. (1897). Cathode rays. *Philos Mag* 44 (269): 293–316.

Thomson, J.J. and Thomson, G.P. (1933). *Ionisation by Collision and the Gaseous Discharge*. 2. Cambridge: University Press.

Tonks, L. (1955). Particle transport,electric currents and pressure balance in a magnetically confined plasma. *Phys Rev* 97: 1143–1145.

Tonks, L. (1967). The birth of plasma. *Am J Phys* 35: 857–858.

Tonks, L. and Langmuir, I. (1929). Oscillations in ionized gases. *Phys Rev* 33: 195–210.

Town, R.P.J., Bell, A.R., and Rose, S.J. (1994). Fokker–Planck simulations of short-pulse-laser-solid experiments. *Phys Rev E* 50: 1413–1421.

Townsend, J.S. (1915). *Electricity in Gases*. Oxford: Oxford University Press.

Townsend, J.S. (1947). *Electrons in Gases*. London: Hutchinson.

van Kampen, N.G. (1956). On the theory of stationary waves in plasmas. *Physica* 21: 949–963.

Vlasov, A.A. (1938). The vibrational properties of an electron gas. *Zh Eksp Teor Fiz* 8: 291. Translation in *Sov Phys Usp* **93**: 721–733, **1968**.

von Engel, A. (1965). *Ionized Gases*, 2e. Oxford: Oxford University Press.

von Engel, A. and Steenbeck, M. (1932). *Elektrische Gasentladungen: Ihre Physik und Technik (German Edition)*. Berlin: Springer-Verlag. A concise account of much of this work is contained in vonEngel (1965).

Watson, G.N. (1944). *A Treatise on the Theory of Bessel Functions*, 2e. Cambridge: Cambridge Uninersity Press.

Wesson, J. (2004). *Tokamaks*, 3e. Oxford: Oxford University press.

Wesson, J. (2011). *Tokamaks*, 4e. Oxford: Oxford University Press.

Weyl, H. (1944). Shock waves in arbitrary fluids. Applied Mathematics Panel Note 12. National defense research committee. Subsequently published in *Commun Pure Appl Maths* 2: 103–122, 1944; Included in Johnson & Chéret (1998, 498–519).

Whittaker, E.T. and Robinson, G. (1944). *Calculus of Observations*, 4e. Glasgow: Blackie.

Yariv, A. (1997). *Optical Electronics in Modern Communications*, 5e. New York: Oxford University Press.

Zel'dovich, Y.B. and Raizer, Y.P. (1966). *Physics of Shock Waves and High Temperature Hydrodynamic Phenomena*. New York: Academic Press.

Zheng, S.B., Wootton, A.J., and Solano, E.R. (1996). Analytic tokamak equilibrium for shaped plasmas. *Phys Plasmas* 3: 1176–1178.

Index